# A MODERN INTRODUCTION TO
# PARTICLE PHYSICS

### Third Edition

# A MODERN INTRODUCTION TO
# PARTICLE PHYSICS

## Third Edition

# Fayyazuddin
# Riazuddin

National Centre for Physics, Pakistan

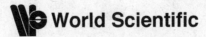

**World Scientific**

NEW JERSEY · LONDON · SINGAPORE · BEIJING · SHANGHAI · HONG KONG · TAIPEI · CHENNAI

*Published by*

World Scientific Publishing Co. Pte. Ltd.

5 Toh Tuck Link, Singapore 596224

*USA office:* 27 Warren Street, Suite 401-402, Hackensack, NJ 07601

*UK office:* 57 Shelton Street, Covent Garden, London WC2H 9HE

**Library of Congress Cataloging-in-Publication Data**
Fayyazuddin, 1930–
  Modern introduction to particle physics / by Fayyazuddin, Riazuddin. -- 3rd ed.
    p. cm.
  Includes bibliographical references and index.
  ISBN-13: 978-981-4338-83-7 (hard cover : alk. paper)
  ISBN-10: 981-4338-83-4 (hard cover : alk. paper)
  1. Particles (Nuclear physics) I. Riazuddin. II. Title.
  QC793.2.F39 2011
  539.7'2--dc23

                        2011015874

**British Library Cataloguing-in-Publication Data**
A catalogue record for this book is available from the British Library.

Printed in Singapore.

Thou seest not in the creation of All-
merciful any imperfection.
Return thy gaze; seest thou any fissure?
Then return thy gaze again, and again and thy
gaze comes back to thee dazzled, aweary
Koran, The Kingdom LXVII.

To Masooda and Mumtaz

# Preface

## Preface for the Third Edition

The main aim in producing the third edition is to bring the book up to date and as such many chapters have been thoroughly revised. In particular the chapters on Heavy Flavors (Chap. 8), Neutrino Physics (Chap. 12), Electroweak Unification (Chap. 13), Weak Decays of Heavy Flavors (Chap. 15), Particle Mixing and $CP$-violation (Chap. 16), Grand Unification, Supersymmetry and Strings (Chap. 17) and Cosmology and Astroparticle Physics (Chap. 18) have gone through major revision with the addition of some new material. To make the book self-contained, appendix A has been extended. An important feature of the 3rd edition is the addition of a substantial number of new problems.

A number of typographical errors have been corrected and a number of figures have been streamlined, using Jaxodraw software.

We wish to express our deep sense of appreciation to Dr. Maqbool Ahmed and Mansoor-ur-Rehman for critically reading Chap. 18 on Cosmology and Astroparticle Physics, making many useful suggestions. We wish to express our deep thank to Aqeel Ahmed (our graduate student) for doing an excellent job in typing, drawing figures and carefully reading some of the chapters; without his help it was difficult to put the manuscript in the final form. Thanks are also due to Ishtiaq Ahmed, Jamil Aslam, M. Junaid, Ali Paracha and Abdur Rehman for assistance in typing the manuscript.

Finally we wish to acknowledge the permission granted by Particle Data Group for reproducing figures indicated in the text which are duely referred. One of us (F) would like to acknowledge the support of Higher Education Commission (HEC), Islamabad.

Fayyazuddin

Riazuddin

## Preface for the Second Edition

Our aim in producing this new edition is to bring the book up to date and as such many chapters have been throughly revised. In particular, the chapters on Neutrino Physics, Partcle Mixing and CP–Violation and Weak Decays of Heavy Flavors have been mostly rewritten incorporating new material and new data. The heavy quark effective field theory has been included and a brief introductory section on supersymmetry and strings has been added. We wish to thank Ansar Fayyazuddin for writing this section.

A number of typographical errors have been corrected. Another change is that we have adopted a metric and notation for gamma matrices commonly used.[1]

Finally we wish to thank Mr. Amjad Hussain Gilani and Dr. Muhammad Nisar who did an excellent job in typing the manuscript; without their help it was difficult to put the manuscript in final shape.

Fayyazuddin

Riazuddin

Jan. 21, 2000

---

[1]See for example, J. D. Bjorken and S.D. Drell, *Relativistic Quantum Mechanics*, McGraw-Hill Book Co., New York (1965).

## Preface for the First Edition

Particle physics has been one of the frontiers of science since J. J. Thompson's discovery of the electron about one hundred years ago. Since then physicists have been concerned with (i) attempts to discover the ultimate constituents of matter, (ii) the fundamental forces through which the fundamental constituents interact, and (iii) seeking a unification of the fundamental forces.

At the present level of experimental resolution, the smallest units of matter appear to be leptons and quarks, which are spin 1/2 fermions. Hadrons (particles which feel the strong force) are composed of quarks. The evidence for this comes from the observed spectrum and static properties of hadrons and from high energy lepton-hadron scattering experiments involving large momentum transfers, which "prove" the actual existence of quarks within hadrons. As originally formulated, the quark model needed three flavors of quarks, up $(u)$, down $(d)$ and strangeness $(s)$ not just $u$ and $d$. The discoveries of the tau leptons and more flavors [charm $(c)$ and bottom $(b)$] were to some extent welcomed and to some extent appeared to be there for no apparent reason since elementary building blocks of an atom are just $u$ and $d$ quarks and electrons. A charm quark was predicted to exist to remove all phenomenological obstacles to a proper and an elegant gauge theory of weak interaction. Without it, nonexistence of strangeness-changing neutral current posed a puzzle. This also restored the quark-lepton symmetry: for each pair of leptons of charges 0 and $-1$ there is a quark pair of charges 2/3 and $-1/3$. The existence of $\tau$-leptons and discovery of the $b$ quark (charge $-1/3$) demand the existence of another quark (charge 2/3), called the top quark, to again restore the quark-lepton symmetry. Indeed, six quark flavors have been proposed to incorporate violation of CP invariance in weak interaction.

Quarks also have a hidden three valued degree of freedom known as color: each quark flavor comes in three colors. The antisymmetry of three-quark wave function of a baryon [e.g. proton] is attributed to color degree of freedom. The three number of colors also manifest themselves in $\pi^0$ decay and in the annihilation of lepton-antilepton into hadrons. We have encountered the following types of charges: gravitational, namely, mass, electric, flavor and color. The fundamental forces through which elementary fermions interact are then simply the forces of attraction or repulsion between these charges. The unification of forces is then sought by searching for a single entity of which the various charges are components in the sense that they can be transformed into one and another. In other words, they form generators of a gauge group $G$ which is taken to be local so that a definite form of interaction between vector fields (which must exist and belong to the adjoint representation of $G$) and elementary fermions (which belong to the fundamental or trivial representation of $G$) is generated with a universal coupling constant. In this respect non-Abelian gauge field theories [Yang-Mills type] have played a major role. Here the field itself is a carrier of "charge" so that there are direct interactions between the field quanta.

Let us first discuss the strong quark interactions. The local gauge group is $SU_C(3)$ generated by three color charges, the field quanta are eight massless spin 1 color carrying gluons. The theory of quark interactions arising from the exchange of gluons is called quantum chromodynamics (QCD). The most striking physical properties of QCD are (i) the concept of a "running coupling constant $\alpha_s(q^2)$", depending on the amount of momentum transfer $q^2$. It goes to zero for high $q^2$ leading to asymptotic freedom and becomes large for low $q^2$, (ii) confinement of quarks and gluons in a hadron so that only color singlets can be produced and observed. Only the property (i) has a rigorous theoretical basis while the property (ii) finds support from hadron spectroscopy and lattice gauge simulations.

Weak and electromagnetic interactions result from a gauge group acting upon flavors. It is $SU_L(2) \times U(1)$ and is spontaneously broken rather than exact as was $SU_C(3)$.

The electroweak theory, together with the quark hypothesis and QCD, form the basis for the so called "Standard Model" of elementary particles. There have been many quantitative confirmations of the predictions of the standard model: existence of neutral weak current mediated by $Z^0$, discovery of weak vector bosons $W^\pm, Z^0$ at the predicated masses, precision determinations of electroweak parameters and coupling constants (e.g. $\sin^2 \theta_W$ which comes out to be the same in all experiments) leading to one loop verification of the theory and providing constraints on the top quark and Higgs masses. Similarly there have been tests of QCD, verifying the running of the coupling constant $\alpha_s(q^2)$, $q^2$-dependence of structure functions in deep-inelastic lepton-nucleon scattering. Other evidences come from hadron spectroscopy and from high energy processes in which gluons play an essential role.

In spite of the above successes, many questions remain: replication of families and how many quarks and leptons are there? QCD does not throw any light on how many quark flavors there should be? Origin of fermion masses, which appear as free parameters since Higgs couplings with fermions contain as many arbitrary coupling constants as there are masses, is another unanswered question. Origin of CP violation at more fundamental level, rigorous basis of confinement and hadronization of quarks are other questions which await answers. Top quark and Higgs boson are still to be discovered.

Symmetry principles have played an important part in our understanding of particle physics. Thus Chapters 2-6 discuss global symmetries and flavor or classifications symmetries like SU(2) and SU(3) and quark model. Chapter 5 provides the necessary group theory and consequences of flavor SU(3). Chapters 2-6 together with Chapters 9, 10 and 11 on neutrino, weak interactions, properties of weak hadronic currents and chiral symmetry comprise mainly what is called old particle physics but include some new topics like neutrino oscillations and solar neutrino problem. These Chapters are included to provide necessary background to new particle physics, comprising mainly the standard model as defined above. The rest of the book is devoted to the standard model and the topics mentioned in paras 2-7 of the preface. Recently there has been an interface of particle physics with cosmology, providing not only an understanding of the history of very early universe but also shedding some light on questions such as dark matter and open or closed universe. Chapter 16 of the book is devoted to this interface.

Particle physics forms an essential part of physics curriculum. This book can be used as a text book, but it may also be useful for people working in the field. The book is so designed as to form one semester course for senior undergraduates (with suitable selection of the material) and one semester course for graduate students. Formal quantum field theory is not used; only a knowledge of non-relativistic quantum mechanics is required for some parts of the book. But for the remaining parts, the knowledge of relativistic quantum mechanics is essential. The familiarity with quantum field theory is an advantage and for this purpose two Appendicess which summarize the Feynman rules and renormalization group techniques, are added.

Initial incentive for this book came from the lectures which we have given at various places: Quaid-e-Azam University, Islamabad, Daresbury Nuclear Physics Laboratory (R), the University of Iowa (R), King Fahd University of Petroleum and Minerals, Dhahran (R) and King Abdulaziz University, Jeddah (F).

We have not prepared a bibliography of the original papers underlying the developments discussed in the book. Remedy for this can be found in the recent review articles and books listed at the end of each Chapter.

We wish to express our deep sense of appreciation to Dr. Ahmed Ali for critically reading the manuscript, for making many useful suggestions and for his help to update the data. We also wish to express our deep thanks to a colleague Mr. El hassan El aaud and a graduate student Mr. F. M. Al-Shamali [of one of us (R)], who drew diagrams and in general assisted in producing the final manuscript. In addition, the typing help provided by Mr. Mohammad Junaid at Research Institute of King Fahd University of Petroleum and Minerals was indispensable in getting the job done. Finally we wish to acknowledge the support of King Fahd University of Petroleum and Minerals for this project under Project No. PH/Particle/123.

We also take this opportunity to express our deep sense of gratitude to Prof. Abdus Salam, who first introduced us to this subject and for his encouragement throughout our work in this field.

Fayyazuddin

Riazuddin

March 4, 1992

# Contents

*Preface*                                                                    vii

1. Introduction                                                                1

   1.1   Fundamental Forces . . . . . . . . . . . . . . . . . .    1
        1.1.1   The Gravitational Force . . . . . . . . . . . . .    2
        1.1.2   The Weak Nuclear Force . . . . . . . . . . . . .    2
        1.1.3   The Electromagnetic Force . . . . . . . . . . .    3
        1.1.4   The Strong Nuclear Force . . . . . . . . . . . .    4
   1.2   Relative Strength of Four Fundamental Forces . . . . . .    4
   1.3   Range of the Three Basic Forces . . . . . . . . . . .    5
   1.4   Classification of Matter . . . . . . . . . . . . . .    7
   1.5   Strong Color Charges . . . . . . . . . . . . . . .    9
   1.6   Fundamental Role of "Charges" in the Unification of Forces   10
   1.7   Strong Quark-Quark Force . . . . . . . . . . . . .   16
   1.8   Grand Unification . . . . . . . . . . . . . . . .   18
   1.9   Units and Notation . . . . . . . . . . . . . . . .   19
   1.10  Problems . . . . . . . . . . . . . . . . . . . .   21
   1.11  References . . . . . . . . . . . . . . . . . . .   21

2. Scattering and Particle Interaction                                        23

   2.1   Introduction . . . . . . . . . . . . . . . . . . .   23
   2.2   Kinematics of a Scattering Process . . . . . . . . . .   26
   2.3   Interaction Picture . . . . . . . . . . . . . . . .   31
   2.4   Scattering Matrix (S-Matrix) . . . . . . . . . . . .   32
   2.5   Phase Space . . . . . . . . . . . . . . . . . . .   36
   2.6   Examples . . . . . . . . . . . . . . . . . . . .   39

|  |  | 2.6.1 | Two-body Scattering | 39 |
|  |  | 2.6.2 | Three-body Decay | 41 |
|  | 2.7 | Electromagnetic Interaction | | 50 |
|  | 2.8 | Weak Interaction | | 52 |
|  | 2.9 | Hadronic Cross-section | | 55 |
|  | 2.10 | Problems | | 56 |
|  | 2.11 | References | | 58 |

3. Space-Time Symmetries    59

|  | 3.1 | Introduction | | 59 |
|  |  | 3.1.1 | Rotation and SO(3) Group | 60 |
|  |  | 3.1.2 | Translation | 62 |
|  |  | 3.1.3 | Lorentz Group | 63 |
|  | 3.2 | Invariance Principle | | 65 |
|  |  | 3.2.1 | U Continuous | 65 |
|  |  | 3.2.2 | $U$ is Discrete (e.g. Space Reflection) | 66 |
|  | 3.3 | Parity | | 66 |
|  | 3.4 | Intrinsic Parity | | 68 |
|  |  | 3.4.1 | Intrinsic Parity of Pion | 70 |
|  | 3.5 | Parity Constraints on S-Matrix for Hadronic Reactions | | 71 |
|  |  | 3.5.1 | Scattering of Spin 0 Particles on Spin $\frac{1}{2}$ Particles | 71 |
|  |  | 3.5.2 | Decay of a Spin $0^+$ Particle into Three Spinless Particles Each Having Odd Parity | 72 |
|  | 3.6 | Time Reversal | | 73 |
|  |  | 3.6.1 | Unitarity | 74 |
|  |  | 3.6.2 | Reciprocity Relation | 75 |
|  | 3.7 | Applications | | 76 |
|  |  | 3.7.1 | Detailed Balance Principle | 76 |
|  | 3.8 | Unitarity Constraints | | 77 |
|  |  | 3.8.1 | Two-Particle Partial Wave Unitarity | 79 |
|  | 3.9 | Problems | | 85 |
|  | 3.10 | References | | 90 |

4. Internal Symmetries    91

|  | 4.1 | Selection Rules and Globally Conserved Quantum Numbers | | 91 |
|  | 4.2 | Isospin | | 97 |
|  |  | 4.2.1 | Electromagnetic Interaction and Isospin | 100 |
|  |  | 4.2.2 | Weak Interaction and Isospin | 101 |

4.3    Resonance Production . . . . . . . . . . . . . . . . . 101

     4.3.1    $\Delta$-resonance . . . . . . . . . . . . . . . . . . 103

     4.3.2    Spin of $\Delta$ . . . . . . . . . . . . . . . . . . . . 103

4.4    Charge Conjugation . . . . . . . . . . . . . . . . . . 107

4.5    G-Parity . . . . . . . . . . . . . . . . . . . . . . . 112

4.6    Problems . . . . . . . . . . . . . . . . . . . . . . . 113

4.7    References . . . . . . . . . . . . . . . . . . . . . . . 117

5.   Unitary Groups and SU(3)                   119

5.1    Unitary Groups and SU(3) . . . . . . . . . . . . . . . 119

5.2    Particle Representations in Flavor SU(3) . . . . . . . . 124

     5.2.1    Mesons . . . . . . . . . . . . . . . . . . . . 126

     5.2.2    Baryons . . . . . . . . . . . . . . . . . . . . 128

5.3    U-Spin . . . . . . . . . . . . . . . . . . . . . . . . 132

5.4    Irreducible Representations of SU(3) . . . . . . . . . . 134

     5.4.1    Young's Tableaux . . . . . . . . . . . . . . . 135

5.5    SU(N) . . . . . . . . . . . . . . . . . . . . . . . . 141

5.6    Applications of Flavor SU(3) . . . . . . . . . . . . . 145

     5.6.1    SU(3) Invariant BBP Couplings . . . . . . . . 145

     5.6.2    VPP Coupling . . . . . . . . . . . . . . . . . 146

5.7    Mass Splitting in Flavor SU(3) . . . . . . . . . . . . 148

5.8    Problems . . . . . . . . . . . . . . . . . . . . . . . 154

5.9    References . . . . . . . . . . . . . . . . . . . . . . . 158

6.   SU(6) and Quark Model                     159

6.1    SU(6) . . . . . . . . . . . . . . . . . . . . . . . . 159

     6.1.1    SU(6) Wave Function for Mesons . . . . . . . 160

6.2    Magnetic Moments of Baryons . . . . . . . . . . . . . 164

6.3    Radiative Decays of Vector Mesons . . . . . . . . . . . 170

6.4    Radiative Decays (Complementary Derivation) . . . . . 176

     6.4.1    Mesonic Radiative Decays $V = P + \gamma$ . . . . . . 176

     6.4.2    Baryonic Radiative Decay . . . . . . . . . . . 177

6.5    Problems . . . . . . . . . . . . . . . . . . . . . . . 179

6.6    References . . . . . . . . . . . . . . . . . . . . . . . 180

7.   Color, Gauge Principle and Quantum Chromodynamics     181

7.1    Evidence for Color . . . . . . . . . . . . . . . . . . . 181

7.2    Gauge Principle . . . . . . . . . . . . . . . . . . . . 184

|      | 7.2.1 | Aharanov and Bohm Experiment . . . . . . . . | 186 |
|      | 7.2.2 | Gauge Principle for Relativistic Quantum Mechanics | 188 |
| 7.3  | Non-Abelion Local Gauge Transformations (Yang-Mills) . | 190 |
| 7.4  | Quantum Chromodynamics (QCD) . . . . . . . . . . . | 194 |
|      | 7.4.1 | Conserved Current . . . . . . . . . . . . . . . | 197 |
|      | 7.4.2 | Experimental Determinations of $\alpha_s(q^2)$ and Asymptotic Freedom of QCD . . . . . . . . . . | 199 |
| 7.5  | Hadron Spectroscopy . . . . . . . . . . . . . . . . | 202 |
|      | 7.5.1 | One Gluon Exchange Potential . . . . . . . . . | 202 |
|      | 7.5.2 | Long Range QCD Motivated Potential . . . . . | 205 |
|      | 7.5.3 | Spin-Spin Interaction . . . . . . . . . . . . . | 209 |
| 7.6  | The Mass Spectrum . . . . . . . . . . . . . . . . | 209 |
|      | 7.6.1 | Meson Mass Relations . . . . . . . . . . . . . | 211 |
|      | 7.6.2 | Baryon Mass Spectrum . . . . . . . . . . . . | 213 |
| 7.7  | Problems . . . . . . . . . . . . . . . . . . . . | 217 |
| 7.8  | References . . . . . . . . . . . . . . . . . . . . | 219 |

8. Heavy Flavors                                                         221

| 8.1  | Discovery of Charm . . . . . . . . . . . . . . . . | 221 |
|      | 8.1.1 | Isospin . . . . . . . . . . . . . . . . . . . . | 223 |
|      | 8.1.2 | SU(3) Classification . . . . . . . . . . . . . . | 223 |
| 8.2  | Charm . . . . . . . . . . . . . . . . . . . . . . | 224 |
|      | 8.2.1 | Heavy Mesons . . . . . . . . . . . . . . . . . | 224 |
|      | 8.2.2 | The Fifth Quark Flavor: Bottom Mesons . . . . | 228 |
|      | 8.2.3 | The Sixth Quark Flavor: The Top . . . . . . . | 228 |
| 8.3  | Strong and Radiative Decays of $D^*$ Mesons . . . . . . | 229 |
| 8.4  | Heavy Baryons . . . . . . . . . . . . . . . . . . | 232 |
| 8.5  | Quarkonium . . . . . . . . . . . . . . . . . . . | 233 |
| 8.6  | Leptonic Decay Width of Quarkonium . . . . . . . . | 237 |
| 8.7  | Hadronic Decay Width . . . . . . . . . . . . . . . | 238 |
| 8.8  | Non-Relativistic Treatment of Quarkonium . . . . . . | 240 |
| 8.9  | Observations . . . . . . . . . . . . . . . . . . . | 245 |
| 8.10 | Tetraquark . . . . . . . . . . . . . . . . . . . | 246 |
| 8.11 | Problems . . . . . . . . . . . . . . . . . . . . | 249 |
| 8.12 | References . . . . . . . . . . . . . . . . . . . . | 254 |

9. Heavy Quark Effective Theory                                          255

| 9.1  | Effective Lagrangian . . . . . . . . . . . . . . . | 255 |

9.2   Spin Symmetry of Heavy Quark . . . . . . . . . . . . . . . 259
9.3   Mass Spectroscopy for Hadrons with One Heavy Quark  .  264
9.4   The P-wave Heavy Mesons: Mass Spectroscopy . . . . . .  269
9.5   Decays of P-wave Mesons . . . . . . . . . . . . . . . . . . 275
9.6   Problems . . . . . . . . . . . . . . . . . . . . . . . . . . . 277
9.7   References . . . . . . . . . . . . . . . . . . . . . . . . . . . 277

10.  Weak Interaction                                              279

10.1   $V - A$ Interaction . . . . . . . . . . . . . . . . . . . . . 279
        10.1.1   Helicity of the Neutrino . . . . . . . . . . . . . . 281
10.2   Classification of Weak Processes . . . . . . . . . . . . . 281
        10.2.1   Purely Leptonic Processes . . . . . . . . . . . . . 281
        10.2.2   Semileptonic Processes . . . . . . . . . . . . . . . 283
        10.2.3   Non-Leptonic Processes . . . . . . . . . . . . . . 287
        10.2.4   $\mu$-Decay . . . . . . . . . . . . . . . . . . . . . 288
        10.2.5   Remarks . . . . . . . . . . . . . . . . . . . . . . . 289
        10.2.6   Semi-Leptonic Processes . . . . . . . . . . . . . . 291
10.3   Baryon Decays . . . . . . . . . . . . . . . . . . . . . . . 292
10.4   Pseudoscalar Meson Decays . . . . . . . . . . . . . . . . 296
        10.4.1   Pion Decay . . . . . . . . . . . . . . . . . . . . . 296
        10.4.2   Strangeness Changing Semi-Leptonic Decays . . . 297
10.5   Hadronic Weak Decays . . . . . . . . . . . . . . . . . . . 299
        10.5.1   Non-Leptonic Decays of Hyperons . . . . . . . . 299
        10.5.2   $\Delta I = 1/2$ Rule for Hyperon Decays . . . . . . . 302
        10.5.3   Non-leptonic Hyperon Decays in Non-Relativistic
                   Quark Model . . . . . . . . . . . . . . . . . . . . 304
10.6   Problems . . . . . . . . . . . . . . . . . . . . . . . . . . . 307
10.7   References . . . . . . . . . . . . . . . . . . . . . . . . . . 310

11.  Properties of Weak Hadronic Currents and Chiral Symmetry    311

11.1   Introduction . . . . . . . . . . . . . . . . . . . . . . . . . 311
11.2   Conserved Vector Current Hypothesis (CVC) . . . . . . . 311
11.3   Partially Conserved Axial Vector Current Hypothesis
        (PCAC) . . . . . . . . . . . . . . . . . . . . . . . . . . . 314
11.4   Current Algebra and Chiral Symmetry . . . . . . . . . . . 317
        11.4.1   Explicit Breaking of Chiral Symmetry . . . . . . . 320
        11.4.2   An Application of Chiral Symmetry to Non-
                   Leptonic Decays of Hyperons . . . . . . . . . . . . 323

11.5    Axial Anomaly . . . . . . . . . . . . . . . . . . . . . . . . . . 325
11.6    QCD Sum Rules . . . . . . . . . . . . . . . . . . . . . . . . . 327
11.7    Problems . . . . . . . . . . . . . . . . . . . . . . . . . . . . . 328
11.8    References . . . . . . . . . . . . . . . . . . . . . . . . . . . . 329

12.   Neutrino      331

12.1    Introduction . . . . . . . . . . . . . . . . . . . . . . . . . . . 331
12.2    Intrinsic Properties of Neutrinos . . . . . . . . . . . . . . 332
12.3    Mass . . . . . . . . . . . . . . . . . . . . . . . . . . . . . . . . 332
     12.3.1   Constraints on Neutrino Mass . . . . . . . . . . 333
     12.3.2   Dirac and Majorana Masses . . . . . . . . . . . 337
     12.3.3   Fermion Masses in the Standard Model (SM) and
              See-saw Mechanism . . . . . . . . . . . . . . . . 339
12.4    Neutrino Oscillations . . . . . . . . . . . . . . . . . . . . . . 343
     12.4.1   Mikheyev-Smirnov-Wolfenstein Effect . . . . . . 345
     12.4.2   Evolution of Flavor Eigenstates in Matter . . . . 349
12.5    Evidence for Neutrino Oscillations . . . . . . . . . . . . . 351
     12.5.1   Disappearance Experiments . . . . . . . . . . . 351
     12.5.2   Appearance Experiments . . . . . . . . . . . . . 351
12.6    Neutrino Mass Models and Mixing Matrix and Symmetries 355
12.7    Neutrino Magnetic Moment . . . . . . . . . . . . . . . . . 360
12.8    Problems . . . . . . . . . . . . . . . . . . . . . . . . . . . . . 362
12.9    References . . . . . . . . . . . . . . . . . . . . . . . . . . . . 363

13.   Electroweak Unification      365

13.1    Introduction . . . . . . . . . . . . . . . . . . . . . . . . . . . 365
13.2    Spontaneous Symmetry Breaking and Higgs Mechanism . 366
     13.2.1   Higgs Mechanism . . . . . . . . . . . . . . . . . 368
     13.2.2   Gauge Symmetry Breaking for Chiral $U_1 \otimes U_2$ Group 369
13.3    Renormalizability . . . . . . . . . . . . . . . . . . . . . . . . 372
13.4    Electroweak Unification . . . . . . . . . . . . . . . . . . . . 374
     13.4.1   Experimental Consequences of the Electroweak
              Unification . . . . . . . . . . . . . . . . . . . . . 381
     13.4.2   Need for Radiative Corrections . . . . . . . . . . 382
     13.4.3   Experiments which Determine $\sin^2 \theta_W$ . . . . . 387
13.5    Decay Widths of $W$ and $Z$ Bosons . . . . . . . . . . . . 389
13.6    Tests of Yang-Mills Character of Gauge Bosons . . . . . 395
13.7    Higgs Boson Mass . . . . . . . . . . . . . . . . . . . . . . . 399

13.8 Upper Bound . . . . . . . . . . . . . . . . . . . . . . . 399
    13.8.1 Unitarity . . . . . . . . . . . . . . . . . . . . . 399
    13.8.2 Finiteness of Couplings . . . . . . . . . . . . . . 400
13.9 Standard Model, Higgs Boson Searches, Production at Decays . . . . . . . . . . . . . . . . . . . . . . . . . . . . 401
    13.9.1 LEP-2 . . . . . . . . . . . . . . . . . . . . . . . 401
    13.9.2 LHC and Tevatron . . . . . . . . . . . . . . . . 402
13.10 Two Higgs Doublet Model (2HDM) . . . . . . . . . . . . 406
13.11 GIM Mechanism . . . . . . . . . . . . . . . . . . . . . . 411
13.12 Cabibbo-Kobayashi-Maskawa Matrix . . . . . . . . . . . 414
13.13 Axial Anomaly . . . . . . . . . . . . . . . . . . . . . . . 416
13.14 Problems . . . . . . . . . . . . . . . . . . . . . . . . . . 421
13.15 References . . . . . . . . . . . . . . . . . . . . . . . . . . 423

14. Deep Inelastic Scattering     425

14.1 Introduction . . . . . . . . . . . . . . . . . . . . . . . . 425
14.2 Deep-Inelastic Lepton-Nucleon Scattering . . . . . . . . 427
14.3 Parton Model . . . . . . . . . . . . . . . . . . . . . . . . 431
14.4 Deep Inelastic Neutrino-Nucleon Scattering . . . . . . . 436
14.5 Sum Rules . . . . . . . . . . . . . . . . . . . . . . . . . 439
14.6 Deep-Inelastic Scattering Involving Neutral Weak Currents 446
14.7 Problems . . . . . . . . . . . . . . . . . . . . . . . . . . 447
14.8 References . . . . . . . . . . . . . . . . . . . . . . . . . . 450

15. Weak Decays of Heavy Flavors     451

15.1 Leptonic Decays of $\tau$ Lepton . . . . . . . . . . . . . . . 451
15.2 Semi-Hadronic Decays of $\tau$ Lepton . . . . . . . . . . . . 453
    15.2.1 Special Cases . . . . . . . . . . . . . . . . . . . . 454
15.3 Weak Decays of Heavy Flavors . . . . . . . . . . . . . . 457
    15.3.1 Leptonic Decays of $D$ and $B$ Mesons . . . . . . 458
    15.3.2 Semileptonic Decays of $D$ and $B$ Mesons . . . . . 459
    15.3.3 (Exclusive) Semileptonic Decays of $D$ and $B$ Mesons 464
    15.3.4 Weak Hadronic Decays of $B$ Mesons . . . . . . . . 471
    15.3.5 Inclusive Hadronic $B$ Decays . . . . . . . . . . . 476
    15.3.6 Radiative Decays of $B_q$ Mesons . . . . . . . . . . 478
15.4 Inclusive Hadronic Decays of $D$-Mesons . . . . . . . . . 479
    15.4.1 Scattering and Annihilation Diagrams . . . . . . 480
15.5 Problems . . . . . . . . . . . . . . . . . . . . . . . . . . 484

15.6    References . . . . . . . . . . . . . . . . . . . . . . .    487

16.   Particle Mixing and $CP$-Violation                                  489

16.1    Introduction . . . . . . . . . . . . . . . . . . . . . . .    489
16.2    $CPT$ and $CP$ Invariance . . . . . . . . . . . . . . . . .    492
16.3    CP-Violation in the Standard Model . . . . . . . . . . .    494
16.4    Particle Mixing . . . . . . . . . . . . . . . . . . . . .    497
16.5    $K^0 - \bar{K}^0$ Complex and $CP$-Violation in $K$-Decay . . . . .    504
16.6    $B^0 - \bar{B}^0$ Complex . . . . . . . . . . . . . . . . . . . .    511
16.7    $CP$-Violation in $B$-Decays . . . . . . . . . . . . . . . . .    515
16.8    $CP$-Violation in Hadronic Weak Decays of Baryons . . . .    518
16.9    Problems . . . . . . . . . . . . . . . . . . . . . . . . .    522
16.10   References . . . . . . . . . . . . . . . . . . . . . . . .    523

17.   Grand Unification, Supersymmetry and Strings                       525

17.1    Grand Unification . . . . . . . . . . . . . . . . . . . .    525
        17.1.1   $q^2$ Evolution of Gauge Coupling Constants and the
                 Grand Unification Mass Scale . . . . . . . . . .    529
        17.1.2   General Consequences of GUTS . . . . . . . . . .    531
17.2    Poincaré Group and Supersymmetry . . . . . . . . . . . .    534
        17.2.1   Introduction . . . . . . . . . . . . . . . . . .    534
        17.2.2   Poincaré Group . . . . . . . . . . . . . . . . .    537
        17.2.3   Two-Component Weyl Spinors . . . . . . . . . . .    539
        17.2.4   Spinor Algebra, Supersymmetry . . . . . . . . .    540
        17.2.5   Supersymmetric Multiplets . . . . . . . . . . .    542
17.3    Supersymmetry and Strings . . . . . . . . . . . . . . . .    544
        17.3.1   Introduction . . . . . . . . . . . . . . . . . .    544
        17.3.2   Supersymmetry . . . . . . . . . . . . . . . . . .    545
17.4    String Theory and Duality . . . . . . . . . . . . . . . .    548
        17.4.1   M-theory . . . . . . . . . . . . . . . . . . . .    550
17.5    Some Important Results . . . . . . . . . . . . . . . . .    552
17.6    Conclusions . . . . . . . . . . . . . . . . . . . . . . .    552
17.7    Problems . . . . . . . . . . . . . . . . . . . . . . . .    552
17.8    References . . . . . . . . . . . . . . . . . . . . . . . .    554

18.   Cosmology and Astroparticle Physics                                557

18.1    Cosmological Principle and Expansion of the Universe . .    557
18.2    The Standard Model of Cosmology . . . . . . . . . . . . .    559

18.3    Cosmological Parameters and the Standard Model Solutions 562

18.4    Accelerating Universe and Dark Energy . . . . . . . . . . . 566

    18.4.1    Evidence from Supernovae . . . . . . . . . . . . . 567

    18.4.2    Evidence from CMB Data . . . . . . . . . . . . . 568

    18.4.3    Quintessence . . . . . . . . . . . . . . . . . . . . 571

    18.4.4    Modified Gravity . . . . . . . . . . . . . . . . . . 573

18.5    Hot Big Bang: Thermal History of the Universe . . . . . . 574

    18.5.1    Thermal Equilibrium . . . . . . . . . . . . . . . . 574

    18.5.2    The Radiation Era . . . . . . . . . . . . . . . . . 576

18.6    Freeze Out . . . . . . . . . . . . . . . . . . . . . . . . . . 581

18.7    Limit on Neutrino Mass . . . . . . . . . . . . . . . . . . . 584

18.8    Primordial Nucleosynthesis . . . . . . . . . . . . . . . . . 585

18.9    Inflation . . . . . . . . . . . . . . . . . . . . . . . . . . . 588

    18.9.1    Horizon Problem . . . . . . . . . . . . . . . . . . 588

    18.9.2    Flatness Problem . . . . . . . . . . . . . . . . . . 590

    18.9.3    Realization of Inflation . . . . . . . . . . . . . . . 591

    18.9.4    Slow-roll Inflation . . . . . . . . . . . . . . . . . 593

18.10   Baryogenesis . . . . . . . . . . . . . . . . . . . . . . . . . 595

    18.10.1   Sakharov's Conditions . . . . . . . . . . . . . . . 597

    18.10.2   Various Scenarios for Baryogenesis . . . . . . . . 598

    18.10.3   Leptogenesis . . . . . . . . . . . . . . . . . . . . 601

18.11   Problems . . . . . . . . . . . . . . . . . . . . . . . . . . . 606

18.12   References . . . . . . . . . . . . . . . . . . . . . . . . . . 607

Appendix A   Quantum Field Theory                                609

A.1    Spin 0 Field . . . . . . . . . . . . . . . . . . . . . . . . . 609

A.2    Spin 1/2 Particle . . . . . . . . . . . . . . . . . . . . . . . 611

    A.2.1    Pauli Representation of $\gamma$ Matrices . . . . . . . . 612

    A.2.2    Weyl Representation of $\gamma$ Matrices . . . . . . . . 613

A.3    Trace of $\gamma$ Matrices . . . . . . . . . . . . . . . . . . . . . 616

A.4    Spin 1 Field . . . . . . . . . . . . . . . . . . . . . . . . . 618

A.5    Massive Spin 1 Particle . . . . . . . . . . . . . . . . . . . 619

A.6    Feynman Rules for S-Matrix in Momentum Space . . . . . 620

A.7    Application of Feynman Rules . . . . . . . . . . . . . . . . 621

    A.7.1    $e^+e^- \rightarrow$ Hadrons . . . . . . . . . . . . . . . . 624

    A.7.2    Electron Scattering and Structureless Spin 1/2 Target 625

A.8    Discrete Symmetries . . . . . . . . . . . . . . . . . . . . . 628

    A.8.1    Charge Conjugation . . . . . . . . . . . . . . . . 628

    A.8.2    Space Reflection . . . . . . . . . . . . . . . . . . 631

       A.8.3   Time Reversal . . . . . . . . . . . . . . . . . . . . .  632

  A.9   Problems . . . . . . . . . . . . . . . . . . . . . . . . . . . .  633

Appendix B   Renormalization Group and Running Coupling Constant  639

  B.1   Feynman Rules for Quantum Chromodynamics . . . . . .  639

  B.2   Renormalization Group, Coupling Constant and Asymptotic Freedom . . . . . . . . . . . . . . . . . . . . . . . . .  640

  B.3   Running Coupling Constant in Quantum Electrodynamics (QED) . . . . . . . . . . . . . . . . . . . . . . . . . . . . .  645

  B.4   Running Coupling Constant for SU(2) Gauge Group . . .  646

  B.5   Renormalization Group and High $Q^2$ Behavior of Green's Function . . . . . . . . . . . . . . . . . . . . . . . . . . . . .  647

       B.5.1   Gluon Propagator . . . . . . . . . . . . . . . . .  649

       B.5.2   Fermion Propagator . . . . . . . . . . . . . . . .  650

  B.6   References for Appendices . . . . . . . . . . . . . . . . . .  652

*Index*                                                     653

# Chapter 1

# Introduction

## 1.1 Fundamental Forces

Particle physics is concerned with the fundamental constituents of matter and the fundamental "forces" through which the fundamental constituents interact among themselves.

Until about 1932, only four particles, namely the proton $(p)$, the neutron $(n)$, the electron $(e)$ and the neutrino $(\nu)$ were regarded as the ultimate constituents of matter. Of these four particles, two, the proton and the electron are electrically charged. The other two are electrically neutral. The neutron and proton form atomic nuclei, the electron and nucleus form atoms while the neutrino comes out in radioactivity, i.e. the neutron decays into a proton, an electron and a neutrino. Each of these particles, called a fermion, spins and exists in two spin (or polarization) states called left-handed (i.e. appears to be spinning clockwise as viewed by an observer that it is approaching) and right-handed (i.e. spinning anti-clockwise) spin states. One may add a fifth particle, the photon to this list. The photon is a quantum of electromagnetic field. It is a boson and carries spin 1, is electrically neutral and has zero mass, due to which it has only two spin directions or it has only transverse polarization. It is a mediator of electromagnetic force. A general feature of quantum field theory is that each particle has its own antiparticle with opposite charge and magnetic moment, but with same mass and spin. Accordingly we have four antiparticles viz., the antiproton $(\overline{p})$, the positron $(e^+)$, the antineutron $(\overline{n})$ and antineutrino $(\overline{\nu})$.

The four particles experience four types of forces:

### 1.1.1   *The Gravitational Force*

This is the force of attraction between two particles and is proportional to their masses and inversely proportional to the square of distance between them. It controls the motion of planets and galaxies and also governs the law of falling bodies. It is a long distance force. It determines the overall structure of the Universe.

Its strength is characterized by the Newton's constant $G_N = 1.3 \times 10^{-54}$ $(\text{GeV}/m)^{-1}$. The gravitational energy between two protons of mass $\sim (1\text{GeV}/c^2)$ at $r = 10^{-15}$m is given by Newton's law

$$V = G_N \frac{m_p^2}{r} \approx 10^{-39} \text{ GeV} = 10^{-30} \text{ eV}. \tag{1.1}$$

Hence on microscopic scale, the gravitational energy is negligibly small as compared with electromagnetic energy. We note that

$$\frac{G_N}{\hbar c} \approx 6.71 \times 10^{-39} \left( \frac{\text{GeV}}{c^2} \right)^{-2} \tag{1.2}$$

and

$$\sqrt{\frac{\hbar c}{G_N}} \approx 10^{19} \frac{\text{GeV}}{c^2} = M_P \tag{1.3a}$$

$$\text{Planck length} = \frac{c\hbar}{M_P c^2} \approx 2 \times 10^{-33} \text{ cm} \tag{1.3b}$$

where $M_P$ is called the Planck mass with the associated Planck length. It is clear from Eqs. (1.3) that gravitational interaction becomes significant at Planck mass or Planck length. Assuming that this interaction is of the same order as the electromagnetic interaction (see below) ($\alpha = e_M^2/\hbar c = 1/137$, $e_M^2 = e^2/4\pi\varepsilon_0$) at Planck mass $M_P$, we conclude that the effective gravitational interaction at 1 GeV is given by

$$\alpha_{G_N} \approx \frac{(1\text{GeV})^2}{M_P^2} \alpha \approx 10^{-38} \alpha \approx 10^{-40}. \tag{1.4}$$

In particle physics, the gravitational interaction may be neglected at the present available energies.

### 1.1.2   *The Weak Nuclear Force*

It is responsible for radioactivity, e.g.

$$n \rightarrow p + e^- + \bar{\nu}_e$$

$$O^{14} \rightarrow N^{14} + e^+ + \nu_e$$

The latter process has half life of 71.4 sec. From the half life we can determine its strength which is given by the Fermi coupling constant [see Chap. 2]

$$\frac{G_F}{(\hbar c)^3} \approx 10^{-5} \text{ GeV}^{-2} \tag{1.5}$$

We note that

$$\frac{(\hbar c)^{3/2}}{\sqrt{G_F}} = 300 \text{ GeV} \tag{1.6a}$$

$$(\hbar c)\sqrt{\frac{G_F}{(\hbar c)^3}} \simeq 0.7 \times 10^{-16} \text{ cm} \tag{1.6b}$$

Eq. (1.6a) gives the energy scale at which the weak interaction becomes significant i.e. of the same order as the electromagnetic interaction. At an energy scale of 1 GeV,

$$\alpha_W = \frac{(1\text{GeV})^2}{(300\text{GeV})^2}\alpha \approx 10^{-5}\alpha. \tag{1.7}$$

It is clear from Eq. (1.6b) that weak force is a short range force effective over a distance of order of $10^{-16}$ cm.

### 1.1.3 *The Electromagnetic Force*

It acts between two electrically charged particles e.g. a negatively charged electron and a positively charged proton attract each other with a force which is proportional to their electric charges and inversely proportional to the square of distance between them. But according to the concept of electromagnetic field introduced by Faraday and Maxwell, the presence of a charged particle produces an electric field and when moving it produces a magnetic field. This field manifests itself in a force between charged particles. It is responsible for the binding of atoms. The interatomic and intermolecular force are all electrical in nature and mainly governs all known phenomena on earth, including life. This force also manifests itself through electromagnetic radiation in the form of light, radio waves and X-rays.

The electromagnetic force is a long range force and its strength is determined by the dimensionless number $\alpha = \frac{1}{137}$. This is because the electromagnetic potential energy in MKS units is given by

$$V = -\frac{e^2}{4\pi\epsilon_0}\frac{1}{r} = -(\frac{e^2}{4\pi\epsilon_0})\frac{1}{\hbar c}\frac{\hbar c}{r}$$

$$= -\alpha\frac{\hbar c}{r} \tag{1.8}$$

where

$$\alpha = \frac{1}{4\pi\epsilon_0}\left(\frac{e^2}{\hbar c}\right) = \frac{1}{137} \tag{1.9}$$

and is called the fine structure constant.

The potential energy between an electron and a proton at a distance $r = 10^{-10}$ m [dimensions of an atom] is given by $V = -14$ eV.

In quantum mechanics, the binding energy between electron and proton due to electromagnetic interaction is given by Bohr's formula

$$|E_1| = \frac{1}{2}\alpha^2(\mu c^2)$$

where $\mu$ is the reduced mass of the system. For hydrogen atom $\mu \approx m_e = 0.51 \frac{\text{MeV}}{c^2}$, which gives $|E_1| \approx 14$ eV.

For a proton ($p$) and antiproton ($\bar{p}$) hypothetical atom ($\mu \simeq \frac{m_p}{2} \simeq 1000\ m_e$) the binding energy provided by the electromagnetic potential is $\left|E_1^{p\bar{p}}\right| \simeq 14$ keV.

### 1.1.4    *The Strong Nuclear Force*

The strong nuclear force is responsible for the binding of protons and neutrons in a nucleus. We have seen that the electromagnetic binding energy for the $p\bar{p}$ atom is of the order of 14 keV, but the binding energy of deuteron (bound n-p system) is about 2 MeV. Thus the strong nuclear force is about 100 times the electromagnetic force. It is a short range force effective over the nuclear dimension of the order of $10^{-13}$ cm.

## 1.2    Relative Strength of Four Fundamental Forces

We conclude from the above discussion that the relative strengths of the four forces are in the order of

$$10^{-40} : 10^{-7} : 10^{-2} : 1 \tag{1.10}$$

The experimental results on the scattering of electron on nuclei can be explained by invoking electromagnetic interaction only. In fact the scattering of $\gamma$-rays on proton at low energy is given by the Thomson formula:

$$\sigma_\gamma = \frac{8\pi}{3}\alpha^2\left(\frac{1}{m_p}\right)^2 \simeq 4\pi\alpha^2\left(\frac{1}{m_p}\right)^2 \simeq 10^{-31}\text{cm}^2. \tag{1.11}$$

The neutrino participates in weak interactions only as reflected by the extreme smallness of the scattering cross-section of neutrino on proton, viz.

$\bar{\nu}_e p \to e^+ n$, which is given by $\sigma_W \simeq 10^{-43} \text{cm}^2$. Comparing the above cross-sections with the one for nucleon-nucleon scattering, which is of the order $\sigma_N \simeq 10^{-24} \text{cm}^2$, we see that the electron and neutrino do not experience strong interaction. Thus we can divide matter into two broad classes: leptons, e.g. $\nu_e$, $e$ and hadrons, e.g. $p$, $n$.

## 1.3  Range of the Three Basic Forces

We now briefly and qualitatively discuss the ranges of the three basic forces. Due to quantum fluctuations, an electron can emit a photon and reabsorb it. Electromagnetic force is mediated by the exchange of photon as depicted in Fig. 1.1.

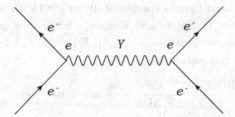

Fig. 1.1   Electromagnetic force mediated by a photon.

Such a photon can exist only for a time

$$\Delta t \sim \frac{\hbar}{\Delta E} = \frac{1}{\omega} \qquad (1.12a)$$

where $\Delta E$ is the energy of the photon. Since the unobserved photon exists for a time $< \frac{1}{\omega}$, it can travel at most

$$R = \frac{c}{\omega}. \qquad (1.12b)$$

Now $\omega$ can be arbitrarily small and therefore $R$ can be arbitrarily large, i.e. the distance over which a photon can transport electromagnetic force is arbitrarily large, i.e. electromagnetic force has infinite range. This is expected from the Coulomb potential $e_M^2/r$.

If we assume that weak interaction is mediated by a vector boson $W$ in analogy with electromagnetic interaction (Fig. 1.2), then since weak interaction is of short range, $W$ must be massive.

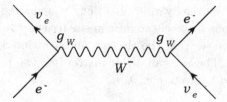

Fig. 1.2   Weak interaction mediated by a vector boson $W$.

The maximum distance to which the virtual W-boson is allowed to travel by the uncertainty relation is

$$R_W \approx \frac{\hbar c}{m_W \, c^2} \tag{1.13a}$$

The W boson has been found experimentally in 1983 with a mass $m_W \approx 80$ GeV/$c^2$ as predicted by Salam and Weinberg when they unified weak and electromagnetic interactions (see Sec. 1.6). Equation (1.13a) then gives the range of the weak interaction as

$$R_W \approx \frac{197 \times 10^{-13} \text{ MeV - cm}}{80 \times 10^3 \text{ MeV}} \approx 2 \times 10^{-16} \text{ cm.} \tag{1.13b}$$

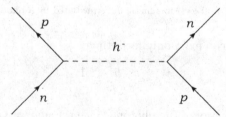

Fig. 1.3   Strong nuclear force mediated by a particle of mass $m_h$.

If the strong nuclear force is mediated by a particle of mass $m_h$, as shown in Fig. 1.3, then its range is given by

$$R_h \approx \frac{\hbar c}{m_h \, c^2} \tag{1.14}$$

Since nuclear force has a range of $10^{-13}$ cm, $m_h \approx 100$ MeV/$c^2$. Yukawa in 1935, predicted the existence of pion by a similar argument. A particle of this mass was discovered in 1938, but it turned out that it was not the Yukawa particle, the pion; it did not interact strongly with matter and therefore is not responsible for strong nucleon force. It was actually the muon, while the pion was discovered in 1947 in the decay

$$\pi^- \rightarrow \mu^- + \nu_\mu,$$

where $\nu_\mu$ is the neutrino corresponding to muon. The mass of $m_\pi$ was found to be 140 MeV/c$^2$. Thus

$$R_h \approx 1.4 \times 10^{-13} \text{cm} \approx \sqrt{2} f, \qquad (1.15)$$

where $f$ is called the Fermi and is equal to $10^{-13}$ cm.

## 1.4  Classification of Matter

The electron and neutrino are just two members of a family called leptons (leptons do not experience strong nuclear force) of which six are presently known to exist, listed below in the table. Similarly, the neutron and proton are members of much larger family called hadrons (hadrons experience strong nuclear force). Hadrons exist in two classes, baryons and mesons; the former carry half integral spin and a quantum number called baryon number while the latter carry zero or integral spin and no baryon number.

| Leptons | Mass[12] | Electric Charge | Life Time[12] |
|---------|----------|-----------------|---------------|
| $\nu_e,\ e^-$ | $m_{\nu_e} < 2$ eV $m_e \approx 0.51$ MeV | $0, -1$ | $\nu_e$ Stable $\tau_e > 4.6 \times 10^{26}$ yrs |
| $\nu_\mu,\ \mu^-$ | $m_{\nu_\mu} < 0.19$ MeV $m_\mu \approx 105.6$ MeV | $0, -1$ | $\nu_\mu$ Stable $\tau_\mu = 2.197 \times 10^{-6}$ s |
| $\nu_\tau,\ \tau^-$ | $m_{\nu_\tau} < 18.2$ MeV $m_\tau \approx 1777$ MeV | $0, -1$ | $\nu_\tau$ Stable $\tau_\tau = 290.6 \times 10^{-15}$ s |

It was found experimentally that the proton or the neutron has a structure; they are not the elementary constituents of matter. The mass spectrum, production and decay characteristics of hadrons can be understood in a much simpler picture if one assumes that the baryons are made up of three constituents called quarks and mesons are made up of quark and antiquark.

The quark carries fractional charges 2/3 or 1/3. They have spin one half and carry different flavors to distinguish them. At present six flavors, called up, down, charm, strange, top and bottom, are known to exist. They are listed in the table below. Quarks play the same role in hadron spectroscopy as neutrons and protons in nuclear spectroscopy and electrons and nuclei in atomic spectroscopy.

| Quark type (Flavor) | Electric charge | Mass [effective mass or constituent mass in a hadron] |
|---|---|---|
| $(u, d)$ | $(2/3, -1/3)$ | 0.33 GeV |
| $(c, s)$ | $(2/3, -1/3)$ | (1.5 GeV, 0.5 GeV) |
| $(t, b)$ | $(2/3, -1/3)$ | (172 GeV, 4.5 GeV) |

Masses of $u$, $d$, $s$ are so-called "constituent" which are effective masses in a hadron; they are not directly measurable but are theoretical estimates. To sum up, according to present view, the six quarks and six leptons form the fundamental constituents of matter.

There are three generations of matter as depicted in the two tables above. The first generation is relevant for the visible universe and the life on earth as the proton $(uud)$, neutron $(udd)$ are composite of up and down quarks. The second and third generations do exist and are produced either in laboratory or in cosmic rays by a collision of particles of first generation.

The table below summarizes the known elementary interactions exhibited by different elementary fermions.

| Interactions | Relativistic length | Leptons $\nu_i, l_i$ | Quarks $u_i, d_i$ |
|---|---|---|---|
| Strong | 1 | | Y |
| Electromagnetic | $10^{-2}$ | N,Y | Y |
| Weak | $10^{-5}$ | Y,Y | Y |
| Gravitational | $10^{-40}$ | Y,Y | Y |

The first generation of quarks $u$ and $d$ form an isodoublet, i.e. they are assigned isospin $I = 1/2$ and $I_3 = \pm 1/2$, [that is why they are called up and down quarks]. The second and third generation of quarks are assigned new quantum numbers as follows: s-quark, strangeness $S = -1$, c-quark, charm, $C = +1$, b-quark, bottomness $B = -1$, t-quark, topness $T = +1$. The second and third generation of quarks account for the strange hadrons, charmed hadrons and $B$-hadrons created in the laboratory in high energy collisions between hadrons of first generation. They are always created in pair, so that final state has $S = 0, C = 0, B = 0$ and $T = 0$ that is to say these quantum numbers are conserved in electromagnetic and strong interactions.

In high energy atomic collisions, we can split an atom into its constituents-atomic nucleus and electrons. In high energy nucleus - nucleus collisions, we can split a nucleus into its constituents viz. neutrons and protons. But in high energy hadron-hadron collisions, a hadron is not split into its constituents viz., quarks. A hadron-hadron collision results not into free quarks but into hadrons. This leads us to the hypothesis of quark confinement, i.e. quarks are always confined in a hadron. The quark-quark force, which keeps the quarks confined in a hadron, is a fundamental strong force on the same level as electromagnetic and weak nuclear forces which are the other two fundamental forces in nature. Its strength is characterized by a dimensionless coupling constant $\alpha_s = g_s^2/4\pi \approx 0.5$ at present energies. It is actually energy dependent. The strong nuclear force between protons and neutrons should then be a complicated interaction derivable from this basic quark-quark force. As for example, the fundamental force for an atomic system is electromagnetic force, the interatomic and intermolecular forces are derivable from the basic electromagnetic force.

## 1.5   Strong Color Charges

We have seen that the quarks form hadrons; the baryons and mesons in the ground state are composites of $(qqq)_{L=0}$ and $(q\bar{q})_{L=0}$. Quarks and (antiquarks) are spin $1/2$ fermions. Now $q$ and $\bar{q}$ spins may be combined to form a total spin S, which is 0 or 1. Total spin for $qqq$ system is $3/2$ or $1/2$. Further as $q$ and $\bar{q}$ have opposite intrinsic parities, the parity of the $q\bar{q}$ system is $P = (-1)(-1)^L = -1$ for the ground state. Thus we have for the ground states [see Chap. 6]:

| Mesons | Baryons |
|---|---|
| $(q\bar{q})_{L=0}$ | $(qqq)_{L=0}$ |
| $S = 0, 1$ | $S = 1/2, 3/2$ |
| $J^P = 0^-, 1^-$ | $J^P = 1/2^+, 3/2^+$ |

Some examples are listed in Table 1.1.

Table 1.1   Some examples of mesons and baryons

| Mesons | Baryons |
|---|---|
| $\pi$, $\rho$ | $p$, $\Delta$ |
| $\pi^+ = (u\bar{d}) \frac{1}{\sqrt{2}} (\uparrow\downarrow - \downarrow\uparrow)$ | $p\,(S = 1/2, S_Z = 1/2)$ <br> $= (uud)\,(\uparrow\uparrow\downarrow)$ |
| $\rho^+\,(S = 1, S_Z = 0):$ <br> $(u\bar{d}) \frac{1}{\sqrt{2}} (\uparrow\downarrow + \downarrow\uparrow)$ <br> $\rho^+\,(S = 1, S_Z = 1):$ <br> $(u\bar{d})\,(\uparrow\uparrow)$ | $\Delta^{++}\,(S = 3/2, S_Z = 3/2)$ <br> $= (uuu)\,(\uparrow\uparrow\uparrow)$ |

There is a difficulty with the above picture; consider, for example the state, $|\Delta^{++}(S_Z = 3/2)\rangle \sim |u^\uparrow u^\uparrow u^\uparrow\rangle$. This state is symmetric in quark flavor and spin indices ($\uparrow$). The space part of the wave function is also symmetric ($L = 0$). Thus, the above state being totally symmetric violates the Pauli principle for fermions. Therefore, another degree of freedom (called color) must be introduced to distinguish the otherwise identical quarks: each quark flavor carries three different strong color charges, red(r), yellow(y) and blue(b), i.e.

$$q = q_a \qquad\qquad a = r, y, b$$

Leptons do not carry color and that is why they do not take part in strong interactions. Including the color, we write, e.g.

$$|\Delta^{++}(S_Z = 3/2)\rangle = \frac{1}{\sqrt{6}} \sum \varepsilon_{abc} \left| u_a^\uparrow u_b^\uparrow u_c^\uparrow \right\rangle,$$

so that the wave function is now antisymmetric in color indices and satisfies the Pauli principle. Other examples, as far as quark content is concerned, are

$$|p\rangle = \frac{1}{\sqrt{6}} \sum \varepsilon_{abc} |u_a u_b d_c\rangle,$$

$$|\pi^+\rangle = \frac{1}{\sqrt{3}} \sum_a |u_a \bar{d}_a\rangle,$$

i.e. these states are color singlets. In fact, all known hadrons are color singlets. Thus, the color quantum number is hidden. This is the postulate of color confinement mentioned earlier and explains the non-existence of free quark ($q$) or such systems as ($qq$), ($q\bar{q}q$) and ($qqqq$). Actually nature has also assigned a more fundamental role to color charges as we briefly discuss below.

## 1.6   Fundamental Role of "Charges" in the Unification of Forces

First thing to note is that the electromagnetic and the strong nuclear forces are each characterized by a dimensionless coupling constant and thus to achieve unification there has to be a "hidden" dimensionless coupling constant associated with the weak nuclear force which is related to the "observed" Fermi coupling constant by a mass scale. That this is so will be clear shortly. Secondly we know that the electromagnetic force is a gauge force describable in terms of electric charge and the current associated with it. This force is mediated by electromagnetic radiation field whose quanta are spin 1 photons. This is generalized: All fundamental forces are gauge forces describable in terms of "charges" and their currents as summarized in Table 1.2.

Note that the coupling constants $\alpha$, $\alpha_2$, $\alpha_s$ in Table 1.2 are dimensionless but they are energy dependent due to quantum effects, a fact which is used in the unification of the forces. Note also that $Q_3$ is not identical with $Q_{em}$; thus the unification of electromagnetic and weak nuclear forces needs another charge, call it $Q_B$ with an associated mediator, call it $B$, which does not change flavor like photon:

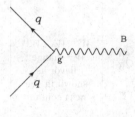

$$\alpha' = \frac{g'^2}{4\pi}$$

Then the photon $\gamma$ associated with the electric charge $Q_{em}$ is a linear combination of the mediators $B$ and $W_3$ bosons, associated with the charges $Q_B$ and $Q_3$ respectively,

$$\gamma = \sin\theta_W W_3 + \cos\theta_W B, \qquad (1.16a)$$

while the second orthogonal combination

$$Z = \cos\theta_W W_3 - \sin\theta_W B, \qquad (1.16b)$$

is associated with a new charge $Q_Z$. $Z$ is the mediator of a new interaction, called neutral weak interaction. The weak mixing angle $\theta_W$ is a fundamental parameter of the theory and in terms of it

$$Q_Z = Q_3 - \sin^2\theta_W Q_{em}. \qquad (1.17)$$

The weak color charges $Q_W, Q_{\bar{W}}$ and $Q_B$ generate the local group $SU_L(2) \times U(1)$ where the subscript $L$ on the weak isospin group SU(2) indicates that we deal with chiral fermions that is to say that the left-handed fermions [i.e. those which appear to be spinning clockwise as viewed by an observer that they are approaching] are doublets under $SU_L(2)$ [required

Table 1.2

| Force | Charges | Mediators of force: Spin 1 gauge particles | Coupling between basic fermions and mediators |
|---|---|---|---|
| Electro-magnetic | $Q_{em}$ | Photon $(\gamma)$ | $\alpha = \frac{e^2}{4\pi}$ |
| Weak Nuclear | $Q_W, Q_{\bar{W}}$ $[Q_W, Q_{\bar{W}}]$ $= Q_3$ $\neq Q_{em}$ | $W^+, W^-, W^0$ $W^{\pm}$ change flavor as shown in the next column | $\alpha_2 = \frac{g_2^2}{4\pi}$ |
| Strong | 3 color charges | 8 color carrying gluons: $G_{ab}$ $a, b$ $= 1, 2, 3$ | $\alpha_s = \frac{g_s^2}{4\pi}$ |

by parity violation in weak interactions] while the right-handed fermions (spinning anticlockwise) are singlets as indicated below:

| | $\begin{pmatrix} \nu_e \\ e^- \end{pmatrix}_L$ | $\begin{pmatrix} u \\ d \end{pmatrix}_L$ | $e_R$ | $u_R$ | $d_R$ |
|---|---|---|---|---|---|
| $I_{3L}$ | $\left(\frac{1}{2}, -\frac{1}{2}\right)$ | $\left(\frac{1}{2}, -\frac{1}{2}\right)$ | | | |
| $Y_W$ | $-1$ | $-1/3$ | $-2$ | $4/3$ | $-2/3$ |

We have [$I_{3L}$ is the same as $Q_3$ and $\frac{1}{2}Y_W$ is identical with $Q_B$]

$$Q_{em} = I_{3L} + \frac{1}{2}Y_W, \tag{1.18a}$$

giving

$$\frac{1}{e^2} = \frac{1}{g_2^2} + \frac{1}{g'^2}, \tag{1.18b}$$

so that we have the unification conditions

$$\sin^2\theta_W = \frac{e^2}{g_2^2}, \qquad \cos^2\theta_W = \frac{e^2}{g'^2}. \tag{1.18c}$$

Unlike photon, which is massless, the weak vector bosons $W^+$, $W^-$ and $Z^0$ must be massive since we know that weak interactions are of short range. This is achieved by spontaneous symmetry breaking (SSB)[see Chap. 13]. For this purpose it is necessary to introduce a self-interacting complex scalar field

$$\phi = \begin{pmatrix} \phi^+ \\ \phi^0 \end{pmatrix},$$

which is a doublet under $SU_L(2)$ and has $Y_W = 1$. This is the so-called Higgs field which also interacts with the chiral fermions introduced earlier as well as with gauge vector bosons, $W^\pm$, $W_3$ and $B$. The scalar field $\phi$ develops a non-zero vacuum expectation value:

$$\langle\phi\rangle = \langle 0|\phi|0\rangle = \begin{pmatrix} 0 \\ \frac{v}{\sqrt{2}} \end{pmatrix},$$

thereby breaking the gauge symmetry of the ground state $|0\rangle$. This amounts to rewriting

$$\phi = \begin{pmatrix} \phi^+ \\ \frac{\phi_1+i\phi_2}{\sqrt{2}} + \frac{v}{\sqrt{2}} \end{pmatrix}$$

where $\phi^+$ and hermitian fields $\phi_1$ and $\phi_2$ have zero vacuum expectation values. In contrast to the gauge invariant vertices shown in Table 1.1 [which are not affected by SSB], one starts with manifestly gauge invariant vertices involving $\phi$ and other fields and then translate them to physical amplitudes after SSB as pictorially shown below [the dotted lines ending in X denotes $\langle\phi\rangle = v/\sqrt{2}$]:

Because of mixing between $W_0$ and $B$, these are not physical particles, the physical particles $\gamma$ and $Z$ are defined in Eq. (1.16). This requires diagonalization of the mass matrix for $W_0 - B$ sector, which on diagonalization gives

$$m_A = 0, \qquad m_Z = \frac{m_W}{\cos\theta_W}, \tag{1.19a}$$

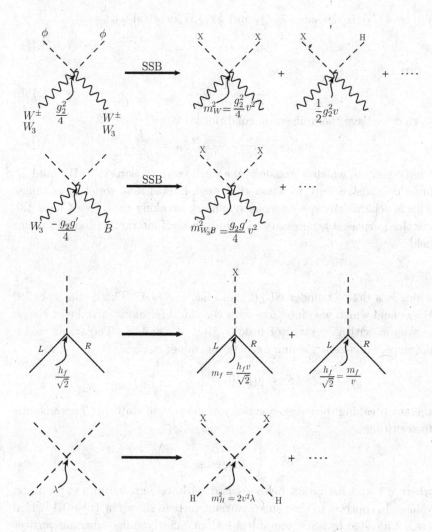

where from the above picture

$$m_W = \frac{1}{2} g_2 v. \tag{1.19b}$$

Further we note from the above picture that the mass of a fermion of flavor $f$ and that of Higgs particle $H$ are respectively given by

$$m_f = \frac{h_f v}{\sqrt{2}}, \qquad m_H = \sqrt{2v^2 \lambda}. \tag{1.19c}$$

What has happened is that $\phi^{\pm}$ and $\phi_2$ have provided the longitudinal degrees of freedom to $W^{\pm}$ and $Z$ which have eaten them up while becoming massive. The remaining electrically neutral scalar field is called the Higgs field and its quantum is called the Higgs particle which we have denoted by $H$ in the above picture. We note from Eq. (1.19a) that

$$\sin^2 \theta_W = 1 - \frac{m_W^2}{m_Z^2} \qquad (1.20a)$$

The directly observed Fermi coupling constant in weak nuclear processes at low energies (i.e. $\ll m_W$) is given by [cf. Eq. (1.18c)]

$$\frac{G_F}{\sqrt{2}} = \frac{g_2^2}{8m_W^2} = \frac{e^2}{8m_W^2 \sin^2 \theta_W}, \qquad (1.20b)$$

or

$$m_W = \left[ \frac{\pi \alpha}{\sqrt{2} G_F \sin^2 \theta_W} \right]^{1/2}, \qquad (1.20c)$$

where $\alpha = \frac{e^2}{4\pi} = \frac{1}{137}$ is the fine structure constant and $G_F$ is the Fermi constant ($\approx 10^{-5}$ GeV$^{-2}$).

The main predictions of the electroweak unification are

(i) existence of a new type of neutral weak interaction mediated by $Z^0$,

(ii) weak vector bosons $W, Z$ whose masses are predicted by the relations (1.20c) and (1.20a), once $\sin^2 \theta_W$ is determined, $\alpha$ and $G_F$ being known,

(iii) existence of the Higgs particle with mass $m_H = \sqrt{2v^2 \lambda}$, which is arbitrary since $\lambda$ is not fixed.

The first prediction was verified more than 30 years ago and the phenomenology of neutral weak interaction gives

$$\sin^2 \theta_W \approx 0.23.$$

One can now use this result to predict $m_W$ and $m_Z$ through the relations (1.20) to get

$$m_W \approx 80 \text{ GeV}, \qquad m_Z \approx 92 \text{ GeV}$$

in agreement with their experimental values. The standard model is in very good shape experimentally. The third prediction is not yet tested and the present lower bound on $m_H$ from Higgs searches at LEP is

$$m_H > 114 \text{ GeV}.$$

We also note that the electroweak unification energy scale is given by

$$\lambda_F \equiv v = 2\frac{m_W}{g_2}$$

$$= \left(\sqrt{2}G_F\right)^{-1/2}$$

$$\approx 250 \text{ GeV}. \tag{1.21}$$

We will briefly discuss the unification of the other two forces with the electroweak force after discussing the origin of the strong force between the two quarks below.

## 1.7   Strong Quark-Quark Force

We have already remarked:

(i) each quark flavor carries 3 colors.
(ii) only color singlets (colorless states) exist as free particles.

Strong color charges are the sources of the strong force between two quarks just as the electric charge is the source of electromagnetic interaction between two electrically charged particles. The analogy is carried further in Table 1.3.

The binding energy provided by one gluon exchange potential of the form mentioned above cannot be sufficient to confine the quarks in a hadron since as one can ionize an atom to knock out an electron, similarly a quark could be separated from a hadron if sufficient energy is supplied. Thus $V^g$, the one gluon exchange potential, can at best provide binding for quarks at short distances and cannot explain their confinement, i.e. impossibility of separating a quark from a hadron. The hope here is that the self interaction of color carrying gluons may give rise to long distance behavior of the potential in QCD completely different from that in QED, where the electrically neutral photon has no self interaction. One hopes that the long range potential in QCD would increase with the distance so that the quarks would be confined in a hadron. Phenomenologically, a potential of form

$$V_{ij}(r) = V_{ij}^g(r) + V^c(r), \tag{1.22a}$$

where

$$V_{ij}^g(r) = -k_s\frac{\alpha_s}{r} + \cdots , \tag{1.22b}$$

| Table 1.3 | |
|---|---|
| **Electromagnetic Force Between 2 Electrically Charged Particles** | **Strong Color Force Between 2 Quarks** |
| We deal with electrically neutral atoms. | We deal with color singlet systems i.e. hadrons. |
| Mediator of the electromagnetic force is electrically neutral massless spin 1 photon, the quantum of the electromagnetic field. | Mediators are eight massless spin 1 color carrying gauge vector bosons, called gluons. |
| Exchange of photon gives the electric potential: | Exchange of gluons gives the color electric potential: |

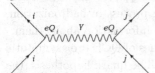 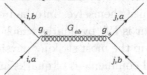

| | |
|---|---|
| $V_{ij}^e = \frac{e^2}{4\pi}\frac{Q_i Q_j}{r}, r = \|\mathbf{r}_i - \mathbf{r}_j\|$ For an electron and proton | $V_{ij}^{q\bar{q}} = -\frac{4}{3}\alpha_s\frac{1}{r}, \alpha_s = \frac{g_s^2}{4\pi},$ for $\bar{q}q$ color singlet system (mesons) while for $qqq$ color singlet system (baryons). |
| $V_{ij} = -\frac{\alpha}{r}, \alpha = \frac{e^2}{4\pi}$ This attractive potential is responsible for the binding of atoms. | $V_{ij}^{qq} = -\frac{2}{3}\alpha_s\frac{1}{r}$ Note the very important fact that in both cases, we get an attractive potential. Without color, $V_{ij}^{qq}$ would have been repulsive. |
| The theory here is called quantum electrodynamics (QED). | The theory here is called quantum chromodynamics (QCD). |
| Due to quantum (radiative) corrections, $\alpha\left(\sqrt{q^2}\right)$ increases with increasing momentum transfer $q^2$, for example | Due to quantum (radiative) corrections, $\alpha_s\left(\sqrt{q^2}\right)$ decreases with increasing $q^2$ [this is brought about by the self interaction of gluons (cf. Table 1.2)], for example |
| $\alpha(m_e) \approx \frac{1}{137},$ $\alpha(m_W) \approx \frac{1}{128}$ | $\alpha_s(m_\tau) \approx 0.35,$ $\alpha_s(m_\Upsilon \simeq 10 \text{ GeV}) \approx 0.16,$ $\alpha_s(m_Z) \approx 0.125.$ That the effective coupling constant decreases at short distances is called the asymptotic freedom property of QCD. |

(... denotes spin dependent terms, (see Chap. 7) and $k_s = 4/3(q\bar{q})$, $2/3(qqq)$) is the single gluon exchange potential while $V^c(r)$ is the confining potential (independent of the quark flavor), has been used in hadron spectroscopy with quite good success. Lattice gauge theories suggest

$$V^c(r) = Kr, \tag{1.22c}$$

with $K \approx 0.25$ $(\text{GeV})^2$, obtained from the quarkonium spectroscopy. This gives the confining potential. Note that as $r$ increases $V^c(r)$ increases and as such it requires an infinite energy to release the quarks from a hadron. However quarks can be produced for example, in electron-positron annihilation

$$e^+e^- \to \gamma \to q\bar{q}$$

In this process, the photon emitted from the electron-positron pair lifts the quark and antiquark pair from the vacuum. The $q\bar{q}$ pair in a very short time hadronize themselves into hadrons. In the process of hadronization, the foot print is left by quarks which provide us information about them.

To sum up the most striking physical properties of QCD are asymptotic freedom and confinement of quarks and gluons. The quark hypothesis, the electroweak theory and QCD form the basis for the "Standard Model" of elementary particles to which most of the book is devoted while Chap. 18 is concerned with the interface of cosmology with particle physics.

We now briefly discuss the attempts to unify the other two forces with the electroweak force.

## 1.8   Grand Unification

The three strong color charges introduced earlier generate the gauge group $SU_C(3)$ while that of the electroweak interaction is $SU_L(2) \times U(1)$. Thus the standard model involves

| $SU_C(3)$ | $\times$ | $SU_L(2)$ | $\times$ | $U(1)$ |
|:---:|:---:|:---:|:---:|:---:|
| $\alpha_s$ | | $\alpha_2$ | | $\alpha'$ |

where the associated coupling constants $\alpha_s$, $\alpha_2$ and $\alpha'$ are very different at the present energies. But these coupling constants are energy dependent due to quantum radiative corrections. Grand unification is an attempt to find a bigger group $G$:

$$G \supset SU_C(3) \times SU_L(2) \times U(1)$$

such that at some energy scale $q^2 = m_X^2$ ,

$$\alpha_s(m_X^2) = \alpha_2(m_X^2) = \alpha'(m_X^2)$$
$$= \alpha_G. \tag{1.23}$$

This merging of coupling at high energy in principle is possible because of the logarithmic dependence of coupling constants with energy scale as calculated in quantum theory (renormalization group analysis, see Appendix

B) and the fact that the two of the three coupling constants, namely, $\alpha'$ and $\alpha_2$ are related at $\sqrt{q^2} = m_W$ through the electroweak unification conditions given in Eq. (1.18c). As will be discussed in Chap. 17, the relation (1.23) holds at $\sqrt{q^2} = m_X \approx 10^{15}$ GeV or less, which gives the grand unification (GUT) scale. In other words, the GUT breaking scale is almost 14 orders of magnitude of $SU(2) \otimes U(1)$ breaking scale. In a way it is good because such a large scale suppresses proton decay, though we now know that it does not work for simple $SU(5)$. The reason is that with the modern values of the couplings, they do not merge at a single scale. They do merge for supersymmetric version of $SU(5)$ (see Chap. 17).

In spite of this GUT scale have some attractive features:

(i) quark-lepton unification
(ii) relationships between quark and lepton masses
(iii) quantization of electric charge, for a simple group it is a consequence of the charge operator being a generator of the group and traceless. So for example, sum of charges in a multiplet containing quarks and leptons = 0, thus giving some relation between quark and lepton charges.

But they still leave arbitrariness in Higgs sector needed to give masses to lepto-quarks and $W^\pm$, $Z$ vector bosons, do not explain number of generations, do not explain fermions mass hierarchy typified by $m_t/m_u \approx 10^5$ and the gauge hierarchy problem $m_W/m_X \approx 10^{-12}$ in a natural way. These mass hierarchies are more naturally accommodated in supersymmetry (see Chap. 17).

## 1.9  Units and Notation

We shall use the natural units:

$$\hbar = c = 1$$

We note that

$$[\hbar] = ML^2T = 6.582 \times 10^{-22} \text{MeV-s}$$
$$[c] = LT^{-1} = 3 \times 10^{10} \text{cm/s}$$
$$[\hbar c] = 197 \times 10^{-13} \text{MeV-cm}$$

If $\hbar = c = 1$, then

$$v = \frac{c^2 p}{E} = \frac{p\,(\mathrm{MeV}/c)}{E\left(\mathrm{MeV}/c^2\right)} \quad \text{(in units of c)}$$

If we take, $M = 1$ GeV

$$L \sim \frac{1}{\mathrm{GeV}} = \frac{\hbar c}{1000\ \mathrm{MeV}} \approx 2 \times 10^{-14}\ \mathrm{cm}$$

$$T \sim \frac{1}{\mathrm{GeV}} = \frac{\hbar}{1000\ \mathrm{MeV}} \approx 6.58 \times 10^{-25}\ \mathrm{s}$$

$$1\ \mathrm{MeV} = 1.6 \times 10^{-6}\mathrm{erg} = 1.6 \times 10^{-13} J$$

$$1\ \mathrm{gm} = 5.61 \times 10^{23}\mathrm{GeV}$$

$$1\ \mathrm{GeV} = 10^{3}\mathrm{MeV}$$

We will denote the position by a 4-vector $x\,(\mu = 0, 1, 2, 3)$ :

$$x^{\mu} = (ct, \mathbf{x}) = (t, \mathbf{x}); \quad \text{contravariant vector}$$

$$x_{\mu} = (ct, -\mathbf{x}) = (t, -\mathbf{x}) = g_{\mu\nu}x^{\nu} \quad \text{covariant vector}$$

$$x^2 = x_{\mu}x^{\mu} = t^2 - \mathbf{x}^2$$

$$\partial_{\mu} = \frac{\partial}{\partial x^{\mu}} = \left(\frac{\partial}{\partial t}, \nabla\right)$$

$$\partial^{\mu} = \frac{\partial}{\partial x_{\mu}} = \left(\frac{\partial}{\partial t}, -\nabla\right)$$

$$\partial_{\mu}\partial^{\mu} = \left(\frac{\partial^2}{\partial t^2}, -\nabla^2\right) = \Box^2$$

with $g_{\mu\nu} = 0, \mu \neq \nu, g_{00} = 1, g_{11} = g_{22} = g_{33} = -1$. On the light cone

$$x^2 = 0, \text{ i.e. } t^2 - \mathbf{x}^2 = 0.$$

The energy $E$ and momentum $\mathbf{p}$ are represented by a 4-vector $p$:

$$p^{\mu} = (E/c, \mathbf{p}) = (E, \mathbf{p}),$$

$$p^2 = p_{\mu}p^{\mu} = p_0^2 - \mathbf{p}^2 = E^2 - \mathbf{p}^2.$$

For a particle on the mass shell

$$E^2 = \mathbf{p}^2 + m^2,$$

i.e.

$$p^2 = p_{\mu}p^{\mu} = m^2.$$

The scalar product

$$p.q = p^{\mu}q_{\mu} = E_p E_q - \mathbf{p}.\mathbf{q}.$$

## 1.10 Problems

(1) Show that the gravitational potential energy for a body of 1 kg on the surface of the earth is $6.250 \times 10^7$ J$= 3.90 \times 10^{26}$ eV. Force of attraction $F = \frac{V}{r} = 9.8$ N, $g = 9.8$ ms$^{-2}$. Take mass of earth $M = 5.974 \times 10^{24}$ kg. Mean equatorial radius of earth is $6.378 \times 10^6$ m.

(2) Newton's gravitational constant

$$G_N = 6.67 \times 10^{-11} \, (\text{kg})^{-1} \, \text{m}^3 \text{s}^{-2}$$
$$= \left(6.67 \times 10^{-11}\right) \left(\text{kgm}^2 \text{s}^{-2}\right) \text{m} \, (\text{kg})^{-2}$$

Show that in GeV:

$$G_N = 1.35 \times 10^{-54} \, (\text{GeV/m}) \left(\text{GeV/c}^2\right)^{-2}.$$

## 1.11 References

1. A. Zee, The unity of forces in the universe, Vol. 1, World Scientific, Singapore (1982).
2. R. N. Mohapatra, Unification and supersymmetry, The frontiers of quark-lepton physics, Springer Verlag, Berlin (1986).
3. I. Hinchliffe, Ann. Rev. Nuclear and Particle Science, 36, 505 (1986).
4. M. B. Green, J. H. Schwarz and E. Witten, Superstring theory I and II, Cambridge University Press (1986).
5. L. Brink and M. Henneaux, Principles of String theory, Plenum (1987).
6. C. H. Llewellyn Smith, Particle Phenomenology: The Standard Model, Proc. of the 1989 Scottish Universities Summer School, Physics of the Early Universe, OUTD-90-160.
7. S. Dimopoulos, S. A. Raby and F. Wilczek, Unification of couplings, Physics Today, 44, 25 (1991).
8. R. E. Marshak, Conceptional foundations of modern particle physics, World Scientific, Singapore (1992).
9. M.E. Peskin, "Beyond standard model" in proceeding of 1996 European School of High Energy Physics CERN 97-03, Eds. N. Ellis and M. Neubert.
10. J. Ellis, "Beyond Standard Model for Hillwalker" CERN-TH/98-329, hep-ph 9812235.
11. E. M. Henley, A. Garcia, Subatomic Physics, World Scientific, 3rd edition (2007).

12. Particle Data Group, N. Nakamura et al., Journal of Physics G, **37**, 075021 (2010).

# Chapter 2

# Scattering and Particle Interaction

## 2.1 Introduction

Most of the information about the properties of particles and their interactions are extracted from the experiments involving scattering of particles. High energy projectiles (electrons, protons, photons and pions) are used to probe the structure of matter at short distances. For high energy particles, the theory of special relativity becomes relevant. As such the Lorentz invariance is assumed in any process involving particles. The transition reaction amplitude is a function of Lorentz invariant variables. We first briefly review of the Lorentz transformation.

In relativity theory, space and time are treated on equal footing and as such we deal with four vectors in four (1+3)-dimensional space:

$$x^\mu = (x^0, x^i) = (ct, \mathbf{x}), \quad \text{contravariant vector}$$
$$x_\mu = (x_0, x_i) = (ct, -\mathbf{x}), \quad \text{covariant vector}$$

Correspondingly

$$\partial_\mu = \frac{\partial}{\partial x^\mu} = \left(\frac{1}{c}\frac{\partial}{\partial t}, \nabla\right)$$

$$\partial^\mu = \frac{\partial}{\partial x_\mu} = \left(\frac{1}{c}\frac{\partial}{\partial t}, -\nabla\right)$$

We introduce metric tensor $g_{\mu\nu}$

$$g_{\mu\nu} = diag(1, -1, -1, -1) = g^{\mu\nu}, g_{00} = 1, \ g_{ij} = -\delta_{ij}$$

which can be used to lower and raise the indices

$$x^\mu = g^{\mu\nu}x_\nu, \quad x_\mu = g_{\mu\nu}x^\nu$$

The Lorentz transformation

$$x'^\mu = \Lambda^\mu_\nu x^\nu \tag{2.1}$$

leaves the length of vector $x$

$$x^2 = x^\mu x_\mu = c^2 t^2 - \mathbf{x}^2$$

invariant so that

$$x^\mu g_{\mu\nu} x^\nu = x'^\alpha g_{\alpha\beta} x'^\beta$$
$$= \Lambda_\mu^\alpha x^\mu \Lambda_\nu^\beta x^\nu g_{\alpha\beta}$$

giving

$$g_{\mu\nu} = \Lambda_\mu^\alpha g_{\alpha\beta} \Lambda_\nu^\beta \tag{2.2}$$

In matrix form the Lorentz transformation is

$$x' = \Lambda x, \ x = \Lambda^T x' \tag{2.3}$$

and the metric

$$g = \left(\Lambda^T\right)_{\mu\alpha} g_{\alpha\beta} \left(\Lambda\right)_{\beta\nu}$$
$$= \Lambda^T g \Lambda \tag{2.4}$$

$\Lambda$ is a $4\times 4$ matrix. It has two sub-matrices, that is, Lorentz transformation has two subgroups

$$\Lambda = \begin{pmatrix} \Lambda(v) & 0 \\ 0 & 1 \end{pmatrix}, \ \Lambda = \begin{pmatrix} 1 & 0 \\ 0 & \Lambda_R \end{pmatrix} \tag{2.5}$$

where

$$\Lambda(v) = \begin{pmatrix} \cosh\zeta & -\sinh\zeta & 0 \\ -\sinh\zeta & \cosh\zeta & 0 \\ 0 & 0 & 1 \end{pmatrix} \tag{2.6}$$

corresponds to the Lorentz velocity transformation. Noting that $\cosh^2\zeta - \sinh^2\zeta = 1$ and putting $\cosh\zeta = \gamma$, $\sinh\zeta = \gamma\frac{v}{c}$ with $\gamma = \frac{1}{\sqrt{1-v^2/c^2}}$, we have the usual Lorentz transformation ($\mathbf{v} = (v,0,0)$)

$$x'^0 = \gamma\left(x^0 - \frac{vx}{c}\right) = \gamma\left(x^0 - \frac{\mathbf{v}\cdot\mathbf{x}}{c}\right)$$
$$x'^1 = \gamma\left(x^1 - \frac{v}{c}x^0\right) = x^1 + (\gamma - 1)\frac{\mathbf{v}\cdot\mathbf{x}}{v^2}\mathbf{v} - \gamma\frac{\mathbf{v}}{c} \tag{2.7}$$
$$x'^2 = x^2$$
$$x'^3 = x^3$$

connecting two inertial frames when one frame moves with velocity $v$ in $x$-direction relative to the other. The matrix

$$\Lambda_R = \begin{pmatrix} \cos\omega & \sin\omega & 0 \\ -\sin\omega & \cos\omega & 0 \\ 0 & 0 & 1 \end{pmatrix} \tag{2.8}$$

gives the rotation in three dimensions along the $x^3$ axis

$$x'^0 = x^0$$
$$x'^1 = \cos\omega \, x^1 + \sin\omega \, x^2$$
$$x'^2 = -\sin\omega \, x^1 + \cos\omega \, x^2 \tag{2.9}$$
$$x'^3 = x^3$$

It is easy to see that a generalization of the Lorentz velocity transformation when $\mathbf{v}$ is not restricted to the $x^1$-direction is

$$x'^0 = \gamma \left[ x^0 - \frac{\mathbf{v} \cdot \mathbf{x}}{c} \right]$$
$$\mathbf{x}' = \mathbf{x} + (\gamma - 1)\frac{\mathbf{v} \cdot \mathbf{x}}{v^2}\mathbf{v} - \gamma\frac{\mathbf{v}}{c}x^0 \tag{2.10}$$

giving $[v_j = -v^j]$

$$\Lambda^0_0 = \gamma, \ \Lambda^i_0 = -\frac{\gamma}{c}v^i, \ \Lambda^0_j = \gamma\frac{v_j}{c}, \ \Lambda^i_j = \delta^i_j - (\gamma - 1)\frac{v^i v_j}{v^2} \tag{2.11}$$

In relativistic mechanics, energy and momentum are treated on equal footing just like the space and time. Accordingly we introduce energy-momentum 4-vector :

$$p^\mu = \left(p^0, \mathbf{p}\right) = (E/c, \mathbf{p})$$
$$p_\mu = (p_0, -\mathbf{p}) = (E/c, -\mathbf{p})$$

so that

$$p^2 = p_\mu p^\mu = E^2/c^2 - \mathbf{p}^2 = m^2 c^2 \tag{2.12}$$

where the particle is on the mass shell.

In terms of $\gamma = \frac{1}{\sqrt{(1-v^2/c^2)}}$, the above relation is identically satisfied with

$$E = \gamma m c^2, \quad \mathbf{p} = \gamma m \mathbf{v}$$

so that

$$\mathbf{v}/c = \frac{c\mathbf{p}}{E} \tag{2.13}$$

Thus in analogy with space-time transformation

$$\mathbf{p}' = \mathbf{p} + \left[(\gamma - 1)\mathbf{v}\frac{\mathbf{v} \cdot \mathbf{p}}{v^2} - \gamma\mathbf{v}\frac{E}{c^2}\right]$$
$$E' = \gamma\left[E - \mathbf{v} \cdot \mathbf{p}\right] \tag{2.14}$$

where the primed quantities are in the frame moving with velocity **v** relative to the one in which $E$ and **p** are measured. It is convenient to write the vector **p** in terms of its longitudinal and transverse components

$$\mathbf{p} = (\mathbf{p}_\parallel, \mathbf{p}_\perp)$$
$$\mathbf{p}'_\parallel = \gamma \left[ \mathbf{p}_\parallel - \frac{\mathbf{v}}{c^2} E \right]$$
$$\mathbf{p}'_\perp = \mathbf{p}_\perp \tag{2.15}$$
$$E' = \gamma \left[ E - \mathbf{v} \cdot \mathbf{p}_\parallel \right]$$

From now on, we will put $c = 1$.

## 2.2    Kinematics of a Scattering Process

Consider a typical 2-body scattering process

$$a + b \rightarrow c + d.$$

We denote the four momenta of particles $a, b, c$ and $d$ by $p_a$, $p_b$, $p_c$, $p_d$ respectively. Energy momentum conservation gives:

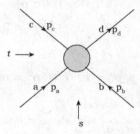

Fig. 2.1    Two-body scattering: $a + b \rightarrow c + d$

$$p_a + p_b = p_c + p_d \tag{2.16a}$$

or in the component form

$$\mathbf{p}_a + \mathbf{p}_b = \mathbf{p}_c + \mathbf{p}_d \tag{2.16b}$$

$$E_a + E_b = E_c + E_d \tag{2.16c}$$

We assume Lorentz invariance in any process involving particles. The reaction transition amplitude is a function of scalars (i.e. Lorentz invariants) formed out of the four vectors $p_a, p_b, p_c$ and $p_d$. The invariants are

$$s = (p_a + p_b)^2 = (p_c + p_d)^2 \qquad (2.17a)$$

$$t = (p_a - p_c)^2 = (p_d - p_b)^2 \qquad (2.17b)$$

$$u = (p_a - p_d)^2 = (p_c - p_b)^2. \qquad (2.17c)$$

But only two of the three scalars are independent:

$$s + t + u = 3p_a^2 + p_b^2 + p_c^2 + p_d^2 + 2p_a \cdot (p_b - p_c - p_d) \qquad (2.18)$$
$$= m_a^2 + m_b^2 + m_c^2 + m_d^2.$$

In an actual scattering experiment, we have a projectile (let it be $a$) and a target ($b$), which is stationary in the laboratory frame. Thus

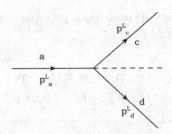

Fig. 2.2   Two-body scattering in the laboratory frame.

$$p_a \equiv \left(E_a^L, \mathbf{p}_a^L\right) = (\nu_L, \mathbf{p}_L)$$
$$p_b \equiv (m_b, 0) \qquad (2.19)$$
$$p_c \equiv \left(E_c^L, \mathbf{p}_c^L\right), \quad p_d \equiv \left(E_d^L, \mathbf{p}_d^L\right).$$

Hence in the laboratory frame:

$$s = (p_a + p_b)^2$$
$$= m_a^2 + m_b^2 + 2m_b\nu_L, \qquad (2.20a)$$
$$t = (p_a - p_c)^2$$
$$= m_a^2 + m_c^2 - 2\nu_L E_c^L + 2\left|\mathbf{p}_L\right|\left|\mathbf{p}_c^L\right|\cos\theta_L. \qquad (2.20b)$$

or

$$\nu_L = \frac{s - m_a^2 - m_b^2}{2m_b} = \frac{p_a \cdot p_b}{m_b} \qquad (2.21a)$$

$$\mathbf{p}_L^2 = -p_a^2 + \nu_L^2 = -m_a^2 + \nu_L^2 \qquad (2.21b)$$

$$\left|\mathbf{p}_L\right| = \frac{\sqrt{\lambda\left(s, m_a^2, m_b^2\right)}}{2m_b}, \qquad (2.21c)$$

where

$$\lambda(x, y, z) = x^2 + y^2 + z^2 - 2xy - 2xz - 2yz. \qquad (2.22)$$

Theoretically, it is convenient to consider a scattering process in the center-of-mass (c.m.) frame. In this frame:

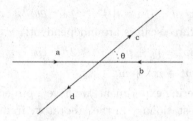

Fig. 2.3   Two-body scattering in the center-of-mass frame.

$$p_a \equiv (E_a, \mathbf{p}), \quad p_b \equiv (E_b, -\mathbf{p}),$$
$$p_c \equiv (E_c, \mathbf{p}'), \quad p_d \equiv (E_d, -\mathbf{p}'). \qquad (2.23)$$

Thus we have

$$s = (p_a + p_b)^2 = (p_c + p_d)^2$$
$$= (E_a + E_b)^2 = (E_c + E_d)^2 \equiv E_{cm}^2 \qquad (2.24a)$$

$$t = m_a^2 + m_c^2 - 2E_a E_c + 2|\mathbf{p}||\mathbf{p}'|\cos\theta$$
$$= m_b^2 + m_d^2 - 2E_b E_d + 2|\mathbf{p}||\mathbf{p}'|\cos\theta. \qquad (2.24b)$$

Now

$$s = E_a^2 + E_b^2 + 2E_a E_b \qquad (2.25)$$
$$= (\mathbf{p}^2 + m_a^2) + (\mathbf{p}^2 + m_b^2) + 2\sqrt{(\mathbf{p}^2 + \mathbf{m}_a^2)}\sqrt{(\mathbf{p}^2 + \mathbf{m}_b^2)}.$$

Solving Eq. (2.25), we get

$$|\mathbf{p}| = \frac{\sqrt{\lambda(s, m_a^2, m_b^2)}}{2\sqrt{s}}. \qquad (2.26a)$$

Similarly by considering, $s = (E_c + E_d)^2$, we get

$$|\mathbf{p}'| = \frac{\sqrt{\lambda(s, m_c^2, m_d^2)}}{2\sqrt{s}}. \qquad (2.26b)$$

We also note that

$$E_a = \frac{s + m_a^2 - m_b^2}{2\sqrt{s}}, \quad E_b = \frac{s + m_b^2 - m_a^2}{2\sqrt{s}}$$

$$E_c = \frac{s + m_c^2 - m_d^2}{2\sqrt{s}}, \quad E_d = \frac{s + m_d^2 - m_c^2}{2\sqrt{s}}. \tag{2.27}$$

For the elastic scattering

$$c \equiv a, \quad d \equiv b$$

and

$$|\mathbf{p}| = |\mathbf{p}'|, \quad E_c = E_a, \quad E_d = E_b$$

$$t = -2\mathbf{p}^2 (1 - \cos\theta) = -4\mathbf{p}^2 \sin^2\frac{\theta}{2}. \tag{2.28}$$

Thus $-t$ is just the square of momentum transfer.

Finally one can derive a relation between the scattering angles $\theta$ and $\theta_L$ using Lorentz transformation. Let us take $\mathbf{p}_L$ and $\mathbf{p}$ along $z$-axis. The c.m. frame is moving relative to the laboratory frame with a velocity:

$$\mathbf{v} = \frac{\mathbf{p}_L}{\nu_L + m_b}. \tag{2.29}$$

Lorentz transformation gives

$$p_c^L \cos\theta_L = \gamma \left[ p' \cos\theta + vE_c \right]$$
$$p_c^L \sin\theta_L = p' \sin\theta \tag{2.30}$$
$$E_c^L = \gamma \left[ E_c + vp' \cos\theta \right].$$

Hence, we get

$$\tan\theta_L = \frac{p' \sin\theta}{\gamma \left[ p' \cos\theta + vE_c \right]}, \tag{2.31a}$$

where

$$\gamma = \frac{1}{\sqrt{1 - v^2}} = \frac{\nu_L + m_b}{E_{cm}}. \tag{2.31b}$$

Equation (2.31b) follows from the relations:

$$p_L = \gamma \left[ p + vE_a \right], \quad \nu_L = \gamma \left[ E_a + vp \right], \quad m_b = \gamma \left[ E_b - vp \right]. \tag{2.32}$$

Another variable which is useful for scattering process, particularly for inclusive reactions, is rapidity. We define the longitudinal rapidity $y$:

$$y = \tanh^{-1}\frac{p_\parallel}{E} = \frac{1}{2} \ln\frac{E + p_\parallel}{E - p_\parallel} \tag{2.33}$$

Now using Eqs. (2.15)

$$E' \pm p'_\parallel = \gamma \left(1 \mp v/c\right) \left(E \pm cp_\parallel\right)$$

one gets

$$y' = y - \frac{1}{2} \ln \frac{1 + v/c}{1 - v/c}$$

and

$$dy' = dy$$

i.e. the shape of rapidity distribution is invariant. It may be noted that by choosing some direction, e.g. beam direction for the $z$-axis, one can write the energy and momentum of the particle as

$$E = m_T \cosh y, \quad p_z \equiv p_\parallel = m_T \sinh y, \quad \mathbf{p}_T = (p_x, p_y)$$

where $m_T$ is called the transverse mass and is given by

$$E^2 - p_\parallel^2 = m_T^2$$

which follows from $\cosh^2 y - \sinh^2 y = 1$

Now

$$E^2 - p_\parallel^2 = E^2 - (\mathbf{p}^2 - \mathbf{p}_T^2) = m^2 + \mathbf{p}_T^2$$

so that

$$E_T^2 = \mathbf{p}_T^2 + m^2 = m_T^2$$

Thus from Eq. (2.33),

$$y = \ln \frac{E + p_\parallel}{m_T}$$

The rapidity $\eta$ is defined as

$$\eta = -\ln(\tan \frac{\theta}{2})$$

where $\theta$ is the scattering angle in the c.m. frame. For high energy scattering ($E \approx |\mathbf{p}|$):

$$y = \frac{1}{2} \ln \frac{E + p_z}{E - p_z} = \frac{1}{2} \ln \frac{E + |\mathbf{p}| \cos\theta}{E - |\mathbf{p}| \cos\theta}$$

$$\approx \frac{1}{2} \ln \left(\frac{\cos^2 \theta/2}{\sin^2 \theta/2}\right) = -\ln(\tan \theta/2) = \eta$$

The scattering cross section is

$$E \frac{d^3\sigma}{d^3p} = E \frac{d^3\sigma}{dp_x dp_y dp_z} = E \frac{d^3\sigma}{p_T dp_T d\phi dp_z}$$

As

$$dp_z = m_T \cosh y \, dy$$
$$= E \, dy$$

and averaging over $\phi$,

$$E\frac{d^3\sigma}{d^3p} = \frac{d^2\sigma}{\pi dy d(p_T^2)}$$

i.e. $E\frac{d^3\sigma}{d^3p}$ is an invariant quantity.

## 2.3   Interaction Picture

In quantum mechanics, the transition rate from initial state $|i\rangle$ to final state $|f\rangle$ is given by

$$W = 2\pi \mid \langle f| \; H_I \; |i\rangle \mid^2 \rho_f(E_f), \qquad (2.34)$$

where $H_I$ is the interaction Hamiltonian viz.

$$H = H_0 + H_I. \qquad (2.35)$$

The above formula is obtained when $H_I$ is treated as small in first order perturbation theory and $|i\rangle$ and $|f\rangle$ are eigenstates of $H_0$. $\rho_f(E_f)$ is the density of final states, i.e. $\rho_f(E_f)dE_f$ = number of final states with energies between $E_f$ and $E_f + dE_f$.

In order to define the transition rate in general, it is convenient to go to interaction picture, which we define below:
The Schrödinger equation is given by

$$i\frac{d}{dt}|\Psi(t)\rangle_S = H|\Psi(t)\rangle_S. \qquad (2.36)$$

We now go to interaction picture by a unitary transformation

$$|\Psi(t)\rangle_I = e^{iH_0t}|\Psi(t)\rangle_S. \qquad (2.37)$$

Then, using Eq. (2.36), we have

$$i\frac{d}{dt}|\Psi(t)\rangle_I = -H_0|\Psi(t)\rangle_I + e^{iH_0t}He^{-iH_0t}|\Psi(t)\rangle_I. \qquad (2.38)$$

Now using (2.35)

$$H_0^I(t) = e^{iH_0t}H_0e^{-iH_0t} = H_0 \qquad (2.39a)$$
$$H_I^I(t) = e^{iH_0t}He^{-iH_0t} = H_0 + H_I^I(t), \qquad (2.39b)$$

where

$$H_I^I(t) = e^{iH_0t} H_I e^{-iH_0t}. \tag{2.39c}$$

Hence we have from Eq. (2.38) [From now on, we will drop the superscript $I$ from $H_I^I(t)$, it will be understood.]

$$i\frac{d}{dt} |\Psi(t)\rangle_I = H_I(t) |\Psi(t)\rangle_I. \tag{2.40}$$

Thus an operator $\widehat{A}$ in Schrödinger picture is related to operator $\widehat{A}_I(t)$ in interaction picture by a unitary transformation

$$\widehat{A}_I(t) = e^{iH_0t} \, \widehat{A} \, e^{-iH_0t} \tag{2.41a}$$

and

$$i\frac{d\,\widehat{A}_I(t)}{dt} = \left[\widehat{A}_I(t), H_0\right]. \tag{2.41b}$$

## 2.4 Scattering Matrix (S-Matrix)

From the general principles of quantum mechanics, the probability of finding the system in state $|b\rangle$, when the system is in state $|\Psi(t)\rangle_I$, is given by $|C_b(t)|^2$ where

$$C_b(t) = \langle b | \Psi(t)\rangle_I. \tag{2.42}$$

Assume that $|\Psi(t)\rangle_I$ is generated from $|\Psi(t_0)\rangle_I$ by a linear operator $U(t, t_0)$:

$$|\Psi(t)\rangle_I = U(t, t_0) |\Psi(t_0)\rangle_I \tag{2.43a}$$

$$U(t_0, t_0) = 1. \tag{2.43b}$$

Substituting Eq. (2.43a) in Eq. (2.40), we get

$$i\frac{\partial U(t, t_0)}{dt} |\Psi(t_0)\rangle_I = H_I(t) U(t, t_0) |\Psi(t_0)\rangle_I \tag{2.44a}$$

so that we obtain

$$i\frac{\partial U(t, t_0)}{dt} = H_I(t) U(t, t_0). \tag{2.44b}$$

We note that $U(t, t_0)$ depends only on the structure of the physical system and not on the particular choice of the initial state $|\Psi(t_0)\rangle_I$. Thus

$$\begin{aligned} |\Psi(t)\rangle_I &= U(t, t_0) |\Psi(t_0)\rangle_I \\ &= U(t, t') |\Psi(t')\rangle_I \\ &= U(t, t') \, U(t', t_0) |\Psi(t_0)\rangle_I \end{aligned} \tag{2.45}$$

Therefore,

$$U(t,\ t')\ U(t',\ t_0) = U(t,\ t_0) \qquad (2.46a)$$

$$I = U(t_0,t_0)\ = U(t_0,\ t)U(t,\ t_0) \qquad (2.46b)$$

$$U(t_0,t)\ = U^{-1}(t,\ t_0) \qquad (2.46c)$$

Thus, the operator $U$ satisfies the group properties.

The formal solution of differential Eq. (2.44a) is given by

$$U(t,t_0)\ = 1 - i\int_{t_0}^{t} H_I(t_1)U(t_1,\ t_0)dt_1. \qquad (2.47)$$

This integral equation can be solved by iteration method, i.e.

$$
\begin{aligned}
U(t,t_0)\ &= 1 - i\int_{t_0}^{t} dt_1 H_I(t_1)\left[1 - i\int_{t_0}^{t_1} dt_2 H_I(t_2)U(t_2,\ t_0)\right] \\
&= 1 - i\int_{t_0}^{t} dt_1 H_I(t_1) + (-i)^2 \int_{t_0}^{t} dt_1 H_I(t_1)\int_{t_0}^{t_1} dt_2 H_I(t_2) \\
&\quad + \cdots
\end{aligned}
\qquad (2.48)
$$

Eq. (2.48) is the basis of perturbation theory.

Now at $t = t_0 \to -\infty$, the system is known to be in an eigenstate $|a\rangle$ of $H_0$. Hence the probability amplitude for transition to an eigenstate $|b\rangle$ of $H_0$ is given by

$$
\begin{aligned}
C_b(t) &= \langle b\,|\Psi(t)\rangle_I \\
&= \lim_{t_0 \to -\infty} \langle b|\,U(t,\ t_0)\,|\Psi(t_0)\rangle_I \\
&= \lim_{t_0 \to -\infty} \langle b|\,U(t,\ t_0)e^{iH_0 t_0}\,|\Psi(t_0)\rangle_S.
\end{aligned}
\qquad (2.49)
$$

Now for $t_0 \to -\infty$,

$$|\Psi(t_0)\rangle_S = |a,t_0\rangle = |a\rangle\,e^{-iE_a t_0}. \qquad (2.50)$$

Hence from Eq. (2.49), we get

$$C_b(t) = \langle b|\,U(t,\ -\infty)\,|a\rangle. \qquad (2.51)$$

Our purpose is to calculate $C_b(t)$ for large $t$ (since for $t \to \infty$, the system is an eigenstate of $H_0$), i.e.

$$\lim_{t \to \infty} C_b(t) = \lim_{t \to \infty} \langle b|\,U(t,\ -\infty)\,|a\rangle = \langle b|\,U(\infty,\ -\infty)\,|a\rangle. \qquad (2.52)$$

The operator

$$S = U(\infty,\ -\infty) \qquad (2.53)$$

with matrix elements

$$S_{ba} = \langle b| U(\infty, -\infty) |a\rangle = \langle b| S |a\rangle \tag{2.54}$$

is called the S-matrix.

An important property of S-matrix is that it is a unitary operator. This follows from the conservation of probability:

$$\sum_b |C_b(\infty)|^2 = 1 \tag{2.55}$$

or

$$\sum_b \langle b| S |a\rangle \langle b| S |a\rangle^* = 1$$

$$\sum_b \langle a| S^\dagger |b\rangle \langle b| S |a\rangle = 1. \tag{2.56}$$

Hence

$$\langle a| S^\dagger S |a\rangle = 1,$$

i.e.

$$S^\dagger S = 1 \tag{2.57}$$

Therefore, $S$ is a unitary operator. It is convenient to introduce "*in*" and "*out*"states [see also Appendix A]. Assume that before time $t_1$ and after time $t_2$, $H = H_0$ and perturbation acts only during the time interval $t_1 < t < t_2$. Then the "*in*" state

$$|a\rangle_{in} = |a, t_1 \to -\infty\rangle \tag{2.58a}$$

Similarly the "*out*" state is

$$|a\rangle_{out} = |a, t_2 \to \infty\rangle \tag{2.58b}$$

The operator $S$ provides the link between the description of the system before $t_1(t_1 \to -\infty)$ and after $t_2$ $(t_2 \to \infty)$. Thus

$$|a\rangle_{in} = S|a\rangle_{out}$$

so that

$$S_{ba} =_{out} \langle b|a\rangle_{in} \tag{2.58}$$

where $a$ and $b$ indicate quantum numbers necessary to specify the initial and final states or in more transparent notation

$$S_{fi} =_{out} \langle f|i\rangle_{in} \tag{2.59}$$

Now using Eq. (2.48)

$$S = U(\infty, -\infty) = 1 - i \int_{-\infty}^{+\infty} dt_1 H_I(t_1)$$

$$+ (-i)^2 \int_{-\infty}^{+\infty} dt_1 \int_{-\infty}^{t_1} dt_2 H_I(t_1) H_I(t_2) + \cdots (2.60)$$

By interchanging the variables of integration $t_1 \to t_2$, we can write the second term as

$$(-i)^2 \int_{-\infty}^{+\infty} dt_2 \int_{-\infty}^{t_2} dt_1 H_I(t_2) H_I(t_1)$$

It can be shown[1] that this term is equally well described as

$$(-i)^2 \int_{-\infty}^{+\infty} dt_1 \int_{t_1}^{\infty} dt_2 H_I(t_2) H_I(t_1)$$

Using this identity we can write $S^{(2)}$ as

$$S^{(2)} = (-i)^2 \frac{1}{2!} \int_{-\infty}^{+\infty} dt_1 \int_{-\infty}^{+\infty} dt_2 T\left[H_I(t_1) H_I(t_2)\right] \qquad (2.61)$$

where the time ordered product

$$\begin{aligned} T\left[H_I(t_1) H_I(t_2)\right] \\ = H_I(t_1) H_I(t_2) \quad t_2 < t_1 \\ = H_I(t_2) H_I(t_1) \quad t_2 > t_1 \end{aligned} \qquad (2.62)$$

In terms of the Hamiltonian density $\mathcal{H}_I(x)$

$$H_I(t) = \int d^3 x \mathcal{H}_I(x)$$

so that

$$S^{(2)} = \frac{(-i)^2}{2!} \int d^4 x_1 \int d^4 x_2 T\left[\mathcal{H}_I(x_1) \mathcal{H}_I(x_2)\right]$$

A generalization of this is

$$S^{(n)} = \frac{(-i)^n}{n!} \int d^4 x_1 \int d^4 x_2 \cdots \int d^4 x_n T\left[\mathcal{H}_I(x_1) \mathcal{H}_I(x_2) \cdots \mathcal{H}_I(x_n)\right] \qquad (2.63)$$

Up to the first order in perturbation

$$S_{fi} = \delta_{fi} - i \int d^4 x \langle f | \mathcal{H}_I(x) | i \rangle \qquad (2.64)$$

---

[1]See J.J. Sakurai, Advanced Quantum Mechanics, Addison-Wesley (1973) p.186.

Using the translational invariance, which is generated by momentum operator one can write

$$\mathcal{H}_I(x) = e^{ip \cdot x}\mathcal{H}_I(0)e^{-ip \cdot x}$$

$$\int d^4x \, \langle f \,|\mathcal{H}_I(x)|\, i\rangle = \int d^4x e^{i(p_f - p_i) \cdot x} \, \langle f \,|\mathcal{H}_I|\, i\rangle$$

$$= (2\pi)^4 \, \delta^4 \left(p_f - p_i\right) \langle f \,|\mathcal{H}_I|\, i\rangle \qquad (2.65)$$

where $\mathcal{H}_I = \mathcal{H}_I(0)$ and $\delta$-function ensures energy-momentum conservation. This suggests that it would be convenient to introduce an operator $T$, called $T$ (transition) matrix with the element $T_{fi}$ so that

$$S_{fi} = \delta_{fi} + i(2\pi)^4 T_{fi} \qquad (2.66)$$

In the first order of perturbation

$$T_{fi} = -\langle f \,|\mathcal{H}_I|\, i\rangle \qquad (2.67)$$

Then, using Eqs. (2.37) and (2.38), the transition probability for large $t$ from a state $|i\rangle$ to state $|f\rangle$ for $i \neq f$ is given by

$$P = \lim_{t \to \infty} |C_f(t)|^2 = |\langle f| \, S \, |i\rangle|^2 = \sum (2\pi)^8 \, \delta^4 \left(p_f - p_i\right) \delta^4\left(0\right) |T_{fi}|^2 \,. \qquad (2.68)$$

Now

$$\delta^4\left(p\right) = \frac{1}{(2\pi)^4} \int e^{-ip^\mu x_\mu} d^4x$$

$$\delta^4\left(0\right) = \frac{1}{(2\pi)^4} \, (\text{Volume}) \, t = \frac{Vt}{(2\pi)^4} \,. \qquad (2.69)$$

Therefore, the transition rate per unit macroscopic volume is given by

$$W_{fi} = \frac{P}{Vt} = (2\pi)^4 \sum \delta^4 \left(p_f - p_i\right) |T_{fi}|^2 \,. \qquad (2.70)$$

To carry out sum over final states, we need to know the density of final states $\rho_f\left(E_f\right)$.

## 2.5   Phase Space

Consider a single particle in one dimension confined in the region $0 \leq x \leq L$. The normalized eigenstate of momentum operator $\widehat{p}$ is given by

$$u_p\left(x\right) = \frac{1}{\sqrt{L}} e^{ipx} \,. \qquad (2.71)$$

The boundary condition that $u_p(x)$ is periodic in the range $L$ gives

$$p = \left(\frac{2\pi}{L}\right) n. \tag{2.72}$$

Thus

$$\frac{dn}{dE} = \left(\frac{L}{2\pi}\right) \frac{dp}{dE} = \rho(E), \tag{2.73}$$

i.e. the number of states within the interval $E$ and $E + dE$ is given by $dn = \rho(E)dE$. In three dimensions, we have

$$\rho(E) = \frac{dn}{dE} = \left(\frac{L}{2\pi}\right)^3 \frac{d}{dE} \int d^3p = \left(\frac{L}{2\pi}\right)^3 p^2 \frac{dp}{dE} \int d\Omega. \tag{2.74}$$

The generalization to the $n$ particles in final state gives:

$$n = \left[\left(\frac{L}{2\pi}\right)^3\right]^{n-1} \int d^3p'_1 \, d^3p'_2 \cdots d^3p'_{n-1}. \tag{2.75}$$

Since

$$\mathbf{p}_i = \mathbf{p}_f = \mathbf{p}'_1 + \mathbf{p}'_2 + \cdots + \mathbf{p}'_n, \tag{2.76}$$

only $(n-1)$ momenta are independent. With the normalization $L = 2\pi$, we can rewrite from Eq. (2.75)

$$n = \int \delta^3 \left[\mathbf{p}_i - (\mathbf{p}'_1 + \mathbf{p}'_2 + \cdots + \mathbf{p}'_n)\right] d^3p'_1 \, d^3p'_2 \cdots d^3p'_n. \tag{2.77}$$

Thus we can write

$$\begin{aligned}
\rho_f(E) = \int & \delta\left[E - (E'_1 + E'_2 + \cdots + E'_n)\right] \\
& \times \delta^3 \left[\mathbf{p}_i - (\mathbf{p}'_1 + \mathbf{p}'_2 + \cdots + \mathbf{p}'_n)\right] \\
& \times d^3p'_1 \, d^3p'_2 \cdots d^3p'_n.
\end{aligned} \tag{2.78}$$

Hence the transition rate [cf. Eqs. (2.70) and (2.78)]

$$\begin{aligned}
W_f = (2\pi)^4 \int & d^3p'_1 \, d^3p'_2 \cdots d^3p'_n \\
& \times \sum_{\text{final spins}} \overline{|T_{fi}|^2} \, \delta^4 \left(p'_1 + p'_2 + \cdots + p'_n - p_i\right),
\end{aligned} \tag{2.79}$$

where bar $(-)$ denotes the average over initial spins if initial particles are unpolarized, otherwise we have to use the density matrix if initial particles are polarized.

**Remarks:**

(1) In the first order perturbation theory

$$T_{fi} = -\langle f| \mathcal{H}_I |i\rangle \cdot \tag{2.80}$$

(2) The normalization of states is

$$\langle \mathbf{p}' |\mathbf{p}\rangle = \int d^3x \, \langle \mathbf{p}' |\mathbf{x}\rangle \, \langle \mathbf{x} |\mathbf{p}\rangle \tag{2.81}$$

$$= \frac{1}{(2\pi)^3} \int e^{-i(\mathbf{p}'-\mathbf{p})\cdot\mathbf{x}} d^3x = \delta(\mathbf{p}'-\mathbf{p}).$$

The phase space $\int d^3p$ is not Lorentz invariant. Thus we consider the Lorentz invariant phase space

$$\int d^4p' \, \delta(p'^2 - m^2) \, \theta(p'_0)$$

$$= \int d^3p' \int dp'_0 \, \frac{1}{2p'_0} \, [\delta(p'_0 - E') + \delta(p'_0 + E')] \, \theta(p'_0)$$

$$= \int \frac{d^3p'}{2E'}. \tag{2.82}$$

Now we write

$$|\mathbf{p}\rangle = \int d^3p' \, |\mathbf{p}'\rangle \, \langle \mathbf{p}' |\mathbf{p}\rangle$$

$$= \int \frac{d^3p'}{2E'} \, |\mathbf{p}'\rangle \, [2E' \, \langle \mathbf{p}' |\mathbf{p}\rangle]$$

$$= \int \frac{d^3p'}{2E'} \, \sqrt{4 \, p_0 \, p'_0} \, \langle \mathbf{p}' |\mathbf{p}\rangle \, |\mathbf{p}'\rangle. \tag{2.83}$$

It is clear from Eq. (2.82), that $\langle p'| T |p\rangle$ is not Lorentz invariant, but $\sqrt{p_0 \, p'_0} \, \langle p'| T |p\rangle$ is. Thus in general we write

$$T_{fi} = N' F_{fi}, \tag{2.84}$$

where $N'$ is a multiple of factors like $1/E_r$. In fact it is convenient to take

$$N' = \left( \prod_r \frac{m_r}{(2\pi)^3 E_r} \prod_s \frac{1}{(2\pi)^3 \, 2E_s} \right)^{1/2}, \tag{2.85}$$

if there are $r$ fermions and $s$ bosons such that

$$r + s = n + m, \quad N' = \left[ \frac{1}{(2\pi)^{3/2}} \right]^{n+m} N, \tag{2.86}$$

$m$ and $n$ being the number of initial and final particles respectively. Hence finally the transition rate is

$$W_f = \frac{(2\pi)^4}{(2\pi)^{3m}} \int \frac{d^3 p_1'}{(2\pi)^3} \frac{d^3 p_2'}{(2\pi)^3} \cdots \frac{d^3 p_n'}{(2\pi)^3}$$

$$\times N^2 \delta^4 \left( p_1' + p_2' + \cdots + p_n' - p_i \right) \sum_{final\ spins} |F_{fi}|^2 . \qquad (2.87)$$

In the first order perturbation theory

$$\left[ \frac{1}{(2\pi)^{3/2}} \right]^{n+m} \prod_r \left( \frac{m_r}{E_r} \right)^{1/2} \prod_s \left( \frac{1}{2E_s} \right)^{1/2} F_{fi} = - \langle f| \mathcal{H}_I |i \rangle . \quad (2.88)$$

For example for $s = 0, r = 4$ (i.e. four fermions process)

$$\frac{1}{(2\pi)^6} \left( \frac{m_a\ m_b\ m_c\ m_d}{E_a\ E_b\ E_c\ E_d} \right)^{1/2} F_{fi} = - \langle f| \mathcal{H}_I |i \rangle . \qquad (2.89)$$

Here $m_a$, $m_b$, $m_c$ and $m_d$ are the masses of four particles $a, b, c$ and $d$ involved in a scattering or a decay process.

## 2.6 Examples

### 2.6.1 Two-body Scattering

Consider the scattering process

$$a + b \to c + d,$$

where $a$ and $c$ are bosons e.g. pions and $b$ and $d$ are fermions, e.g. nucleons. The scattering cross section is given by

$$d\sigma = \frac{dW}{(\text{Flux})_{in}}, \qquad (2.90)$$

where $(\text{Flux})_{in}$ is the incident flux defined as

$$(\text{Flux})_{in} = \rho_1 \rho_2 v_{in} = \frac{v_{in}}{(2\pi)^6}. \qquad (2.91)$$

$$v_{in} = \left| \frac{\mathbf{p}_a}{E_a} - \frac{\mathbf{p}_b}{E_b} \right|. \qquad (2.92)$$

We calculate the scattering cross section in the c.m. frame. In this frame:

$$\mathbf{p}_a = -\mathbf{p}_b = \mathbf{p}_{ab} = \mathbf{p}, \quad \mathbf{p}_c = -\mathbf{p}_d = \mathbf{p}_{cd} = \mathbf{p}' \qquad (2.93a)$$

$$E_{cm} = E_a + E_b = E_c + E_d. \qquad (2.93b)$$

$$v_{in} = |\mathbf{p}| \frac{E_{cm}}{E_a E_b}. \tag{2.94}$$

Now from Eq. (2.87)

$$dW = \frac{(2\pi)^4}{(2\pi)^6} \int \frac{d^3 p_c}{(2\pi)^3} \frac{d^3 p_d}{(2\pi)^3} \left( \frac{m_b m_d}{4 E_b E_d E_c E_a} \right) \delta^4 \left( p_c + p_d - p_a - p_b \right) \overline{\sum_{\text{spin}} |F_{fi}|^2}.$$
$$\tag{2.95}$$

We can write

$$\delta^4 \left( p_c + p_d - p_a - p_b \right) = \delta^3 \left( \mathbf{p}_c + \mathbf{p}_d - \mathbf{p}_a - \mathbf{p}_b \right) \delta \left( E_c + E_d - E_a - E_b \right). \tag{2.96}$$

The integration over $d^3 p_d$ in Eq. (2.95) can be removed by the three-dimensional $\delta$-function. Writing

$$d^3 p_c = |\mathbf{p}'|^2 d |\mathbf{p}'| d\Omega', \tag{2.97}$$

Eq. (2.95) gives

$$dW = \frac{m_b m_d}{4 E_b E_a} \frac{1}{(2\pi)^8} \int |\mathbf{p}'|^2 d |\mathbf{p}'| d\Omega' \delta \left( E_{cm} - \sqrt{\mathbf{p}'^2 + m_c^2} - \sqrt{\mathbf{p}'^2 + m_d^2} \right)$$
$$\frac{1}{\sqrt{\mathbf{p}'^2 + m_c^2}} \frac{1}{\sqrt{\mathbf{p}'^2 + m_d^2}} \overline{\sum_{\text{spins}} |F_{fi}|^2}. \tag{2.98}$$

Now using the formula

$$\int dx \, \delta \left[ E - Y(x) \right] F(x) = \left[ F(x) \frac{1}{Y'(x)} \right]_{E = Y(x)}, \tag{2.99}$$

we have from Eqs. (2.90) and (2.98)

$$d\sigma = \frac{dW \, (2\pi)^6}{v_{in}}$$

$$= \frac{1}{64\pi^2} \frac{|\mathbf{p}'|}{|\mathbf{p}|} \frac{1}{E_{cm}^2} |M|^2 d\Omega', \tag{2.100}$$

where

$$|M|^2 = \begin{cases} 4 m_b m_d \overline{\sum_{\text{spins}} |F_{fi}|^2}, \text{ if particles a,c bosons and b, d fermions.} \\ 16 m_a m_b m_c m_d \overline{\sum_{\text{spins}} |F_{fi}|^2}, \text{ if all particles are fermions} \\ \overline{\sum_{\text{spins}} |F_{fi}|^2}, \text{ if all particles are bosons} \end{cases} \tag{2.101}$$

In the c.m. frame,

$$dt \equiv -dQ^2$$

$$= -\frac{|\mathbf{p}| |\mathbf{p}'|}{\pi} d\Omega'$$

$$\frac{d\sigma}{dQ^2} = \frac{1}{E_{cm}^2} \frac{1}{|\mathbf{p}|^2} \frac{|M|^2}{64\pi} \qquad (2.102)$$

In the Lab frame we note from Eqs. (2.29), (2.31b) and (2.32)

$$p = \gamma(p_L - v\nu_L)$$
$$= \frac{\gamma p_L m_b}{\nu_L + m_b} = \frac{p_L m_b}{E_{cm}}$$

so in the Lab frame

$$\frac{d\sigma}{dQ^2} = \frac{1}{m_b^2} \frac{1}{|\mathbf{p}_L|^2} \frac{|M|^2}{64\pi}$$

From Eq. (2.20b),

$$dQ^2 = \frac{1}{\pi} |\mathbf{p}_L| |\mathbf{p}_L'| d\Omega_L$$

one gets

$$\frac{d\sigma}{d\Omega_L} = \frac{1}{m_b^2} \frac{|\mathbf{p}_L'|}{|\mathbf{p}_L|} \frac{|M|^2}{64\pi^2}. \qquad (2.103)$$

## 2.6.2   Three-body Decay

### 2.6.2.1   Three-body Phase Space

Consider a three-body decay

$$a \rightarrow b + c + d$$
$$m \rightarrow m_1 + m_2 + m_3$$
$$k = p_1 + p_2 + p_3.$$

The decay rate [cf. Eq. (2.87)] is given by [$\rho_{in} = \frac{1}{(2\pi)^3}$]

$$d\Gamma = \frac{dW}{\rho_{in}}$$
$$= (2\pi)^4 \int \frac{d^3 p_1}{(2\pi)^3} \int \frac{d^3 p_2}{(2\pi)^3} \int \frac{d^3 p_3}{(2\pi)^3} \left( \frac{m m_1 m_2 m_3}{E \, E_1 E_2 E_3} \right)$$
$$\times \delta^3 (\mathbf{p}_1 + \mathbf{p}_2 + \mathbf{p}_3 - \mathbf{k}) \delta (E_1 + E_2 + E_3 - E) |M|^2, \qquad (2.104)$$

where for definiteness, we have taken all the particles to be fermions.

We evaluate Eq. (2.104) in the rest frame of particle $m$. In this frame $\mathbf{k} = 0$ and $E = m$. Hence we have

$$\mathbf{p}_1 + \mathbf{p}_2 + \mathbf{p}_3 = 0$$
$$E_1 + E_2 + E_3 = m. \qquad (2.105)$$

From Eq. (2.104), removing the integration over $d^3p_3$ due to three-dimensional $\delta$-function, we get

$$d\Gamma = \frac{4\pi}{(2\pi)^5} (m_1 m_2 m_3) \int p_1^2 dp_1 \, p_2^2 dp_2 \, d\Omega_{12} \frac{1}{E_1 E_2 E_3}$$

$$\times \delta \left[ E_1 + E_2 + \sqrt{(\mathbf{p}_1 + \mathbf{p}_2)^2 + m_3^2} - m \right] |M|^2 . \qquad (2.106)$$

After performing the angular integration over $\Omega_{12}$, we obtain

$$d\Gamma = \frac{2(2\pi)^2}{(2\pi)^5} m_1 m_2 m_3 \int \frac{|\mathbf{p}_1| \, |\mathbf{p}_2| \, E_1 E_2 \, dE_1 \, dE_2}{E_1 E_2 E_3} \frac{E_3}{|\mathbf{p}_1| \, |\mathbf{p}_2|} |\overline{M}|^2$$

$$= \frac{2 m_1 m_2 m_3}{(2\pi)^3} \int dE_1 \, dE_2 \, |\overline{M}|^2 , \qquad (2.107)$$

where $|\overline{M}|^2$ is the value $|M|^2$ after the angular integration has been performed. In order to evaluate the integral in Eq. (2.107), it is convenient to define the invariants:

$$s_{12} = (k - p_3)^2 = (p_1 + p_2)^2$$

$$s_{13} = (k - p_2)^2 = (p_1 + p_3)^2$$

$$s_{23} = (k - p_1)^2 = (p_2 + p_3)^2 . \qquad (2.108)$$

In the rest frame of particle $m$, we have

$$s_{12} = m^2 + m_3^2 - 2mE_3$$

$$s_{13} = m^2 + m_2^2 - 2mE_2$$

$$s_{23} = m^2 + m_1^2 - 2mE_1 \qquad (2.109)$$

$$s_{12} + s_{13} + s_{23} = m^2 + m_1^2 + m_2^2 + m_3^2. \qquad (2.110)$$

On the other hand, in the c.m. frame of particles 1 and 2, we put

$$\mathbf{p}_1 = -\mathbf{p}_2 = \mathbf{p} \quad \text{and} \quad \mathbf{p}_3 = \mathbf{q}. \qquad (2.111)$$

In this frame, we denote the energies of particles 1, 2 and 3, by $\omega_1$, $\omega_2$, $\omega_3$ respectively. Thus in this frame

$$s_{13} = (\omega_1 + \omega_3)^2 - (\mathbf{p} + \mathbf{q})^2 = m_1^2 + m_3^2 - 2\mathbf{p}.\mathbf{q} + 2\omega_1 \omega_3$$

$$s_{23} = (\omega_2 + \omega_3)^2 - (\mathbf{p} - \mathbf{q})^2 = m_2^2 + m_3^2 + 2\mathbf{p}.\mathbf{q} + 2\omega_2 \omega_3$$

$$s_{12} = (\omega_1 + \omega_2)^2 . \qquad (2.112)$$

For fixed $s_{12}$, the range of $s_{23}$ is determined by letting $\mathbf{q}$ to be parallel or antiparallel to $\mathbf{p}$. Thus

$$(s_{23})_{\min}^{\max} = (\omega_2 + \omega_3)^2 - \left[ \sqrt{\omega_3^2 - m_3^2} \mp \sqrt{\omega_2^2 - m_2^2} \right]^2 . \qquad (2.113)$$

We also note that one can express $\omega_1$, $\omega_2$ and $\omega_3$ in terms of $s_{12}$.

$$\omega_1 = \frac{s_{12} + m_1^2 - m_2^2}{2\sqrt{s_{12}}}$$

$$\omega_2 = \frac{s_{12} - m_1^2 + m_2^2}{2\sqrt{s_{12}}}$$

$$\omega_3 = \frac{m^2 - m_3^2 - s_{12}}{2\sqrt{s_{12}}}. \tag{2.114}$$

In terms of the invariants $s_{13}$ and $s_{23}$, Eq. (2.107) can be written as

$$d\Gamma = \frac{2m_1 m_2 m_3}{(2\pi)^3 (4m^2)} \int ds_{23}\, ds_{12}\, \left|\overline{M}\right|^2. \tag{2.115}$$

The scatter plot in $s_{23}$ and $s_{12}$ is called a Dalitz plot (Fig. 2.4). If $\left|\overline{M}\right|^2$ is a constant, we have uniform distribution of events. Non-uniform distribution of events over Dalitz plot will indicate a structure in $\left|\overline{M}\right|^2$ and would provide an important information about the dynamics underlying the process concerned.

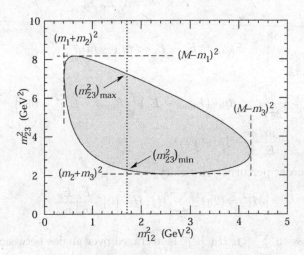

Fig. 2.4   Dalitz plot for a three body final state [ref. 6].

### 2.6.2.2   $\beta$-decay

$$A \to B + e^- + \overline{\nu}_e, \qquad p = p_B + p_e + p_\nu$$

e.g.

$$O^{14} \rightarrow N^{14} + e^+ + \nu_e.$$

We obtain from Eq. (2.104)

$$d\Gamma = \frac{(4\pi)^2}{(2\pi)^5} \int p_e^2 dp_e \, p_\nu^2 dp_\nu \, d\Omega_{e\nu} \, \delta \left( E_B + E_e + E_\nu - m_A \right)$$
$$\times \frac{m_A \, m_B \, m_e \, m_\nu}{m_A \, E_B \, E_e \, E_\nu} \, |M|^2 , \tag{2.116}$$

where

$$p_\nu \, dp_\nu = E_\nu \, dE_\nu. \tag{2.117}$$

It is a very good approximation to neglect the recoil of the particle $B$, so that $\mathbf{p}_B \approx 0$ and $E_B = m_B$. For this case, the $\delta$-function removes the integration over $dE_\nu$ and we get

$$d\Gamma = \frac{(4\pi)^2}{(2\pi)^5} \, p_e^2 \, dp_e \, d\Omega_{e\nu} \, (m_A - m_B - E_e)$$
$$\times \left( (m_A - m_B - E_e)^2 - m_\nu^2 \right)^{1/2} \left( \frac{m_e \, m_\nu}{E_e \, E_\nu} \, |M|^2 \right). \tag{2.118}$$

Let us write

$$E_{\max} = E_e + E_\nu \approx m_A - m_B. \tag{2.119}$$

Then one gets

$$d\Gamma = \frac{(4\pi)^2}{(2\pi)^5} \, p_e^2 \, dp_e \, (E_{\max} - E_e) \left( (E_{\max} - E_e)^2 - m_\nu^2 \right)^{1/2}$$
$$\times \left( \frac{m_e \, m_\nu}{E_e \, E_\nu} \, |M|^2 \right) d\Omega_{e\nu}. \tag{2.120}$$

In the first order perturbation theory,

$$|M|^2 = (2\pi)^{12} \sum_{\text{spin}} |\langle f | H_W | i \rangle|^2 \frac{E_e \, E_\nu}{m_e \, m_\nu}. \tag{2.121}$$

If the expression $\overline{\sum_{\text{spin}} |\langle f | H_W | i \rangle|^2}$ is averaged over angles between electron and neutrino, $d\Gamma$ can be integrated over $d\Omega_{e\nu}$ and we obtain

$$dΓ = \frac{(4\pi)^2}{(2\pi)^5} \, p_e^2 \, dp_e \, (E_{\max} - E_e)$$
$$\times \left( (E_{\max} - E_e)^2 - m_\nu^2 \right)^{1/2} (2\pi)^{12} \overline{\sum_{\text{spin}} |\langle f | H_W | i \rangle|}. \tag{2.122}$$

Let us make the simplest assumption that the averaged expression is independent of electron energy $E_e$. In this case

$$\frac{1}{p_e^2}\left(\frac{d\Gamma}{dp_e}\right) \propto (E_{\max} - E_e)\left[(E_{\max} - E_e)^2 - m_\nu^2\right]^{1/2}. \tag{2.123}$$

If we neglect the mass of the neutrino, then

$$K \equiv \left(\frac{d\Gamma}{p_e^2\, dp_e}\right)^{1/2} \propto (E_{\max} - E_e). \tag{2.124}$$

From Eq. (2.124), we see that plot of $\left(d\Gamma/p_e^2\, dp_e\right)^{1/2}$ versus $E_e$ should be a straight line. This is called Fermi or Kurie plot. Figure 2.5 shows that it is indeed a straight line. Therefore, the assumption that the matrix elements $\langle f|\, H_W\, |i\rangle$ are independent of energy is correct.

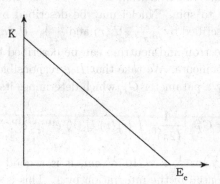

Fig. 2.5   Fermi or Kurie plot.

From Eq. (2.123), we get

$$\Gamma_\beta = \frac{1}{2\pi^3}\left[(2\pi)^{12}\, \overline{|\langle f|\, H_W\, |i\rangle|^2}\right]\int_0^{p_e^{\max}}(E_{\max} - E_e)^2\, p_e^2\, dp_e$$

$$= \frac{1}{2\pi^3}\, m_e^5\left[(2\pi)^{12}\, \overline{|\langle f|\, H_W\, |i\rangle|^2}\right]f(\rho_0), \tag{2.125}$$

where

$$f(\rho_0) = \int_0^{\rho_0}\rho^2\left(\sqrt{\rho_0^2 + 1} - \sqrt{\rho^2 + 1}\right)^2 d\rho \tag{2.126}$$

$$\rho = \frac{p_e}{m_e}, \qquad \rho_0 = \frac{p_e^{\max}}{m_e}.$$

This does not take into account Coulomb corrections due to Coulomb force which the electron experiences with the nucleus of charge $Ze$ once it has left the nucleus. This can be taken into account in the integral $f(\rho_0)$ and the formula (2.126) remains valid. The lifetime for $\beta$-decay $\tau_\beta = 1/\Gamma_\beta$, but it is the half life $t_{1/2} = \tau_\beta(\ln 2)$ which is experimentally measured, while $f$ is computed. $ft_{1/2}$ is called the $ft$ value. It is assumed that $H_W$ is universal, i.e. the same for all decays and this assumption is supported by experiments. $ft$ values for $\beta$-decay vary from about $10^3$ to $10^{23}$ seconds. This variation is due to the phase space available in the final state characterized by $E_{max}$ and hence by $f(\rho_0)$. Other cause of variation is due to the nuclear wave functions that enter into the calculation of matrix elements $\langle f| H_W |i\rangle$. Without the universality of $H_W$, an understanding of weak interaction would be hopeless. Some characteristic $ft$ values are shown in Table 2.1.

We now consider the transition $O^{14} \to N^{14}$ so that we do not have complications due to spin. Nuclei may be described by highly localized wave functions described by $\frac{1}{(2\pi)^{3/2}}U_i(r)$ and $\frac{1}{(2\pi)^{3/2}}U_f(r)$ which vanish for $r > 10^{-13}$ cm. Electron and neutrino can be described by plane waves as they carry large momenta. We take that $H_W$ responsible for $\beta$-transitions is characterized by a parameter $G_F$ which determines its strength. Thus

$$|\langle f| H_W |i\rangle|^2 = G_F^2 \left| \frac{1}{(2\pi)^6} \int U_f^*(r) \, U_i(r) \, e^{i\mathbf{p_e \cdot r}} \, e^{i\mathbf{p_\nu \cdot r}} d^3r \right|^2 . \qquad (2.127)$$

Since $p_e/h \sim 10^{11}$ cm$^{-1}$, $r \approx 10^{-13}$ cm, it is a good approximation to replace the exponential in the integration by 1. This is called the allowed approximation. Thus one gets from Eq. (2.128)

$$(2\pi)^{12} \; \overline{|\langle f| H_W |i\rangle|^2} = G_F^2. \qquad (2.128)$$

Hence we obtain from Eq. (2.126)

$$\Gamma_\beta = \frac{G_F^2 \, m_e^5}{2\pi^3} \, f(\rho_0)$$

$$= \frac{1}{\tau_\beta} = \frac{\ln 2}{t_{1/2}} \qquad (2.129\text{a})$$

or

$$G_F^2 = \frac{(2\pi^3 \, \ln 2)}{ft} \, \frac{1}{m_e^5} \qquad (2.129\text{b})$$

or

$$(G_F \, m_N^2)^2 = \frac{(2\pi^3 \, \ln 2)}{ft} \left( \frac{m_N}{m_e} \right)^5 \frac{1}{m_N}. \qquad (2.129\text{c})$$

Using $\frac{1}{m_N} \approx (0.7) \, 10^{-24}$ sec and $ft = 3100$ sec we get

$$G_F \, m_N^2 \approx 1.5 \times 10^{-5} \tag{2.130}$$

to be compared with its present accurate determination

$$G_F = 1.166 \times 10^{-5} \text{ GeV}^2.$$

We rewrite Eq. (2.130b) in a more transparent form

$$f \, t_{1/2} = \frac{2\pi^3 (\ln 2)}{|M_F|^2} \left[ \frac{(\hbar c)^3}{G_F} \right]^2 \frac{1}{(m_e c^2)^4} \left( \frac{\hbar}{m_e c^2} \right) \tag{2.131}$$

where $M_F$ is called Fermi matrix element. We note that $f \, t_{1/2}$ is constant. With

$$\frac{G_F}{(\hbar c)^3} = 1.166 \times 10^{-5} \text{ GeV}^{-2} \tag{2.132}$$

one gets

$$f \, t_{1/2} = \frac{5972}{|M_F|^2} \text{ sec} \tag{2.133}$$

In $\beta$-decay, isobars are involved, i.e. it is a transition with $\Delta I_3 = \pm 1$. In particular for

$$O^{14} \to N^{14} : \quad \left[ J^P = 0^+, \, I = 1, \, I_3 = 0,1 \right]$$
$$M_F = \langle \psi_f : 0^+; 1,0 \, | I_- | \, \psi_i = 0^+, 1, 1 \rangle \tag{2.134}$$
$$= \sqrt{2}.$$

Also in the Fermi transition, $\Delta J = 0$. Thus for the decay $O^{14} - N^{14}$

$$f \, t_{1/2} = 2986 \text{ sec} \tag{2.135}$$

to be compared with the experimental value 3100 sec; a discrepancy of only 1%. In view of the simplification used, it is in good agreement with the experimental value. However, the decay $He^6 \to Li^6$ is forbidden in the Fermi theory as it involves a transition $\Delta J = \pm 1$, whereas in Fermi theory, the selection rule is $\Delta J = 0$.

But the decay $_2He^6 \to \, _3Li^6 + e^- + \bar{\nu}_e$ does occur. Gamow and Teller, then introduced an additional matrix element

$$M_{GT} = C_A \langle \psi_f | \sigma | \psi_i \rangle \Rightarrow \int U_f^* \langle \chi_f | \sigma | \chi_i \rangle U_i \, d^3x \tag{2.136}$$

which allows the selection rule $\Delta J = 0, \pm 1$. Summing over the spins and taking the average for the initial spin **s**

$$|M_{GT}|^2 = \frac{1}{2} \sum_i \sum_f \langle \chi_f | \sigma | \chi_i \rangle \cdot \langle \chi_f | \sigma | \chi_i \rangle^*$$

$$= \frac{1}{2} \sum_i \sum_f \langle \chi_i | \sigma | \chi_f \rangle \cdot \langle \chi_f | \sigma | \chi_i \rangle$$

$$= \frac{3}{2} \sum_i \langle \chi_i | \sigma^2 | \chi_i \rangle = 3 \qquad (2.137)$$

where we have used $\sigma^2 = 3$ and there is no interference term as $Tr(\sigma) = 0$. Hence we replace $|M_F|^2$ in Eq. (2.131) given by

$$|M|^2 = g_V^2 |M_F|^2 + g_A^2 |M_{GT}|^2$$

where $g_V = 1$ and $g_A = 1.261$. The results are summarized in Table 2.1.

Table 2.1

|  |  | $t_{1/2}$ | $T_e^{max}$ | $|M_F|^2$ | $|M_{GT}|^2$ | $f\, t_{1/2}(s)$ |
|---|---|---|---|---|---|---|
| $n \to p$ | $\frac{1}{2}^+ \to \frac{1}{2}^+$ | 10.6 min | 0.7821 | 1 | 3 | 1100 |
| $He^6 \to {}^6 Li$ | $0^- \to 1^-$ | 0.813 s | 3.50 | 0 | 3 | 810 |
| $O^{14} \to {}^{14} N$ | $0^+ \to 0^+$ | 71.4 s | 1.812 | 2 | 0 | 3100 |
| $H^3 \to He^3$ | $\frac{1}{2} \to \frac{1}{2}$ | 12.33 Yr | 0.0186 | 1 | 3 | 1131 |

Finally we note from Eq. (2.124) that a non-vanishing neutrino mass reveals itself as a downward deviation from a straight Kurie plot as the energy approaches its nominal ($m_\nu = 0$) kinematically allowed maximum $T_e^{max}$. We can write Eq. (2.124):

$$\left( \frac{d\Gamma}{dT_e} \right) \propto T_e^{3/2} \frac{(T_e + 2m_e)^{3/2}}{T_e + 2m_e} (T_e^{max} - T_e) \left[ (T_e^{max} - T_e)^2 - m_\nu^2 \right]^{1/2},$$
$$(2.138)$$

where

$$T_e = E_e - m_e = \sqrt{\mathbf{p}_e^2 + m_e^2} - m_e. \qquad (2.139)$$

We note that the effect of $m_\nu$ is near $T_e = T_e^{max}$, otherwise $(T_e^{max} - T_e)^2 \gg m_\nu^2$. If we put

$$\frac{T_e}{T_e^{max}} = x, \quad x_e = \frac{m_e}{T_e^{max}}, \qquad (2.140)$$

then

$$\Gamma_{\nu=0} \propto (T_e^{max})^5 \int_0^1 x^{3/2} (1-x)^2 \frac{(x + 2x_e)^{3/2}}{(x + x_e)} dx. \qquad (2.141)$$

Hence it follows from Eqs. (2.138) and (2.141), that if $m_\nu \neq 0$, then the fraction of events $G(m_\nu)$ which will be absent at the end point is given by

$$G(m_\nu) = \frac{1}{\Gamma} \int_{T_e^{max}-m_\nu}^{T_e^{max}} \left(\frac{d\Gamma}{dT_e}\right)_{m_\nu=0} dT_e$$

$$\propto \frac{(T_e^{max})^2 \, m_\nu^3}{(T_e^{max})^5} = g\left(\frac{m_\nu}{T_e^{max}}\right)^3, \qquad (2.142)$$

where $g$ is some constant. Hence it follows from Eq. (2.142), that in order to have $G(m_\nu)$ as large as possible, $T_e^{max}$ has to be as small as possible. Thus we see from Table 2.1, that tritium ($H^3$) is most suitable to determine the mass $m_\nu$ of neutrino experimentally, since electrons from this decay have very low end-point energy (18.6 keV).

The distortion at the extreme end of the Kurie plot due to $m_\nu \neq 0$ is shown in Fig. 2.6. Thus in order to determine $m_\nu$ one has to look for such a distortion, but note that the deviation is in fact quite small. Moreover, the fraction of the events in the energy range of 18.5 keV $\leq Ee \leq$ 18.6 keV is only $3 \times 10^{-7}$. The experiment is hence quite difficult and even then it would be extremely difficult to determine $m_\nu$ better than 2 eV by this method. We shall come back to this point in Chap. 12.

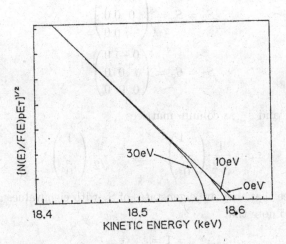

Fig. 2.6   The distortion at the extreme end of the Kurie plot due to $m_\nu \neq 0$.

## 2.7  Electromagnetic Interaction

A monochromatic electromagnetic wave is composed of N monoenergetic photons, each having energy and momentum, $E = \hbar\omega$ and $\mathbf{p} = \hbar\mathbf{k}$. The electromagnetic field is described by a vector potential $\mathbf{A}$ with polarization vector $\varepsilon$. Electromagnetic waves are transverse waves so that $\mathbf{k} \cdot \varepsilon = 0$ and thus have two independent states of polarization. We can conveniently describe it as left-circularly or right-circularly polarized photon or we can say that a photon has two helicity states $\pm 1$. Such a photon can be described by a polarization vector

$$\varepsilon_\pm = \frac{1}{\sqrt{2}}\,(\mp 1,\ -i,\ 0), \quad \varepsilon^0 = 0, \tag{2.143}$$

where we have taken the propagation vector $\mathbf{k}$ along $z$-axis.

The spin 1 matrices $\mathbf{S}$ are given by

$$(S_i)_{jk} = -i\epsilon_{ijk}. \tag{2.144}$$

Writing them explicitly, we have

$$S_1 = S_x = \begin{pmatrix} 0 & 0 & 0 \\ 0 & 0 & -i \\ 0 & i & 0 \end{pmatrix}$$

$$S_2 = S_y = \begin{pmatrix} 0 & 0 & i \\ 0 & 0 & 0 \\ -i & 0 & 0 \end{pmatrix}$$

$$S_3 = S_z = \begin{pmatrix} 0 & -i & 0 \\ i & 0 & 0 \\ 0 & 0 & 0 \end{pmatrix} \tag{2.145}$$

If we write $\varepsilon^+$ and $\varepsilon^-$ as column matrices

$$\varepsilon_+ = \frac{1}{\sqrt{2}}\begin{pmatrix} -1 \\ -i \\ 0 \end{pmatrix}, \quad \varepsilon_- = \frac{1}{\sqrt{2}}\begin{pmatrix} 1 \\ -i \\ 0 \end{pmatrix}, \tag{2.146}$$

it is easy to see that they are eigenstates of $S_z$ with eigenvalues $\pm 1$ respectively. We also note that

$$\varepsilon_+^* \cdot \varepsilon_+ = 1 = \varepsilon_-^* \cdot \varepsilon_-$$

$$\varepsilon_+^* \cdot \varepsilon_- = 0 = \varepsilon_-^* \cdot \varepsilon_+ \tag{2.147}$$

$$\varepsilon_\lambda^* \cdot \varepsilon_\lambda = \delta_{\lambda\lambda'}, \lambda, \lambda' = \pm 1. \tag{2.148}$$

For a real photon, if we sum over polarizations (spin), we have

$$\sum_{\lambda=\pm 1} \varepsilon_{i\lambda}^* \varepsilon_{j\lambda} = \delta_{ij} - \frac{k_i k_j}{\mathbf{k}^2}. \tag{2.149}$$

In quantum field theory, electromagnetic force between two electrons (or any charged particles) is assumed to be mediated by photons, the quanta of electromagnetic field. The simplest case is the exchange of a single photon as shown in Fig. 2.7.

The Coulomb potential between two charged particles is $e^2/4\pi r$ in rationalized Gaussian units. This is the Fourier transform of an amplitude $M(\mathbf{q})$ corresponding to the diagram shown in Fig. 2.7. Thus we write

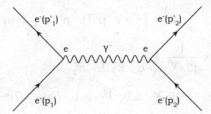

Fig. 2.7   Electron-electron scattering through exchange of a photon.

$$\frac{e^2}{4\pi r} = \frac{1}{(2\pi)^3} \int_{-\infty}^{\infty} e^{i\mathbf{q}\cdot\mathbf{r}} M(\mathbf{q}) \, d^3q. \tag{2.150}$$

In order to find $M(\mathbf{q})$, we note that

$$\int_{-\infty}^{\infty} \frac{e^{i\mathbf{q}\cdot\mathbf{r}}}{\mathbf{q}^2} d^3q = 2\pi \int_0^{\infty} \int_0^{\pi} e^{i|\mathbf{q}|r\cos\theta} \frac{1}{\mathbf{q}^2} |\mathbf{q}|^2 \, d|\mathbf{q}| \sin\theta \, d\theta$$

$$= \frac{4\pi}{r} \int_0^{\infty} \frac{\sin |\mathbf{q}| r}{|\mathbf{q}|} \, d|\mathbf{q}|$$

$$= \frac{4\pi}{r} \int_0^{\infty} \frac{\sin x}{x} \, dx = \frac{4\pi}{r} \frac{\pi}{2} = \frac{2\pi^2}{r}. \tag{2.151}$$

Hence we have

$$M(\mathbf{q}) = \frac{e^2}{\mathbf{q}^2}. \tag{2.152}$$

This gives the matrix elements of the above diagram (Fig. 2.7) in momentum space in non-relativistic limit. A relativistic generalization of this is

$$\langle T \rangle = -\langle M \rangle = \frac{e^2}{(2\pi)^6} \sqrt{\frac{m_1 \, m_2 \, m_1' \, m_2'}{E_1 \, E_2 \, E_1' \, E_2'}} \frac{\langle J^\mu \rangle_1 \, \langle J_\mu \rangle_2}{q^2}, \tag{2.153}$$

i.e.

$$F = \frac{g_{\mu\nu} \, \langle J^{\mu}\rangle_1 \, \langle J^{\nu}\rangle_2}{q^2}, \tag{2.154}$$

where $\langle J^{\mu}\rangle$ is the expectation value of the electromagnetic current. $g_{\mu\nu}/q^2$ is called the Feynman propagator of the photon. $J_{\mu}$ is given by

$$J^{\mu} = e \, \overline{\psi} \, \gamma^{\mu} \, \psi,$$

so that in free particle approximation

$$\langle J^{\mu}\rangle_i = e\overline{u}\left(\mathbf{p}_i'\right) \gamma^{\mu} u\left(\mathbf{p}_i\right), \quad i = 1, 2. \tag{2.155}$$

Thus

$$T = \frac{1}{q^2} \, \overline{u}\left(\mathbf{p}_2'\right) e\gamma^{\mu} u\left(\mathbf{p}_2\right) \overline{u}\left(\mathbf{p}_1'\right) e\gamma_{\mu} u\left(\mathbf{p}_1\right)$$

$$\times \frac{1}{(2\pi)^6} \frac{m^2}{\sqrt{E_1 \, E_2 \, E_1' \, E_2'}}. \tag{2.156}$$

In the non-relativistic limit $\frac{\mathbf{p}^2}{m} \approx 0$, $\frac{\mathbf{p}'^2}{m} \approx 0$, $E_1 = E_2 = E_1' = E_2' \approx m$,

$$\overline{u}\left(\mathbf{p}\right) \gamma^0 u\left(\mathbf{p}\right) \approx 1, \quad \overline{u}\left(\mathbf{p}\right) \gamma u\left(\mathbf{p}\right) \approx 0,$$

$$q^2 = \left(p_1' - p_1\right)^2 = \left(E_1' - E_1\right)^2 - 4\mathbf{p}^2 \approx -4\mathbf{p}^2,$$

and $q^2 \rightarrow -\mathbf{q}^2$ so that we have from Eq. (2.153)

$$-T = M\left(\mathbf{q}\right) = \frac{e^2}{\mathbf{q}^2}. \tag{2.157}$$

## 2.8   Weak Interaction

If weak nuclear force is mediated by exchange of some particle, then this particle must have a finite mass, since weak nuclear force is a short range force. We assume that mediator of this force is a vector particle of finite mass. It, therefore, has three directions of polarization or it is a spin 1 particle with $M_z = \pm 1, 0$. These spin states can be expressed as

$$\varepsilon_{\pm} = \frac{1}{\sqrt{2}} \left(\mp 1, \, -i, \, 0\right), \; \varepsilon_{\pm}^0 = 0$$

$$\varepsilon_0 = \left(0, \, 0, \, \frac{q_0}{m_W}\right), \; \varepsilon_0^0 = \frac{|\mathbf{q}|}{m_W}. \tag{2.158}$$

In this representation

$$q = (q_0, 0, \, 0, \, |\mathbf{q}|), \; q^2 = m_W^2 \tag{2.159}$$

so that

$$q \cdot \varepsilon = 0. \tag{2.160}$$

It is easy to see that $\varepsilon_\pm$, $\varepsilon_0$ are eigenstates of $S_z$ with eigenvalues $\pm 1$, 0 respectively. For a spin 1 particle on the mass-shell

$$\sum_{\lambda = \pm 1, 0} \varepsilon_\lambda^\mu \varepsilon_\lambda^\sigma = \varepsilon_+^{\mu^*} \varepsilon_+^\sigma + \varepsilon_-^\mu \varepsilon_-^{\sigma^*} + \varepsilon_0^\mu \varepsilon_0^{\sigma^*} = -g^{\mu\sigma} + \frac{q^\mu \, q^\sigma}{m_W^2}. \tag{2.161}$$

In order to estimate the strength of weak interaction, we evaluate the matrix elements of the scattering process

$$\nu_e + e^- \rightarrow \nu_e + e^-$$

as given by the diagram in Fig. 2.8. In analogy with Eq. (2.154), the

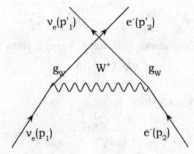

Fig. 2.8   Neutrino-electron scattering through exchange of vector boson.

scattering amplitude $F$ is given by [the propagator $1/\left(q^2\right)$ is replaced by $1/\left(q^2 - m_W^2\right)$ as W-boson is massive]

$$F = \frac{\langle J^{W\mu} \rangle_1 \, \langle J_\mu^W \rangle_2}{(q^2 - m_W^2)} \tag{2.162a}$$

Now in contrast to electron, neutrino is a two-component object and its wave function is $(1 - \gamma_5)u(p)$. Thus in analogy with Eq. (2.155)

$$\langle J^{W\mu} \rangle_i = g_W \, \overline{u}(\mathbf{p}_i') \, \gamma^\mu \, (1 - \gamma_5) \, u(\mathbf{p}_i), \quad i = 1, \, 2, \tag{2.162b}$$

where $g_W$ is the strength of weak interaction just as $e$ is the strength of the electromagnetic interaction. Thus for $q^2 << m_W^2$

$$F = -\frac{g_W^2}{m_W^2} \left[ \overline{u}(\mathbf{p}_2') \, \gamma^\mu \, (1 - \gamma_5) \, u(\mathbf{p}_2) \right] \left[ \overline{u}(\mathbf{p}_1') \, \gamma_\mu \, (1 - \gamma_5) \, u(\mathbf{p}_1) \right] \tag{2.162c}$$

$$= -\frac{g_W^2}{m_W^2} \left( A^\mu B_\mu \right)$$

Using Eq. (A.48), we get

$$|F|^2 = \frac{g_W^4}{m_W^4} \sum_{\text{spin}} A^\mu B_\mu A^{*\nu} B_\nu^* \tag{2.163}$$

$$= \frac{g_W^4}{m_W^4} \frac{1}{m_e m_\nu} \left[ p_2'^\mu \, p_2^\nu + p_2'^\nu \, p_2^\mu - p_2 \cdot p_2' g^{\mu\nu} + i\varepsilon^{\mu\nu\rho\sigma} p_{2\rho} \, p_{2\sigma}' \right]$$

$$\times \frac{2}{m_e m_\nu} \left[ p_{1\mu}' \, p_{1\nu} + p_{1\nu}' \, p_{1\mu} - p_1 \cdot p_1' g_{\mu\nu} + i\varepsilon_{\mu\nu\alpha\beta} \, p_1^\alpha \, p_1'^\beta \right]$$

$$= \frac{g_W^4}{m_W^4} \frac{2}{m_e^2 m_\nu^2} \left( s - m_e^2 - m_\nu^2 \right)^2, \tag{2.164}$$

where

$$s = (p_1 + p_2)^2 = (p_1' + p_2')^2 = E_{cm}^2. \tag{2.165}$$

From Eqs. (2.103) and (2.164), we get

$$\sigma = \frac{2g_W^4}{\pi m_W^4} \frac{1}{s} \left( s - m_e^2 - m_\nu^2 \right)^2. \tag{2.166}$$

If we neglect the lepton masses (viz for $s \gg m_e^2$), then we have

$$\sigma = \left( \frac{g_W^2}{4\pi} \right)^2 \left( \frac{8}{m_W^4} \right) 4\pi s. \tag{2.167}$$

Now $G_F/\sqrt{2} = g_W^2/m_W^2$ so that

$$\sigma = G_F^2 \frac{s}{\pi}. \tag{2.168}$$

Taking $\sigma \approx 10^{-38}$ cm$^2$ at $s = (1 \text{ GeV})^2$, we get

$$\left( \frac{10^{-38}}{4 \times 10^{-28}} \right) \pi \text{ GeV}^{-4} = G_F^2$$

$$G_F \approx 10^{-5} \text{ GeV}^{-2} \tag{2.169}$$

to be compared with Eq. (2.130). This shows the universality of the weak interaction since $G_F$ is the same as obtained from the $\beta$-decay or from the scattering of neutrinos on leptons.

In unified electroweak theory [see Chap. 13]

$$g_W \, \sin \theta_W = \frac{e}{2\sqrt{2}}, \tag{2.170}$$

where $\sin \theta_W$ is a parameter of the theory. Experimentally, $\sin^2 \theta_W \approx 1/4$, thus we get

$$\frac{G_F}{\sqrt{2}} = \frac{e^2}{8m_W^2 \sin^2 \theta_W} = 4\pi \frac{\alpha}{8m_W^2 \sin^2 \theta_W} \tag{2.171}$$

or

$$m_W = \left[ \frac{\pi\alpha}{\sqrt{2}} \left( \sin^2 \theta_W \, G_F \right)^{-1} \right]^{1/2}. \tag{2.172}$$

Using $\sin^2 \theta_W \approx 1/4$, and Eq. (2.172), we get

$$m_W \approx 80 \text{ GeV}. \tag{2.173}$$

## 2.9    Hadronic Cross-section

Consider the $N - N$ scattering through the pion exchange. In particular consider the diagram (Fig. 2.9).

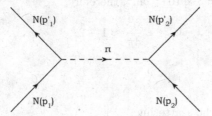

Fig. 2.9    Nucleon-nucleon scattering through pion exchange.

Neglecting the spin of the nucleon

$$F \approx g_s^2 \frac{1}{q^2 - m_\pi^2}, \qquad q^2 = (p_1' - p_1)^2. \tag{2.174}$$

From Eqs. (2.103) and (2.174), we get

$$\frac{d\sigma}{d\Omega} = g_s^4 \frac{1}{(q^2 - m_\pi^2)^2} m_N^4 \frac{1}{4\pi^2} \frac{|\mathbf{p}'|}{|\mathbf{p}|} \frac{1}{E_{cm}^2}. \tag{2.175}$$

For elastic scattering $|\mathbf{p}'| = |\mathbf{p}|$, so that

$$\begin{aligned}
\sigma &= \frac{g_s^4}{4\pi^2} \frac{m_N^4}{m_\pi^4} \frac{1}{s} \int_0^\pi \frac{2\pi \sin\theta d\theta}{\left[1 + \frac{2|\mathbf{p}|^2}{m_\pi^4}(1 - \cos\theta)\right]^2} \\
&= \left(\frac{g_s^2}{4\pi}\right)^2 \left(\frac{m_N}{m_\pi}\right)^2 \left(\frac{4\pi}{m_\pi^2}\right) \left(\frac{4m_N^2}{s}\right) \frac{1}{1 + \frac{s - 4m_N^2}{m_\pi^2}}. 
\end{aligned} \tag{2.176}$$

Now

$$s = 4m_N^2 \left(1 + \frac{E_L^K}{2m_N}\right), \tag{2.177}$$

where $E_L^K$ is the incident kinetic energy of the nucleon. Now $\frac{1}{m_\pi^2} \approx 2 \times (10^{-13})^2$ cm$^2$, $\left(\frac{m_N}{m_\pi}\right)^2 \approx 50$, thus

$$\sigma = \left(\frac{g_s^2}{4\pi}\right)^2 (1.3 \times 10^{-23}\text{cm}^2) \frac{1}{\left[1 + 14\frac{E_L^K}{m_N}\right]}. \tag{2.178}$$

For $E_L^K \ll \frac{m_\pi}{14} \approx 10$ MeV,

$$\sigma = \left(\frac{g_s^2}{4\pi}\right)^2 \left(1.3 \times 10^{-23} \text{cm}^2\right). \tag{2.179}$$

Experimentally $\sigma \approx 5 \times 10^{-23} \text{cm}^2$, therefore,

$$\frac{g_s^2}{4\pi} \approx 1 - 2. \tag{2.180}$$

## 2.10    Problems

(1) Show that for the scattering

$$e^- e^+ \to \gamma \to \pi^+ (k_1) \; \pi^- (k_2)$$

the differential and total cross sections are given by ($s \gg m_e^2$):

$$\frac{d\sigma}{d\Omega} = \frac{\alpha^2}{s} |F(s)|^2 \frac{\left(\frac{s}{4} - m_\pi^2\right)^{3/2}}{s^{3/2}} \left(1 - \cos^2 \theta\right)$$

$$\sigma = \frac{8\pi}{3s} \alpha^2 |F(s)|^2 \frac{\left(\frac{s}{4} - m_\pi^2\right)^{3/2}}{s^{3/2}}$$

where $s = q^2 = (k_1 + k_2)^2$ and $F(s)$ is the electromagnetic form factor of the pion, defined by $\langle 0| J_\mu^{em} |\pi^+(k_1)\pi^-(k_2)\rangle = F(s)(k_1 + k_2)_\mu$.
**Hint:** See Appendix A.

(2) Consider the decay

$$\omega \to \pi^+ \pi^- \pi^0.$$

Discuss the Dalitz plot for this decay.
**Hint:** From Lorentz invariance, the decay amplitude

$$F_\lambda \sim \varepsilon_{\lambda\mu\nu\rho} \, p_1^\mu \, p_2^\nu \, p_3^\rho,$$

where $p_1$, $p_2$ and $p_3$ are four momenta of pions.

(3) Consider a process in which one proton is at rest and the other collides with it, as a result of collision a particle of rest mass $M$ is produced, in addition to the two protons.

$$p + p \to M + p + p.$$

   (a) Find the minimum energy the moving proton must have in order to make this reaction possible.

   (b) What would be the corresponding energy if both the protons are moving.

(4) (a) A K-meson decay into a muon and a neutrino

$$K^- \to \mu^- \bar{\nu}_e$$

Find the energy and velocity of $\mu^-$. Using the above result show that whether the reaction

$$\bar{\nu}_\mu + p \to \mu^+ + n$$

is allowed energetically.

$$m_K = 494 \; \frac{\text{MeV}}{c^2}, \quad m_\mu = 106 \; \frac{\text{MeV}}{c^2}$$

(b) In c.m frame

$$s = (E_a^{cm} + E_b^{cm})^2 = E_{cm}^2$$
$$p_a = (Ea, \; \mathbf{p}), \quad p_b = (Eb, \; -\mathbf{p})$$

In the lab-frame

$$p_a = (E_L, \; \mathbf{p}_L)$$
$$p_b = (m_b, 0)$$

Show that in the limit $E_L > m_a, \; m_b$

$$E_{cm} = \sqrt{2 m_b E_L}$$

(5) Consider the decay

$$W^- \to \mu^- + \bar{\nu}_\mu$$
$$m_W \approx 80.4 \text{ GeV}/c^2, \quad m_\mu \approx 106 \text{ MeV}/c^2$$
$$\tau = \gamma \tau_0, \quad \tau_0 = 2.2 \times 10^{-6} \text{ s}$$

What distance $\mu^-$ travels, before its decays?

(6) To explore a structure of size of linear dimension $d$, we need a beam of particles of de Broglie wavelength $\lambda \le d$. What is the momentum of beam particles required for $d = 0.01$ fm?

(7) For the decay

$$a \to b + c,$$

show that the decay width is given by

$$\Gamma = \frac{1}{32\pi^2} \frac{|\mathbf{p}|}{m_a^2} \int |M|^2 \, d\Omega$$

$$\Gamma = \frac{1}{8\pi} \frac{|\mathbf{p}|}{m_a^2} |M|^2,$$

if $|M|^2$ is independent of angles and $|\mathbf{p}|$ is the momentum in the rest frame of particle $a$ and

$$|M|^2 = \begin{cases} \overline{\sum_{\text{spin}} |F_{fi}|^2}, \text{ if all particles are bosons.} \\ (4m_b m_c) \overline{\sum_{\text{spin}} |F_{fi}|^2} \text{ if particles } b \text{ and } c \text{ are fermions.} \\ (4m_a m_b) \overline{\sum_{\text{spin}} |F_{fi}|^2} \text{ if particles } a \text{ and } b \text{ are fermions.} \end{cases}$$

(8) Consider the process

$$e^-(p_1) + e^+(p_2) = \bar{q}(k_1) + q(k_2) + g(k_3)$$

Define $s = (p_1 + p_2)^2 = q^2$ , $q = p_1 + p_2 = k_1 + k_2 + k_3$

$$x_i = \frac{2k_i \cdot q}{q^2}$$

Show that the phase space integral is

$$\int \frac{d^3k_1 d^3k_2 d^3k_3}{(2\pi)^9} \frac{1}{2E_1} \frac{1}{2E_2} \frac{1}{2E_3} (2\pi)^4 \delta^4 (q - k_1 - k_2 - k_3)$$

$$= \frac{q^2}{128\pi^3} \int dx_1 \int dx_2.$$

## 2.11    References

1. G. Källen, Elementary Particle Physics, Addison - Wesley, Reading, Massachusetts (1964)
2. S. Gasirowicz, Elementary Particle Physics, Wiley, New York (1966).
3. H. M. Pilkuhn, Relativistic Particle Physics, Springer - Verlag, New York (1979).
4. D. H. Perkins, Introduction to High Energy Physics (3rd Edition), Addison - Wesley, Reading, Massachusetts (1987).
5. E. M. Henley, A. Garcia, Subatomic Physics, World Scientific, 3rd edition (2007).
6. Particle Data Group, K. Nakamura et al., Journal of Physics, G **37**, 0750212 (2010).

# Chapter 3

# Space-Time Symmetries

## 3.1 Introduction

Symmetries have played an important role in the progress of physics. There is a close connection between a symmetry and a conservation law. This is stated in the form of Noether's theorem. This can be illustrated by a simple example. Consider a single particle of mass $m$ moving in a time independent potential $V(x^i)$. Such a system is described by the action integral

$$I = \int_{t_1}^{t_2} dt L(x^i, t) \tag{3.1}$$

where the Lagrangian

$$L(x^i, t) = \frac{1}{2} m \left( \frac{dx^i}{dt} \right)^2 - V(x^i). \tag{3.2}$$

Let us subject this system to a small change : $x^i(t) \rightarrow x^i(t) + \delta x^i(t)$ then

$$I \rightarrow I + \delta I = \int_{t_1}^{t_2} dt \left[ \frac{1}{2} m \frac{d(x^i + \delta x^i)}{dt} - V(x^i + \delta x^i) \right]$$

To the order $\delta(x^i)$

$$V(x^i + \delta x^i) = V(x^i) + \delta x^i \frac{\partial V}{\partial x^i} \tag{3.3}$$

and

$$\frac{d(x^i + \delta x^i)}{dt} \frac{d(x^i + \delta x^i)}{dt} = \left( \frac{dx^i}{dt} \right)^2 + 2 \frac{dx_i}{dt} \frac{d}{dt} (\delta x^i)$$

$$= \left( \frac{dx^i}{dt} \right)^2 + 2 \left[ \frac{d}{dt} \left( \delta x^i \frac{dx^i}{dt} \right) - \delta x^i \left( \frac{d^2 x^i}{dt^2} \right) \right] \tag{3.4}$$

Thus

$$\delta I = \int_{t_1}^{t_2} dt \delta x^i \left[ -m\frac{d^2 x^i}{dt^2} - \frac{\partial V(x^i)}{\partial x^i} \right] + m \int_{t_1}^{t_2} \frac{d}{dt} \left( \delta x^i \frac{dx^i}{dt} \right) \qquad (3.5)$$

The second integral is known as the surface term and can be eliminated if we assume that the variations to the path vanish at the end points $\delta x^i(t_1) = 0 = \delta x^i(t_2)$. Then we get the classical equation of motion as a result of extermization of $I$:

$$m\frac{d^2 x^i}{dt^2} = -\frac{\partial V(x^i)}{\partial x^i}$$

### 3.1.1   *Rotation and SO(3) Group*

To establish the connection between symmetry of the action and existence of conserved quantities, assume that $V(x^i)$ is a function of length $x^2 = x^i x^i$ only and consider rotation of coordinates

$$x'^i = R^{ij} x^j \qquad (3.6)$$

which leaves the length of a vector **x** invariant

$$x'^i x'^i = R^{ik} x^k R^{ij} x^j = x^i x^i$$
$$R^{ik} R^{ij} = \delta^{kj} \qquad (3.7)$$

For an infinitesimal rotation

$$R^{ij} = \delta^{ij} + \epsilon^{ij} \qquad (3.8)$$
$$\epsilon^{ij} = -\epsilon^{ji} \qquad (3.9)$$
$$x^i = \epsilon^{ij} x^j \qquad (3.10)$$

Thus $I$ is manifestly invariant under rotation since it depends on the length. Thus $\delta I = 0$ but now one cannot put surface integral equal to zero since the boundary condition $\delta x^i(t_1) = 0 = \delta x^i(t_2)$ will destroy the rotational invariance. So the invariance of $I$, together with equation of motion gives

$$0 = \int_{t_1}^{t_2} dt \frac{d}{dt} \left( \delta x_i m \frac{dx^i}{dt} \right) = \delta x_i m \frac{dx^i}{dt} \Big|_{t_1}^{t_2} \qquad (3.11)$$

But

$$\delta x^i = \epsilon^{ij} x^j \qquad (3.12)$$

so that

$$\delta x^i \left( m\frac{dx^i}{dt} \right) = \epsilon^{ij} x^j \left( m\frac{dx^i}{dt} \right)$$

$$= \frac{1}{2} \epsilon^{ij} L^{ij} \qquad (3.13)$$

$$L^{ij} = m\left(x^i\frac{dx^j}{dt} - x_j\frac{dx^i}{dt}\right) \qquad (3.14)$$

where $L^{ij}$ are components of angular momentum and from Eq. (3.11)

$$L^{ij}(t_1) = L^{ij}(t_2) \qquad (3.15)$$

i.e. $L^{ij}$ are conserved. Thus the conservation of angular momentum is a consequence of the rotational invariance of the action. The same thing holds in Quantum Mechanics where $L^{ij}$ are operators. They generate $SO(3)$ group which have just three generators corresponding to three independent parameters $\epsilon^{ij}$. One can easily identify these generators $[\hbar = 1]$ as

$$L^{ij} = i\left(x^i\frac{\partial}{\partial x^j} - x^j\frac{\partial}{\partial x^i}\right) \qquad (3.16)$$

because of the identity

$$\begin{aligned}
\delta x^i &= \frac{i}{2}\epsilon^{jk}L^{jk}x^i = \frac{i^2}{2}\epsilon^{jk}\left(x^j\frac{\partial}{\partial x^k} - x^k\frac{\partial}{\partial x^j}\right)x^i \\
&= -\frac{1}{2}\epsilon^{jk}\left(x^j\delta^{ik} - x^k\delta^{ij}\right) \\
&= -\frac{1}{2}\left(\epsilon^{ji}x^j - \epsilon^{ik}x^k\right) \\
&= \epsilon^{ij}x^j \qquad (3.17)
\end{aligned}$$

It is easy to see that $L^{ij}$ satisfy the commutation relation

$$\left[L^{ij}, L^{mn}\right] = i\left[\delta^{in}L^{jm} - \delta^{jn}L^{im} + i \leftrightarrow j\right] \qquad (3.18)$$

which forms the Lie algebra of $SO(3)$. The most general representation of the generators of $SO(3)$ is given by

$$J^{ij} = L^{ij} + S^{ij} \qquad (3.19)$$

where the hermitian $S^{ij}$ satisfies the same commutation relation as $L^{ij}$ and commute with them and refer to internal degree of freedom. For spin-$\frac{1}{2}$ particle,

$$S^{ij} = \frac{1}{2}\sigma^{ij} = \frac{1}{2}\epsilon^{ijk}\sigma^k \qquad (3.20)$$

where $\sigma$'s are Pauli matrices. Consider

$$L^{ij} = \epsilon^{ijk}L^k \qquad (3.21)$$

$$J^{12} = J^3 = L^3 + \frac{1}{2}\sigma^3 \qquad (3.22)$$

$$\left[J^i, J^j\right] = i\epsilon^{ijk}J^k \qquad (3.23)$$

In Quantum Mechanics a transformation is associated with unitary operator $U$, which corresponding to the rotation

$$x'^i = x^i + \epsilon^{ij} x^j \tag{3.24}$$

is

$$U_R = 1 - \frac{i}{2} \epsilon^{ij} J^{ij} \tag{3.25}$$

which on exponentiation becomes

$$U_R = e^{-\frac{i}{2} \epsilon^{ij} J^{ij}}$$
$$= e^{-i\omega \cdot \mathbf{J}} \tag{3.26}$$

where

$$\epsilon^{ij} = \epsilon^{ijk} \omega^k \tag{3.27}$$

### 3.1.2  *Translation*

For space-time translation

$$x'^\mu = x^\mu + a^\mu \tag{3.28}$$

or

$$\delta x^\mu = x'^\mu - x^\mu = \epsilon^\mu \tag{3.29}$$

and

$$P^\mu = -i\partial_\mu = -i \frac{\partial}{\partial x^\mu} \tag{3.30}$$

is corresponding generator since

$$\delta x^\mu = i\epsilon^\nu (-i\partial_\nu) x^\mu = \epsilon^\nu \delta_\nu^\mu = \epsilon^\mu \tag{3.31}$$

By Noether's theorem, the invariance of action under translation gives conservation of $P^\mu$, i.e. energy momentum.

The corresponding unitary operator is

$$U_T = 1 - i\epsilon^\mu P_\mu \tag{3.32}$$

which on exponentiation gives

$$U_T = e^{-i\epsilon^\mu P_\mu} \tag{3.33}$$

The above discussion can be extended to Lorentz group.

### 3.1.3 *Lorentz Group*

As discussed in Sec. 2.1, a Lorentz transformation is

$$x'^\mu = \Lambda^\mu_\nu x^\nu \tag{3.34}$$

$$g_{\mu\nu} = \Lambda^\alpha_\mu g_{\alpha\beta} \Lambda^\beta_\gamma \tag{3.35}$$

An infinitesimal Lorentz transformation can be written as

$$x'^\mu = \delta^\mu_\nu + g^{\mu\alpha}\epsilon_{\alpha\nu}x^\nu \tag{3.36}$$

so that

$$\Lambda^\mu_\nu = \delta^\mu_\nu + g^{\mu\alpha}\epsilon_{\alpha\nu} \tag{3.37}$$

and the above condition (3.35) gives

$$\epsilon_{\mu\nu} = -\epsilon_{\nu\mu} \tag{3.38}$$

Corresponding to six parameters $\epsilon_{\mu\nu}$, the Lorentz group has six generators

$$L^{\mu\nu} = i\left(x^\mu \partial^\nu - x^\nu \partial^\mu\right), \quad L_{\mu\nu} = i\left(x_\mu \partial_\nu - x_\nu \partial_\mu\right) \tag{3.39}$$

since in terms of these we have the identity

$$\begin{aligned}
\delta x^\lambda = x'^\lambda - x^\lambda &= \frac{i^2}{2}\epsilon_{\mu\nu}\left(x^\mu \partial^\nu - x^\nu \partial^\mu\right)x^\lambda \\
&= -\frac{1}{2}\epsilon_{\mu\nu}\left(x^\mu g^{\nu\lambda} - x^\nu g^{\mu\lambda}\right) \\
&= g^{\lambda\mu}\epsilon_{\mu\nu}x^\nu
\end{aligned} \tag{3.40}$$

It is easy to check that $L^{\mu\nu}$'s satisfies the commutation relations

$$[L^{\mu\nu}, L^{\rho\sigma}] = i\left\{g^{\mu\rho}L^{\nu\sigma} - g^{\nu\sigma}L^{\mu\rho} + \mu \leftrightarrow \nu\right\} \tag{3.41}$$

which form Lie algebra of the Lorentz group. The most general representations of the generators of this group that obey the above commutation relations are given by

$$M^{\mu\nu} = L^{\mu\nu} + S^{\mu\nu} \tag{3.42}$$

where the hermitian $S^{\mu\nu}$ satisfies the same commutation relation as $L^{\mu\nu}$ and commute with them so that

$$[M^{\mu\nu}, M^{\rho\sigma}] = i\left[g^{\mu\rho}M^{\nu\sigma} - g^{\nu\sigma}M^{\mu\rho} + \mu \leftrightarrow \nu\right] \tag{3.43}$$

For a Dirac spin $\frac{1}{2}$ field

$$S^{\mu\nu} = \frac{1}{2}\Sigma^{\mu\nu} = \frac{i}{2}\left[\gamma^\mu, \gamma^\nu\right], \tag{3.44}$$

since this is the only combination of $\gamma$-matrices which have six independent components and is antisymmetric in $\mu, \nu$. The unitary transformation corresponding to the infinitesimal Lorentz transformation is

$$U_\Lambda = 1 + \frac{i}{2}\epsilon_{\mu\nu}M^{\mu\nu} \tag{3.45}$$

which on exponentiation is

$$U_\Lambda = e^{\frac{i}{2}\epsilon_{\mu\nu}M^{\mu\nu}} \tag{3.46}$$

Just as the Lorentz transformation has two subgroups as discussed in Sec. 2.1, the six generators of the Lorentz group spilt into three generators $M^{ij}$ which belong to $SO(3)$

$$M^{ij} = -\epsilon^{ijk}J_k = \epsilon^{ijk}J^k \tag{3.47}$$

where $\epsilon^{123} = 1$, $\epsilon_{123} = -1$, $\epsilon^{oijk} = \epsilon^{ijk}$

$$\left[J^i, J^j\right] = i\epsilon^{ijk}J^k \tag{3.48}$$

satisfying the commutation relations of $SO(3)$. The other three generators $K^i = M^{0i}$ give the Lorentz boosts with commutation relations

$$\left[J^j, K^l\right] = i\epsilon^{jlm}K^m \tag{3.49}$$

$$\left[K^i, K^j\right] = -\left[M^{0i}, M^{oj}\right]$$
$$= -ig^{00}M^{ij} = -i\epsilon^{ijk}J^k \tag{3.50}$$

Note that the minus sign in the last equation which is manifestation of the non-compactness of the Lorentz group. Note also $K$'s are antihermitian. It is useful to introduce hermitian combination

$$M^i = \frac{1}{2}\left(J^i + iK^i\right) \tag{3.51}$$

$$N^i = \frac{1}{2}\left(J^i - iK^i\right) \tag{3.52}$$

which satisfy

$$\left[M^i, M^j\right] = i\epsilon^{ijk}M^k \tag{3.53}$$

$$\left[N^i, N^j\right] = i\epsilon^{ijk}N^k \tag{3.54}$$

$$\left[N^i, M^j\right] = 0 \tag{3.55}$$

This algebra is identical to the Lie algebra of $SU_M(2)\otimes SU_N(2)$ with the Casimir operator

$$M^2 = M^iM^i \tag{3.56}$$

$$N^2 = N^iN^i \tag{3.57}$$

$$\left[M^2, M^i\right] = 0 = \left[N^2, N^i\right] \tag{3.58}$$

Thus in analogy with angular momentum, one can use the eigenvalues of $M^2, M^3, N^2, N^3$ to label the irreducible representations of the Lorentz group, but this is beyond the scope of this book.

## 3.2  Invariance Principle

We now formulate an invariance principle in a general way. Consider a transition from an initial state $|i\rangle$ to a final state $|f\rangle$. This transition is described by a matrix element $\langle f | S | i \rangle$. The invariance means:

$$\langle f | S | i \rangle = \langle f^u | S | i^u \rangle = \langle f | U^\dagger S U | i \rangle \tag{3.59}$$

or

$$S = U^\dagger S U \tag{3.60a}$$

or

$$[S, \, U] = 0. \tag{3.60b}$$

Here

$$|i^u\rangle = U |i\rangle$$
$$|f^u\rangle = U |f\rangle \tag{3.61}$$

are the transformed states. We see that the invariance under unitary transformation means that S-matrix commutes with it. Since S-matrix is related to the Hamiltonian of the system, it follows that $[H, U] = 0$.

We consider two cases when $U$ is continuous and discrete, as discussed below.

### 3.2.1  *U Continuous*

$U$ can be built out of infinitesimal transformations. Thus we need to consider an infinitesimal transformation:

$$U = 1 - i \, \varepsilon \, \widehat{F}, \tag{3.62}$$

where $\widehat{F}$ is a hermitian operator. $\widehat{F}$ can often be identified with an observable of the system, for example, the energy-momentum $P_\mu$ or the angular momentum $J$. $\widehat{F}$ is called the generator of the transformation represented by U. From Eq. (3.60b), we get

$$\left[ S, \, \widehat{F} \right] = 0. \tag{3.63}$$

This means that $\widehat{F}$ is conserved. To see this, let $|i\rangle$ and $|f\rangle$ be eigenstates of $\widehat{F}$ :

$$\widehat{F} |i\rangle = F_i |i\rangle$$
$$\widehat{F} |f\rangle = F_f |f\rangle. \tag{3.64}$$

From Eq. (3.63), we have

$$\langle f | \left[ S, \cdot \widehat{F} \right] | i \rangle = 0 \tag{3.65a}$$

or

$$(F_i - F_f) \langle f | S | i \rangle = 0. \tag{3.65b}$$

Hence, we get

$$F_i = F_f \quad \text{if} \quad \langle f | S | i \rangle \neq 0, \tag{3.66}$$

i.e. $\widehat{F}$ is conserved (eigenvalue of $\widehat{F}$ is conserved) in the transition $|i\rangle$ to $|f\rangle$. $\widehat{F}$ is then said to be a constant of motion. We have already discussed some of the common transformations and their generators in Sec. 3.1. Invariance under these transformations means that the corresponding generators are conserved. There is no evidence that space-time symmetries are violated by the fundamental laws of nature. The translation and rotational symmetries implies that space is homogeneous and isotropic.

### 3.2.2   *U is Discrete (e.g. Space Reflection)*

$$U^2 = 1. \tag{3.67a}$$

Eigenvalues of $U$ are

$$U' = \pm 1. \tag{3.67b}$$

Thus $U$ is both unitary and hermitian. $U$ can be regarded as an observable.

### 3.3   Parity

Consider a transformation corresponding to space reflection:

$$\mathbf{x} \rightarrow \mathbf{x}' = -\mathbf{x}. \tag{3.68}$$

The corresponding unitary operator is denoted by $\widehat{P}$, which acts on a wave function gives

$$\widehat{P} \, \Psi (\mathbf{x}, \, t) = \Psi (-\mathbf{x}, \, t). \tag{3.69}$$

Now

$$\widehat{P}^2 = 1, \tag{3.70}$$

so that $\widehat{P}$ has two eigenvalues $\pm 1$. If

$$\left[ S, \, \widehat{P} \right] = 0 \quad \text{or} \quad \left[ H, \, \widehat{P} \right] = 0 \tag{3.71}$$

then we say that parity is conserved. $\widehat{P}$ does not commute with all types of $H$. In particular, the weak interaction Hamiltonian $H_W$ does not commute with $\widehat{P}$ :

$$\left[ H_W, \ \widehat{P} \right] \neq 0 \tag{3.72}$$

i.e. parity is not conserved in weak processes.

Under parity operator $\widehat{P}$

$$\mathbf{x} \to -\mathbf{x}, \quad \mathbf{p} \to -\mathbf{p} \tag{3.73}$$

but the orbital angular momentum

$$\mathbf{L} = \mathbf{x} \times \mathbf{p} \to \mathbf{L}, \tag{3.74a}$$

so that

$$\mathbf{J} \to \mathbf{J}, \quad \sigma \to \sigma. \tag{3.74b}$$

Such vectors are called axial vectors. Also under parity, the scalars transform as:

$$\mathbf{x} \cdot \mathbf{p} \to \mathbf{x} \cdot \mathbf{p} \tag{3.75a}$$

$$(\mathbf{p_1} \times \mathbf{p_2}) \cdot \mathbf{p_3} \to - (\mathbf{p_1} \times \mathbf{p_2}) \cdot \mathbf{p_3} \tag{3.75b}$$

$$\mathbf{J} \cdot \mathbf{p} \to - \mathbf{J} \cdot \mathbf{p}. \tag{3.75c}$$

The scalars which change sign under parity are called pseudoscalars. All the three quantities are rotational invariant, but the last two have different behavior under $\widehat{P}$.

A particle when it is in an orbital angular momentum state $l$ has an orbital parity associated with it. In polar co-ordinates $\mathbf{x} \equiv (r, \theta, \phi)$, so that $\mathbf{x} \to -\mathbf{x}$ implies

$$r \to r, \quad \theta \to \pi - \theta, \quad \phi \to \pi + \phi. \tag{3.76}$$

Now we can write the wave function of a particle as

$$\Psi(\mathbf{x}) = R(r) Y_{lm}(\theta, \phi) \tag{3.77a}$$

$$Y_{lm}(\theta, \phi) = (-1)^m \left[ \frac{(2l+1)(l-m)!}{4\pi (l+m)!} \right]^{1/2} P_l^m(\cos\theta) \, e^{im\phi}. \tag{3.77b}$$

Under space inversion

$$P_l^m(\cos\theta) \to P_l^m(-\cos\theta) = (-1)^{l+m} P_l^m(\cos\theta) \tag{3.78a}$$

$$e^{im\phi} \to e^{im(\phi+\pi)} = (-1)^m e^{im\phi}, \tag{3.78b}$$

so that

$$Y_{lm}(\theta, \phi) \to (-1)^l \, Y_{lm}(\theta, \phi). \tag{3.78c}$$

We see that the orbital parity of a particle in an angular momentum state $l$ is $(-1)^l$.

## 3.4   Intrinsic Parity

As far as orbital parity is concerned, it is independent of the species of
particles and depends only on orbital angular momentum state of system
of particles. When creation or annihilation of particles takes place, we
have to assign an intrinsic parity to each particle. Consider, for example,
a photon, the quantum of electromagnetic field represented by a vector
potential :

$$\mathbf{A}(\mathbf{x}) = \varepsilon\, f(x), \tag{3.79}$$

where $\varepsilon$ is the polarization vector and $f(x)$ is a scalar function. Now the
interaction of a charged particle with electromagnetic field is introduced by
the gauge invariant substitution:

$$\mathbf{p} \to \mathbf{p} - e\,\mathbf{A}(\mathbf{x}). \tag{3.80}$$

Since $\mathbf{x}$ and $\mathbf{p}$ change sign under $\widehat{P}$, it follows that

$$\mathbf{A}(\mathbf{x}) \to -\mathbf{A}(-\mathbf{x}) \tag{3.81a}$$

i.e.

$$\widehat{P}\,\mathbf{A}(\mathbf{x})\,\widehat{P}^{-1} = -\mathbf{A}(-\mathbf{x}). \tag{3.81b}$$

This means that under parity

$$\varepsilon \to -\varepsilon. \tag{3.82}$$

The behavior of the polarization vector $\varepsilon$ characterizes what we call the
intrinsic parity of a photon. Thus we say that intrinsic parity of a photon
is odd. Similarly for any particle $a$ represented by a state vector $|a, \mathbf{p}\rangle$,

$$\widehat{P}\,|a, \mathbf{p}\rangle = \eta_a^P\,|a, -\mathbf{p}\rangle, \tag{3.83}$$

where $\eta_a^P$ is called the intrinsic parity of particle $a$. Note that $\eta_a^P = \pm 1$.
We now show that the conservation of parity leads to multiplicative conservation law. Consider a reaction

$$a + b \to c + d. \tag{3.84}$$

We can write the initial state

$$|i\rangle = |a\rangle\,|b\rangle\,|\text{ relative motion }\rangle. \tag{3.85}$$

Here $|a\rangle$ and $|b\rangle$ describe the internal states of $a$ and $b$, while the third
factor describes their relative motion. This state can be described by a
wave function $R(r)\,Y_{lm}(\theta, \phi)$. Since, we assume that parity is conserved in

the reaction (3.84), it follows that the states $|i\rangle$ and $|f\rangle$ are eigenstates of $\widehat{P}$, with eigenvalues $\eta_i^P$ and $\eta_f^P$ respectively. Now

$$\eta_i^P = \eta_a^P \, \eta_b^P \, (-1)^l \tag{3.86a}$$

$$\eta_f^P = \eta_c^P \, \eta_d^P \, (-1)^{l'}, \tag{3.86b}$$

where $\eta_a^P$, $\eta_b^P$, $\eta_c^P$, and $\eta_d^P$ are intrinsic parities of $a$, $b$, $c$ and $d$ respectively and $(-1)^l$ and $(-1)^{l'}$ are their orbital parities in the initial and final states.

Parity conservation for the reaction (3.84) gives

$$\eta_i^P = \eta_f^P \tag{3.87a}$$

or

$$\eta_a^P \, \eta_b^P \, (-1)^l = \eta_c^P \, \eta_d^P \, (-1)^{l'} \tag{3.87b}$$

i.e. parity is conserved as a multiplicative quantum number.

However, the law of parity conservation is not universal, in particular it does not hold for weak interactions. Then it follows from Eq. (3.72) that it is not possible to find simultaneous eigenstates of $H_W$ and $\widehat{P}$. Thus if parity is not conserved, the energy eigenstates $|\Psi\rangle$ are not expected to be eigenstates of parity. In this case, we can write

$$|\Psi\rangle = |\Psi_{\text{regular}}\rangle + y\,|\Psi_{\text{irregular}}\rangle, \tag{3.88}$$

where $|\Psi_{\text{regular}}\rangle$ and $|\Psi_{\text{irregular}}\rangle$ have opposite parities. $y$ is called the parity mixing amplitude and is a measure of the degree of parity non-conservation. Parity violation is maximum if $|y|^2 = 1$. Several experiments involving hadrons show that in hadronic interactions

$$|y|^2 < 10^{-13}.$$

Experiments involving atomic transitions show that parity is conserved to a high degree in electromagnetic interaction and that $|y|^2 < 10^{-14}$. For weak interactions, the parity violation is maximum viz $|y|^2 = 1$. It follows that in order to determine the intrinsic parity of a particle, one cannot use weak interactions. Only by considering reactions involving hadronic or electromagnetic interactions, one can determine the intrinsic parity of a particle. Even then the intrinsic parity cannot be fixed uniquely and we have to use a convention viz the intrinsic parity of a proton is $+1$, i.e.

$$\eta\,(\text{proton}) = +1. \tag{3.89}$$

Since proton and neutron form an isospin doublet, we also take

$$\eta\,(\text{neutron}) = +1. \tag{3.90}$$

### 3.4.1  *Intrinsic Parity of Pion*

We shall assume that the spin of pion is zero (we shall show later, how it comes out to be zero). Consider first the decay $\pi^0 \to 2\gamma$. Here we have two polarization vectors $\varepsilon_1$ and $\varepsilon_2$ corresponding to two $\gamma$-rays, whose momenta we take as $\mathbf{k}_1$ and $\mathbf{k}_2$, such that (gauge invariance) $\mathbf{k}_1 \cdot \varepsilon_1 = 0$, $\mathbf{k}_2 \cdot \varepsilon_2 = 0$. We also note that $\varepsilon_1 \cdot \varepsilon_2 = 0$. Now only the momentum $\mathbf{k} = \mathbf{k}_1 - \mathbf{k}_2$ is independent as $\mathbf{K} = \mathbf{k}_1 + \mathbf{k}_2 = 0$ in the rest frame of $\pi^0$. It is clear that the only invariant which we can form is $\mathbf{k} \cdot (\varepsilon_1 \times \varepsilon_2)$, which is a pseudoscalar, showing that intrinsic parity of $\pi^0$ is $-1$.

Consider the capture of $\pi^-$ at rest by deuteron. The dominant processes are

$$\pi^- + d \to n + n \tag{3.91}$$
$$\to n + n + \gamma.$$

Parity conservation for the first reaction gives

$$\eta_\pi \, \eta_d \, (-1)^l = \eta_n \, \eta_n \, (-1)^{l'} = (-1)^{l'}, \tag{3.92}$$

where $l$ is the relative orbital angular momentum of $\pi^- d$ and $l'$ is that of two neutrons. There is evidence that $\pi^-$ is captured in $l = 0$ orbital state. Thus from Eq. (3.92), we get

$$\eta_\pi \, \eta_d = (-1)^{l'}. \tag{3.93}$$

The deuteron is a bound state of a proton and neutron and has spin 1. The relative angular momentum of the two nucleons in deuteron is predominantly zero. Thus deuteron is a predominantly $^3S_1$ state, i.e. for a deuteron $J^P = 1^+$. It follows that the total angular momentum of the initial state is $J = 1$. Conservation of angular momentum gives $J_{\text{final}} = 1$. The spin $S$ of the two neutron system is either 0 or 1. Thus for $J = 1$, we have two possibilities: Triplet spin state ($S = 1$): $l' = 2, 1, 0$, i.e. the final state is $^3D_1$ or $^3P_1$ or $^3S_1$. For the singlet spin state ($S = 0$): $l' = 1$ and the final state is $^1P_1$. Now the Pauli exclusion principle requires that the final state must be antisymmetric. Since the triplet spin state is symmetric, the orbital state must be antisymmetric, i.e. $l' = 1$ and allowed final state is $^3P_1$. For the spin singlet state, since it is antisymmetric, $l'$ should be even. Thus $^1P_1$ state is not allowed by the Pauli exclusion principle. Hence we have the result that the final state must be $^3P_1$ so that from Eq. (3.93), we get

$$\eta_{\pi^-} = (-1)^{l'} = -1 \tag{3.94}$$

since $\eta_d = +1$. Thus for a pion $J^P = 0^-$ and it is called a pseudoscalar particle.

## 3.5   Parity Constraints on S-Matrix for Hadronic Reactions

### 3.5.1   *Scattering of Spin 0 Particles on Spin $\frac{1}{2}$ Particles*

Consider two-body elastic scattering of a spin 0 particle on a spin $\frac{1}{2}$ particle

$$
\begin{array}{ccccc}
a & + & b & \rightarrow & c & + & d \\
\mathbf{p}_1 & & (\mathbf{p}_2, \sigma) & & \mathbf{p}'_1 & & (\mathbf{p}'_2, \sigma)
\end{array}
$$

In the center-of-mass frame

$$
\mathbf{p}_1 = -\,\mathbf{p}_2 = \mathbf{p}_i
$$
$$
\mathbf{p}'_1 = -\,\mathbf{p}'_2 = \mathbf{p}_f. \tag{3.95}
$$

For the elastic scattering $|\mathbf{p}_i| = |\mathbf{p}_f| = |\mathbf{p}| = \mathbf{p}$. The initial and final states can be labeled as $|i\rangle = |\mathbf{p}_i, \ \sigma\rangle, |f\rangle = |\mathbf{p}_f, \ \sigma\rangle$. Under parity

$$
\hat{P}|i\rangle = \eta_i^p |-\mathbf{p}_i, \sigma\rangle, \quad \hat{P}|f\rangle = \eta_f^p |-\mathbf{p}_f, \sigma\rangle. \tag{3.96}
$$

The transition matrix elements

$$
\langle \mathbf{p}_f, \sigma| T |\mathbf{p}_i, \sigma\rangle = \langle \mathbf{p}_f, \sigma| \hat{P}^\dagger \hat{P}\, T\, \hat{P}^\dagger \hat{P} \, |\mathbf{p}_i, \sigma\rangle
$$
$$
= \eta_f^p \eta_i^p \, \langle -\mathbf{p}_f, \sigma| \hat{P}\, T\, \hat{P}^\dagger \, |-\mathbf{p}_i, \sigma\rangle. \tag{3.97}
$$

Now invariance under $\hat{P}$ implies

$$
\hat{P}\, T\, \hat{P}^\dagger = T. \tag{3.98}
$$

Because of elastic scattering

$$
\eta_i^p = \eta_f^p. \tag{3.99}
$$

Therefore, we have from Eqs. (3.97), (3.98) and (3.99)

$$
\langle -\mathbf{p}_f, \sigma| T |-\mathbf{p}_i, \sigma\rangle = \langle \mathbf{p}_f, \sigma| T |\mathbf{p}_i, \sigma\rangle. \tag{3.100}
$$

If we assume rotational invariance, then $\langle T \rangle$ can depend only on the rotational invariant quantities $p$, $\ \mathbf{p}_f \cdot \mathbf{p}_i$, $\ \sigma \cdot \mathbf{p}_i$, $\ \sigma \cdot \mathbf{p}_f$, $\ \sigma \cdot (\mathbf{p}_i \times \mathbf{p}_f)$. We need not consider $\sigma^2$ or higher powers of it, because $\sigma^2 = 3$ and $(\sigma \cdot \mathbf{a})(\sigma \cdot \mathbf{b}) = \mathbf{a} \cdot \mathbf{b} + i\sigma \cdot (\mathbf{a} \times \mathbf{b})$. Thus these quantities can be reduced to either a constant or $\sigma$. In other words, assuming rotational invariance only, we can write in spin space

$$
\langle \mathbf{p}_f, \sigma| T |\mathbf{p}_i, \sigma\rangle = [A(p, \theta) + A_1(p, \theta)\, \sigma \cdot \mathbf{p}_i + A_2(p, \theta)\, \sigma \cdot \mathbf{p}_f + B(p, \theta)\, \sigma \cdot (\mathbf{p}_i \times \mathbf{p}_f)].
$$
$$
\tag{3.101}
$$

This is a $2 \times 2$ matrix in spin space. It is understood that the above matrix elements are to be taken between spin wave functions $\chi_f^\dagger$ and $\chi_i$ for the final and initial states. Thus using rotational invariance alone, we have $2^2 = 4$ independent amplitudes. If in addition we assume invariance under parity, then Eqs. (3.100) and (3.101) imply $A_1 = 0 = A_2$. Therefore, invariance under rotation and space-inversion gives

$$\langle \mathbf{p}_f, \sigma | T | \mathbf{p}_i, \sigma \rangle = \chi_f^\dagger \left[ A\left(p, \theta\right) + B\left(p, \theta\right) \sigma \cdot \left(\mathbf{p}_i \times \mathbf{p}_f\right) \right] \chi_i. \tag{3.102}$$

This is an example which shows how a symmetry principle restricts the form of a transition matrix.

### 3.5.2 *Decay of a Spin $0^+$ Particle into Three Spinless Particles Each Having Odd Parity*

Consider the decay

$$A \to P_1 + P_2 + P_3$$

where all the particles have spin 0. Consider the decay in the rest frame of particle $A$. We have

$$0 = \mathbf{p}_1 + \mathbf{p}_2 + \mathbf{p}_3, \tag{3.103}$$

where $\mathbf{p}_1$, $\mathbf{p}_2$ and $\mathbf{p}_3$ are momenta of particles $P_1$, $P_2$ and $P_3$ respectively. The transition matrix elements for the decay is given by

$$M\left(\mathbf{p}_1, \ \mathbf{p}_2, \ \mathbf{p}_3\right) = \langle \ P_1\left(\mathbf{p}_1\right) \ P_2\left(\mathbf{p}_2\right) \ P_3\left(\mathbf{p}_3\right) | T \ | A\left(0\right) \ \rangle. \tag{3.104}$$

Under parity

$$\hat{P} | \ A\left(0\right) \ \rangle = | \ A\left(0\right) \ \rangle$$
$$\hat{P} | P_i\left(\mathbf{p}_i\right) \ \rangle = - | P_i\left(-\mathbf{p}_i\right) \ \rangle, \qquad i = 1, 2, 3. \tag{3.105}$$

Now

$$M\left(\mathbf{p}_1, \ \mathbf{p}_2, \ \mathbf{p}_3\right) = \langle \ P_1\left(\mathbf{p}_1\right) \ P_2\left(\mathbf{p}_2\right) \ P_3\left(\mathbf{p}_3\right) | \widehat{P}^\dagger \widehat{P} \ T \ \widehat{P}^\dagger \widehat{P} | A\left(0\right) \ \rangle$$
$$= (-1)^3 \langle \ P_1\left(-\mathbf{p}_1\right) \ P_2\left(-\mathbf{p}_2\right) \ P_3\left(-\mathbf{p}_3\right) | \widehat{P} \ T \ \widehat{P}^\dagger | A\left(0\right) \ \rangle. \tag{3.106}$$

If parity is conserved

$$\hat{P} \ T \ \hat{P}^\dagger = T \tag{3.107}$$

and we have from Eqs. (3.106) and (3.107)

$$M\left(\mathbf{p}_1, \ \mathbf{p}_2, \ \mathbf{p}_3\right) = -M\left(-\mathbf{p}_1, \ -\mathbf{p}_2, \ -\mathbf{p}_3\right). \tag{3.108}$$

Because of the rotational invariance, $M$ can be a function of rotational invariant quantities $\mathbf{p}_1 \cdot \mathbf{p}_2$, $\mathbf{p}_2 \cdot \mathbf{p}_3$, $\mathbf{p}_3 \cdot \mathbf{p}_1$ and $\mathbf{p}_1 \cdot \left(\mathbf{p}_2 \times \mathbf{p}_3\right)$. But the last invariant is zero, since $\mathbf{p}_3 = -(\mathbf{p}_1 + \mathbf{p}_2)$. Hence the rotational and space-inversion invariance implies

$$M\ (\mathbf{p}_1 \cdot \mathbf{p}_2,\ \mathbf{p}_2 \cdot \mathbf{p}_3,\ \mathbf{p}_3 \cdot \mathbf{p}_1) = -M\ (\mathbf{p}_1 \cdot \mathbf{p}_2,\ \mathbf{p}_2 \cdot \mathbf{p}_3,\ \mathbf{p}_3 \cdot \mathbf{p}_1)$$

or

$$M = 0.$$

Thus we have the result that the decay of a spinless particle with even parity to three pseudoscalar particles is forbidden if we assume invariance under space-inversion. On the other hand, decay of a spinless particle with odd parity to three pseudoscalar particles will be allowed under space-inversion invariance.

## 3.6   Time Reversal

Under time reversal

$$t \to -t, \quad \mathbf{x} \to \mathbf{x}. \tag{3.109a}$$

Therefore,

$$\mathbf{p} \to -\mathbf{p}, \quad \mathbf{L} \to -\mathbf{L}, \quad \sigma \to -\sigma. \tag{3.109b}$$

Let $\Pi$ denote the operation which transforms quantum mechanical states and operators under the above transformation, i.e. under $t \to -t$. First we show that $\Pi$ cannot be a unitary operator. Under $\Pi$, the commutation relation

$$[\widehat{q}_i, \ \widehat{p}_j] = i\hbar\delta_{ij} \to -i\hbar\delta_{ij}, \tag{3.110}$$

is not invariant. Hence the transformation generated by $\Pi$ cannot be unitary. But we want the above commutation relation to be invariant under $\Pi$. A way out of this difficulty is as follows: All $c$-numbers are simultaneously transformed into their complex conjugates. Such a transformation is called antiunitary. Then under $\Pi$,

$$\widehat{q}_i \to \Pi\,\widehat{q}_i\,\Pi^{-1} = \widehat{q}_i, \quad \widehat{p}_j \to \Pi\,\widehat{p}_j\,\Pi^{-1} = -\widehat{p}_j \tag{3.111}$$

$$i \to -i$$

and the commutation relation (3.110) remains invariant. Also, we note that

$$\Pi\,\mathbf{J}\,\Pi^{-1} = -\mathbf{J} \tag{3.112}$$

and the commutation relation

$$[J_i, \; J_j] = i\varepsilon_{ijk} \, J_k \tag{3.113}$$

is preserved.

If $H_0$ and $V$ are invariant under time reversal, then

$$\Pi \, H_0 \, \Pi^{-1} = H_0$$

$$\Pi \, H_{int} \, \Pi^{-1} = H_{int}. \tag{3.114}$$

Now [cf. Eqs. (2.60) and (2.61)], under time reversal we have to follow the rule , $i \to -i$ and as such

$$\Pi S \Pi^{-1} = S^\dagger$$
$$\Pi T \Pi^{-1} = T^\dagger$$

Now referring to Eq. (2.58)

$$\Pi |a\rangle_{in} = \Pi |a, t_0 \to -\infty\rangle$$
$$= |a^t, t_0 \to \infty\rangle$$
$$= |a^t\rangle_{out}.$$

Invariance under time reversal implies

$$\langle f \,|\, T \,|\, i \rangle = \langle f \,|\, \Pi^{-1} \Pi \, T \, \Pi^{-1} \Pi \,|\, i \rangle$$
$$= \langle f^t \,|\, T^\dagger \,|\, i^t \rangle^*$$
$$= \langle i^t \,|\, T \,|\, f^t \rangle. \tag{3.115}$$

From time reversal invariance, we derive some important results:

### 3.6.1  *Unitarity*

Consider the weak decay $B \to f$. The decay amplitude is

$$A_f =_{out} \langle f \,|H|\, B \rangle$$

Time reversal invariance gives

$$A_f =_{in} \langle f^t \,|H|\, B^t \rangle^*$$
$$=_{out} \langle f^t \,|S^\dagger H|\, B^t \rangle^* \tag{3.116}$$

The superscript $t$ on states represents that we have to reverse momenta and spins. If we use the fact that spins are to be summed and final states are to be integrated and work in the rest frame of $B$, then we can remove superscript $t$ and Eq. (3.116) gives

$$A_f^* = \sum_n {}_{\text{out}} \langle f | S^\dagger | n \rangle_{\text{out}} {}_{\text{out}} \langle n | H | B \rangle \qquad (3.117)$$

$$A_f^* = \sum_n S_{nf}^* A_n$$

$$= \sum_n (\delta_{nf} - iT_{nf}^*) A_n \qquad (3.118)$$

$$A_f^* = A_f - i \sum_n T_{nf}^* A_n \qquad (3.119)$$

$$A_f^* - A_f = -i \sum_n T_{nf}^* A_n \qquad (3.120)$$

$$\Im A_f = \frac{1}{2} \sum_n T_{nf}^* A_n \qquad (3.121)$$

where $T_{nf}$ denotes the scattering amplitude for $f - n$ scattering.

### 3.6.2   Reciprocity Relation

Let us specify the initial and final states as

$$| i \rangle = | \alpha, \mathbf{p}_i, m_i \rangle$$
$$| f \rangle = | \beta, \mathbf{p}_f, m_f \rangle. \qquad (3.122)$$

Then

$$| i^t \rangle = | \alpha, -\mathbf{p}_i, -m_i \rangle$$
$$| f^t \rangle = | \beta, -\mathbf{p}_f, -m_f \rangle. \qquad (3.123)$$

where $m_i$ and $m_f$ denote the $z$-component of spin and $\alpha$ and $\beta$ denote the all other quantum numbers which may be necessary to specify the states.

Therefore, Eq. (3.115) gives

$$\langle \beta, \mathbf{p}_f, m_f | T | \alpha, \mathbf{p}_i, m_i \rangle = \langle \alpha, -\mathbf{p}_i, -m_i | T | \beta, -\mathbf{p}_f, -m_f \rangle. \qquad (3.124)$$

This expresses the equality of two scattering processes obtained by reversing the momenta and spin-components and interchanging the initial and final states. This is known as reciprocity relation and is a consequence of invariance under time reversal. Since $\Pi$ is not a unitary operator, therefore, it does not have observable eigenvalues. The states cannot be labeled by such eigenvalues. Therefore, invariance under $\Pi$ cannot be tested by searching for time-parity forbidden decays. It can be tested by using the relation of the form given in Eq. (3.124). No violation of time reversal has been found in hadronic and electromagnetic interactions.

## 3.7   Applications

### 3.7.1   *Detailed Balance Principle*

#### 3.7.1.1   *Determination of Spin of the Pion*

If we assume invariance under time reversal, we get Eq. (3.124). In addition, if we assume parity conservation, we have from Eq. (3.124).

$$\langle \beta, \, \mathbf{p}_f, \, m_f | T \, | \alpha, \, \mathbf{p}_i, \, m_i \rangle = \langle \, \alpha, \, -\mathbf{p}_i, \, -m_i | \, \widehat{P}^\dagger T \, \widehat{P} \, \, | \beta, \, -\mathbf{p}_f, \, -m_f \rangle$$
$$= \langle \alpha, \, \mathbf{p}_i, \, -m_i | T \, | \beta, \, \mathbf{p}_f, \, -m_f \rangle. \quad (3.125)$$

If the spins are summed, then we can write

$$\sum_{\text{spin}} |\langle \beta, \, \mathbf{p}_f, \, m_f | T \, | \alpha, \, \mathbf{p}_i, \, m_i \rangle|^2 = \sum_{\text{spin}} |\langle \alpha, \, \mathbf{p}_i, \, m_i | T \, | \beta, \, \mathbf{p}_f, \, m_f \rangle|^2.$$
$$(3.126)$$

This is called the "semi detailed balance principle". We now apply the above result to two-body scattering

$$a + b \to c + d, \quad e.g. \quad p + p \to \pi^+ + d.$$

Then we get [cf. Eq. (2.100)]

$$\frac{d\,\sigma}{d\Omega}(a + b \to c + d) = \frac{1}{16\pi^2} \frac{m_N^2}{E_{cm}^2} \frac{p_{cd}}{p_{ab}} \frac{1}{(2s_a + 1)(2s_b + 1)} \sum_{\text{spin}} |F_{ab \to cd}|^2$$
$$(3.127)$$

and

$$\frac{d\,\sigma}{d\Omega}(c + d \to a + b) = \frac{1}{16\pi^2} \frac{m_N^2}{E_{cm}^2} \frac{p_{ab}}{p_{cd}} \frac{1}{(2s_c + 1)(2s_d + 1)} \sum_{\text{spin}} |F_{cd \to ab}|^2.$$
$$(3.128)$$

But Eq. (3.126) gives

$$\sum_{\text{spin}} |F_{ab \to cd}|^2 = \sum_{\text{spin}} |F_{cd \to ab}|^2. \quad (3.129)$$

Hence we have

$$\frac{d\,\sigma}{d\Omega}(a + b \to c + d) = \frac{(2s_c + 1)(2s_d + 1)}{(2s_a + 1)(2s_b + 1)} \frac{p_{cd}^2}{p_{ab}^2} \frac{d\sigma}{d\Omega}(c + d \to a + b).$$
$$(3.130)$$

This is known as the principle of detailed balance. We now apply the above result to the reaction

$$p + p \to \pi^+ + d.$$

Then from Eq. (3.130), we get

$$\frac{d\sigma}{d\Omega}\left(p+p \to \pi^+ + d\right) = \frac{3\left(2s_\pi + 1\right)}{4}\frac{p_\pi^2}{p_p^2}\frac{d\sigma}{d\Omega}\left(\pi^+ + d \to p + p\right), \quad (3.131)$$

where we have used the result that the proton spin $s_p = \frac{1}{2}$ and that the deuteron spin $s_d = 1$. For the total cross sections, we get

$$\sigma\left(p+p \to \pi^+ + d\right) = \frac{3}{4}\left(2s_\pi + 1\right)\frac{p_\pi^2}{p_p^2}\sigma\left(\pi^+ + d \to p + p\right). \quad (3.132)$$

From the experimentally measured cross sections, we find $s_\pi = 0$, i.e. the spin of the pion is zero.

## 3.8 Unitarity Constraints

So far, assuming rotational invariance, we have discussed the constraints on the $T$-matrix imposed by space reflection and time reversal invariance. In this section, we discuss the constraints on the $T$-matrix due to the unitarity of the $S$-matrix.

Unitarity of the $S$-matrix gives

$$S\,S^\dagger = 1 \quad (3.133a)$$

or

$$\langle j \mid S\,S^\dagger \mid i\rangle = \langle j \mid i\rangle = \delta_{ji}, \quad (3.133b)$$

where $|i\rangle$ and $|j\rangle$ are initial and final states. Introduce a complete set of states $|k\rangle$,

$$\sum_k \langle j \mid S \mid k\rangle \langle k \mid S^\dagger \mid i\rangle = \delta_{ji} \quad (3.134a)$$

or

$$\sum_k \langle j|\left[1 + i\left(2\pi\right)^4 \delta^4\left(P_j - P_k\right)T\right]|k\rangle \langle k|\left[1 - i\left(2\pi\right)^4 \delta^4\left(P_k - P_i\right)T^\dagger\right]|i\rangle$$

$$= \delta_{ji}, \quad (3.134b)$$

which gives

$$- i\left(2\pi\right)^4 \sum_k \left[\delta_{ki}\,\delta^4\left(P_j - P_k\right)T_{jk} - \delta_{jk}\delta^4\left(P_k - P_i\right)\left(T_{ki}^\dagger\right)\right]$$

$$= \left(2\pi\right)^8 \sum_k \delta^4\left(P_j - P_i\right) \langle j \mid T \mid k\rangle \langle k \mid T^\dagger \mid i\rangle\,\delta^4\left(P_i - P_k\right) \quad (3.135a)$$

or

$$-i\left[T_{ji} - T_{ij}^*\right] = (2\pi)^4 \sum_k \langle j \, | T \, | k \rangle \, \langle k | \, T^\dagger \, | i \rangle \, \delta^4 \left(P_i - P_k\right). \qquad (3.135b)$$

In Eq. (3.135b), $\sum_k$ means integration over momenta and sum over other quantum numbers. Only those states will contribute which are allowed by energy-momentum conservation implied by the $\delta$-function in Eq. (3.135b). For forward elastic scattering and no spin flip, $i = j$ and we get

$$2\Im T_{ii} = (2\pi)^4 \sum_k \langle i \, | T \, | k \rangle \, \langle k \, | T^\dagger \, | i \rangle \, \delta^4 \left(P_k - P_i\right). \qquad (3.136)$$

For two-body scattering viz

$$a + b \to 1 + 2 + \cdots$$

the right-hand side of Eq. (3.136) is the transition rate $W_i$ [cf. Eq. (2.99)], where $W_i(i = a + b)$ is given by

$$W_i = \sigma_i \, (\text{Flux})_{\text{in}} = \sigma_i \, \frac{1}{(2\pi)^6} \frac{|\mathbf{p}| \, E_{cm}}{E_a \, E_b}. \qquad (3.137)$$

Expressing the T-matrix, in terms of the amplitude $F$, we have

$$T_{ii} = \left[\frac{1}{(2\pi)^{3/2}}\right]^4 N \, F_{ii}, \qquad (3.138a)$$

where

$$N = \begin{cases} \frac{m_a \, m_b}{E_a E_b}, & a \text{ and } b \text{ both fermions} \\ \frac{m_b}{2E_a E_b}, & a \text{ a boson, } b \text{ fermion} \\ \frac{1}{4E_a E_b}, & a \text{ and } b \text{ both bosons} \end{cases} \qquad (3.138b)$$

Hence, we have from Eq. (3.136):

$$2n\Im F_{ii} = E_{cm} \, |\mathbf{p}| \, \sigma_{ab} = \frac{1}{2}\sqrt{\lambda \, \left(s, m_a^2, m_b^2\right)} \, \sigma_{ab}, \qquad (3.139)$$

where $F_{ii}$ is the forward elastic scattering amplitude, $\sigma_i = \sigma_{ab}$ is the total cross section for the reaction $a + b \to 1 + 2 + \cdots$ and

$$n = \begin{cases} m_a m_b \\ \frac{m_b}{2} \\ \frac{1}{4} \end{cases} \qquad (3.140)$$

depending upon the nature of particles $a$ and $b$. Equation (3.139) is known as the optical theorem. As a simple example, consider $a$ and $b$ to be spinless particles. Then we can express

$$F_{ij}(s, \theta) = 8\pi \, s^{1/2} \sum_{L=0}^{\infty} (2L + 1) \, F_{ij,L}(s) \, P_L(\cos\theta)$$

$$\equiv 8\pi \, s^{1/2} \, f_{ij}(s, \theta). \qquad (3.141)$$

If we put $f_{ii}(s, 0) = f(0)$ we have from Eqs. (3.139), (3.140), and (3.141)

$$\Im f(0) = \frac{|\mathbf{p}|}{4\pi} \, \sigma_{ab} \tag{3.142}$$

the usual form of optical theorem in potential scattering.

### 3.8.1 Two-Particle Partial Wave Unitarity

Assume that for each channel $k$, three or more particles states can be neglected. We work in the center-of-mass frame, with initial state $i = a + b$, so that $\mathbf{p}_a = \mathbf{p}_b = \mathbf{p}$. We take $\mathbf{p}$ along $z$-axis. Two-body Lorentz invariant phase space is given by

$$n_k \, (2\pi)^4 \int \frac{d^3 p_{1k}}{(4\pi)^3 E_{1k}} \frac{d^3 p_{2k}}{(2\pi)^3 E_{2k}} \delta^4 \left( p_{1k} + p_{2k} - P_i \right). \tag{3.143}$$

In the center-of-mass frame, $\mathbf{p}_{1k} = -\mathbf{p}_{2k} = \mathbf{p}_k$, where

$$p_k = |\mathbf{p}_k| = \frac{\sqrt{\lambda. (s, \, m_{1k}^2, \, m_{2k}^2)}}{s^{1/2}} \tag{3.144}$$

$$n_k = \begin{cases} \frac{1}{4}, & \text{both bosons} \\ \frac{m_{2k}}{2}, & \text{1st particle boson, 2nd one fermion} \\ m_{1k} \, m_{2k}, & \text{both fermions} \end{cases} \tag{3.145}$$

Then working out the integral (3.143), we get

$$n_k \, \frac{1}{4\pi^2} \frac{p_k}{s^{1/2}} \, d\Omega', \tag{3.146}$$

where $\Omega' \equiv (\theta', \phi')$ is the solid angle between $\mathbf{p}$ and $\mathbf{p}_k$. $\Omega \equiv (\theta, \phi)$ is the solid angle between $\mathbf{p}$ and $\mathbf{p}_{1j}$ where $\mathbf{p}_{1j}$ is the momentum of first particle in the state $j$. $\Omega'' \equiv (\theta'', \phi'')$ is the solid angle between $\mathbf{p}_k$ and $\mathbf{p}_{1j}$. For the two-particle states in channel $k$, the unitarity relation (3.135b) becomes, on using Eq. (3.146)

$$-i \left[ F_{ji}(\Omega) - F_{ij}^*(-\Omega) \right] = \sum_k \frac{n_k \, p_k}{4\pi^2 \, s^{1/2}} \int F_{jk}(\Omega'') \, F_{ik}^*(-\Omega') \, d\Omega'. \tag{3.147}$$

We use the general relation (3.147) for two-particle unitarity for three important cases:

**Case (i)**: Collision between spinless particles. In this case $i, j,$ and $k$ are simply channel indices. For this case, we can expand $F_{ij}(\theta)$ in terms of the Legendre polynomials of $\cos\theta$ [this is a consequence of rotational invariance; there can be no dependence on the magnetic quantum number $m$

and hence no dependence on $\phi$]. This expansion is given in Eq. (3.141). Similarly $F_{ik}(\Omega')$ is independent of $\phi'$ for spinless particles and can be expanded in terms of $P_L(\cos\theta')$. Likewise $F_{jk}(\Omega'')$ can be expanded in terms of $P_L(\cos\theta'')$. Hence, we have from Eq. (3.147)

$$- i8\pi \, s^{1/2} \sum_L \left(F_{ji,\,L}(s) - F^*_{ij,\,L}(s)\right) (2L+1) \, P_L(\cos\theta)$$

$$= \frac{1}{4}\sum_k \frac{p_k}{4\pi^2 \, s^{1/2}} \, 64\pi^2 \, s$$

$$\times \sum_{L''}\sum_{L'} (2L''+1) \, (2L'+1) \, F_{jk,\,L''}(s) \, F^*_{ik,\,L'}(s)$$

$$\times \int P_{L''}(\cos\theta'') \, P_{L'}(\cos\theta') \, \sin\theta' \, d\theta' \, d\phi'. \tag{3.148}$$

In order to evaluate the integral on the right-hand side of Eq. (3.148), we use the following formulae:

$$P_L(\cos\theta'') = \frac{4\pi}{(2L+1)}\sum_M Y^*_{LM}(\theta,0) \, Y_{LM}(\theta',\phi') \tag{3.149a}$$

$$\int_0^{2\pi} P_L(\cos\theta'') \, d\phi' = \frac{4\pi}{(2L+1)}\sum_M \int_0^{2\pi} Y^*_{LM}(\theta,0) \, Y_{LM}(\theta',\phi') \, d\phi'$$

$$= \frac{8\pi^2}{4\pi} P_L(\cos\theta) \, P_L(\cos\theta') \tag{3.149b}$$

$$\int_{-1}^1 d\,(\cos\theta') \, P_L(\cos\theta') \, P_{L'}(\cos\theta') = \frac{2}{2L+1}\delta_{L'L}. \tag{3.149c}$$

We get from Eq. (3.148), using Eqs. (3.149)

$$\frac{1}{2i}\sum_L (2L+1)\left(F_{ji,\,L}(s) - F^*_{ij,\,L}(s)\right) P_L(\cos\theta)$$

$$= \sum_k p_k \sum_{L'} (2L'+1) \, F_{jk,\,L'}(s) \, F^*_{ik,\,L'}(s) \, P_{L'}(\cos\theta) \tag{3.150}$$

Since the Legendre polynomials are linearly independent, we get the desired 2-body partial-wave unitarity relation

$$\frac{1}{2i}\left(F_{ji,\,L}(s) - F^*_{ij,\,L}(s)\right) = \sum_k p_k \, F_{jk,\,L}(s) \, F^*_{ik,\,L}(s). \tag{3.151}$$

If we are interested only in elastic scattering, we may drop indices $i$ and $j$ and we obtain

$$\Im F_L(s) = \sum_k p_k \, |F_{k,\,L}|^2. \tag{3.152}$$

Occasionally all channels except the elastic one are closed at low energies. Then $p_k = p$ and we have

$$\Im F_L = p \, |F_L|^2 \, , \tag{3.153}$$

so that we can put

$$F_L = \frac{1}{p} \, e^{i\delta_L} \sin \delta_L, \tag{3.154}$$

where $\delta_L$ is a real function of $s$. We can also express

$$F_L = \frac{1}{2ip} \left( e^{2i\delta_L} - 1 \right) = p^{-1} \left( \cot \delta_L - i \right)^{-1}. \tag{3.155}$$

The differential cross section is given by

$$\frac{d \, \sigma_{ij}}{d\Omega} = \frac{1}{4s} \frac{p'}{p} \frac{1}{16\pi^2} \, |F_{ij} \, (s, \theta)|^2$$

$$= \frac{p'}{p} \left| \sum_L (2L + 1) F_{ij,L} \, (s) \, P_L \, (\cos \theta) \right|^2, \tag{3.156}$$

where we have used Eq. (3.141). Using the orthogonality of Legendre polynomials, we get

$$\sigma_{ij} = 4\pi \frac{p'}{p} \sum_L (2L + 1) \, |F_{ji,L} \, (s)|^2 \equiv \sum_L \sigma_{ji,L}, \tag{3.157a}$$

where

$$\sigma_{ji,L} = 4\pi \frac{p'}{p} \, (2L + 1) \cdot |F_{ji,L}|^2. \tag{3.157b}$$

For "purely elastic" region, where Eq. (3.154) applies, we have from Eq. (3.157b)

$$\sigma_L = \frac{4\pi}{p^2} \, (2L + 1) \sin^2 \delta_L. \tag{3.158}$$

**Case (ii):** Particles $a$ and $b$ carry spin. Here it is convenient to introduce helicity. Let $\lambda_1$ and $\lambda_2$ be helicities of particle $a$ and $b$ respectively and let $\lambda = \lambda_1 - \lambda_2$. In the center-of-mass frame $\mathbf{p}_a = -\mathbf{p}_b = \mathbf{p}$. Let us take the vector $\mathbf{p} = (p, \theta, \phi)$. In the center-of-mass frame we represent the two-particle state as $[\Omega = (\theta, \phi)]$

$$|p, \lambda_1, \lambda_2, \Omega\rangle = |\mathbf{p}, \lambda_1\rangle \, |-\mathbf{p}, \lambda_2\rangle \, (-1)^{s_2 - \lambda_2} \tag{3.159}$$

The last factor in Eq. (3.159) is due to phase convention. Noting that $(\mathbf{J} = \mathbf{J}_1 + \mathbf{J}_2)$, $\mathbf{J} \cdot \mathbf{p} = \mathbf{J}_1 \cdot \mathbf{p} - \mathbf{J}_2 \cdot (-\mathbf{p})$, we have

$$\frac{\mathbf{J} \cdot \mathbf{p}}{|\mathbf{p}|} |p, \lambda_1, \lambda_2, \Omega\rangle = \lambda \, |p, \lambda_1, \lambda_2, \Omega\rangle \tag{3.160}$$

Now

$$R \, |0,0\rangle = |\theta, \phi\rangle \tag{3.161}$$

where $R$ is the rotation operator $e^{-i\theta \mathbf{n} \cdot \mathbf{J}/\hbar}$ $[\mathbf{n} = (-\sin\phi, \cos\phi, 0)]$ and

$$
\begin{aligned}
R \, |J\,M\rangle &= \sum_{M'} |J\,M'\rangle \, \langle J\,M' \,|\, R \,|J\,M\rangle \\
&= \sum_{M'} |J\,M'\rangle \, d^J_{M'M}\,(\Omega)
\end{aligned} \tag{3.162}
$$

where $d^J_{M'M}\,(\Omega)$ are rotation matrices. Thus

$$
\begin{aligned}
\langle J\,M\,, \lambda |\, \theta\phi\rangle &= \langle J\,M, \lambda \,|\, R \,|\, 0\,0\rangle \\
&= \sum_{M'} \langle J\,M, \lambda \,|\, R \,|J\,M', \lambda\rangle \, \langle J\,M', \lambda \,|\, 0\,0\rangle \\
&= \sum_{M'} d^J_{MM'}\,(\Omega) \, \langle J\,M'\,, \lambda |\, 0\,0\rangle \\
&= \sum_{M'} d^J_{MM'}\,(\Omega) \, \sqrt{\frac{2J+1}{4\pi}} \delta_{M'\lambda} \\
&= \sqrt{\frac{2J+1}{4\pi}} d^J_{M\lambda}\,(\Omega)
\end{aligned} \tag{3.163}
$$

Hence

$$
\begin{aligned}
|\theta, \phi, \lambda\rangle &= \sum_{JM} |JM, \lambda\rangle \, \langle JM, \lambda |\, \theta\phi\rangle \\
&= \sum_{JM} \sqrt{\frac{2J+1}{4\pi}} |JM, \lambda\rangle \, d^J_{M\lambda}\,(\Omega)\,.
\end{aligned} \tag{3.164}
$$

Thus we can write

$$|\theta, \phi, \lambda_1, \lambda_2\rangle = \sum_{JM} |JM\lambda_1\lambda_2\rangle \, \sqrt{\frac{2J+1}{4\pi}} d^J_{M\lambda}\,(\Omega)\,. \tag{3.165}$$

We now consider the scattering process $a+b = c+d$. Let $\lambda_1$ and $\lambda_2$ be initial helicities and $\lambda_1'$ and $\lambda_2'$ be final helicities. $\lambda = \lambda_1 - \lambda_2$ and $\lambda' = \lambda_1' - \lambda_2'$. We take initial momentum $\mathbf{p} = (p, 0, 0)$ and final momentum $\mathbf{p}' = (p', \theta, \phi)$.

We can write the scattering amplitude on using Eq. (3.165)

$$
\begin{aligned}
F_{ji}\left(\lambda_1'\lambda_2', \lambda_1\lambda_2, \Omega\right) &= \langle \theta\, \phi\, \lambda_1'\lambda_2', j|\, F\,|0\, 0\, \lambda_1\lambda_2,\, i\rangle \\
&= \sum_{J'M'}\sum_{JM} d_{M'\lambda'}^{*J'}\left(\Omega\right)\sqrt{\frac{2J'+1}{4\pi}}\, d_{M\lambda}^{J}\left(0\right)\sqrt{\frac{2J+1}{4\pi}} \\
&\quad\times\langle J'M'\,\lambda_1'\lambda_2',\, j|\,F\,|JM\,\lambda_1\lambda_2, i\rangle \\
&= \sum_{J'M'}\sum_{JM} d_{M'\lambda'}^{*J'}\left(\Omega\right)\sqrt{\frac{(2J'+1)(2J+1)}{(4\pi)^2}} \\
&\quad\times\delta_{M\lambda}n\delta_{JJ'}\delta_{MM'}\delta_{M\lambda}F_{ji}^{J}\left(\lambda_1'\lambda_2', \lambda_1\lambda_2, s\right) \\
&= n\sum_{J}\frac{(2J+1)}{4\pi}F_{ji}^{J}\left(\lambda_1'\lambda_2', \lambda_1\lambda_2, s\right)d_{\lambda\lambda'}^{J*}\left(\theta,\phi\right).
\end{aligned}
$$

$$(3.166)$$

Note that $n = 8\pi^2\dfrac{s^{1/2}}{\sqrt{m_a\,m_b\,m_c\,m_d}}$ when all the particles are fermions. For spinless particles $n = 32\pi^2\,s^{1/2}$, $\lambda = \lambda' = 0$, $J = L$ and $d_{00}^{J*}\left(\theta,\phi\right) = \sqrt{\frac{4\pi}{2J+1}}Y_{L0}^{*}\left(\theta,\phi\right) = P_L\left(\cos\theta\right)$, we get back Eq. (3.141). The differential scattering cross section for the process $a + b \to c + d$ is given by

$$
\frac{d\sigma}{d\Omega} = \frac{|\mathbf{p}'|}{|\mathbf{p}|}\frac{1}{(2s_1+1)(2s_2+1)}
$$

$$
\times\sum_{\lambda_1'\lambda_2'\lambda_1\lambda_2}\left|\sum_{J}(2J+1)\,F^{J}\left(\lambda_1'\lambda_2', \lambda_1\lambda_2, s\right)d_{\lambda\lambda'}^{J}\left(\theta\right)\right|^2 \tag{3.167}
$$

where we have used

$$
d_{\lambda\lambda'}^{J}\left(\theta,\phi\right) = e^{i(\lambda-\lambda')\phi}\,d_{\lambda\lambda'}^{J}\left(\theta\right). \tag{3.168}
$$

To proceed further we note the following properties of rotation matrices

$$
d_{\lambda'\lambda}^{J}\left(-\Omega\right) = \left(d^{J^{-1}}\left(\Omega\right)\right)_{\lambda'\lambda} = \left(d^{J\dagger}\left(\Omega\right)\right)_{\lambda'\lambda} = d_{\lambda\lambda'}^{J*}\left(\Omega\right) \tag{3.169}
$$

$$
\int d_{\lambda\lambda'}^{J*}\left(\Omega\right)\,d_{M\lambda'}^{J'}\left(\Omega\right)\,d\Omega = \frac{4\pi}{(2J+1)}\delta_{JJ'}\,\delta_{\lambda M} \tag{3.170}
$$

$$
d^{J}\left(\Omega''\right) = d^{J}\left(-\Omega'\right)\,d^{J}\left(\Omega\right) = d^{J\dagger}\left(\Omega'\right)\,d^{J}\left(\Omega\right)
$$

or

$$
\begin{aligned}
d_{\lambda\lambda'}^{J}\left(\Omega''\right) &= \sum_{M}\left(d^{J\dagger}\left(\Omega'\right)\right)_{\lambda M}\cdot\left(d^{J}\left(\Omega\right)\right)_{M\lambda'} \\
&= \sum_{M}d_{M\lambda}^{J*}\left(\Omega'\right)\,d_{M\lambda'}^{J}\left(\Omega\right)
\end{aligned} \tag{3.171}
$$

Note that in Eq. (3.171) $d^J(\Omega'')$ has been expressed as product of two rotation matrices corresponding to $-\Omega'$, $\Omega$. Then using Eqs. (3.166), (3.169), (3.170) and (3.171), we get from Eq. (3.151), for the two-particle partial wave unitarity relation.

$$\frac{1}{2i}\left[F_{ji}^J\left(\lambda_1'\lambda_2';\lambda_1\lambda_2;s\right) - F_{ij}^{J^*}\left(\lambda_1\lambda_2;\lambda_1'\lambda_2';s\right)\right]$$
$$= \sum_k p_k \sum_{\lambda_{k_1}}\sum_{\lambda_{k_2}} F_{jk}^J\left(\lambda_1'\lambda_2';\lambda_{k_1}\lambda_{k_2};s\right) F_{ik}^{J^*}\left(\lambda_1\lambda_2;\lambda_{k_1}\lambda_{k_2};s\right). \qquad (3.172)$$

The two-particle elastic unitarity gives

$$\frac{1}{2i}\left[F^J\left(\lambda_1'\lambda_2';\lambda_1\lambda_2;s\right) - F^{J^*}\left(\lambda_1\lambda_2;\lambda_1'\lambda_2';s\right)\right]$$
$$= p\sum_{\lambda_1''\lambda_2''} F^J\left(\lambda_1'\lambda_2';\lambda_1''\lambda_2'';s\right) F^{J^*}\left(\lambda_1\lambda_2;\lambda_1''\lambda_2'';s\right). \qquad (3.173)$$

Assuming parity conservation, we get

$$F^J\left(-\lambda_1',-\lambda_2';-\lambda_1,-\lambda_2;s\right)$$
$$= \eta\left(-1\right)^{s_1'+s_2'-s_1-s_2} F^J\left(\lambda_1'\lambda_2';\lambda_1\lambda_2;s\right) \qquad (3.174)$$

where $s_1$, $s_2$, $s_1'$ and $s_2'$ are the spins of particle $a$ and $b$ in the initial and final states and $\eta$ is the product of their intrinsic parities. Equation (3.174) shows that not all the amplitudes are independent. Time reversal invariance puts additional restrictions on the amplitudes $F^J$'s namely

$$F_{ji}^J\left(\lambda_1'\lambda_2';\lambda_1\lambda_2;s\right) = F_{ij}^J\left(\lambda_1\lambda_2;\lambda_1'\lambda_2';s\right) \qquad (3.175)$$

For the elastic scattering:

$$F^J\left(\lambda_1'\lambda_2';\lambda_1\lambda_2;s\right) = F^J\left(\lambda_1\lambda_2;\lambda_1'\lambda_2';s\right) \qquad (3.176)$$

Finally using the orthogonality of d-matrices, we get integrated cross section for elastic scattering from Eq. (3.167)

$$\sigma = \sum_J \sigma_J$$

where

$$\sigma_J = 4\pi\frac{(2J+1)}{(2s_1+1)(2s_2+1)}\sum_{\lambda_1'\lambda_2'\lambda_1\lambda_2}\left|F^J\left(\lambda_1'\lambda_2';\lambda_1\lambda_2;s\right)\right|^2 \qquad (3.177)$$

In particular, when all the particles have spin 1/2, the $S$-wave unitarity gives $\left(s = 4p^2\right)$

$$\sigma_S = \frac{4\pi}{4}\frac{\sin^2\delta_0}{p^2} < \frac{\pi}{p^2} \qquad (3.178)$$

For special case of the elastic scattering of $a+b \to a+b$, where $a$ carries spin $s$ and $b$ is spinless, we have $\lambda = \lambda_1 - \lambda_2 = \lambda_1$ and $\lambda' = \lambda'_1 - \lambda'_2 = \lambda'_1$. For this case we have from Eqs. (3.166), (3.167), (3.173), (3.174) and (3.176) (for $a$ to be fermion)

$$F_{\lambda'\lambda}(\Omega) = \frac{4\pi\sqrt{s}}{m_a} \sum_J (2J+1) F_{\lambda'\lambda}^J(s) d_{\lambda\lambda'}^{J*}(\theta,\phi) \tag{3.179}$$

$$\frac{d\sigma}{d\Omega} = \frac{1}{(2s+1)} \sum_{\lambda'\lambda} \left| \sum_J (2J+1) F_{\lambda'\lambda}^J(s) d_{\lambda\lambda'}^{J*}(\theta) \right|^2 \tag{3.180}$$

$$\sigma_J = 4\pi \left( \frac{2J+1}{2s+1} \right) \sum_{\lambda'\lambda} \left| F_{\lambda'\lambda}^J(s) \right|^2 \tag{3.181}$$

$$\frac{1}{2i} \left[ F_{\lambda'\lambda}^J(s) - F_{\lambda\lambda'}^{J*}(s) \right] = p \sum_{\lambda''} F_{\lambda'\lambda''}^J(s) F_{\lambda'\lambda''}^{J*}(s) \tag{3.182}$$

$$F_{\lambda'\lambda}^J = F_{-\lambda'-\lambda}^J \tag{3.183}$$

We end this chapter with the following remarks. We have shown how the symmetry principles put restrictions on the $S$-matrix. In this way, we get the minimum set of observables to describe the experimental data. This approach is especially rewarding, when the underlying dynamics is not known, which is the case for the hadronic interactions.

## 3.9 Problems

(1) Nucleon-nucleon scattering

$$N_a + N_b \to N_a + N_b$$

In the center-of-mass frame

$$\mathbf{p}_a = -\mathbf{p}_b = \mathbf{p}$$
$$\mathbf{p}'_a = -\mathbf{p}'_b = \mathbf{p}'$$

Introduce three orthogonal unit vectors $\mathbf{l}, \mathbf{m}, \mathbf{n}$

$$\mathbf{l} = \frac{\mathbf{p} \times \mathbf{p}'}{|\mathbf{p} \times \mathbf{p}'|}, \qquad \mathbf{m} = \frac{\mathbf{p}' - \mathbf{p}}{|\mathbf{p}' - \mathbf{p}|} \qquad \mathbf{n} = \frac{\mathbf{p}' + \mathbf{p}}{|\mathbf{p}' + \mathbf{p}|}$$

$$l_i l_j + m_i m_j + n_i n_j = \delta_{ij} \qquad i = 1, 2, 3$$

Then $T$-matrix can be written as

$$\langle T \rangle = \langle \alpha,\ \mathbf{p}',\ \sigma_1,\ \sigma_2 | T\ |\alpha,\ \mathbf{p},\ \sigma_1,\ \sigma_2 \rangle$$

$$= [A_1 + B_1\ \sigma_1 \cdot \mathbf{l} + C_1\ \sigma_1 \cdot \mathbf{m} + D_1\ \sigma_1 \cdot \mathbf{n}]$$

$$\times [A_2 + B_2\ \sigma_2 \cdot \mathbf{l} + C_2\ \sigma_2 \cdot \mathbf{m} + D_2\ \sigma_2 \cdot \mathbf{n}]$$

It is understood that the matrix elements are to be taken between the spin states $\chi_{af}^\dagger\ \chi_{bf}^\dagger$ and $\chi_{ai}\ \chi_{bi}$. Then using parity conservation and time reversal invariance, show that $T$ can be written as

$$\langle T \rangle = \left( \frac{m}{E} \right) [H_1 + H_2\ \sigma_1 \cdot \mathbf{l}\ \sigma_2 \cdot \mathbf{l} + iH_3\ (\sigma_1 + \sigma_2) \cdot \mathbf{l}$$

$$+ iH_3'\ (\sigma_1 - \sigma_2) \cdot \mathbf{l} + H_4\ \sigma_1 \cdot \mathbf{m}\ \sigma_2 \cdot \mathbf{m} + H_5\ \sigma_1 \cdot \mathbf{n}\ \sigma_2 \cdot \mathbf{n}]$$

For identical nucleons like $p - p$ and $n - n$ scattering, $\langle T \rangle$ has to be symmetric under $1 \leftrightarrow 2$; hence $H_3' = 0$. Show that

$$\sigma_1 \cdot \sigma_2 = \sigma_1 \cdot \mathbf{l}\ \sigma_2 \cdot \mathbf{l} + \sigma_1 \cdot \mathbf{m}\ \sigma_2 \cdot \mathbf{m} + \sigma_1 \cdot \mathbf{n}\ \sigma_2 \cdot \mathbf{n}.$$

By eliminating $\sigma_1 \cdot \mathbf{n}\ \sigma_2 \cdot \mathbf{n}$, using the above relation, express $\langle T \rangle$ as

$$\langle T \rangle = \left( \frac{m}{E} \right) [G_1 + G_2\ \sigma_1 \cdot \sigma_2 + G_3\ \sigma_1 \cdot \mathbf{m}\ \sigma_2 \cdot \mathbf{m}$$

$$+ G_4\ i\,(\sigma_1 + \sigma_2) \cdot \mathbf{l} + G_5\ \frac{1}{2}\,(\sigma_1 \cdot \mathbf{l}\ \sigma_2 \cdot \mathbf{l} + \sigma_2 \cdot \mathbf{l}\ \sigma_1 \cdot \mathbf{l})$$

$$+ G_6\ i\,(\sigma_1 - \sigma_2) \cdot \mathbf{l}]$$

Using

$$\frac{1}{(2\pi)^3} \int_{-\infty}^{\infty} e^{i\ \mathbf{q} \cdot \mathbf{r}}\ G\left(\mathbf{q}^2\right) d^3q = V\,(r),$$

show the most general form of 2-nucleon potential can be written in the form

$$V_{12} = V_c + V_s\ \sigma_1 \cdot \sigma_2 + V_T\ S_{12} + V_{LS}\ \sigma \cdot \mathbf{L} + V_5\ Q_{12} + V_6\ (\sigma_1 - \sigma_2) \cdot \mathbf{L}$$

where

$$S_{12} = 3\ \sigma_1 \cdot \widehat{\mathbf{r}}\ \sigma_2 \cdot \widehat{\mathbf{r}} - \sigma_1 \cdot \sigma_2$$

$$Q_{12} = \frac{1}{2}\,[\sigma_1 \cdot \mathbf{L}\ \sigma_2 \cdot \mathbf{L} + \sigma_2 \cdot \mathbf{L}\ \sigma_1 \cdot \mathbf{L}]$$

$$\sigma = \sigma_1 + \sigma_2 = 2\mathbf{S}$$

$$\mathbf{L} = (\mathbf{r} \times \mathbf{p})$$

(2) Consider the elastic scattering $a + b \to a + b$, where $a$ is spin half particle and $b$ is spinless. For this case $J = L \pm 1/2$. Expressing the two independent amplitudes $F^J_{1/2\,1/2}$ and $F^J_{1/2,\,-1/2}$ as

$$F^J_{1/2\,\pm 1/2} = \frac{1}{2}\left(f_{L+} \pm f_{(L+1)_-}\right),$$

where $L_\pm$ correspond to $J = L \pm 1/2$, and then using Eq. (3.153), show that

$$\Im f_{L\pm} = p\left|f_{L\pm}\right|^2$$

Hence one can write

$$f_{L+} = \frac{1}{p}\, e^{i\delta_{L+}} \sin \delta_{L+}$$

$$f_{L-} = \frac{1}{p}\, e^{i\delta_{L-}} \sin \delta_{L-}$$

The scattering matrix [cf. Eq. (3.102)] can be written as

$$F_{M'M}(\theta, \phi) = \frac{4\pi s^{1/2}}{m_a}\, \chi^\dagger_{M'}\, [f + ig\, \sigma \cdot \mathbf{n}]\, \chi_M$$

where $\mathbf{n} = \frac{\mathbf{p} \times \mathbf{p}'}{|\mathbf{p} \times \mathbf{p}'|}$. If $\mathbf{p}$ is along $z$-axis, then

$$\chi_{M'} = e^{-i\theta\, \sigma \cdot \mathbf{n}}\, \chi_M, \qquad \mathbf{n} = (-\sin\phi, \cos\phi, 0).$$

Using the relations (where the prime denotes differentiation with respect to $\cos\theta$),

$$d^J_{1/2\,1/2}(\theta, \phi) = d^J_{1/2\,1/2} = \frac{1}{J+1/2}\cos\frac{\theta}{2}\left(P'_{J+1/2} - P'_{J-1/2}\right)$$

$$= d^J_{-1/2\,-1/2}(\theta)$$

$$d^J_{1/2\,-1/2}(\theta, \phi) = e^{-i\phi}\, d^J_{1/2\,-1/2}(\theta)]$$

$$= \frac{e^{-i\phi}}{J+1/2}\sin\frac{\theta}{2}\left(P'_{J+1/2} + P'_{J-1/2}\right)$$

$$d^J_{-1/2\,1/2}(\theta, \phi) = e^{-i\phi}\, d^J_{1/2\,-1/2}(\theta)$$

show that

$$f(\theta) = \frac{m_a}{4\pi s^{1/2}}\left[F_{1/2\,1/2}\cos\frac{\theta}{2} + e^{i\phi}\, F_{1/2\,-1/2}\sin\frac{\theta}{2}\right]$$

$$g(\theta) = \frac{m_a}{4\pi s^{1/2}}\left[e^{i\phi}\, F_{1/2\,-1/2}\cos\frac{\theta}{2} - F_{1/2\,1/2}\sin\frac{\theta}{2}\right]$$

Now, using Eqs. (3.179) and (3.180), show that

$$f(\theta) = \sum_{L=0}^{\infty} \left[ (L+1)\, f_{L_+} + L f_{L_-} \right] P_L (\cos\theta)$$

$$g(\theta) = \sum_{L=0}^{\infty} \left[ (f_{L_+} - L f_{L_-})\, \sin\theta\, P'_L (\cos\theta) \right]$$

$$\frac{d\sigma}{d\Omega} = \frac{m_a^2}{16\pi^2 s} \frac{1}{2} \sum_{MM'} |F_{MM'}|^2 = |f|^2 + |g|^2$$

(3) Consider the decay

$$a_1 \rightarrow \rho\pi$$

$$p^\mu = k^\mu + q^\mu, \quad \mathbf{p} = \mathbf{k} + \mathbf{q}$$

$$p^2 = m_{a_1}^2, \quad k^2 = m_\rho^2, \quad q^2 = m_\pi^2 \approx 0$$

List all the $\ell$-values allowed by the conservation of angular momentum and parity. The decay width is given by

$$\Gamma = \frac{1}{8\pi} \frac{|\mathbf{k}|}{m_{a_1}^2} |M|^2$$

where

$$|M|^2 = \sum_{Polarization} |F|^2$$

Take

$$F = f_{a\rho\pi}\, m_{a_1}\, \eta \cdot \varepsilon$$

where

$$\eta^\mu : \text{Polarization of axial-vector particle } a_1$$

$$\varepsilon^\mu : \text{Polarization of } \rho$$

Find $\Gamma$.

(4) Consider the decay

$$V \rightarrow P_1 P_2$$

$$V(1^-) \rightarrow P_1(0^-) P_2(0^-)$$

(a) Show that it is a $p$-wave decay, if it is a strong decay.

(b) $p^\mu$ : 4-momentum of $V$

$\quad$ $\varepsilon^\mu$ : polarization vector of $V$

$\quad$ $p \cdot \varepsilon = 0$

Using, Lorentz invariance, show that the amplitude has the form

$$F = g_{Vpp} q \cdot \varepsilon, \quad q^\mu = (p_1 - p_2)^\mu$$

(c) Show that the decay width is given by

$$\Gamma = \frac{g_{VPP}^2}{4\pi} \left(\frac{2}{3}\right) \frac{|\mathbf{p}|^3}{m_V^2}$$

$|\mathbf{p}|$ is the momentum in the rest frame of $V$

$$\sum_\lambda \varepsilon_\mu^\lambda \varepsilon_\nu^{*\lambda} = -g_{\mu\nu} + \frac{p_\mu p_\nu}{m_V^2}$$

(5) Consider the decay

$$\pi^0 \to \gamma\gamma$$

(a) Show that it is a $p$-wave decay.

(b) Show that the decay amplitude can only have the form:

$$F = e^2 F_{\pi\gamma\gamma} \mathbf{k} \cdot \left(\varepsilon^\lambda \times \varepsilon^{\lambda'}\right)$$

Show that

$$\Gamma\left(\pi^0 \to \gamma\gamma\right) = 4\pi\alpha^2 F_{\pi^0\gamma\gamma}^2 \frac{|\mathbf{k}|^3}{m_{\pi^0}^2}$$

$$= \frac{\pi\alpha^2}{2} F_{\pi^0\gamma\gamma}^2 m_{\pi^0} \text{ (in the rest frame of } \pi^0)$$

Find $F_{\pi^0\gamma\gamma}$, using the experimental value

$$\tau = (8.4 \pm 0.5) \times 10^{-17} \text{s}.$$

(6) Consider the decay

$$A \to X + V$$

$$J^P(X) = 0^-, \quad J^P(V) = 1^-, \quad J^P(A) = 1^+, 1^-$$

List allowed $l$-values for the final state. If the parity is conserved, which $l$-values are excluded for $J^P(A) = 1^+$ and $J^P(A) = 1^-$.

(7) (a) For the decay

$$X \to V_1 + V_2, \quad J^P(X) = 0^-$$

the decay amplitude is given by

$$A = A_1 \varepsilon_1 \cdot \varepsilon_2 + A_2 \varepsilon_1 \cdot \mathbf{p} \, \varepsilon_2 \cdot \mathbf{p} + A_3 \mathbf{p} \cdot (\varepsilon_1 \times \varepsilon_2)$$

where $\varepsilon_1$ and $\varepsilon_2$ are polarization vectors of $V_1$ and $V_2$ and $\mathbf{p}$ is the momentum in the rest frame of $X$. If the parity is conserved which of the above amplitudes are zero.

(b) List all the allowed $l$-values for the final state. If the parity is conserved which $l$-values are excluded.

## 3.10   References

1. S. Gasirowicz, Elementary particle physics, Wiley, New York (1966).
2. H. M. Pilkuhn, Relativistic particle physics, Springer-Verlag, New York (1979).
3. P. Ramond, Field Theory: A Modern Primer, Westview Press; 2nd edition (2001).
4. E. M. Henley, A. Garcia, Subatomic Physics, World Scientific, 3rd edition (2007).

# Chapter 4

# Internal Symmetries

Hadrons found in nature are not fundamental constituents of matter. There are hundreds of them. They can be divided into two classes: (a) baryons: they are fermions with half integer spin, i.e. $J = 3/2, 1/2$; (b) mesons: they are bosons with integral spin, i.e. $J = 0, 1, 2$. Some of the low lying mesons with $J^P = 0^-$ and $J^P = 1^-$ are shown in Figs. 4.1 and 4.2. Low lying baryons with $J^P = 1/2^+, 3/2^+$ are shown in Figs. 4.3 and 4.4. Hadrons with the same $J^P$ are distinguished from each other by some internal quantum numbers. The assignment of these quantum numbers is meaningful, since these quantum numbers are additively conserved in hadronic interactions.

## 4.1 Selection Rules and Globally Conserved Quantum Numbers

A particle would decay into two or more lighter ones if the decay is allowed by energy-momentum conservation. The reason is that the entropy $S = k_B \ln$ (phase space). Since phase space for the lightest particles is largest and the entropy $S$ tends to increase, the system tends to decay into the lightest particles, unless there is some selection rule to forbid that decay. But we know that certain decays, although allowed by energy-momentum and angular momentum conservation, do not take place. Thus there must be selection rules or conservation laws which forbid these decays.

We now list these "global" conservation laws:

(i) Electric charge conservation: The decay $e^- \rightarrow \nu + \gamma$ is not seen ($\tau_e > 4.3 \times 10^{23}$ years). This is a consequence of electric charge conservation: "Electric charge is additively conserved in any process". This in turn is a consequence of the invariance of Hamiltonian under the global gauge

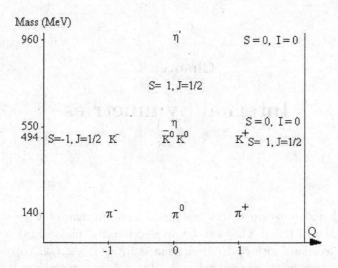

Fig. 4.1   Lowest lying pseudoscalar mesons $J^P = 0^-$.

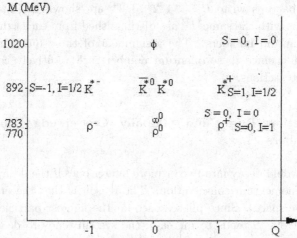

Fig. 4.2   Lowest lying vector mesons $J^P = 1^-$.

transformation $U_Q(1)$ :

$$|\Psi\rangle \rightarrow e^{i\hat{Q}\Lambda} \, |\Psi\rangle \qquad (4.1a)$$

so that

$$\left[\hat{Q}, \, H\right] = 0. \qquad (4.1b)$$

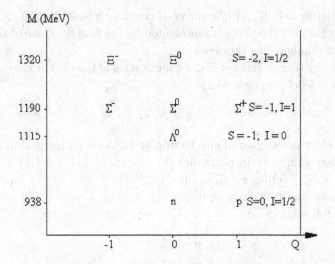

Fig. 4.3   Lowest lying $1/2^+$ baryons.

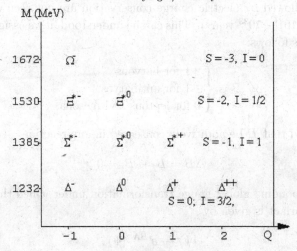

Fig. 4.4   Lowest lying $3/2^+$ baryons.

The electric charge $\hat{Q}$ is a generator of $U_Q(1)$ global gauge group. If $\Lambda$ is a function of space-time viz $\Lambda = \Lambda(\mathbf{r}, t)$, then the gauge transformation is called local. Actually, electric charge has a dual rule; it is also a generator of the local gauge group, $U = e^{i\hat{Q}\Lambda(\mathbf{r},t)}$. It is a feature of local gauge group that corresponds to this transformation, there is a vector field $A_\mu$ coupled to the matter field $\Psi$, with a universal coupling whose strength is just the electric charge of the particle represented by the field $\Psi$. None of the other quantum numbers has this feature.

A closely related concept is the quantization of the electric charge, which at particle level is expressed as

$$q = N_q \, e, \tag{4.2}$$

i.e. the electric charge $q$ of any hadron or lepton is an integral multiple of elementary charge $e$. In particular $N_n = 0$ $[q_n = (-0.4 \pm 1.1) \times 10^{-21} \, e]$ and $N_e + N_p = 0$ $[|(q_e + q_p)| < 1.0 \times 10^{-21} \, e]$.

(ii) Baryon charge conservation:

The following decays

$$p \to e^+ + \gamma$$
$$p \to e^+ + \pi^0$$

although allowed by electric charge conservation are not seen experimentally ($\tau_p > 10^{31} - 10^{33}$ years). This can be understood, if we assign a baryon charge $B$ as follows:

$$B = \begin{cases} +1 \text{ for baryons} \\ -1 \text{ for antibaryos} \\ 0 \text{ for leptons and mesons} \end{cases} \tag{4.3}$$

and demand that $B$ be additively conserved in any reaction

$$\Delta B = B_f - B_i = 0. \tag{4.4}$$

The corresponding global gauge transformation under which the Hamiltonian is invariant is given by

$$|\Psi\rangle \to e^{i\hat{B}\Lambda} \, |\Psi\rangle. \tag{4.5}$$

(iii) Lepton charge conservation:

Some decay modes of leptons are not seen. The absence of these decay modes is a consequence of non-conservation of lepton charge which is assigned as follows:

$$L = \begin{cases} +1 \text{ for leptons} \\ -1 \text{ for antileptons} \\ 0 \text{ for all other particles.} \end{cases} \tag{4.6}$$

Any reaction in which $L$ is additively conserved ($\Delta L = 0$) is allowed; otherwise it is forbidden. Some examples are given below:

| | | | | |
|---|---|---|---|---|
| $n \rightarrow p + e^- + \bar{\nu}_e$ | | | Allowed | |
| $L$  0 | 0 | 1 | $-1$ $\Delta L = L_f - L_i = 0$ | |
| $\bar{\nu}_e + (Z, A) \rightarrow (Z + 1, A) + e^-$ Not allowed | | | | |
| $L$ $-1$ | 0 | 0 | 1 | $\Delta L = 2$ |
| $\bar{\nu}_e + (Z, A) \rightarrow (Z - 1, A) + e^+$ Allowed | | | | |
| $L$ $-1$ | 0 | 0 | $-1$ $\Delta L = 0$ | |

Further, the reaction [antineutrinos obtained from the decay of pile neutrons in a fission reactor $(n \rightarrow p + e^- + \bar{\nu}_e)$]

$$\bar{\nu}_e + {}^{37}Cl \rightarrow e^- + {}^{37}Ar$$

for which $\Delta L = 2$ is not seen, but the reaction using solar neutrinos,

$$\nu_e + {}^{37}Cl \rightarrow e^- + {}^{37}Ar$$

has been seen and is allowed by lepton charge conservation. Also the allowed reaction

$$\bar{\nu}_e + p \rightarrow e^+ + n$$

has been observed with expected cross section. The global gauge transformation, under which the Hamiltonian is invariant is given by

$$|\Psi\rangle \rightarrow e^{i\hat{L}\Lambda} |\Psi\rangle. \tag{4.7}$$

It was later discovered that the neutrino produced in the decay $\pi^+ \rightarrow \mu^+ \nu$ was not the same as $\nu_e$ since if it were so, a reaction of the type

$$\nu + (Z, A) \rightarrow (Z + 1, A) + e^-$$

would have been observed. Instead what was observed was $\mu^-$ replacing $e^-$. This clearly shows that the neutrino accompanying $\mu^+$ in $\pi^+$ decay is different from $\nu_e$ and is denoted by $\nu_\mu$. The muon number defined as

$$L_\mu = \begin{cases} +1 \text{ for } \mu^-, \ \nu_\mu \\ -1 \text{ for } \mu^+, \ \bar{\nu}_\mu \\ 0 \text{ for all other particles} \end{cases}$$

is conserved in processes involving $\mu^\pm, \nu_\mu, \bar{\nu}_\mu$. The best limits are

$$\frac{\Gamma(\mu \rightarrow e\gamma)}{\Gamma(\mu \rightarrow all)} < 1.2 \times 10^{-11}$$

and

$$\frac{\Gamma(\mu \rightarrow 3e)}{\Gamma(\mu \rightarrow all)} < 1.0 \times 10^{-12}$$

(iv) Strangeness and Hypercharge:

It is clear from Figs. 4.1-4.4, that hadrons with the same spin and parity occur in nature as multiplets. Consider, for example, $J^P = 0^-$ mesons. We distinguish the triplet of pions $(\pi^\pm, \pi^0)$, the doublets $(K^+, K^0)$ and $(\bar{K}^0, K^-)$ by assigning a new quantum number, called strangeness: $S(\pi) = 0$, $S(K) = +1$ and $S(\bar{K}) = -1$. The singlets $\eta$ and $\eta'$ have strangeness $S = 0$. Similarly the baryons with $J^P = 1/2^+$ are assigned the strangeness quantum number as follows: For the doublet $(p, n)$, $S = 0$, for the triplet $(\Sigma^\pm, \Sigma^0)$, $S = -1$, for the singlet $(\Lambda^0)$, $S = -1$, and for the doublet $(\Xi^0, \Xi^-)$, $S = -2$. Sometimes, it is convenient to write $Y = B + S$, where $Y$ is called the hypercharge.

The quantum number $S$ is additively conserved in hadronic interactions. In any process, involving hadronic interactions, $\Delta S$ must be zero. This immediately leads to the result that in hadronic collisions, the strange particles are produced in pairs:

$$
\pi^- + p
\begin{cases}
\to K^0 + \Lambda^0 & \Delta S = 0 \\
\nrightarrow K^- + \Sigma^+ & \Delta S = -2 \\
\nrightarrow K^- + p & \Delta S = -1 \\
\to n + K^+ + K^- & \Delta S = 0.
\end{cases}
\tag{4.8}
$$

Experimentally, only the first and the last reactions are seen and the cross section for these reactions is typical of strong interactions. On the other hand, strange particles decay into ordinary particles by weak interactions:

$$
\Lambda \to p\,\pi^-
$$
$$
K^0 \to \pi^+\pi^-.
$$

These decays have lifetimes of the order $10^{-10}$ seconds, characteristics of weak interactions. Thus strangeness is not conserved in weak interactions.

In strong interactions, since both quantum numbers $B$ and $S$ are conserved, it is clear that hypercharge is also conserved. The gauge transformation under which the Hamiltonian is invariant is given by

$$
|\Psi\rangle \to e^{i\hat{Y}\Lambda} |\Psi\rangle.
\tag{4.9}
$$

It is interesting to note that the hypercharge of a multiplet is just equal to twice the average charge of that multiplet, i.e.

$$
Y = 2\,\langle Q \rangle = 2\,\langle q/e \rangle.
\tag{4.10}
$$

For example for the triplet of pions $(\pi^\pm, \pi^0)$, $\langle Q \rangle = 0$ and $Y = 0$, for the doublet $(p, n)$, $\langle Q \rangle = 1/2$ and $Y = 1$, whereas for the doublet $(\bar{K}^0, K^-)$ or $(\Xi^0, \Xi^-)$, $\langle Q \rangle = 1/2$ and $Y = -1$.

It is tempting to assign another quantum number, called isospin to each multiplet. For example we can assign $I = 1, I_3 = +1, 0, -1$ to the triplet of pions $(\pi^+, \pi^0, \pi^-)$ and $I = 1/2$, $I_3 = 1/2$ and $-1/2$ to the doublet $(p, n)$. We will discuss isospin in the next section. Here we summarize the conservation laws for internal quantum numbers $Q, B, S$ and $I$ for the three basic interactions.

| Quantum Number | Hadronic | Electromagnetic | Weak |
|:---:|:---:|:---:|:---:|
| $Q$ | Yes | Yes | Yes |
| $B$ | Yes | Yes | Yes |
| $S$ or $Y$ | Yes | Yes | No |
| Isospin | Yes | No | No |

## 4.2  Isospin

We now introduce isospin. From Figs. 4.1 and 4.2, it is apparent that particles occur in nature as multiplets. In analogy with ordinary spin, we can regard proton and neutron as an isospin doublet (nucleon) $N = \begin{pmatrix} p \\ n \end{pmatrix}$, with $I = 1/2$ and $I_3 = \pm 1/2$.

The concept of isospin is meaningful only if in hadronic interactions isospin is conserved. This is indeed the case. Experiments on nucleon-nucleon scattering show that after subtracting the effect of Coulomb force in $pp$ scattering, $pp$, $np$ and $nn$ hadronic forces are equal in strength and have the same range. That is nuclear forces do not depend on the charge of the particle and are thus charge independent. It is now known that all hadronic forces, not just the one between nucleons are charge independent.

The two states of nucleon $N$ viz $p$ and $n$ will have similar properties as far as hadronic forces are concerned. Without electromagnetic interaction, proton and neutron will have the same mass, but its presence makes their masses slightly different. This is supported by the fact that $(m_p - m_n) = -1.2$ MeV only, i.e. about 0.1% of $m_p$.

Like ordinary angular momentum, we introduce a quantity isospin $\mathbf{I} \equiv (I_1, I_2, I_3)$ in isospin space. The operator $\hat{\mathbf{I}}$ satisfies the commutation relations of angular momentum $J$ viz.

$$\left[\hat{I}_i, \ \hat{I}_j\right] = i\varepsilon_{ijk} \ \hat{I}_j, i = 1, 2, 3. \tag{4.11}$$

As a consequence of these commutation relations, it is possible to find a complete set of simultaneous eigenstates $|I \ I_3\rangle$ of $\hat{\mathbf{I}}^2$, and $\hat{I}_3$ with eigenvalues

$I(I+1)$ and $I_3$:

$$\hat{\mathbf{I}}^2 \, |I \, I_3\rangle = I(I+1) \, |I \, I_3\rangle \tag{4.12a}$$

$$\hat{I}_3 \, |I \, I_3\rangle = I_3 \, |I \, I_3\rangle . \tag{4.12b}$$

$\hat{I}_3$ has $(2I+1)$ eignenvalues

$$-I, \cdots, +I. \tag{4.13a}$$

The possible eigenvalues of $I$ are

$$I = 0, 1/2, 1, 3/2, 2, \cdots \tag{4.13b}$$

Thus, all the multiplets in Figs. 4.1 and 4.2 belong to an irreducible representation of the isospin group, i.e. they have any of the possible eigenvalues of $I$ given in Eq. (4.13b). For example, the proton and neutron states can be written as far as the isospin is concerned as

$$|p\rangle = |1/2 \; 1/2\rangle$$
$$|n\rangle = |1/2 \; -1/2\rangle , \tag{4.14}$$

and the pions can be represented as

$$|\pi^+\rangle = |1 \; 1\rangle$$
$$|\pi^0\rangle = |1 \; 0\rangle$$
$$|\pi^-\rangle = |1 \; -1\rangle \tag{4.15}$$

The charge of a state is given by the relation

$$Q = \left(\frac{q}{e}\right) = I_3 + \langle Q \rangle = I_3 + \frac{1}{2}Y. \tag{4.16}$$

This is called the Gell-Mann-Nishijima relation.

Charge independence of hadronic force implies that this force does not distinguish any direction in isospin space that is to say that hadronic interactions are invariant under a rotation in isospin space in complete analogy with ordinary angular momentum. This means that the $S$-matrix or the hadronic part of the Hamiltonian $H_h$ commutes with the rotation operator

$$U_I = e^{-i\alpha \cdot \hat{\mathbf{I}}}, \qquad \alpha = \omega \, \mathbf{n} \tag{4.17}$$

in isospin space, i.e.

$$[S, \; U_I] = 0, \; \text{or} \; [H_h, \; U_I] = 0. \tag{4.18}$$

$\hat{\mathbf{I}}$ is the generator of a rotation group in the isospin space. For an infinitesimal rotation

$$U_I = 1 - i\alpha \cdot \hat{\mathbf{I}}. \tag{4.19}$$

Hence, we have

$$\left[S,\, \hat{\mathbf{I}}\right] = 0, \text{ or } \left[H_h,\, \hat{\mathbf{I}}\right] = 0, \tag{4.20}$$

i.e. isospin is conserved in any process involving hadronic interactions. Thus we have the selection rules

$$\Delta\, |I|^2 = 0, \qquad \Delta\, I_3 = 0. \tag{4.21}$$

Since in the absence of electromagnetic interaction, the mass Hamiltonian $H_M$ commutes with $\hat{\mathbf{I}}$, the eignestates of $H_M$ with the same $I$, i.e. $(2I + 1)$ states with different values of $I_3$, are degenerate in mass.

As an illustration of isospin conservation, we consider the $\pi - N$ scattering.

$$\pi^+ p \to \pi^+ p$$
$$\pi^- p \to \pi^- p$$
$$\to \pi^0 n$$

We can write

$$|\pi^+\, p\rangle = |1\ 1\rangle\ |1/2\ 1/2\rangle = |1\ 1/2\ 1\ 1/2\rangle$$
$$|\pi^-\, p\rangle = |1\ 1/2\ -1\ 1/2\rangle$$
$$|\pi^0\, n\rangle = |1\ 1/2\ 0\ -1/2\rangle. \tag{4.22}$$

Now the scattering amplitude $F$ is given by

$$\langle \pi^-\, p\, |F|\, \pi^-\, p\rangle$$
$$= \sum_{II_3} \sum_{I'I'_3} \langle \pi^-\, p|\, I'\, I'_3\, 1\, 1/2\rangle$$
$$\times \langle I'\, I'_3\, 1\, 1/2\, |F|\, I\, I_3\, 1\, 1/2\rangle\ \langle I\, I_3\, 1\, 1/2\, |\pi^-\, p\rangle$$
$$= \sum_{II_3} \sum_{I'I'_3} \langle \pi^-\, p|\, I'\, I'_3\, 1\, 1/2\rangle\ F_I\, \delta_{II'}\, \delta_{I_3 I'_3}\ \langle I\, I_3\, 1\, 1/2\, |\pi^-\, p\rangle$$
$$= \sum_{I} \langle \pi^-\, p|\, I\, -1/2\, 1\, 1/2\rangle\ F_I\, \langle I\, -1/2\, 1\, 1/2\, |\pi^-\, p\rangle. \tag{4.23}$$

Using the Clebsch-Gordan coefficients, we have

$$\langle \pi^-\, p\, |F|\, \pi^-\, p\rangle = \frac{1}{3}\, F_{\frac{3}{2}} + \frac{2}{3}\, F_{\frac{1}{2}}. \tag{4.24a}$$

Similarly, we get

$$\langle \pi^0\, n\, |F|\, \pi^-\, p\rangle = \frac{\sqrt{2}}{3}\, F_{\frac{3}{2}} - \frac{\sqrt{2}}{3}\, F_{\frac{1}{2}} \tag{4.24b}$$

$$\langle \pi^+\, p\, |F|\, \pi^+\, p\rangle = F_{\frac{3}{2}}. \tag{4.24c}$$

Without using isospin invariance we have three independent amplitudes. With its use we have only two independent amplitudes. Thus

$$\sigma_{\pi^+} = \rho \left| F_{\frac{3}{2}} \right|^2 = \sigma^{(+)} \tag{4.25a}$$

$$\sigma_{\pi^-} \equiv \left[ \sigma \left( \pi^- \, p \to \pi^- \, p \right) + \sigma \left( \pi^- \, p \to \pi^0 \, n \right) \right] \equiv \sigma^{(-)} + \sigma^{(0)}$$

$$= \rho \left[ \frac{1}{3} \left| F_{\frac{3}{2}} \right|^2 + \frac{2}{3} \left| F_{\frac{1}{2}} \right|^2 \right]. \tag{4.25b}$$

Here $\rho$ is the kinematical factor. If $F_{3/2} \gg F_{1/2}$, then from Eq. (4.24)

$$\sigma^{(+)} : \sigma^{(-)} : \sigma^{(0)} = 9 : 1 : 2.$$

Experimentally, the cross-sections are in the ratio $(122 \pm 8) : (12.8 \pm 1.10) :$ $(25.6 \pm 1.3)$ for the kinetic energy of the pion from 120 MeV to 300 MeV. Thus it is clear that the scattering takes place predominantly in the $I = 3/2$ state for the above energy range.

Finally, we note that since the electric charge is always conserved, the conservation of $I_3$ implies $Y$-conservation and vice versa. To summarize, for hadronic interactions

$$\Delta |\mathbf{I}|^2 = 0$$

$$\Delta(Q, \, B, \, Y) = 0. \tag{4.26}$$

### 4.2.1    *Electromagnetic Interaction and Isospin*

Because of Eq. (4.16), electromagnetic interaction breaks the rotational symmetry in the isospin space:

$$\left[ H_{em}, \, \hat{I} \right] \neq 0 \tag{4.27a}$$

but

$$\left[ H_{em}, \, \hat{I}_3 \right] = 0. \tag{4.27b}$$

Hence $H_{em}$ is invariant under an isospin rotation about the 3rd axis, i.e. $I_3$ is still conserved by the electromagnetic interaction.

We can say that the isospin symmetry is broken by the electromagnetic interaction and a small mass difference between the members of an isospin multiplet may arise due to the electromagnetic interaction. Since

$$\left[ H_{em}, \, \hat{Q} \right] = 0, \tag{4.28}$$

therefore, it follows from Eq. (4.27b) that

$$\left[ H_{em}, \, \hat{Y} \right] = 0. \tag{4.29}$$

Hence for electromagnetic interaction, we have the selection rules:
$$\Delta I_3 = 0, \qquad \Delta Y = 0, \qquad \Delta B = 0, \qquad (4.30a)$$
but
$$\Delta |\mathbf{I}|^2 \neq 0. \qquad (4.30b)$$

### 4.2.2 *Weak Interaction and Isospin*

Consider the weak processes
$$\Lambda \rightarrow p + \pi^-$$
$$n \rightarrow p + e^- + \bar{\nu}_e.$$
Clearly $I_3$ is not conserved in weak interactions and hence $I^2$ is also not conserved. It follows that $Y$ is also not conserved, since $Q$ is conserved. Thus for weak interactions, we have the selection rules:
$$\Delta |\mathbf{I}|^2 \neq 0, \qquad \Delta Y \neq 0, \qquad \Delta B = 0. \qquad (4.31)$$

## 4.3 Resonance Production

We now consider the reaction shown in Fig. 4.5. We have three particles in the final state, produced incoherently. Let us consider the pair of particles $(n\pi^+)$, $(n\pi^-)$ and $(\pi^+ \pi^-)$. We define the invariant mass of each system designated by Eq. (4.22) and $|1,1,1,-1\rangle$:
$$s_{12} = (E_1 + E_2)^2 - (\mathbf{p}_1 + \mathbf{p}_2)^2 \qquad (4.32a)$$
$$s_{13} = (E_1 + E_3)^2 - (\mathbf{p}_1 + \mathbf{p}_3)^2 \qquad (4.32b)$$
$$s_{23} = (E_2 + E_3)^2 - (\mathbf{p}_2 + \mathbf{p}_3)^2 \qquad (4.32c)$$

If the reaction proceeds as in Fig. 4.5, $n$, $\pi^+$ and $\pi^-$ will have energy and momentum statistically distributed. The number of $(n\ \pi^+)$ pairs with an invariant mass $\sqrt{s_{12}}$, $N(s_{12})$ can also be calculated. $N(s_{12})$ can be plotted as a function of $\sqrt{s_{12}}$ and the result is called a phase space spectrum as shown in Fig. 4.7. If the reaction takes place as shown in Fig. 4.6, i.e. with $\pi^+$'s strongly correlated with the $n$'s, then energy-momentum conservation demands
$$E_\Delta = E_1 + E_2$$
$$\mathbf{p}_\Delta = \mathbf{p}_1 + \mathbf{p}_2$$
$$m_\Delta = \left[ E_\Delta^2 - p_\Delta^2 \right]^{1/2} = \sqrt{s_{12}}. \qquad (4.33)$$

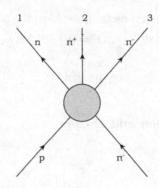

Fig. 4.5   The reaction $\pi^- p \to n\pi^+ \pi^-$, $\pi^- p \to p\pi^0 \pi^-$.

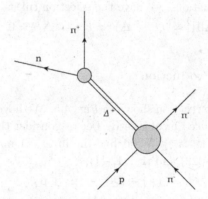

Fig. 4.6   The pion production through resonance $\pi^- p \to \Delta^+ \pi^- \to n\pi^+ \pi^-$.

In this case the final $n\,\pi^+$ results from the decay of a quasi-stable particle $\Delta^+$, called a resonance. In this situation, $N(s_{12})$ shows a strong peak at $\sqrt{s_{12}} = m_\Delta$ (Fig. 4.7). The finite width of the peak shows that the particle is very short lived, the life time $\tau = \frac{1}{\Gamma}$, $\Gamma$ being the width of the resonance. Actually a broad peak is seen experimentally at $\sqrt{s_{12}} = m_\Delta = 1238$ MeV with the full width at half maximum $\Gamma_\Delta \approx 120$ MeV [see Fig. 4.8].

Similarly, if we consider the pair $n\,\pi^-$ one finds a peak due to $\Delta^-$. The $(\pi^+ \pi^-)$ invariant mass distribution, $N(s_{23})$, also shows a broad peak at about $\sqrt{s_{23}} = 750$ MeV, due to the $\rho^0$ resonance.

### 4.3.1 $\Delta$-resonance

We now discuss the quantum numbers of the $\Delta$-resonance. We first determine its isospin. The resonance $\Delta$ is seen both in $\pi^- p$ and $\pi^+ p$ scattering. Since for $\pi^+ p$, $I = 3/2$ is the only possibility, it follows that its isospin must be 3/2. This is confirmed in the $\pi^+ p$ and $\pi^- p$ scattering experiments at energies at which multiple mesons production is insignificant viz the processes:

$$\pi^+ p \to \pi^+ p$$
$$\pi^- p \to \pi^- p$$
$$\to \pi^0 n.$$

If the $I = 3/2$ channel dominates in the above processes, we then have from Eq. (4.25) $\sigma_{\pi^+}/\sigma_{\pi^-} = 3$, at the resonance energy. This is what is borne out experimentally, showing unambiguously that the resonance channel is $I = 3/2$ (see Fig. 4.8).

### 4.3.2 *Spin of $\Delta$*

We first consider two-body scattering

$$a + b \to R \to a' + b' \tag{4.34}$$

through a resonance $R$. Suppose the spin of $R$ is $J$. Consider the decay

$$R \to a + b.$$

Let **p** be the momentum in the center-of-mass frame of particles $a$ and $b$. Let $\lambda_1$ and $\lambda_2$ be their helicities. Now $|\mathbf{p}| = p$ and its direction is given by $\omega \equiv (\theta, \phi)$. We can write the helicity state [cf. Eq. (3.165)]

$$|\lambda_1 \lambda_2 \, \omega\rangle = \sum_{J'M'} \sqrt{\frac{2\,J' + 1}{4\pi}} d_{M'\lambda}^{J'} (\theta, \phi) |J'M', \lambda\rangle, \tag{4.35}$$

where $\lambda = \lambda_1 - \lambda_2$. Therefore, the decay amplitude is given by

$$F_\lambda(\omega) = \sum_{J'M'} \sqrt{\frac{2\,J' + 1}{4\pi}} d_{M\lambda}^{J'*} \langle J'M', \lambda |F| JM\rangle. \tag{4.36}$$

We now take $R$ and $a$ to be fermions and $b$ a boson. Now

$$\langle J'M', \lambda |F| JM\rangle = \delta_{JJ'}\, \delta_{MM'}\, F_\lambda^J (s)\sqrt{4\pi}. \tag{4.37}$$

Therefore,

$$\begin{aligned} F_\lambda (\omega) &= \sqrt{2J + 1} d_{M\lambda}^{J*} (\theta, \phi)\, F_\lambda^J (s) \\ &= \sqrt{2J + 1} e^{i(\lambda - M)\phi}\, F_\lambda^J (s)\, d_{M\lambda}^{J*}(\theta). \end{aligned} \tag{4.38}$$

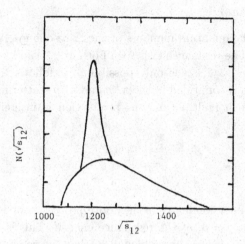

Fig. 4.7   Phase space plot for $(n\pi^+)$ pairs.

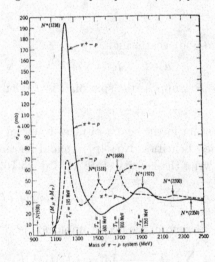

Fig. 4.8   The resonance scattering for $\pi^+ p$ and $\pi^- p$ channels.

Now

$$d\Gamma = (2\pi)^4 \int \frac{d^3 p_a}{(2\pi)^3} \frac{d^3 p_b}{(2\pi)^3} \frac{m_R \, m_a}{2 E_R \, E_a \, E_b} \overline{\sum_{\text{spin}}} |F_\lambda(\omega)|^2, \qquad (4.39a)$$

or

$$\frac{d\Gamma}{d(\cos\theta)} = \frac{4\,m_a}{16\pi\,s^{1/2}}\,|\mathbf{p}|\overline{\sum_{\text{spin}}}\,(2J+1)\,\left|F_\lambda^J(s)\right|^2\left|d_{M\lambda}^J(\theta)\right|^2. \tag{4.39b}$$

Therefore

$$\Gamma = \frac{m_a}{2\pi\,s^{1/2}}\,|\mathbf{p}|\overline{\sum_{\text{spin}}}\,\left|F_\lambda^J(s)\right|^2, \tag{4.40}$$

where we have used the orthogonality of $d$-functions. When $R$, $a$ and $b$ all are bosons, we get

$$\Gamma = \frac{1}{8\pi\,s}\,|\mathbf{p}|\overline{\sum_{\text{spin}}}\,\left|F_\lambda^J(s)\right|^2. \tag{4.41}$$

For a resonance scattering as in Eq. (4.34), the invariant scattering amplitude is given by

$$F(ab \to R \to a'b') = \sum_M F(ab \to R)\,F(R \to a'b')\,\phi_R(s), \tag{4.42}$$

where $\phi_R(s)$ is the resonance factor. Now using Eq. (4.38), we have

$$F(ab \to R \to a'b') = \sum_M d_{M\lambda}^J(\omega')\,d_{M\lambda'}^{J*}(\omega'')\,(2J+1)$$

$$\times F_\lambda^J\,(ab \to R)\,F_{\lambda'}^J\,(R \to a'b')\,\phi_R(s), \tag{4.43}$$

where $\omega' \equiv (\theta',\phi')$ and $\omega'' \equiv (\theta'',\phi'')$ are the polar and azimuthal angles of particles $a$ and $a'$ with respect to some fixed direction. Using the group property of $d$-functions

$$\sum_M d_{M\lambda}^J(\omega')\,d_{M\lambda'}^{J*}(\omega'') = d_{\lambda\lambda'}^{J*}(\theta,\phi), \tag{4.44}$$

where $\theta$ and $\phi$ are the polar and azimuthal angles of the particle $a'$ relative to $a$. Hence we have

$$F(ab \to R \to a'b') = (2J+1)\,d_{\lambda\lambda'}^{J*}(\theta,\phi)$$

$$\times F_\lambda^J\,(ab \to R)\,F_{\lambda'}^J\,(R \to a'b')\,\phi_R(s). \tag{4.45}$$

Now comparing it with [cf. Eq. (3.179) for the $J$th partial wave]

$$F_{\lambda'\lambda}(\omega) = \frac{4\pi\sqrt{s}}{\sqrt{m_a\,m_a'}}\,(2J+1)\,F_{\lambda'\lambda}^J(s)\,d_{\lambda\lambda'}^{J*}(\theta,\phi), \tag{4.46}$$

we have

$$F_{\lambda'\lambda}^J(s) = \frac{\sqrt{m_a\,m_a'}}{4\pi\sqrt{s}}F_\lambda^J\,(ab \to R)\,F_{\lambda'}^J\,(R \to a'b')\,\phi_R(s). \tag{4.47}$$

Now the partial wave cross section in the angular momentum state $J$ is given by

$$\sigma_J = 4\pi \frac{2J+1}{(2S_a+1)(2S_b+1)} \frac{|\mathbf{p}'|}{|\mathbf{p}|} \sum_{\lambda\lambda'} \left| F^J_{\lambda'\lambda}(s) \right|^2. \qquad (4.48)$$

Using

$$\left| F^J_\lambda(ab \to R) \right|^2 = \left| F^J_\lambda(R \to ab) \right|^2 \qquad (4.49)$$

and Eqs. (4.40), (4.47) and (4.48), we get

$$\sigma_J = \frac{4\pi}{|\mathbf{p}|^2} \frac{2J+1}{(2S_a+1)(2S_b+1)} \left[ \frac{\Gamma(R \to ab)\,\Gamma(R \to a'b')}{4} \, |\phi_R(s)|^2 \right]. \qquad (4.50)$$

The resonance factor is given in the Breit-Wigner form:

$$|\phi_R(s)|^2 = \left[ \frac{1}{\left(\sqrt{s}-m_R\right)^2 + \frac{\Gamma^2}{4}} \right]. \qquad (4.51)$$

Hence we have

$$\sigma_J = \frac{\pi}{|\mathbf{p}|^2} \frac{2J+1}{(2S_a+1)(2S_b+1)} \left[ \frac{\Gamma(R \to ab)\,\Gamma(R \to a'b')}{\left(\sqrt{s}-m_R\right)^2 + \frac{\Gamma^2}{4}} \right]. \qquad (4.52)$$

Consider now the process

$$\pi^+ p \to \Delta^{++} \to \pi^+ p. \qquad (4.53)$$

From Eq. (4.52), we get

$$\sigma_J(m_\Delta) = \frac{2\pi}{|\mathbf{p}|^2}(2J+1).$$

Experimentally, near the resonance

$$\sigma_J \approx \frac{8\pi}{|\mathbf{p}|^2}, \qquad (4.54)$$

giving $J = 3/2$.

It is also possible to determine the spin of a resonance by angular distribution of its decay products. This we illustrate by considering the $\Delta$-resonance viz $\Delta^{++} \to \pi^+ p$. Take the z-axis along the direction of the nucleon (or pion) in their center-of-mass frame, so that $l^i_z = 0$ (i-refers to $\pi^+ p$ in the initial state and $l$ refers to orbital angular momentum). Since pion is spinless, $M^i = \pm 1/2$. If $J$ is the spin of $\Delta$-resonance, then $M = \pm 1/2$, by angular momentum conservation. Now from Eq. (4.39b), the angular distribution of $p\pi^+$ in the final state is given by

$$I(\theta) \propto \sum_{M,\lambda} \left| F^J_\lambda(s) \right|^2 \left| d^J_{M\lambda}(\theta) \right|^2. \qquad (4.55)$$

Thus for $J = 1/2$, $M = +1/2, -1/2$,

$$I(\theta) \propto \left( \left| F_{1/2}^{1/2}(s) \right|^2 + \left| F_{-1/2}^{1/2}(s) \right|^2 \right) \left( \left| d_{1/2\ 1/2}^{1/2}(\theta) \right|^2 + \left| d_{1/2\ -1/2}^{1/2}(\theta) \right|^2 \right).$$

$$(4.56)$$

Using [Problem 3.2], we have

$$I(\theta) \propto \left[ \cos^2 \frac{\theta}{2} + \sin^2 \frac{\theta}{2} \right]. \tag{4.57}$$

Thus the angular distribution is isotropic.

For $J = 3/2$ and $M = \pm 1/2$, we have

$$I(\theta) \propto \left( \left| F_{1/2}^{3/2}(s) \right|^2 + \left| F_{-1/2}^{3/2}(s) \right|^2 \right) \left( \left| d_{1/2\ 1/2}^{3/2}(\theta) \right|^2 + \left| d_{1/2\ -1/2}^{3/2}(\theta) \right|^2 \right).$$

$$(4.58)$$

Again using [Problem 3.2], we have

$$I(\theta) \propto \left( 1 + 3 \cos^2 \theta \right). \tag{4.59}$$

We note that $I(-\theta) = I(\theta)$. The observed angular distribution of the protons or the pions at the resonance agrees with the prediction of Eq. (4.59), showing that $J = 3/2$ for the $\Delta$. The above derivation clearly shows that the angular distribution depends only on the value of $J$, and not on the parity, i.e. orbital angular momentum which never enters in the helicity representation used above.

## 4.4 Charge Conjugation

It is a general feature of relativistic quantum mechanics that corresponding to a particle, there is an antiparticle which has the same mass and spin as its particle. We treat particle and antiparticle on equal footing. We, therefore, postulate an operator $U_c$, which changes a particle into its antiparticle. The operator $U_c$ is a unitary operator. Thus, for example

$$U_c \left| \pi^+ \right\rangle = \left| \pi^- \right\rangle \tag{4.60a}$$
$$U_c \left| p \right\rangle = \left| \bar{p} \right\rangle. \tag{4.60b}$$

In general, for a charged particle

$$U_c \left| Q, \mathbf{p}, \mathbf{s} \right\rangle = \left| -Q, \mathbf{p}, \mathbf{s} \right\rangle, \tag{4.61}$$

where $|Q, \mathbf{p}, \mathbf{s}\rangle$ represents a single particle state with charge $Q$, momentum $\mathbf{p}$ and spin $\mathbf{s}$. Now

$$\hat{Q} \, |Q, \, \mathbf{p}, \, \mathbf{s}\rangle = Q \, |Q, \, \mathbf{p}, \, \mathbf{s}\rangle \tag{4.62a}$$

$$U_c \, \hat{Q} \, |Q, \, \mathbf{p}, \, \mathbf{s}\rangle = Q \, |{-Q}, \, \mathbf{p}, \, \mathbf{s}\rangle \tag{4.62b}$$

$$\hat{Q} \, U_c \, |Q, \, \mathbf{p}, \, \mathbf{s}\rangle = \hat{Q} \, |{-Q}, \, \mathbf{p}, \, \mathbf{s}\rangle$$

$$= -Q \, |{-Q}, \, \mathbf{p}, \, \mathbf{s}\rangle . \tag{4.62c}$$

Therefore, we have

$$U_c \, \hat{Q} + \hat{Q} \, U_c = 0, \tag{4.63a}$$

$$\left[U_c \, \hat{Q}\right]_+ = 0, \tag{4.63b}$$

i.e. $U_c$ and $Q$ do not commute. Hence it is not possible to find simultaneous eigenstates of $U_c$ and $\hat{Q}$. In general, for any additive internal quantum number, such as $Q$, $I_3$, $B$, $Y$ and $L$,

$$U_c \, |Q, \, I_3, \, B, \, Y, \, L\rangle = |{-Q}, \, {-I_3}, \, {-B}, \, {-Y}, \, {-L}\rangle \tag{4.64}$$

and consequently,

$$[U_c, \, Q_i] \neq 0, \tag{4.65}$$

where

$$Q_i = \hat{I}_3, \, \hat{B}, \, \hat{Y}, \, \text{or} \, \hat{L}.$$

Now

$$U_c \, |B\rangle = |{-B}\rangle$$

$$U_c^2 \, |B\rangle = U_c \, |{-B}\rangle = |B\rangle . \tag{4.66}$$

Therefore,

$$U_c^2 = 1 \tag{4.67}$$

and eigenvalues of $U_c$ are $\pm 1$, i.e. $U_c$ is a discrete transformation.

It follows from Eq. (4.64) that states with $Q \neq 0$, $B \neq 0$, $Y \neq 0$, etc. cannot be eigenstates of $U_c$. Only states with $Q = 0$, $B = 0$, $Y = 0$, $I_3 = 0$ can be eigenstates of $U_c$. For them it is possible to define the charge conjugation parity $\eta_c$:

$$U_c \, |B = 0\rangle = \eta_c \, |B = 0\rangle , \tag{4.68}$$

where

$$\eta_c^2 = 1 \quad \text{or} \quad \eta_c = \pm 1. \tag{4.69}$$

$\eta_c$ is a multiplicatively conserved quantum number in any process which conserves $C$-parity. The C-parity is either $+1$ or $-1$.

Charge conjugation is an internal symmetry. If

$$[U_c, H] = 0, \quad \text{or} \quad [U_c, S] = 0, \tag{4.70}$$

we say that the corresponding interaction is invariant under charge conjugation $U_c$. While strong and electromagnetic interactions are invariant under $U_c$, weak interactions are not

$$[U_c, H_{\text{weak}}] \neq 0 \tag{4.71}$$

This is clear from the fact that neutrinos and antineutrinos which come out in $\beta$-decay of nuclei have opposite polarizations or helicities $[H = 2\mathbf{s} \cdot \mathbf{p}/|\mathbf{p}|]$. If charge conjugation were conserved in weak interactions, neutrino and antineutrino would have the same helicity.

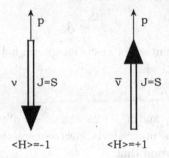

Fig. 4.9   The neutrino with helicity $-1$ and antineutrino with helicity $+1$.

How to test charge conjugation in hadronic interactions? Consider for example, the reactions

$$\bar{p} + p \to \pi^+ + h$$
$$\to \pi^- + \bar{h},$$

where $h$ $(\bar{h})$ denote all other hadrons with $B = 0$ and with positive (negative) electric charge. Now

$$\langle \bar{p}\, p|S|\pi^+\, h \rangle = \langle \bar{p}\, p|U_c^{-1}\, U_c\, S\, U_c^{-1}\, U_c|\pi^+\, h \rangle$$
$$= \langle p\, \bar{p}|S|\pi^-\, \bar{h} \rangle, \tag{4.72}$$

where we have assumed that $S$ is invariant under $U_c$:

$$U_c\, S\, U_c^{-1} = S. \tag{4.73}$$

Thus $C$-invariance requires that positive and negative pions have same energy spectrum. Comparison of $\pi^+$ and $\pi^-$ distributions show no difference, the result is stated as

$$\left| \frac{C - \text{nonconserving amplitude}}{C - \text{conserving amplitude}} \right| \leq 0.01.$$

As we have discussed, $\gamma$, $\pi^0$ and $\eta^0$ can be eigenstates of $U_c$. We now determine the $C$-parity of these states. Now under $U_c$, the electromagnetic current $j_\mu^{em}$:

$$j_\mu^{em} \xrightarrow{U_c} - j_\mu^{em}. \tag{4.74}$$

But the electromagnetic field $A_\mu$ satisfies the equation

$$\Box^2 A_\mu = j_\mu^{em}. \tag{4.75}$$

Thus from Eq. (4.75), it follows that

$$A_\mu \xrightarrow{U_c} - A_\mu. \tag{4.76}$$

Since a photon is a quantum of electromagnetic field, it follows that the $C$-parity of photon is $-1$ viz.

$$\eta_c\,(\gamma) = -1. \tag{4.77}$$

The decays $\pi^0 \to 2\gamma$ and $\eta^0 \to 2\gamma$ and $\omega^0 \to \pi^0\gamma$ have been observed. Hence if these reactions proceed via electromagnetic interaction, it then follows from $C$-conservation that

$$\eta_c\,\left(\pi^0\right) = +1$$
$$\eta_c\,\left(\eta^0\right) = +1 \tag{4.78}$$
$$\eta_c\,\left(\omega^0\right) = -1. \tag{4.79}$$

Since $\pi^0 \to 3\gamma$ and $\eta^0 \to 3\gamma$ can proceed via electromagnetic interaction, but have never been seen, these decays are strictly forbidden due to $C$-conservation in electromagnetic interaction. We conclude that the electromagnetic interaction is invariant under $U_c$.

$$[U_c,\ H_{\text{em}}] = 0. \tag{4.80}$$

Consider now the positronium, the bound states of $e^-$ and $e^+$. Let us consider $e^- - e^+$ in definite $(l,\ s)$ state. Now $e^-$ and $e^+$ are identical fermions which differ only in their electric charges. We can use a generalized Pauli principle for the positronium viz "under total exchange of particles (which consists of changing simultaneously $Q$, $\mathbf{r}$ and $s$ labels), the state should change sign or be antisymmetric". Under exchange of space co-ordinates, we get a factor $(-1)^l$, under spin co-ordinate exchange, we get a factor $(-1)^{s+1}$ ($s = 0$ for spin singlet state and $s = 1$, for spin triplet state), exchange of electric charge gives a factor $\eta_c$. We require the state to be antisymmetric, i.e.

$$(-1)^l \, (-1)^{s+1} \, \eta_c = -1 \tag{4.81}$$

or

$$\eta_c = (-1)^{l+s} \tag{4.82}$$

which gives the charge conjugation parity of the positronium in $(l, s)$ state.

The positronium $(e^- - e^+)$ can decay into $n \; \gamma$ by electromagnetic in-teraction. $C$-parity conservation gives

$$(-1)^{l+s} = (-1)^n. \tag{4.83}$$

From Eq. (4.82), we get the following selection rules:

| $l = 0 = s$ | $^1S_0 \to 2\gamma$ | Allowed |
|---|---|---|
| | $^1S_0 \to 3\gamma$ | strictly forbidden |
| $l = 0$ | $^3S_1 \to 2\gamma$ | strictly forbidden |
| $s = 1$ | $^3S_1 \to 3\gamma$ | Allowed |

Similarly for $(p - \bar{p})$ and quark-antiquark systems: $\eta_c = (-1)^{l+s}$.

Now for $(\pi^+ - \pi^-)$ system for which $B = 0$, $Y = 0$, $Q = 0$, generalized Pauli principle requires that the state should be symmetric (even) under total exchange of pions that is

$$(-1)^l \, \eta_c = 1 \tag{4.84}$$

or

$$\eta_c = (-1)^l. \tag{4.85}$$

Similarly for $\pi^0 - \pi^0$ system we get $\eta_c = (-1)^l$. For this case, since two $\pi^0$'s are identical particle, ordinary Pauli principle requires that $(-1)^l =$ even, i.e. they must be in an orbital state with $l$ even. Thus $\eta_c$ must be $+1$ for $\pi^0 - \pi^0$ system, whereas $\eta_c$ depends upon $l$ value for $\pi^+\pi^-$ system.

## 4.5  G-Parity

For strong interactions, both isospin and $C$-parity are conserved. For hadrons, it is convenient to define a new operator $\hat{G}$ = charge conjugation $+180^\circ$ rotation around 2nd axis in isospin space. It follows that strong interactions are invariant under $G$, but

$$[\hat{G},\ H_{\text{em}}] \neq 0 \tag{4.86a}$$

$$[\hat{G},\ H_{\text{weak}}] \neq 0 \tag{4.86b}$$

i.e. electromagnetic and weak interactions are not invariant under $G$.

Under $180^\circ$ rotation around the 2nd axis in isospin space, we have

$$|\pi_1\rangle \rightarrow -|\pi_1\rangle$$
$$|\pi_2\rangle \rightarrow |\pi_2\rangle$$
$$|\pi_3\rangle \rightarrow -|\pi_3\rangle. \tag{4.87}$$

Therefore, we get

$$\left|\pi^{\pm,0}\right\rangle \rightarrow -\left|\pi^{\mp,0}\right\rangle. \tag{4.88}$$

Under charge conjugation

$$\left|\pi^{\pm}\right\rangle \xrightarrow{U_c} \left|\pi^{\mp}\right\rangle \tag{4.89a}$$

$$\left|\pi^{0}\right\rangle \xrightarrow{U_c} \left|\pi^{0}\right\rangle. \tag{4.89b}$$

Thus we have

$$\left|\pi^{\pm}\right\rangle \xrightarrow{\hat{G}} -\left|\pi^{\pm}\right\rangle \tag{4.90a}$$

and

$$\left|\pi^{0}\right\rangle \xrightarrow{\hat{G}} -\left|\pi^{0}\right\rangle \tag{4.90b}$$

Thus the $G$-parity of pions is $G(\pi) = -1$. The nucleon state $|N\rangle$, under $180^\circ$ rotation about 2nd axis in isospin space transforms as

$$|N_R\rangle = e^{i\ \tau_2\ \pi/2}\ |N\rangle$$
$$= \left(\cos\frac{\pi}{2} + i\ \tau_2 \sin\frac{\pi}{2}\right)|N\rangle$$
$$= i\ \tau_2\ |N\rangle \tag{4.91}$$

i.e.

$$|p\rangle \xrightarrow{I_{2(\pi)}} |n\rangle$$
$$|n\rangle \xrightarrow{I_{2(\pi)}} -|p\rangle. \tag{4.92}$$

But

$$|p\rangle \overset{U_c}{\to} |\bar{p}\rangle$$

$$|n\rangle \overset{U_c}{\to} |\bar{n}\rangle. \tag{4.93}$$

Therefore,

$$|p\rangle \overset{G}{\to} |\bar{n}\rangle \tag{4.94a}$$

$$|n\rangle \overset{G}{\to} -|\bar{p}\rangle. \tag{4.94b}$$

Only states with $B = 0$ and $Y = 0$ for which isospin $I$ is integer can be eigenstates of $\hat{G}$. Only for such states we can define G-parity $G$. In general G-parity of a state with isospin $I$ is given by

$$G = \eta_c (-1)^I \tag{4.95}$$

Thus for $\eta^0$ and $\omega^0$, $G = +1$ and $-1$ respectively. Thus for fermion-antifermion e.g. $q - \bar{q}$ system, the $G$-parity is given by

$$G = (-1)^{l+s+I} = \eta_c (-1)^I \tag{4.96}$$

For $(\pi^+\pi^-)$ system

$$G = (-1)^l (-1)^I = (-1)^{l+I} = 1 \tag{4.97}$$

The concept of $G$-parity is particularly useful to see whether a decay which does not involve $\gamma$ is strong or electromagnetic (problems 4.3 and 4.4).

## 4.6   Problems

(1) Consider pion nucleon scattering

$$\pi^i + N \to \pi^j + N$$

where $i$ and $j$ are isospin indices of the incoming and outgoing pions respectively. Using isospin invariance, show that in isospin space, the scattering amplitude $A$ can be written

$$A_{ji} = \left\{ A^{(+)} \, \delta_{ji} + A^{(-)} \frac{1}{2} [\tau_j, \tau_i] \right\}.$$

Show that isospin $\frac{3}{2}$ and $\frac{1}{2}$ projection operators are given by

$$P_{3/2} = \frac{2 + \mathbf{t} \cdot \tau}{3}, \qquad P_{1/2} = \frac{1 - \mathbf{t} \cdot \tau}{3}$$

where $\tau$ are Pauli matrices and $\mathbf{t}$ are isospin matrices for $I = 1$. Further show that

$$\left(P_{3/2}\right)_{ji} = \frac{1}{3}\left\{2\,\delta_{ji} - \frac{1}{2}\,[\tau_j, \tau_i]\right\}$$

$$\left(P_{1/2}\right)_{ji} = \frac{1}{3}\left\{1\delta_{ji} + \frac{1}{2}\,[\tau_j, \tau_i]\right\}.$$

**Hint:** $(t_k)_{ji} = i\varepsilon_{jki}$.

(2) Show that for the decay

$$i \to f + \gamma,$$

either $\Delta I = 0$ or $|\Delta I| = 1$.

Hence show that for the decay $\eta \to \pi^+\pi^-\gamma$, pions are in $I = 1$ state and $l$ is odd, but for the decay $\omega \to \pi^+\pi^-\gamma$, pions are in $I = 0$ or 2 state and $l$ is even.

(3) Show that the decay $\omega \to \pi^+\pi^-$ is forbidden in strong interaction, but is allowed by electromagnetic interaction. What are the values of isospin $I$ and orbital angular momentum for the pions ?

(4) Show that $\eta \to \pi^+\pi^-\pi^0$ is forbidden in strong interaction but is allowed by electromagnetic interaction. Determine the possible values of isospin for the final pions.

(5) A meson $f$ has been seen in the invariant mass plot of $(\pi\pi)$ in the reaction:

$$\pi^- p \to n\pi^+\pi^-$$

but not in the reaction

$$\pi^+ p \to p\pi^+\pi^0$$

What can you say about isospin of $f$? What is its electric charge?

(6) Let us consider the following particles:

$$X^0 \quad : I = 0, \quad J^{PC} = 0^{-\,+}$$
$$Y^{\pm,0} \quad : I = 1, \quad J^{PG} = 1^{-\,+}$$

Which of the following decays are allowed by strong interaction? Also for what values of orbital angular momentum and isospin of the final states the decays is allowed.

(a) $X^0 \to \pi^+\pi^-$

(b) $Y^- \to X^0\pi^-$

(c) $Y^0 \to \pi^+\pi^-$

(d) $Y^+ \rightarrow X^0 \pi^+$

(e) $Y^- \rightarrow \pi^- \eta^0$

(7) (a) List all the decay channels allowed by strong interaction for $\Omega^-$.

    i. Why strong decays of $\Omega^-$ are not seen?

    ii. Why radiative decay of $\Omega^-$ is not possible?

(b) List all the possible weak decays (Semi-leptonic and Hadronic) for $\Omega^-$.

(8) A particle with spin angular momentum **J** has the following Hamiltonian in an external magnetic field **B** $(\mathbf{B} = \nabla \times \mathbf{A})$

$$H_{mag} = -\left(\frac{ge\hbar}{2mc}\right)\left(\frac{\mathbf{J}}{\hbar}\right) \cdot \mathbf{B}$$

How $H_{mag}$ transform under **C**, **P** and **T** respectively?

(9) For $(\pi^+ \pi^-)_{\ell'}$ :

$$\eta_C\left(\pi^+ \pi^-\right) = (-1)^{\ell'}$$

$K \rightarrow 3\pi$ is parity conserving decay, i.e. parity of three pions in the final state is negative.

Show that CP-conservation implies

$$K_2^0 \rightarrow \pi^0 \pi^0 \pi^0 \quad \text{is allowed}$$
$$K_1^0 \rightarrow \pi^0 \pi^0 \pi^0 \quad \text{is forbiden}$$
$$K_1^0 \rightarrow \pi^+ \pi^- \pi^0 \quad \text{is allowed if } \ell' \text{ is odd}$$
$$K_2^0 \rightarrow \pi^+ \pi^- \pi^0 \quad \text{is allowed if } \ell' \text{ is even}$$

where

$$CP|K_1^0\rangle = |K_1^0\rangle \quad \text{and} \quad CP|K_2^0\rangle = -|K_2^0\rangle$$

G-parity of $\pi^+ \pi^- \pi^0$ is

$$\eta_G = \eta_C\left(-1\right)^I = (-1)^{\ell'+I} = -1$$

Hence show that

$$\text{for } \ell' \text{ even} \rightarrow I = 1, 3$$
$$\text{for } \ell' \text{ odd} \rightarrow I = 0, 2$$

(10) G-parity is defined as

$$\eta_G = \eta_C(-1)^I$$

where $I$ is the isospin

|       | $J^{PC}$ | $I^G$ |
|-------|----------|-------|
| $\omega$ | $1^{--}$ | $0^-$ |
| $\rho$   | $1^{--}$ | $1^+$ |
| $\pi$    | $0^{-+}$ | $1^-$ |
| $a_1$    | $1^{++}$ | $1^-$ |

C-parity and G-parity are conserved in the strong decays. For $(\pi^+\pi^-)$ system

$$\eta_C = (-1)^{\ell'}, \quad \eta_G = \eta_C(-1)^{I'} = (-1)^{I'+\ell'}$$

where $\ell'$ and $I'$ are the orbital angular momentum and isospin of the $(\pi^+\pi^-)$. Which of the following decays are allowed and which are forbidden?

$$\omega^0 \to \pi^+\pi^-\pi^0$$
$$\rho^0 \to \pi^+\pi^-\pi^0$$
$$a_1^0 \to \rho^0\pi^0, \ \rho^+\pi^-$$
$$a_1^+ \to \rho^0\pi^+, \ \omega\pi^+$$

(11) For the radiative decay

$$|i\rangle \to |f\rangle + \gamma$$

Show that either

$$\Delta|I| = 0, \quad G(f) = -G(i)$$

or

$$\Delta|I| = \pm 1, \quad \Delta G = 0$$

Photon behaves either like "$\rho$-mesons", i.e. it carries "$I = 1 \ G = 1$" or like "$\omega$-mesons", i.e. it carries "$I = 0, \ G = -1$"

$$\rho^0 \to \pi^0\gamma \quad \Delta|I| = 0, \ G(f) = -1, \ G(i) = 1$$
$$\omega^0 \to \pi^0\gamma, \quad \Delta|I| = 1, \quad \Delta G = 0$$

(12) Which selection rule forbids, the decay

$$\omega \to \pi^+\pi^-$$

as a strong decay. Show that this decay has

$$\Delta|I| = 1, \Delta G \neq 0, G(f) = -G(i),$$

i.e. this decay is allowed by electromagnetic interaction as second order in electromagnetism.

(13) Show that the decay

$$\omega \to \pi^+ \pi^- \pi^0$$

is allowed in strong interaction. What are the allowed $l$ and $I$ values for $(\pi^+ \pi^-)$ system?

(14) By using the similar arguments as used in $\Delta$-resonance, show that $\rho(770$ MeV) meson has $I = 1$ and $J = 1$.

## 4.7 References

1. S. Gasirowicz, Elementary particle physics, Wiley, New York (1966).

2. H. M. Pilkuhn, Relativistic particle physics, Springer-Verlag, New York (1979).

3. T. D. Lee, Particle physics and introduction to field theory, Harwood, New York (1981).

4. E. M. Henley, A. Garcia, Subatomic Physics, World Scientific, 3rd edition (2007).

5. Particle Data Group, K. Nakamura et al., J. Phys. **G 37**, 075021 (2010).

# Chapter 5

# Unitary Groups and SU(3)

## 5.1 Unitary Groups and SU(3)

Consider a vector $\phi_i$, $i = 1, 2, \cdots, N$ in an $N$-dimensional vector space. An arbitrary transformation in this space is

$$\phi_i' = a_i^j \, \phi_j \qquad i, j = 1, 2, \cdots, N. \tag{5.1}$$

For unitary group $U(N)$ in $N$ dimensions,

$$a_i^{*k} \, a_j^k \equiv \left(a^\dagger\right)_k^i \, a_j^k = \delta_j^i. \tag{5.2}$$

For the group $SU(N)$, we also have

$$\det \ a = 1. \tag{5.3}$$

The basic assumption of all the group theoretical approaches to classification of hadrons is that particles belong to an irreducible representation of some group (in our case $SU(N)$) form a multiplet and thus have the same space-time properties, especially the mass, spin and parity. The basic mathematical problem is the investigation of the representations of a group. There are two approaches to this investigation (i) global way, (ii) infinitesimal way. For continuous groups it is convenient to restrict to (ii). For the general infinitesimal transformation:

$$\phi_i' = \left[\delta_i^j + \varepsilon_i^j\right] \phi_j. \tag{5.4}$$

Then conditions (5.2) and (5.3) give

$$\varepsilon_i^{*j} = -\varepsilon_j^i$$
$$\varepsilon_i^i = 0. \tag{5.5}$$

The unitary transformation corresponding to (5.4) may be written as

$$U(a) = 1 - \varepsilon_i^j \, A_j^i + O\left(\varepsilon^2\right). \tag{5.6}$$

$A_j^i$ are called the generators of the group $U(N)$ and characterize the group completely. The $N \times N$ unitary complex matrices $U(a)$ form the representation of $U(N)$. Hence there are $N^2$ arbitrary real parameters and thus there are $N^2$ generators of the group $U(N)$. For $SU(N)$ we have $N^2 - 1$ generators because of the unimodularity condition.

The matrices $U(a)$ have the group property:

$$U(b) \ U(a) = U(c)$$
$$U(a) = U(a) \ U(1), \ U(1) = 1$$
$$U^{-1}(b) \ U(a) \ U(b) = U(b^{-1}ab)$$
$$U^\dagger(a) \ U(a) = 1. \tag{5.7}$$

It is easy to see that Eqs. (5.5) and (5.6) give

$$\left(A_j^i\right)^\dagger = A_i^j. \tag{5.8}$$

By taking $a$ to be an infinitesimal transformation, it is easy to derive (see problem 5) the commutation relations

$$\left[A_i^j, \ A_k^l\right] = \delta_k^j \ A_i^l - \delta_i^l \ A_k^j. \tag{5.9}$$

For the transformation (5.4), we have

$$\phi_k' \equiv U^{-1}(a) \ \phi_k \ U(a) = \phi_k + \varepsilon_i^j \left[A_j^i, \phi_k\right]. \tag{5.10}$$

But we can write

$$\phi_k = U_k^l \ \phi_l.$$
$$\phi_k' = \left(\delta_k^l + \varepsilon_i^j \left(M_j^i\right)_k^l\right) \phi_l. \tag{5.11}$$

Comparing Eqs. (5.11) and (5.4), we get

$$\left(M_j^i\right)_k^l = \delta_k^i \delta_j^l. \tag{5.12}$$

Hence from Eq. (5.10) one has

$$\left[A_j^i, \phi_k\right] = \left(M_j^i\right)_k^l \phi_l$$
$$= \delta_k^i \delta_j^l \ \phi_l = \delta_k^i \ \phi_j. \tag{5.13}$$

The matrices $M_j^i \left[\left(M_j^i\right)^\dagger = M_i^j\right]$ give the representation of group $U(N)$ for the fundamental representation $\phi_i$. Let us define a vector

$$\phi^i = \phi_i^*.$$

It belongs to the representation $\bar{N}$ of $U(N)$, whereas vector $\phi_i$ belongs to the representation $N$ of $U(N)$. Thus $\phi^i$ transforms as

$$\phi^i \rightarrow \phi'^i \equiv \phi'^{*}_i = \left(\delta^j_i - \varepsilon^{*j}_i\right)\phi^{*}_j$$
$$= \left(\delta^i_j - \varepsilon^i_j\right)\phi^j. \tag{5.14}$$

Hence it follows that

$$\left[A^i_j, \phi^k\right] = -\delta^k_j \, \phi^i. \tag{5.15}$$

Now if we consider a tensor $T^k_l$, it transforms as $\phi^k \phi_l$, so that

$$\left[A^i_j, T^k_l\right] = \delta^i_l \, T^k_j - \delta^k_j \, T^i_l. \tag{5.16}$$

Thus the tensor $T^i_j$ transforms in the same way as the generator $A^i_j$.

Let us now restrict ourselves to $SU(N)$. The generators of $SU(N)$ must be traceless. Hence we can write its generators as

$$F^i_j = A^i_j - \frac{1}{N}\delta^i_j \, A^k_k, \tag{5.17}$$

so that

$$U(a) = 1 - \varepsilon^i_j \, F^i_j$$
$$\left(F^i_j\right)^\dagger = F^j_i$$
$$F^i_i = 0. \tag{5.18}$$

Since $A^k_k$ is a $U(N)$ invariant, the commutation relation for $F^i_j$ remains the same as in Eq. (5.9) viz.

$$\left[F^i_j, \, F^k_l\right] = \delta^i_l \, F^k_j - \delta^k_j \, F^i_l. \tag{5.19}$$

The matrices $M^i_j$ must now be traceless, hence

$$\left(M^i_j\right)^k_l = \delta^i_l \, \delta^k_j - \frac{1}{N}\delta^i_j \, \delta^k_l. \tag{5.20}$$

Thus instead of Eqs. (5.13) and (5.15), one has

$$\left[F^i_j, \, \phi_k\right] = \delta^i_k \, \phi_j - \frac{1}{N}\delta^i_j \, \phi_k$$
$$\left[F^i_j, \, \phi^k\right] = -\delta^k_j \, \phi^i + \frac{1}{N}\delta^i_j \, \phi^k. \tag{5.21}$$

Let's now confine to $SU(3)$. It is convenient to express eight generators $F^i_j$ $(i,j = 1,2,3)$ in terms of hermitian operators $F_A$, $(A = 1, \cdots, 8)$ introduced by Gell-Mann. The relationship between $F_A$ and $F^i_j$ is as follows:

$$F^1_2 = F_1 - i\,F_2, \quad F^2_1 = F_1 + i\,F_2, \quad \frac{1}{2}\left(F^1_1 - F^2_2\right) = F_3,$$
$$F^1_3 = F_4 - i\,F_5, \quad F^3_1 = F_4 + i\,F_5, \quad F^2_3 = F_6 - i\,F_7,$$
$$F^3_2 = F_6 - i\,F_7, \quad F^3_3 = -\frac{2}{\sqrt{3}}F_8.$$

From the commutation relation (5.19) for $F_j^i$, one can show that $F_A$'s satisfy the standard commutation relation of a Lie group:

$$[F_A, \ F_B] = C_{AB}^D F_D$$
$$= if_{ABC} \ F_C, \tag{5.22}$$

where the structure constants $f_{ABC}$ are real and antisymmetric. $F_A$'s also satisfy the Jaccobi identity

$$[F_A, \ [F_B, \ F_C]] + [F_B, \ [F_C, \ F_A]] + [F_C, \ [F_A, \ F_B]] = 0. \tag{5.23}$$

Infinitesimal unitary transformation generated by $F_A$ is

$$U = 1 - i \ \varepsilon_A \ F_A,$$

$\varepsilon_A$ being infinitesimal real parameters. For an inifinitesinal transformation, the vectors $\phi_i$ and $\phi^i$ transform as

$$\phi_i' = U_i^j \ \phi_j$$
$$= \left[ \delta_i^j + \frac{i}{2} \varepsilon_A \left( \lambda_A \right)_i^j \right] \phi_j, \tag{5.24}$$

$$\phi'^{\ i} = U_i^{*\ j} \phi^j = \left[ \delta_i^j - \frac{i}{2} \varepsilon_A \left( \lambda_A^* \right)_i^j \right] \phi^j$$
$$= \left[ \delta_i^j - \frac{i}{2} \varepsilon_A \left( \lambda_A \right)_i^j \right] \phi^j, \tag{5.25}$$

since the matrices $\lambda_A$ are hermitian. The matrices $\lambda_A$ are related to $M_j^i$ in the same way as $F_A$ are related to $F_j^i$ . Thus

$$M_2^1 = \frac{1}{2} \left( \lambda_1 - i\lambda_2 \right), \ M_1^2 = \frac{1}{2} \left( \lambda_1 + i\lambda_2 \right), \cdots, M_3^3 = -\frac{1}{\sqrt{3}} \lambda_8. \tag{5.26}$$

Now for $SU(3)$, the matrix elements $M_j^i$ are given by Eq. (5.20)

$$\left( M_j^i \right)_l^k = \delta_l^i \ \delta_j^k - \frac{1}{3} \delta_j^i \ \delta_l^k. \tag{5.27}$$

Using Eqs. (5.26) and (5.27), we can explicitly write $3 \times 3$ matrices $\lambda_A$. They are

$$\lambda_1 = \begin{pmatrix} 0 & 1 & 0 \\ 1 & 0 & 0 \\ 0 & 0 & 0 \end{pmatrix}, \ \lambda_2 = \begin{pmatrix} 0 & -i & 0 \\ i & 0 & 0 \\ 0 & 0 & 0 \end{pmatrix}, \ \lambda_3 = \begin{pmatrix} 1 & 0 & 0 \\ 0 & -1 & 0 \\ 0 & 0 & 0 \end{pmatrix},$$

$$\lambda_4 = \begin{pmatrix} 0 & 0 & 1 \\ 0 & 0 & 0 \\ 1 & 0 & 0 \end{pmatrix}, \ \lambda_5 = \begin{pmatrix} 0 & 0 & -i \\ 0 & 0 & 0 \\ i & 0 & 0 \end{pmatrix}, \ \lambda_6 = \begin{pmatrix} 0 & 0 & 0 \\ 0 & 0 & 1 \\ 0 & 1 & 0 \end{pmatrix},$$

$$\lambda_7 = \begin{pmatrix} 0 & 0 & 0 \\ 0 & 0 & -i \\ 0 & i & 0 \end{pmatrix}, \ \lambda_8 = \begin{pmatrix} \frac{1}{\sqrt{3}} & 0 & 0 \\ 0 & \frac{1}{\sqrt{3}} & 0 \\ 0 & 0 & -\frac{2}{\sqrt{3}} \end{pmatrix}. \tag{5.28}$$

Obviously the matrices $\frac{\lambda_A}{2}$ satisfy the same commutation relations as the generators $F_A$, so that

$$[\lambda_A, \lambda_B] = 2i \, f_{ABC} \, \lambda_C. \tag{5.29a}$$

They are traceless and have the following properties:

$$Tr(\lambda_A \lambda_B) = 2\delta_{AB} \tag{5.29b}$$

$$[\lambda_A, \lambda_B]_+ = 2 \, d_{ABC} \, \lambda_C + \frac{4}{3}\delta_{AB}, \tag{5.30}$$

where $d_{ABC}$ are real and are totally symmetric. Defining $\lambda_0 = \sqrt{\frac{2}{3}}I$, the commutation and anticommutation relations can be written as

$$[\lambda_A, \lambda_B] = 2i \, f_{ABC} \, \lambda_C$$
$$[\lambda_A, \lambda_B]_+ = 2 \, d_{ABC} \, \lambda_C$$
$$Tr(\lambda_A \lambda_B) = 2\delta_{AB}$$
$$d_{0BC} = \sqrt{\frac{2}{3}} \, \delta_{BC}, \qquad f_{0BC} = 0, \tag{5.31}$$

where $A, B, C = 0, 1, \cdots, 8$. Thus $\lambda_A$ are closed both under commutation and anticommutation. We also note that $\lambda_2$, $\lambda_5$, $\lambda_7$ are antisymmetric while the rest of them are symmetric. We express this fact by writing

$$\lambda_A^T = \eta_A \, \lambda_A \qquad \text{(not summed)}, \tag{5.32}$$

where $\eta_A = -1$, for $A = 2, 5, 7$ and $+1$ otherwise. The following identities follow from Eqs. (5.31) and (5.32):

$$\eta_A \, \eta_B \, \eta_C \, f_{ABC} = -f_{ABC}$$
$$\eta_A \, \eta_B \, \eta_C \, d_{ABC} = d_{ABC} \quad \text{(repeated indices not summed)}$$

$$\tag{5.33}$$

i.e. $f_{ABC}$ ($d_{ABC}$) is zero if even (odd) number of indices take the value 2, 5 or 7. The values of $f_{ABC}$ and $d_{ABC}$ have been tabulated by Gell-Mann and are reproduced in Table 5.1. The role of $F_A$ is the same in $SU(3)$ as that of isospin **I** in $SU(2)$ and for this reason $F_A$'s are sometimes called component of F-spin.

Table 5.1   Values of $f_{ABC}$ and $d_{ABC}$

| ABC | $f_{ABC}$ | ABC | $d_{ABC}$ |
|-----|-----------|-----|-----------|
| 123 | 1 | 118 | $1/\sqrt{3}$ |
| 147 | 1/2 | 146 | 1/2 |
| 156 | $-1/2$ | 157 | 1/2 |
| 246 | 1/2 | 228 | $1/\sqrt{3}$ |
| 257 | 1/2 | 247 | $-1/2$ |
| 345 | 1/2 | 256 | 1/2 |
| 367 | $-1/2$ | 338 | $1/\sqrt{3}$ |
| 458 | $\sqrt{3}/2$ | 344 | 1/2 |
| 678 | $\sqrt{3}/2$ | 355 | 1/2 |
| | | 366 | $-1/2$ |
| | | 377 | $-1/2$ |
| | | 448 | $-1/\left(2\sqrt{3}\right)$ |
| | | 558 | $-1/\left(2\sqrt{3}\right)$ |
| | | 668 | $-1/\left(2\sqrt{3}\right)$ |
| | | 778 | $-1/\left(2\sqrt{3}\right)$ |
| | | 888 | $-1/\sqrt{3}$ |
| | | $d_{0AB}$ | $\sqrt{2/3}\delta_{AB}$ |

## 5.2   Particle Representations in Flavor SU(3)

Out of the eight tensor generators $F_j^i$ of $SU(3)$, the set $F_1^1$, $F_2^1$, $F_1^2$ and $F_2^2$ form the generators of the subgroup $SU(2) \times U(1)$. We have $SU(3) \supset SU(2) \times U(1) \supset SU(2)$. It is convenient to classify states in an $SU(3)$ representation by making use of this fact. The generators of the $SU(2) \times U(1)$ subgroup which are conveniently taken to correspond to isospin and hypercharge are

$$I_+ = F_1^2, \ I_- = F_2^1, \ I_3 = \frac{1}{2}\left(F_1^1 - F_2^2\right)$$
$$Y = F_1^1 + F_2^2 = -F_3^3, \tag{5.34}$$

in the case of $SU(3)$ group. There are thus two diagonal operators in $SU(3)$, namely $I_3$ and $Y$. $SU(3)$ is, therefore, a group of rank 2. Further if we define the electric charge as

$$Q = F_1^1 \text{ in } SU(3), \tag{5.35}$$

Eq. (5.34) give the Gell-Mann-Nishijima relation

$$Q = I_3 + \frac{Y}{2}. \tag{5.36}$$

The fundamental representation is a vector which we write as $q_i$. Let us take

$$q_i \equiv \begin{bmatrix} q_1 \\ q_2 \\ q_3 \end{bmatrix} = \begin{bmatrix} \bar{u} \\ \bar{d} \\ \bar{s} \end{bmatrix}, \tag{5.37}$$

as the field operator which creates a u-quark, a d-quark or a s-quark viz.

$$\bar{u}\,|0\rangle = |u\rangle\,,\ \bar{d}\,|0\rangle = |d\rangle\,,\ \bar{s}\,|0\rangle = |s\rangle\,. \tag{5.38}$$

The field operators $q_i$ belong to the representation **3** of $SU(3)$, whereas the field operators

$$q^i = q_i^* = \begin{bmatrix} q^1 \\ q^2 \\ q^3 \end{bmatrix} \equiv \begin{bmatrix} u \\ d \\ s \end{bmatrix} \tag{5.39}$$

belong to the representation $\bar{\mathbf{3}}$ of $SU(3)$. $q^i$ create antiquarks or annihilate quarks. From Eq. (5.21), we have

$$\left[F_j^i,\ q_k\right] = \delta_k^i\, q_j - \frac{1}{3}\delta_j^i\, q_k. \tag{5.40}$$

In the matrix notation, we can write the field operators $q_i$ and $q^i$ as row and column matrix respectively viz.

$$\bar{q} = \begin{pmatrix} \bar{u} & \bar{d} & \bar{s} \end{pmatrix}$$

$$q = \begin{pmatrix} u \\ d \\ s \end{pmatrix}. \tag{5.41}$$

Then it follows from Eqs. (5.24) and (5.25):

$$[F_A,\ q] = -\frac{\lambda_A}{2}q$$

$$[F_A,\ \bar{q}] = \bar{q}\frac{\lambda_A}{2}. \tag{5.42}$$

Hence we see from Eqs. (5.36), (5.38), (5.40) or (5.42), that the quark states or simply quarks belong to the triplet representation of $SU(3)$ and have the following quantum numbers:

|          |         | $I_3$ | $Y$ | $Q$ |
|----------|---------|-------|------|------|
| $|u\rangle$ |         | $1/2$ | $1/3$ | $2/3$ |
|          | $I = 1/2$ |       |      |      |
| $|d\rangle$ |         | $-1/2$ | $1/3$ | $-1/3$ |
| $|s\rangle$ | $I = 0$ | $0$ | $-2/3$ | $-1/3$ |

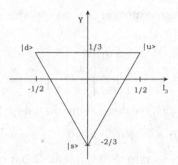

Fig. 5.1   Weight diagram for **3**.

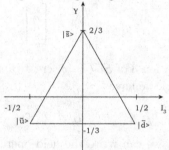

Fig. 5.2   Weight diagram for **3̄**.

It is convenient to plot each state of the triplet representation on a $I-Y$ plot as shown in Fig. 5.1. Such a diagram is called the weight diagram.

The **3̄** representation of $SU(3)$ is not equivalent to **3**; it transforms as $q^i = q_i^*$. It is the hypercharge which distinguishes **3** and **3̄**. Antiquarks belong to the **3̄** representation of $SU(3)$ and the weight diagram is shown in Fig. 5.2.

### 5.2.1   *Mesons*

Quarks are taken to be spin 1/2 particles. To build observed particles from quarks, it is convenient to assign a baryon number 1/3 to quarks. Thus

$$q_i \left| 0 \right\rangle : B = 1/3$$
$$q^i \left| 0 \right\rangle : B = -1/3.$$

Consider

$$q^i q_j = \left( q^i q_j - \frac{1}{3}\delta_j^i q^k q_k \right) + \frac{1}{3}\delta_j^i q^k q_k$$

$$= P_j^i \text{ (octet)} \qquad \text{Singlet}$$

$$\mathbf{\bar{3} \otimes 3} = \qquad \mathbf{8} \qquad \oplus \qquad \mathbf{1}$$

$P_j^i$ can be regarded as a field operator for pseudoscalar mesons. Thus

$$P_j^i |0\rangle \equiv |P_j^i\rangle = \left( q^i q_j - \frac{1}{3}\delta_j^i q^k q_k \right) |0\rangle \tag{5.43}$$

has baryon number zero and is an octet. It may be taken to represent octet of pseudoscalar mesons $\pi$, $K$ and $\eta$. We write (in our notation upper index is row index and lower index is column index)

$$P_j^i = \frac{1}{\sqrt{2}} (\lambda_A)_j^i \, \pi_A \tag{5.44}$$

where identification is shown in Table 5.2.

Table 5.2   Pseudoscalar Mesons $J^P = 0^-$ [cf. Eqs. (5.43) and (5.44)]

| State and its quark content | $|D^1, Y, I, I_3\rangle$ |
|---|---|
| $\left\| P_1^2 \right\rangle = \left\| \pi^+ \right\rangle = \left\| u\,\bar{d} \right\rangle$ | $\frac{\pi_1 + i\pi_2}{\sqrt{2}}\Big\rangle : -\left\|8, 0, 1, 1\right\rangle$ |
| $\left\| \frac{P_1^1 - P_2^2}{\sqrt{2}} \right\rangle = \left\| \pi^0 \right\rangle = \left\| \frac{u\,\bar{u} - d\,\bar{d}}{\sqrt{2}} \right\rangle$ | $\left\| \pi_3 \right\rangle : \left\|8, 0, 1, 0\right\rangle$ |
| $\left\| P_2^1 \right\rangle = \left\| \pi^- \right\rangle = \left\| d\,\bar{u} \right\rangle$ | $\frac{\pi_1 - i\pi_2}{\sqrt{2}}\Big\rangle : \left\|8, 0, 1, -1\right\rangle$ |
| $\left\| P_1^3 \right\rangle = \left\| K^+ \right\rangle = \left\| u\,\bar{s} \right\rangle$ | $\frac{\pi_4 + i\pi_5}{\sqrt{2}}\Big\rangle : \left\|8, 1, 1/2, 1/2\right\rangle$ |
| $\left\| P_2^3 \right\rangle = \left\| K^0 \right\rangle = \left\| d\,\bar{s} \right\rangle$ | $\frac{\pi_6 + i\pi_7}{\sqrt{2}}\Big\rangle : \left\|8, 1, 1/2, -1/2\right\rangle$ |
| $\left\| P_3^2 \right\rangle = \left\| \bar{K}^0 \right\rangle = \left\| s\,\bar{d} \right\rangle$ | $\frac{\pi_6 - i\pi_7}{\sqrt{2}}\Big\rangle : \left\|8, -1, 1/2, 1/2\right\rangle$ |
| $\left\| P_3^1 \right\rangle = \left\| K^- \right\rangle = \left\| s\,\bar{u} \right\rangle$ | $\frac{\pi_4 - i\pi_5}{\sqrt{2}}\Big\rangle : \left\|8, -1, 1/2, -1/2\right\rangle$ |
| $\left\| -\frac{3}{\sqrt{6}} P_3^3 \right\rangle = \left\| \eta_8 \right\rangle = \left\| \frac{u\,\bar{u} + d\,\bar{d} - 2s\bar{s}}{\sqrt{6}} \right\rangle$ | $= \left\| \pi_8 \right\rangle : \left\|8, 0, 0, 0\right\rangle$ |

Hence in a matrix notation, the octet of pseudoscalar mesons $J^P = 0^-$ can be represented by a matrix:

$$P = \begin{pmatrix} \frac{1}{\sqrt{6}}\eta_8 + \frac{1}{\sqrt{2}}\pi^0 & \pi^+ & K^+ \\ \pi^- & \frac{1}{\sqrt{6}}\eta_8 - \frac{1}{\sqrt{2}}\pi^0 & K^0 \\ K^- & \overline{K^0} & -\frac{2}{\sqrt{6}}\eta_8 \end{pmatrix}. \tag{5.45}$$

The singlet pseudoscalar meson $\eta_1$ is given by

$$|\eta_1\rangle = \left| \frac{u\bar{u} + d\bar{d} + s\bar{s}}{\sqrt{3}} \right\rangle = \frac{u\bar{u} + d\bar{d} + s\bar{s}}{\sqrt{3}} |0\rangle : |1, 0, 0, 0\rangle .$$

Another possible set of candidates for the octet of bosons is vector mesons $J^P = 1^-$ :

$$\rho^+ \rho^0 \rho^- \qquad I = 1,\ Y = 0$$
$$K^{*^+} K^{*^0} \qquad I = 1/2,\ Y = 1$$
$$K^{*^0} K^{*^-} \qquad I = 1/2,\ Y = -1$$
$$\omega_8 \qquad I = 0,\ Y = 0$$

A singlet vector boson is denoted as $\omega_1$. In broken $SU(3)$, a singlet meson can mix with the eighth component of an octet. For example, $\omega_8$ and $\omega_1$ can mix and physical particles are mixtures of them and are denoted by $\omega$ and $\phi$. The weight diagram for mesons is given in Fig. 5.3.

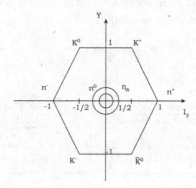

Fig. 5.3   Weight diagram for pseudoscalar meson octet.

### 5.2.2   *Baryons*

We now consider baryons. Baryons have $B = 1$ and they must be constructed out of 3 quarks. For this purpose let us proceed as follows. Writing

$$q_j\, q_k = \frac{1}{2}\left(q_j\, q_k + q_k\, q_j\right) + \frac{1}{2}\left(q_j\, q_k - q_k\, q_j\right)$$

$$= \frac{1}{\sqrt{2}} S_{jk} + \frac{1}{\sqrt{2}} A_{jk}, \tag{5.46}$$

where the symmetric tensor

$$S_{jk} = \frac{1}{\sqrt{2}}\left(q_j\, q_k + q_k\, q_j\right) \tag{5.47}$$

has six independent components. The antisymmetric tensor

$$A_{jk} = \frac{1}{\sqrt{2}}\left(q_j\, q_k - q_k\, q_j\right) \tag{5.48}$$

has three independent components. Now a vector $T^i$ belonging to the representation $\bar{\mathbf{3}}$ can be written in terms of $A_{lm}$ as

$$T^i = \varepsilon^{ilm} A_{lm}$$

or

$$A_{jk} = \frac{1}{2}\varepsilon_{ijk} T^i. \tag{5.49}$$

Hence we have the result

$$\mathbf{3} \otimes \mathbf{3} = \mathbf{6} \oplus \bar{\mathbf{3}}$$

and

$$\mathbf{3} \otimes \mathbf{3} \otimes \mathbf{3} = (\mathbf{6} \otimes \mathbf{3}) + (\bar{\mathbf{3}} \otimes \mathbf{3}).$$

First consider $\bar{\mathbf{3}} \otimes \mathbf{3}$ :

$$T^i q_j = \left(T^i q_j - \frac{1}{3} \delta^i_j T^k q_k\right) + \frac{1}{3} \delta^i_j T^k q_k \tag{5.50}$$

viz.

$$\bar{\mathbf{3}} \otimes \mathbf{3} = \mathbf{8} \oplus \mathbf{1}.$$

The octet operator for baryons can be written as

$$\bar{B}^i_j = \frac{1}{2}\left(T^i q_j - \frac{1}{3} \delta^i_j T^k q_k\right), \tag{5.51a}$$

where

$$T^i = \varepsilon^{ilm} A_{lm} = \frac{1}{2\sqrt{6}}\varepsilon^{ilm} (q_l q_m - q_m q_l). \tag{5.51b}$$

For the singlet representation,

$$\frac{1}{2}\frac{1}{\sqrt{3}}T^k q_k = \frac{1}{2\sqrt{3}}\varepsilon^{klm} A_{lm} q_k$$

$$= \frac{1}{2\sqrt{6}}\varepsilon^{klm} (q_l q_m - q_m q_l) q_k. \tag{5.52}$$

Now let's consider $\mathbf{6} \otimes \mathbf{3}$: It is given by

$$S_{ij} q_k = S_{ij} q_k + S_{jk} q_i + S_{ki} q_j - S_{jk} q_i - S_{ki} q_j$$
$$= \tilde{T}_{\{ijk\}} - S_{jk} q_i - S_{ki} q_j, \tag{5.53a}$$

where

$$\tilde{T}_{\{ijk\}} = S_{ij} q_k + S_{jk} q_i + S_{ki} q_j \tag{5.53b}$$

is completely symmetric tensor and has 10 independent components. Now we show that

$$- (S_{jk}\, q_i + S_{ki}\, q_j) + 2 S_{ij}\, q_k = \varepsilon_{kjl}\, \varepsilon^{lmn}\, S_{in}\, q_m + \varepsilon_{kil}\, \varepsilon^{lmn}\, S_{jn}\, q_m. \quad (5.54)$$

**Proof:**

$$R.H.S = \left( \delta_k^m\, \delta_j^n - \delta_k^n\, \delta_j^m \right) S_{in}\, q_m + \left( \delta_k^m\, \delta_i^n - \delta_k^n\, \delta_i^m \right) S_{jn}\, q_m$$

$$= S_{ij}\, q_k - S_{ik}\, q_j + S_{ji}\, q_k - S_{jk}\, q_i$$

$$= - (S_{jk}\, q_i + S_{ki}\, q_j) + 2 S_{ij}\, q_k = L.H.S.$$

Hence from Eqs. (5.53a) and (5.54), one gets

$$S_{ij}\, q_k = \frac{1}{3}\tilde{T}_{\{ijk\}} + \frac{1}{3}\left[ \varepsilon_{kjl}\, \varepsilon^{lmn}\, S_{in}\, q_m + \varepsilon_{kil}\, \varepsilon^{lmn}\, S_{jn}\, q_m \right]$$

$$= \frac{1}{3}\tilde{T}_{\{ijk\}} + \frac{1}{3}\left[ \varepsilon_{kjl}\, \delta_i^r + \varepsilon_{kil}\, \delta_j^r \right] \varepsilon^{lmn}\, S_{rn}\, q_m. \quad (5.55)$$

Hence we have

$$\mathbf{6 \otimes 3 = 10 \oplus 8}.$$

We write the decouplet representation:

$$\bar{T}_{\{ijk\}} = \sqrt{3}\, \tilde{T}_{\{ijk\}}$$

$$= \frac{1}{\sqrt{3}}\left[ S_{ij}\, q_k + S_{jk}\, q_i + S_{ki}\, q_j \right] \quad (5.56)$$

and the octet ($8'$) representation:

$$\bar{B}_r^{\prime l} = \frac{1}{\sqrt{3}}\varepsilon^{lmn}\, S_{rn}\, q_m$$

$$\bar{B}_l^{\prime l} = 0. \quad (5.57)$$

Hence the final result becomes

$$\mathbf{3 \otimes 3 \otimes 3 = (6 \oplus \bar{3}) \otimes 3 = (6 \otimes 3) \oplus (\bar{3} \otimes 3)}$$

$$= \mathbf{10 \oplus 8 \oplus\ 8' \ \oplus\ 1}.$$

### 5.2.2.1  *Baryon States*

(i) **Octet Representation 8**

From Eqs. (5.51) and (5.57), we have

$$\bar{B}_j^i\, |0\rangle = |B_j^i\rangle$$

$$= \frac{1}{2\sqrt{2}}\left[ \varepsilon^{ilm}\, (q_l q_m - q_m q_l)\, q_j - \frac{1}{3}\delta_j^i \varepsilon^{klm}\, (q_l q_m - q_m q_l)\, q_k \right] |0\rangle \quad (5.58)$$

Table 5.3 Baryons $J^P = \frac{1}{2}^+$

| State: **8** | Quark content | $Q$ | $I$ | $I_3$ | $Y$ |
|---|---|---|---|---|---|
| $\|p\rangle = \bar{B}_1^3 \|0\rangle$ | $\frac{1}{\sqrt{2}} \|[u,d]\,u\rangle$ | 1 | $\frac{1}{2}$ | $\frac{1}{2}$ | 1 |
| $\|n\rangle = \bar{B}_2^3 \|0\rangle$ | $\frac{1}{\sqrt{2}} \|[u,d]\,d\rangle$ | 0 | $\frac{1}{2}$ | $\frac{-1}{2}$ | 1 |
| $\|\Sigma^+\rangle = \bar{B}_1^2 \|0\rangle$ | $\frac{1}{\sqrt{2}} \|[u,s]\,u\rangle$ | 1 | 1 | $+1$ | 0 |
| $\|\Sigma^0\rangle =$ $\frac{1}{\sqrt{2}}(\bar{B}_1^1 - \bar{B}_2^2)\|0\rangle$ | $\frac{1}{2} \|[d,s]\,u + [u,s]\,d\rangle$ | 0 | 1 | 0 | 0 |
| $\|\Sigma^-\rangle = \bar{B}_2^1 \|0\rangle$ | $\frac{1}{\sqrt{2}} \|[d,s]\,d\rangle$ | $-1$ | 1 | $-1$ | 0 |
| $\|\Lambda^0\rangle = -\frac{3}{\sqrt{6}} \bar{B}_3^3 \|0\rangle$ | $\frac{1}{\sqrt{12}} \|2\,[u,d]\,s$ $- [d,s]\,u - [s,u]\,d\rangle$ | 0 | 0 | 0 | 0 |
| $\|\Xi^-\rangle = \bar{B}_3^1 \|0\rangle$ | $\frac{1}{\sqrt{2}} \|[d,s]\,s\rangle$ | $-1$ | $1/2$ | $-1/2$ | $-1$ |
| $\|\Xi^0\rangle = \bar{B}_3^2 \|0\rangle$ | $\frac{1}{\sqrt{2}} \|[s,u]\,s\rangle$ | 0 | $1/2$ | $1/2$ | $-1$ |

| State : **8′** | Quark content |
|---|---|
| $\bar{B}_1'^3 \|0\rangle$ | $\frac{1}{\sqrt{6}} \left\| \left([u,d]_+\,u - 2uud\right)\right\rangle$ |
| $\bar{B}_2'^3 \|0\rangle$ | $-\frac{1}{\sqrt{6}} \left\| \left([u,d]_+\,d - 2ddu\right)\right\rangle$ |
| $\bar{B}_1'^2 \|0\rangle$ | $\frac{1}{\sqrt{6}} \left\| \left([u,s]_+\,u - 2uus\right)\right\rangle$ |
| $\frac{1}{\sqrt{2}}(\bar{B}_1'^1 - \bar{B}_2'^2)\|0\rangle$ | $\frac{1}{\sqrt{12}} \left\{ \left\|(-2\,[u,d]_+\,s \right.\right.$ $\left.\left. + [u,s]_+\,d + [d,s]_+\,u\right)\right\rangle \right\}$ |
| $\bar{B}_2'^1 \|0\rangle$ | $\frac{1}{\sqrt{6}} \left\| \left([d,s]_+\,d - 2dds\right)\right\rangle$ |
| $-\frac{3}{\sqrt{6}}\bar{B}_3'^3 \|0\rangle$ | $-\frac{1}{2} \left\| \left([s,d]_+\,u - [s,u]_+\,d\right)\right\rangle$ |
| $\bar{B}_3'^1 \|0\rangle$ | $\frac{1}{\sqrt{6}} \left\| \left(2ssd - [d,s]_+\,s\right)\right\rangle$ |
| $\bar{B}_3'^2 \|0\rangle$ | $\frac{1}{\sqrt{6}} \left\| \left([s,u]_+\,s - 2ssu\right)\right\rangle$ |

and for representation **8′** :

$$\bar{B}_j'^{\,i} \,|0\rangle = \left| B_j'^{\,i} \right\rangle = \frac{1}{2\sqrt{3}}\varepsilon^{ikl}\, S_{jl}\, q_k \,|0\rangle. \qquad (5.59)$$

The octet of baryons are then identified as given in Table 5.3. Hence from Eq. (5.58) and Table 5.3, we see that known eight $J^P = 1/2^+$ baryons can be represented as $3 \times 3$ matrices

$$B_j^i = \begin{pmatrix} \frac{1}{\sqrt{6}}\Lambda^0 + \frac{1}{\sqrt{2}}\Sigma^0 & \Sigma^+ & p \\ \Sigma^- & \frac{1}{\sqrt{6}}\Lambda^0 - \frac{1}{\sqrt{2}}\Sigma^0 & n \\ \Xi^- & \Xi^0 & -\frac{2}{\sqrt{6}}\Lambda^0 \end{pmatrix} \qquad (5.60a)$$

$$\bar{B}_j^i = \begin{pmatrix} \frac{1}{\sqrt{6}}\bar{\Lambda}^0 + \frac{1}{\sqrt{2}}\bar{\Sigma}^0 & \bar{\Sigma}^- & \bar{\Xi}^- \\ \bar{\Sigma}^+ & \frac{1}{\sqrt{6}}\bar{\Lambda}^0 - \frac{1}{\sqrt{2}}\bar{\Sigma}^0 & \bar{\Xi}^0 \\ \bar{p} & \bar{n} & -\frac{2}{\sqrt{6}}\bar{\Lambda}^0 \end{pmatrix}. \qquad (5.60b)$$

Note that

$$\bar{B}_j^i = B_i^{*j}\, \gamma^0, \qquad (5.61)$$

where the symbol $*$ denotes complex conjugation with respect to $SU(3)$ but hermitian conjugation for the field operators. The weight diagram for the octet representation is shown in Fig. 5.4.

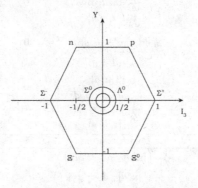

Fig. 5.4   Weight diagram for $\frac{1}{2}^{+}$ baryon octet.

**Singlet representation 1:**
From Eq. (5.56), one gets

$$
\begin{aligned}
\Lambda_1^0 &= \frac{1}{2}\frac{1}{\sqrt{6}}\; \varepsilon^{klm}\; (q_l\, q_m - q_m\, q_l)\, q_k\; |0\rangle \\
&= \frac{1}{2}\frac{2}{\sqrt{6}}\; \left\{ [\bar{d},\, \bar{s}]\; \bar{u} + [\bar{s},\, \bar{u}]\; \bar{d} + [\bar{u},\, \bar{d}]\; \bar{s} \right\} |0\rangle . \\
&= \frac{1}{\sqrt{6}}\; |[d,\, s]\; u + [s,\, u]\; d + [u,\, d]\; s\rangle . \tag{5.62}
\end{aligned}
$$

**(ii) Decuplet representation 10:**
From Eq. (5.56), we have

$$
|T_{ijk}\rangle = \frac{1}{\sqrt{3}}\left\{ S_{ij}\, q_k + S_{jk}\, q_i + S_{ki}\, q_j \right\} |0\rangle . \tag{5.63}
$$

The detailed identification of the states of decuplet representation are given in Table 5.4. The weight diagram for the decuplet of baryons is shown in Fig. 5.5.

## 5.3   U-Spin

We have labeled the states within an irreducible representation of $SU(3)$ uniquely by the eigenvalues of $\mathbf{I}^2$, $I_3$ and $Y$. The reason is that $SU(3)$

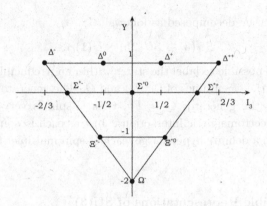

Fig. 5.5   Weight diagram for $\frac{3}{2}^+$ baryon decuplet.

Table 5.4   Decuplet $J^P = 3/2^+$ [cf. Eq. (5.63)]

| State | Quark content | $Q$ | $I$ | $I_3$ | $Y$ |
|---|---|---|---|---|---|
| $\lvert\Delta^{++}\rangle \equiv \frac{1}{\sqrt{6}}\lvert T_{111}\rangle$ | $\lvert uuu\rangle$ | 2 | $\frac{3}{2}$ | $\frac{3}{2}$ | 1 |
| $\lvert\Delta^{+}\rangle \equiv \frac{1}{\sqrt{2}}\lvert T_{112}\rangle$ | $\frac{1}{\sqrt{3}}\lvert udu + duu + uud\rangle$ | 1 | $\frac{3}{2}$ | $\frac{1}{2}$ | 1 |
| $\lvert\Delta^{0}\rangle \equiv \frac{1}{\sqrt{2}}\lvert T_{122}\rangle$ | $\frac{1}{\sqrt{3}}\lvert udd + ddu + dud\rangle$ | 0 | $\frac{3}{2}$ | $\frac{-1}{2}$ | 1 |
| $\lvert\Delta^{-}\rangle \equiv \frac{1}{\sqrt{2}}\lvert T_{222}\rangle$ | $\lvert ddd\rangle$ | $-1$ | $\frac{3}{2}$ | $\frac{-3}{2}$ | 1 |
| $\lvert\Sigma^{*+}\rangle \equiv \frac{1}{\sqrt{2}}\lvert T_{113}\rangle$ | $\frac{1}{\sqrt{3}}\lvert uus + usu + suu\rangle$ | 1 | 1 | 1 | 0 |
| $\lvert\Sigma^{*0}\rangle \equiv \frac{1}{\sqrt{2}}\lvert T_{123}\rangle$ | $\frac{1}{\sqrt{6}}\left\lvert \begin{array}{c} uds + dus + dsu \\ +sdu + sud + usd \end{array}\right\rangle$ | 0 | 1 | 0 | 0 |
| $\lvert\Sigma^{*-}\rangle \equiv \frac{1}{\sqrt{2}}\lvert T_{322}\rangle$ | $\frac{1}{\sqrt{3}}\lvert sdd + dds + dsd\rangle$ | $-1$ | 1 | $-1$ | 0 |
| $\lvert\Xi^{*0}\rangle \equiv \frac{1}{\sqrt{2}}\lvert T_{133}\rangle$ | $\frac{1}{\sqrt{3}}\lvert uss + ssu + sus\rangle$ | 0 | $\frac{1}{2}$ | $\frac{1}{2}$ | $-1$ |
| $\lvert\Xi^{*-}\rangle \equiv \frac{1}{\sqrt{2}}\lvert T_{233}\rangle$ | $\frac{1}{\sqrt{3}}\lvert dss + ssd + sds\rangle$ | $-1$ | $\frac{1}{2}$ | $\frac{-1}{2}$ | $-1$ |
| $\lvert\Omega^{-}\rangle \equiv \frac{1}{\sqrt{6}}\lvert T_{333}\rangle$ | $\lvert sss\rangle$ | $-1$ | 0 | 0 | $-2$ |

contains the direct product of $SU(2)_I \times U(1)_Y$, i.e.

$$SU(3) \supset SU(2)_I \times U(1)_Y.$$

The generators of $SU(2)_I$ and $U(1)_Y$ are identified with the generators of $SU(3)$ as given in Eq. (5.34). However, we can take a different decomposition. For example the generators

$$U_+ = F_2^3, \; U_- = F_3^2, \; U_3 = \frac{1}{2}\left(F_2^2 - F_3^3\right) \tag{5.64}$$

are the generators of group $SU(2)_U$ . These generators commute with the generator

$$Q = F_1^1. \tag{5.65}$$

Thus $SU(3)$ can be decomposed as follows

$$SU(3) \supset SU(2)_U \times U(1)_Q.$$

Therefore, it is possible to label the states within an irreducible representation $SU(3)$ by the eigenvalues of $U^2$, $U_3$ and $Q$. The generators of $SU(2)_U$ commute with the generator $Q = F_1^1$, thus $U$-spin is very useful when dealing with electromagnetic interactions. Just as each isospin multiplet is associated with a definite hypercharge, each $U$-spin multiplet has a definite charge.

## 5.4   Irreducible Representations of SU(3)

We have already encountered two irreducible representations:

$$\text{triplet} = q_i$$

$$\text{octet} = P_j^i = q^i q_j - \frac{1}{3} \delta_j^i \, q^k q_k.$$

The octet representation is a regular or an adjoint representation of $SU(3)$ because $P_j^i$ transforms in the same way as the generators $F_j^i$.

Now we look at more general representations of $SU(3)$. The general prescription for finding the basic tensors $T_{i_1 \cdots i_p}^{j_1 \cdots j_q}$ for an irreducible representation of $SU(3)$ is:

1. Construct tensors $T_{i_1 \cdots i_p}^{j_1 \cdots j_q}$.
2. Symmetrize among $i_1 \cdots i_p$ and $j_1 \cdots j_q$ indices.
3. Subtract traces so that all contractions give zero, e.g.

$$T_{i\ i_2 \cdots i_p}^{i\ j_2 \cdots j_q} = 0, \text{ etc.}$$

The linearly independent components of tensor $T$ then supply an irreducible representation of $SU(3)$ which is designated as $(p, q)$. The dimensionality of such a representation can be easily computed. First let us calculate the number of independent components for a symmetric tensor with $p$ lower (or $q$ upper) indices. We note that each index can take only the value 1, 2 or 3. Thus the number of independent components are the same as the number of ways of separating $p$ identical objects with two identical partitions:

$$\frac{(p+2)!}{p!\, 2!} = \frac{(p+2)\,(p+1)}{2}.$$

Thus a tensor which is symmetric in $p$ lower and $q$ upper indices has

$$B(p,q) = \frac{(p+2)\,(p+1)\,(q+2)\,(q+1)}{4}$$

independent components. But the trace condition shows that a symmetric object with $p-1$ lower indices and $q-1$ upper indices vanishes. This gives $B(p-1,\ q-1)$ conditions. Hence a symmetric traceless tensor has dimensions

$$D(p,q) = B(p,q) - B(p-1,\ q-1)$$
$$= (p+1)\,(q+1)\left(\frac{p+q}{2}+1\right). \tag{5.66}$$

Thus we have for example:

| Representation | Dimensionality |
|:---:|:---:|
| $(p,q)$ | $D(p,q)$ |
| $(0,0)$ | $1$ : Singlet |
| $(1,0)$ | $3$ : Triplet |
| $(0,1)$ | $3$ : Triplet |
| $(1,1)$ | $8$ : Octet |
| $(3,0)$ | $10$ : Decuplet |
| $(2,2)$ | $27$ : 27 plet |

### 5.4.1 *Young's Tableaux*

By taking the direct product of basic representation **3** with itself, we can generate the representations of higher dimensions. These representations are however reducible. Let us now discuss a general method to decompose these reducible representations into irreducible representations. We have already discussed some simple examples.

We represent the fundamental representation **3** by a box, i.e. associate index $i$ with a box.

$$\square \ : \ \mathbf{3} : q_i. \tag{5.67}$$

We note that the representation $\bar{\mathbf{3}}$ is antisymmetric combination of two **3**'s viz.

$$T^i = \varepsilon^{ijk}(q_j\ q_k - q_k\ q_j)\frac{1}{\sqrt{2}}$$
$$= \varepsilon^{ijk}\,A_{jk},$$
$$A_{ij} = \varepsilon_{ijk}\,T^k. \tag{5.68}$$

This can be represented by a column of two boxes

$$\begin{array}{c}\square\\\square\end{array} \ : \ \bar{\mathbf{3}} : T^i. \tag{5.69}$$

Since a tensor index takes only three values $i = 1, 2, 3$, a column in Young's tableau can have at most three boxes

$$\begin{array}{|c|} \hline i \\ \hline j \\ \hline k \\ \hline \end{array} : \varepsilon_{ijk}. \tag{5.70}$$

It is completely antisymmetric and it corresponds to the trivial singlet representation. We note that

$$1 = \begin{array}{|c|}\hline\\\hline\\\hline\end{array} = \begin{array}{|c|c|}\hline&\\\hline&\\\hline\end{array} = \begin{array}{|c|c|c|}\hline&&\\\hline&&\\\hline\end{array} = \cdots \tag{5.71}$$

Consider the $(p, q)$ representation. It is a tensor whose components are

$$T^{j_1 \cdots j_q}_{i_1 \cdots i_p} \tag{5.72}$$

symmetric in lower and upper indices and traceless. We can lower the upper indices with $\varepsilon$ tensors, obtaining an object with $p + 2q$ lower indices:

$$t_{i_1 \ldots i_p \, k_1 l_1 \ldots k_q l_q} = \varepsilon_{j_1 k_1 l_1} \ldots \varepsilon_{j_q k_q l_q} \, T^{j_1 \cdots j_q}_{i_1 \cdots i_p} \tag{5.73}$$

It is clear that $t$ is antisymmetric in each pair $k_x \longleftrightarrow l_x$, $x = 1, \cdots, q$. Since there are $p + 2q$ indices, we arrange $p + 2q$ boxes as follows:

$$\tag{5.74}$$

| $k_1$ | $\cdots$ | $k_q$ | $i_1$ | $\cdots$ | $i_p$ |
|-------|----------|-------|-------|----------|-------|
| $l_1$ | $\cdots$ | $l_q$ | | | |

This is a Young Tableau. The most general Young Tableau has only two rows. An irreducible representation is completely specified by two integers $(p, q)$. Note that $(\overline{p, q}) = (q, p)$. It is clear from Eq. (5.71) that tableaux of the form

$$\tag{5.75}$$

| | | | $k_1$ | $\cdots$ | $k_q$ | $i_1$ | $\cdots$ | $i_p$ |
|--|--|--|-------|----------|-------|-------|----------|-------|
| | | | $l_1$ | $\cdots$ | $l_q$ | | | |
| | | | | | | | | |

still corresponds to the representation $(p, q)$. Comparison with the Young Tableau gives the following rule for preparing a tensor with the right symmetry properties to give a state in $(p, q)$: First, symmetrize indices in each row of the tableau. Then antisymmetrize the indices in each column. If we have more general tableau with columns of more than two boxes, the rules for forming a tensor are the same as before. Assign an index to each

Table 5.5   Irreducible representations of $SU(3)$.

| $(p,q)$ | $D(p,q)$ | Tableau | Tensor |
|---|---|---|---|
| | 1 | $\boxed{i}$ $\boxed{j}$ $\boxed{k}$ | $\mathbf{1}: \varepsilon_{ijk}$ |
| $(1,0)$ | 3 | $\boxed{i}$ | $\mathbf{3}: q_i$ |
| $(0,1)$ | 3 | $\boxed{j}$ $\boxed{k}$ | $\mathbf{\bar{3}}: T^i = \varepsilon^{ijk}\, t_{jk}$ |
| $(2,0)$ | 6 | $\boxed{i\,\,j}$ | $\mathbf{6}: T_{ij}$ |
| $(0,2)$ | 6 | $\boxed{k_1\,k_2}$ $\boxed{l_1\,l_2}$ | $\mathbf{\bar{6}}: T^{ij} = \varepsilon^{ik_1l_1}\varepsilon^{jk_2l_2}\, t_{k_1l_1k_2l_2}$ |
| $(1,1)$ | 8 | $\boxed{k\,\,j}$ $\boxed{l}$ | $\mathbf{8}: T^i_j = e^{ikl}\, t_{klj}$ |
| $(3,0)$ | 10 | $\boxed{i\,\,j\,\,k}$ | $\mathbf{10}: T_{ijk}$ |
| $(0,3)$ | 10 | $\boxed{l_1\,l_2\,l_3}$ $\boxed{m_1\,m_2\,m_3}$ | $\mathbf{\overline{10}}: \begin{aligned}T^{ij} &= \varepsilon^{il_1m_1}\varepsilon^{jl_2m_2}\varepsilon^{kl_3m_3}\\&\times t_{l_1m_1l_2m_2l_3m_3}\end{aligned}$ |
| $(2,1)$ | 15 | $\boxed{l\,\,k\,\,j}$ $\boxed{m}$ | $\mathbf{15}: T^i_{kj} = e^{ilm}\, t_{lmkj}$ |
| $(2,2)$ | 27 | $\boxed{m_1\,m_2\,l\,\,k}$ $\boxed{n_1\,n_2}$ | $\mathbf{27}: \begin{aligned}T^{ij}_{lk} &= \varepsilon^{il_1m_1}\varepsilon^{jl_2m_2}\\&\times t_{m_1n_1m_2n_3lk}\end{aligned}$ |

box. Then symmetrize the indices in each row and finally antisymmetrize the indices in each column.

Some of the common irreducible representations of $SU(3)$ are shown in Table 5.5.

### Decomposition of Product Representations

We now consider the decomposition of the direct product of irreducible representations $(p,q)$ and $(r,s)$ corresponding to tableaux $A$ and $B$.

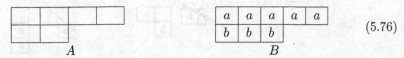

$$(5.76)$$

$$A \qquad\qquad B$$

We now give a recipe for the decomposition of the direct product of $(p,q)$ and $(r,s)$ with the aid of Young Tableaux. Put $a$'s in the top row of $B$ and $b$'s in the second row. Take boxes with $a$ from $B$ and add them to $A$, each in a different column, to form new tableaux. Then, take the boxes with $b$ and add them to form tableaux, again each box in a different column, with one additional restriction given below. On reading the added symbols from right to left and from the top to bottom, the number of $a$'s must be greater than or equal to that of $b$'s, i.e. forget all tableaux which concave upwards or towards the lower left. This avoids double counting of tensors. The tableaux formed in this way correspond to irreducible representations in $(p, q) \otimes (r, s)$. We now give several examples to illustrate how this recipe works.

**Examples**

(i)

$$\square \otimes \boxed{a} = \boxed{\phantom{x}\,a} \oplus \begin{array}{c}\square\\\boxed{a}\end{array} \qquad (5.77)$$

$$\mathbf{3} \quad \otimes \quad \mathbf{3} \quad = \quad \mathbf{6} \quad \oplus \quad \mathbf{\bar{3}}$$

(ii)

$$\begin{array}{c}\square\\\square\end{array} \otimes \boxed{a} = \boxed{\phantom{x}\,a}\; \oplus \; \begin{array}{c}\square\\\boxed{a}\end{array} \qquad (5.78)$$

$$\mathbf{\bar{3}} \quad \otimes \quad \mathbf{3} \quad = \quad \mathbf{8} \quad \oplus \quad \mathbf{1}$$

(iii)

$$\square \otimes \begin{array}{c}\boxed{a}\\\boxed{b}\end{array} = \boxed{\phantom{x}\,a} \qquad \oplus \qquad \begin{array}{c}\square\\\boxed{a}\end{array}$$

$$= \boxed{\cancel{b}\,a}\; \boxed{b} \qquad \qquad \boxed{\cancel{b}}\;\begin{array}{c}\square\\\boxed{a}\\\boxed{b}\end{array}$$

$$\mathbf{3} \quad \otimes \quad \mathbf{\bar{3}} \quad = \quad \mathbf{8} \quad \oplus \quad \mathbf{1}$$

$$(5.79)$$

We discard the first and the third tableaux in Eq. (5.79) as they do not satisfy the constraint that number of $a$'s greater than or equal to the number of $b$'s as we go from right to left or top to bottom.

(iv)

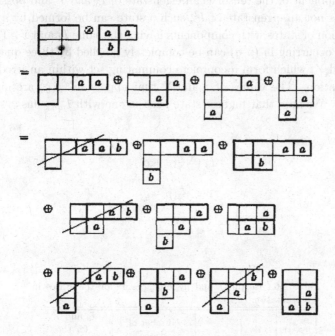

$$8 \otimes 8 = 10 \oplus 27 \oplus 8 \oplus \overline{10} \oplus 8 \oplus 1$$

or

$$(1,1) \otimes (1,1) = (3,0) \oplus (2,2) \oplus (1,1) \oplus (0,3) \oplus (1,1) \oplus (0,0). \qquad (5.80)$$

The slashed tableaux are discarded because they do not satisfy the constraint $a$'s $\geq b$'s.

(v)

$$8 \otimes 3 = 15 \oplus \overline{6} \oplus 3 \qquad (5.81)$$

$$(1,1) \otimes (1,0) = (2,1) \oplus (0,2) \oplus (1,0). \qquad (5.82)$$

To summarize, an arbitrary irreducible representation of $SU(3)$ is denoted by two integers, each positive or zero: $(p,q)$. The corresponding irreducible tensor is denoted by $T^{j_1 \cdots j_q}_{i_1 \cdots i_p}$. It transforms as $\phi^*_{j_1} \cdots \phi^*_{j_q} \; \phi_{i_1} \cdots \phi_{i_p}$.

Each component of the tensor is an eigenstate of $I_3$ and $Y$ and possibly of $I^2$. If it is not an eigenstate of $I^2$, such a state can be formed by a linear combination of states with components having the same $I_3$ and $Y$. The basic states occurring in $(p,q)$ can be completely labelled by three quantities $I$, $I_3$, and $Y$, which form a complete commuting set within an irreducible representation. The values of $I$ and $Y$ that appear in $(p,q)$ are given in Table 5.6. We note that highest state i.e. the one with $I_{max}$ has

$$I_3 = \frac{1}{2}(p+q)$$
$$Y = \frac{1}{3}(p-q). \tag{5.83}$$

Table 5.6  Isospin $I$ and hypercharge $Y$ for the states in representation $(p,q)$.

| $(p,q)$ | $Y$ | $I$ | Number of states | $I_3$ and $Y$ for the highest state |
|---|---|---|---|---|
| $(1,0)$ | $\frac{1}{3}$ $-\frac{2}{3}$ | $\frac{1}{2}$ $0$ | $2$ $1$ | $\frac{1}{2}$, $\frac{1}{3}$ |
| $(0,1)$ | $\frac{2}{3}$ $-\frac{1}{3}$ | $0$ $\frac{1}{2}$ | $1$ $2$ | $\frac{1}{2}$, $\frac{-1}{3}$ |
| $(1,1)$ | $1$ $0$ $-1$ | $\frac{1}{2}$ $1,0$ $\frac{1}{2}$ | $2$ $3+1=4$ $2$ | $1$, $0$ |
| $(3,0)$ | $1$ $0$ $-1$ $-2$ | $\frac{3}{2}$ $1$ $\frac{1}{2}$ $0$ | $4$ $3$ $2$ $1$ | $\frac{3}{2}$, $1$ |
| $(2,1)$ | $\frac{4}{3}$ $\frac{1}{3}$ $-\frac{2}{3}$ $-\frac{5}{3}$ | $1$ $\frac{3}{2},\frac{1}{2}$ $1,0$ $\frac{1}{2}$ | $3$ $4+2=6$ $3+1=4$ $2$ | $\frac{3}{2}$, $\frac{1}{3}$ |
| $(2,2)$ | $2$ $1$ $0$ $-1$ $-2$ | $1$ $\frac{3}{2},\frac{1}{2}$ $2,1,0$ $\frac{3}{2},\frac{1}{2}$ $1$ | $3$ $4+2=6$ $5+3+1=9$ $4+2=6$ $3$ | $2$, $0$ |

## 5.5   SU(N)

We now discuss Young's tableaux for $SU(N)$. Again we assign an index to a box. Thus fundamental representation $N(\phi_i,\ i = 1, \cdots, N)$ is represented by a box:

$$\square\ : N \tag{5.84}$$

The tensor $\varepsilon_{i_1 i_2 \cdots i_N}$ is represented by a column of $N$ boxes

$$: 1. \tag{5.85}$$

It describes the singlet representation $\mathbf{1}$ of $SU(N)$. Now $\varepsilon_{i_1 i_2 \cdots i_N}$ is a completely antisymmetric tensor:

$$\varepsilon_{i_1 i_2 \cdots i_N} = \begin{cases} 0, & \text{if any of the two indices are equal} \\ \pm 1, & \text{if } i_1 \cdots i_N \text{ is an even (odd)} \\ & \text{permutation of } 1, 2, \cdots, N. \end{cases} \tag{5.86}$$

The $N$-dimensional representation $\bar{N}$ is described by a column of $(N - 1)$ boxes.

$$: \bar{N} \tag{5.87}$$

Hence we see that for $SU(2)$, $\mathbf{2}$ and $\bar{\mathbf{2}}$ are equivalent representation and both will be represented by $\square$. Only for $N \geq 3$, $N$ and $\bar{N}$ are distinct representations.

We now discuss the decomposition of the product of representation $N$

by itself into irreducible representations of $SU(N)$.

$$N \qquad\qquad N \quad = \quad \tfrac{1}{2}N\,(N+1) \qquad\qquad \tfrac{1}{2}N(N-1).$$
$$\tag{5.88}$$

Thus $N^2$ components decompose into two irreducible representations of dimensions $\frac{N\,(N+1)}{2}$ and $\frac{N\,(N-1)}{2}$ viz.

$$\phi_i\,\phi_j = \frac{1}{\sqrt{2}}\left(S_{ij} + A_{ij}\right), \tag{5.89}$$

where

$$S_{ij} = \frac{1}{\sqrt{2}}\left(\phi_i\,\phi_j + \phi_j\,\phi_i\right). \tag{5.90a}$$

$$A_{ij} = \frac{1}{\sqrt{2}}\left(\phi_i\,\phi_j - \phi_j\,\phi_i\right). \tag{5.90b}$$

We can regard $S_{ij}$ as an $N \times N$ matrix, but since it is a symmetric matrix, it has only $\frac{N^2-N}{2} + N = \frac{1}{2}N\,(N+1)$ independent elements and this gives the dimension of symmetric representation $S_{ij}$. Again if we regard $A_{ij}$ as $N \times N$ matrix, we can easily see that it has $\frac{N^2-N}{2} = \frac{1}{2}N\,(N-1)$ independent elements and this gives the dimension of antisymmetric representation $A_{ij}$.

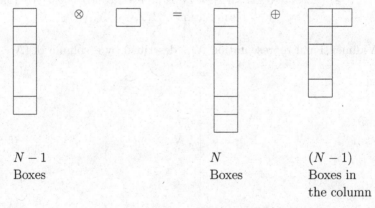

$N-1$                 $N$                $(N-1)$

Boxes                 Boxes             Boxes in

                                                           the column

$$\bar{N} \otimes N = 1 \oplus \text{Adjoint representation of dimension } N^2 - 1. \tag{5.91}$$

Thus

$$\phi^i\,\phi_j = T^i_j + \frac{1}{N}\,\delta^i_j\,\phi^k\,\phi_k, \tag{5.92a}$$

where

$$T_j^i = \phi^i \, \phi_j - \frac{1}{N} \, \delta_j^i \, \phi^k \, \phi_k. \qquad (5.92b)$$

The adjoint representation has the same dimension as the number of generators of $SU(N)$. For example for $SU(6) : \bar{6} \otimes 6 = 1 \oplus 35$.

We now give a general recipe to calculate the dimension of irreducible representations in the decomposition of the product of representation $N$ by itself. To calculate the dimension of an array of boxes there is a recipe which involves calculation of $\frac{\text{Numerator}}{\text{Denominator}}$.

**Numerator:** Insert $N$ in each of the diagonal boxes starting from the top left-hand corner of the tableaux.

$$\begin{array}{|c|c|c|}
\hline
N & N+1 & N+2 \\
\hline
N-1 & N & N+1 \\
\hline
N-2 & N-1 & N \\
\hline
\end{array} \qquad (5.93)$$

Along the diagonals immediately above and below insert $N + 1$ and $N - 1$ respectively. In the next diagonals insert $N + 2$ and so on. The numerator is equal to the product of all these numbers. For example for the tableaux

$$\begin{array}{|c|c|}
\hline
N & N+1 \\
\hline
N-1 & N \\
\hline
\end{array} \qquad (5.94)$$

the numerator $= N^2 \, (N + 1) \, (N - 1) = N^2 \, (N^2 - 1)$.

**Denominator:** The denominator is given by the "product of hooks". We associate each box with a value of the hook. To find it, draw a line entering the row in which the box lies from the right. On entering the box, this line turns downwards through an angle of $90^o$ and then proceeds along the column until it leaves the diagram. The value of hook associated with that box is then the total number of boxes that the line has passed through, including the box in question. The product of hooks is the denominator. We illustrate this by the following example. Consider the tableau (5.94). The hooks associated with each box are shown in Eq. (5.95).

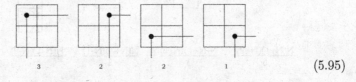

$$\begin{array}{cccc}
3 & 2 & 2 & 1
\end{array} \qquad (5.95)$$

We see that the denominator $= 3 \times 2 \times 2 \times 1 = 12$. Hence the tableau (5.94) corresponds to an irreducible representation of dimension $N^2 \left(N^2 - 1\right)/12$. Let us now consider some more examples:

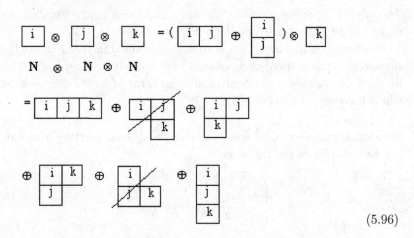

$$(5.96)$$

To avoid double counting, we discard the slashed tableaux. Thus we can write

$$\phi_i \, \phi_j \, \phi_k \sim T_{\{ijk\}} + T_{[ij]k} + T_{[ik]j} + T_{[ijk]}. \qquad (5.97)$$

Note further that

$$T_{[ij]k} + T_{[ki]j} + T_{[jk]i} = 0. \qquad (5.98)$$

In order to find the dimension of these representations, we note that

$$\frac{N(N+1)(N+2)}{6} \qquad \frac{N(N+1)(N-1)}{3} \qquad \frac{N(N+1)(N-1)}{3} \qquad \frac{N(N-1)(N-2)}{6} \qquad (5.99)$$

For example for SU(6) : $\mathbf{6} \otimes \mathbf{6} \otimes \mathbf{6} = \mathbf{56} \oplus \mathbf{70} \oplus \mathbf{70} \oplus \mathbf{20}$.

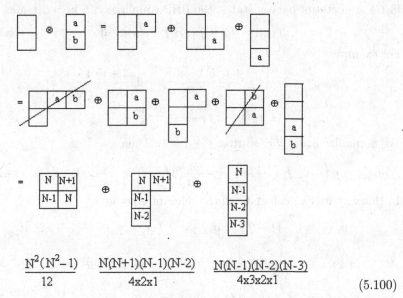

$$\frac{N^2(N^2-1)}{12} \qquad \frac{N(N+1)(N-1)(N-2)}{4 \times 2 \times 1} \qquad \frac{N(N-1)(N-2)(N-3)}{4 \times 3 \times 2 \times 1} \tag{5.100}$$

Hence we have

$$\frac{N(N-1)}{2} \quad \frac{N(N-1)}{2} \quad \frac{N^2(N^2-1)}{12} \quad \frac{N(N^2-1)(N-2)}{8} \quad \frac{N(N-1)(N-2)(N-3)}{24} \tag{5.101}$$

For example for SU(3) : $\mathbf{\bar{3}} \otimes \mathbf{\bar{3}} = \mathbf{\bar{6}} \oplus \mathbf{3}$ and for SU(5) : $\mathbf{10} \otimes \mathbf{10} = \mathbf{50} \oplus \mathbf{\overline{45}} \oplus \mathbf{\bar{5}}$.

## 5.6  Applications of Flavor SU(3)

### 5.6.1  *SU(3) Invariant BBP Couplings*

If $O_A$ is an octet operator, the matrix elements of this operator between the states $|8, B\rangle$ and $|8, C\rangle$ can be written as

$$\langle 8, C | O_A | 8, B \rangle = i\, f_{ABC}\, F + d_{ABC}\, D. \tag{5.102}$$

In particular if $O_A$ is pseudoscalar meson octet operator $P_A$, and $|8, B\rangle$, $|8, C\rangle$ are octet of baryon states, the BBP couplings can be written as

$$g_{ABC} = 2g \left[ i \, f_{ABC} \, f + d_{ABC} \, d \right]. \qquad (5.103)$$

For example

$$g_{\pi^0 pp} = 2g \left[ i \, f_3 \frac{4 + i5}{\sqrt{2}} \frac{4 - i5}{\sqrt{2}} f + d_3 \frac{4 + i5}{\sqrt{2}} \frac{4 - i5}{\sqrt{2}} d \right]$$

$$= g \, (f + d) = -g_{\pi^0 nn} = \frac{g_{\pi^- pn}}{\sqrt{2}}. \qquad (5.104)$$

We normalize $g_{\pi^0 pp} = g$, so that $f + d = 1$. Then

$$g_{Kp\Lambda} = g \left( -\sqrt{3} f - \frac{1}{\sqrt{3}} d \right) = -\frac{g}{\sqrt{3}} \, (3f + d) = -\frac{g}{\sqrt{3}} \, (1 + 2f). \qquad (5.105)$$

In this way we can calculate all the relevant couplings:

$$g_{\pi\Lambda\Sigma} = \frac{2}{\sqrt{3}} (1 - f) \, g, \quad g_{\pi\Sigma\Sigma} = 2 \, f \, g, \quad g_{\pi\Xi\Xi} = -(1 - 2f) \, g$$

$$g_{KN\Lambda} = -\frac{1}{\sqrt{3}} (1 + 2f) \, g, \quad g_{KN\Sigma} = -(1 - 2f) \, g$$

$$g_{K\Lambda\Xi} = -\frac{1}{\sqrt{3}} (1 - 4f) \, g, \quad g_{K\Sigma\Xi} = - \, g$$

$$g_{\eta_8 NN} = -\frac{1}{\sqrt{3}} (1 - 4f) \, g, \quad g_{\eta_8 \Lambda\Lambda} = -\frac{2}{\sqrt{3}} (1 - f) \, g$$

$$g_{\eta_8 \Sigma\Sigma} = \frac{2}{\sqrt{3}} (1 - f) \, g, \quad g_{\eta_8 \Xi\Xi} = -\frac{1}{\sqrt{3}} (1 + 2f) \, g. \qquad (5.106)$$

Experimentally

$$\frac{g_{\pi NN}^2}{4\pi} \equiv \frac{g^2}{4\pi} \approx 14$$

$$f \geq 0.35, \qquad r \equiv \frac{d}{f} = \frac{1 - f}{f} \leq 1.85. \qquad (5.107)$$

### 5.6.2  VPP Coupling

Here we take $O_A = V_A$, the vector meson octet and $|8, B\rangle$ and $|8, C\rangle$ are octets of pseudoscalar meson states. Now under charge conjugation $C$:

$$V_A \to -\eta_A V_A, \text{ (no summation over } A)$$

$$|8, B\rangle \to \eta_A |8, B\rangle, \qquad (5.108)$$

where

$$\eta_B = \left( \begin{array}{ll} +1, & B = 1, 3, 4, 6, 8 \\ -1, & B = 2, 5, 7 \end{array} \right). \qquad (5.109)$$

Hence the invariance under charge conjugation gives

$$\langle 8, C \,|V_A|\, 8, B \rangle \to -\eta_A\, \eta_B\, \eta_C\; \langle 8, C \,|V_A|\, 8, B \rangle$$
$$= -\eta_A\, \eta_B\, \eta_C\; [\gamma_F\, i\, f_{ABC} + \gamma_D\, d_{ABC}]$$
$$= [\gamma_F\, i\, f_{ABC} + \gamma_D\, d_{ABC}]. \tag{5.110}$$

But [cf. Eq. (5.33)]

$$\eta_A\, \eta_B\, \eta_C\, f_{ABC} = -f_{ABC}$$
$$\eta_A\, \eta_B\, \eta_C\, d_{ABC} = d_{ABC}. \tag{5.111}$$

Therefore, we have

$$\gamma_D = -\gamma_D \quad \text{or} \quad \gamma_D = 0.$$

Hence $VPP$ has only $F$-type coupling. Thus

$$\langle 8, C \,|V_A|\, 8, B \rangle = i\, f_{ABC}\, 2\gamma, \tag{5.112}$$

where we have put $\gamma_F = 2\gamma$. For example $V_3 = \rho^0$:

$$\left\langle 8, \frac{1 - i\,2}{\sqrt{2}} \,|V_3|\, 8, \frac{1 + i\,2}{\sqrt{2}} \right\rangle = i\, f_3\, \frac{1 + i\,2}{\sqrt{2}}\, \frac{1 - i\,2}{\sqrt{2}}\, 2\gamma = 2\gamma. \tag{5.113}$$

Thus $\gamma_{\rho\pi\pi} = 2\gamma$. It is straightforward to calculate the $VPP$ coupling for other members of the octet, which are given below:

$$\gamma_{\rho\pi\pi} = 2\gamma, \;\; \gamma_{K^{*+}\pi^0 K^+} = \gamma = \frac{1}{\sqrt{2}}\, \gamma_{K^{*+}\pi^+ K^0}, \;\; \gamma_{\omega_8 K\bar{K}} = \sqrt{3}\gamma. \tag{5.114}$$

The decay width of decay $V \to PP$ is given by

$$\Gamma(V \to PP) = \frac{\gamma^2_{VPP}}{4\pi}\, \frac{2}{3}\, \left(\frac{p^3_{cm}}{m^2_V}\right). \tag{5.115}$$

Hence we have

$$\Gamma(\rho \to \pi\pi) = \frac{\gamma^2}{4\pi}\, \frac{8}{3}\, \left(\frac{p^3_{\pi\pi}}{m^2_V}\right) = 149.1 \pm 2.9 \text{ MeV}. \tag{5.116}$$

This gives $\frac{\gamma^2}{4\pi} \approx 0.74$. Now

$$\Gamma_{\text{tot}}\left(K^{*+} \to K\pi\right) = \Gamma\left(K^{*+} \to \pi^0 K^+\right) + \Gamma\left(K^{*+} \to \pi^+ K^0\right)$$
$$= \frac{\gamma^2}{4\pi}\, (1 + 2)\, \frac{2}{3}\, \frac{p^3_{K\pi}}{m^2_{K^*}},$$

and

$$\frac{\Gamma_{\text{tot}}\left(K^{*+} \to K\pi\right)}{\Gamma(\rho \to \pi\pi)} = \frac{3}{4}\, \frac{p^3_{K\pi}/m^2_{K^*}}{p^3_{\pi\pi}/m^2_\rho} \approx 0.29. \tag{5.117}$$

This gives $\Gamma_{\text{tot}}\left(K^{*^+} \to K\pi\right) \approx 44.5$ MeV to be compared with the experimental value $49.8 \pm 0.8$ MeV.

In broken $SU(3)$, $\omega_8$ can mix with the singlet $\omega_1$, so that the physical particles $\omega$ and $\phi$ are linear combinations of $\omega_8$ and $\omega_1$:

$$\phi = \omega_8 \cos\theta - \omega_1 \sin\theta$$
$$\omega = \omega_8 \sin\theta + \omega_1 \cos\theta$$
$$\begin{pmatrix} \phi \\ \omega \end{pmatrix} = \begin{pmatrix} \cos\theta & -\sin\theta \\ \sin\theta & \cos\theta \end{pmatrix} \begin{pmatrix} \omega_8 \\ \omega_1 \end{pmatrix}. \tag{5.118}$$

Let us now show that $\omega_1 \to PP$ is forbidden by charge conjugation invariance. The invariant coupling in this case is

$$\omega_{1\mu} \, P_j^i \, \partial_\mu \, P_i^j$$

which changes sign under charge conjugation. Hence

$$\Gamma\left(\phi \to K^+ K^-\right) = \cos^2\theta \, \Gamma\left(\omega_8 \to K\bar{K}\right)$$

Therefore,

$$\Gamma\left(\phi \to K^+ K^-\right) = \cos^2\theta \, \frac{\gamma^2}{4\pi}\left(3 \times \frac{2}{3}\right)\left(\frac{p_{KK}^3}{m_\phi^2}\right), \tag{5.119}$$

and

$$\frac{\Gamma\left(\phi \to K^+ K^-\right)}{\Gamma\left(\rho \to \pi\pi\right)} = \frac{3}{4}\cos^2\theta \left(\frac{p_{KK}^3/m_\phi^2}{p_{\pi\pi}^3/m_\rho^2}\right) = 0.013, \tag{5.120}$$

where we have used $\cos^2\theta = 2/3$ [see Eq. (5.153) below]. This gives $\Gamma\left(\phi \to K^+ K^-\right) = 1.95$ MeV to be compared with the experimental value 2.1 MeV.

## 5.7  Mass Splitting in Flavor SU(3)

In exact $SU(3)$, the particles belonging to an irreducible representation of $SU(3)$ must have the same mass. But we note that all members of a supermultiplet do not have the same mass. This means that $SU(3)$ is not an exact symmetry of strong interactions, but is a broken symmetry of these interactions. This means that the interaction Hamiltonian consists of two parts viz.

$$H = H_0 + H_1, \tag{5.121a}$$

where

$$[F_j^i, H_0] = 0 \tag{5.121b}$$

$$[F_j^i, H_1] \neq 0 \tag{5.121c}$$

i.e. $H_0$ is $SU(3)$ invariant, but $H_1$ breaks the $SU(3)$ symmetry. If we take $H_1$ such that

$$[\mathbf{I}, H_1] = 0, \qquad [Y, H_1] = 0, \tag{5.122}$$

$H_1$ still preserves the isospin symmetry and hypercharge is conserved in its presence. The first of Eqs. (5.122) holds only in the absence of electromagnetic interaction. In order that $SU(3)$ to be meaningful, $H_1$ must be at least an order of magnitude weaker than $H_0$.

The simplest general form of $H_1$ in $SU(3)$ which satisfies Eqs. (5.121c) and (5.122) is

$$H_1 \sim T_3^3 \text{ or } \lambda_8. \tag{5.123}$$

To get $H_1$ from the quark model, the mass Hamiltonian for quarks is given by

$$H_q = m_u \, \bar{u} \, u + m_d \, \bar{d} \, d + m_s \, \bar{s} \, s,$$

where $m_u$, $m_d$ and $m_s$ are masses of u-quark, d-quark and s-quark respectively. In the exact $SU(3)$ limit, $m_u = m_d = m_s$. If $SU(3)$ is broken but isospin symmetry $SU(2)$ is still exact, then $m_u = m_d \neq m_s$. Now one can write

$$
\begin{aligned}
H_q &= \bar{m} \, (\bar{u} \, u + \bar{d} \, d) + m_s \, \bar{s} \, s + (m_u - m_d) \frac{\bar{u} \, u - \bar{d} \, d}{2} \\
&= \frac{2\bar{m} + m_s}{3} \, (\bar{u} \, u + \bar{d} \, d + \bar{s} \, s) \\
&\quad + \frac{\bar{m} - m_s}{\sqrt{3}} \frac{1}{\sqrt{3}} \, (\bar{u} \, u + \bar{d} \, d - 2s\bar{s}) \\
&\quad + (m_u - m_d) \frac{1}{2} \, (\bar{u} \, u - \bar{d} \, d),
\end{aligned} \tag{5.124}
$$

where

$$\bar{m} = \frac{m_u + m_d}{2}, \quad \frac{2\bar{m} + m_s}{3} = \frac{m_u + m_d + m_s}{3} \equiv m_q$$

$$\frac{(\bar{m} - m_s)}{\sqrt{3}} = \frac{m_u + m_d - 2m_s}{2\sqrt{3}} \equiv \frac{\lambda}{2}. \tag{5.125}$$

$H_q$ becomes

$$H_q = m_q \, \bar{q} \, q + \lambda \, \bar{q} \, \frac{\lambda_8}{2} q + (m_u - m_d) \, \bar{q} \, \frac{\lambda_3}{2} q. \tag{5.126}$$

This also shows that $SU(3)$ symmetry breaking term transforms as $\lambda_8$ under $SU(3)$.

It was shown by Okubo that for any irreducible representation $(p,q)$ of $SU(3)$, the matrix element of tensor $T_3^3$ is given by

$$\langle (p,q)\ I,\ Y\ |T_3^3|\ (p,q)\ I,\ Y \rangle = a + bY + c\left[\frac{Y^2}{4} - I(I+1)\right], \quad (5.127)$$

where $a$, $b$, $c$ are independent of quantum numbers $I$ and $Y$ but in general depend on $(p,q)$. Thus we can write the mass formula for particles in a multiplet of $SU(3)$ as

$$m = m_0 + \Delta m = a + bY + c\left[\frac{Y^2}{4} - I(I+1)\right]. \quad (5.128)$$

Let us apply this formula to baryon octet. Then from Eq. (5.128), we get

$$\frac{m_N + m_\Xi}{2} = \frac{3m_\Lambda + m_\Sigma}{4} \quad (5.129)$$

whereas for pseudoscalar meson octet, we get

$$m_K^2 = \frac{3\ m_{\eta_8}^2 + m_\pi^2}{4}. \quad (5.130)$$

In Eq. (5.130), we have used squared masses, as in the Lagrangian for bosons, the square of boson masses appear. Equations (5.129) and (5.130) are well known Gell-Mann-Okubo mass formulae.

For the decuplet

$$I = \frac{Y}{2} + 1, \quad (5.131)$$

and Eq. (5.128) reduces to

$$m = a' + b'Y \quad (5.132)$$

and we obtain the equal spacing rule for the decuplet.

$$m_\Omega - m_{\Xi^*} = m_{\Xi^*} - m_{\Sigma^*} = m_{\Sigma^*} - m_\Delta. \quad (5.133)$$

The mass relations (5.129) and (5.133) are well satisfied experimentally and are regarded as a great success of $SU(3)$. Similarly for vector bosons we get

$$m_{K^*}^2 = \frac{3\ m_{\omega_8}^2 + m_\rho^2}{4}. \quad (5.134)$$

Since due to mixing between $\omega_8$ and the singlet $\omega_1$, the physical particles are $\phi$ and $\omega$, the formula (5.134) is not directly applicable. We will come to this formula later. Similar remarks are applicable to the mass formula (5.130).

For octet and decuplet representations of $SU(3)$, one can easily derive the mass formula as follows. We note from Eq. (5.126) that

$$H_1 \sim \bar{q} \, \frac{\lambda_8}{2} \, q = q_i \left(\frac{\lambda_8}{2}\right)^i_j q^j$$

$$= \left(\frac{\lambda_8}{2}\right)^i_j O^j_i = (T_8)^i_j \, O^j_i \qquad (5.135)$$

where $O^i_j$ is an octet operator viz.

$$O^i_j = q^i \, q_j - \frac{1}{3}\delta^i_j \, q^k \, q_k \qquad (5.136)$$

and

$$T_8 = \frac{\lambda_8}{2}. \qquad (5.137)$$

Hence we see that $H_1$ transforms under $SU(3)$ as

$$H_1 \sim (T_8)^i_j \, O^j_i = \left(M^3_3\right)^i_j \, O^j_i = \left(\delta^3_j \, \delta^i_3 - \frac{1}{3}\delta^3_3 \, \delta^i_j\right) O^j_i$$

$$= O^3_3. \qquad (5.138)$$

Thus to first order in $\lambda$, the mass splitting for the state $|A\rangle$ of an $SU(3)$ multiplet is given by

$$\Delta m = \lambda \, \langle A \,|H_I|\, A \rangle$$

$$= \lambda \, \langle A \,|O^3_3|\, A \rangle. \qquad (5.139)$$

Let us apply it to baryon octet:

$$O^3_3 = O_F \left(\bar{B}^i_3 \, B^3_i - \bar{B}^3_j \, B^j_3\right) + O_D \left(\bar{B}^i_3 \, B^3_i + \bar{B}^3_j \, B^j_3\right)$$

$$= O_F \left(\Xi^-\Xi^- + \Xi^0\Xi^0 - \bar{p}p - \bar{n}n\right)$$

$$+ O_D \left[\Xi^-\Xi^- + \Xi^0\Xi^0 + \bar{p}p + \bar{n}n + 2\left(-\frac{2}{\sqrt{6}}\bar{\Lambda}\right)\left(-\frac{2}{\sqrt{6}}\Lambda\right)\right]. \qquad (5.140)$$

Hence we have

$$m_p = m_0 + \lambda\left(-O_F + O_D\right) = m_n$$

$$m_\Sigma = m_0$$

$$m_\Lambda = m_0 + \frac{4}{3}\lambda \, O_D$$

$$m_\Xi = m_0 + \lambda\left(O_F + O_D\right). \qquad (5.141)$$

This gives the Gell-Mann-Okubo mass formula (5.129) for baryons.

For the decuplet, we have

$$\lambda \, O_3^3 = \lambda \, \bar{T}^{ij3} T_{ij3}. \tag{5.142}$$

This is the only possibility as $T_{ijk}$ is a completely symmetric tensor.

$$\begin{aligned}
\lambda \, O_3^3 &= [\bar{T}^{113} T_{113} + 2\bar{T}^{123} T_{123} + 2\bar{T}^{133} T_{133} \\
&\quad + 2\bar{T}^{233} T_{233} + \bar{T}^{223} T_{223} + \bar{T}^{333} T_{333}] \\
&= \lambda \left[ 2 \, \bar{\Sigma}^{*+} \, \Sigma^{*+} + 2 \, \bar{\Sigma}^{*0} \, \Sigma^{*0} + 2 \, \bar{\Sigma}^{*-} \, \Sigma^{*-} \right. \\
&\quad \left. + 4\bar{\Xi}^{*-} \Xi^{*-} + 4 \, \bar{\Xi}^{*0} \Xi^{*0} + 6\bar{\Omega}^- \Omega^- \right].
\end{aligned} \tag{5.143}$$

This gives

$$\begin{aligned}
m_\Delta &= m_0 \\
m_{\Sigma^*} &= m_0 + 2\lambda \\
m_{\Xi^*} &= m_0 + 4\lambda \\
m_{\Omega^-} &= m_0 + 6\lambda,
\end{aligned} \tag{5.144}$$

and hence this gives the mass relation (5.133) for the decuplet.

For octet of vector mesons

$$\begin{aligned}
O_3^3 &= O_D \left[ V_3^i \, V_i^3 + V_j^3 \, V_3^j \right] \\
&= O_D \left[ \bar{K}^{*+} K^{*+} + \bar{K}^{*-} K^{*-} + \bar{K}^{*0} K^{*0} + 2 \left( -\frac{2}{\sqrt{6}} \, \omega_8 \right) \left( -\frac{2}{\sqrt{6}} \, \omega_8 \right) \right]
\end{aligned} \tag{5.145}$$

Hence one gets

$$\begin{aligned}
m_\rho^2 &= m_0^2 \\
m_{K^*}^2 &= m_0^2 + \lambda \, O_D \\
m_{\omega_8}^2 &= m_0^2 + \frac{4}{3}\lambda \, O_D.
\end{aligned} \tag{5.146}$$

This gives the octet formula (5.134) for vector mesons. Now $\omega_8$ and $\omega_1$ mix, when $SU(3)$ is broken, the mass matrix in $\omega_8$ and $\omega_1$ basis can be written

$$M^2 = \begin{pmatrix} m_8^2 & m_{18}^2 \\ m_{18}^2 & m_1^2 \end{pmatrix}. \tag{5.147}$$

Using Eq. (5.118), we can diagonalize it:

$$U^T M^2 U = \begin{pmatrix} m_\phi^2 & 0 \\ 0 & m_\omega^2 \end{pmatrix}, \tag{5.148}$$

where

$$U = \begin{pmatrix} \cos\theta & -\sin\theta \\ \sin\theta & \cos\theta \end{pmatrix}. \tag{5.149}$$

This gives

$$m_\phi^2 + m_\omega^2 = m_8^2 + m_1^2 \tag{5.150a}$$

$$m_\phi^2 - m_\omega^2 = \left(\cos^2\theta - \sin^2\theta\right)\left(m_8^2 - m_1^2\right)$$
$$+ 4\sin\theta\cos\theta\, m_{18}^2, \tag{5.150b}$$

$$\tan 2\theta = \frac{2\, m_{18}^2}{m_8^2 - m_1^2} \tag{5.151a}$$

$$\tan^2\theta = \frac{m_\phi^2 - m_8^2}{m_8^2 - m_\omega^2} = \frac{3\, m_\phi^2 - 4\, m_{K^*}^2 + m_\rho^2}{4\, m_{K^*}^2 - m_\rho^2 - 3\, m_\omega^2}. \tag{5.151b}$$

Now using $m_{K^*} = 892$ MeV, $m_\rho = 770$ MeV, $m_\omega = 783$ MeV and $m_\phi = 1020$ MeV, we have $m_8 \equiv m_\omega = 930$ MeV, $m_1 \equiv m_\omega = 880$ MeV and

$$\tan\theta \approx 0.84, \qquad \theta \approx 40°. \tag{5.152}$$

It is tempting to take

$$\tan\theta \approx \frac{1}{\sqrt{2}} \approx 0.71, \qquad \theta \approx 35.3°. \tag{5.153}$$

For this case $\sin\theta \approx \frac{1}{\sqrt{3}}$, $\cos\theta \approx \frac{\sqrt{2}}{\sqrt{3}}$ and

$$|\phi\rangle = \frac{\sqrt{2}}{\sqrt{3}}\, |\omega_8\rangle - \frac{1}{\sqrt{3}}\, |m_1\rangle = -|s\,\bar{s}\rangle \tag{5.154a}$$

$$|\omega\rangle = \frac{1}{\sqrt{3}}\, |\omega_8\rangle + \frac{\sqrt{2}}{\sqrt{3}}\, |\omega_1\rangle = \left|\frac{u\,\bar{u} + d\,\bar{d}}{\sqrt{2}}\right\rangle. \tag{5.154b}$$

Hence we have

$$m_\omega^2 = m_\rho^2 \tag{5.155a}$$

and from Eqs. (5.151b) and (5.153)

$$m_\phi^2 - m_\omega^2 = 2\left(m_{K^*}^2 - m_\rho^2\right). \tag{5.155b}$$

Equation (5.153) gives the "ideal mixing". With this mixing $\phi$ is made up of $s\bar{s}$, i.e. of strange quarks only. Experimentally it is observed that $\phi \to \rho\pi$ or $3\pi$ is very much suppressed as compared with $\phi \to K\bar{K}$. Note that $\rho\pi$ or $3\pi$ does not contain any strange quark. The suppression of $\phi$ decay into non-strange particles is explained by the so-called Okubo-Zweig-Iizuka rule (OZI rule): "**The decays which correspond to disconnected quark diagrams are forbidden**". Thus the decay in Fig. 5.6a is allowed but the decay in Fig. 5.6b is forbidden. There is no theoretical basis for the OZI rule. No strong interaction selection rule forbids the decay of $\phi \to \rho\pi$ or $3\pi$. But experimentally this rule seems to be well satisfied. The small decay width for $\phi \to \rho\pi$ ($3\pi$) can be explained by some deviation from the "ideal mixing" which allows small admixture of non-strange quarks in $\phi$.

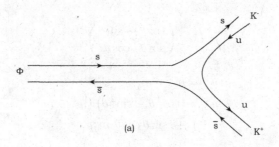

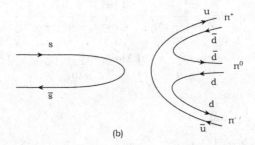

Fig. 5.6   (a) $\phi \to K\bar{K}$ decay allowed by OZI rule, (b) $\phi \to \pi^+\pi^-\pi^0$ decay suppressed by OZI rule.

## 5.8   Problems

(1) Show that for a vector operator $O_i (i = 1, 2)$ under $SU(2)$

$$[I_A, O_i] = O_j \left(\frac{\tau_A}{2}\right)^j_i, \qquad A = 1, 2, 3.$$

Given

$$\langle \alpha, 3/2, -1/2 | O_1 | \beta, 1, -1 \rangle = F,$$

find

$$\langle \alpha, 3/2, -3/2 | O_2 | \beta, 1, -1 \rangle.$$

The states are labelled as $|\alpha, I, I_3\rangle$.

(2) Suppose that $(\tau_A)^i_j \equiv (\tau_A)_{ij}$ and $(\sigma_A)^\alpha_\beta \equiv (\sigma_A)_{\alpha\beta}$ are Pauli matrices in two different two-dimensional spaces. In the four-dimensional product space, define the basis vectors

$$|\mu = 1\rangle = |i = 1\rangle \, |\alpha = 1\rangle, \; |\mu = 2\rangle = |i = 1\rangle \, |\alpha = 2\rangle$$
$$|\mu = 3\rangle = |i = 2\rangle \, |\alpha = 1\rangle, \; |\mu = 4\rangle = |i = 2\rangle \, |\alpha = 2\rangle.$$

Define
$$T_{AB} = \tau_A \otimes \sigma_B; \quad (T_{AB})_\nu^\mu = (\tau_A)_j^i \, (\sigma_B)_\beta^\alpha,$$
$$\mu, \nu = 1, \cdots, 4, \qquad A, B = 1, 2, 3.$$
Evaluate
$$T_{21} = (\tau_2 \otimes \sigma_1), \qquad \text{as a } 4 \times 4 \text{ matrix.}$$

(3) A second ranked mixed tensor $T_j^i$ transforms as $\phi_i^* \, \phi_j$, under the unitary transformation

$$\phi_i' = a_i^j \, \phi_j,$$

show that

$$\left[ F_j^i, \, T_l^k \right] = \delta_l^i \, T_j^k - \delta_j^k \, T_l^i.$$

(4) (a). Using the following relation for $SU(3)$:

$$\left[ F_j^i, \, q_k \right] = \delta_k^i \, q_j - \frac{1}{3} \delta_j^i \, q_k,$$

show that

$$F_1^2 \, |d\rangle = |u\rangle, \qquad F_3^1 \, |u\rangle = |s\rangle.$$

(b). Using the relation

$$\left[ F_j^i, \, \bar{B}_l^k \right] = \delta_l^j \, \bar{B}_j^k - \delta_j^k \, \bar{B}_l^i$$

and Eq. (5.60b) of the text, show that

$$F_2^1 \, |\Sigma^+\rangle = -\sqrt{2} \, |\Sigma^0\rangle, \qquad F_3^2 \, |p\rangle = - \, |\Sigma^+\rangle.$$

(5) From the group property

$$U^{-1}(b) \, U(a) \, U(b) = U(b^{-1}ab),$$

derive the commutation relation for the generators of the unitary group $U(N)$:

$$\left[ A_i^j, \, A_k^l \right] = \delta_k^j \, A_i^l - \delta_i^l \, A_k^j.$$

(6) Show that $\lambda_1$, $\lambda_2$ and $\lambda_3$ generate an $SU(2)$ subalgebra of $SU(3)$. Show that the representations generated by the remaining $\lambda$'s or their linear combinations transform as doublets and singlet representations of $SU(2)$.

(7) Show that $\lambda_2$, $\lambda_5$ and $\lambda_7$ generate an $SU(2)$ subalgebra of $SU(3)$. Show that the representation generated by the linear combinations of remaining $\lambda$'s transform as 5-dimensional representation of $SU(2)$.
   **Hint:** $\lambda_5 \mp i\lambda_2$ act as raising and lowering operators.

(8) Find the matrix generators $\lambda_A$ $(A = 1, \cdots, 15)$ for the group $SU(4)$.

(9) The following assignments for 3 quarks are given instead of the usual ones:

|     | $B$ | $S$ | $I$   | $I_3$ |
|-----|-----|-----|-------|-------|
| $u'$ | 1   | $-2$ | $1/2$ | $1/2$  |
| $d'$ | 1   | $-2$ | $1/2$ | $-1/2$ |
| $s'$ | 1   | $-3$ | 0     | 0      |

Find the charge $Q$ and hypercharge $Y$ for each quark in this case. Mesons can be constructed as $\bar{q}'\, q$ as before. If baryons are constructed as $q'\, q'\, q'$, can the above assignment of quarks work? If not, discuss the difficulties encountered.

(10) Find the $U$-spin eigenstates for the baryon octet and decuplet. Plot them on $Q$ versus $U_3$ plot.

(11) As far as $SU(3)$ is concerned, magnetic moment operator transforms as $T_1^1$ which is singlet under $U$-spin. Using this fact and the $U$-spin multiplets found above, show that the baryon octet magnetic moments are related as follows:

$$\mu_{\Sigma^+} = \mu_p, \quad \mu_{\Sigma^-} = \mu_{\Xi^-}, \quad \mu_{\Xi^0} = \mu_n = \frac{1}{2}\left(3\mu_\Lambda - \mu_{\Sigma^0}\right),$$

$$\mu_{\Sigma^0 - \Lambda^0} = -\frac{\sqrt{3}}{2}\left(\mu_\Lambda - \mu_\Sigma\right)$$

(12) In $SU(3)$, find

$$10 \otimes 8, \qquad \overline{10} \otimes 10, \qquad 8 \otimes 3.$$

(13) In $SU(5)$, show that

$$5 \otimes 5 = 24 \oplus 1, \qquad 10 \otimes 10 = \bar{5} \oplus 50 \oplus 45$$

$$\bar{5} \otimes 10 = 5 \oplus \overline{45}, \qquad \overline{10} \otimes 10 = 1 \oplus 24 \oplus 75.$$

(14) Consider the representation 6 of $SU(3)$. Write down the particle content of this representation in terms of quarks. If $SU(3)$ breaking Hamiltonian $H_I$ transforms as $O_3$ or $T_8$, write down the mass formula for these particles.

(15) Draw the weight diagrams for the 15 plet and 27 vlet representations of $SU(3)$.

(16) Consider the $O^-$ nonet. Experimental masses are

$$m_\pi = 137 \text{ MeV}, \quad m_K = 496 \text{ MeV},$$

$$m_\eta = 549 \text{ MeV}, \quad m_{\eta'} = 958 \text{ MeV}.$$

From the octet mass formula, find $m_{\eta_8}$. Compare it with $m_\eta$. Assuming that discrepancy between the two values is entirely due to $\eta_1 - \eta_8$ mixing in broken $SU(3)$, so that

$$|\eta'\rangle = \cos\theta \,|\eta_1\rangle + \sin\theta \,|\eta_8\rangle, \quad |\eta\rangle = -\sin\theta \,|\eta_1\rangle + \cos\theta \,|\eta_8\rangle,$$

find from the experimental masses and $m_{\eta_8}$, the values of $m_{\eta_1}$ and the mixing angle $\theta$. If we write

$$|\eta\rangle = \cos\phi \,|\eta_{ns}\rangle - \sin\phi \,|\eta_s\rangle, \quad |\eta'\rangle = \sin\phi \,|\eta_{ns}\rangle + \cos\phi \,|\eta_s\rangle,$$

where

$$|\eta_{ns}\rangle = \frac{1}{\sqrt{2}} \,|\bar{u}\,u + \bar{d}\,d\rangle, \quad |\eta_s\rangle = \frac{1}{\sqrt{2}} \,|\bar{s}\,s\rangle,$$

show that

$$\phi = \tan^{-1}\sqrt{2} + \theta.$$

(17) You are given an octet operator

$$O = \cos\theta \, O_{1+i2} + \sin\theta \, O_{4+i5},$$

determine the $SU(3)$ matrix elements for the transitions:

$$n \to p \quad , \quad \Sigma^- \to n \quad , \quad \Sigma \to \Lambda \quad , \quad \Sigma^0 \to p$$
$$\Xi^- \to \Xi^0 \quad , \quad \Xi^0 \to \Lambda \quad , \quad \Xi^0 \to \Sigma \quad , \quad \Xi^- \to \Sigma^+$$

in terms of $F$, $D$ and $\theta$.

(18) Write down the $D\,B\,P$ couplings in the $SU(3)$ limit for the process

$$D \to B\,p$$

where

$$D : \text{Bayron decuplet} \quad J^P = 3/2^+$$
$$B : \text{Baryon octet} \quad J^P = 1/2^+$$
$$P : \text{Meson octet} \quad J^P = 0^-.$$

Hence show that for the energetically allowed decays, they are in the following ratios:

| $\Delta^{++}$ | | $\Sigma^{*+}$ | | $\Sigma^*$ | | $\Xi^{*0}$ | | $\Xi^{*-}$ |
|---|---|---|---|---|---|---|---|---|
| $\to p\pi^+$ | : | $\to \Lambda\pi$ | : | $\to \Sigma\pi$ | : | $\to \Xi^-\pi^+$ | : | $\to \Xi^{*0}\pi^0$ |
| $-\sqrt{6}$ | : | $\sqrt{3}$ | : | $1$ | : | $\sqrt{2}$ | : | $-1$ |

(19) Show that why in Eq. (5.102), only two reduced matrix elements $F$ and $D$ are possible?

**Hint:** From field operators $\bar{B}^i_j$, $B^i_j$ and $P^i_j$ one can form only two independent $SU(3)$ scalars: $\bar{B}^i_j i\gamma_5 B^j_k P^k_i$ and $\bar{B}^i_j i\gamma_5 P^j_k B^k_i$, giving Yukawa coupling between pseudoscalar mesons and $J^0 = \frac{1}{2}^+$ baryons. The above two scalars can be arranged antisymmetric $H_F$ and symmetric $H_D$ combinations.

## 5.9   References

1. M. Gell-Mann and Y. Ne'eman, The eightfold way, Benjamin, New York (1964).
2. S. Okubo, Lectures on Unitary Symmetry, (unpublished Univ. of Rochester Rep.). For SU(3), we have drawn heavily on this reference.
3. P. Carruthers, Introduction to unitary symmetry, Interscience, New York (1966).
4. D. B. Lichtenberg, Unitary symmetry and elementary particles (2nd edition), Academic Press, New York (1978).
5. F. E. Close, An introduction to quarks and partons, Academic Press, New York (1979).
6. R. Slansky, Group theory for unified model building, Physics Report 79c, 1 (1981).
7. H. Georgi, Lie algebra in particle physics, Benjamin Cummings, Reading Massachusetts (1982).

# Chapter 6

# SU(6) and Quark Model

## 6.1   SU(6)

Quarks have spin 1/2. The well-known baryons with spin 1/2 and spin 3/2 are in the octet and decuplet representations of flavor SU(3). We note that within each representation the mass splitting between adjacent members is of the same order. For example

$$m_\Lambda - m_N \approx 170 \text{ MeV}, \quad m_\Xi - m_\Sigma = 125 \text{ MeV};$$

$$m_{\Sigma^*} - m_\Delta \approx 153 \text{ MeV}, \quad m_\Omega - m_{\Xi^*} = 142 \text{ MeV}.$$

It is tempting to put these two representations in an irreducible representation of a group higher than SU(3). But octet and decuplet representations have different spins. This means that the proposed group cannot commute with angular momentum (spin). The proposed group must contain $SU(3) \otimes SU\sigma(2)$ as its subgroup. This might cause some trouble, since we are combining an internal symmetry with a space-time symmetry. It does cause trouble but this does not show up until one tries to make the theory relativistic [see Chap. 17].

We note that spin 3/2 baryon decuplet has $(10 \times 4)$ states and spin 1/2 baryon octet has $(8 \times 2)$ states. Thus, we look for an irreducible representation with 56 dimensions. Such a representation occurs in the decomposition of the product of representation **6** of SU(6) by itself viz.

$$\mathbf{6} \otimes \mathbf{6} \otimes \mathbf{6} = \mathbf{56} \oplus \mathbf{70} \oplus \mathbf{70} \oplus \mathbf{20}. \tag{6.1}$$

The representation **56** is completely symmetric irreducible representation of SU(6). The six quark states (in this section we will not write $| \ \rangle$ explicitly) $(u\uparrow u\downarrow \ d\uparrow d\downarrow \ s\uparrow s\downarrow)$ can be put in the fundamental representation **6**. We denote such a state as $\Psi_{i\alpha} : \alpha = 1, 2; i = 1, 2, 3$. In matrix notation.

$$\Psi \equiv \begin{pmatrix} u\uparrow & d\uparrow & s\uparrow \\ u\downarrow & d\downarrow & s\downarrow \end{pmatrix}. \tag{6.2}$$

Now SU(3), SU(2) and SU(3)⊗SU(2) are subgroups of SU(6). The representation **6** splits under these subgroups as shown in the table below:

| Subgroups of SU(6) | Quarks Representation | | Generators |
|---|---|---|---|
| $SU(3)$ | $(u \uparrow d \uparrow s \uparrow)$ $(u \downarrow d \downarrow s \downarrow)$ $(\mathbf{3, 1})$ , $(\mathbf{3, 1})$ | | $\frac{1}{2}\lambda_A \otimes 1$ $A = 1 \cdots 8$ |
| $SU(2)$ | $(u \uparrow u \downarrow)$ $(d \uparrow d \downarrow)$ $(s \uparrow s \downarrow)$ $(\mathbf{1, 2})$ , $(\mathbf{1, 2})$ , $(\mathbf{1, 2})$ | | $1 \otimes \frac{1}{2}\sigma_n$ $n = 1, 2, 3$ |
| $SU(3) \times SU(2)$ | | $(\mathbf{3, 2})$ | $\left(\frac{\lambda_A}{2} \otimes \frac{\sigma_n}{2}\right)$ |

Thus, one can see that SU(6) has 35 generators. Hence the adjoint representation of SU(6) has dimension 35 and is given in the following decomposition:

$$\mathbf{6} \otimes \mathbf{\bar{6}} = \mathbf{35} \oplus \mathbf{1}. \tag{6.3}$$

The representations **56** and **35** split under the subgroup SU(3)⊗SU(2) as follows:

$$\mathbf{56} \quad : \quad [(\mathbf{3, 2}) \otimes (\mathbf{3, 2}) \otimes (\mathbf{3, 2})]_{\text{symmetric}}$$

$$= \quad \underset{\substack{\text{baryon} \\ \text{decuplet}}}{(\mathbf{10, 4})} \quad + \quad \underset{\substack{\text{baryon} \\ \text{octet}}}{(\mathbf{8, 2})} \tag{6.4}$$

$$\mathbf{35} \quad : \quad [(\mathbf{3, 2}) \otimes (\mathbf{\bar{3}, 2})]$$

$$= \quad \underset{\substack{\text{nonet} \\ \text{of vector} \\ \text{mesons}}}{(\mathbf{8, 3}) \oplus (\mathbf{1, 3})} \quad \oplus \quad \underset{\substack{\text{octet of} \\ \text{pseudoscalar} \\ \text{mesons}}}{(\mathbf{8, 1})} \quad \oplus \quad \underset{\substack{\text{singlet} \\ \text{pseudoscalar} \\ \text{meson}}}{(\mathbf{1, 1})} \tag{6.5}$$

### 6.1.1  *SU(6) Wave Function for Mesons*

The mesons are composite of $q\bar{q}$. The lowest lying mesons have

$$(q\bar{q})_{L=0} \quad \text{and} \quad P = (-1)(-1)^0 = -1.$$

The spin wave functions are given by:

$$\text{Spin singlet state}: \quad \chi_A = \frac{1}{\sqrt{2}} |\uparrow\downarrow - \downarrow\uparrow\rangle \tag{6.6a}$$

Spin triplet states: $\quad \chi_S^{1,0,-1} = |\uparrow\uparrow\rangle, \dfrac{1}{\sqrt{2}} |\uparrow\downarrow + \downarrow\uparrow\rangle, |\downarrow\downarrow\rangle.$ \hfill (6.6b)

The spin singlet state is antisymmetric, it gives $J^P = 0^-$, whereas the spin triplet states are symmetric and gives $J^P = 1^-$. Thus one can write $SU(6)$ states for

$$J^P = 0^- \; (^1S_0):$$

$$(q_i \bar{q}_j)_{L=0} \, \chi_A = (q_i \bar{q}_j) \frac{1}{\sqrt{2}} \left( \uparrow_{(i)} \downarrow_{(j)} - \downarrow_{(i)} \uparrow_{(j)} \right)$$

For example

$$\pi^+ = \left( u^\uparrow \bar{d}^\downarrow - u^\downarrow \bar{d}^\uparrow \right)$$

$$\pi^0 = \frac{1}{2} \left( u^\uparrow \bar{u}^\downarrow - u^\downarrow \bar{u}^\uparrow - d^\uparrow \bar{d}^\downarrow + d^\downarrow \bar{d}^\uparrow \right)$$

Similarly one can write $SU(6)$ states for other members of nonent. For $J^P : 1^- \left( ^3S_1 \right)$, $SU(6)$ states are as follows:

| | $J_z = 1$ | $J_z = 0$ | $J_z = -1$ |
|---|---|---|---|
| | $(q_i \bar{q}_j)_{L=0}\, \chi_S^1$ | $(q_i \bar{q}_j)_{L=0}\, \chi_S^0$ | $(q_i \bar{q}_j)_{L=0}\, \chi_S^{-1}$ |
| | $(q_i \bar{q}_j) \left( \uparrow_{(i)} \uparrow_{(j)} \right)$ | $(q_i \bar{q}_j) \frac{1}{\sqrt{2}} \left( \uparrow_{(i)} \downarrow_{(j)} + \downarrow_{(i)} \uparrow_{(j)} \right)$ | $(q_i \bar{q}_j) \left( \downarrow_{(i)} \downarrow_{(j)} \right)$ |
| $\rho^+ :$ | $\left( u^\uparrow \bar{d}^\uparrow \right)$ | $\frac{1}{\sqrt{2}} \left( u^\uparrow \bar{d}^\downarrow + u^\downarrow \bar{d}^\uparrow \right)$ | $\left( u^\downarrow \bar{d}^\downarrow \right)$ |
| $\omega^0 :$ | $\frac{1}{\sqrt{2}} \left( u^\uparrow \bar{d}^\uparrow + u^\uparrow \bar{d}^\uparrow \right)$ | $\frac{1}{2} \left( u^\uparrow \bar{u}^\downarrow + u^\downarrow \bar{u}^\uparrow + d^\uparrow \bar{d}^\downarrow + d^\downarrow \bar{d}^\uparrow \right)$ | $\frac{1}{\sqrt{2}} \left( u^\downarrow \bar{d}^\downarrow + u^\downarrow \bar{d}^\downarrow \right)$ |

Similarly one can write for other states.

We now briefly discuss, the $p$-wave, $i.e.$ $L = 1$ mesons:

$$(q\bar{q})_{L=1}, \quad \text{Parity } P = (-1)(-1)^1 = +1$$

For $L = 1$ mesons, we have the following $SU(6)$ wave functions

$$(q\bar{q})_{L=1}\, \chi_A^0 : \; ^1P_1 \text{ state } 1^+$$
$$(q\bar{q})_{L=1}\, \chi_S^{1,0,-1} : \; ^3P_0 \text{ state } 0^+$$
$$: \; ^3P_1 \text{ state } 1^+$$
$$: \; ^3P_2 \text{ state } 2^+$$

Now C-parity for the mesons composite of identical flavor quark and anti-quark $C = (-1)^{L+S}$ and G-parity $G = C (-1)^I$. Hence for

$$
\begin{array}{ccccc}
L = 0 & S = 0 & C = +1 & G = 1^- & \pi \\
L = 0 & S = 0 & C = +1 & G = 1^+ & \eta,\, \eta' \\
L = 0 & S = 1 & C = -1 & G = 1^+ & \rho \\
L = 0 & S = 1 & C = -1 & G = 1^- & \omega,\, \phi
\end{array}
$$

For

$$L = 1 \quad S = 0 \quad C = -1 \quad {}^1P_1 \qquad\qquad\qquad 1^{+-}$$
$$L = 1 \quad S = 1 \quad C = 1 \quad {}^3P_0, \ {}^3P_1, {}^3P_2 \quad 0^{++}, \ 1^{++}, \ 2^{++}$$

Hence in the quark model we have nonent of these states. Some of these states can be identified in the particle data book.

Lowest lying baryons are made up of three quarks: $(qqq)_{L=0}$, $P = (-1)^0(1)^3 = 1$. Here we have to combine three spin 1/2's. In this case we have the following decomposition:

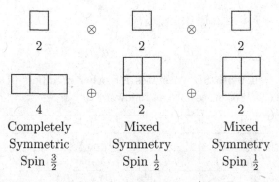

| Completely | Mixed | Mixed |
| Symmetric | Symmetry | Symmetry |
| Spin $\frac{3}{2}$ | Spin $\frac{1}{2}$ | Spin $\frac{1}{2}$ |

It is convenient to combine first two spin 1/2's. For this case we have $S = 0$ with spin wave function $\chi_A$ [Eq. (6.6a)] and $S = 1$ with spin wave functions $\chi_S$ [Eq. (6.6b)]. Let us now combine spin 0 with the remaining spin 1/2 and we get the spin $S = 1/2$ and the following wave function $\chi_{MA}$:

$$\chi_{MA}^{1/2} = \frac{1}{\sqrt{2}} |(\uparrow\downarrow - \downarrow\uparrow) \uparrow\rangle, \quad \chi_{MA}^{-1/2} = \frac{1}{\sqrt{2}} |(\uparrow\downarrow - \downarrow\uparrow) \downarrow\rangle. \qquad (6.7)$$

Now combine spin 1 with the remaining spin 1/2. For this case one gets $S = 3/2$ and $S = 1/2$. The spin wave functions for this case are given in Table 6.1.

In table 6.1, the numerical coefficients are Clebsch-Gordon coefficients in combining spin 1 and spin 1/2.

The state function for the completely symmetric representation **56** of SU(6) can be written:

$$\Phi_S \, \chi_S + \frac{1}{\sqrt{2}} \left[ \Phi_{MS} \, \chi_{MS} + \Phi_{MA} \, \chi_{MA} \right], \qquad (6.8)$$

where [cf. Eqs. (5.63), (5.58) and (5.59)]

$$\Phi_S = |T_{ijk}\rangle \qquad\qquad\qquad\qquad (6.9a)$$

Table 6.1   Spin wave functions for $S = 3/2$ and $S = 1/2$ resulting in the combination of spin 1 and spin 1/2.

| $S_z = 3/2$ | $S_z = 1/2$ | $S_z = -1/2$ | $S_z = -3/2$ |
|---|---|---|---|
| $\chi_S^{3/2}$ Symmetric:<br><br>$\lvert\uparrow\uparrow\uparrow\rangle$, | $\frac{1}{\sqrt{3}}\lvert\uparrow\uparrow\downarrow + \uparrow\downarrow\uparrow$ $+ \downarrow\uparrow\uparrow\rangle$, | $\frac{1}{\sqrt{3}}\lvert\downarrow\downarrow\uparrow + \downarrow\uparrow\downarrow$ $+ \uparrow\downarrow\downarrow\rangle$, | $\lvert\downarrow\downarrow\downarrow\rangle$ |
| $\chi_{MS}^{1/2}$ (Mixed symmetry: symmetric in 1 and 2) | $\frac{1}{\sqrt{3}}\left\lvert -\frac{(\uparrow\downarrow+\downarrow\uparrow)\uparrow}{\sqrt{2}}\right.$ $+\sqrt{2}\uparrow\uparrow\downarrow\bigr\rangle$, | $\frac{1}{\sqrt{3}}\left\lvert \frac{(\uparrow\downarrow+\downarrow\uparrow)\downarrow}{\sqrt{2}}\right.$ $-\sqrt{2}\downarrow\downarrow\uparrow\bigr\rangle$ | |

$$\Phi_{MA} = \overline{B}^i_j \lvert 0\rangle, \quad \Phi_{MS} = -\overline{B'}^i_j \lvert 0\rangle. \tag{6.9b}$$

The spin state functions $\chi_S$, $\chi_{MS}$ and $\chi_{MA}$ are given in Table 6.1 and Eq. (6.7). Using Tables 5.4 and 6.1, we can write the state function $\Phi_S \chi_S$ for the decuplet. For example,

$$
\begin{aligned}
&\lvert\Delta^0,\ S_z = 1/2\rangle \\
&= \frac{1}{\sqrt{3}}\,(u\,u\,d + u\,d\,u + d\,u\,u) \\
&\quad \times \frac{1}{\sqrt{3}}\,(\uparrow\uparrow\downarrow + \uparrow\downarrow\uparrow + \downarrow\uparrow\uparrow) \\
&= \frac{1}{3}\left[
\begin{array}{l}
u^\uparrow u^\uparrow d^\downarrow + u^\uparrow d^\uparrow u^\downarrow + d^\uparrow u^\uparrow u^\downarrow \\
+u^\uparrow u^\downarrow d^\uparrow + u^\uparrow d^\downarrow u^\uparrow + d^\uparrow u^\downarrow u^\uparrow \\
+u^\downarrow u^\uparrow d^\uparrow + u^\downarrow d^\uparrow u^\uparrow + d^\downarrow u^\uparrow u^\uparrow
\end{array}
\right].
\end{aligned}
\tag{6.10}
$$

$$
\left\lvert\Delta^+,\ S_z = \frac{3}{2}\right\rangle = \frac{1}{\sqrt{3}}\,(u\,u\,d + u\,d\,u + d\,u\,u)\lvert\uparrow\uparrow\uparrow\rangle \tag{6.11}
$$

$$
= \frac{1}{\sqrt{3}}\,(u^\uparrow u^\uparrow d^\uparrow + u^\uparrow d^\uparrow u^\uparrow + d^\uparrow u^\uparrow u^\uparrow)
$$

$$
\left\lvert\Omega^-,\ S_z = \frac{3}{2}\right\rangle = s\,s\,s\lvert\uparrow\uparrow\uparrow\rangle = \lvert s^\uparrow s^\uparrow s^\uparrow\rangle. \tag{6.12}
$$

Similarly one can calculate all the other states.

For baryon octet $1/2^+$ states, we explicitly calculate the state

$|p, \ S_z = 1/2\rangle$. It is given by

$$|p, \ S_z = 1/2\rangle = \frac{-1}{\sqrt{2}} \left\{ [(u \ d + d \ u) \ u - 2u \ u \ d] - \frac{1}{\sqrt{6}} [(\uparrow\downarrow + \downarrow\uparrow) \uparrow - 2 \uparrow\uparrow\downarrow] \right.$$

$$\left. + \left(1/\sqrt{2}\right) \left[ (u \ d - d \ u) u \ \frac{1}{\sqrt{2}} [(\uparrow\downarrow - \downarrow\uparrow) \uparrow] \right] \right\}$$

$$= \frac{1}{6\sqrt{2}} \begin{bmatrix} \left(u^\uparrow d^\downarrow + u^\downarrow d^\uparrow\right) u^\uparrow - 2u^\uparrow d^\uparrow u^\downarrow \\ + \left(d^\uparrow u^\downarrow + d^\downarrow u^\uparrow\right) u^\uparrow - 2d^\uparrow u^\uparrow u^\downarrow - 2u^\uparrow u^\downarrow d^\uparrow \\ -2u^\downarrow u^\uparrow d^\uparrow + 4u^\uparrow u^\uparrow d^\downarrow + 3u^\uparrow d^\downarrow u^\uparrow \\ -3u^\downarrow d^\uparrow u^\uparrow - 3d^\uparrow u^\downarrow u^\uparrow + 3d^\downarrow u^\uparrow u^\uparrow \end{bmatrix}$$

$$= \frac{1}{\sqrt{18}} \begin{bmatrix} 2u^\uparrow d^\downarrow u^\uparrow + 2u^\uparrow u^\uparrow d^\downarrow + 2d^\downarrow u^\uparrow u^\uparrow \\ -u^\uparrow u^\downarrow d^\uparrow - u^\uparrow d^\uparrow u^\downarrow - u^\downarrow d^\uparrow u^\uparrow \\ -d^\uparrow u^\downarrow u^\uparrow - d^\uparrow u^\uparrow u^\downarrow - u^\downarrow u^\uparrow d^\uparrow \end{bmatrix}. \qquad (6.13)$$

The rest of the states can also be calculated in a similar way. (See problem 6.3.)

Finally we give the state functions for the representations **70**, **70** and **20**. They are as follows:

Representation **70** : MS

$\Phi_S \ \chi_{MS} :$      **(10, 2)**      : 20

$\Phi_{MS} \ \chi_S :$      **(8, 4)**      : 32

$\frac{1}{\sqrt{2}} \left(-\Phi_{MS} \ \chi_{MS} + \Phi_{MA} \ \chi_{MA}\right) : (\mathbf{8, 2}) : 16$

$\Phi_A \ \chi_{MA} : (\mathbf{1, 2}) : 2, \qquad \Phi_A = |\Lambda_1^0\rangle [\text{cf. Eq. (5.62)}].$

Representation **70** : MA

$\Phi_S \ \chi_{MA} :$      **(10, 2)**      : 20

$\Phi_{MA} \ \chi_S :$      **(8, 4)**      : 32

$\frac{1}{\sqrt{2}} \left(\Phi_{MS} \ \chi_{MA} + \Phi_{MA} \ \chi_{MS}\right) : (\mathbf{8, 2}) : 16$

$\Phi_A \ \chi_{MA} : (\mathbf{1, 2}) : 2$

Representation **20**

$\Phi_A \ \chi_S : (\mathbf{1, 4}) : 4$

$\frac{1}{\sqrt{2}} \left(\Phi_{MS} \ \chi_{MA} - \Phi_{MA} \ \chi_{MS}\right) : (\mathbf{8, 2}) : 16$

We will not give the detailed identification for these states.

## 6.2   Magnetic Moments of Baryons

Magnetic moment operator is given by

$$\widehat{\mu} = g\mu_0 \ \mathbf{J} \ / \ \hbar. \qquad (6.14)$$

Define the magnetic moment $\mu$ of a particle of mass $m$:

$$\mu = g\mu_0 J, \qquad (6.15)$$

where

$$\mu_0 = e\hbar/2mc \tag{6.16}$$

and $J$ is the angular momentum which appears in the eigenvalue of $J^2$ which is $J(J+1)\hbar^2$. For electron, $J = 1/2$, $g = -2$, i.e.

$$\mu_e = -e\hbar \,/\, 2m_e c. \tag{6.17}$$

For a spin $1/2$ particle, $\mathbf{J} = 1/2\hbar\sigma$. Thus for a quark, the magnetic moment operator is given by

$$\widehat{\mu}_q = 2Q_q \left( \frac{e\hbar}{2m_q c} \right) \frac{1}{2}\sigma_q$$

$$= \mu_q \sigma_q, \tag{6.18}$$

where

$$\mu_q = Q_q \left( \frac{e\hbar}{2m_q c} \right) \tag{6.19}$$

is the magnetic moment of the quark.

The magnetic moment operator for a baryon of $J^P = 1/2^+$ in the quark model is given by

$$\widehat{\mu}_B = \sum_q \mu_q \sigma_q. \tag{6.20}$$

We need to calculate the expectation value of $\widehat{\mu}_{Bz}$ viz.

$$\mu_B = \langle \widehat{\mu}_{Bz} \rangle = \sum_q \mu_q \langle \sigma_{qz} \rangle. \tag{6.21}$$

Let us explicitly calculate the magnetic moment of the proton. For the proton

$$\widehat{\mu}_{pz} = \mu_u \sigma_{uz} + \mu_d \sigma_{dz} + \mu_u \sigma_{uz}. \tag{6.22}$$

Using the proton state $|p, \sigma_z = 1\rangle$ as given in Eq. (6.13), we have (we will not write $\sigma_z = 1$ explicitly in the state)

$$\widehat{\mu}_{pz} |p\rangle = \frac{1}{\sqrt{18}} \left\{ \begin{array}{l} 2\left(\mu_u - \mu_d + \mu_u\right) u^\uparrow d^\downarrow u^\uparrow + 2\left(\mu_u + \mu_u - \mu_d\right) u^\uparrow u^\uparrow d^\downarrow \\ +2\left(-\mu_d + \mu_u + \mu_u\right) d^\downarrow u^\uparrow u^\uparrow - \left(\mu_u - \mu_u + \mu_d\right) u^\uparrow u^\downarrow d^\uparrow \\ -\left(\mu_u + \mu_d - \mu_u\right) u^\uparrow d^\uparrow u^\downarrow - \left(-\mu_u + \mu_d + \mu_u\right) u^\downarrow d^\uparrow u^\uparrow \\ -\left(\mu_d - \mu_u + \mu_u\right) d^\uparrow u^\downarrow u^\uparrow - \left(\mu_d + \mu_u - \mu_u\right) d^\uparrow u^\uparrow u^\downarrow \\ -\left(-\mu_u + \mu_u + \mu_d\right) u^\downarrow u^\uparrow d^\uparrow \end{array} \right\} \tag{6.23}$$

Hence

$$\mu_p = \langle p | \widehat{\mu}_{pz} | p \rangle$$
$$= \frac{1}{18} \left[ 12 \left( 2\mu_u - \mu_d \right) + 6\mu_d \right]$$
$$= \frac{4}{3}\mu_u - \frac{1}{3}\,\mu_d. \tag{6.24}$$

Similarly, we can also calculate the magnetic moments for the rest of the baryons in the octet.

However, one can use simplified state functions to calculate the magnetic moments. In this calculation the order in which quarks appear is important. For the proton, write the state function

$$|p\rangle = |u\ u\ d\rangle\ \chi_{MS}^{1/2} = |u\ u\ d\rangle \left( -\frac{1}{\sqrt{6}} \right) |[(\uparrow\downarrow + \downarrow\uparrow)\uparrow - 2\uparrow\uparrow\downarrow]\rangle \tag{6.25}$$

$$\sigma_z\,(1)\ |[(\uparrow\downarrow + \downarrow\uparrow)\uparrow - 2\uparrow\uparrow\downarrow]\rangle = |[(\uparrow\downarrow - \downarrow\uparrow)\uparrow - 2\uparrow\uparrow\downarrow]\rangle \tag{6.26a}$$

$$\sigma_z\,(2)\ |[(\uparrow\downarrow + \downarrow\uparrow)\uparrow - 2\uparrow\uparrow\downarrow]\rangle = |[(-\uparrow\downarrow + \downarrow\uparrow)\uparrow - 2\uparrow\uparrow\downarrow]\rangle \tag{6.26b}$$

$$\sigma_z\,(3)\ |[(\uparrow\downarrow + \downarrow\uparrow)\uparrow - 2\uparrow\uparrow\downarrow]\rangle = |[(\uparrow\downarrow + \downarrow\uparrow)\uparrow + 2\uparrow\uparrow\downarrow]\rangle. \tag{6.26c}$$

Hence

$$\widehat{\mu}_{pz}\,\chi_{MS}^{1/2} = \left[ \mu_u\sigma_z\,(1) + \mu_u\sigma_z\,(2) + \mu_d\sigma_z\,(3) \right] \chi_{MS}^{1/2}$$
$$= \left( -\frac{1}{\sqrt{6}} \right) \left\{ -4\mu_u\ |\uparrow\uparrow\downarrow\rangle + \mu_d\ |[(\uparrow\downarrow + \downarrow\uparrow)\uparrow + 2\uparrow\uparrow\downarrow]\rangle \right\}. \tag{6.27}$$

Therefore,

$$\mu_p = \langle\widehat{\mu}_{pz}\rangle_p = \frac{1}{6} \left[ 8\mu_u + (1 + 1 - 4)\,\mu_d \right]$$
$$= \frac{4}{3}\mu_u - \frac{1}{3}\,\mu_d$$

For $|\Lambda^0\rangle$, the simplified state function is given by

$$|\Lambda^0\rangle = -|u\ d\ s\rangle\ \chi_{MA}^{1/2} = -|u\ d\ s\rangle\ \frac{1}{\sqrt{2}}\,|(\uparrow\downarrow - \downarrow\uparrow)\uparrow\rangle \tag{6.28}$$

$$\widehat{\mu}_{\Lambda z} = \mu_u\sigma_z\,(1) + \mu_d\sigma_z\,(2) + \mu_s\sigma_z\,(3) \tag{6.29}$$

$$\widehat{\mu}_{\Lambda z}\chi_{MA}^{1/2} = \frac{1}{\sqrt{2}} \left[ \begin{array}{c} \mu_u\ |(\uparrow\downarrow + \downarrow\uparrow)\uparrow\rangle + \mu_d\ |(-\uparrow\downarrow - \downarrow\uparrow)\uparrow\rangle \\ + \mu_s\ |(\uparrow\downarrow - \downarrow\uparrow)\uparrow\rangle \end{array} \right]. \tag{6.30}$$

Therefore,

$$\mu_\Lambda = \langle \widehat{\mu}_{\Lambda z} \rangle_\Lambda = \frac{1}{2} \left[ 0 + 2 \, \mu_s \right] = \mu_s. \tag{6.31}$$

For $|\Sigma^0\rangle$, the simplified state function is

$$|\Sigma^0\rangle = |u \, d \, s\rangle \, \chi_{MS}^{1/2} = |u \, d \, s\rangle \left( -\frac{1}{\sqrt{6}} \right) |[(\uparrow\downarrow + \downarrow\uparrow) \uparrow -2 \uparrow\uparrow\downarrow]\rangle \tag{6.32}$$

$$\widehat{\mu}_{\Sigma^0 z} \chi_{MS}^{1/2} = -\frac{1}{\sqrt{6}} \{ \mu_u \, |[(\uparrow\downarrow - \downarrow\uparrow) \uparrow -2 \uparrow\uparrow\downarrow]\rangle$$
$$+ \mu_d \, |[(-\uparrow\downarrow + \downarrow\uparrow) \uparrow -2 \uparrow\uparrow\downarrow]\rangle$$
$$+ \mu_s \, |[(\uparrow\downarrow + \downarrow\uparrow) \uparrow +2 \uparrow\uparrow\downarrow]\rangle \}. \tag{6.33}$$

Therefore,

$$\mu_{\Sigma^0} = \langle \widehat{\mu}_{\Sigma^0 z} \rangle_{\Sigma^0} = \frac{1}{6} \left[ \mu_u \, (4) + \mu_d \, (4) + \mu_s \, (1 + 1 - 4) \right]$$
$$= \frac{2}{3} \mu_u + \frac{2}{3} \mu_d - \frac{1}{3} \mu_s. \tag{6.34}$$

From Eqs. (6.30) and (6.32), we get

$$\mu_{\Sigma^0 - \Lambda^0} = \langle \Sigma^0 | \, \widehat{\mu}_{\Lambda z} \, | \Lambda \rangle$$
$$= \frac{1}{\sqrt{12}} \left[ 2\mu_u - 2\mu_d \right]$$
$$= \frac{1}{\sqrt{3}} \left[ \mu_u - \mu_d \right] = \mu_{\Lambda^0 - \Sigma^0}. \tag{6.35}$$

The magnetic moments for the rest of the baryons in the octet, can be written from Eq. (6.23) as follows:

$$\mu_n : \qquad (\mu_u \longleftrightarrow \mu_d) = \frac{4}{3} \, \mu_d - \frac{1}{3} \, \mu_u. \tag{6.36}$$

$$\mu_{\Sigma^+} : \qquad (\mu_d \longleftrightarrow \mu_s) = \frac{4}{3} \, \mu_u - \frac{1}{3} \, \mu_s. \tag{6.37}$$

$$\mu_{\Sigma^-} : \qquad (\mu_u \longleftrightarrow \mu_d \text{ in } \mu_{\Sigma^+}) = \frac{4}{3} \, \mu_d - \frac{1}{3} \, \mu_s. \tag{6.38}$$

$$\mu_{\Xi^0} : \qquad (\mu_d \longleftrightarrow \mu_s \text{ in } \mu_n) = \frac{4}{3} \, \mu_s - \frac{1}{3} \, \mu_u. \tag{6.39}$$

$$\mu_{\Xi^-} : \qquad (\mu_u \longleftrightarrow \mu_d \text{ in } \mu_{\Xi^0}) = \frac{4}{3} \, \mu_s - \frac{1}{3} \, \mu_d. \tag{6.40}$$

In order to compare these magnetic moments with their experimental values, we introduce the following quantities:

$$\mu_0 = \frac{e\hbar}{2\overline{m}c}, \qquad \overline{m} = \frac{m_u + m_d}{2}. \tag{6.41}$$

We can write

$$\mu_0 = \mu_N \left( \frac{m_p}{\overline{m}} \right), \tag{6.42a}$$

where

$$\mu_N = \frac{e\hbar}{2m_p c}. \tag{6.42b}$$

Here $\mu_N$ is the nucleon magneton. Thus we can write the magnetic moments of $u$, $d$ and $s$ quarks in terms of $\mu_N$ :

$$\mu_u = \frac{2}{3} \left( \frac{m_p}{m_u} \right) \mu_N \tag{6.43a}$$

$$\mu_d = -\frac{1}{3} \left( \frac{m_p}{m_d} \right) \mu_N \tag{6.43b}$$

$$\mu_s = -\frac{1}{3} \left( \frac{m_p}{m_s} \right) \mu_N. \tag{6.43c}$$

We will now assume isospin symmetry, i.e. will take $m_u = m_d = \overline{m}$. We see that there are two unknown numbers $\overline{m}$ and $m_s$. These numbers will be fixed from the experimental values of $\mu_p$ and $\mu_\Lambda$. From Eqs. (6.24), (6.31) and (6.42b), we obtain

$$\mu_p = \frac{m_p}{\overline{m}} \, \mu_N = 2.793 \, \mu_N. \tag{6.44}$$

$$\mu_\Lambda = -\frac{1}{3} \frac{m_p}{m_s} \, \mu_N = -0.613 \, \mu_N. \tag{6.45}$$

On the right-hand side of Eqs. (6.44) and (6.45), we have put their experimental values. From Eqs. (6.44) and (6.45), we get

$$\overline{m} = m_u = m_d \approx 336 \text{ MeV} \tag{6.46a}$$

$$m_s \approx 510 \text{ MeV}. \tag{6.46b}$$

It is interesting to compare these values with those obtained from the naive quark model. Now proton is made up of $uud$ quarks and $\Lambda$ is made up of $uds$ quarks:

$$3\,\overline{m} = m_p, \qquad \overline{m} \approx 313 \text{ MeV} \tag{6.47a}$$

$$2\,\overline{m} + m_s = m_\Lambda, \qquad m_s = \frac{3m_\Lambda - 2m_p}{3} = 490 \text{ MeV}. \tag{6.47b}$$

The masses of $u$, $d$ and $s$ quarks given in Eqs. (6.46a) and (6.46b) are called the constituent quark masses. These are effective masses of the quarks confined in a hadron. The constituent quark masses are quite different from those appearing in the Hamiltonian or the Lagrangian. These masses are called current quark masses [see Chap. 11].

Using Eqs. (6.43) and (6.44), we get

$$\mu_u \approx 1.862 \; \mu_N \tag{6.48a}$$

$$\mu_d \approx -0.991 \; \mu_N \tag{6.48b}$$

$$\mu_s \approx -0.613 \; \mu_N. \tag{6.48c}$$

Using Eqs. (6.48) the predictions of quark model for the baryon magnetic moments as given in Eqs. (6.24), (6.31) and (6.34)-(6.40) are tabulated in Table 6.2 along with their experimental values. If we put $m_u = m_d = m_s$, in Eqs. (6.24), (6.31) and (6.34)-(6.40), one gets the SU(6) predictions

$$\mu_p = \mu_{\Sigma^+} = -\frac{3}{2} \, \mu_n = -3 \, \mu_\Lambda = -3 \, \mu_{\Sigma^-} = 3 \, \mu_{\Sigma^0}$$

$$= -\frac{3}{2} \, \mu_{\Xi^0} = -3 \, \mu_{\Xi^-} = \sqrt{3} \, \mu_{\Sigma^0 - \Lambda^0}. \tag{6.49}$$

We conclude this section by the following observations:
1) The quark model is simpler than SU(6).
2) It is more predictive than SU(6). It gives information about the scale of magnetic moments.
3) It gives good account of some corrections to SU(6) relations.
From Table 6.2, we see the agreement between quark model values of baryon magnetic moments and their experimental values is not bad.

Table 6.2 Magnetic moments of baryons: Quark model predictions and comparison with their experimental values.

| Magnetic moment | Quark model values (in $\mu_N$) | Experimental values (in $\mu_N$) |
|---|---|---|
| $\mu_p$ | input | 2.793 |
| $\mu_n$ | $-1.862 : \frac{4}{3}\mu_d - \frac{1}{3}\mu_u$ | $-1.913$ |
| $\mu_\Lambda$ | input | $-0.613 \pm 0.004$ |
| $\mu_{\Sigma^+}$ | $2.687 : \frac{4}{3}\mu_u - \frac{1}{3}\mu_s$ | $2.458 \pm 0.010$ |
| $\mu_{\Sigma^-}$ | $-1.037 : \frac{4}{3}\mu_d - \frac{1}{3}\mu_s$ | $-1.160 \pm 0.025$ |
| $\mu_{\Sigma^0}$ | $0.785 : \frac{2}{3}\mu_u + \frac{2}{3}\mu_d - \frac{1}{3}\mu_s$ | – |
| $\mu_{\Xi^0}$ | $-1.438 : \frac{4}{3}\mu_s - \frac{1}{3}\mu_u$ | $-1.250 \pm 0.014$ |
| $\mu_{\Xi^-}$ | $-0.507 : \frac{4}{3}\mu_s - \frac{1}{3}\mu_d$ | $-0.6507 \pm 0.0025$ |
| $\mu_{\Sigma^0 - \Lambda^0}$ | $1.647 : \frac{1}{\sqrt{3}}(\mu_u - \mu_d)$ | $1.61 \pm 0.08$ |

## 6.3    Radiative Decays of Vector Mesons

For a quark and antiquark system, the Hamiltonian is given by

$$
H = \frac{\widehat{\mathbf{p}}_1^2}{2m_1} + \frac{\widehat{\mathbf{p}}_2^2}{2m_2} + V(\mathbf{r}_1, \mathbf{r}_2)
$$
$$
= \frac{(\sigma_1 \cdot \widehat{\mathbf{p}}_1)(\sigma_1 \cdot \widehat{\mathbf{p}}_1)}{2m_1} + \frac{(\sigma_2 \cdot \widehat{\mathbf{p}}_2)(\sigma_2 \cdot \widehat{\mathbf{p}}_2)}{2m_2} + V(\mathbf{r}_1, \mathbf{r}_2). \qquad (6.50)
$$

To introduce electromagnetic interaction, we make the gauge invariant replacement

$$
\widehat{\mathbf{p}} \to \widehat{\mathbf{p}} - eQ\mathbf{A}(\mathbf{r}, t), \qquad (6.51)
$$

where $\mathbf{A}(\mathbf{r}, t)$ is the electromagnetic field, $eQ$ is the electric charge of the quark and $\widehat{\mathbf{p}} = -i\nabla$. From Eqs. (6.50) and (6.51), we get

$$
H = \sum_{i=1}^{2} \left[ \begin{array}{c} \frac{1}{2m_i}\widehat{\mathbf{p}}_i^2 - \frac{e\,Q_i}{2m_i}(\sigma_i \cdot \widehat{\mathbf{p}}_i)(\sigma_i \cdot \mathbf{A}(\mathbf{r}_i, t)) \\ -\frac{e\,Q_i}{2m_i}(\sigma_i \cdot \mathbf{A}(\mathbf{r}_i, t))\sigma_i \cdot \widehat{\mathbf{p}}_i + \frac{e^2\,Q_i^2}{2m_i}\mathbf{A}^2(\mathbf{r}_i, t) \end{array} \right] + V(\mathbf{r}_1, \mathbf{r}_2).
$$
$$
(6.52)
$$

Using the identities

$$
(\sigma \cdot \widehat{\mathbf{p}})(\sigma \cdot \mathbf{A}) + (\sigma \cdot \mathbf{A})(\sigma \cdot \widehat{\mathbf{p}}) = \widehat{\mathbf{p}} \cdot \mathbf{A} + \mathbf{A} \cdot \widehat{\mathbf{p}} + i\sigma \cdot (-i\nabla \times \mathbf{A}) \quad (6.53a)
$$

$$
\widehat{\mathbf{p}} \cdot \mathbf{A} = \mathbf{A} \cdot \widehat{\mathbf{p}} - \nabla \cdot \mathbf{A} \qquad (6.53b)
$$

and the gauge condition

$$
\nabla \cdot \mathbf{A} = 0, \qquad (6.53c)
$$

Eq. (6.52) becomes

$$
H = H_0 + H_{\text{int}}, \qquad (6.54)
$$

where

$$
H_0 = \sum_{i=1}^{2} \frac{1}{2m_i}\mathbf{p}_i^2 + V(\mathbf{r}_1, \mathbf{r}_2) \qquad (6.55)
$$

$$
H_{\text{int}} = -e\sum_i \frac{Q_i}{2m_i} \left[ 2\mathbf{A}(\mathbf{r}_i, t) \cdot \widehat{\mathbf{p}}_i + i\sigma_i(-i\nabla_i \times \mathbf{A}(\mathbf{r}_i, t)) \right]. \qquad (6.56)
$$

In Eq. (6.56), the second order term $e^2$ has been neglected. Now [see Appendix A]

$$
\mathbf{A}(\mathbf{r}, t) = \frac{1}{\sqrt{2V}} \sum_{\mathbf{k}'} \sum_{\lambda'} \frac{1}{\sqrt{\omega'}} \left[ \begin{array}{c} \varepsilon^{\lambda'} a_{\lambda'}(\mathbf{k}')\, e^{i\mathbf{k}' \cdot \mathbf{r}} e^{-i\omega' t} \\ +\varepsilon^{*\lambda'} a_{\lambda'}^{\dagger}(\mathbf{k}')\, e^{-i\mathbf{k}' \cdot \mathbf{r}} e^{i\omega' t} \end{array} \right] \qquad (6.57)
$$

where $\varepsilon^{\lambda'}$ is the polarization vector, $a_{\lambda'}(\mathbf{k}')$ and $a_{\lambda'}^{\dagger}(\mathbf{k}')$ are the annihilation and creation operators for the photon respectively. They satisfy the commutation relation

$$\left[a_\lambda(\mathbf{k})\ a_{\lambda'}^{\dagger}(\mathbf{k}')\right] = \delta_{\lambda\lambda'}\ \delta(\mathbf{k} - \mathbf{k}').\tag{6.58}$$

Let us now consider the emission of a photon viz the process

$$a \to b + \gamma.\tag{6.59}$$

We note that

$$a_\lambda(\mathbf{k})\,|a\rangle = 0\tag{6.60a}$$

$$\langle b\ \gamma| = \langle b|\,a_\lambda(\mathbf{k})\tag{6.60b}$$

$$\langle b\ \gamma|\,a_{\lambda'}^{\dagger}(\mathbf{k}') = \langle b\ |\,a_\lambda(\mathbf{k})\ a_{\lambda'}^{\dagger}(\mathbf{k}')$$
$$= \langle b\ |\left[\delta_{\lambda\lambda'}\ \delta(\mathbf{k} - \mathbf{k}') - a_{\lambda'}^{\dagger}(\mathbf{k}')\ a_\lambda(\mathbf{k})\right].$$
$$\tag{6.60c}$$

It is clear from Eqs. (6.60a), (6.60b), and (6.60c) that only second half of Eq. (6.57) contributes and the matrix elements for the process (6.59) are given by

$$H_{ba} = -e\sum_i \langle b\ |\,\frac{Q_i}{2m_i}\,\frac{1}{\sqrt{2V\omega}}e^{-i\mathbf{k}\cdot\mathbf{r}_i}$$
$$\times \left[2\varepsilon^{*\lambda}\cdot\widehat{\mathbf{p}}_i - i\sigma_i\cdot(\mathbf{k}\times\varepsilon^{*\lambda})\right]|a\rangle\ e^{i\omega t},\tag{6.61}$$

where we have used

$$-\nabla\times\mathbf{A} \propto (-i)^2\ \mathbf{k}\times\varepsilon^{\lambda*}.\tag{6.62}$$

In Eq. (6.61), the term with $2\,\varepsilon^{*\lambda}\cdot\widehat{\mathbf{p}}_i$ gives the electric transition and the term $\sigma_i\left(\mathbf{k}\times\varepsilon^{*\lambda'}\right)$ gives the magnetic transition.

Making the dipole approximation so that in the expansion

$$e^{-i\mathbf{k}\cdot\mathbf{r}_i} = 1 - i\,\mathbf{k}\cdot\mathbf{r}_i + \cdots,\tag{6.63}$$

we retain only the first term. Then

$$H_{ba}^{E_1} = -e\sum_i \frac{1}{\sqrt{2V\omega}}\,\langle b|\,Q_i\frac{\widehat{\mathbf{p}}_i}{m_i}\,|a\rangle\cdot\varepsilon^{\lambda*}\ e^{i\omega t}.\tag{6.64}$$

Now

$$i\frac{\widehat{\mathbf{p}}_i}{m_i} = [\mathbf{r}_i,\ H_0] + 0(e).\tag{6.65}$$

We now choose the center-of-mass (c.m.) frame and introduce

$$\mathbf{r} = \mathbf{r}_1 - \mathbf{r}_2 \tag{6.66a}$$

$$\mathbf{R} = \frac{m_1\mathbf{r}_1 + m_2\mathbf{r}_2}{m_1 + m_2} \tag{6.66b}$$

$$\frac{1}{\mu} = \frac{1}{m_1} + \frac{1}{m_2}. \tag{6.66c}$$

In the c.m. frame $\mathbf{R} = 0$, so that

$$[\mathbf{R},\ H] = 0. \tag{6.67}$$

Therefore, Eqs. (6.64)-(6.67):

$$H_{ba}^{E_1} = \frac{ie\mu\omega}{\sqrt{2V\omega}} \left[ \langle b| \left( \frac{Q_1}{m_1} - \frac{Q_2}{m_2} \right) \mathbf{r} |a\rangle \cdot \varepsilon^{\lambda^*} \right] e^{i\omega t}, \tag{6.68}$$

where we have used the fact that $|a\rangle$ and $|b\rangle$ are eigenstates of $H_0$ with eigenvalues $E_a$ and $E_b$:

$$H_0 |a\rangle = E_a |a\rangle, \tag{6.69a}$$

$$H_0 |b\rangle = E_b |a\rangle, \tag{6.69b}$$

$$E_a - E_b = \omega. \tag{6.69c}$$

We shall make use of Eq. (6.68) later. Here we consider the magnetic transition in dipole approximation, i.e. allowed M1 transition. For M1 transition we get from Eq. (6.61)

$$H_{ba}^{M_1} = \frac{i\,e}{\sqrt{2V\omega}}\ \langle b| \sum_i \frac{Q_i}{2m_i}\sigma_i \cdot \left( \mathbf{k} \times \varepsilon^{\lambda^*} \right) |a\rangle\ e^{i\omega t}. \tag{6.70}$$

We consider the decays of the form

$$V \to P + \gamma$$

$$^3S_1 \to {}^1S_0 + \gamma. \tag{6.71}$$

For the transition $^3S_1 \to {}^1S_0$, $\Delta L = 0$ and there is no change in parity. Therefore, it is M1 transition and the Hamiltonian given in Eq. (6.70) is relevant for the decay (6.71). Now we can write

$$\sigma \cdot \left( \mathbf{k} \times \varepsilon^{\lambda^*} \right) = \sigma_z \left( \mathbf{k} \times \varepsilon^{\lambda^*} \right)_z + \sqrt{2}\,s_+ \left( \mathbf{k} \times \varepsilon^{\lambda^*} \right)_-$$

$$+\sqrt{2}\,s_- \left( \mathbf{k} \times \varepsilon^{\lambda^*} \right)_+, \tag{6.72a}$$

where

$$s_+ = \frac{1}{2} \left( \sigma_x + i\sigma_y \right), \quad s_- = \frac{1}{2} \left( \sigma_x - i\sigma_y \right), \tag{6.72b}$$

$$\left(\mathbf{k} \times \varepsilon^{\lambda^*}\right)_{\pm} = \frac{1}{\sqrt{2}} \left[\left(\mathbf{k} \times \varepsilon^{\lambda^*}\right)_x \pm i \left(\mathbf{k} \times \varepsilon^{\lambda^*}\right)_y\right]. \qquad (6.72c)$$

If we take the matrix elements between $V(S_z = 0)$ and $P(S_z = 0)$, we need to consider $\sigma_{iz} \left(\mathbf{k} \times \varepsilon^{\lambda^*}\right)_z$, i.e. we have to calculate the matrix elements of the operator

$$\widehat{\mu}_z = \sum_i (Q_i/2m_i)\,\sigma_{iz} \qquad (6.73a)$$

between the states

$$|V,\ S_z = 0\rangle \quad \text{and} \quad |P\rangle. \qquad (6.73b)$$

$$|V,\ S_z = 0\rangle : |q_1 \bar{q}_2\rangle \chi_S^0 = \frac{1}{\sqrt{2}} |q_1 \bar{q}_2\rangle \left|\left(\uparrow_{(1)}\downarrow_{(2)} + \downarrow_{(1)}\uparrow_{(2)}\right)\right\rangle. \qquad (6.74)$$

Now

$$\sigma_{1z} |\chi_S^0\rangle = |\chi_A^0\rangle, \qquad \sigma_{2z} |\chi_S^0\rangle = -|\chi_A^0\rangle. \qquad (6.75)$$

We explicitly calculate $\langle\widehat{\mu}_z\rangle$ for the transition $\omega^0 \to \pi^0$

$$|\omega,\ S_z = 0\rangle = \frac{1}{\sqrt{2}} |u\bar{u} + d\bar{d}\,\rangle \chi_S^0.$$

Using Eq. (6.75); one gets

$$\widehat{\mu}_z |\omega,\ S_z = 0\rangle = \frac{2}{3m_u} \frac{1}{\sqrt{2}} |u\bar{u} + d\bar{d}\,\rangle \chi_A^0 - \frac{1}{3m_d} \frac{1}{\sqrt{2}} |d\bar{d}\,\rangle \chi_A^0. \qquad (6.76)$$

Now

$$|\pi^0\rangle = \frac{1}{\sqrt{2}} |\,u\bar{u} - d\bar{d}\,\rangle \chi_A^0. \qquad (6.77)$$

Hence

$$\langle\pi^0|\,\widehat{\mu}_z\,|\omega^0,\ S_z = 0\rangle = \frac{1}{6} \left(\frac{2}{m_u} + \frac{1}{m_d}\right). \qquad (6.78)$$

Similarly one can calculate $\langle\widehat{\mu}_z\rangle$ for other members of the octet. They are given in Table 6.3.

We now calculate the decay rate for $V \to P\gamma$. According to Fermi Golden Rule, the decay rate is given by

$$\Gamma = 2\pi \left|\langle P|\, H_{\text{int}}^{M1} \,|V\rangle\right|^2 \rho(E). \qquad (6.79)$$

If we consider the decay of the vector meson at rest, then

$$E_V = m_V\,; \qquad 0 = \mathbf{k} + \mathbf{k}_P, \quad |\mathbf{k}| = \omega$$

Table 6.3   The matrix elements $\langle P|\, \hat{\mu}_z\, |V,\ S_z = 0\rangle$ for M1 transition for the decay $V \to P + \gamma$.

| Transition | Matrix elements $\langle P|\hat{\mu}_z|V,\ S_z = 0\rangle$ | Transition | Matrix elements $\langle P|\hat{\mu}_z|V,\ S_z = 0\rangle$ |
|---|---|---|---|
| $\omega^0 \to \pi^0$ | $\frac{1}{6}\left(\frac{2}{m_u} + \frac{1}{m_d}\right)$ | $\rho^0 \to \pi^0$ | $\frac{1}{6}\left(\frac{2}{m_u} - \frac{1}{m_d}\right)$ |
| $\rho^\pm \to \pi^\pm$ | $\frac{1}{6}\left(\frac{2}{m_u} - \frac{1}{m_d}\right)$ | $\omega^0 \to \eta_{ns}$ | $\frac{1}{4}\left(\frac{4}{3m_u} - \frac{2}{3m_d}\right)$ |
| $\rho^0 \to \eta_{ns}$ | $\frac{1}{4}\left(\frac{4}{3m_u} + \frac{2}{3m_d}\right)$ | $\phi \to \eta_s$ | $-\frac{1}{3m_s}$ |
| $K^{*+} \to K^+$ | $\frac{1}{6}\left(\frac{2}{m_u} - \frac{1}{m_s}\right)$ | $K^{*0} \to K^0$ | $-\frac{1}{6}\left(\frac{1}{m_s} + \frac{1}{m_d}\right)$ |

$$E_P = \sqrt{\omega^2 + m_P^2}$$

and

$$\rho\,(E) = \int \delta\,(m_V - E_P - \omega)\,\frac{V}{(2\pi)^3}\omega^2 d\omega d\Omega$$

$$= \frac{V\omega^2}{(2\pi)^3}\frac{E_P}{m_V}d\Omega,\quad [m_V = E_P + \omega]. \tag{6.80}$$

Now $\langle P|\, H_{\text{int}}^{M1}\, |V\rangle$ is given in Eq. (6.70) with $a = V$ and $b = P$. In order to calculate $\Gamma$, we have to average over the initial spins of vector meson $V$ and sum over the final spins of the photon. The vector meson has three spin orientations $S_z = +1, 0, -1$. Instead of calculating $\langle H_{\text{int}}^{M1}\rangle$ for $S_z = \pm 1$, $0$ and then taking the average, it is more convenient to calculate $\langle H_{\text{int}}^{M1}\rangle$ for $S_z = 0$ and forget about the spin average. Thus from Eqs. (6.70) and (6.74), we get

$$\left|\langle P|\, H_{\text{int}}^{M1}\, |V\rangle\right|^2 = \sum_{\lambda=1,\,2}\frac{e^2}{2V\omega}\left|\langle P|\,\hat{\mu}_z\,|V,\ S_z = 0\rangle\right|^2\left|\left(\mathbf{k}\times\varepsilon_z^{\lambda*}\right)\right|^2. \tag{6.81}$$

From now on we will not write $S_z = 0$ explicitly in $|V\rangle$. We note the following properties of the polarization vector $\varepsilon^\lambda$:

$$\varepsilon^\lambda \cdot \varepsilon^{\lambda'} = \delta_{\lambda\lambda'}$$

$$\mathbf{k}\cdot\varepsilon^\lambda = 0,\qquad \lambda = 1, 2$$

$$\sum_\lambda \varepsilon_n^{*\lambda}\varepsilon_{n'}^\lambda = \left(\delta_{nn'} - \frac{k_n\,k_{n'}}{k^2}\right),\quad n, n' = 1, 2, 3. \tag{6.82}$$

Using Eq. (6.81), we have

$$\sum_{\lambda=1,\,2}\left|\left(\mathbf{k}\times\varepsilon^{*\lambda}\right)_z\right|^2 = k^2\left(1 - \cos^2\theta\right) \tag{6.83}$$

and

$$\int d\Omega \; k^2 \left(1 - \cos^2 \theta\right) = \frac{8\pi}{3} k^2. \tag{6.84}$$

Hence from Eqs. (6.79), (6.80) and (6.83), one gets

$$\Gamma = \frac{4\alpha}{3} \left|\langle P| \widehat{\mu}_z |V\rangle\right|^2 k^3 \frac{E_P}{m_V}. \tag{6.85}$$

For the decay

$$P \rightarrow V + \gamma, \tag{6.86}$$

we only sum over the spin of vector meson and do not take the average. Hence for this decay, one has

$$\Gamma \left(P \rightarrow V + \gamma\right) = 4\alpha \; \left|\langle V| \widehat{\mu}_z |P\rangle\right|^2 k^3 \frac{E_V}{m_P}. \tag{6.87}$$

We note that a relativistic treatment of the phase space gives the expressions (6.85) and (6.87) without the factor $E_P/m_V$ and $E_V/m_P$ respectively. Thus we can write Eq. (6.85):

$$\Gamma = \frac{4\alpha}{3} \left|\langle P| \widehat{\mu}_z |V\rangle\right|^2 k^3 \Omega^2, \tag{6.88}$$

where $\Omega$ is the overlap integral. It is of order 1, but it may differ from 1, if we take into account the distortion of wave function due to symmetry breaking introduced by the quark mass differences. $\Omega$ may vary from process to process. We assume that this variation is not large. Then we can fix $\Omega$ by using one decay, which we take $\rho^{\pm} \rightarrow \pi^{\pm} + \gamma$. Using Eq. (6.87), Table 6.7, $m_u = m_d = 336$ MeV and $k = 372$ MeV, we get

$$\Gamma \left(\rho^{\pm} \rightarrow \pi^{\pm} + \gamma\right) = (123 \; KeV) \; \Omega^2 \tag{6.89}$$

But

$$\Gamma_{\text{exp}} \left(\rho^{\pm} \rightarrow \pi^{\pm} + \gamma\right) = (67 \pm 7) \; \text{keV}. \tag{6.90}$$

which gives

$$\Omega \approx 0.735. \tag{6.91}$$

Using this value of $\Omega$ and $m_u/m_s = 0.66$, we can compare the predictions of quark model using Table 6.3 and Eq. (6.88) with their experimental values. This is given in Table 6.4.

We notice from Table 6.4 that agreement between the predictions and experiment is only fair. This is understandable since the relativistic corrections become important for hadrons involving light quarks (see, for instance [6]).

Table 6.4   Quark model prediction for $V \to P + \gamma$ with $\Omega = 0.735$.

| Decay | $k$ (in keV) | $\Gamma$ (in keV) | $\Gamma$ (Experimental) (in keV) |
|---|---|---|---|
| $\omega^0 \to \pi^0 + \gamma$ | 380 | $(9.6)\ (123)\,\Omega^2$ $= (638)$ | $703 \pm 23$ |
| $K^{*\pm} \to K^\pm + \gamma$ | 307 | $124\,\Omega^2 = (67)$ | $50.3 \pm 4.4$ |
| $K^{*0} \to K^0 + \gamma$ | 309 | $190\,\Omega^2 = (103)$ | $116 \pm 16$ |

## 6.4   Radiative Decays (Complementary Derivation)

### 6.4.1   *Mesonic Radiative Decays $V = P + \gamma$*

The decay

$$V = P + \gamma$$

is a parity conserving decay. It is a transition from $^3S_1 \to {}^1S_0$; thus it is a M1 transition. The angular momentum and parity conservation implies $l = 1$ in the final state, i.e. it is a $p$-wave decay. Now

$$p = p' + k.$$

Let $\eta$ and $\epsilon$ be the polarization vectors of $V$ and $\gamma$ respectively.

$$p \cdot \eta = 0, \quad k \cdot \epsilon = 0, \quad \mathbf{k} \cdot \epsilon = 0.$$

From Lorentz invarience, the only invariant one can form for this decay is $\epsilon^{\mu\nu\rho\sigma} p_\mu p'_\nu \eta_\rho \epsilon_\sigma$. Hence the T-matrix

$$T = \frac{1}{\left[(2\pi)^{3/2}\right]^3} \frac{F}{\sqrt{2p_0 2p'_0 2k_0}},$$

where

$$F = 2eg_{VP\gamma}\epsilon^{\mu\nu\rho\sigma} p_\mu p'_\nu \eta_\rho \epsilon_\sigma.$$

The decay width

$$\Gamma = \frac{1}{8\pi} \frac{|\mathbf{k}|}{m_V^2} |M|^2$$

$$|M|^2 = \overline{\sum_{pol} |F|^2}$$

In the rest frame of vector meson $V$:   $p = (m_V, 0)$, $\mathbf{p} = -\mathbf{k}$, $p \cdot \eta = 0$, give $\eta^0 = 0$, $\eta_\mu = (0, \eta)$.

Thus in the rest frame of $V$ :

$$\sum_\lambda \eta_\mu^{(\lambda)} \eta_\nu^{*(\lambda)} = -g_{\mu\nu} + \frac{p_\mu k_\nu}{m_V^2}$$

$$\Rightarrow \sum_\lambda \eta_i^{(\lambda)} \eta_j^{*(\lambda)} = \delta_{ij}$$

Hence in the rest frame of $V$ :

$$F = -2m_V e g_{VP\gamma} \epsilon^{ijn} k_i \eta_j \epsilon_n$$

$$= 2m_V e g_{VP\gamma} \epsilon_{ijn} k_i \eta_j \epsilon_n$$

$$= 2m_V e g_{VP\gamma} \mathbf{k} \cdot (\eta \times \epsilon)$$

$$\overline{\sum_{pol}} |F|^2 = \frac{4}{3} e^2 m_V^2 g_{VP\gamma}^2 \epsilon_{ijn} k_i \left( \sum_\lambda \eta_j^{(\lambda)} \eta_{j'}^{(\lambda^*)} \right) \epsilon_{i'j'n'} k_{i'} \left( \sum_\lambda \epsilon_n^{(\lambda)} \epsilon_{n'}^{(\lambda^*)} \right)$$

$$= \frac{4}{3} e^2 m_V^2 g_{VP\gamma}^2 \left( 2\mathbf{k}^2 \right)$$

$$\Gamma = \frac{e^2 g_{VP\gamma}^2}{8\pi} \frac{|\mathbf{k}|}{m_V^2} \left( \frac{8}{3} m_V^2 \mathbf{k}^2 \right)$$

$$= \frac{4\alpha}{3} g_{VP\gamma}^2 |\mathbf{k}|^3 \text{, where } \alpha = \frac{e^2}{4\pi}$$

Since it is an M1 transition, we can put

$$g_{VP\gamma} = \mu = \langle P| \widehat{\mu}_z |V \rangle \,,$$

$$\Gamma = \frac{4\alpha}{3} \mu^2 |\mathbf{k}|^3$$

In quark model:

$$\widehat{\mu}_z = \sum_i \left( \frac{Q_i}{2m_i} \right) \sigma_{ij}$$

## 6.4.2  *Baryonic Radiative Decay*

The radiative decay

$$B\left( \frac{1}{2}^+ \right) \to B'\left( \frac{1}{2}^+ \right) + \gamma$$

is an M1 transition. The transition matrix for this decay is

$$T = \frac{1}{(2\pi)^{a/2}} \sqrt{\frac{mm'}{2k_0 p_0 p_0'}} F$$

where
$$F = \bar{u}(p') \left[ \frac{F_2}{m + m'} i\epsilon_\mu \sigma^{\mu\nu} k_\nu \right] u(p)$$
This follows from Lorentz invariance and electromagnetic current conservation.

The decay width is given by [see Chap. 2]
$$\Gamma = \frac{1}{2\pi} \left( \frac{m'}{m} \right) |\mathbf{k}| |M|^2$$
where
$$|M|^2 = \overline{\sum_{spin}} \sum_{polarization} |F|^2$$
In the rest frame of $B$ :
$$p = (m_B, 0), \qquad \mathbf{p}' = -\mathbf{k} = |\mathbf{k}| \, \mathbf{n}$$
$$F \approx \chi_f^\dagger \frac{iF_2}{2m'} \sigma \cdot (\mathbf{k} \times \epsilon) \chi_i$$
where we have put
$$E' = \sqrt{\mathbf{k}^2 + m'^2} \approx m'$$
Now
$$|M|^2 = \left( \frac{F_2}{2m'} \right)^2 \frac{1}{2} \sum_{pol} Tr \left[ (\sigma \cdot (\mathbf{k} \times \epsilon)) (\sigma \cdot (\mathbf{k} \times \epsilon)) \right]$$
$$= \left( \frac{F_2}{2m'} \right)^2 2 |\mathbf{k}|^2$$
so
$$\Gamma = \frac{1}{\pi} \frac{F_2^2}{4mm'} |\mathbf{k}|^3$$
Since it is an M1 transition
$$\frac{F_2^2}{4mm'} = \left( \mu_{\Sigma^0 \to \Lambda^0}^0 \right)^2 \frac{e^2}{(2m_N)^2}$$
Hence
$$\Gamma = 4\alpha \frac{1}{4m_N^2} \left( \mu_{\Sigma^0 \to \Lambda^0}^0 \right)^2 |\mathbf{k}|^3$$
Using the quark model values: $\mu_{\Sigma^0 \to \Lambda^0}^0 = 1.647$, we get
$$\Gamma \approx 9.08 \times 10^{-3} \quad \text{MeV}$$
$$\tau = \frac{\hbar}{\Gamma}$$
$$\approx 7.3 \times 10^{-20} \quad \text{sec}$$
to be compared with experimental value $(7.4 \pm 0.4) \times 10^{-20}$ sec.

To conclude that quark model gives results compatible with experiments.

## 6.5   Problems

(1) In quark model, using SU(6) wave functions, show that the Fermi matrix element for $n \rightarrow p$ transition:

$$\left\langle p, \; S_z = \frac{1}{2} \middle| \sum_q \tau_q^+ \middle| n, \; S_z = \frac{1}{2} \right\rangle = 1.$$

Find the Gamow-Teller matrix element

$$\left\langle p, \; S_z = \frac{1}{2} \middle| \sum_q \tau_q^+ \sigma_{qz} \middle| n, \; S_z = \frac{1}{2} \right\rangle.$$

(2) Show that the transition moment between $\Delta^+$ and $p$ is given by

$$\left\langle p, \; S_z = \frac{1}{2} \middle| \widehat{\mu}_z \middle| \Delta^+, \; S_z = \frac{1}{2} \right\rangle = \frac{2\sqrt{2}}{3} \mu_p.$$

(3) Write all the SU(6) states for the octet $J^P = \frac{1}{2}^+$. You have to calculate $\left| \Sigma^0, \; S_z = \frac{1}{2} \right\rangle$ and $\left| \Lambda^+, \; S_z = \frac{1}{2} \right\rangle$ the other states can be calculated as follows:

$$\left| n^0, \; S_z = \frac{1}{2} \right\rangle : \quad \text{Change } u \rightarrow d \text{ and overall sign in } \left| p^+ \right\rangle$$

$$\left| -\Sigma^+, \; S_z = \frac{1}{2} \right\rangle : \quad \text{Change } d \rightarrow s \text{ in } \left| p^+ \right\rangle$$

$$\left| \Sigma^-, \; S_z = \frac{1}{2} \right\rangle : \quad \text{Change } u \rightarrow d \text{ and overall sign in } \left| -(\Sigma^+) \right\rangle$$

$$\left| \Xi^0, \; S_z = \frac{1}{2} \right\rangle : \quad \text{Change } d \rightarrow s \text{ and overall sign in } \left| n^0 \right\rangle$$

$$\left| \Xi^-, \; S_z = \frac{1}{2} \right\rangle : \quad \text{Change } u \rightarrow d \text{ in } \left| \Xi^0 \right\rangle$$

(4) Consider $M1$ transition decay

$$\Sigma^0 \rightarrow \Lambda^0 + \gamma.$$

Calculate its decay rate in the non-relativistic quark model and compare it with its experimental value

$$\tau = (7.4 \pm 0.4) \times 10^{-20} \text{ sec.}$$

**Hint:** M1 transition operator is

$$\sum_q \frac{Q_q}{2m_q} \sigma_q \cdot (\mathbf{k} \times \varepsilon^{*\lambda}) = \widehat{\mu}_z \left( \mathbf{k} \times \varepsilon^{*\lambda} \right)_z + \sqrt{2} \, \widehat{\mu}_+ \left( \mathbf{k} \times \varepsilon^{*\lambda} \right)_-$$

$$+ \sqrt{2} \, \widehat{\mu}_- \left( \mathbf{k} \times \varepsilon^{*\lambda} \right)_+ ,$$

where

$$\widehat{\mu}_z = \sum_q \frac{Q_q}{2m_q}\sigma_{qz}, \quad \widehat{\mu}_\pm = \sum_q \frac{Q_q}{2m_q}\sigma_{q\pm}$$

$$\left|\Lambda,\ S_z = \pm\frac{1}{2}\right\rangle = -|u\ d\ s\rangle\ \chi_{MA}^{\pm 1/2}$$

$$\left|\Sigma^0,\ S_z = \pm\frac{1}{2}\right\rangle = |u\ d\ s\rangle\ \chi_{MS}^{\pm 1/2}.$$

(5) Calculate the decay rates for the following decays in quark model:

$$\phi \to \eta + \gamma$$

$$\eta' \to \rho^0 + \gamma$$

$$\to \omega^0 + \gamma$$

and compare them with their experimental values $(54.9 \pm 6.5)$ keV, $(72 \pm 13)$ keV, $(72 \pm 13)$ keV and $(6.5 \pm 1.0)$ keV respectively. [You may take $\eta_8 - \eta_1$ mixing angle as $\theta = -10°$.]

## 6.6   References

1. M. Gell-Mann and Y. Ne'eman, The eightfold way, Benjamin, New York (1964).
2. J. J. J. Kokkedee, The quark model, Benjamin, New York (1969).
3. O. W. Greenberg, Ann. Rev. Nucl. Part. Science 28, 327 (1978).
4. F. E. Close, An introduction to quarks and partons, Academic Press, New York (1979).
5. S. Godfrey, N. Isgur, Phys. Rev. D32, 189 (1985); V. O. Galkin and R. N. Faustauve, Sov. J. Nucl. Phys. 44, 1023 (1986) and Proceedings of the International Seminar "QUARKS' 88", USSR May (1988) p. 264.
6. Particle Data Group, K. Nakamura et al., Journal of Physics, G **37**, 0750212 (2010).

# Chapter 7

# Color, Gauge Principle and Quantum Chromodynamics

## 7.1 Evidence for Color

As we have discussed in the introduction in order that 3 quark wave function of lowest lying baryons satisfy the Pauli principle, each quark flavor carries three color charges, red $(r)$, yellow $(y)$ and blue $(b)$, i.e.

$$q_a \qquad\qquad a = r, y, b.$$

Leptons do not carry color and that is the reason why they do not experience strong interactions. Thus each quark belongs to a triplet representation of color SU(3), which we write as $SU_C(3)$. Now SU(3) has the remarkable property that $\mathbf{3} \otimes \mathbf{3} \otimes \mathbf{3} = \mathbf{10} \oplus \mathbf{8} \oplus \mathbf{8} \oplus \mathbf{1}$ and $\bar{\mathbf{3}} \otimes \mathbf{3} = \mathbf{8} \oplus \mathbf{1}$, so that baryons which are bound states of 3 quarks belong to the singlet representation, which is totally antisymmetric as required by the Pauli principle and mesons which are bound states of $q\bar{q}$ belong to the singlet representation which is totally symmetric. This assignment takes into account the fact that all known hadrons are color singlets. Thus the color is hidden. This is the postulate of color confinement and explains the non-existence of free quarks.

Evidence for color also comes from $\pi^0 \to 2\gamma$ decay. Since $\pi^0$ is bound state of $q\bar{q}$, i.e. $|\pi^0\rangle = \frac{1}{\sqrt{2}} |u\bar{u} - d\bar{d}\rangle$, one can imagine that the decay takes place as shown in Fig. 7.1. The matrix elements $M$ for the $\pi^0$-decay, without and with color [where we have to sum over the 3 colors for the quarks in the above diagrams] are respectively proportional to

$$M \propto \frac{1}{\sqrt{2}} \left[ \left(\frac{2}{3}\right)^2 - \left(-\frac{1}{3}\right)^2 \right] e^2 = \frac{1}{\sqrt{2}\,3} e^2$$

$$M \propto \frac{1}{\sqrt{2}} \left[ \left(\frac{2}{3}\right)^2 - \left(-\frac{1}{3}\right)^2 \right] 3e^2 = \frac{1}{\sqrt{2}} e^2.$$

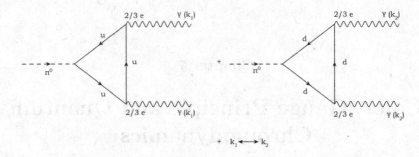

Fig. 7.1   Triangle diagrams for $\pi^0 \to 2\gamma$ through its constituents.

In fact the above quark triangle diagrams predict

$$M = e^2 F = \frac{e^2}{2\pi^2}\frac{S_\pi}{f_\pi} \tag{7.1a}$$

where

$$S_\pi = \begin{cases} \frac{1}{3\sqrt{2}} & \text{, without color} \\ \frac{1}{\sqrt{2}} & \text{with color} \end{cases}, \tag{7.1b}$$

and $f_\pi$ is the pion decay constant and is determined from the decay $\pi^+ \to \mu^+ + \nu_e$ [see Chap. 10]; its value is $130.41 \pm 0.231$ MeV. Hence the decay rate is given by

$$\Gamma(\pi^0 \to 2\gamma) = 4\pi\alpha^2|F|^2\frac{m_{\pi^0}^3}{16}$$

$$= \frac{\alpha^2}{16\pi^3 f_\pi^2}S_\pi^2 m_{\pi^0}^3. \tag{7.2}$$

With $S_\pi = \frac{1}{\sqrt{2}}$, this gives $\Gamma(\pi^0 \to 2\gamma) = 7.77$ eV in very good agreement with the experimental value $\Gamma_{exp} = 7.82 \pm 0.31$. Without color $\Gamma_{th}$ will be a factor of 9 less in complete disagreement with the experimental value.

Furthermore another evidence for color comes from measuring the ratio of $e^-e^+$ annihilation processes

$$R = \frac{\sigma(e^-e^+ \to \text{hadrons})}{\sigma(e^-e^+ \to \mu^-\mu^+)} \tag{7.3}$$

in the large center-of-mass energy $\sqrt{s} = \sqrt{(p_1 + p_2)^2}$ limit, where $p_1$ and $p_2$ are the momenta of $e^-$ and $e^+$ respectively. To the lowest order in electromagnetic interaction, Eq. (A.78) gives in the asymptotic region ($s \gg m_e^2, m_\mu^2$)

$$\sigma(e^-e^+ \to \mu^-\mu^+) = \frac{4\pi}{3}\alpha^2\frac{1}{s}. \tag{7.4}$$

Now for the inclusive process $e^- e^+ \to$ hadrons, we expect this to take place via $e^- e^+ \to q\bar{q}$ and quarks (antiquarks) fragment into hadrons [see Fig. 7.2], so that

$$(e^- e^+ \to \text{hadrons}) = \sum_q \sigma(e^- e^+ \to q\bar{q})$$

where the analogue of Eq. (7.4) gives in the asymptotic region $[s \gg m_e^2, m_q^2]$

$$\sigma(e^- e^+ \to q\bar{q}) = \frac{4\pi}{3}\alpha[3e_q^2]\frac{1}{s}, \qquad (7.5)$$

where $e_q$ (in units of e) are the electric charges of the quarks which enter the photon-$q\bar{q}$ vertex [see Fig. 7.2] and the factor 3 arises because we have to sum over 3 colors for each quark flavor $q$. This gives in the asymptotic region

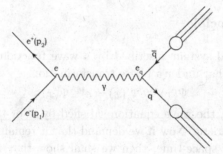

Fig. 7.2   One photon exchange diagram for hadron production in $e^- e^+$ annihilation.

$$R = 3\sum_q e_q^2. \qquad (7.6)$$

For example, above the bottom quark threshold (see Chap. 8) i.e. for $\sqrt{s}$ in the range $2m_b < \sqrt{s} \ll m_Z$ [so that weak interaction effects can be neglected],

$$3\sum_q e_q^2 = 3\left(\frac{4}{9} + \frac{1}{9} + \frac{1}{9} + \frac{4}{9} + \frac{1}{9}\right) = \frac{11}{3},$$

which is confirmed by experimental measurement of $R$ above $\sqrt{s} > 2m_b$ [see Fig. 7.3]. Actually nature has also assigned a more fundamental role to color charges. We know that electromagnetic force is a gauge force; here we postulate that strong force is also a gauge force. In order to discuss the gauge force, we first state the gauge principle.

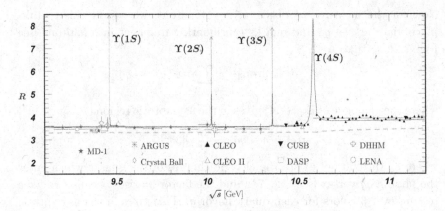

Fig. 7.3   Compilation of R-values from different $e^- e^+$ experiments [29].

## 7.2   Gauge Principle

Suppose a physical system described by a wave function $\Psi(x)$, $x \equiv (t, \mathbf{r})$ has the property that under a phase transformation

$$\Psi(x) \rightarrow \Psi'(x) = e^{ie\Lambda}\Psi(x) \tag{7.7}$$

(with $\Lambda$ constant), the wave equation satisfied by $\Psi$ or the corresponding Lagrangian is invariant. Now if we demand that it remains invariant when $\Lambda$ is a function of space-time, then we shall show that it is necessary to introduce a vector boson which is coupled to a vector current with universal coupling $e$. We call such a phase transformation local gauge transformation and the vector boson associated with it is a mediator of force whose strength is determined by the charge $e$.

This is best illustrated by considering a non-relativistic particle of charge $e$ and mass $m$ described by a complex wave function $\Psi(x)$. Consider a space-time dependent phase transformation given in Eq. (7.7), with $\Lambda$ as a function of $x$ and $e$ the electric charge. For this case the physical law is given by the Schrödinger equation

$$-\frac{1}{2m}\nabla^2\Psi = i\frac{\partial\Psi}{\partial t}. \tag{7.8}$$

This is not invariant under the local gauge transformation (7.7). In order to restore gauge invariance, it is necessary to postulate a vector field $A_\mu \equiv (\phi, \mathbf{A})$ and make the substitutions

$$\nabla \rightarrow \nabla - ie\mathbf{A}$$
$$\frac{\partial}{\partial t} \rightarrow \frac{\partial}{\partial t} + ie\phi$$

or
$$\partial_\mu \to \partial_\mu + ieA_\mu. \tag{7.9}$$
Equation (7.8) now becomes
$$-\frac{1}{2m}(\nabla - ie\mathbf{A})^2 \Psi = i\left(\frac{\partial}{\partial t} + ie\phi\right)\Psi. \tag{7.10}$$
This equation is invariant under the transformation (7.7), provided that $\mathbf{A}$ and $\phi$ simultaneously undergo the transformations:
$$\mathbf{A} \to \mathbf{A} + \nabla\Lambda$$
$$\phi \to \phi - \frac{\partial}{\partial t}\Lambda$$
or
$$A_\mu \to A_\mu - \partial_\mu\Lambda. \tag{7.11}$$
$A_\mu \equiv (\phi, -\mathbf{A})$ are the electromagnetic potentials. From the present point of view, the necessity for the existence of the electromagnetic potential $A_\mu(x)$ is a consequence of assuming invariance under the local gauge transformation. The electromagnetic fields $\mathbf{E}$ and $\mathbf{B}$ are related to the vector potential $A_\mu$ as follows:
$$\mathbf{E} = -\frac{\partial\mathbf{A}}{\partial t} - \nabla\phi$$
$$\mathbf{B} = \nabla \times \mathbf{A}. \tag{7.12}$$
They are clearly invariant under the gauge transformations (7.11).

The Lagrangian density which gives Eq. (7.10) is given by
$$\mathcal{L} = -\frac{1}{2m}\nabla\Psi^* \cdot \nabla\Psi + \frac{1}{2i}\left(\Psi^*\frac{\partial\Psi}{\partial t} - \Psi\frac{\partial\Psi^*}{\partial t}\right)$$
$$-e(\rho\phi - \mathbf{j}\cdot\mathbf{A}) + \frac{1}{2}(\mathbf{E}^2 - \mathbf{B}^2), \tag{7.13}$$
where
$$\rho = \Psi^*\Psi,$$
$$\mathbf{j} = \frac{1}{2im}(\Psi^*\nabla\Psi - (\nabla\Psi^*)\Psi) - \frac{e}{2m}\mathbf{A}\Psi^*\Psi. \tag{7.14a}$$
$\mathcal{L}$ is clearly invariant under the gauge transformations (7.7) and (7.11). $\rho$ and $\mathbf{j}$ satisfy the equation of continuity
$$\frac{\partial\rho}{\partial t} + \nabla\cdot\mathbf{j} = 0. \tag{7.14b}$$
This implies that the charge
$$Q = \int \rho(x)d^3x \tag{7.14c}$$
is conserved. Note also that the last term in Eq. (7.13) can be written in manifestly covariant form $-\frac{1}{4}F_{\mu\nu}F^{\mu\nu}$, where $F_{\mu\nu} = \partial_\mu A_\nu - \partial_\nu A_\mu$ is the electromagnetic field tensor. The term $-\frac{1}{4}F_{\mu\nu}F^{\mu\nu}$ is the Lagrangian density for pure electromagnetic field.

### 7.2.1   *Aharanov and Bohm Experiment*

We now discuss the question of testing the applicability of the gauge principle in electromagnetism. Taking the vector potential $\mathbf{A}$ to be independent of time and putting $V = e\phi$, we try solution of Eq. (7.10) in the following form

$$\Psi(\mathbf{r},\,t) = \Psi^0(\mathbf{r},\,t)e^{i\gamma(\mathbf{r})} \tag{7.15a}$$

where

$$\gamma(\mathbf{r}) = e \int^{\mathbf{r}} \mathbf{A}(\mathbf{r}') \cdot d\ell'. \tag{7.15b}$$

Here $\Psi$ can be regarded as a wave function of a particle that goes from one place to another along a certain route where a field $\mathbf{A}$ is present while $\Psi^0$ is the wave function for the same particle along the same route but with $\mathbf{A} = 0$. It is easy to see that $[\mathbf{A} \to \mathbf{A} + \nabla\gamma]$

$$\mathbf{D}\Psi \equiv (\nabla - ie\mathbf{A})\Psi$$
$$= e^{i\gamma(\mathbf{r})}\nabla\Psi^0$$
$$\mathbf{D}^2\Psi = e^{i\gamma(\mathbf{r})}\nabla^2\Psi^0$$

Thus (7.15a) is a solution of Eq. (7.10) when $\mathbf{A}(\mathbf{r}) \neq 0$ if $\Psi^0(\mathbf{r},t)$ satisfies

$$-\frac{1}{2m}\nabla^2\Psi^0 + V\Psi^0 = i\frac{\partial\Psi^0}{\partial t}. \tag{7.16}$$

The solution (7.15a) has some striking physical consequences as shown in the two-slit electron interferometer experiment proposed by Aharanov and Bohm [Fig. 7.4].

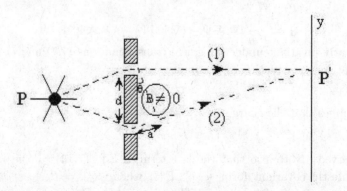

Fig. 7.4   Double slit electron interferometer to test Aharanov-Bohm effect.

In this experiment the magnetic field $\mathbf{B}$ (pointing in a horizontal direction out of the paper) is produced by a long solenoid of small cross-section and is confined to the interior of the solenoid so that the two electron beams (1) and (2) can go above and below the $\mathbf{B} \neq 0$ region but stay within the $\mathbf{B} = 0$ region and finally meet in the interference region $P'$. In the interference region, the wave function for the electron is

$$\Psi = \Psi_1 + \Psi_2$$

so that

$$|\Psi|^2 = |\Psi_1^0|^2 + |\Psi_2^0|^2 + 2|\Psi_1^0||\Psi_2^0|\cos[\gamma_1(\mathbf{r}) - \gamma_2(\mathbf{r})] \qquad (7.17a)$$

where

$$\gamma_1 = \gamma_1^0 + e\int_{(1)P}^{P'} \mathbf{A}(\mathbf{r}') \cdot d\ell'. \qquad (7.17b)$$

$$\gamma_2 = \gamma_2^0 + e\int_{(2)P}^{P'} \mathbf{A}(\mathbf{r}') \cdot d\ell'. \qquad (7.17c)$$

Here $\gamma_1^0$ and $\gamma_2^0$ are the phases of the wave functions $\Psi_1^0$ and $\Psi_2^0$ in the absence of $\mathbf{A}$. The interference pattern is determined by the phase difference

$$\delta(B \neq 0) = \gamma_1 - \gamma_2$$

$$= \gamma_1^0 - \gamma_2^0 + e\oint_C \mathbf{A}(\mathbf{r}') \cdot d\ell'.$$

$$= \delta(B = 0) + \Delta, \qquad (7.18a)$$

where $C$ is the closed path P P' P and

$$\Delta = e\oint_C \mathbf{A}(\mathbf{r}') \cdot d\ell' = e\int_S \mathbf{B} \cdot d\sigma = e\Phi. \qquad (7.18b)$$

In Eq. (7.18a) we have used Stokes theorem and put $\mathbf{B} = \nabla \times \mathbf{A}$ and $\Phi$ is the magnetic flux through the surface $S$ bounded by the closed path $C$. Note the important fact that the phase difference $\Delta$ is gauge invariant since

$$\int \nabla\Lambda . d\mathbf{l} = \int d\Lambda = 0$$

while the individual phases $\gamma_1$ and $\gamma_2$ are not. Note also the remarkable fact that the amount of interference can be controlled by varying magnetic flux even though in the idealized experimental arrangement, electrons never enter the region $B \neq 0$.

Now referring to Fig. 7.4

$$\frac{\text{Phase difference}}{2\pi} = \frac{\text{Path difference}}{\lambda}$$

$$\frac{\delta}{2\pi} = \frac{a}{\lambda} = \frac{1}{\lambda}d\sin\theta \approx \frac{d}{\lambda}\frac{y}{L}$$

where $L$ is the distance of the screen from the slits. Thus from Eq. (7.18a) we see that the diffraction maximum of the interference pattern for $B \neq 0$ is shifted from that for $B = 0$ by the amount $\Delta y$ given by

$$\Delta y = e\Phi \left( \frac{L}{d} \frac{\lambda}{2\pi} \right). \tag{7.19}$$

This shift in the diffraction maximum, being gauge invariant, should be measurable. In fact the existence and magnitude of Aharanov-Bohm effect has been confirmed to within 5% of the theoretical prediction (7.19) by two qualitatively different experimental arrangements - one involving an electron biprism interferometer while the second used a Josephson-junction interferometer.

The following comments are in order.

(i) Measurement of Aharanov-Bohm effect not only verifies the gauge principle in electromagnetism but also quantum mechanics itself since classically the dynamical behavior of electrons is controlled by Lorentz force which is zero when the electrons go through magnetic field free region; yet in quantum mechanics observable effects are seen and depend on the magnetic field in a region inaccessible to the electrons.

(ii) The vector potential **A** rather than the fields plays a crucial role as the basic dynamical variable in quantum mechanics.

(iii) By varying **B** (and hence $\Phi$) we change the relative phase between the contributions from the two paths and move interference pattern up and down. When $\Delta = 2n\pi$ or $\Phi = n\phi_0$ [$\phi_0 = 2\pi/e = 4.135 \times 10^{-7}$ gauss cm$^2$], the interference pattern will return to its initial form, as if there were no field. In other words, an integral multiple of the flux quantum $\phi_0$ will not make any observable difference to the quantum mechanics of the particle.

### 7.2.2  *Gauge Principle for Relativistic Quantum Mechanics*

We now discuss the gauge principle for relativistic  quantum mechanics. The spin 1/2 particle is described by the Dirac equation with the Lagrangian density:

$$\mathcal{L} = \bar{\Psi}(x) i\gamma^\mu \partial_\mu \Psi(x) - m\bar{\Psi}(x)\Psi(x). \tag{7.20}$$

In order that the Lagrangian density $\mathcal{L}$ be invariant under the gauge transformation (7.7), we must introduce a vector field $A_\mu(x)$ satisfying Eq. (7.11) and replace in Eq. (7.20) $\partial_\mu \Psi$ by

$$\partial_\mu \Psi(x) \rightarrow (\partial_\mu + ieA_\mu)\Psi \equiv D_\mu \Psi. \tag{7.21}$$

$D_\mu$ is called the covariant derivative. The gauge invariant Lagrangian density is given by

$$\mathcal{L} = \bar{\Psi}(x)i\gamma^\mu(\partial_\mu + ieA_\mu)\Psi - m\bar{\Psi}(x)\Psi(x) - \frac{1}{4}F^{\mu\nu}F_{\mu\nu} \qquad (7.22)$$

$$F_{\mu\nu} = \partial_\mu A_\nu - \partial_\nu A_\mu. \qquad (7.23)$$

It is easy to see that under the transformation (7.11), $F_{\mu\nu}$ is invariant. Under the transformations (7.7) and (7.11),

$$D_\mu\Psi \to e^{ie\Lambda(x)}D_\mu\Psi, \qquad (7.24)$$

so that $\bar{\Psi}D_\mu\Psi$ is gauge invariant, and so is $m\bar{\Psi}\Psi$. From Eq. (7.22), we see that the interaction of matter field $\Psi$ with the electromagnetic field $A_\mu$ is given by

$$\mathcal{L}_{int} = e\bar{\Psi}\gamma^\mu\Psi A_\mu = -eJ^\mu_{em}A_\mu, \qquad (7.25a)$$

where

$$J^\mu_{em} = e\bar{\Psi}\gamma^\mu\Psi, \ \partial_\mu J^\mu_{em} = 0 \qquad (7.25b)$$

is the electromagnetic current. We conclude that the gauge principle viz the invariance of fundamental physical law under the gauge transformation gives correctly the form of interaction of a charged particle with electromagnetic field. To sum up the consequences of the electromagnetic force as a gauge force are as follows:

(i) It is universal viz any charged particle is coupled with the electromagnetic field A with a universal coupling strength given by e, the electric charge of the particle.

(ii) $J^\mu_{em}$ is conserved.

(iii) The electromagnetic field is a vector and hence the associated quantum, the photon, has spin 1.

(iv) The photon must be massless, since the mass term $\mu^2 A^\mu A_\mu$ is not invariant under the gauge transformation. Thus unbroken gauge symmetry gives rise to long range force mediated by a massless gauge boson i.e. photon.

(v) The covariant derivative $D_\mu$ is an operator whose commutator is

$$[D_\mu, D_\nu] = ieF_{\mu\nu}$$
$$F_{\mu\nu} = \partial_\mu A_\nu - \partial_\nu A_\mu. \qquad (7.26)$$

## 7.3 Non-Abelion Local Gauge Transformations (Yang-Mills)

We now generalize the idea of local gauge transformation when there are more than one type of state. We first extend it to an isospin doublet $\Psi$, a two-component object:

$$\Psi = \begin{pmatrix} \Psi_1 \\ \Psi_2 \end{pmatrix}$$

$\Psi$ belongs to the fundamental representation of of isospin group characterized by SU(2). As discussed in Chap. 5, there exist transformations between different states (cf. Eq. (5.24))

$$\psi_a \rightarrow \psi'_a = U_a^b \psi_b$$
$$= \left[ \delta_a^b + \frac{i}{2} \Lambda_A (\tau_A)_a^b \right] \psi_b, \qquad a, b = 1, 2, \dots \qquad (7.27)$$

where

$$UU^\dagger = 1, \ \det U = 1.$$

Correspondingly there exists a unitary operator

$$U = 1 - i\mathbf{\Lambda} \cdot \mathbf{I} = 1 - i\Lambda_A I_A, \qquad A = 1, 2, 3 \qquad (7.28)$$

$I_A$ are generators of the group SU(2). They are hermitian and traceless and satisfy the commutation relations

$$[I_A, I_B] = i\epsilon_{ABC} I_C \qquad (7.29)$$

For the fundamental representation,

$$\Psi_a \rightarrow \Psi'_a = U\Psi_a U^\dagger$$
$$= \Psi_a - i\Lambda_A [I_A, \Psi_a] \qquad (7.30)$$

so that

$$[I_A, \Psi_a] = -\frac{1}{2} (\tau_A)_a^b \Psi_b$$

$\tau_A$ are Pauli matrices

$$\tau_1 = \begin{pmatrix} 0 & 1 \\ 1 & 0 \end{pmatrix}, \ \tau_2 = \begin{pmatrix} 0 & -i \\ i & 0 \end{pmatrix}, \ \tau_3 = \begin{pmatrix} 1 & 0 \\ 0 & -1 \end{pmatrix} \qquad (7.31)$$

In general exponentiating (7.28),

$$U = e^{-i\Lambda_A I_A}$$

whose matrix representation is

$$U = e^{i\Lambda_A T_A}$$

where for the fundamental representation

$$T_A = t_A = \frac{1}{2}\tau_A$$

$$Tr\,[t_A t_B] = \frac{1}{2}\delta_{AB}$$

$$[t_A, t_B] = i\epsilon_{ABC} t_C$$

For the regular or adjoint representation, the matrix $T_A^G$ associated with each generator is

$$\left(T_A^G\right)_{BC} = -i\epsilon_{ABC}$$

$$[I_A, T_B^G] = i\epsilon_{ABC} T_C^G$$

In contrast to the Abelian case where the gauge field is $A_\mu$, we now have a triplet of gauge fields, $\mathbf{W}_\mu$ which form a basis for 3-dimensional irreducible adjoint representation for $SU(2)$, with the transformation law

$$[I_A, W_{B\mu}] = -\left(T_A^G\right)_{BC} W_{C\mu} = i\epsilon_{ABC} W_{C\mu}, \qquad (7.32)$$

i.e. it transforms in the same way as generators of the group. Thus under SU(2), $W_{A\mu}$ transforms as

$$W_{A\mu} \rightarrow W'_{A\mu} = W_{A\mu} - \epsilon_{ABC}\Lambda_B W_{C\mu}$$

$$\mathbf{W}_\mu \rightarrow \mathbf{W}'_\mu = \mathbf{W}_\mu - \mathbf{\Lambda} \times \mathbf{W}_\mu \qquad (7.33)$$

$$\mathbf{W}_{\mu\nu} \rightarrow \mathbf{W}'_{\mu\nu} = \mathbf{W}_{\mu\nu} - \mathbf{\Lambda} \times \mathbf{W}_{\mu\nu} \qquad (7.34)$$

The last transformation follows since $\mathbf{W}_{\mu\nu} = \partial_\mu \mathbf{W}_\nu - \partial_\nu \mathbf{W}_\mu$ is a vector in SU(2). It is convenient to define the matrices $\Lambda$ and $W_\mu$

$$\frac{1}{2}\tau \cdot \mathbf{\Lambda} = \frac{1}{2}\tau_A \Lambda_A = \Lambda; \qquad \frac{1}{2}\tau \cdot \mathbf{W}_\mu = W_\mu \qquad (7.35)$$

For the local SU(2) gauge transformation, $\Lambda_A$ is a function of $x$.

Thus the gauge invariant Lagrangian, under the infinitesimal local gauge transformation

$$\Psi(x) \rightarrow (1 + \frac{1}{2}i\,\tau \cdot \mathbf{\Lambda}(\mathbf{x}))\Psi(\mathbf{x})$$

$$= (1 + i\,\Lambda(x))\Psi(x) \qquad (7.36)$$

is given by

$$\mathcal{L} = i\bar{\Psi}\gamma^\mu D_\mu \Psi - m\bar{\Psi}\Psi - \frac{1}{2}Tr(W^{\mu\nu}W_{\mu\nu}) \qquad (7.37)$$

provided that $W_\mu$ transforms as

$$\mathbf{W}_\mu \to \mathbf{W}_\mu - \mathbf{\Lambda} \times \mathbf{W}_\mu - \frac{1}{g}\partial_\mu\mathbf{\Lambda}$$

$$W_\mu \to W_\mu + i[\Lambda, \ W_\mu] - \frac{1}{g}\partial_\mu\Lambda \qquad (7.38)$$

Note the appearance of $\frac{1}{g}\partial_\mu\Lambda$ compared to (7.33). Here the covariant derivative is

$$D_\mu = \partial_\mu + igW_\mu = \partial_\mu + \frac{i}{2}g\ \tau.\mathbf{W}_\mu$$

and

$$\mathbf{W}_{\mu\nu} = \partial_\mu\mathbf{W}_\nu - \partial_\nu\mathbf{W}\mu - g\mathbf{W}_\mu \times \mathbf{W}_\nu$$

$$W_{\mu\nu} = \partial_\mu W_\mu - \partial_\nu W_\mu + ig[W_\mu, W_\nu]$$

$$= D_\mu W_\nu - D_\nu W_\mu \qquad (7.39)$$

$$[D_\mu, D_\nu] = igW_{\mu\nu} \qquad (7.40)$$

$$Tr\,[W^{\mu\nu}W_{\mu\nu}] = \frac{1}{2}\mathbf{W}^{\mu\nu} \cdot \mathbf{W}_{\mu\nu} \qquad (7.41)$$

For finite gauge transformation

$$U = e^{i\Lambda(\mathbf{x})}$$

$$\Psi(x) \to U\Psi(x)$$

It follows from Eq. (7.40) that

$$W_{\mu\nu} \to W'_{\mu\nu} = \frac{i}{g}\left[UD_\mu U^\dagger, UD_\nu U^\dagger\right]$$

$$= \frac{i}{g}U\,[D_\mu, D_\nu]\,U^\dagger$$

$$= UW_{\mu\nu}U^\dagger$$

Further

$$Tr\,[W^{\mu\nu}W_{\mu\nu}] \to Tr\,[UW^{\mu\nu}U^\dagger UW_{\mu\nu}U^\dagger]$$

$$= Tr\,[UW^{\mu\nu}W_{\mu\nu}U^\dagger]$$

$$= Tr\,[W^{\mu\nu}U^\dagger UW_{\mu\nu}]$$

$$= Tr\,[W_{\mu\nu}W^{\mu\nu}]$$

$$= Tr\,[W^{\mu\nu}W_{\mu\nu}]$$

where we have used

$$Tr\,[AB] = Tr\,[BA]$$

The mass term $\bar{\Psi}\Psi$ is obviously invariant. Now all we have to show is that $\bar{\Psi}\gamma^\mu D_\mu \Psi$ is invariant.

$$\bar{\Psi}\gamma^\mu D_\mu \Psi \to \bar{\Psi}(x) U^\dagger (x) \gamma^\mu \left[ \partial_\mu U (x) \Psi (x) + U (x) \partial_\mu \Psi (x) \right]$$
$$+ig \left[ \bar{\Psi}(x) U^\dagger (x) \gamma^\mu W'_\mu U (x) \Psi (x) \right]$$

Using

$$\partial_\mu \left[ U^\dagger (x) U (x) \right] = 0,$$

we see that it is invariant provided that

$$W_\mu \to W'_\mu = U W_\mu U^\dagger - \frac{i}{g} U (x) \left( \partial_\mu U^\dagger (x) \right)$$

Under $SU(2)$ gauge transformation the Lagrangian,

$$\mathcal{L} = i\bar{\Psi}\gamma^\mu \partial_\mu \Psi - m\bar{\Psi}\Psi - \frac{1}{4}\mathbf{W}^{\mu\nu} \cdot \mathbf{W}_{\mu\nu} \qquad (7.42)$$

which is invariant under the global guage transformations (7.27) and (7.33), is transformed to $\mathcal{L} + \delta\mathcal{L}$, when $\mathbf{\Lambda}$ is a function of $x$, where

$$\delta\mathcal{L} = -\frac{1}{2}\bar{\Psi}\gamma^\mu \, \tau \, \Psi \cdot \partial_\mu \mathbf{\Lambda} - \partial_\mu \mathbf{\Lambda} \cdot (\mathbf{W}^{\mu\nu} \times \mathbf{W}_\nu)$$
$$= -[\frac{1}{2}\bar{\Psi}\gamma^\mu \, \tau \, \Psi + (\mathbf{W}^{\mu\nu} \times \mathbf{W}_\nu)] \cdot \partial_\mu \mathbf{\Lambda} \qquad (7.43)$$

Now under SU(2) gauge transformation, we have

$$\delta\mathcal{L} = \frac{\partial \mathcal{L}}{\partial \mathbf{\Lambda}} \cdot \delta\mathbf{\Lambda} + \frac{\partial \mathcal{L}}{\partial(\partial_\mu \mathbf{\Lambda})} \cdot \delta(\partial_\mu \mathbf{\Lambda}) \qquad (7.44)$$

Comparison with Eq. (7.43) gives

$$\mathbf{J}^\mu = -\frac{\partial L}{\partial(\partial_\mu \mathbf{\Lambda})} = \frac{1}{2}\bar{\Psi}\gamma^\mu \, \tau \, \Psi + \mathbf{W}^{\mu\nu} \times \mathbf{W}_\nu$$

and invariance of the Lagrangian under the global gauge transformation $(\partial_\mu \mathbf{\Lambda} = 0)$, $\delta\mathcal{L} = 0$ gives the Noether Theorem:

$$\partial_\mu \mathbf{J}^\mu = -\frac{\partial \mathcal{L}}{\partial \mathbf{\Lambda}} = 0 \qquad (7.45)$$

Hence the interaction part of the Lagrangian (7.37) can be written as

$$\mathcal{L}_{int.} = g \, \mathbf{J}^\mu \cdot \mathbf{W}_\mu$$
$$= \frac{1}{2}g \, \bar{\Psi}\gamma^\mu \, \tau \cdot \mathbf{W}_\mu \, \Psi + g \, \mathbf{W}^{\mu\nu} \cdot (\mathbf{W}_\nu \times \mathbf{W}_\mu) \qquad (7.46)$$

## 7.4   Quantum Chromodynamics (QCD)

We now generalize the ideas of Secs. 7.2 and 7.3 to the case where there is more than one type of states, e.g. $q_a$ ($a = 1, 2, 3$) and where there exist transformations [$\mathrm{SU}_C(3)$] between the different states

$$q_a \to q'_a = U_a^b q_b, \tag{7.47a}$$

with

$$U(x) = \exp\left[\frac{i}{2}\lambda_A \Lambda_A(x)\right], \tag{7.47b}$$

$$UU^\dagger = 1, \qquad det\, U = 1$$

and repeated indices imply summation. Here $q_a$ ($a = 1, 2, 3$) for a particular quark flavor $q$ form the fundamental representation of the color SU(3) group and $\lambda_A$, $A = 1 \cdots 8$, are the eight matrix generators of the group $\mathrm{SU}_C(3)$ [see Chap. 5 for the form of these matrices. Although in Chap. 5 we discussed flavor SU(3) but the mathematics is the same].

Each quark flavor carries three color charges: red ($r$), yellow ($y$) and blue ($b$), i.e.

$$q_a; \; a = 1, 2, 3$$

$q_a$ belongs to the triplet representation **3** of $SU_C(3)$. Under infinitesimal $SU_C(3)$ gauge transformation

$$U = 1 - i\Lambda_A F_A, \; A = 1 \cdots 8, \; \Lambda_A = \Lambda_A(x)$$

Now $q_a$ transforms as

$$\begin{aligned}
q_a \to q'_a &= U q_a U^\dagger \\
&= (1 - i\Lambda_A F_A) q_a (1 + i\Lambda_A F_A) \\
&= q_a - i\Lambda_A[F_A, \, q_a] \\
&= q_a + i\Lambda_A(T_A)_a^b q_b
\end{aligned}$$

Hence

$$\begin{aligned}
[F_A, \, q_a] &= -(T_A)_a^b q_b \\
&= -(\frac{\lambda_A}{2})_a^b q_b
\end{aligned}$$

where $\lambda_A$ are Gell-Mann matrices discussed in Chap. 5. Gluons, the gauge vector bosons of $SU_C(3)$ belong to the octet representation, i.e. to the adjoint representation of $SU_C(3)$ :

$$\begin{aligned}
[F_A, \, G_{B\mu}] &= -(T_A)_B^C G_{C\mu} \\
&= i f_{ABC} G_{C\mu}
\end{aligned}$$

Thus

$$UG_{B\mu}U^\dagger = G_{B\mu} + i\Lambda_A (T_A)^C_B G_{C\mu}$$
$$= G_{B\mu} - \Lambda_A f_{BAC} G_{C\mu}$$

Quarks are spin 1/2 particles. The Lagrangian density for free quarks is

$$\mathcal{L} = \bar{q}^a i\gamma^\mu \partial_\mu q_a - \bar{q}^a m q_a, \tag{7.48a}$$

where

$$q_a = \begin{pmatrix} u_a \\ d_a \\ s_a \end{pmatrix} \text{ and } m = \begin{pmatrix} m_u & & \\ & m_d & \\ & & m_s \end{pmatrix} \tag{7.48b}$$

is clearly invariant under the SU(3) transformation (7.47a) with $\Lambda$ constant. If we now require that the Lagrangian density (7.48a) be invariant under the gauge transformation (7.47a) , with $\Lambda(x)$ as function of space-time, then as we have seen in Sec. 7.3, we must replace $\partial_\mu$ by its covariant derivative which in the present case takes the form

$$D_\mu = \left( \partial_\mu - \frac{i}{2} g_s \lambda \cdot \mathbf{G}_\mu \right) = \left( \partial_\mu - \frac{i}{2} g_s \lambda_A G_{A\mu} \right) \tag{7.49}$$

where $g_s$ is a scale parameter, the coupling constant and $\mathbf{G}_{A\mu}$ are vector gauge fields, their number being equal to the generators of $SU_C(3)$ group, namely 8. Then we note the important fact that the covariant derivatives satisfy the commutation relation

$$[D_\mu, D_\nu] = -ig_s \frac{\lambda_B}{2} [\partial_\mu, G_{B\nu}] - ig_s \frac{\lambda_A}{2} [G_{A\nu}, \partial_\nu]$$
$$+ (-ig_s)^2 \left[ \frac{\lambda_A}{2}, \frac{\lambda_B}{2} \right] G_{A\mu} G_{B\nu}$$
$$= -ig_s \frac{\lambda_C}{2} \{ \partial_\mu G_{C\nu} - \partial_\nu G_{C\mu} + g_s f_{ABC} G_{A\mu} G_{B\nu} \}$$
$$= -ig_s \{ \partial_\mu G_\nu - \partial_\nu G_\mu - ig_s [G_\mu, G_\nu] \}$$
$$= -ig_s G_{\mu\nu}, \tag{7.50}$$

where in the matrix notation

$$\frac{1}{2}\lambda \cdot \mathbf{G}_\mu = \frac{1}{2}\lambda_A G_{A\mu} \equiv G_\mu, \tag{7.51a}$$

$$\frac{1}{2}\lambda \cdot \mathbf{\Lambda}(x) = \frac{1}{2}\lambda_A \Lambda_A \equiv \Lambda, \tag{7.51b}$$

$$G_{\mu\nu} \equiv \frac{1}{2}\lambda \cdot \mathbf{G}_{\mu\nu} = \partial_\mu G_\nu - \partial_\nu G_\mu - ig_s \left[G_\mu, G_\nu\right]$$

$$= D_\mu G_\nu - D_\nu G_\mu \tag{7.51c}$$

$$G_{A\mu\nu} = G_{A\nu} - \partial_\nu G_{A\mu} + g_s f_{ABC} G_{B\mu} G_{C\nu} \tag{7.51d}$$

$$Tr(G_{\mu\nu}G_{\mu\nu}) = Tr\left(\frac{1}{2}\lambda_A G_{A\mu\nu}\frac{1}{2}\lambda_B G_{B\mu\nu}\right)$$

$$= \frac{1}{4}Tr(\lambda_A \lambda_B)G_{A\mu\nu}G_{B\mu\nu}$$

$$= \frac{1}{2}\mathbf{G}_{\mu\nu} \cdot \mathbf{G}_{\mu\nu} \tag{7.51e}$$

Note the important fact that $\mathbf{G}_{\mu\nu}$ in Eq. (7.50) provides the generalization of $F_{\mu\nu}$ [cf. Eq. (7.26) in Abelian case] for the present non-Abelian case. The two differ in the appearance of the last term in Eqs. (7.51c) or (7.51d). This is because the gauge fields themselves carry color charges in contrast to photons which are electrically neutral in the electromagnetic case. Now if we replace the Lagrangian density (7.48a) by

$$\mathcal{L} = \bar{q}^a i\gamma^\mu \left(\partial_\mu - \frac{i}{2}g_s\lambda_A G_{A\mu}\right)^b_a q_b - \bar{q}^a m q_a - \frac{1}{4}G_A^{\mu\nu}G_{A\mu\nu} \tag{7.52a}$$

or in the matrix notation by

$$\mathcal{L} = \bar{q}i\gamma^\mu \left(\partial_\mu - ig_s G_\mu\right) q - \bar{q}mq - \frac{1}{2}Tr(G^{\mu\nu}G_{\mu\nu}), \tag{7.52b}$$

then the Lagrangian density (7.52a) is invariant under the infinitesimal gauge transformation [cf. Eq. (7.47b)]

$$q \to \left(1 + \frac{i}{2}\lambda \cdot \Lambda(x)\right) q, \tag{7.53}$$

provided that the vector fields $G_{A\mu}$ undergo the simultaneous transformation

$$G_\mu \to G_\mu + i\left[\Lambda, G_\mu\right] + \frac{1}{g_s}\partial_\mu\Lambda \tag{7.54a}$$

or

$$G_{A\mu} \to G_{A\mu} - f_{ABC}\Lambda_B G_{C\mu} + \frac{1}{g_s}\partial_\mu\Lambda_A. \tag{7.54b}$$

To see this, we note that under these transformations

$$D_\mu q \rightarrow \left(1 + \frac{i}{2}\lambda \cdot \Lambda(x)\right) D_\mu q \qquad (7.55a)$$

$$G_{A\mu\nu} \rightarrow G_{A\mu\nu} - f_{ABC}\Lambda_B G_{C\mu\nu}. \qquad (7.55b)$$

It is then trivial to show that the Lagrangian (7.52a) is gauge invariant.

For the finite gauge transformation (7.47a), we have the gauge invariance provided that the gauge fields $G_\mu$ simultaneously undergo the transformation

$$G_\mu \rightarrow UG_\mu U^\dagger + \frac{i}{g_s} U\partial_\mu U^\dagger \qquad (7.56)$$

Under these transformations:

$$D_\mu q \rightarrow U(D_\mu q) \qquad (7.57a)$$

$$G_{\mu\nu} \rightarrow UG_{\mu\nu}U^\dagger \qquad (7.57b)$$

and hence the Lagrangian (7.52a) is gauge invariant. [see Sec. 7.3.]

The eight gauge vector bosons $G_{A\mu}$ are called gluons. They are mediators of strong interaction between quarks just as photons are mediators of electromagnetic force between electrically charged particles. The gauge transformation given in Eq. (7.47a) is called the non-Abelian gauge transformation, whereas the gauge transformation (7.7) is called the Abelian gauge transformation. The non-Abelian gauge transformation was first considered by Yang and Mills and gauge bosons are sometimes called Yang-Mills fields.

## 7.4.1 *Conserved Current*

In order to discuss the conserved current associated with gauge fields, we discuss a general method. Suppose we have a set of fields which we denote by $\phi_a(x)$. The Lagrangian is a function of these fields $\phi_a$ and $\partial_\mu \phi_a$:

$$\mathcal{L} = \mathcal{L}(\phi_a, \partial_\mu \phi_a). \qquad (7.58)$$

Consider an infinitesimal gauge transformation

$$\phi_a(x) \rightarrow \phi_a(x) + i\Lambda_A(x)(T_A)_a^b \phi_b. \qquad (7.59)$$

$T_A$ are matrices corresponding to the non-Abelian gauge group and the representation to which the fields $\phi_a(x)$ belong. From Eq. (7.58),

$$\delta\mathcal{L} = \sum_\phi \frac{\partial \mathcal{L}}{\partial \phi_a}\delta\phi_a + \sum_\phi \frac{\partial \mathcal{L}}{\partial(\partial_\mu \phi_a)}\delta(\partial_\mu \phi_a). \qquad (7.60)$$

Using the Euler-Lagrange equations

$$\frac{\partial \mathcal{L}}{\partial \phi_a} - \partial_\mu \left( \frac{\partial \mathcal{L}}{\partial(\partial_\mu \phi_a)} \right) = 0 \qquad (7.61)$$

and the fact that $\delta(\partial_\mu \phi_a) = \partial_\mu \delta(\phi_a)$, we have

$$\delta \mathcal{L} = \sum_\phi \left[ \partial_\mu \left( \frac{\partial \mathcal{L}}{\partial(\partial_\mu \phi_a)} \right) \delta \phi_a + \frac{\partial \mathcal{L}}{\partial(\partial_\mu \phi_a)} \delta(\partial_\mu \phi_a) \right]$$

$$= \sum_\phi \partial_\mu \left[ \frac{\partial \mathcal{L}}{\partial(\partial_\mu \phi_a)} \delta \phi_a \right]. \qquad (7.62)$$

On using Eq. (7.59) so that $\delta \phi_a = i\Lambda_A (T_A)_a^b \phi_b$,

$$\delta \mathcal{L} = \sum_\phi \partial_\mu \left[ \frac{\partial \mathcal{L}}{\partial(\partial_\mu \phi_a)} i\Lambda_A (T_A)_a^b \phi_b \right]. \qquad (7.63)$$

If we take $\Lambda_A$ as constant, i.e. independent of $x$, then we can rewrite Eq. (7.63) as

$$\delta \mathcal{L} = \partial_\mu \sum_\phi i \left[ \frac{\partial \mathcal{L}}{\partial(\partial_\mu \phi_a)} (T_A)_a^b \phi_b \right] \Lambda_A \equiv -\partial_\mu F_A^\mu \Lambda_A, \qquad (7.64)$$

where

$$F_A^\mu = -\sum_\phi i \left( \frac{\partial \mathcal{L}}{\partial(\partial_\mu \phi_a)} \right) (T_A)_a^b \phi_b. \qquad (7.65)$$

Hence we have the Noether's theorem. If the Lagrangian is invariant under the gauge transformation (7.59) with constant $\Lambda_A$, i.e. $\delta \mathcal{L} = 0$, then the current given in Eq. (7.65) is conserved.

Let us apply this to the QCD Lagrangian (7.52a). Here $\phi_a$ corresponds to $G_{B\mu}$ and $q_a$. Now, for the gauge vector bosons which belong to the adjoint representation of $SU_C(3)$, we have $i(T_A)_B^C = -f_{BAC}$ and for the quarks which belong to the triplet representation of $SU_C(3)$, $T_A = \frac{1}{2}\lambda_A$. Then from the Lagrangian (7.52a):

$$\frac{\partial \mathcal{L}}{\partial(\partial_\mu q)} = i\bar{q}\gamma^\mu, \qquad \frac{\partial \mathcal{L}}{\partial(\partial_\mu G_{A\nu})} = -G_A^{\mu\nu}$$

Hence Eq. (7.65) gives

$$F_A^\mu = \frac{1}{2}\bar{q}^a \gamma^\mu (\lambda_A)_a^b q_b - f_{ABC} G_{B\nu} G_C^{\mu\nu} \qquad (7.66)$$

$$= \frac{1}{2}\bar{q}\gamma^\mu \lambda_A q + f_{ABC} G_B^{\mu\nu} G_{C\nu}.$$

The current $F_A^\mu$ is universally coupled to the gauge fields $G_{A\mu}$ with universal coupling $g_s$. Now the interaction part of the Lagrangian (7.52a) is given by

$$\mathcal{L}_{int} = g_s F_A^\mu G_{A\mu}$$

$$= g_s G_{A\mu} \bar{q}^a \gamma^\mu \left(\frac{\lambda_A}{2}\right)_a^b q_b - g_s f_{ABC} G_{A\mu} G_{B\nu}$$

$$\times \left[\frac{1}{2}\left(\partial^\mu G_C^\nu - \partial^\nu G_C^\mu\right) + \frac{1}{4} g_s f_{CDE} G_D^\mu G_E^\nu\right] \qquad (7.67)$$

The last term of Eq. (7.67) represents the self interaction of gauge bosons among themselves as they carry the color charges. This term is very important in QCD and is responsible for the asymptotic freedom of QCD.

From Eq. (7.67), the $qqG$, $GGG$ and $GGGG$ vertices in the momentum space can be represented graphically as shown in Fig. 7.5.

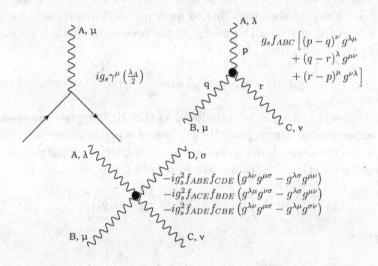

Fig. 7.5   Graphic representation of $qqG$, $GGG$ and $GGGG$ vertices.

The Feynman rules for the QCD Lagrangian are discussed in Appendix B.

## 7.4.2 *Experimental Determinations of $\alpha_s(q^2)$ and Asymptotic Freedom of QCD*

The important physical properties of QCD are

(i) the gluons, being mediators of strong interaction between quarks, are vector particles and carry color; both of these properties are supported by hadron spectroscopy discussed in the next section,

(ii) asymptotic freedom which implies that the effective coupling constant $\alpha_s = g_s^2/4\pi$ decreases logarithmically at short distances or high momentum transfers, a property which has a rigorous theoretical basis. This is the basis for perturbative QCD which is relevant for processes involving large momentum transfers,

(iii) confinement which implies that potential energy between color charges increases linearly at large distances so that only color singlet states exist, a property not yet established but find support from lattice simulations and qualitative pictures (see next section) and from quarkonium spectroscopy to be discussed in Chap. 8.

In this section, we discuss the present evidence for QCD being asymptotic free. First we note that due to quantum radiative corrections, $\alpha_s$ evolves with the characteristic energy of the process in which it appears. Actually these corrections give

$$g_s(Q^2) = g_{s0} \left[ 1 + g_{s0}^2 b_0 \ln \frac{\lambda^2}{Q^2} + \cdots \right], \qquad (7.68a)$$

where $\lambda^2 \gg Q^2$ and must be introduced so that the integrals involved in these corrections are convergent. Here $\cdots$ denotes higher order corrections and $\sqrt{Q^2}$ is the momentum carried by a gluon at quark-quark-gluon vertex which defines $g_s(Q^2)$. It is convenient to rewrite Eq. (7.68a) as

$$\frac{1}{g_s^2(Q^2)} = \frac{1}{g_{s0}^2} \left[ 1 - 2g_{s0}^2 b_0 \ln \frac{\lambda^2}{Q^2} + O(g_s^4) \right]. \qquad (7.68b)$$

This gives

$$\alpha_s^{-1}(q^2) - \alpha_{s0}^{-1} = -8\pi b_0 \ln \frac{\lambda^2}{Q^2}. \qquad (7.68c)$$

We now eliminate the unobserved "bare" coupling constant $\alpha_{s0}$ and the cut-off $\lambda^2$ by making a subtraction at $Q^2 = \mu^2$. Thus we obtain

$$\alpha_s^{-1}(Q^2) - \alpha_s^{-1}(\mu^2) = b \ln \frac{Q^2}{\mu^2} \qquad (7.68d)$$

with $b = 8\pi b_0$. Or

$$\alpha_s(Q^2) = \frac{1}{\alpha_s^{-1}(\mu^2) + b \ln Q^2/\mu^2}. \qquad (7.68e)$$

The constant $b$ is evaluated in Appendix B and is given by

$$b = \frac{1}{4\pi}\left(11 - \frac{2}{3}n_f\right) \tag{7.68f}$$

where $n_f$ is the number of effective quark flavors. Another way of writing Eq. (7.68e) is

$$\alpha_s^{-1}(Q^2) = b\ln\frac{Q^2}{\Lambda_{QCD}^2} \tag{7.68g}$$

where

$$\alpha_s^{-1}(\mu) - b\ln\mu^2 = -b\ln\Lambda_{QCD}^2.$$

Thus finally we have

$$\alpha_s(Q^2) = \frac{4\pi}{\left(11 - \frac{2}{3}n_f\right)\ln\frac{q^2}{\Lambda_{QCD}^2}} \tag{7.68h}$$

and we see the running of $\alpha_s(Q^2)$ with $Q^2$. $\Lambda_{QCD}$ is the QCD scale factor which effectively defines the energy scale at which the running coupling constant attains its maximum value. $\Lambda_{QCD}$ can be determined from experiment. For $\frac{2}{3}n_f < 11$, it is clear from Eq. (7.68b) or (7.68h) that $\alpha_s(Q^2)$ decreases as $Q^2$ increases and approaches zero as $Q^2 \to \infty$ or $r \to 0$. This is known as the asymptotic freedom property of QCD. This is due to the factor 11 in Eq. (7.68f) or (7.68h) and arises due to the self-interaction of gluons (see Appendix B).

We now discuss the experimental determination of the coupling constant $\alpha_s(Q^2)$ at various values of $Q^2$ from different reactions, starting from the lowest value of $\sqrt{Q^2}$.

(1) From Eq. (7.5)

$$R = \frac{\sigma(e^+ e^- \to q\,\bar{q} \to hadrons)}{\sigma(e^+ e^- \to \mu^+ \mu^-)}$$

$$= 3\sum_q e_q^2[1 + \alpha_s\frac{\sqrt{s}}{\pi} + 1.411(\alpha_s\frac{\sqrt{s}}{\pi})^2 \cdots]$$

For $s > 2m_b^2$, $3\sum_q e_q^2 = \frac{11}{3}$.

By fitting the values of R at $\sqrt{s} = 34$ GeV, shown in Fig. 7.3, one obtains

$$\alpha_s(34\ \text{GeV}) = 0.142 \pm 0.03$$

(2) From Eq. (8.71) and Eq. (8.69):

$$\frac{\alpha_s^2(m_\Psi)}{\alpha^2} = \frac{2}{3}\frac{m_\eta^2}{m_\Psi^2}\frac{\Gamma(\eta_c \to hadrons)}{\Gamma(\Psi \to e^+ e^-)}$$

$$\frac{\alpha_s^3(m_V)}{\alpha^2} = \frac{81\pi}{10(\pi^2-9)}e_q^2\frac{\Gamma(V \to hadrons)}{\Gamma(V \to e^+ e^-)}$$

where $\eta_c$ is $(q\,\bar{q})^1 S_1$ state while $V$ is $(q\,\bar{q})^3 S_1$ state, $e_q$ corresponds to charge of the quark $q$.

Using the experimental values for the leptonic and hadronic decay widths,

$$\alpha_s(m_{J/\psi}) = 0.217 \pm 0.009, \quad \alpha_s(m_\Upsilon) = 0.163 \pm 0.005 \qquad (7.69a)$$

The value of $\alpha_s$ obtained from the scaling violations in deep inelastic lepton-nucleon scattering [see Chap. 14] gives

$$\alpha_s\left(\sqrt{q^2} = 2.6 \text{ GeV}\right) = 0.264 \pm 0.101. \qquad (7.70)$$

Finally from the semi-leptonic branching ratio $R_\tau$ for the inclusive decay $\tau \to \nu_\tau$ + hadrons, one obtains

$$\alpha_s(m_\tau) = 0.35 \pm 0.03$$

Figure 7.6 shows the values of $\alpha(m_z)$ deduced from the various experiments. Figure 7.7 clearly shows the experimental evidence for the running of $\alpha_s(q)$ i.e. decrease of the coupling constant as $q$ increases as indicated by Eq. (7.68a). An average of the values in Fig. 7.6 gives

$$\alpha_s(m_Z) = 0.119 \pm 0.002$$

which coresponds to

$$\Lambda_{QCD} = 219^{+25}_{-23} \text{ MeV}. \qquad (7.71)$$

The LEP / SLAC value for $\alpha_s(m_Z)$ is $0.124 \pm 0.004$.


## 7.5  Hadron Spectroscopy

### 7.5.1  *One Gluon Exchange Potential*

All known hadrons are color singlets. Just as an exchange of photon gives force of repulsion between like charges and force of attraction between unlike charges, the exchange of gluon gives force of attraction between color singlet

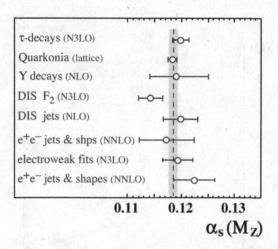

Fig. 7.6  Summary of the values of $\alpha_s(m_z)$ from various processes. The values shown indicate the process and the measured value of $\alpha_s$ extrapolated upto $\mu = m_z$. The error shown is the total error including theoretical uncertainties [29].

states. The exchange of gluons can provide binding between quarks in a hadron.

For $q\bar{q}$ system (meson), the color electric potential due to one gluon exchange diagram [see Fig. 7.8] is given by:

$$V_{ij} = -g_s^2 \frac{1}{4\pi r} \sum_{A=1}^{8} \left(\frac{\lambda_A}{2}\right)_b^a \left(\frac{\lambda_A}{2}\right)_c^d \frac{1}{\sqrt{3}}\delta_a^c \frac{1}{\sqrt{3}}\delta_d^b. \tag{7.72}$$

The factors $\frac{1}{\sqrt{3}}\delta_a^c$ and $\frac{1}{\sqrt{3}}\delta_d^b$ in the initial and final states arise due to normalized color singlet totally symmetric wave function for the $q\bar{q}$ system. The minus sign arises due to the coupling of a vector particle to the antiquark. Here $i$, $j$ are flavor indices and $a$, $b$, $c$, $d$ are color indices. Since $Tr(\lambda_A\lambda_B) = 2\delta_{AB}$, $Tr(\lambda_A\lambda_A) = 16$,

$$V_{ij} = -\frac{4}{3}\frac{\alpha_s}{r}, \alpha_s = \frac{g_s^2}{4\pi}. \tag{7.73}$$

For three quarks system (baryon), one gluon exchange diagram (Fig. 7.9) gives the following two-body potential

$$V_{ij} = g_s^2 \frac{1}{4\pi r} \frac{\varepsilon_{eac}}{\sqrt{6}} \frac{\varepsilon^{ebd}}{\sqrt{6}} \left(\frac{\lambda_A}{2}\right)_d^c \left(\frac{\lambda_A}{2}\right)_b^a. \tag{7.74a}$$

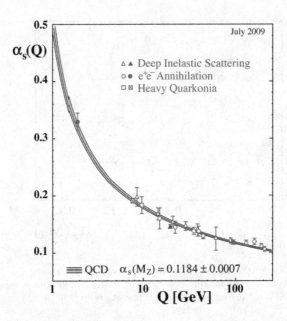

Fig. 7.7   Summary of the values of $\alpha_s(\mu)$ at the values of $\mu$ where they are measured. The figure clearly shows the decrease in $\alpha_s(\mu)$ with increasing $\mu$ [29].

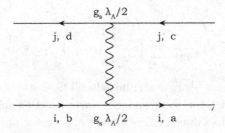

Fig. 7.8   Diagram generating one–gluon exchange potential for $q\bar{q}$ system.

The factors $\frac{\varepsilon_{eac}}{\sqrt{6}}$ and $\frac{\varepsilon_{ebd}}{\sqrt{6}}$ arise due to the fact that three-quark color wave function is totally antisymmetric in color indices. Using $\varepsilon_{eac}\varepsilon^{ebd} = \delta_a^b\delta_c^d - \delta_a^d\delta_c^b$, and $Tr\lambda_A = 0$,

$$V_{ij} = -\frac{2}{3}\frac{\alpha_s}{r}. \tag{7.74b}$$

Note the important fact that in both cases, we get an attractive potential. We also note that $V_{ij}^{q\bar{q}} = 2V_{ij}^{qq}$ for color singlet states. Thus we can write

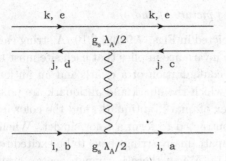

Fig. 7.9 Diagram generating one-gluon exchange two-body potential for three quarks (baryon) system.

the two-body one-gluon exchange potential as

$$V_{ij} = k_s \frac{\alpha_s}{r}, k_s = \left\{ \begin{array}{cc} -\frac{4}{3} & q\bar{q} \\ -\frac{2}{3} & qq \end{array} \right\}. \tag{7.75}$$

Since the running coupling constant $\alpha_s$ becomes smaller as we decrease the distance, the effective potential $V_{ij}$ approaches the lowest order one-gluon exchange potential given in Eq. (7.74a) as $r \to 0$. Now in momentum space, we can write the potential in QCD perturbation theory for small distances ($r < 0.1$ fm) as

$$V(\mathbf{q}^2) = k_s 4\pi \alpha_s(\mathbf{q}^2)/\mathbf{q}^2, \tag{7.76}$$

where $V(r)$ is the Fourier transform of $V(\mathbf{q}^2)$ and $\mathbf{q}^2$ is the momentum conjugate to $r$. The running coupling $\alpha_s(\mathbf{q}^2)$ in QCD is given by Eq. (7.68h).

We conclude that for short distances, one can use the one gluon exchange potential, taking into account the running coupling constant $\alpha_s(\mathbf{q}^2)$.

### 7.5.2 *Long Range QCD Motivated Potential*

The second regime, i.e. for large $r$, QCD perturbation theory breaks down and we have the confinement of the quarks. Thus unlike the short range part of the potential, the long range part cannot be calculated on perturbative QCD as the QCD constants become large in this region. Perturbative QCD gives no hint of intrinsically nonperturbative phenomena such as color confinement. One may look for the origin of this yet unsatisfactorily explained phenomena. There are many pictures which support the existence of a linear confining term. One of these is discussed below:

### 7.5.2.1　*The string picture of hadrons*

This picture is depicted in Figs. 7.9 and 7.10. A string carries color indices at its ends. Gauge invariance implies that each site must be a color-singlet. Thus, an allowed configuration of a quark and an antiquark on adjacent sites is the one in which the quark and antiquark are linked by a string so that the color index of quark (antiquark) and the color index of the string at that end are contracted to form a color singlet. When a quark and an antiquark are far apart, many strings have to be excited to connect the two sites [see Fig. 7.11]. When there is enough energy available to create a new $q\bar{q}$ pair, the system breaks up permitting the formation of two color singlets. Calculation based on this theory shows that the energy stored in this configuration is:

Fig. 7.10　String picture of $q\bar{q}$.

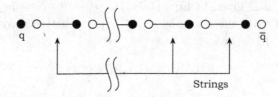

Fig. 7.11　String separation of a quark-antiquark pair.

$$E = T_0 \frac{L}{a} \qquad \text{for } L \gg a,$$

where $L$ is the quark-antiquark separation and $T_0$ is the string tension. To isolate a quark for example, the antiquark in the above illustration has to be removed to infinity; it clearly takes an infinite amount of energy to do this. This is the basis of color confinement. The confining potential is of the form:

$$V(r) \sim \text{constant} \times r,$$

for $r > 1/M$, where $M$ is a typical hadronic mass scale. Thus $\frac{1}{M}$ is of order of the hadron size of 1 fm = 5 GeV$^{-1}$ so that $M \approx 200$ MeV. The confining potential is spin and flavor independent. This picture is supported by the observation that hadrons of a given internal symmetry quantum number but different spins obey a simple spin (J)-mass (M) straight line relation, i.e. we say that they lie on linear Regge trajectories, an example of which is displayed in Fig. 7.12.

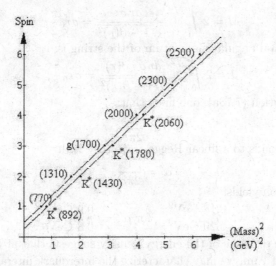

Fig. 7.12   Regge trajectories for non-strange ($I = 1$) and strange ($I = 1/2$) bosons.

For the families of hadrons composed entirely of light quarks, the above mentioned relation between $J$ and $M^2$ for Regge trajectories is given by:

$$J(M^2) = \alpha_0 + \alpha' M^2, \qquad (7.77a)$$

with

$$\alpha \approx 0.8 - 0.9 (\text{GeV}/c^2)^{-2}. \qquad (7.77b)$$

The connection between linear energy density and the linear Regge trajectory is provided by the string model formulated by Nambu. We consider a massless (and for simplicity spinless) quark and antiquark connected by a string of length $r_0$, which is characterized by an energy per unit length $\sigma$. The situation is sketched below:

For a given value of length $r_0$, the largest achievable angular momentum $J$ occurs when the ends of the string move with the velocity of light. In these circumstances, the speed at any point along the string at a distance $r$ from the center will be: $(\beta = v/c)$

$$\beta(r) = 2r/r_0.$$

The total mass of the system is then:

$$M = 2 \int_0^{r_0/2} \frac{dr\,\sigma}{\sqrt{1 - \beta(r)^2}} = \sigma r_0 \frac{\pi}{2}, \qquad (7.78a)$$

while the orbital angular momentum of the string is:

$$J = 2 \int_0^{r_0/2} \frac{dr\,\sigma\,r\beta(r)}{\sqrt{1 - \beta(r)^2}} = \sigma r_0^2 \frac{\pi}{8}, \qquad (7.78b)$$

Using the relation (7.78a), one finds that:

$$J = \frac{M^2}{2\pi\sigma}, \qquad (7.79a)$$

which corresponds to a linear Regge trajectory with

$$\alpha' = \frac{1}{2\pi\sigma}. \qquad (7.79b)$$

This connection yields:

$$\sigma = \begin{matrix} 0.18 \text{ GeV}^2 \\ 0.20 \text{ GeV}^2 \end{matrix} \quad \text{for } \alpha' = \begin{matrix} 0.9 \text{ GeV}^{-2} \\ 0.8 \text{ GeV}^{-2} \end{matrix} \qquad (7.80)$$

This heuristic estimate of the energy density suggests that at a separation of the order of 1 fm, we may characterize the interquark interaction by the linear potential

$$V(r) = \sigma r. \qquad (7.81)$$

The lattice gauge theory calculations also support the linear form for the long range part of the QCD potential.

Thus phenomenological potential of the form

$$V_{ij}(r) = V_{ij}^G(r) + V_{ij}^C(r) \qquad (7.82)$$

can be used for heavy quarks. The Cornell potential

$$V(r) = -\frac{K}{r} + \frac{r}{a^2} + C, \qquad (7.83a)$$

where

$$K = 0.48, \; a = 2.34(\text{GeV})^{-1} \quad \text{and} \quad C = -0.25 \qquad (7.83b)$$

has been used successfully to describe mass spectrum of charmonium and bottomonium systems [see Chap. 8]. Note that value of $\left(a \equiv \frac{1}{\sqrt{\sigma}}\right)$ in Eq. (7.83b) is consistent with the value of $\sigma$ stated above [cf. Eq. (7.80)]. The purely phenomenological potentials of the form and

$$V(r) = a + br^{0.1} \qquad (7.84a)$$

and

$$V(r) = C \ln r \qquad (7.84b)$$

have also been used successfully for $c\bar{c}$ and $b\bar{b}$ systems.

### 7.5.3  Spin-Spin Interaction

Finally, we note that a spin $1/2$ charged particle of charge $eQ_i$ has a magnetic momentum $\mu_i = \frac{eQ_i}{2m_i}\sigma_i$. In quantum mechanics, the energy splitting between S-states (zero orbital angular momentum) is given by two-particle operator (Fermi contact term)

$$H_{ij}^M = \frac{1}{4\pi}\left[-\frac{8\pi}{3}\mu_i \cdot \mu_j \delta^3(\mathbf{r}_i - \mathbf{r}_j)\right]$$  (7.85)

$$= -\frac{8\pi}{3}\alpha\frac{Q_iQ_j}{2m_im_j}\sigma_i \cdot \sigma_j \delta^3(\mathbf{r})$$

Similarly in QCD, we have eight color-magnetic moments

$$\mu_A^i = \frac{g_s}{2m_i}\left(\frac{\lambda_A}{2}\right)\sigma, \qquad A = 1, \cdots, 8.$$  (7.86)

The analogous two-particle interaction for QCD is then given by

$$H_{ij} = \left[-\frac{8\pi}{3}\mu_A^{(i)} \cdot \mu_A^{(j)}\delta^3(\mathbf{r}_i - \mathbf{r}_j)\right]\frac{1}{4\pi}.$$  (7.87)

Again for a color singlet system $\alpha \to k_s\alpha_s$ (cf. Eq. (7.75))

$$H_{ij} = -\frac{8\pi}{3}\alpha_s k_s \frac{\sigma_i \cdot \sigma_j}{4m_im_j}\delta^3(r),$$  (7.88)

Eq. (7.88) gives $m(^3S_1) > m(^1S_0)$ [for example $m_\rho > m_\pi$] in agreement with the experimental result. This supports the fact that gluons are spin 1 particles.

## 7.6  The Mass Spectrum

The one gluon exchange potential is obtained by summing over all possible quark indices in $V_{ij}^G$ in a multiquark system like $q\bar{q}$ and $qqq$. Thus

$$V^G = \frac{1}{2}\sum_{i \neq j}V_{ij}^G$$

$$= \frac{1}{2}\left[\sum_{i>j}V_{ij}^G + \sum_{i<j}V_{ij}^G\right]$$

$$= \frac{1}{2}\left[\sum_{i>j}\left(V_{ij}^G + V_{ji}^G\right)\right]$$

$$= \sum_{i>j}V_{ij}^G$$  (7.89)

The potential $V_G$ for S-states is found to be [in non-relativistic limit keeping terms up to $(p^2/m^2)$]

$$V_G = k_s \alpha_s \sum_{i>j} \left[ \frac{1}{r} - \frac{1}{2m_i m_j} \left( \frac{\mathbf{p}_i \cdot \mathbf{p}_j}{r} + \frac{\mathbf{r}(\mathbf{r} \cdot \mathbf{p}_i) \cdot \mathbf{p}_j}{r^3} \right) \right.$$
$$\left. - \frac{\pi}{2} \delta^3(\mathbf{r}) \left( \frac{1}{m_i^2} + \frac{1}{m_j^2} + \frac{16\mathbf{s}_i \cdot \mathbf{s}_j}{3m_i m_j} \right) \right]. \tag{7.90}$$

The second term in the bracket will be ignored so that one gluon potential is velocity independent. The first term on the right-hand side is the potential in the extreme non-relativistic limit $\left( \frac{v}{c} \approx 0 \right)$; spin dependent term is due to the color magnetic moments interaction as mentioned previously.

For S–states,

$$\left\langle \Psi_s \left| \delta^3(\mathbf{r}) \right| \Psi_s \right\rangle$$
$$= \int \Psi_s^*(\mathbf{r}) \delta^3(\mathbf{r}) \Psi_s(\mathbf{r}) d^3 r$$
$$= |\Psi_s(0)|^2. \tag{7.91}$$

Now our Hamiltonian, including the rest masses of the quarks can be written as

$$H(\mathbf{r}) = \sum_i m_i + \sum_i \frac{\hat{\mathbf{p}}_i^2}{2m_i} + V_C(r) + V_G(r), \tag{7.92}$$

where

$$\hat{\mathbf{p}}_i^2 = -\frac{\hbar^2}{2m_i} \nabla_i^2. \tag{7.93}$$

Here $V_C(r)$ is the confining potential, $V_G(r)$ is the one gluon exchange potential given in Eq. (7.90), i is the quark flavor index, i.e. i = u, d, s for ordinary hadrons. We will take $m_u = m_d$. In order to discuss the mass spectrum of hadrons, we have to take the expectation value of the Hamiltonian $H(\mathbf{r})$ with respect to the relevant wave functions of the hadrons. The wave function is the product of three parts viz unitary spin, spin and space parts. For s-wave, we write the space function as $\Psi_s(\mathbf{r})$. Let us first take the expectation value of $H(\mathbf{r})$ with respect to $\Psi_s(\mathbf{r})$, we have

$$m \equiv \langle \Psi_s | H | \Psi_s \rangle$$
$$= \sum_i m_i + \sum_i \frac{1}{m_i} \left\langle \Psi_s \left| \hat{\mathbf{p}}_i^2 \right| \Psi_s \right\rangle + \left\langle \Psi_s \left| V_C(r) + k_s \alpha_s \frac{1}{r} \right| \Psi_s \right\rangle$$
$$- k_s \alpha_s \frac{\pi}{2} \left( \frac{1}{m_i^2} + \frac{1}{m_j^2} + \frac{16\mathbf{s}_1 \cdot \mathbf{s}_2}{3m_i m_j} \right) |\Psi_s(0)|^2 \tag{7.94}$$

Note that the mass operator $m$ is still an operator in unitary spin and spin space. We first apply the mass formula (7.94) to pseudoscalar meson system.

### 7.6.1  *Meson Mass Relations*

From Eq. (7.94), the mass operator for S-wave mesons can be written as

$$m = m_0 + m_1 + m_2 + a \left[ \frac{1}{m_1} + \frac{1}{m_2} \right]$$

$$+ \bar{d} \left[ \frac{1}{m_1^2} + \frac{1}{m_2^2} + \frac{16 \mathbf{s}_1 \cdot \mathbf{s}_2}{3 m_1 m_2} \right], \tag{7.95a}$$

$$m_0 = A_0 + k_s \alpha_s b \tag{7.95b}$$

where

$$a = \left\langle \Psi_s \left| \hat{\mathbf{p}}_i^2 \right| \Psi_s \right\rangle \tag{7.96a}$$

$$A_0 = \left\langle \Psi_s \left| V_C(r) \right| \Psi_s \right\rangle \tag{7.96b}$$

$$b = \left\langle \Psi_s \left| \frac{1}{r} \right| \Psi_s \right\rangle \tag{7.96c}$$

$$\bar{d} = -\frac{4}{3} \alpha_s \frac{\pi}{2} \left\langle \Psi_s \left| \delta^3(\mathbf{r}) \right| \Psi_s \right\rangle \tag{7.96d}$$

For $(q\bar{q})_{L=0}$ the indices $i = 1$ and $j = 2$ refer to the constituent antiquark and quark respectively

$$\frac{1}{4} \sigma_1 \cdot \sigma_2 = \mathbf{s}_1 \cdot \mathbf{s}_2 = \left\{ \begin{array}{ll} \frac{1}{4} & \text{spin triplet state } S = 1: \text{ vector meson} \\ -\frac{3}{4} & \text{spin singlet state } S = 0: \text{ pseudoscalar meson.} \end{array} \right.$$

Thus we have,

$$m(^3S_1) - m(^1S_0) = \frac{32}{9} \pi \alpha_s \frac{1}{m_1 m_2} |\psi_s(0)|^2$$

Hence we have the following mass relations, $m_u \approx m_d$

$$m_\rho = m_\omega$$

$$m_\phi = 2 m_{K^*} - m_\rho$$

$$(m_{K^*} - m_K) = \frac{m_u}{m_s} (m_\rho - m_\pi)$$

These relations are well satisfied experimentally with $m_u = m_d \approx 336$ MeV, $m_s \approx 510$ MeV (cf. Chap. 6).

If gluons were scalar particles, then $\mathbf{s}_1 \cdot \mathbf{s}_2$ term would be absent so that $m(^3S_1) = m(^1S_0)$ in disagreement with the experimental observation. For pseudoscalar gluons, $k_s = \frac{4}{3}$, since pseudoscalar coupling is the same for antiquarks. In this case we would have $m(^3S_1) < m(^1S_0)$, again in disagreement with the experimental result. We conclude that the experimental results about meson spectrum support the fact that gluons are vector particles and are thus quanta of QCD.

For pseudoscalar mesons $\eta_{ns}$ and $\eta_s$, we get

$$m_{\eta_{ns}} = m_\pi \tag{7.97a}$$

$$m_{\eta_s} = (2m_K - m_\pi) + O(\lambda^2). \tag{7.97b}$$

These formulae are badly broken. Thus the above analysis breaks down for $J = 0$ mesons, $\eta$ and $\eta'$. The reason for this is that our Hamiltonian does not take into account quark-antiquark annihilation into gluons. The lowest order annihilation diagram is shown in Fig. 7.13. This diagram contributes only to $^1S_0$ state, because of charge conjugation conservation. Since gluons do not carry any flavor, therefore it contributes to $I = Y = 0$, $^1S_0$ states only. This diagram is relevant only for $\eta$ and $\eta'$ mesons, and is of order $O(\alpha_s^2)$. For $I = Y = 0$ vector bosons, the diagram with three-gluon exchange contributes, which is of order $O(\alpha_s^3)$ and hence can be neglected.

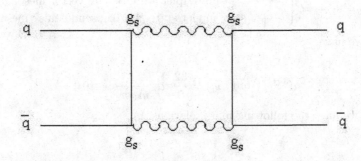

Fig. 7.13   The $q\bar{q}$ annihilation diagram for $^1S_0$ state through two gluons.

We now take into account the diagram of Fig. 7.13 for pseudoscalar mesons. If $u\bar{u}$, $d\bar{d}$ and $s\bar{s}$ can annihilate with an amplitude $A$, which we assume to be SU(3) invariant, then there will be an additional contribution to the mass matrix, which in the $u\bar{u}$, $d\bar{d}$ and $s\bar{s}$ basis is given by

$$M_{ann} = \begin{pmatrix} A & A & A \\ A & A & A \\ A & A & A \end{pmatrix}. \tag{7.98}$$

Taking into account Eq. (7.96) and the fact that $|\eta_s\rangle = |s\bar{s}\rangle$, $|\eta_{ns}\rangle = \frac{1}{\sqrt{2}}|(u\bar{u} + d\bar{d})\rangle$, $|\pi_0\rangle = \frac{1}{\sqrt{2}}|(u\bar{u} - d\bar{d})\rangle$, we get in $\pi^0$, $\eta_{ns}$ and $\eta_s$ basis, the mass matrix

$$\begin{pmatrix} m_\pi & 0 & 0 \\ 0 & m_\pi + 2A & \sqrt{2}A \\ 0 & \sqrt{2}A & 2m_K - m_\pi + A \end{pmatrix}. \tag{7.99}$$

From Eq. (7.98), we note that we have to diagonalize the mass matrix

$$M \to M + M_{ann} = \begin{pmatrix} m_\pi + 2A & \sqrt{2}A \\ \sqrt{2}A & 2m_K - m_\pi + A \end{pmatrix}. \tag{7.100}$$

For this purpose, we define the physical states as (see Problem 5.15)

$$|\eta\rangle = \cos\phi\,|\eta_{ns}\rangle - \sin\phi\,|\eta_s\rangle$$
$$|\eta'\rangle = \sin\phi\,|\eta_{ns}\rangle + \cos\phi\,|\eta_s\rangle \tag{7.101}$$

Then the mass eigenvalues are given by

$$m_\eta m_{\eta'} = m_\pi(2m_K - m_\pi) + A(4m_K - m_\pi)$$
$$m_\eta + m_{\eta'} = m_{\eta_{ns}} + m_{\eta_s} = 2m_K + 3A. \tag{7.102}$$

Using the experimental values for $\eta$ and $\eta'$, we can determine $A$. The mass scale $A$ comes out to be $\approx 172$ MeV, a rather low value compared to $m_\eta$ and $m_{\eta'}$ which is both interesting and reasonable.

To conclude, we have shown that mass spectrum of vector mesons can be explained successfully. With the addition of annihilation diagram, the pseudoscalar meson mass spectrum can also be understood.

### 7.6.2 Baryon Mass Spectrum

In order to discuss the mass spectrum of the baryons, it is convenient to first calculate the matrix elements of the spin operator

$$\Omega_{ss} = \sum_{i>j} \frac{1}{m_i m_j} \mathbf{s}_i \cdot \mathbf{s}_j \tag{7.103}$$

between spin states. The eigenvalues of $\mathbf{s}_i \cdot \mathbf{s}_j$ are $1/4$ and $-3/4$ for spin triplet and singlet states respectively. Therefore,

$$\mathbf{s}_i \cdot \mathbf{s}_j \left|\uparrow\uparrow\right\rangle = \frac{1}{4}\left|\uparrow\uparrow\right\rangle$$

$$\mathbf{s}_i \cdot \mathbf{s}_j \left[\frac{1}{\sqrt{2}}\left|(\uparrow\downarrow + \downarrow\uparrow)\right\rangle\right] = \frac{1}{4\sqrt{2}}\left|(\uparrow\downarrow + \downarrow\uparrow)\right\rangle$$

$$\mathbf{s}_i \cdot \mathbf{s}_j \left|\downarrow\downarrow\right\rangle = \frac{1}{4}\left|\downarrow\downarrow\right\rangle \tag{7.104a}$$

$$\mathbf{s}_i \cdot \mathbf{s}_j \left[\frac{1}{\sqrt{2}}\left|(\uparrow\downarrow - \downarrow\uparrow)\right\rangle\right] = -\frac{3}{4\sqrt{2}}\left|(\uparrow\downarrow - \downarrow\uparrow)\right\rangle \tag{7.104b}$$

From Eqs. (7.104a) and (7.104b), we get

$$\mathbf{s}_i \cdot \mathbf{s}_j \left|i^\uparrow j^\downarrow\right\rangle = -\frac{1}{4}\left|i^\uparrow j^\downarrow\right\rangle + \frac{1}{2}\left|i^\downarrow j^\uparrow\right\rangle$$

$$\mathbf{s}_i \cdot \mathbf{s}_j \left|i^\downarrow j^\uparrow\right\rangle = -\frac{1}{4}\left|i^\downarrow j^\uparrow\right\rangle + \frac{1}{2}\left|i^\uparrow j^\downarrow\right\rangle \tag{7.105}$$

The spin wave functions for baryons are given in Table 6.1 and Eq. (6.7). Using these wave functions, we get with the help of Eqs. (7.104a), (7.104b) and (7.105) for $\frac{1}{2}^+$ baryons with $s_z = \frac{1}{2}$.

$$\Omega_{ss}\left|p\right\rangle$$

$$= \sum_{i>j} \frac{1}{m_i m_j} \mathbf{s}_i \cdot \mathbf{s}_j \left|uud\right\rangle \left(-\frac{1}{\sqrt{6}}\right)\left|(\uparrow\downarrow + \downarrow\uparrow) - 2\left|\uparrow\uparrow\downarrow\right\rangle\right.$$

$$= \frac{1}{m_u^2}\left|uud\right\rangle \left(-\frac{1}{\sqrt{6}}\right)\left\{\frac{1}{4}\left|(\uparrow\downarrow + \downarrow\uparrow) - 2\left|\uparrow\uparrow\downarrow\right\rangle\right|\right.$$

$$+ \left[\frac{1}{4}\left|\uparrow\downarrow\uparrow\right\rangle - \frac{1}{4}\left|\downarrow\uparrow\uparrow\right\rangle + \frac{1}{2}\left|\uparrow\uparrow\downarrow\right\rangle - 2\left(-\frac{1}{4}\left|\uparrow\uparrow\downarrow\right\rangle + \frac{1}{2}\left|\downarrow\uparrow\uparrow\right\rangle\right)\right]$$

$$+ \left[-\frac{1}{4}\left|\uparrow\downarrow\uparrow\right\rangle + \frac{1}{2}\left|\uparrow\uparrow\downarrow\right\rangle + \frac{1}{4}\left|\downarrow\uparrow\uparrow\right\rangle - 2\left(-\frac{1}{4}\left|\uparrow\uparrow\downarrow\right\rangle + \frac{1}{2}\left|\uparrow\downarrow\uparrow\right\rangle\right)\right]\right\}$$

$$= \left|uud\right\rangle \frac{1}{m_u^2}\left(-\frac{3}{4}\right)\frac{1}{\sqrt{6}}\left|(\uparrow\downarrow + \downarrow\uparrow)\uparrow - 2\uparrow\uparrow\downarrow\right\rangle$$

$$= -\frac{3}{4m_u^2}\left|p\right\rangle. \tag{7.106a}$$

Similarly we get

$$\Omega_{ss}\left|\Lambda\right\rangle = -\frac{3}{4m_u^2}\left|\Lambda\right\rangle \tag{7.106b}$$

$$\Omega_{ss}\left|\Sigma^0\right\rangle = \frac{1}{4}\left(\frac{1}{m_u^2} - \frac{4}{m_u m_s}\right)\left|\Sigma^0\right\rangle \tag{7.106c}$$

$$\Omega_{ss}\left|\Xi^0\right\rangle = \frac{1}{4}\left(\frac{1}{m_u^2} - \frac{4}{m_u m_s}\right)\left|\Xi^0\right\rangle, \tag{7.106d}$$

where we have used

$$
\begin{aligned}
|\Lambda\rangle &= -|uds\rangle \, \chi_{MS}^{1/2} \\
|\Sigma^0\rangle &= |uds\rangle \, \chi_{MS}^{1/2} \\
|\Xi^0\rangle &= |ssu\rangle \, \chi_{MS}^{1/2}.
\end{aligned}
\tag{7.107}
$$

For $\frac{3}{2}^+$ baryons, we take $s_z = 3/2$ and calculate the matrix elements of $\Omega_{ss}$. Now

$$
\begin{aligned}
\Omega_{ss} |\Delta^{++}\rangle &= \sum_{i>j} \frac{1}{m_i m_j} \mathbf{s}_i \cdot \mathbf{s}_j \, |uuu\rangle \, |\uparrow\uparrow\uparrow\rangle \\
&= \frac{3}{4} \frac{1}{m_u^2} |\Delta^{++}\rangle
\end{aligned}
\tag{7.108a}
$$

Similarly we get

$$
\Omega_{ss} |\Sigma^{*+}\rangle = \frac{1}{4}\left( \frac{2}{m_u m_s} + \frac{1}{m_u^2} \right) |\Sigma^{*+}\rangle
\tag{7.108b}
$$

$$
\Omega_{ss} |\Xi^{*0}\rangle = \frac{1}{4}\left( \frac{2}{m_u m_s} + \frac{1}{m_u^2} \right) |\Xi^{*0}\rangle
\tag{7.108c}
$$

$$
\Omega_{ss} |\Omega^-\rangle = \frac{3}{4} \frac{1}{m_u^2} |\Omega^-\rangle,
\tag{7.108d}
$$

where we have used

$$
|\Sigma^{*+}\rangle = |uus\rangle \, |\uparrow\uparrow\uparrow\rangle
\tag{7.109a}
$$

$$
|\Xi^{*0}\rangle = |ssu\rangle \, |\uparrow\uparrow\uparrow\rangle
\tag{7.109b}
$$

$$
|\Omega^-\rangle = |sss\rangle \, |\uparrow\uparrow\uparrow\rangle.
\tag{7.109c}
$$

Since the spin-spin interaction term from Eqs. (7.90), (7.103) and (7.105) is,

$$
\frac{16\pi}{9} \alpha_s \, |\psi_s(0)|^2 \, \Omega_{ss}
\tag{7.110}
$$

we have from Eqs. (7.105) and (7.107):

$$
m\left( J = \frac{3}{2} \right) > m\left( J = \frac{1}{2} \right)
$$

in agreement with experimental observations for gluons with color, $k_s = -2/3$. If gluons do not carry color, then $k_s = 1$ instead of $-2/3$ and we would get results in contradiction with experimental values. This supports that the vector gluons carry color.

The spin dependent term $\Omega_{ss}$ splits the masses of baryons with the same quark content, but with different spin. Thus, we get from Eqs. (7.106a), (7.108a) and (7.110):

$$m_\Delta - m_p = \frac{8\pi}{3} \frac{1}{m_u^2} |\psi_s(0)|^2 \tag{7.111a}$$

$$m_\Sigma - m_\Lambda = \frac{16\pi\alpha_s}{9m_u^2} \left(1 - \frac{m_u}{m_s}\right) |\psi_s(0)|^2 \tag{7.111b}$$

$$m_{\Sigma^*} - m_\Sigma = \frac{8\pi}{3} \frac{\alpha_s}{m_u m_s} |\psi_s(0)|^2 \tag{7.111c}$$

$$m_{\Xi^*} - m_\Xi = \frac{8\pi}{3} \frac{\alpha_s}{m_u m_s} |\psi_s(0)|^2 \tag{7.111d}$$

From Eqs. (7.110),

$$\frac{m_{\Xi^*} - m_\Xi}{m_{\Sigma^*} - m_\Sigma} = 1 \qquad \text{(Exp. value 1.12)} \tag{7.112a}$$

$$\frac{2m_{\Sigma^*} + m_\Sigma - 3m_\Lambda}{2(m_\Delta - m_p)} = 1 \qquad \text{(Exp. value 1.04)} \tag{7.112b}$$

$$\frac{m_\Sigma - m_\Lambda}{m_\Delta - m_p} = \frac{2}{3}\left(1 - \frac{m_u}{m_s}\right) = 0.23 \quad \text{(Exp. value 0.26)} \tag{7.112c}$$

In the above derivation, the effects of wave function distortion due to symmetry breaking by quark effective masses have been neglected. These effects will give slight deviations from unity in the relations (7.112a, 7.112b).

We now discuss the baryon masses of same spin, using Eqs. (7.94) and (7.105). We can write the baryon mass formula:

$$m = (m_1 + m_2 + m_3) + \left(\frac{1}{m_1} + \frac{1}{m_2} + \frac{1}{m_3}\right) \langle\psi_s| \hat{\mathbf{p}}_i^2 |\psi_s\rangle$$

$$+ \langle\psi_s| \left(\left(V_C - \frac{2}{3}\alpha_s \frac{1}{r}\right)\right) |\psi_s\rangle + \sum_{i>j} \left(\frac{1}{m_i^2} + \frac{1}{m_j^2}\right) \frac{\pi}{3}\alpha_s |\psi_s(0)|^2$$

$$+ \frac{16\pi}{3}\alpha_s |\psi_s(0)|^2 \langle B |\Omega_{ss}| B\rangle \tag{7.113}$$

From Eq. (7.113), we get the Gell-Mann-Okubo mass formula

$$\frac{m_p + m_\Xi}{2} = \frac{m_\Sigma + 3m_\Lambda}{2} \tag{7.114}$$

We conclude that both the meson and baryon mass spectra can be explained quite well in QCD. In this simple picture, we have used non-relativistic quantum mechanics for u, d and s quarks. Although this approximation is not so good for these quarks (as their masses are less than 1/2 GeV) and at this energy scale QCD perturbation theory may not be a good approximation, even then the results are good.

## 7.7 Problems

(1) Show that the Lagrangian

$$\mathcal{L} = i\bar{\Psi}\gamma^\mu D_\mu \Psi - m\bar{\Psi}\Psi - \frac{1}{2}Tr(W^{\mu\nu}\,W_{\mu\nu})$$

where

$$D_\mu = \partial_\mu + igW_\mu$$
$$W_{\mu\nu} = \partial_\mu W_\nu - \partial_\nu W_\mu + ig[W_\mu, W_\nu]$$

is invariant under the infinitesimal gauge transformation

$$\Psi(x) \to (1 + i\Lambda(x))\Psi(x)$$
$$W_\mu \to W_\mu + i[\Lambda,\ W_\mu] - \frac{1}{g}\partial_\mu\Lambda$$

First show that under the infinitesimal gauge transformation

$$D_\mu\Psi \to (1 + i\Lambda)D_\mu\Psi$$
$$W_{\mu\nu} \to W_{\mu\nu} + i[\Lambda,\ W_{\mu\nu}]$$

(2) Consider the Lagrangian

$$\mathcal{L} = i\bar{\Psi}\gamma^\mu(\partial_\mu + \frac{i}{2}g\,\tau\cdot\mathbf{W}_\mu)\Psi - m\bar{\Psi}\Psi - \frac{1}{4}\mathbf{W}^{\mu\nu}\cdot\mathbf{W}_{\mu\nu}$$
$$\mathbf{W}_{\mu\nu} = \partial_\mu\mathbf{W}_\nu - \partial_\nu\mathbf{W}_\mu - g\mathbf{W}_\mu\times\mathbf{W}_\nu$$
$$W_{A\mu\nu} = \partial_\mu W_{A\nu} - \partial_\nu W_{A\mu} - g\epsilon_{ABC}W_{B\mu}\times W_{C\nu}$$
$$\tau\cdot\mathbf{W} = \tau_A W_{A\mu}$$

Use Euler-Lagrange equations to show that

$$\partial_\mu W_A^{\mu\nu} = g\bar{\Psi}\gamma^\nu\frac{\tau_A}{2}\Psi + g\epsilon_{ABC}W_B^{\mu\nu}W_{C\nu}$$
$$= gJ_A^\nu$$
$$J_A^\nu = \bar{\Psi}\gamma^\nu\frac{\tau_A}{2}\Psi + \epsilon_{ABC}W_B^{\mu\nu}W_{C\nu}$$
$$\mathcal{L}_{int.} = gJ_A^\nu W_{A\nu}$$

Write the Lagrangian in detail for

$$\Psi = \begin{pmatrix} \nu \\ e \end{pmatrix}_L$$

(3) In QED, for

$$e^- \mu \to e^- \mu$$

scattering, the cross section is given

$$\frac{d\sigma}{dQ^2} = \frac{2\pi\alpha^2}{s^2}\left(\frac{s^2 + u^2}{t^2}\right)$$

Show that in QCD for

$$u\bar{d} \to u\bar{d}$$

or

$$ud \to ud$$
$$\frac{d\sigma}{dQ^2} = \frac{4\pi\alpha_s^2}{s^2}\left(\frac{9s^2 + u^2}{t^2}\right)$$

To take into account the non-Abelion aspect of QCD, show that

$$\alpha\,|F|^2 \to \alpha_s \sum_{color} |F|^2 \Rightarrow [\frac{1}{12}Tr(\lambda_A\lambda_B)][\frac{1}{12}Tr(\lambda_A\lambda_B)]\alpha_s\,|F|^2$$

(4) In QED, for Bhaba scattering

$$e^+e^- \to e^+e^-$$
$$\frac{d\sigma}{dQ^2} = \frac{2\pi\alpha^2}{s^2}[\frac{t^2 + u^2}{s^2} + \frac{s^2 + u^2}{t^2} + \frac{2u^2}{st}]$$

Show that in QCD for

$$u\bar{u} \to u\bar{u}$$
$$\frac{d\sigma}{dQ^2} = \frac{4\pi\alpha_s^2}{9s^2}[\frac{t^2 + u^2}{s^2} + \frac{s^2 + u^2}{t^2} - \frac{2}{3}\frac{u^2}{st}]$$

Note for the interference term, we get

$$Tr(\lambda_A\lambda_B\lambda_C\lambda_D)$$

**Hint:**

$$f_{ABC}f_{ABC'} = 3\delta_{CC'}$$
$$d_{ABC}d_{ABC'} = \frac{5}{3}\delta_{CC'}$$

Interference term will give $-\frac{2}{27}$

Draw the Feynman diagram for $uu \to uu$, then using the crossing symmetry, show that

$$\frac{d\sigma}{dQ^2} = \frac{4\pi\alpha_s^2}{9s^2}\left[\frac{t^2 + s^2}{u^2} + \frac{s^2 + u^2}{t^2} - \frac{2}{3}\frac{s^2}{ut}\right].$$

## 7.8 References

1. E. Abers and B. W. Lee, Gauge Theories, Phys. Rep. 9c, 1 (1973).
2. Y. Nambu, Phys. Rev. D10, 4262 (1974).
3. B. W. Lee, "Particle Physics", in Physics and Contemporary Needs, Vol. 1 (Ed. Riazuddin), 321, Plenum Press, New York (1977).
4. D. Bohm and H. J. Hiley, Il Nuovo Cimento 52A, 295 (1979).
5. S. Mandelstam, In Proc. 1979 Int. Sym. on Lepton and Photon Interaction at High Energies (Ed. T. B. W. Kirk and H. D. I. Abarbanel). Fermi Lab., Batavia, Illinois (1979).
6. A. De Rujula, H. Georgi and S. L. Glashow, Phys. Rev. D12, 147 (1976). 2. B. W. Lee, Ref. 2 in A above. 3. O. W. Greenberg, Ann. Rev. Nucl. Part. Sci. 28, 327 (1978). 4. F. E. Close, An Introduction to Quarks and Partons, Academic Press, New York, 1979. 5. C. Quigg, "Models for Hadrons", in Gauge Theories in High Energy Physics, edited by M. K. Gaillard and R. Stora (Les Houches, 1981), North-Holland, Amsterdam, 1983, p. 645. 6. J. L. Rosner, "Quark Models", in Techniques and Concepts of High Energy Physics (St. Croix, 1980), edited by T. Ferbel, Plenum, New York, 1981.
7. S. Gasiorowicz and J. L. Rosner, Am. J. Phys. 49, 954 (1981).
8. E. Reya, Perturbative Quantum Chromodynamics Phys. Rep. 69 C, 195 (1981).
9. A. H. Mueller, Perturbative QCD at High Energies, Phys. Rep. 73 C, 237 (1981).
10. G. Altarelli, Partons in Quantum Chromodynamics, Phys. Rep. 81 C, 1 (1982)
11. F. Wilczek, Quantum Chromodynamics: the Modern Theory of the Strong Interaction, Ann. Rev. Nucl. and Part. Sci. 32, 177 (1982).
12. K. Huang, Quarks, Leptons and Gauge Fields, World Scientific, Singapore (1982).
13. K. Moriyasu, An Elementary Primer for Gauge Theory, World Scientific, Singapore (1983).
14. C. Quigg, Gauge Theories of the Strong, Weak and Electromagnetic Interactions, Benjamin/Cummings, Reading Massachusetts, (1983).
15. M. Creutz, Quarks, Gluons and Lattices, Cambridge University Press (1983).
16. Ta-Pei Cheng and Ling-Fong Li, Gauge Theory of Elementary Particle Physics, Clarendon Press, Oxford (1984).
17. M. Chaichian and N. F. Nelipa, Introduction to Gauge Field Theories,

Springer-Verlag (1984).

18. C. Quigg, Quantum Chromodynamics near the Confinement Limit, FERMILAB-Conf-85/126-T (1985).

19. D. W. Duke and R. G. Roberts, Phys. Rep. 120, 275 (1985).

20. T. Muta, Foundation of Quantum Chromodynamics, World Scientific, Singapore (1987).

21. T. D. Lee, Particle Physics and Introduction to Field Theory (revised edition), Harwood Academic, New York (1988).

22. J. J. R. Aichison and A. J. G. Hey, Gauge Theories in Particle Physics (2nd edition), Adam Hilger, Bristol, England (1988).

23. R. D. Field, Applications of Perturbative QCD, Addison-Wesley (1989).

24. Perturbative Quantum Chromodynamics, Editor: A. H. Mueller, World Scientific, Singapore (1989).

25. G. Altarelli, Experimental Tests of Perturbative QCD, Ann. Rev. Nucl. and Part. Sci, 39, 357 (1989).

26. C. H. Llewellyn Smith, Particle Phenomenology: The Standard Model, OUTP-90-16P, The Proceedings of the 1989 Scottish Universities Summer School: Physics of the Early Universe.

27. R. E. Marshak, Conceptual Foundations of Modern Particle Physics, World Scientific (1992).

28. M.E. Peskin and D.V. Schroeder, An Introduction to Quantum Field Theory, Addison–Wesley, Reading, Mass (1995).

29. Particle Data Group, K. Nakamura et al., Journal of Physics, G **37**, 0750212 (2010).

# Chapter 8

# Heavy Flavors

## 8.1 Discovery of Charm

The $J/\Psi$ was discovered in 1974 in the reaction

$$p + Be \longrightarrow e^+ e^- + X$$

at $\sqrt{s} = 7.6$ GeV. A narrow peak at $m(e^+ e^-) = 3.1$ GeV was found. It was also seen in $e^+ e^-$ collision at $\sqrt{s} = 3.105$ GeV in the following reactions

$$e^- e^+ \longrightarrow e^- e^+$$
$$e^- e^+ \longrightarrow \mu^- \mu^+$$
$$e^- e^+ \longrightarrow \text{hadrons}.$$

The width of the resonance was very narrow. It was less than the energy spread of the beam, $\Gamma \leq 3$ MeV. For this reason, the width cannot be read off directly from resonance curve. The resonant cross section for any final state $f$ :

$$e^- e^+ \longrightarrow J/\Psi \longrightarrow f$$

is given by the Breit-Wigner formula [cf. Eq. (4.52)]:

$$\sigma_{ef} = \frac{\pi}{k^2} \frac{2J+1}{(2s_1+1)(2s_2+1)} \frac{\Gamma_e \Gamma_f}{(\sqrt{s}-m)^2 + \frac{\Gamma^2}{4}} \qquad (8.1)$$

where $J$ is the spin of the resonance, $m$ is its mass, $s_1 = s_2 = 1/2$ is the spin of electron or positron and

$$s = E_{cm}^2 = 4(k^2 + m_e^2) \approx 4k^2 \qquad (8.2)$$

Here $k = |\mathbf{k}|$ is the center-of-mass momentum. $\Gamma$ is the total width, $\Gamma_e$ and $\Gamma_f$ are the partial widths into $e^- e^+$ and $f$ respectively. We can write Eq. (8.1)

$$\sigma_{ef} = \frac{\pi}{s}(2J+1)\frac{\Gamma_e \Gamma_f}{(\sqrt{s}-m)^2 + \frac{\Gamma^2}{4}} \tag{8.3}$$

Since the resonance is very narrow, $\Gamma$ is very small and it is a good approximation to replace the denominator in Eq. (8.3) by the $\delta$-function $\frac{2\pi}{\Gamma}\delta(\sqrt{s}-m)$ and then integration of Eq. (8.3) gives

$$\int \sigma_{ef} d\sqrt{s} = \pi(2J+1)\frac{2\pi\Gamma_e\Gamma_f}{m^2\Gamma} \tag{8.4}$$

Now $\sum_f \sigma_{ef} = \sigma_{tot}$, $\sum_f \Gamma_f = \Gamma$ and assuming $\Gamma_e = \Gamma_\mu$, we have for the process

$$e^- e^+ \longrightarrow \Psi \longrightarrow \mu^- \mu^+$$

$$\int \sigma_{ef} d\sqrt{s} = 2\pi^2(2J+1)\frac{\Gamma_e^2}{m^2\Gamma} \tag{8.5}$$

We also have for the total cross section

$$\int (\Sigma\sigma_{ef})d\sqrt{s} = 2\pi^2(2J+1)\frac{\Gamma_e}{m^2} \tag{8.6}$$

Assuming the spin of the resonance $J/\Psi$, $J=1$, we determine the widths $\Gamma_e = \Gamma_\mu$ and the total decay width $\Gamma$. Since $\Gamma = \Gamma_e + \Gamma_\mu + \Gamma_h$, we can also determine the hadronic decay width $\Gamma_h$. The experimental values for these decay widths are given below:

$$m(J/\Psi) = 3096.916 \pm 0.011 \text{ MeV},$$

$$\Gamma_e = \Gamma_\mu = 5.55 \pm 0.14 \pm 0.02 \text{ keV},$$

$$\Gamma = 93.2 \pm 2.1 \text{ keV}.$$

The $J/\Psi$ spin-parity can be determined from a study of the interference between $e^- e^+ \longrightarrow \gamma \longrightarrow \mu^- \mu^+$ and $e^- e^+ \longrightarrow \Psi \longrightarrow \mu^- \mu^+$. The cross section for the QED process $e^- e^+ \longrightarrow \gamma \longrightarrow \mu^- \mu^+$ is well known [Eq. (78) of Appendix A] and is given by ($s \gg m_e^2, m_\mu^2$)

$$\frac{d\sigma}{d\Omega} = \frac{\alpha^2}{4s}(1+\cos^2\theta)$$

$$\sigma = \frac{4\pi\alpha^2}{s}. \tag{8.7}$$

If the spin-parity of $J/\Psi$ is that of photon viz $1^-$, then the angular distribution would not change by the interference between QED amplitude and the resonant amplitude (See problem 1). In fact, experimentally, it was found to be $(1+\cos^2\theta)$ near the resonance, clearly establishing the spin-parity of $J/\Psi$ to be $1^-$ .

### 8.1.1  *Isospin*

Experimentally the decay $J/\Psi \to p\bar{p}$ occurs with a branching ratio $\Gamma_{p\bar{p}}/\Gamma = (0.214 \pm 0.010)\%$, which is too large to be explained by the electromagnetic effects. Now $p\bar{p}$ can have only $I = 0$ or $I = 1$. Thus the isospin of $J/\Psi$ is either 0 or 1. If $J/\Psi$ has $I = 1$, then the decay $J/\Psi \to \rho^0 \pi^0$ is forbidden while for $I = 0$ (see problem 8.2), we have

$$\frac{\Gamma\left(J/\Psi \to \rho^0 \pi^0\right)}{\Gamma\left(J/\Psi \to \rho^- \pi^+\right) + \Gamma\left(J/\Psi \to \rho^+ \pi^-\right)} = \frac{1}{2} \tag{8.8}$$

to be compared with the experimental value of $0.494 \pm 0.068$. Thus the isospin of $J/\Psi$ is 0. Now $G$-parity is given by $G = (-1)^I C$, where $C$ is the charge conjugation parity of $J/\Psi$. Since $I = 0$, therefore, $G = C$. The allowed decay $J/\Psi \to \rho^0 \pi^0$ fixes its $C$–parity to be $C = (-1)(+1) = -1$. Hence $G = -1$ for $J/\Psi$.

### 8.1.2  *SU(3) Classification*

Due to C-invariance, the VPP coupling is F-type (see Sec. 5.6.2) which is not possible if V is SU(3) singlet. Thus SU(3) singlet vector meson cannot decay into two pseudoscalar mesons belonging to the same SU(3) multiplet. Thus in particular if $J/\Psi$ is SU(3) singlet, then $J/\Psi \to K\bar{K}$ is forbidden while $J/\Psi \to K^*\bar{K}$ or $J/\Psi \to \bar{K}^*K$ is allowed by C-invariance. Experimentally one finds

$$\frac{\Gamma\left(J/\Psi \to K^+ K^-\right)}{\Gamma\left(J/\Psi \to K\bar{K}^*\right)} \approx 2.6 \times 10^{-2} \tag{8.9}$$

which shows that $J/\Psi$ is SU(3) singlet. If $J/\Psi$ is SU(3) singlet, then its invariant coupling with PV is given by

$$\Psi Tr(PV) = \Psi \left[ \rho^0 \pi^0 + \rho^- \pi^+ + \rho^+ \pi^- + K^{*-} K^+ + K^{*+} K^- \right.$$
$$\left. + \bar{K}^{*0} K^0 + K^{*0} \bar{K}^0 + \frac{2}{6} \omega_8 \eta_8 + \frac{4}{6} \omega_8 \eta_8 \right]. \tag{8.10}$$

Hence

$$\frac{\Gamma\left(J/\Psi \to \rho\pi\right)}{\Gamma\left(J/\Psi \to K\bar{K}^*\right)} = 1 \text{ (phase space correction)} \tag{8.11}$$
$$= 1.2$$

to be compared with the experimental value $1.39 \pm 0.12$. To summarize, the $J/\Psi$ resonance is SU(3) singlet with $J^{PC} = 1^{--}$, $G = -1$ and $I = 0$.

## 8.2   Charm

Although $J/\Psi$ itself does not carry any new quantum number, its unusually narrow width in spite of large available phase space suggests that it is a bound state of $c\bar{c}$, where $c$ is a quark with a flavor which is outside the three flavors $u$, $d$ and $s$ of SU(3). This new flavor is called charm. The quark $c$ is assigned a new quantum number $C = 1$ and $C = 0$ for $u$, $d$ and $s$ quarks. Thus to take this quantum number into account, the Gell-Mann-Nishijima relation would be modified to

$$Q = I_3 + \frac{1}{2}(Y + C). \qquad (8.12)$$

For the charmed quark $c$, $C = 1$, $I_3 = 0$, $Y = B = 1/3$. Thus the charge of charmed quark is $2/3$ and its mass $m_c \approx \frac{1}{2}m_{J/\Psi} = 1.55$ GeV.

The narrow width of $J/\Psi$ (87 keV compared to 100 MeV for $\rho$) can be qualitatively understood by the $OZI$ rule, just as the suppression of $\phi \rightarrow 3\pi$ compared to $\phi \rightarrow K\bar{K}$ is explained (see Sec. 5.7) by this rule. Thus the decay depicted in Fig. 8.1 is allowed but that shown in Fig. 8.2 is suppressed by OZI rule. But the decay $J/\psi \rightarrow D\bar{D}$ shown in Fig. 8.1 is not allowed energetically since $m_{J/\Psi} < 2m_D$.

### 8.2.1   *Heavy Mesons*

The heavy quark $Q$ $(\bar{Q})$ can form bound states with light quark $\bar{q}(q)$, where $Q = c$, $b$ and $q = u$, $d$, $s$ :

$$
\begin{array}{llllll}
 & : & (^1S_0) : (Q\bar{q})_{L=0}\chi_A^0 & : & (^3S_1) : (Q\bar{q})_{L=0}\chi_S^{+1,\,0,\,-1} \\
Q = c & : & (c\bar{q}) = 0^- & : & (c\bar{q}) = 1^- \\
C = 1 & : & D^+,\ D^0,\ D_s^+ & : & D^{*+},\ D^{*0},\ D_s^{*+} \\
C = -1 & : & (q\bar{c}) : \bar{D}^0,\ D^-, D_s^- & : & \bar{D}^{*0}, D^{*-}, D_s^{*-}
\end{array}
$$

For $L = 0$, i.e. for s-wave

$$\mathbf{J} = \mathbf{S}_q + \mathbf{S}_Q$$
$$J^2 = S_q^2 + S_Q^2 + 2\mathbf{S}_q \cdot \mathbf{S}_Q \qquad (8.13)$$

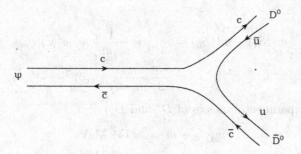

Fig. 8.1   $J/\psi$ allowed by OZI.

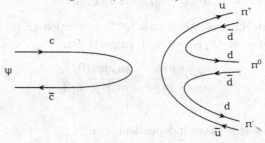

Fig. 8.2   $J/\psi$ suppressed by $OZI$ rule.

We have $[\hbar = 1]$

$$\langle \mathbf{S}_q \cdot \mathbf{S}_Q \rangle_{J=0,1} = \frac{1}{2}\left[ J(J+1) - \frac{3}{2}\right]$$

$$= -\frac{3}{4} \text{ for singlet state: } \sigma_q \cdot \sigma_Q = -3$$

$$= \frac{1}{4} \text{ for triplet state: } \sigma_q \cdot \sigma_Q = 1$$

$$m(^3S_1) > m(^1S_0) \tag{8.14}$$

The mass spliting between $^3S_1$ and $^1S_0$ is given by the Fermi contact term (see Chap. 7) viz

$$H_{ij} = \frac{8\pi}{3}\alpha_s k_s \frac{\sigma_i \cdot \sigma_j}{4m_i m_j}\delta^3(r)$$

$$k_s = 4/3 \text{ for } q\bar{q}$$

$$= 2/3 \text{ for } qq \tag{8.15}$$

Hence

$$m(^3S_1) - m(^1S_0) = \frac{8\pi}{3}\frac{4}{3}\alpha_s \frac{1}{m_q m_Q}\langle\psi_s|\delta^3(r)|\psi_s\rangle$$

$$= \frac{32\pi}{9}\frac{\alpha_s}{m_q m_Q}|\psi_s(0)|^2 \qquad (8.16)$$

From the experimental masses of $D^*$ and $D_s^*$,

$$m_{D^{*+}} - m_{D^+} \approx 141 \text{ MeV}$$

$$m_{D^{*0}} - m_{D^0} \approx 142 \text{ MeV}$$

$$m_{D_s^{*\pm}} - m_{D_s^\pm} \approx 144 \text{ MeV}$$

These values are in agreement with QCD prediction (8.16) for the mass difference between $m(^3S_1) - m(^1S_0)$. In particular

$$\frac{m_{D_s^*} - m_{D_s}}{m_{D^*} - m_D} \approx \frac{m_d}{m_s}\frac{|\psi_s(0)|^2_{D_s}}{|\psi_s(0)|^2_D} \qquad (8.17)$$

$$\approx 1$$

i.e. right-hand side is flavor independent.

We now discuss $L = 1$ (p-wave) heavy mesons: $(Q\bar{q})_{L=1}$. As a first approximation, we take heavy quark as stationary and its spin is decoupled. It is natural to couple orbital angular momentum $\mathbf{L}$ with $\mathbf{S}_q$ in heavy quark limit:

$$\mathbf{j} = \mathbf{L} + \mathbf{S}_q \qquad (8.18)$$

The total angular momentum $\mathbf{J}$ of the bound state $Q\bar{q}$ is given by

$$\mathbf{J} = \mathbf{j} + \mathbf{S}_Q \qquad (8.19)$$

Now

$$j^2 = S_q^2 + L^2 + 2\mathbf{S}_q.\mathbf{L}$$

Eigenvalues of $\mathbf{S}_q.\mathbf{L}$ are given by $[\hbar = 1]$

$$\langle\mathbf{S}_q \cdot \mathbf{L}\rangle = \frac{1}{2}\left[j(j+1) - l(l+1) - \frac{3}{4}\right] \qquad (8.20)$$

Now

$$j = l - 1/2, l + 1/2.$$

Thus for $L = 1$

$$j = 1/2, 3/2$$

$$\langle \mathbf{S}_q \cdot \mathbf{L} \rangle_{j=1/2} = -1$$

$$\langle \mathbf{S}_q \cdot \mathbf{L} \rangle_{j=3/2} = \frac{1}{2} \tag{8.21}$$

Hence

$$m(j = 3/2) > m(j = 1/2) \tag{8.22}$$

Since

$$J = j + 1/2, \; j - 1/2,$$

for $j = 3/2$, we have $J = 2, 1$: $D_2^*$, $D_1$ degenerate $J^P = 2^+$, $1^+$

for $j = 1/2$, we have $J = 1, 0$: $D_1^*$, $D_0$ degenerate $J^P = 1^+$, $0^+$

$$m(j = 3/2) = \frac{5m_{D_2^*} + 3m_{D_1}}{8}$$

$$m(j = 1/2) = \frac{3m_{D_1^*} + m_{D_0}}{4} \tag{8.23}$$

The degeneracy between $(D_2^*, D_1)$ and $(D_1^*, D_0)$ is lifted by spin orbital coupling (see Chap. 9). All the charmed mesons are listed in Table 8.1. We note that $D^+$, $D^0$, $D_s^+$ $(\bar{D}^0, D^-, D_s^-)$ can decay only weakly, whereas $J^P = 1^-$ $D$-meson decay strongly and radiatively.

Table 8.1

| Charmed meson | Quark content | Mass (MeV) | Lifetime ($10^{-12}$ sec) /Width | $J^P$ |
|---|---|---|---|---|
| $D^0$ | $c\bar{u}$ | $1864.54 \pm 0.17$ | $\tau = 0.4101 \pm 0.0015$ | $0^-$ |
| $D^+$ | $c\bar{d}$ | $1869.62 \pm 0.2$ | $\tau = 1.040 \pm 0.007$ | $0^-$ |
| $D_s^+$ | $c\bar{s}$ | $1968.49 \pm 0.34$ | $\tau = 0.5 \pm 0.007$ | $0^-$ |
| $D^{*0}$ | $c\bar{u}$ | $2006.97 \pm 0.19$ | $\Gamma < 2.1$ MeV | $1^-$ |
| $D^{*+}$ | $c\bar{d}$ | $2010.27 \pm 0.17$ | $\Gamma = (96 \pm 22)$ KeV | $1^-$ |
| $D_s^{*+}$ | $c\bar{s}$ | $2112.3 \pm 0.7$ | $\Gamma \le 1.9$ MeV | $1^-$ |
| $D_0^{*0}$ | $c\bar{u}$ | $2318 \pm 29$ | $\Gamma \le 267 \pm 40$ MeV | $1^-$ |
| $D_1^0$ | $c\bar{u}$ | $2422.3 \pm 1.3$ | $\Gamma = 20.4 \pm 1.7$ MeV | $1^+$ |
| $D_1^+$ | $c\bar{d}$ | – | – | $1^+$ |
| $D_2^{*0}$ | $c\bar{u}$ | $2461 \pm 1.6$ | $\Gamma = 43 \pm 4$ MeV | $2^+$ |
| $D_2^{*+}$ | $c\bar{d}$ | $2460.1^{+2.6}_{-3.5} \pm 1.6$ | $\Gamma = 37 \pm 6$ MeV | $2^+$ |
| $D_{s0}^{*\pm}$ | $c\bar{s}$ | $2317.8 \pm 0.6$ | $\Gamma < 3.8$ MeV | $0^+$ |
| $D_{s1}^{*\pm}$ | $c\bar{s}$ | $2459.6 \pm 0.6$ | $\Gamma < 3.5$ MeV | $1^+$ |
| $D_{s1}^{\pm}$ | $c\bar{s}$ | $2535.35 \pm 0.34 \pm 0.5$ | $\Gamma < 2.3$ MeV | $1^+$ |
| $D_{s2}^*$ | $c\bar{s}$ | $2572.6 \pm 0.9$ | $\Gamma = 20 \pm 5$ MeV | $2^+$ |

## 8.2.2  *The Fifth Quark Flavor: Bottom Mesons*

Fifth quark was discovered, when in 1977 the upsilon meson $\Upsilon(J^{PC} = 1^{--})$ was found experimentally as a narrow resonance at Fermi Lab. with mass $\sim 9.5$ GeV. This was later confirmed in $e^+e^-$ experiments at DESY and CESR which determined its mass to be $9460 \pm 10$ MeV and also its width. The updated parameters of this resonance [from the Particle Data Group Tables] are mass $9460.37 \pm 0.21$ MeV and width $52.5 \pm 1.8$ keV. Again the narrow width in spite of large phase space available suggests the existence of a fifth quark flavor called beauty, with a new quantum number $B = -1$ for the bottom ($b$) quark. With this assignment the formula $Q = I_3 + 1/2(Y + B + C)$ would give the charge of $b$ quark the value $-1/3(I_3 = 0)$. The mass of $b$ quark is expected to be around 4.9 GeV as suggested by the $\Upsilon$ mass which is regarded as a $^3S_1$ bound state of $b\bar{b}$.

Thus one would expect particles with $B = \pm 1$, such as $b\bar{q}$ or $q\bar{b}$. The lowest lying bound states $b\bar{q}$ and $q\bar{b}$ have been found experimentally. The $B = -1$ states $(\bar{B}^0, B^-)\bar{B}^0_s$ form an SU(3) triplet ($\bar{3}$) and $B = +1$ states $(B^+, B^0)B^0_s$ form another triplet (3). For p-wave multiplets

$$(q\bar{b})_{L=1} \quad J^P = 2^+, 1^+ \quad \left[ \begin{array}{c} (B_2^{*+,0}, B_1^{+,0}) \\ (B_{s_2}^{*0}, B_{s_1}^{0}) \end{array} \right]_{j=3/2}$$

$$J^P = 1^+, 0^+ \quad \left[ \begin{array}{c} (B_1^{*+,0}, B_0^{+,0}) \\ (B_{s_1}^{*0}, B_{s_0}^{0}) \end{array} \right]_{j=1/2}$$

The masses and decay time of B-mesons are given below

$$B^\pm = 5279.16 \pm 0.31 \text{ MeV}, \quad \tau = (1.638 \pm 0.11) \times 10^{-12} \text{ sec}$$

$$B^0 = 5279.53 \pm 0.33 \text{ MeV}, \quad \tau = (1.530 \pm 0.069) \times 10^{-12} \text{ sec}$$

The masses and decay widths of other $B$-mesons can be found in particle data book.

## 8.2.3  *The Sixth Quark Flavor: The Top*

The top quark $t$ with $Q = 2/3$ and new flavor $T = 1$ was expected on theoretical grounds. It was first found experimentally in 1996; its mass is $m_t = 172.0 \pm 0.9 \pm 1.3$ GeV. Since $(t, b)$ form a weak doublet, it decays weakly to $W^+ + b$, i.e.

$$t \rightarrow W^+ + b$$

The predicted decay rate is [see Chap. 13]

$$\Gamma(t \rightarrow W^+ + b) = \frac{G_F}{8\pi\sqrt{2}} m_t^3 \left(1 - \frac{m_W^2}{m_t^2}\right)^2 \left(1 + 2\frac{m_W^2}{m_t^2}\right) \tag{8.24}$$

where we have neglected the $b$ quark mass compared to $m_W$ and $m_t$. Taking $m_t = 172$ GeV and $m_W = 80$ GeV, $G_F = 1.166 \times 10^{-5}$ GeV$^{-2}$, we get

$$\Gamma \approx 1.48 \text{ GeV}.$$

If $QCD$ correction is taken into account, then

$$\Gamma \approx 1.36 \text{ GeV}.$$

which gives, the life time of $\tau$ to be

$$\tau = 4.84 \times 10^{-25} \text{ sec}$$

Thus $t$ quark decays before it can form bound states such as $t\bar{t}$ and $t\bar{q}$.

## 8.3 Strong and Radiative Decays of $D^*$ Mesons

$$D^* \to D + \pi$$
$$D^* \to D + \gamma$$

These are p-wave decays. The decay width for the strong decay $D^* \to D\pi$ is given by

$$\Gamma(D^{*+} \to D^+ \pi^0) = \frac{g^2}{4\pi} \frac{2}{3} \frac{|\mathbf{P}|^3_{D\pi}}{m^2_{D^*}} \tag{8.25}$$

Now $D^+$, $D^0$ form an isospin doublet. SU(2) gives

$$\begin{aligned}
D^{*+} &\to D^+ \pi^0 : -g \\
D^{*+} &\to D^0 \pi^+ : \sqrt{2}g \\
D^{*0} &\to D^0 \pi^0 : g \\
D^{*0} &\to D^+ \pi^- : \sqrt{2}g
\end{aligned} \tag{8.26}$$

The decay $D^{*0} \to D^+ \pi^-$ is not energetically allowed. The experimental value for the total decay width of $D^{*+}$ is

$$\begin{aligned}
\Gamma_{D^{*+}} &= 96 \pm 22 \text{ keV} \\
Br(D^{*+} &\to D^0 \pi^+) = (67.7 \pm 0.5)\% \\
Br(D^{*+} &\to D^+ \pi^0) = (30.7 \pm 0.5)\% \\
Br(D^{*+} &\to D^+ \gamma) = (1.6 \pm 0.4)\%
\end{aligned} \tag{8.27}$$

From the above equations, we get

$$\begin{aligned}
\Gamma(D^{*+} &\to D^0 \pi^+) = (65 \pm 15) \text{ keV} \\
\Gamma(D^{*+} &\to D^+ \pi^0) = (36 \pm 8) \text{ keV} \\
\Gamma(D^{*+} &\to D^+ \gamma) = (1.5 \pm 0.5) \text{ keV}
\end{aligned} \tag{8.28}$$

Thus we get

$$\frac{g^2}{4\pi} = (3.32 \pm 0.76) \tag{8.29}$$

The decay width for the radiative decay is given by

$$\Gamma = \frac{4\alpha}{3}\mu^2|\mathbf{k}|^3 \tag{8.30}$$

$$\mu_{D^+} = \langle D^+| \sum_i \frac{Q_i}{2m_i}\sigma_{iz}|D^{*+}\rangle$$

$$= \frac{1}{6}\left[-\frac{1}{m_d} + \frac{2}{m_c}\right]$$

$$\mu_{D^0} = \frac{1}{3}\left[\frac{1}{m_u} + \frac{1}{m_c}\right]$$

$$\mu_{D_s} = \frac{1}{6}\left[-\frac{1}{m_s} + \frac{2}{m_c}\right] \tag{8.31}$$

Using $m_d = m_u \approx 336$ MeV, $m_s \approx 490$ MeV, $m_c \approx 1500$ MeV,

$$\Gamma(D^{*+} \to D^+\gamma) = 1.8 \text{ keV}$$

$$\Gamma(D^{*0} \to D^0\gamma) = 37 \text{ keV}$$

$$\Gamma(D_s^{*+} \to D_s\gamma) = 0.36 \text{ keV} \tag{8.32}$$

Now

$$\Gamma(D^{*0} \to D^0\pi^0) = \frac{|\mathbf{p}|_{D^0\pi^0}^3}{|\mathbf{p}|_{D^+\pi^0}^3}\frac{m_{D^*}^2}{m_{D^{*0}}^2}\Gamma(D^{*+} \to D^+\pi^0)$$

$$= (52 \pm 10) \text{ keV}$$

$$\Gamma(D^{*0} \to D^0\gamma) = \left(\frac{38.1 \pm 0.29}{61.9 \pm 0.29}\right)\Gamma(D^{*0} \to D^0\pi^0) \tag{8.33}$$

$$= (32 \pm 6) \text{ keV}$$

consistent with the quark model value for the radiative decay given in Eq. (8.32). The strong decays of

$$D_s^{*+} \to D^0 K^+$$

$$\to D^+ K^0$$

are not energetically allowed. Thus decay channel allowed are radiative decay, $D_s^{*+} \to D_s\gamma$ and the decay $D_s^{*+} \to D_s^+\pi^0$ is isospin violating decay, hence it is also electromagnetic decay. The full width of $D_s^{*+}$ is $\Gamma < 1.9$ MeV, and

$$Br(D_s^{*+} \to D_s^+\gamma) = (94 \pm 0.7)\% $$

$$Br(D_s^{*+} \to D_s^+\pi^0) = (5.8 \pm 0.7)\% \tag{8.34}$$

We now briefly discuss the strong decays of p-wave $D$-mesons. Parity and angular momentum selection rules imply the following allowed decay modes

| | | | |
|---|---|---|---|
| $D_2^* \to D^*\pi$ | $l = 2$ | d-wave decay |
| $\to D\pi$ | $l = 2$ | d-wave decay |
| $D_1 \to D^*\pi$. | $l = 0, 2$ | s and d-wave decays |
| $D_1 \nrightarrow D\pi$ | $l = 1$ | not allowed due to parity |
| $D_1^* \to D^*\pi$ | $l = 0, 2$ | s and d-wave decays |
| $D_0^* \nrightarrow D^*\pi$ | | not allowed |
| $D_0^* \to D\pi$ | $l = 0$ | s-wave decays |

The allowed s-wave decays are very broad, since $D_2^*$, $D_1$ belong to one multiplet viz $j = 3/2$; it is reasonable to assume that the s-wave decay of $D_1$ is suppressed, this is also born out by the experimental values of decay widths

$$\Gamma_{D_1^0} = (20.4 \pm 1.7) \text{ MeV}$$
$$\Gamma_{D_2^{*0}} = (43 \pm 4) \text{ MeV}$$
$$\Gamma_{D_2^{*\pm}} = (37 \pm 6) \text{ MeV} \tag{8.35}$$

The resonances $D_1^*$, $D_0^*$ have not been seen experimentally; their decay widths are expected to be of the order of few hundred MeV. However, a resonance $D_0^*$ at $(2138 \pm 28)$ MeV with width $(267 \pm 40)$ MeV has been experimentally discovered. Same selection rules hold for p-wave, $D_{sJ}$ mesons; $J = 2, 1, 1, 0$.

The decay channels and the widths for the multiplet, $j = 3/2$ are given below

$$D_{s_1}^+ \to D^{*+}K^0$$
$$\to D^{*0}K^+$$
$$D_{s_2}^+ \to D^0 K^+$$
$$\to D^+ K^0$$
$$\to D^{*0}K^+$$
$$\to D^{*+}K^0$$

$$\Gamma_{D_{s_1}} < 2.3 \text{ MeV} \tag{8.36}$$

$$\Gamma_{D_{s_2}^*} = 20 \pm 5 \text{ MeV} \tag{8.37}$$

For the multiplet, $j = 1/2$ $(D^*_{s_1}, D^*_{s_0})$, the above decays are not energetically allowed, only channels energetically allowed are radiative decays and isospin violating decays

$$D^*_{s_1} \rightarrow D^{*+}_s \pi^0$$
$$\rightarrow D^+_s \gamma$$
$$\rightarrow D^{*+}_s \gamma$$
$$D^*_{s_0} \rightarrow D^+_s \pi^0$$

They are narrow resonances, since strong decays are energetically not allowed. Experimentally it is more likely to find a narrow resonance than a very broad resonance.

## 8.4   Heavy Baryons

Since $u$, $d$, $s$, belong to the triplet representation of SU(3), the charmed and bottom baryons with spin parity $\frac{1}{2}^+$ belong to either triplet representation $\mathbf{3}$ or sextet representation $\mathbf{6}$ of SU(3). Using the Pauli principle, the unitary spin and spin wave functions of spin $\frac{1}{2}^+$ baryons can be written as

$$A_{ij} = \frac{1}{\sqrt{2}} (q_i q_j - q_j q_i) Q \chi_{MA} \tag{8.38}$$

$$S_{ij} = \frac{1}{\sqrt{2}} (q_i q_j + q_j q_i) Q \chi_{MS}, \tag{8.39}$$

where $i, j = 1, 2, 3$ ($q_1 = u$, $q_2 = d$, $q_3 = s$, $Q = c$ or $b$) and the spin wave functions $\chi_{MA}$ and $\chi_{MS}$ are given in Eq. (6.8) and Table 6.3 respectively. Note that $A_{ij}$ belongs to triplet representation $\mathbf{\bar{3}}$ of SU(3): In particular we have an isospin singlet and isospin doublet: For $Q = c$

| $J^P = \frac{1}{2}^+$ | Mass(GeV) | Mean life($10^{-15}$ sec) |
|---|---|---|
| $A_{12} = \frac{1}{\sqrt{2}} (ud - du) c : \Lambda_c^+$ | 2.286 | 200±6 |
| $I = \frac{1}{2} \begin{bmatrix} A_{13} = \frac{1}{\sqrt{2}} (us - su) c : \Xi_c^+ \\ A_{23} = \frac{1}{\sqrt{2}} (ds - sd) c : \Xi_c^0 \end{bmatrix}$ | 2.468<br>2.471 | 442±26<br>$112^{+13}_{-10}$ |

These baryons decay weakly.

For $Q = b$

| | Mass | Mean life($10^{-12}$ sec) |
|---|---|---|
| $A_{12} : \Lambda_b^0$ | 5.620 | 1.383 |
| $\begin{bmatrix} A_{13} : \Xi_b^0 \\ A_{23} : \Xi_b^- \end{bmatrix}$ | 5.792 | 1.42 |

$S_{ij}$ belongs to sextet representation of SU(3). In particular, we have an isospin triplet, an isospin doublet and an isospin singlet:

| $J^P = \frac{1}{2}^+$ | | Mass(GeV) | Decay channel |
|---|---|---|---|
| $I = 1$ | $\begin{cases} S_{11} = \sqrt{2}\Sigma_{cc}^+ \\ S_{12} = \Sigma_c^+ \\ S_{22} = \sqrt{2}\Sigma_c^0 \end{cases}$ | 2.455 $\rightarrow$ | $\Lambda_c^+\pi^0(\Gamma \sim 2.23 MeV)$ |
| $I = 1/2$ | $\begin{cases} S_{13} = \Xi_c'^+ \\ S_{23} = \Xi_c'^0 \end{cases}$ | 2.576 | $\Xi_c\gamma(-)$ |
| $I = 0$ | $S_{33} = \frac{1}{\sqrt{2}}(2ss)c : \sqrt{2}\Omega_c^0$ | 2.697 | $-(\tau = 69 \times 10^{-15} \sec)$ |
| $I = 1$ | $\begin{cases} S_{11} = \sqrt{2}(\Sigma_b^+), \\ S_{12} = (\Sigma_b^0), \\ S_{22} = \sqrt{2}(\Sigma_b^-), \end{cases}$ | | |
| $I = 1/2$ | $\begin{cases} S_{13} = (\Xi_b'^0) \\ S_{23} = (\Xi_b'^-) \end{cases}$ | | |
| $I = 0$ | $S_{33} = \sqrt{2}(\Omega_b^-)$ | | |

The spin $\frac{3}{2}^+$ baryons also belong to the sextet representation of $SU(3)$. They are given in Eq. (8.39), with $\chi_{MS}$ replaced by $\chi_A$ where the spin wave functions $\chi_s$ are given in Table 6.1. The six spin $\frac{3}{2}^+$ baryons are labelled as

$$\Sigma_c^{*++}(\Sigma_b^{*+}), \ \Sigma_c^{*+}(\Sigma_b^{*0}), \ \Sigma_c^{*0}(\Sigma_b^{*-}),$$

$$\Xi_c^{*+}(\Xi_b^{*0}), \ \Xi_c^{*0}(\Xi_b^{*-}), \ \Omega_c^{*+}(\Omega_b^{*0})$$

In addition to $C = +1$ and $B = -1$ baryons considered above, we also have the following baryons with $C = 2$ and $B = -2$ belonging to the triplet representation of $SU(3)$ with spin parity $(3/2)^+$:

$$\Xi_{cc}^{*++} = ccu\chi_s, \ \Xi_{cc}^{*+} = ccd\chi_s, \ \Omega_{cc}^{*+} = ccs\chi_s$$
$$\Xi_{bb}^{*0} = bbu\chi_s, \ \Xi_{bb}^{*-} = bbu\chi_s, \ \Omega_{bb}^{*-} = bbs\chi_s \quad (8.40)$$

Finally we have singlets with $C = 3$ and $B = -3$, namely

$$\Omega_{ccc}^{*++} = ccc\chi_s, \ \Omega_{bbb}^{*-} = bbb\chi_s \quad (8.41)$$

## 8.5  Quarkonium

The bound system of heavy quarks $Q\bar{Q}$, $Q = c, b$, is called quarkonium e.g. charmonium $c\bar{c}$ and bottomonium $b\bar{b}$. Since quarks are fermions with spin 1/2, their bound system can be written as $(Q\bar{Q})_{L,S}$. Now spin $S$ can have two values 0 and 1 with spin wave function antisymmetric and symmetric respectively. If we regard $Q$ and $\bar{Q}$ as identical fermions which differ only in their charges, then we can state generalized Pauli principle: The wave function is antisymmetric with the exchange of particles $Q$ and $\bar{Q}$. Under particle exchange, we get with space coordinates exchange, a factor $(-1)^L$, with spin coordinates exchange, a factor $(-1)^{S+1}$ and with charge exchange, a factor $C$ ($C$ is called $C$-parity). Hence Pauli principle gives

$$(-1)^{L+S+1}C = -1. \tag{8.42}$$

Therefore,

$$C = (-1)^{L+S}. \tag{8.43}$$

Hence we have the result

$$C = \begin{cases} -1 & L+S \quad \text{odd} \\ +1 & L+S \quad \text{even.} \end{cases} \tag{8.44}$$

Also for $(Q\bar{Q})$ system, the parity

$$P = (-1)(-1)^L = (-1)^{L+1}. \tag{8.45}$$

Let us now use the spectroscopic notation,

$$L \; = \; \begin{matrix} 0, & 1, & 2, & 3, & \cdots \\ S, & P, & D, & E, & \cdots \end{matrix} .$$

A state is completely specified as

$$n \; {}^{2S+1}L_J,$$

where $n$ is the principal quantum number and $J$ is the total angular momentum. Thus for $L = 0$, we have the following states

$$J^{PC}$$

$$\begin{matrix} n & {}^1S_0 & C = +1, & n = 1, 2, \cdots & 0^{-+} \\ n & {}^3S_1 & C = -1, & n = 1, 2, \cdots . & 1^{--} \end{matrix}$$

The ground state is therefore a hyperfine doublet $1\,{}^1S_0(0^{-+})$ and $1\,{}^3S_1(1^{--})$. For $L = 1$, we have the following states

$$\begin{matrix} n & {}^1P_J & J = +1, & C = -1, & 1^{+-} \\ n & {}^3P_J & J = 0, 1, 2 & C = 1, & 0^{++}, 1^{++}, 2^{++}. \end{matrix}$$

Finally, we note that for $L = 2$, we have the following states

$$\begin{matrix} n & {}^1D_J & J = 2, & C = +1, & 2^{-+} \\ n & {}^3D_J & J = 1, 2, 3 & C = -1, & 1^{--}, 2^{--}, 3^{--}. \end{matrix}$$

The low lying states for $L = 0, 1$ are listed below (Masses in GeV and decay widths in MeV).

| | | | |
|---|---|---|---|
| $n = 1$ | $^1S_0$ | $[\eta_c(2.980)$ | $26.7 \pm 3.0$ |
| | $^3S_1$ | $\big[\ J/\Psi(3.097)$ | $(93.2 \pm 2.1) \times 10^{-3}$ |
| | | $\Upsilon(9.460)$ | $(54.02 \pm 1.25) \times 10^{-3}$ |
| $n = 2$ | | $\big[\ \acute{\eta}_c(3.637)$ | $14 \pm 7$ |
| | | $\acute{\psi}(3.686)$ | $(317 \pm 9) \times 10^{-3}$ |
| | | $\acute{\Upsilon}(10.023)$ | $(31.98 \pm 2.63) \times 10^{-3}$ |
| | $^1P_1(1^{+-})$ | $h_c(3.526)$ | $< 1$ |
| | $^3P_0(0^{++})$ | $\chi_{c0}(3.415)$ | $10.2 \pm 0.7$ |
| | $^3P_1(1^{++})$ | $\chi_{c1}(3.510)$ | $0.89 \pm 0.05$ |
| $n = 1$ | $^3P_2(2^{++})$ | $\chi_{c2}(3.556)$ | $2.03 \pm 0.12$ |
| | $^1P_1(1^{+-})$ | $h_b$ | - |
| | $^3P_0(0^{++})$ | $\chi_{b0}(9.859)$ | - |
| | $^3P_1(1^{++})$ | $\chi_{b1}(9.893)$ | - |
| | $^3P_2(2^{++})$ | $\chi_{b2}(9.912)$ | - |

It is interesting to see that the state $^3D_1$ has the same quantum number as $^3S_1$. They can therefore mix, but the mixing is expected to be small.

The states $^3P_J$ and $^1P_1$ is a hyperfine quartet (degenerate), but this degeneracy is removed due to hyperfine splitting. The low lying charmonium states listed above are shown in Fig. 8.3. Most of these states have been discovered experimentally. The transitions and decays of charmonium states are shown in Fig. 8.3. Similar transitions and decays occur for bottomonium bound states.

From Fig. 8.3, we note that both $M1$ and $E1$ radiative transitions are possible:

$$J/\Psi \rightarrow \eta_c + \gamma$$
$$\Psi' \rightarrow \eta_c + \gamma$$
$$\Psi' \rightarrow \eta_c' + \gamma \ \ \text{(no parity change)} \quad \text{M1 transitions}$$
$$\rightarrow \Psi + \gamma$$
$$\left.\begin{array}{l} \Psi' \rightarrow \chi_c + \gamma \\ \chi \rightarrow \eta_c + \gamma \end{array}\right] \ \text{(parity changes)} \ \ \text{E1 transitions}$$

From Eq. (6.88) and Table 6.3, we get (for example)

$$\Gamma(\Psi' \rightarrow \eta_c\gamma) = \frac{4\alpha}{3}\left[\frac{2}{3}\frac{1}{m_c}\right]^2 k^3\Omega^2$$

$$= 2.7 \ \text{keV} \, \Omega \ \ (1.21 \pm 0.37 \ \text{expt. value}) \qquad (8.46)$$

where $\Omega$ is the overlap integral defined as

$$\Omega_{n'n} = \int_0^\infty e^{i\mathbf{q}\cdot\mathbf{r}} \phi_{n'00}(\mathbf{r})\phi_{n00}(\mathbf{r})\, d^3\mathbf{r}. \tag{8.47}$$

and $\mathbf{q} = (m_{sp}/m)\mathbf{k}$, $\mathbf{k}$ is the momentum carried by photon, $m_{sp}$ is the mass of the spectator quark and $m$ is the mass of the bound state. For $\Omega = 1$, $\Gamma$ is about a factor of two larger than the experimental value.

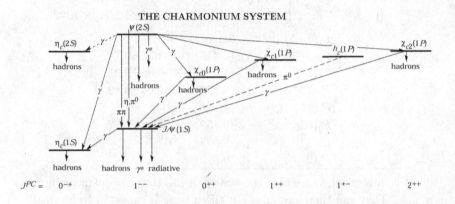

Fig. 8.3   The charmonium spectrum ($c\bar{c}$ bound state) [16].

For $E1$ transitions $nS_1 \to n'P_J$ and $nP_J \to n'S_1$ ($J = 0, 1, 2$) the decay widths can be written (cf. Eq. (6.68))

$$\Gamma_{nS_1 \to n'P_J} = \frac{4\alpha}{3}\left[\frac{2J+1}{3}\right]|M_{n'n}|^2 k^3 \tag{8.48}$$

$$\Gamma_{nP_J \to n'S_1} = \frac{4\alpha}{3}|M_{n'n}|^2 k^3 \tag{8.49}$$

where

$$M_{n'n} = \,<Q>\Omega_{n'n},\ \Omega_{n'n} = (1/\sqrt{3})$$
$$\times \int_0^\infty [j_0(qr) - 2j_2(qr)]R_{n'0}(r)R_{n1}(r)r^3 dr \tag{8.50}$$

Note that $j_0$ and $j_2$ are spherical Bessel functions and $R_{nl}$ are radial wave functions. In order to predict these decay widths one needs to know the radial wave functions, i.e. some potential model is needed.

Finally, we note that there are 22 states below $B$ threshold as compared with eight states below charm threshold. This is a consequence of the fact that interquark potential is flavor independent (as expected in QCD) so that $E_{n2} - E_{n1}$ is the same for $c\bar{c}$ and $b\bar{b}$. (Note that charm threshold is at about 3.74 GeV whereas $B$ threshold is at about 10.55 GeV.)

## 8.6   Leptonic Decay Width of Quarkonium

The decays of $^3S_1(Q\bar{Q})$ state $(V)$ into charged leptons proceeds through the virtual photon as shown in Fig. 8.4.

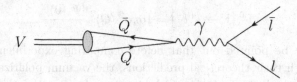

Fig. 8.4   The decay of $V$ in to charged leptons through virtual photon.

The scattering cross section for the $Q\bar{Q} \to l\bar{l}$ is given by Eq. (A.77)

$$\sigma = \frac{4\pi\alpha^2}{3}\,\langle Q \rangle^2\,\frac{1}{s}\,\frac{\beta_l}{\beta_Q}$$
$$\times\left[1 + \frac{2 - \beta_Q^2 - \beta_l^2}{2} + \frac{\left(1 - \beta_Q^2\right)\left(1 - \beta_l^2\right)}{4}\right] \qquad (8.51)$$

where

$$\beta_l = \frac{\sqrt{s - 4m_l^2}}{\sqrt{s}}, \quad \beta_Q = \frac{\sqrt{s - 4m_Q^2}}{\sqrt{s}}$$

$$s = E_{cm}^2 \qquad (8.52)$$

and $Q$ is the charge of the quark $Q$. Now the cross section $\sigma$ can be written as

$$\sigma = \frac{3}{4}\sigma_t + \frac{1}{4}\sigma_s, \qquad (8.53)$$

where $\sigma_t$ is the cross section for $^3S_1$ state and $\sigma_s$ is the cross-section for $^1S_0$ state. Since the photon is coupled to a conserved vector current, therefore it contributes only to spin triplet state. Thus $\sigma_s = 0$. Hence the decay rate in the limit $\beta_l \to 1$ $(s = 4m_Q^2 \gg 4m_l^2)$ is given by

$$\Gamma = (\text{incident flux})\,\sigma_t$$
$$= 2\beta_Q|\Psi_s(0)|^2\frac{4}{3}\sigma, \qquad (8.54)$$

where the incident flux $= \rho_{in}(2\beta_Q) = 2|\Psi_s(0)|^2\beta_Q$. Hence from Eqs. (8.51) and (8.54),

$$\Gamma\left[^3S_1(V) \to l^+l^-\right] = \frac{16\pi\alpha^2}{3}\,\langle Q \rangle^2\,\frac{|\Psi_s(0)|^2}{m_V^2}, \qquad (8.55)$$

where we have put $s = 4m_Q^2 \approx m_V^2$ and $\beta_Q \approx 0$ (in the non-relativistic limit).

Taking into account the color $|V\rangle = \frac{1}{\sqrt{3}} \sum_a |\bar{Q}_a Q_a\rangle$, we multiply Eq. (8.55) by a factor of three. Hence we have

$$\Gamma^0 \left[ V \to l^+ l^- \right] = 16\pi\alpha^2 \langle Q\rangle^2 \frac{|\Psi_s(0)|^2}{m_V^2}. \tag{8.56}$$

It may be pointed out that before comparing experimental leptonic widths with their theoretical predictions, the vacuum polarization contributions to the leptonic decay width have to be removed so that

$$\Gamma^0 = \Gamma^{exp}(1 - \Pi)^2$$

where $(1 - \Pi)^2 = 0.958, 0.932$ for charmonium and bottomonium respectively and then it is $\Gamma^0$ which is to be compared with the theoretical predictions.

In the quark model, the electromagnetic current

$$\begin{aligned} J_\mu^{e.m} &= \frac{2}{3}\bar{u}\gamma_\mu u - \frac{1}{3}\bar{d}\gamma_\mu d - \frac{1}{3}\bar{s}\gamma_\mu s + \frac{2}{3}\bar{c}\gamma_\mu c - \frac{1}{3}\bar{b}\gamma_\mu b \\ &\sim \frac{1}{\sqrt{2}}\rho_\mu^0 + \frac{1}{3}\frac{1}{\sqrt{2}}\omega_\mu^0 - \frac{1}{3}\phi_\mu^0 + \frac{2}{3}\psi_\mu - \frac{1}{3}\Upsilon_\mu \end{aligned} \tag{8.57}$$

For

$$V_\mu \to e^+ e^-$$

$$\langle 0 | J_\mu^{e.m} | V \rangle = f_V \frac{m_V}{(2\pi)^{3/2}} \frac{\epsilon_\mu}{\sqrt{2q_0}}$$

$$\Gamma \left[ V \to e^+ e^- \right] = \frac{4\pi\alpha^2}{3} \frac{f_V^2}{m_V} \tag{8.58}$$

where (cf Eq. (8.57))

$$f_V = \frac{1}{\sqrt{2}}f_\rho, \quad \frac{1}{3}\frac{1}{\sqrt{2}}f_\omega, \quad -\frac{1}{3}f_\phi, \quad \frac{2}{3}f_\psi, \quad -\frac{1}{3}f_\Upsilon \tag{8.59}$$

for $V = \rho, \omega, \phi, \psi$ and $\Upsilon$ respectively.

## 8.7   Hadronic Decay Width

The decays of quarkonium states $^3S_1$ and $^1S_0$ to ordinary hadrons are suppressed by the *OZI* rule. The narrowness of their decay widths can be explained as follows. By $C$-conservation $^3S_1$ state can decay in the lowest order to three gluons and thus its hadronic decay width is proportional to $\alpha_s^3 \times$ (probability of conversion of gluons into hadrons). Since color is confined, this probability is unity. Similarly the decay of $^1S_0$ into hadrons is proportional to $\alpha_s^2$, since by $C$-conservation it can decay into two gluons. Here analogy with positronium is in order. Positronium in $^1S_0$ state (para positronium) decay into two photons via the diagram (Fig. 8.5).

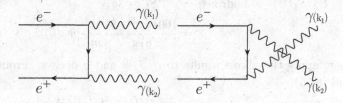

Fig. 8.5   Positronium ($^1S_0$ state) decay into two photons.

In the low energy limit the cross section for the above process is given by

$$\sigma = \frac{\pi}{\beta}\left(\frac{\alpha}{m_e}\right)^2. \tag{8.60}$$

Since $\sigma_t = 0$, we get using Eq. (8.53)

$$\sigma_s = 4\sigma = \frac{4\pi}{\beta}\left(\frac{\alpha}{m_e}\right)^2. \tag{8.61}$$

Hence the decay rate

$$\Gamma\left[^1S_0(e^-e^+) \to 2\gamma\right] = |\beta\Psi_s(0)|^2\, 4\,\sigma = 16\pi\frac{\alpha^2}{4m_e^2}|\Psi_s(0)|^2. \tag{8.62}$$

For $(Q\bar{Q})$ $^1S_0$ state decaying into $2\gamma$, replace

$$e^4 \to \left[\sqrt{3}Q^2 e^2\right]^2 = 3Q^4 e^4 \tag{8.63}$$

and $4m_e^2 \to 4m_Q^2 \approx m_P^2$. Hence

$$\Gamma\left[^1S_0(m_P) \to 2\gamma\right] = \frac{16\pi\alpha^2}{m_P^2}3Q^4|\Psi_s(0)|^2. \tag{8.64}$$

For $\eta_c \to 2$ gluons, replace $\alpha^2$ by $\frac{2}{3}\alpha_s^2$ in Eq. (8.62) [see problem 8.2], so that the hadronic decay rate is

$$\Gamma\left[\eta_c \to \text{hadrons}\right] = \frac{32\pi}{3}\frac{\alpha_s^2(m_{\eta_c})}{m_{\eta_c}^2}|\Psi_s(0)|^2. \tag{8.65}$$

The decay rate for $^3S_1(e^-e^+)$ system going to $3\gamma$ is given by

$$\Gamma\left[^3S(e^-e^+) \to 3\gamma\right] = \frac{64\pi}{9\pi}(\pi^2 - 9)\frac{\alpha^3}{4m_e^2}|\Psi_s(0)|^2. \qquad (8.66)$$

For the decay of $^3S_1(\bar{Q}Q) \to 3g$, we replace $\alpha^3$ by $5\alpha_s^3/18$ [see problem 8.5] and $(2m_e)^2 = (2m_Q)^2 \approx m_V^2$ in Eq. (8.66). Hence

$$\Gamma\left[^3S_1(V) \to \text{hadrons}\right] = \Gamma\left[^3S_1 \to 3g\right]$$

$$= \frac{160\pi(\pi^2 - 9)}{81\pi}\frac{\alpha_s^3}{m_V^2}|\Psi_s(0)|^2. \qquad (8.67)$$

We now apply the above results to $\phi$, $J/\Psi$ and $\gamma$ decays. From Eqs. (8.67) and (8.55),

$$\alpha_s^3(m_V) = \frac{81\pi\alpha^2 <Q>^2}{10(\pi^2 - 9)}\frac{\Gamma(V \to \text{hadrons})}{\Gamma(V \to e^-e^+)} \qquad (8.68)$$

From Eq. (8.68),

$$\alpha_s(m_\phi) \approx 0.45, \quad \alpha_s(m_\Psi) \approx 0.22, \quad \alpha_s(m_\Upsilon) \approx 0.19,$$

where we have used $\Gamma(\phi \to \text{non-strange mesons}) \approx 653$ keV, $\Gamma(J/\Psi \to \text{hadrons}) \approx 82$ keV, $\Gamma(\Upsilon \to \text{hadrons}) \approx 54$ keV, $\Gamma(\phi \to e^+e^-) \approx 1.26$ keV, $\Gamma(J/\Psi \to e^+e^-) \approx 5.55$ keV, $\Gamma(\Upsilon \to e^+e^-) \approx 1.34$ keV. From this we see a realization of the asymptotic freedom of $QCD$, the coupling $\alpha_s(q^2)$ falls with the increase of $q^2$.

Finally from Eqs. (8.65), (8.67) and (8.55), we have [with $\alpha_s(m_{\eta_c}) = \alpha_s(m_\Psi)$]

$$\Gamma(\eta_c \to \text{hadrons}) = \frac{27\pi}{5(\pi^2 - 9)}\frac{m_\Psi^2}{m_{\eta_c}^2}\frac{1}{\alpha_s(m_\Psi)}\Gamma(J/\Psi \to \text{hadrons}) \qquad (8.69)$$

$$= \frac{3}{2}\left[\frac{\alpha_s(m_\Psi)}{\alpha}\right]^2\frac{m_\Psi^2}{m_{\eta_c}^2}\Gamma(J/\Psi \to e^+e^-)$$

$$\approx 7.6 \text{ MeV} \qquad (8.70)$$

where we have used $\alpha_s(m_\Psi) \approx 0.22$. This value is lower than the experimental value $\Gamma_{tot} \approx 13.2^{+3.8}_{-3.2}$ MeV for $\eta_c$.

## 8.8 Non-Relativistic Treatment of Quarkonium

From a theoretical point of view, heavy quark system (quarkonium) is interesting because this is a relatively simple system. To a good approximation, the quark motion in this bound state should be non-relativistic. Thus we can use the Schrödinger equation for $Q\bar{Q}$ system:

$$-\frac{\hbar^2}{2\mu}\nabla^2\Psi(\mathbf{r}) + [V(r) - E]\Psi(\mathbf{r}) = 0. \tag{8.71}$$

$\mu$ is the reduced mass of $Q\bar{Q}$ system, i.e. $\mu = \frac{1}{2}m_Q$. For central potential, we can use the wave function:

$$\Psi(\mathbf{r}) = R(r)\,Y_{lm}(\theta,\phi). \tag{8.72}$$

The radial wave function $R(r)$ satisfies the equation

$$-\frac{\hbar^2}{2\mu}\left[\frac{d^2}{dr^2} + \frac{2}{r}\frac{d}{dr}\right]R(r) - \left[E - V(r) - \frac{l(l+1)\hbar^2}{2\mu r^2}\right]R(r) = 0. \tag{8.73}$$

If we define a radial function

$$\chi(r) = r\,R(r), \tag{8.74}$$

then $\chi(r)$ satisfies the equation

$$\frac{d^2\chi}{dr^2} + \left[\frac{2\mu}{\hbar^2}(E - V(r)) - \frac{l(l+1)}{r^2}\right]\chi = 0. \tag{8.75}$$

The wave function $\chi(r)$ is normalized as

$$\int_0^\infty [\chi(r)]^2 dr = \int_0^\infty [R(r)]^2 r^2 dr$$

$$= 1 \tag{8.76}$$

with the boundary conditions

$$\chi(0) = 0 \tag{8.77}$$

For S-waves:

$$\chi(r) \to 0 \quad \text{as} \quad r \to \infty$$

$$\tag{8.78}$$

$$R(r) \to 0$$

For $S$-waves:

$$\chi'(0) = R(0) = \sqrt{4\pi}\Psi_s(0). \tag{8.79}$$

We now prove two important results:

**1.**

$$|\Psi_s(0)|^2 = \frac{\mu}{2\pi\hbar^2}\left\langle\frac{dV}{dr}\right\rangle. \tag{8.80}$$

**Proof.**

From Eq. (8.75) for $l = 0$

$$\frac{\chi''}{\chi} = -\frac{\mu}{2\pi\hbar^2}(E - V). \tag{8.81}$$

Therefore,

$$\frac{d}{dr}\left[\frac{\chi''}{\chi}\right] = \frac{2\mu}{\hbar^2}\frac{dV}{dr} \tag{8.82}$$

Taking the expectation value [note that $\chi(r)$ is real], we get

$$\int_0^\infty \chi\frac{d}{dr}\left[\frac{\chi''}{\chi}\right]\chi dr = \frac{2\mu}{\hbar^2}\int_0^\infty \chi\frac{dV}{dr}\chi dr. \tag{8.83}$$

Integrating left-hand side by parts, we get

$$l.h.s. = \left[\frac{\chi''}{\chi}\chi^2\right]_0^\infty - \int_0^\infty \frac{\chi''}{\chi}2\chi\chi' dr$$

$$= \left[\frac{\chi''}{\chi}\chi^2 - \chi'^2\right]_0^\infty$$

$$= [\chi'(0)]^2 = [R(0)]^2, \tag{8.84}$$

where we have used the boundary conditions (8.77), (8.78). Hence Eqs. (8.79) and (8.83) gives

$$|\Psi_s(0)|^2 = \frac{\mu}{2\pi\hbar^2}\left\langle\frac{dV}{dr}\right\rangle. $$

## 2. Virial Theorem

$$\langle T\rangle = \frac{1}{2}\left\langle r\frac{dV}{dr}\right\rangle \tag{8.85}$$

**Proof.**

From Eq. (8.82), we have

$$\int_0^\infty \chi\, r\frac{d}{dr}\left(\frac{\chi''}{\chi}\right)\chi dr = \frac{2\mu}{\hbar^2}\left\langle r\frac{dV}{dr}\right\rangle \tag{8.86}$$

Integrating left-hand side by parts and using Eqs. (8.77) and (8.78), we get

$$l.h.s. = 2\int_0^\infty \chi'^2 dr \tag{8.87}$$

Therefore,

$$2\int_0^\infty \chi'^2 dr = \frac{2\mu}{\hbar^2}\left\langle r\frac{dV}{dr}\right\rangle \tag{8.88}$$

Now from Eq. (8.81),

$$\left\langle \frac{\chi''}{\chi} \right\rangle = -\frac{2\mu}{\hbar^2} [E- <V>].\tag{8.89}$$

But

$$\left\langle \frac{\chi''}{\chi} \right\rangle = \int_0^\infty \chi\chi'' dr$$

$$= |-\chi\chi'|_0^\infty - \int_0^\infty \chi'^2 dr$$

$$= -\int_0^\infty \chi'^2 dr\tag{8.90}$$

Hence from Eqs.(8.88) – (8.90), we get

$$E - \langle V \rangle = \frac{1}{2}\left\langle r\frac{dV}{dr} \right\rangle\tag{8.91}$$

or

$$\langle T \rangle = \frac{1}{2}\left\langle r\frac{dV}{dr} \right\rangle.$$

Let us apply Eq. (8.85) to one gluon exchange potential $V(r) = -\frac{2}{3}\alpha_s\frac{1}{r}$. For this case

$$\left\langle \frac{p^2}{2\mu} \right\rangle = \frac{2}{3}\alpha_s\left\langle \frac{1}{r} \right\rangle = \frac{2}{3}\alpha_s\frac{1}{a},\tag{8.92}$$

where $a = 3/4\mu\alpha_s$ is the Bohr radius. Thus, we get

$$\frac{v}{c} = \frac{2}{3}\alpha_s\tag{8.93}$$

As $\alpha_s$ decreases with mass, for sufficiently high mass $v/c \ll 1$ and one can treat dynamics non-relativistically. For the special case of power law potential

$$V(r) = A + \lambda r^\nu,\tag{8.94}$$

one can obtain interesting results by studying the scaling of Schrödinger equation (8.75). Put $\rho = \beta r$, where $\beta$ is some parameter such that it makes $\rho$ dimensionless. Let us put $\chi(r) = u(\rho)$ and $\bar{E} = E - A$. Then from Eq. (8.75)

$$-\frac{d^2}{d\rho^2}u(\rho) = \left[\frac{2\mu}{\hbar^2}\frac{1}{\beta^2}\bar{E} - \frac{2\mu}{\hbar^2}\lambda\frac{1}{\beta^{2+\nu}}\rho^\nu - \frac{l(l+1)}{\rho^2}\right]u(\rho)\tag{8.95}$$

Put $|\lambda| = \frac{2\mu}{\hbar^2}\beta^{2+\nu}$, this gives

$$\beta = \left[\frac{2\mu|\lambda|}{\hbar^2}\right]^{1/2+\nu}.\tag{8.96}$$

Then put

$$\varepsilon = \frac{2\mu}{\hbar^2} \left[ \frac{2\mu |\lambda|}{\hbar^2} \right]^{1/2+\nu} \bar{E}, \qquad (8.97)$$

where $\varepsilon$ is dimensionless. If we write $sgn(\lambda) = \lambda/|\lambda|$, Eq. (8.95) gives

$$\frac{d^2}{d\rho^2} u(\rho) + \left[ \varepsilon - sgn(\lambda)\rho^\nu - \frac{l(l+1)}{\rho^2} \right] u(\rho) = 0 \qquad (8.98)$$

which depends only upon pure numbers. We now study the consequences of Eq. (8.98).

(i) Lengths and quantities with the dimensions of length depend upon constituent quark mass $m = 2\mu$ and coupling strength $|\lambda|$ as

$$L \sim \frac{1}{\beta} \sim (\mu |\lambda|)^{-1/2+\nu}. \qquad (8.99)$$

Particle density at the origin of coordinates

$$|\Psi_s(0)|^2 \sim L^{-3} \propto (\mu |\lambda|)^{-1/2+\nu}. \qquad (8.100)$$

(ii) Level spacing between energy levels depends on $\mu$ and $|\lambda|$ as

$$\Delta E \sim \frac{1}{\mu} (\mu |\lambda|)^{2/2+\nu} \propto \mu^{-\nu/2+\nu} |\lambda|^{2/2+\nu} \qquad (8.101)$$

The "power law" potential corresponding to the limiting value $\nu \to 0$ is simply the logarithmic potential.

$$V(r) = C \ln \frac{r}{r_0}. \qquad (8.102)$$

We summarize these results for the power law potentials in Table 8.2.

Table 8.2

|  | Coulomb like $\nu = -1$ | Simple harmonic oscillator $\nu = 2$ | linear $\nu = 1$ | log $\nu = 0$ | Power law $\nu = 0.1$ |
|---|---|---|---|---|---|
| $|\Psi_s(0)|^2$ | $\mu^3$ | $\mu^{3/4}$ | $\mu$ | $\mu^{3/2}$ | $\mu^{1.43}$ |
| $\Delta E$ | $\mu$ | $\mu^{-1/2}$ | $\mu^{-1/3}$ | constant | $\mu^{-0.048}$ |

## 8.9 Observations

$$m_{\Upsilon'} - m_{\Upsilon} = m_{\Psi'} - m_{\Psi} \tag{8.103}$$

implies either $\nu = 0$ or $\nu$ is very small. In fact Martin has shown that the potential

$$V(r) = -8.04\,\text{GeV} + 6.870(r/1\,\text{GeV}^{-1})^{0.1} \tag{8.104}$$

gives a good fit to quarkonium mass spectrum.

The logarithmic potential

$$V(r) = (0.71\,\text{GeV}) \ln\left(\frac{r}{r_0}\right) \tag{8.105}$$

gives good fit to the data. The two forms are numerically indistinguishable for $0.1\,\text{fm} \leq r \leq 1\,\text{fm}$.

If we plot $|\Psi_s(0)|^2$ for the vector bosons $\rho$, $\omega$, $\phi$, $\Psi$, $\Upsilon$ versus $\mu$ in a log-log plot, a straight line fit is possible, i.e.

$$|\Psi_s(0)|^2 \sim \mu^p$$

with $p \sim 1.6$. Again this supports the power law potential with $\nu$ very small, i.e. $\nu \sim 0.1$.

Both for charmonium and bottomonium, the low lying bound state energy spectrum satisfies the rule

$$E_{1s} < E_{1p} < E_{2s}. \tag{8.106}$$

In particular we find for $c\bar{c}$

$$\bar{m}(1\,^3P) - \bar{m}(1\,S) = 457\,\text{MeV}$$

$$m(2\,^3S_1) - m(1\,^3S_1) \equiv \Psi' - J/\Psi = 589\,\text{MeV} \tag{8.107}$$

$$m(1\,^3S_1) - m(1\,^1S_0) = J/\Psi - \eta_c = 117\,\text{MeV}$$

$$m(2\,^3S_1) - m(2\,^1S_0) = \Psi' - \eta'_c = 82\,\text{MeV},$$

where

$$\bar{m}(S) = \frac{3m(^3S_1) + m(^1S_0)}{4} \tag{8.108}$$

$$\bar{m}(^3P) = \frac{5m(^3P_2) + 3m(^3P_0) + m(^3P_0)}{9}. \tag{8.109}$$

For Coulomb potential, the energy spectrum satisfies the rule

$$E_{1s} < E_{2s} = E_{2p} < E_{3s} = E_{3p} = E_{3d} \tag{8.110}$$

and for the harmonic oscillator potential

$$E_{1s} < E_{1p} < E_{2s} = E_{1d} < E_{2p}. \tag{8.111}$$

Further, the harmonic oscillator potential gives the level spacing as follows:

$$E_{1p} - E_{1s} = E_{2s} - E_{1p} = \frac{1}{2}(E_{2s} - E_{1s}) \tag{8.112}$$

Thus although oscillator potential is a confining potential, the level spacing is not in agreement with the experimental results.

The QCD inspired Cornell potential [cf. Eq. (7.82a)]

$$V(r) = C - \frac{K}{r} + \frac{r}{a^2}$$

reproduces the mass spectrum for $c\bar{c}$ and $b\bar{b}$ bound states quite well (see the problem 8.9).

Thus we see that the quarkonium spectroscopy is consistent with a potential that increases linearly at large distances, thereby supporting the color confinement. We also saw in this chapter (as well as in Chap. 7) a realization of other striking property of QCD, namely the running of the QCD coupling constant $\alpha_s(q^2)$ with $q^2$.

## 8.10   Tetraquark

Tetraquark mesons are composite of diquark-diantiquark. A diquark is either in symmetric color state $6_c$ or color antisymmetric state $\bar{3}_c$. The antidiquark is either in symmetric color state $\bar{6}_c$ or color antisymmetric state $3_c$. Now

$$6_c \otimes \bar{6}_c = 35_c \oplus 1_c$$
$$\bar{3}_c \otimes 3_c = 8_c \oplus 1_c$$
$$\bar{3}_c \otimes \bar{6}_c = \overline{10}_c \oplus 8_c$$
$$6_c \otimes 3_c = 10_c \oplus 8_c$$

Hence only $6_c \otimes \bar{6}_c$ and $\bar{3}_c \otimes 3_c$ give color singlet state. However for both diquark and antidiquark in color symmetric state, the one gluon exchange potential is repulsive unlike attractive one gluon exchange potential for diquark and antidiquark in color antisymmetric state. In any case, we will confine ourselves to the color singlet tetraquark composite of diquark and antidiquark in color triplet states $\bar{3}_c$ and $3_c$ respectively.

Now diquarks are either antisymmetric or symmetric in flavor:

$$[qq] = \frac{1}{\sqrt{2}}(q_i q_j - q_j q_i) \quad i, \, j = u, \, d, \, s, \, c$$

$$\{qq\} = \frac{1}{\sqrt{2}}(q_i q_j + q_j q_i)$$

Similarly for antidiquark flavor states. For antisymmetric color state $\bar{3}_c$ or $3_c$, Pauli principle requires overall wave function of diquark or antidiquark to be symmetric in flavor, in space and spin: (s, denote the spin of diquark or antidiquark)

$$[qq]_{L=0, \, s=0} \quad ; P = 1 \qquad [qq]_{L=1, \, s=1} \quad ; P = -1$$

$$\{qq\}_{L=0, \, s=1} \quad ; P = 1 \qquad \{qq\}_{L=1, \, s=0} \quad ; P = -1$$

$$[\bar{q}\bar{q}]_{L=0, \, s=0} \quad ; P = 1 \qquad [\bar{q}\bar{q}]_{L=1, \, s=1} \quad ; P = -1$$

$$\{\bar{q}\bar{q}\}_{L=0, \, s=1} \quad ; P = 1 \qquad \{\bar{q}\bar{q}\}_{L=1, \, s=0} \quad ; P = -1$$

For $u$, $d$, $s$ quarks, the underlying flavor symmetry is SU(3) (S denotes strangeness quantum number)

For this case

$$[qq] \, , \quad \begin{cases} I = 0 & I = 1/2 \\ S = 0 & , \quad S = -1 \end{cases}$$

$$[\bar{q}\bar{q}] \, , \quad \begin{cases} I = 0 & I = 1/2 \\ S = 0 & , \quad S = -1 \end{cases}$$

Hence we have a nonet of low lying scalar mesons $0^+$, composite of tetraquark viz

$$[qq]_{L=0, \, s=0} \, [\bar{q}\bar{q}]_{L=0, \, s=0}$$

Nonet of scalar mesons $0^+$ have inverse mass spectrum:

Isosinglet: $\sigma(600)$

Two isodoublet: $K(800)$ $\begin{cases} S = 1 \\ S = -1 \end{cases}$

degenerate $\begin{bmatrix} \text{Isosinglet:} & f_0(980) \\ \text{Isotriplet:} & a_0(980) \end{bmatrix}$

This mass spectrum does not fit $0^+$ as bound states $(q\bar{q})_{L=1}$.

The following tetraquark states are of interest.

$$\left( [qq]_{L=0, \, s=0} \{\bar{q}\bar{q}\}_{L=0, \, s=1} \pm \{qq\}_{L=0, \, s=1} [\bar{q}\bar{q}]_{L=0, \, s=0} \right)$$

These states have $J^{PC} = 1^{++}, 1^{+-}$. The following states have $J^{PC} = 0^{++}$, $1^{++}, 2^{++}$

$$\{qq\}_{L=0,\ s=1}\ \{\bar{q}\bar{q}\}_{L=0,\ s=1}$$
$$\{qq\}_{L=1,\ s=0}\ \{\bar{q}\bar{q}\}_{L=1,\ s=0}$$

Both these states give the tetraquark mesons with $J^{PC} = 0^{++}, 1^{++}, 2^{++}$. We see that there are a large number of tetraquark states. It is hard to formulate selection rules to select the states which can exist as composite of tetraquark.

As an example, we consider the following two tetraquark charmonium states, with $L = 0$:

$$\left([cq]_{L=0,\ s=0}\ \{\bar{c}\bar{q}\}_{L=0,\ s=1} \pm \{cq\}_{L=0,\ s=1}\ [\bar{c}\bar{q}]_{L=0,\ s=0}\right)$$

The two states have $J^{PC} = 1^{++}$ and $1^{+-}$. The $J^{PC} = 1^{++}$ meson is identified with the state $X(3872)$. This state was discovered in the $J/\Psi$ $\pi^+\pi^-$ distribution. However, this state decay in two different final states:

$$X \rightarrow J/\Psi\pi^+\pi^-$$
$$\rightarrow J/\Psi\pi^+\pi^-\ \pi^0$$

with equal probability:

$$R = \frac{Br(X(3872) \rightarrow J/\Psi\pi^+\pi^-\ )}{Br(X(3872) \rightarrow J/\Psi\pi^+\pi^-\ \pi^0)}$$

The $C$-parity of $(\pi^+\pi^-)$ is $C = (-1)^l$; $C$-conservation requires $l = 1$, $I(\pi^+\pi^-) = 1$ as required by Bose statistics. Thus the $G$-parity of $(\pi^+\pi^-)$ system $(-1)^{l+I} = +1$, i.e. $(\pi^+\pi^-)$ has the same $C$ and $G$ parities as $\rho^0$ meson.

Hence for the decay channel

$$X \rightarrow J/\Psi\pi^+\pi^-,\ \ \Delta G = 0,\ \Delta|I| = 1$$

For the decay channel

$$X \rightarrow J/\Psi\pi^+\pi^-\ \pi^0$$

we note that $G$-parity of the $(\pi^+\pi^-\ \pi^0)$ is $-1$ and $C$-conservation requires the $C(\pi^+\pi^-\ \pi^0) = -1$. Hence the $(\pi^+\pi^-\pi^0)$ system has the same quantum number as $\omega^0$. Hence for this decay channel $\Delta G \neq 0$, $\Delta|I| = 0$. We conclude that both decay channels are electromagnetic decay, for the channel $X \rightarrow$

$J/\Psi\pi^+\pi^-$, $\Delta G = 0$, $\Delta|I| = 1$, whereas for the channel $X \to J/\Psi\pi^+\pi^-\pi^0$, $\Delta G \neq 0$, $\Delta|I| = 0$. Experimentally, the branching ratio is

$$Br(X(3872) \to J/\Psi\pi^+\pi^-) \approx Br(X(3872) \to J/\Psi\pi^+\pi^-\pi^0)$$

For the open charm the tetraquark, $D$-states

$$\begin{cases} (cq)_{L=1,\ s=0}(\bar{q}\bar{q}')_{L=0,\ s=1} \\ (cq)_{L=0,\ s=1}(\bar{q}\bar{q}')_{L=1,\ s=0} \end{cases}$$

have $J^{PC} = 2^-, 1^-, 0^-$. The $D$-states

$$\begin{cases} (cq)_{L=1,\ s=0}(\bar{q}\bar{q}')_{L=1,\ s=0} \\ (cq)_{L=0,\ s=1}(\bar{q}\bar{q}')_{L=0,\ s=1} \end{cases}$$

have $J^{PC} = 2^+, 1^+, 0^+$.

In particular, for $q = d$, $\acute{q} = s$, $D_s^+$ : states have $J^{PC} = 2^+, 1^+, 0^+$ out of the tetraquark states

$$(cd)(\bar{d}\bar{s}) \qquad \begin{cases} L = 0,\ s = 1 \\ L = 1,\ s = 0 \end{cases}$$

have been considered as possible tetraquark mesons.

## 8.11   Problems

(1) Write down the amplitude F for the Feynman diagrams shown in Fig. 8.6, Show that

$$\frac{d\sigma}{d\Omega} = \frac{\alpha^2}{4s}\left[(1 + \cos^2\theta) + \frac{g^4}{e^4}\frac{s^2}{(m_{res}^2 - s)^2 + m_{res}^2\Gamma^2}\right]_{J^P=0^-,0^+}$$

$$\frac{d\sigma}{d\Omega} = \frac{\alpha^2}{4s}(1 + \cos^2\theta)\left[1 - 2\frac{g^2}{e^2}\frac{s(m_{res}^2 - s)}{(m_{res}^2 - s)^2 + m_{res}^2\Gamma^2}\right.$$

$$\left.+\frac{g^4}{e^4}\frac{s^2}{(m_{res}^2 - s)^2 + m_{res}^2\Gamma^2}\right]_{J^P=1^-}$$

$$\frac{d\sigma}{d\Omega} = \frac{\alpha^2}{4s}\left\{(1 + \cos^2\theta) + \left[1 + \frac{g^4}{e^4}\frac{s^2}{(m_{res}^2 - s)^2 + m_{res}^2\Gamma^2}\right]\right.$$

$$\left.-4\cos\theta\frac{g^2}{e^2}\frac{s(m_{res}^2 - s)}{(m_{res}^2 - s)^2 + m_{res}^2\Gamma^2}\right\}_{J^P=1^+}$$

In writing above equations, the finite width of resonance has been taken care of. Hence we conclude: experimental angular distribution excludes $J^P = 0^-, 0^+, 1^+$ and higher spin.

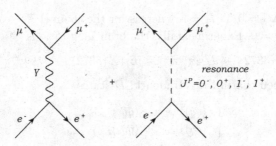

Fig. 8.6   $e^- e^+ \to \mu^- \mu^+$ through $\gamma$ or resonances.

(2) Show that if $J/\Psi$ has isospin $I = 0$,

$$\frac{\Gamma \left( \Psi \to \rho^0 \pi^0 \right)}{\Gamma \left( \Psi \to \rho^- \pi^+ \right) + \Gamma \left( \Psi \to \rho^+ \pi^- \right)} = \frac{1}{2}.$$

**Hint:** The $\rho\pi$ final state has $I = 0$, 1 or 2. But we are interested in $I = 0$. Using C.G. coefficients

$$\left| \rho^- \pi^+ \right\rangle = \frac{1}{\sqrt{3}} \left| 0, 0 \right\rangle + \cdots$$

$$\left| \rho^0 \pi^0 \right\rangle = \frac{1}{\sqrt{3}} \left| 0, 0 \right\rangle + \cdots$$

(3) Show that $D^* D \pi$ couplings in $SU(3)$ are given by

$$D^{*0} \to D^0 \pi^0 : g$$
$$D^{*0} \to D^+ \pi^- : \sqrt{2} g$$
$$D^{*+} \to D^0 \pi^+ : \sqrt{2} g$$
$$D^{*+} \to D^+ \pi^0 : -g$$

$$D^{*0} \to D^0 \eta_8 : \tfrac{1}{\sqrt{3}} g$$
$$D^{*+} \to D^+ \eta_8 : \tfrac{1}{\sqrt{3}} g$$
$$D_s^{*+} \to D^0 K^+ : \sqrt{2} g$$
$$D_s^{*+} \to D^+ K^0 : \sqrt{2} g$$
$$D_s^{*+} \to D_s^+ \eta_8 : -\tfrac{2}{\sqrt{3}} g$$

(4) Consider the decays

$$\acute{\psi} = J/\Psi \eta$$
$$= J/\Psi \pi^0$$

Show that above decays are p-wave decays. The second decay is isospin violating decay. Obtain the decay widths

$$\Gamma(\acute{\psi} \to J/\Psi \eta)$$

and

$$\Gamma(\acute{\psi} \to J/\Psi \pi^0)$$

in terms of the coupling constants $g_{\acute{\psi} J/\Psi \eta}$ and $g_{\acute{\psi} J/\Psi \pi^0}$. Obtain the values of these couplings. Hence show that isospin violating interaction is of the same order as electromagnetic interaction from the experimental branching ratios.

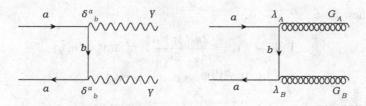

Fig. 8.7 $(q\bar{q}) \to 2\gamma$ or 2 gluons.

(5) For $(q\bar{q})_{\text{color singlet}} \to 2\gamma$ or 2 gluons, as shown in Fig. 8.7, where $a, b = 1, 2, 3$ are 3 colors of quarks, $A, B = 1, \cdots, 8$ are eight colors of gluons, show that

$$\frac{M(2g)}{M(2\gamma)} = \frac{\alpha_s}{\alpha Q_q^2} \frac{\frac{1}{2}\delta_{AB}}{3}$$

and hence show that

$$\frac{\Gamma(2g)}{\Gamma(2\gamma)} = \frac{2}{9} \frac{\alpha_s^2}{\alpha^2 Q_q^4}$$

For $(q\bar{q})_{\text{color singlet}} \to 3\gamma$ or 3 gluons coupled symmetrically in gluon color, show that

$$\frac{\Gamma(3g)}{\Gamma(3\gamma)} = \frac{5}{54} \frac{\alpha_s^3}{\alpha^3 Q_q^6}$$

**Hint:** Use $\frac{1}{16} d_{ABC} d_{ABC} = \frac{5}{6}$

(6) Using Eqs. (8.59) and (8.60) in the text and the experimental table below

| Vector Meson | $m_V$ (MeV) | $\Gamma$ (keV) |
|---|---|---|
| $\rho$ | 770 | $7.04 \pm 0.02$ |
| $\omega$ | 783 | $0.60 \pm 0.02$ |
| $\phi$ | 1020 | $1.26 \pm 0.04$ |
| $\Psi$ | 3097 | $5.55 \pm 0.14$ |
| $\Upsilon$ | 9460 | $1.340 \pm 0.018$ |

show that

$$f_\rho = 221 \text{ MeV}, \quad f_\omega = 194 \text{ MeV}, \quad f_\phi = 228 \text{ MeV},$$
$$f_\psi = 416 \text{ MeV}, \quad f_\Upsilon = 715 \text{ MeV}$$

(7) $J/\Psi$ can also decay into hadrons via electromagnetic interaction through one photon exchange

Show that

$$\Gamma(\Psi \to h)_\gamma = \frac{2\pi}{m_\Psi} \frac{(4\pi)^2 \alpha^2}{3m_\Psi^2} f_\Psi^2 .3m_\Psi^2 \; \rho(m_\Psi^2)$$

$$= 12\pi^2 \Gamma(\Psi \to e^- e^+) \; \rho(m_\Psi^2)$$

Using Fig. 8.8,

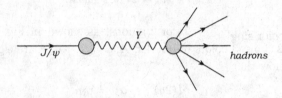

Fig. 8.8   Decay of $J/\Psi$ into hadrons via electromagnetic interaction through one photon exchange

$$\rho(m_\Psi^2) = \frac{1}{12\pi^2} \sum_i Q_i^2$$

$$= \frac{1}{12\pi^2} 3(\frac{4}{9} + \frac{1}{9} + \frac{1}{9}) \; \text{below charm threshold}$$

$$= \frac{1}{12\pi^2} (2)$$

Show that

$$\Gamma(\Psi \to h)_\gamma = 2\Gamma(\Psi \to e^- e^+)$$

$$= 11.10 \pm 0.28 \; \text{KeV}$$

Exp: $12.60 \pm 0.28$ keV

(8) Using the formula for $^3S_1$ bound state

$$\frac{\Gamma[(Q\bar{Q})(2^3S_1) \to e^- e^+]}{\Gamma[(Q\bar{Q})(1^3S_1) \to e^- e^+]} = \frac{m(Q\bar{Q})_{2^3S_1}}{m(Q\bar{Q})_{1^3S_1}} \frac{|\Psi_{2s}(0)|_{Q\bar{Q}}^2}{|\Psi_{1s}(0)|_{Q\bar{Q}}^2}$$

and the experimental values for the leptonic decay width and the masses, show that

$$\frac{|\Psi_{2s}(0)|_{c\bar{c}}^2}{|\Psi_{1s}(0)|_{c\bar{c}}^2} \approx 0.61, \quad \frac{|\Psi_{2s}(0)|_{b\bar{b}}^2}{|\Psi_{1s}(0)|_{b\bar{b}}^2} \approx 0.65.$$

(9) By writing

$$E = <H> = <\Psi|H|\Psi> = 2m + \frac{1}{2\mu} <\mathbf{p}^2> + <V(r)>$$

where $\mu = m/2$ and $V(r) = C - \frac{K}{r} + \frac{1}{a^2}r$, evaluate the energy eigen-values $E_{1s}$, $E_{1p}$ and $E_{2s}$ by variational principle.

**Hint:** Write $\bar{\varepsilon} = (2\mu b^4)^{1/3}\bar{E} = (2\mu b^4)^{1/3}[<H> -2m - C]$, $(E = \bar{E} + C)$ and $\Psi = \frac{u(r)}{r}Y_{lm}(\theta,\phi)$ and express

$$\bar{\varepsilon} = \frac{\int \left[-u\frac{d^2u}{dy^2} + \left(y - \frac{\eta}{y} + \frac{l(l+1)}{y^2}\right) u^2\right] dy}{\int u^2 dy}$$

where $y = (2\mu/b^2)^{1/3}r$ and $\eta = (4\mu^2 b^2)^{1/3}K$. [All these quantities are dimensionless and are therefore also suitable for numerical solution on a computer.]

Using the trial wave functions

$$1S : u = Nye^{-1/2\beta^2 y^2},$$
$$2S : Ny\left(1 - \frac{2}{3}\beta^2 y^2\right)e^{-1/2\beta^2 y^2}$$
$$1P : u = Ny^2 e^{-1/2\beta^2 y^2},$$

minimize $\bar{\varepsilon}$ in order to determine the parameter $\beta$ for each wave func-tion. Then find $\bar{\varepsilon}$. For numerical purpose, use $m_c = 1.52$ GeV, $K = 0.48$, $a = 2.34$ GeV$^{-1}$. Compare your results with the experi-mental values.

Using the equation

$$|\Psi_{ns}(0)|^2 = \frac{\mu}{2\pi}\left\langle\frac{dV}{dr}\right\rangle_{ns} = \frac{\mu}{2\pi}\left[\frac{1}{a^2} + K\left\langle\frac{1}{r^2}\right\rangle_{ns}\right],$$

evaluate $|\Psi_{ns}(0)|^2$ by using the above wave functions for $1S$ or $2S$ states in order to determine $<1/r^2>_{ns}$ with the parameters $\beta$'s determined in the first part of the problem. Hence evaluate the mass difference $\Psi - \eta_c$ and $\Psi' - \eta_c'$ and compare your results with the experimental values.

(10) From Eq. (8.76), using similar proceedure used in deriving Eq. (8.81), show that for p-wave

$$\left\langle\frac{1}{r^3}\right\rangle_p = \frac{2\mu}{4}\left\langle\frac{dV}{dr}\right\rangle.$$

## 8.12 References

1. C. Quigg and J. L. Rosner, Phys. Rep. 56, 167 (1979).
2. H. Grosse and A. Martin, Phys. Rep. 60, 341 (1980).
3. M. E. Peskin, "Aspects of the Dynamics of Heavy Quark Systems", in Dynamics and Spectroscopy at High Energy, Proceedings of the 1983 SLAC Summer Institute in Particle Physics, edited by Patricia M. McDonough, Stanford Linear Accelerator Report SLAC-267, p. 151.
4. E. Eichten, "The Last Hurrah for Quarkonium Physics: The Top System", The Sixth Quark, Proceedings of the 1984 SLAC Summer Institute in Particle Physics, edited by Patricia M. McDonough, Stanford Linear Accelerator Center Report SLAC-281, January, 1985, p. 1.
5. R. N. Cahn (editor), $e^+e^-$ Annihilation: New Quarks and Leptons (Annual Reviews Special Collections Program), Benjamin/Cummings, Menlo Park, California, 1985.
6. C. Quigg, Quantum Chromodynamics near the Confinement Limit, Fermi Lab.-Conf.-85/126-T (1985).
7. J. Lee-Franzini, Nucl. Phys. B3, 139 (1988).
8. F.E.Close, An Introduction to Quarks and Patrons, Academic press, London (1979); Reports on Progress in Physics, 51, 583 (1989).
9. High Energy Electron-Positron Physics, Eds. A. Ali and P. Söding, World Scientific, Singapore (1988).
10. J. L. Rosner, Heavy Quarks, Quark Mixing and C. P. Violation, EF1 90-63, to be published in Proc. TAS1-90 Boulder, Colorado, 1990 (World Scientific, Singapore).
11. W.Lucha, F.F.Schöberl and D.Gromes, Phys. Rep. 200, 127 (1991).
12. Fayyazuddin and Riazuddin, Phys. Rev. D69, 114008 (2004); arXiv:hep-ph/0309283.
13. R.L. Jaffe, 'Exotica', arXiv:hep-ph/04090605 (2004).
14. F.E. Close, Hadron Spectoscopy (theory): Di-quark, Tetra-quark, Penta-quarks and no quarks arXiv: hep-ph/0411396 (2004).
15. L. Maiani, F. Piccinini, A.D. Polosa and V. Riquer, Phys. Rev. D71, 014028 (2005).
16. Particle Data Group, K. Nakamura et al., Journal of Physics, G 37, 0750212 (2010).
17. K. Teraski, Tetra-quark systems in Heavy Mesons $D_{s0}^+(2317)$ and $X(3872)$ and related, arXiv:1005.5573 (2010).
18. A.H. Fariborz, R. Jora and J. Schechter, Light Scalar Puzzle in QCD, arXiv:1005.4614 (2010).

# Chapter 9

# Heavy Quark Effective Theory

## 9.1 Effective Lagrangian

The QCD Lagrangian (7.52)

$$\mathcal{L} = \bar{q} i \gamma^\mu \left( \partial_\mu - i g_s G_\mu \right) q - \bar{q} m q - \frac{1}{2} Tr(G^{\mu\nu} G_{\mu\nu})$$

for light quarks is chiral invariant in the limit $m_{u,d,s} \to 0$ as discussed in Chap. 11. For a heavy quark $c$, $b$, or $t$, the chiral symmetry does not hold. However, QCD has asymptotic freedom which implies that the effective coupling constant $\alpha_s$ decreases logarithmically at short distances or high momentum transfers. This is the basis for perturbative QCD, i.e. above a certain mass scale $\mu$, the perturbative QCD is applicable. The size of a hadron is of the order of $\Lambda_{QCD}$, where $\Lambda_{QCD} \sim 0.2$ GeV (see Chap. 7). Thus for a bound state of quarks or (quark-antiquark), we are in the nonperturbative regime, i.e. in the confinement region.

Consider for example a bound state of light-heavy quark-antiquark, viz $q\bar{Q}$ or $Q\bar{q}$. In the limit $m_Q \to \infty$, heavy quark (antiquark) can be taken as a static source of field in which light antiquark (quark) moves. The situation is like hydrogen atom. In the limit $m_Q \to \infty$, the Hamiltonian for the light degrees of freedom in analogy with H atom can be written to order $v^2/c^2$

$$H = \frac{\hat{p^2}}{2m_q} + V_c(r) - \frac{\hat{p^4}}{8m_q^2} - \frac{1}{4m_q^2} \sigma_q.(\mathbf{E}^c \times \hat{\mathbf{p}}) - \frac{1}{8m_q^2} \nabla \cdot \mathbf{E}^c, \qquad (9.1)$$

where $\mathbf{E}^c$ is the color electric field, $V_c(r)$ is related to $\mathbf{E}^c$ by $\mathbf{E}^c = -\frac{dV_c}{dr}\frac{\mathbf{r}}{r}$, and $\hat{\mathbf{p}} = -i\nabla$. Although $\lambda_{QCD}/m_Q$ is small, it is still finite. The effective heavy quark theory provides a framework to take into account $1/m_Q$ corrections.

The starting point is to define a four-velocity

$$v_\mu = \frac{dx_\mu}{d\tau}, \qquad \mathbf{u} = \frac{d\mathbf{x}}{d\tau}\frac{d\tau}{dt} = \frac{v}{\gamma} \tag{9.2}$$

$$v_o = \gamma = v^o$$

so that

$$v^2 = v^\mu v_\mu = \gamma^2 - \gamma^2 \mathbf{u}^2 = \gamma^2(1 - \mathbf{u}^2) = 1 \tag{9.3}$$

It is convenient to define

$$\slashed{v} = \gamma^\mu v_\mu, \qquad \slashed{v}^2 = v^2 = 1 \tag{9.4}$$

The Dirac equation for a heavy quark is given by

$$(i\gamma^\mu D_\mu - m)\Psi = 0 \tag{9.5}$$

where $D_\mu$ is the covariant derivative

$$D_\mu = \frac{\partial}{\partial x_\mu} - ig_s G_\mu \tag{9.6}$$

$$G_\mu = \frac{\lambda_A}{2} G_\mu^A$$

We define the projection operators

$$P_\pm = \frac{1}{2}\left(1 \pm \slashed{v}\right) \tag{9.7}$$

Note that in the rest frame $v = 0$

$$P_\pm = \frac{1}{2}(1 \pm \gamma^0),$$

i.e. it projects out upper and lower components of $\Psi$. Write

$$\begin{aligned}\Psi &= P_+\Psi + P_-\Psi \\ &= e^{-imv.x}[h_{+v} + h_{-v}] \\ &= e^{-im\slashed{v}v.x}h_{+v} + e^{im\slashed{v}v.x}h_{-v}\end{aligned} \tag{9.8}$$

where

$$h_{\pm v} = e^{imv\cdot x} P_\pm \Psi$$

We note that

$$\psi h_{+v} = h_{+v}, \qquad \psi h_{-v} = -h_{-v} \tag{9.9}$$

$$P_+\gamma^\mu = \gamma^\mu P_- + v^\mu$$

$$P_-\gamma^\mu = \gamma^\mu P_+ - v^\mu \tag{9.10}$$

Using Eqs. (9.8), (9.9), and (9.10), we obtain from Eq. (9.5)

$$(i\gamma \cdot D + iv \cdot D)h_{-v} + iv \cdot D h_{+v} = 0 \tag{9.11}$$

$$(i\gamma \cdot D - iv \cdot D)h_{+v} - (2m + iv \cdot D)h_{-v} = 0 \tag{9.12}$$

Note that $h_{+v}$ and $h_{-v}$ are not decoupled. The next step is to show that to order $1/m^2$, the equations for $h_{+v}$ and $h_{-v}$ are decoupled. From Eq. (9.12), one obtain

$$
\begin{aligned}
h_{-v} &= \frac{i\gamma \cdot D - iv \cdot D}{2m + iv \cdot D}h_{+v} \\
&= \frac{1}{2m}[1 - \frac{iv \cdot D}{2m} + ....][i\gamma \cdot D - iv \cdot D]h_{+v}
\end{aligned} \tag{9.13}
$$

Thus from Eqs. (9.11) and (9.13) to order $1/m$, we get

$$[i\gamma \cdot D + iv \cdot D]\frac{i\gamma \cdot D - iv \cdot D}{2m}h_{+v} + iv \cdot D h_{+v} = 0 \tag{9.14}$$

Now

$$
\begin{aligned}
\gamma \cdot D \, \gamma \cdot D &= \gamma^\mu \gamma^\nu D_\mu D_\nu \\
&= D^2 - \frac{i}{2}\sigma^{\mu\nu}[D_\mu, D_\nu] \\
&= D^2 - \frac{g_s}{2}\sigma^{\mu\nu}G_{\mu\nu}
\end{aligned} \tag{9.15}
$$

Hence from Eq. (9.14), we obtain

$$[iv \cdot D - \frac{D^2}{2m} + \frac{g_s}{4m}\sigma^{\mu\nu}G_{\mu\nu} - \frac{(iv \cdot D)^2}{2m}]h_{+v} = 0 \tag{9.16}$$

This is the Pauli form of Dirac equation to order $1/m$. The corresponding Lagrangian for the field $h_{+v}$ is given by

$$\mathcal{L}_{eff} = \bar{h}_{+v}[iv \cdot D - \frac{D^2 + (iv \cdot D)^2}{2m} + \frac{g_s}{4m}\sigma^{\mu\nu}G_{\mu\nu}]h_{+v} \tag{9.17a}$$

$$= \bar{h}_{+v}[iv \cdot D - \frac{D_\perp^2}{2m} + \frac{g_s}{4m}\sigma^{\mu\nu}G_{\mu\nu}]h_{+v} \tag{9.17b}$$

Note that $h_{+v}$ annihilates a heavy quark. In the limit $m \to \infty$

$$\mathcal{L}_{eff} = \bar{h}_{+v} iv \cdot Dh_{+v} \tag{9.17c}$$

Now from the relation

$$-i\partial_\mu \Psi(x) = [\hat{P}_\mu, \Psi(x)] \tag{9.18}$$

it follows through the transformation (9.8) that

$$-i\partial_\mu h_{+v} = mv_\mu h_{+v} + [\hat{P}_\mu, h_{+v}] \tag{9.19}$$

This shows that a derivative acting on $h_{+v}$ corresponds to a factor of the residual momentum $k_\mu$ carried by the heavy quark

$$-k_\mu = mv_\mu - p_\mu \tag{9.20}$$

so that $k_\mu$ indicates how much heavy quark is of-mass shell. In the limit $m \to \infty$ (no recoil limit) with $v_\mu$ and $k_\mu$ fixed

$$v_\mu^{quark} \equiv \frac{p_\mu}{m} = v_\mu + \frac{k_\mu}{m} \to v_\mu \tag{9.21}$$

One would expect the heavy quark to carry most of the momentum of the $\bar{q}Q$ bound state, but not all:

$$p_\mu^B = p_\mu + l_\mu = mv_\mu^{quark} + l_\mu \tag{9.22}$$

where $p_\mu^B = m_B v_\mu$ is the momentum of the bound system and $l_\mu$ is that which is carried by the light degree of freedom. Now $m_B = m + m_q - B$ where $B$ is the binding energy supplied by the interaction through gluon. Thus from Eq. (9.22)

$$v_\mu^{quark} = \frac{m_B}{m} v_\mu - \frac{l_\mu}{m} \tag{9.23}$$

so that again $v_\mu^{quark} \to v_\mu$ as $m \to \infty$ and a comparison of Eq. (9.23) with Eq. (9.21) shows that in the limit $m \to \infty$, the interaction with gluons can change $k_\mu$ but not $v_\mu^{quark} = v_\mu$; the velocity of the heavy quark can be altered only by an external current which absorbs "infinite momentum". Thus in a hadron, the light degrees of freedom are independent of the heavy quark mass, i.e. residual motion of the heavy quark in a hadron can be taken into account by adding the effective Hamiltonian for heavy quark $Q$ from the $\mathcal{L}_{eff}$ given in Eq. (9.17) to the Hamiltonian for the light quark given in Eq. (9.1). We will come to this point later, when we discuss the masses of heavy hadrons.

The third term in the Lagrangian (9.17a) undergoes short-distance QCD corrections and as such is multiplied by the renormalization factor

$$Z_Q(\mu) = \left[\frac{\alpha_s(\mu)}{\alpha_s(m)}\right]^{-9/25} \tag{9.24}$$

with $Z_Q(\mu = m) = 1$.

The heavy quark propagator in QCD can be written in HQET using Eq. (9.20):

$$i\delta_b^a \frac{m + \gamma \cdot p}{p^2 - m^2 + i\epsilon} = i\delta_b^a \frac{1 + \not v}{2(v \cdot k + i\epsilon)} \tag{9.25}$$

where we have neglected the term $k^2/m \to 0$. The gluon heavy quark vertex can be written from the Lagrangian (9.17) and is given by

$$ig_s v^\mu (T_s)_b^a \tag{9.26}$$

The following relations are useful

$$\not v \gamma^\mu + \gamma^\mu \not v = 2v^\mu$$
$$\not v \gamma_\mu \not v = 2v_\mu \not v - \gamma_\mu \tag{9.27}$$

From Eq. (9.27), one can write

$$\bar h_{+v} \gamma^\mu h_{+v} = \bar h_{+v} v^\mu h_{+v} \tag{9.28}$$

## 9.2   Spin Symmetry of Heavy Quark

In the limit $m \to \infty$, the Lagrangian $\mathcal{L}_{eff}$ given in Eq. (9.17) has additional symmetries not present in the full QCD Lagrangian. One such symmetry, namely the spin symmetry of heavy quark is reflected in the fact that the first term in the Lagrangian Eq. (9.17b) makes no reference to the Dirac structure at all which can couple to the spin degrees of $h_{+v}$.

More explicitly define the spin:

$$s^i = -s_i = -\gamma_5 \not v \gamma \cdot e_i \tag{9.29}$$

where

$$e_i \cdot v = v_\mu e_i^\mu = 0,$$
$$e_{j\mu} e_k^\mu = -\delta_{jk} \tag{9.30}$$

In the rest frame of $h_v$

$$\begin{pmatrix} v = 0 \\ v_0 = 1 \end{pmatrix}, e_i^\mu = \delta_i^\mu$$

thus

$$s_i = \gamma_5 \left( \gamma^0 v_0 \right) \gamma_\mu \delta_i^\mu$$
$$= v_0 \gamma^0 \gamma_5 \gamma_i = \begin{pmatrix} \sigma_i & 0 \\ 0 & \sigma_i \end{pmatrix} = -s^i, \qquad (9.31)$$

i.e. we get the usual definition of the spin. We note that the Lagrangian $\mathcal{L}_{eff}$ given in Eq. (9.17a) is invariant under the infinitesimal transformation

$$\delta h_{+v} = i\theta \cdot \mathbf{s} h_{+v}$$
$$\delta \bar{h}_{+v} = -i\theta \cdot \mathbf{s} \bar{h}_{+v} \qquad (9.32)$$

Now the Noether current is given by

$$\mathbf{J}^\mu = -i \frac{\partial \mathcal{L}}{\partial \left( D_\mu h_{+v} \right)} \mathbf{s} h_{+v}$$
$$= \bar{h}_{+v} v^\mu \mathbf{s} h_{+v} \qquad (9.33)$$

Hence the spin operator is given by

$$\mathbf{S} = \int \mathbf{J}^0 \left( \mathbf{x}, t \right) d^3 x$$
$$= v_0 \int \bar{h}_{+v} \mathbf{s} h_{+v} d^3 x. \qquad (9.34)$$

Note that

$$[S_i, h_{+v}] = -s_i h_{+v} \qquad (9.35)$$

We conclude that the Lagrangian $\mathcal{L}_{eff}$ in Eq. (9.17a) is invariant under SU(2) of heavy quark spin symmetry. It means that pseudoscalar and vector meson states $|P(v)\rangle$ and $|V(v, \epsilon)\rangle$, containing the same heavy quark, with momentum $p_\mu^B = m_B v_\mu$ can be related to each other:

$$S_3(v) |P(v)\rangle = - |V(v, \epsilon)\rangle, \quad S^3(v) |P(v)\rangle = |V(v, \epsilon)\rangle \qquad (9.36)$$

Thus their masses are degenerate in this limit. This degeneracy is lifted by the third term in the Lagrangian (9.17) giving e.g. for B* $(1^-)$ and B$(0^-)$ mesons, the mass difference $(m_B^* - m_B)$ which scales like $1/m_b$.

The second symmetry of the Lagrangian (9.17) in the limit $m \to \infty$ arises when we introduce two distinct flavors $h_1$ (e.g.b) and $h_2$ (e.g.c). Since the first term in Eq. (9.17) makes no reference to masses $m_i$ $(i = 1, 2)$, and since mass is the only property which can distinguish between quarks of different flavors in QCD, the effective theory has a symmetry under which $h_1(v) \leftrightarrow h_2(v)$. It may be emphasized that this symmetry does not in any way depend on $m_1 = m_2$ but only on $m_i \gg \Lambda$, where $\Lambda$ is a scale parameter such that $1/\Lambda$ determines the size of light degrees of freedom in the bound state and is a few hundred $MeV$; it may vary from process to process. Note also that the flavor symmetry holds between heavy quark fields of the same velocity and not with the same momentum. This flavor symmetry together with the spin symmetry mentioned above gives rise to SU(4) symmetry for the system $\begin{pmatrix} h_1(v) \\ h_2(v) \end{pmatrix}$ and has been used to relate the matrix elements of flavor changing effective currents which mediate weak decays of mesons containing heavy quarks. Since we will be dealing with $h_{+v}$ only, we will drop the subscript $+$ in what follows. We first note that

$$\left[ S_i^v, \bar{h}_{v'} \Gamma h_v \right] = -\bar{h}_{v'} \Gamma s_i h_v$$

$$\left[ S_i^{v'}, \bar{h}_{v'} \Gamma h_v \right] = \bar{h}_{v'} s_i \Gamma h_v \tag{9.37}$$

These equations follow from Eq. (9.35). Their use is as follows:

Consider the transition $B^-(v) \to D\left(v'\right)$. Then from Eq. (9.37), we obtain

$$i \left\langle D^0 \left(v'\right) \middle| \left[ S_i^{v'}, \bar{c}_{v'} \Gamma b_v \right] \middle| B^-(v) \right\rangle = \left\langle D^0 \left(v'\right) \middle| \left[ \bar{c}_{v'} s_i \Gamma b_v \right] \middle| B^-(v) \right\rangle \tag{9.38}$$

Now using Eq. (9.36),

$$i \left\langle D^{*0} \left(v'\right) \middle| \bar{c}_{v'} \Gamma b_v \middle| B^-(v) \right\rangle = \left\langle D^0 \left(v'\right) \middle| \left[ \bar{c}_{v'} s_3 \Gamma b_v \right] \middle| B^-(v) \right\rangle \tag{9.39}$$

where $s_3 = \gamma_5 \psi' \gamma^\rho \epsilon_\rho^*$, $\epsilon_\rho$ is polarization vector for $D^{*0}$. For $\Gamma = \gamma^\mu (1 - \gamma_5)$, we have

$$s_3 \gamma^\mu (1 - \gamma_5) = \gamma_5 \; \slashed{v}' \gamma^\rho \epsilon_\rho^* \; \gamma^\mu (1 - \gamma_5) \tag{9.40}$$

Now using

$$\gamma^\lambda \gamma^\rho \gamma^\mu = g^{\lambda\rho}\gamma^\mu - g^{\lambda\mu}\gamma^\rho + g^{\rho\mu}\gamma^\lambda + i\epsilon^{\lambda\rho\mu\sigma}\gamma_5\gamma_\sigma \tag{9.41}$$

$$i \left\langle D^{*0} \left(v'\right) \middle| \left[ \bar{c}_{v'} \gamma^\mu (1 - \gamma_5) b_v \right] \middle| B^-(v) \right\rangle$$

$$= -\left( v'^\mu \epsilon^{*\sigma} - v'^\sigma \epsilon^{*\mu} - i\epsilon^{\mu\lambda\rho\sigma} v'_\lambda \epsilon_\rho^* \right)$$

$$\times \left\langle D^0 \left(v'\right) \middle| \left[ \bar{c}_{v'} \gamma_\sigma (1 - \gamma_5) b_v \right] \middle| B^-(v) \right\rangle$$

$$= -\left( v'^\mu \epsilon^{*\sigma} - v'^\sigma \epsilon^{*\mu} - i\epsilon^{\mu\lambda\rho\sigma} v'_\lambda \epsilon_\rho^* \right)$$

$$\times \frac{1}{\sqrt{4v_0 v'_0}} \left[ \xi_0 \left(v \cdot v'\right) \left(v + v'\right)_\sigma \right] \tag{9.42}$$

where we have used the fact that $\bar{c}\gamma^\mu b$ is a symmetry current, so that

$$\left\langle D^0 \left(v'\right) \middle| [\bar{c}_{v'} \gamma_\mu b_v] \middle| B^- (v) \right\rangle = \frac{1}{\sqrt{4v_0 v_0'}} \left[ \xi_0 \left(v \cdot v'\right) \left(v + v'\right)_\mu \right] \quad (9.43)$$

with

$$\xi_0 \left(v^2\right) = \xi_0 \left(1\right) = 1 \quad (9.44)$$

Another application of Eq. (9.37) is for the matrix elements of the current $\bar{q}\gamma^\mu (1 - \gamma_5) b$, $(q = u, d, s)$ between the vacuum and $B$ meson state viz

$$\langle 0 | [S_i^v, \bar{q}\Gamma b] | B_q \rangle = - \langle 0 | \bar{q}\Gamma s_i b | B_q \rangle \quad (9.45)$$

Hence we get

$$i \langle 0 | \bar{q}\Gamma b | B_q^* \rangle = - \langle 0 | \bar{q}\Gamma \left(\gamma_5 \slashed{v} \gamma \cdot \epsilon\right) b | B_q \rangle \quad (9.46)$$

Thus for $\Gamma = \gamma^\mu (1 - \gamma_5)$ on using Eqs. (9.40) and (9.41), one obtains

$$i\sqrt{\frac{1}{m_{B_q^*}}} \left( f_{B_q^*} \epsilon_\mu \right) = i f_{B_q} m_{B_q} v^2 \epsilon_\mu = i f_{B_q} \sqrt{m_{B_q}} \epsilon_\mu$$

$$f_{B_q^*} = \sqrt{m_{B_q} B_q^*} f_{B_q} = m_{B_q} f_{B_q} \quad (9.47)$$

where we have used

$$\langle 0 | \bar{q}\gamma^\mu \gamma_5 b | B_q (p) \rangle = \sqrt{\frac{1}{2p_0}} i p^\mu f_{B_q} = \sqrt{\frac{m_B}{2v_0}} i f_{B_q} v^\mu \quad (9.48)$$

and

$$\langle 0 | \bar{q}\gamma^\mu b | B_q^* (p) \rangle = \sqrt{\frac{1}{2p_0}} \epsilon^\mu f_{B_q^*} = \sqrt{\frac{1}{2m_{B^*} v_0}} \epsilon^\mu f_{B_q^*} \quad (9.49)$$

The results obtained in Eqs. (9.42) and (9.47) will be used in Chap. 15, where we will discuss the semileptonic decays of $B$ mesons involving the vector and axial form factors in the transitions $B \to D, D^*$.

Similar results can also be derived by the following procedure (called trace technique). For vector and pseudoscalar meson fields, we can write $(P = B \text{ or } D)$

$$H_a = \frac{1 + \slashed{v}}{2} \left[ i\gamma^\mu P_{a\mu}^* - \gamma_5 P_a \right], \quad (9.50)$$

where $a = 1, 2, 3$ for $u$, $d$ and $s$ quarks and $P_{a\mu}^*$ and $P_a$ are annihilation operators normalized as

$$\langle 0 | P_{a\mu}^* | Q \bar{q}_a (1^-) \rangle = \epsilon_\mu \quad (9.51)$$

$$\langle 0 | P_a | Q \, \bar{q}_a \left( 0^- \right) \rangle = 1 \tag{9.52}$$

We define the adjoint field

$$\overline{H}_a = \gamma^0 H_a^\dagger \gamma^0 = \left[ -\overline{P}_{a\mu}^* i\gamma^\mu + \overline{P}_a \gamma_5 \right] \frac{1 + \not{v}}{2} \tag{9.53}$$

We note that

$$\not{v} H_a = H_a$$
$$H_a \not{v} = -H_a \tag{9.54}$$

The spin symmetry which relates $P_a$ and $P_{a\mu}^*$ is automatically incorporated in Eq. (9.50).

In case of spinor field $\overline{\Psi}(x)$, the wave function can be written

$$\langle 0 | \overline{\Psi}(x) | p \rangle \sim e^{-ipx} u(p) \tag{9.55}$$

Then in view of Eqs. (9.50)-(9.52), we can write for mesons containing a heavy quark

$$\langle 0 | H_a | P_a (v) \rangle = -\frac{1 + \not{v}}{2} \gamma_5 \tag{9.56}$$

$$\langle 0 | H_a | P_a^* (v, \epsilon) \rangle = \frac{1 + \not{v}}{2} i\gamma . \epsilon \tag{9.57}$$

We now apply the above considerations for the matrix elements $\langle 0^- (v') | J_\lambda | 0^- (v) \rangle$ and $\langle 1^- (v', \epsilon^*) | J_\lambda | 0^- (v) \rangle$ where $J_\lambda = V_\lambda - A_\lambda$. Using Eqs. (9.56), (9.57) and (9.50)-(9.53), we get

$$\langle 0^- (v') | J^\lambda | 0^- (v) \rangle$$

$$= \frac{1}{\sqrt{4v_0 v_0'}} [-\xi (v \cdot v')] Tr \left[ -\gamma^5 \frac{1 + \not{v}'}{2} \gamma^\lambda \frac{1 + \not{v}}{2} \gamma^5 \right]$$

$$= \frac{1}{\sqrt{4v_0 v_0'}} \xi (v \cdot v') (v + v')^\lambda \tag{9.58}$$

$$\langle 1^- (v', \epsilon_\mu^*) | J^\lambda | 0^- (v) \rangle$$

$$= \frac{1}{\sqrt{4v_0 v_0'}} [-\xi (v \cdot v')] \epsilon_\mu^* \times$$

$$Tr \left[ -i\gamma^\mu \frac{1 + \not{v}'}{2} \gamma^\lambda (1 - \gamma_5) \times \frac{1 + \not{v}}{2} \gamma_5 \right]$$

$$= \frac{-i}{\sqrt{4v_0 v_0'}} \xi (v \cdot v') \left[ i\epsilon^{\lambda\rho\mu\sigma} v_\rho' \epsilon_\mu^* v_\sigma + (1 + v' \cdot v) \epsilon^{*\lambda} - v \cdot \epsilon^* v_\lambda' \right] \tag{9.59}$$

Note that since only vector current contributes in Eq. (9.58), the form factor $\xi (-v \cdot v')$ is normalized as $\xi (1) = 1$. Comparing Eqs. (9.58) and (9.59) with Eqs. (9.40) and (9.46), we see that trace technique gives exactly the same results for $B \to D(D^*)$ transitions as previously obtained.

## 9.3   Mass Spectroscopy for Hadrons with One Heavy Quark

We now discuss the effective Lagrangian (9.17a) in relation to hadronic masses containing one heavy quark. We introduce the following notation: Write a pseudoscalar (vector) heavy meson as $P_q(P_q^*)$, $P = B$ or $D$, $q = u$, $d$ or $s$ and we take $m_u = m_d$. The heavy baryon is written as $B_Q$, $Q = b$ or $c$, B$=\Lambda, \Xi, \Sigma, \Xi'$ or $\Omega$ ; the light quark content and spin configurations are contained in these symbols.

Define

$$\bar{a}(P) = \langle P| \bar{h}_v D^2 h_v |P \rangle \tag{9.60a}$$

$$\bar{d}(P) = Z_Q(\mu) \langle P| g_s \bar{h}_v \sigma^{\mu\nu} G_{\mu\nu} h_v |P \rangle \tag{9.60b}$$

We take $\bar{a}(P)$ and $\bar{d}(P)$ independent of the light quark flavor $q$. We also assume

$$\bar{a}(B) = \bar{a}(D) = \bar{a} \tag{9.61a}$$

$$\frac{\overline{m}_B}{m_b} Z_b^{-1}(\mu) \bar{d}(B) = \frac{\overline{m}_D}{m_c} Z_c^{-1}(\mu) \bar{d}(D) = \bar{d} \tag{9.61b}$$

These assumptions imply that interquark interactions are flavor independent. To understand the physical meaning of these terms we go to the rest frame of $Q$ (i.e. $v = 0$). In this frame

$$iv \cdot D = g_s G_\circ \tag{9.62}$$

$$-D^2 = (\nabla - ig_s \mathbf{G}) \cdot (\nabla - ig_s \mathbf{G}) = \mathbf{D}^2 \tag{9.63}$$

$$\sigma^{\mu\nu} G_{\mu\nu} = 2 \begin{pmatrix} 0 & -i\sigma \cdot \mathbf{E}^c \\ -i\sigma \cdot \mathbf{E}^c & 0 \end{pmatrix} + 2 \begin{pmatrix} \sigma \cdot \mathbf{B}^c & 0 \\ 0 & \sigma \cdot \mathbf{B}^c \end{pmatrix} \tag{9.64a}$$

where

$$G_{0j} = -E_j^c , \qquad B_i^c = \frac{1}{2} \varepsilon_{ijk} G_{jk} \tag{9.64b}$$

are the color electric and magnetic fields respectively. Thus $\bar{h}_v \mathbf{D}^2 h_v$ is gauge invariant extension of kinetic energy term representing the residual motion of heavy quark in a hadron and term $\bar{h}_v \sigma_Q \cdot \mathbf{B}^c h_v$ describes the color magnetic coupling of the heavy–quark spin to the gluon field ($\mathbf{S} = \frac{1}{2}\sigma$)[we have exhibited the subscript $Q$ with $\sigma$].

Matching Eq. (9.62) with $V_c$ in Eq. (9.1), we can write the effective Hamiltonian for a bound hadron containing one heavy quark $Q$ as

$$H = H_q + H_Q \tag{9.65}$$

where $H_Q$ takes care of the residual motion of the heavy quark. Note that

$$g_s G_0 = -V_c(r)$$

has been absorbed in $H_q$. In view of Eqs. (9.63) and (9.64), $H_Q$ is obtained from Eq. (9.16) or from $\mathcal{L}_{eff}$ in Eq. (9.17a) to order $1/m_Q$ as follows:

$$H_Q = -\frac{\mathbf{D}^2}{2m_Q} - \frac{\sigma_Q \cdot \mathbf{B}^c}{2m_Q} Z_Q(\mu) g_s \qquad (9.66)$$

Note that the second term on the right-hand side of Eq. (9.66) represents the interaction of color magnetic field $\mathbf{B}^c$ with color magnetic moment of the heavy quark $\mu_Q = \frac{\sigma_Q}{2m_Q}(Z_Q(\mu) g_s)$.

We note that the term $\sigma_Q \cdot \mathbf{B}^c$ gives rise to color magnetic moment interaction of the type $\mu_q \cdot \mu_Q$ which induces a term of the type ($k_s = \frac{4}{3}$)

$$\frac{8\pi}{3} \alpha_s k_s \frac{\sigma_Q \cdot \sigma_q}{4 m_Q m_q} \delta^3(\mathbf{r})$$

Now

$$\langle \sigma_q \cdot \sigma_Q \rangle = \begin{cases} -3 : & \text{for spin singlet } (^1S_0) \text{ state} \\ 1 : & \text{for spin triplet } (^3S_1) \text{ state} \end{cases}$$

Then the Hamiltonian (9.65) gives the masses of the heavy meson $P_q$ and $P_q^*$

$$m_{P_q} = m_Q + \bar{\Lambda}_q + \frac{\bar{a}}{2m_Q} - \frac{\bar{d}(P)}{2m_Q} \qquad (9.67)$$

$$m_{P_q^*} = m_Q + \bar{\Lambda}_q + \frac{\bar{a}}{2m_Q} + \frac{\bar{d}(P)}{6m_Q} \qquad (9.68)$$

where we have put

$$\bar{\Lambda}_q = \langle H_q \rangle + m_q \qquad (9.69)$$

and

$$\bar{d} = \frac{Z_Q(\mu)}{2m_Q} (8\pi \alpha_s k_s) |\Psi_s(0)|^2$$

From Eqs. (9.67) and (9.69) the following mass relations are obtained

$$m_{B_d^*} - m_{B_d} = m_{B_s^*} - m_{B_s} = \frac{2}{3} \frac{\bar{d}(B)}{m_b} \qquad (9.70a)$$

$$m_{D_d^*} - m_{D_d} = m_{D_s^*} - m_{D_s} = \frac{2}{3} \frac{\bar{d}(D)}{m_c} \qquad (9.70b)$$

$$m_{B_s} - m_{B_d} = (\overline{\Lambda}_s - \overline{\Lambda}_d) + O\left(\frac{1}{m_b}\right) \qquad (9.71a)$$

$$m_{D_s} - m_{D_d} = (\overline{\Lambda}_s - \overline{\Lambda}_d) + O\left(\frac{1}{m_c}\right) \qquad (9.71b)$$

$$\overline{m}_B - \overline{m}_D = (m_b - m_c)\left[1 - \frac{\overline{a}}{2m_b m_c}\right] \qquad (9.71c)$$

where

$$\overline{m}_P = \frac{3m_{P^*} + m_P}{4} \qquad (9.72)$$

Experimentally (in MeV)

$$m_{D_d^{\pm *}} - m_{D_d^{\pm}} = 140.64 \pm 0.10, \qquad m_{D_s^{\pm *}} - m_{D_s^{\pm}} = 143.8 \pm 0.4$$

$$m_{B_d^*} - m_{B_d} = 45.78 \pm 0.35, \qquad m_{B_s^*} - m_{B_s} = 46.5 \pm 1.2$$

$$m_{B_s} - m_{B_d} = 87.5 \pm 0.6, \qquad m_{D_s} - m_{D_d} = 98.87 \pm 0.30$$

$$\overline{m}_B = 5313, \qquad \overline{m}_D = 1971$$

which are compatible with equalities obtained in Eqs. (9.70) and (9.71). From Eqs. (9.70) and (9.61b), we also obtain

$$\frac{\overline{m}_B\,(m_{B^*} - m_B)}{\overline{m}_D\,(m_{D^*} - m_D)} = \frac{\frac{\overline{m}_B \overline{d}(B)}{m_b}}{\frac{\overline{m}_D \overline{d}(D)}{m_c}} = \frac{Z_c(\mu)}{Z_b(\mu)} = \left[\frac{\alpha_s(m_b)}{\alpha_s(m_c)}\right]^{9/25} \qquad (9.73a)$$

Using the experimental values for the masses, we obtain

$$\frac{\alpha_s(m_b)}{\alpha_s(m_c)} \simeq 0.69 \qquad (9.73b)$$

Eq. (9.73b) is compatible with $\alpha_s(m_b) \simeq 0.22$, $\alpha_s(m_c) \simeq 0.32$ [see Chap. 15] used in discussing the decays of D and B mesons. If we use $m_b = 4.9$, $m_c = 1.5$ (in GeV), then from Eq. (9.70), we get

$$\frac{\overline{d}(B)}{m_b} \simeq 0.07 \text{ GeV}, \ \overline{d}(B) \simeq 0.34 \text{ GeV}^2 \qquad (9.74a)$$

$$\frac{\overline{d}(D)}{m_c} \simeq 0.213 \text{ GeV}, \ \overline{d}(D) \simeq 0.32 \text{ GeV}^2 \qquad (9.74b)$$

It may be noted that we cannot use Eq. (9.71c) to determine $\overline{a}$, in a meaningful way since it is very sensitive to quark masses $m_b$, $m_c$ which are not well known. Finally we note from Eqs. (9.67), (9.69) and (9.72) that

$$\overline{m}_B - m_b = \overline{\Lambda}_d + O\left(\frac{1}{m_b}\right) \qquad (9.75a)$$

which gives for $m_b \simeq 4.9$ GeV,

$$\overline{\Lambda}_d \simeq 0.41 \text{ GeV} \tag{9.75b}$$

With the help of Eqs. (7.105) and (7.106), we can derive the mass formulae for heavy baryons using Eqs. (9.65) and (9.66). The masses for the baryons $\Lambda_Q$, $\Sigma_Q$ and $\Sigma_Q^\star$, can be obtained as

$$m_{\Lambda_Q} = m_Q + \widetilde{\Lambda}_d - \frac{3}{4m_u m_d}\widetilde{\lambda} + \frac{\widetilde{a}}{2m_Q} \tag{9.76}$$

$$m_{\Sigma_Q} = m_Q + \widetilde{\Lambda}_d - \frac{3}{4m_u m_d}\widetilde{\lambda} + \frac{\widetilde{a}}{2m_Q} - \frac{\widetilde{d}(\Sigma_Q)}{m_Q} \tag{9.77a}$$

$$m_{\Sigma_Q^\star} = m_Q + \widetilde{\Lambda}_d + \frac{3}{4m_u m_d}\widetilde{\lambda} + \frac{\widetilde{a}}{2m_Q} + \frac{\widetilde{d}(\Sigma_Q)}{2m_Q} \tag{9.77b}$$

where $\widetilde{a}$ and $\widetilde{d}$ are given in Eqs. (9.60) with $P$ replaced by $B_Q$ and

$$\widetilde{\Lambda}_d = \langle H_q \rangle + m_d + m_u \tag{9.78}$$

and parameter $\widetilde{\lambda}$ arises from the color magnetic moment interaction of light quarks in a heavy baryon [cf. Eq. (7.86)]:

$$H_{q_1 q_2} = -\frac{8\pi}{3}\left(-\frac{2}{3}\alpha_s\right)\frac{\mathbf{S}_{q_1} \cdot \mathbf{S}_{q_2}}{m_{q_1} m_{q_2}}\delta^3(\mathbf{r}) \tag{9.79}$$

From Eq. (9.79), one can write

$$\langle H_{q_1 q_2} \rangle = \widetilde{\lambda}\left\langle \frac{\mathbf{S}_{q_1} \cdot \mathbf{S}_{q_2}}{m_{q_1} m_{q_2}}\right\rangle \tag{9.80a}$$

where

$$\widetilde{\lambda} = \frac{16}{3}\left(\frac{2}{3}\alpha_s\right)\frac{\pi}{2}\langle \overline{\Psi}|\delta^3(\mathbf{r})|\overline{\Psi}\rangle \tag{9.80b}$$

The masses for other baryons can be obtained from $m_{\Lambda_Q}$, $m_{\Sigma_Q}$ and $m_{\Sigma_Q^\star}$ by appropriate replacement in the light flavor index. From Eqs. (9.77), (9.78), we get

$$m_{\Sigma_Q^\star} - m_{\Sigma_Q} = m_{\Xi_Q^\star} - m'_{\Xi_Q} = m_{\Omega_Q^\star} - m_{\Omega_Q} = \frac{3\widetilde{d}(\Sigma_Q)}{2m_Q} \tag{9.81}$$

$$\frac{1}{2}\left(m_{\Sigma_Q} + m_{\Omega_Q}\right) - m'_{\Xi_Q} = \frac{1}{2}\left(m_{\Sigma_Q^\star} + m_{\Omega_Q^\star}\right) - m_{\Xi_Q^\star} = 0 \tag{9.82}$$

$$\left(m_{\Lambda_b} - m_{\Lambda_c}\right) = (m_b - m_c)\left[1 - \frac{\widetilde{a}}{2m_b m_c}\right] \tag{9.83}$$

With the present experimental values for the charmed baryons. [see Chap. 8 and Ref. [11]]

$$m_{\Sigma_c^\star} - m_{\Sigma_c} = m_{\Xi_c^\star} - m'_{\Xi_c} \simeq 65 \text{ MeV} \tag{9.84}$$

Thus the mass relation (9.82) is well satisfied. From Eq. (9.82) then we obtain

$$m_{\Omega_c^\star} = m_{\Omega_c} + 65 \text{ MeV} \simeq 2765 \text{ MeV} \tag{9.85}$$

We also note from Eqs. (9.74b), (9.82) and (9.84) that

$$\frac{\widetilde{d}(\Sigma_c)}{\overline{d}(D)} \simeq 0.2 \tag{9.86}$$

Further from Eqs. (9.71) and (9.83),

$$\frac{m_b - m_c}{2m_b m_c} [\overline{a} - \widetilde{a}] = \left[ \left( m_{\Lambda_b} - m_{\Lambda_c} \right) - (\overline{m}_B - \overline{m}_D) \right] \tag{9.87}$$

Now $m_{\Lambda_b} - m_{\Lambda_c} \simeq 3.34 \text{ GeV} \simeq \overline{m}_B - \overline{m}_D$, therefore $\widetilde{a} = \overline{a}$. Using $\widetilde{a} = \overline{a}$, one gets from Eqs. (9.77), (9.78), (9.71), and (9.72):

$$\overline{m}_{\Lambda_b} - \overline{m}_B = \overline{m}_{\Lambda_c} - \overline{m}_D = \widetilde{\Lambda}_d - \overline{\Lambda}_d \tag{9.88}$$

where

$$\overline{m}_{\Lambda_Q} = \frac{1}{4} \left[ m_{\Lambda_Q} + m_{\Sigma_Q} + 2m_{\Sigma_Q^\star} \right] \tag{9.89}$$

Thus using $\overline{m}_{\Lambda_c} = 2.443 \text{ GeV}$,

$$\widetilde{\Lambda}_d - \overline{\Lambda}_d = 0.47 \text{ GeV} \tag{9.90}$$

Also from Eqs. (9.77) and (9.78)

$$m_{\Sigma_Q} - m_{\Lambda_Q} = \frac{\widetilde{\lambda}}{m_u m_d} - \frac{\overline{d}(\Sigma_Q)}{m_Q} \tag{9.91}$$

Thus in particular for charmed baryons,

$$\frac{\widetilde{\lambda}}{m_u m_d} = \left( m_{\Sigma_c} - m_{\Lambda_c} \right) + \frac{2}{3} \left( m_{\Sigma_c}^\star - m_{\Sigma_c} \right)$$

$$\simeq 168 + 43 \simeq 211 \text{ MeV} \tag{9.92}$$

which is not very much different from $\frac{\overline{d}(D)}{m_c} \simeq 213 \text{ MeV}$ (see Eq. (9.74b)).

The success of the mass formulae Eqs. (9.70) and (9.82) cannot be taken as verification of HQET, since similar formulae also hold for light hadrons [see Sec. 7.6.2]. If we follow the approach of Chap. 7 for baryons and put

$$\widetilde{d} = \frac{\widetilde{\lambda}}{m_q} \tag{9.93}$$

then Eq. (9.83) remains unchanged. Using $m_{\Sigma_c} - m_{\Lambda_c} = 168 \text{ MeV}$, we obtain

$$m_{\Sigma_c^\star} - m_{\Sigma_c} \simeq 67 \text{ MeV}$$

$$m_{\Sigma_b^\star} - m_{\Sigma_b} \simeq 20 \text{ MeV} \tag{9.94}$$

$$m_{\Sigma_b} - m_{\Lambda_b} \simeq 199 \text{MeV}, \tag{9.95}$$

i.e. the results similar to the ones obtained in HQET.

## 9.4 The P-wave Heavy Mesons: Mass Spectroscopy

So far we have discussed only S-wave heavy mesons. We now discuss P-wave mesons for which experimental evidence is available. Since the spin of heavy quark is decoupled, it is natural to combine

$$\mathbf{j} = \mathbf{L} + \mathbf{S}_q \tag{9.96}$$

with $\mathbf{S}_Q$ to give

$$\mathbf{J} = \mathbf{j} + \mathbf{S}_Q \tag{9.97}$$

for the bound $Q\bar{q}$ system. Thus for the P states we get two multiplets, one with $j = 3/2$ and the other with $j = 1/2$. Hence for $\ell = 1$ we have four multiplets

$$\left[D_2^* \left(2^+\right)\right], \ \left[D_1 \left(1^+\right)\right]_{j=3/2}$$

and

$$\left[D_1^* \left(1^+\right)\right], \ \left[D_0 \left(0^+\right)\right]_{j=1/2}$$

It is useful to write down the angular momentum part of the wave functions for the four P-states. According to the angular momentum scheme outlined above, P-state can be labeled as $|JMjs_Q\rangle$. We can write these states for $D_J$ mesons:

$$|D_2^\star, M = \pm 1\rangle = \begin{cases} \frac{1}{\sqrt{2}} \left[Y_{10}\chi_+^{+1} + Y_{11}\chi_+^0\right] \\ \frac{1}{\sqrt{2}} \left[Y_{10}\chi_+^{-1} + Y_{1-1}\chi_+^0\right] \end{cases}$$

$$|D_2^\star, M = 0\rangle = \frac{1}{\sqrt{6}} \left[Y_{1-1}\chi_+^{+1} + 2Y_{10}\chi_+^0 + Y_{11}\chi_+^{-1}\right] \tag{9.98a}$$

$$|D_1, M = \pm 1\rangle = \begin{cases} \frac{1}{\sqrt{6}} \left[-Y_{10}\chi_+^{+1} + Y_{11}\left(\chi_+^0 + 2\chi_-^0\right)\right] \\ \frac{1}{\sqrt{6}} \left[Y_{10}\chi_+^{-1} + Y_{1-1}\left(-\chi_+^0 + 2\chi_-^0\right)\right] \end{cases}$$

$$|D_1, M = 0\rangle = \frac{1}{\sqrt{6}} \left[-Y_{1-1}\chi_+^{+1} + 2Y_{10}\chi_-^0 + Y_{11}\chi_+^{-1}\right] \tag{9.98b}$$

$$|D_1^\star, M = \pm 1\rangle = \begin{cases} \frac{1}{\sqrt{3}} \left[-Y_{10}\chi_+^{+1} + Y_{11}\left(\chi_+^0 - \chi_-^0\right)\right] \\ \frac{1}{\sqrt{3}} \left[Y_{10}\chi_+^{-1} + Y_{1-1}\left(-\chi_+^0 - \chi_-^0\right)\right] \end{cases}$$

$$|D_1^\star, M = 0\rangle = \frac{1}{\sqrt{3}} \left[-Y_{1-1}\chi_+^{+1} - Y_{10}\chi_-^0 + Y_{11}\chi_+^{-1}\right] \tag{9.99a}$$

$$|D_0, M = 0\rangle = \frac{1}{\sqrt{3}} \left[Y_{1-1}\chi_+^{+1} - Y_{10}\chi_+^0 + Y_{11}\chi_+^{-1}\right] \tag{9.99b}$$

The degeneracy between $j = 3/2$ and $j = 1/2$ states is removed by the spin-orbit coupling term $\mathbf{S}_q \cdot \mathbf{L}$ in the Hamiltonian $H_q$ given in Eq. (9.1). viz the term

$$\frac{1}{2m_q^2} \mathbf{S}_q \cdot \mathbf{L} \times \left[ \frac{1}{r} \frac{dV_c}{dr} \right]$$

where

$$V_c = -\frac{4}{3} \alpha_s \frac{1}{r} = -\frac{K'}{r} \tag{9.100}$$

is one gluon potential. Thus using [cf. Eq. (8.21)]

$$\langle \mathbf{S}_q \cdot \mathbf{L} \rangle_{j=3/2,1/2} = \frac{1}{2}, \ -1 \tag{9.101}$$

we have

$$m_{j=3/2} - m_{j=1/2} = \left( \frac{1}{2} + 1 \right) \bar{\lambda}_1 = \frac{3}{2} \bar{\lambda}_1 \tag{9.102a}$$

where

$$m_{j=3/2} = \frac{5 m_{D_2^*} + 3 m_{D_1}}{8} \tag{9.102b}$$

$$m_{j=1/2} = \frac{3 m_{D_1^*} + m_{D_0}}{4} \tag{9.102c}$$

$$\bar{\lambda}_{1q} = \frac{1}{4m_q^2} \left\langle \frac{1}{r} \frac{dV_c}{dr} \right\rangle = \frac{K'}{4m_q^2} \left\langle \frac{1}{r^3} \right\rangle_{1p} \tag{9.102d}$$

(subscript 1 on $\bar{\lambda}$ refers to $l = 1$ state). The degeneracy between the doublet $D_2^*$ and $D_1$ and the doublet $D_1^*$ and $D_0$ is removed by the term $\sigma_Q \cdot \mathbf{B}^c$ in the Hamiltonian (9.66). For P-wave this term induces the color magnetic moment interaction of the type

$$S_{12} = [12 \, (\mathbf{S}_q \cdot \mathbf{n}) \, (\mathbf{S}_Q \cdot \mathbf{n}) - 4\mathbf{S}_q \cdot \mathbf{S}_Q] \tag{9.103a}$$

where $\mathbf{n}$ is a unit vector $\frac{\mathbf{r}}{r}$. To see this we note the interaction Hamiltonian for dipole-dipole interaction is given by

$$H_{int} = -\mu_Q \cdot \mathbf{B}^c \tag{9.103b}$$

where the field $\mathbf{B}^c$ produced by the magnetic dipole $\mu_q$ is given by

$$\mathbf{B(r)} = \left[ \frac{3\mathbf{n}(\mathbf{n} \cdot \mu_q) - \mu_q}{r^3} + \frac{8\pi}{3} \, \mu_q \delta^3(\vec{r}) \right] \tag{9.103c}$$

Hence

$$H_{int} = \frac{1}{r^3} \left[ (\mu_Q \cdot \mu_q) - 3(\mathbf{n} \cdot \mu_Q)(\mathbf{n} \cdot \mu_q) \right] - \frac{8\pi}{3} \, \mu_Q \cdot \mu_q \delta^3(\vec{r}) \tag{9.104}$$

This term induces a term

$$\frac{1}{12 m_q m_Q} S_{12} \left[ \frac{1}{r} \frac{dV_c}{dr} - \frac{d^2 V_c}{dr^2} \right] = \frac{3}{12 m_q m_Q} S_{12} \left( \frac{K'}{r^3} \right) \tag{9.105}$$

in Eq. (9.1). Then using the angular wave functions for the states $D_2^*$, $D_1$, $D_1^*$, and $D_0$ given in Eqs. (9.98) and (9.99),

$$\langle S_{12} \rangle_{D_2^*} = -2/5, \qquad \langle S_{12} \rangle_{D_1} = \frac{2}{3} \tag{9.106a}$$

$$\langle S_{12} \rangle_{D_1^*} = \frac{4}{3}, \qquad \langle S_{12} \rangle_{D_0} = -4 \tag{9.106b}$$

Hence the masses for these states, can be written as

$$m_{D_{q2}^*} = m_c + \bar{\Lambda}_{1q} + \frac{1}{2} \bar{\lambda}_{1q} + \frac{\bar{a}_1}{2 m_c} - \frac{2}{5} \frac{\bar{d}_1 (D)}{2 m_c} \tag{9.107}$$

$$m_{D_{q1}} = m_c + \bar{\Lambda}_{1q} + \frac{1}{2} \bar{\lambda}_{1q} + \frac{\bar{a}_1}{2 m_c} + \frac{2}{3} \frac{\bar{d}_1 (D)}{2 m_c} \tag{9.108}$$

$$m_{D_{q1}^*} = m_c + \bar{\Lambda}_{1q} - \bar{\lambda}_{1q} + \frac{\bar{a}_1}{2 m_c} + \frac{4}{3} \frac{\bar{d}_1 (D)}{2 m_c} \tag{9.109}$$

$$m_{D_{q0}} = m_c + \bar{\Lambda}_{1q} - \bar{\lambda}_{1q} + \frac{\bar{a}_1}{2 m_c} - 4 \frac{\bar{d}_1 (D)}{2 m_c} \tag{9.110}$$

where

$$\bar{d}_1 (D) = \frac{K'}{2 m_q} \left\langle \frac{1}{r^3} \right\rangle_{1p}$$

and the parameters $\bar{a}_1$ and $\bar{d}_1$ refer to P-state, similar to $\bar{a}$ and $\bar{d}$ for $S$-state. From Eqs. (9.105) and (9.110),

$$m_{D_{q2}^*} - m_{D_{q1}} = -\frac{16}{15} \frac{\bar{d}_1 (D)}{2 m_c} \tag{9.111}$$

$$m_{D_{q1}^*} - m_{D_{q0}} = \frac{16}{3} \frac{\bar{d}_1 (D)}{2 m_c}$$

$$= -5 \left( m_{D_{2q}^*} - m_{D_{1q}} \right) \tag{9.112}$$

Needless to say that for $b$-flavor P-states, replace $D$ by $B$ and $m_c$ by $m_b$. Using the experimental values for the masses, one finds $m_{D_2^*} - m_{D_1} \approx 39$ MeV; $m_{D_{s2}^*} - m_{D_{s1}} \approx 38$ MeV. Thus relation which gives $m_{D_2^*} - m_{D_1} = m_{D_{s2}^*} - m_{D_{s1}}$ is well satisfied. From Eqs. (9.107) and (9.108)

$$\frac{\bar{d}_1 (D)}{2 m_c} \approx -36 \text{ MeV} \tag{9.113}$$

Using the values of $m_{D_2^*}$, $m_{D_1}$; $m_{D_{s2}^*}$, and $m_{D_{s1}}$, Eq. (9.112) gives

$$m_{D_1^*} - m_{D_0} \approx -190 \approx m_{D_{s1}^*} - m_{D_{s0}} \tag{9.114}$$

Now, Eq. (9.112) gives $m_{D^*_{1q}} < m_{D_{0q}}$, which is against the hierarchy in which we expect the P-wave states, i.e. $D_{1q}$ to lie above $D_{0q}$. In deriving Eq. (9.112), i.e. $m_{D^*_{1q}} < m_{D_{0q}}$, we have used the same $\bar{d}_1$ for $j = 3/2$ and $j = 1/2$ states. However, to arrive at the result $m_{D^*_{1q}} > m_{D_{0q}}$, $\bar{d}_1(j = 3/2)$ must have opposite sign to that of $\bar{d}_1(j = 1/2)$; in that case there is no reason to believe that they have the same magnitude. This is highly unlikely; in fact it is impossible since the sign and magnitude is determined by $S_{12}$ given in Eq. (9.105), with the one gluon exchange potential given in Eq. (9.100). However the above deficiency can be removed by taking into account the relative motion of heavy quark to give the spin orbit coupling $\frac{\mathbf{S}\cdot\mathbf{L}}{2m_q m_Q}$. This term induces a term

$$\frac{1}{m_q m_Q}(\mathbf{S}\cdot\mathbf{L})\left[\frac{1}{r}\frac{dV_c}{dr}\right] = \frac{2}{2m_q m_Q}(\mathbf{S}\cdot\mathbf{L})\left(\frac{K'}{r^3}\right) \tag{9.115}$$

Then the wave functions given in Eqs. (9.98) and (9.99), give

$$\langle\mathbf{S}\cdot\mathbf{L}\rangle_{D^*_2} = 1, \qquad\qquad \langle\mathbf{S}\cdot\mathbf{L}\rangle_{D_1} = -\frac{1}{3} \tag{9.116}$$

$$\langle\mathbf{S}\cdot\mathbf{L}\rangle_{D^*_1} = -\frac{2}{3}, \qquad\qquad \langle\mathbf{S}\cdot\mathbf{L}\rangle_{D_0} = -2 \tag{9.117}$$

These terms give an additional contribution to the last two terms in Eqs. (9.107)-(9.110) which are modified to,

$$\frac{9}{5}\frac{\bar{d}_1(D)}{m_c}, \; -\frac{1}{3}\frac{\bar{d}_1(D)}{m_c}, \; -\frac{2}{3}\frac{\bar{d}_1(D)}{m_c}, \; -6\frac{\bar{d}_1(D)}{m_c} \tag{9.118}$$

Thus instead of Eq. (9.111) and Eq. (9.112),

$$m_{D^*_{q2}} - m_{D_{q1}} = \frac{32}{15}\frac{\bar{d}_1(D)}{m_c} \tag{9.119}$$

$$m_{D^*_{q1}} - m_{D_{q0}} = \frac{16}{3}\frac{\bar{d}_1(D)}{m_c} \tag{9.120}$$

$$m_{D^*_{q1}} - m_{D_{q0}} = \frac{5}{2}\left(m_{D^*_{q2}} - m_{D_{q1}}\right) \tag{9.121}$$

From Eqs. (9.119) and (9.70b):

$$\frac{\bar{d}_1(D)}{\bar{d}(D)} = \frac{5}{16}\frac{m_{D^*_{q2}} - m_{D_{q1}}}{m_{D^*_q} - m_{D_q}} \approx \begin{cases} 0.086 & \text{for} \quad q = d \\ 0.082 & \text{for} \quad q = s \end{cases} \tag{9.122}$$

i.e. $\bar{d}_1(D)$ and $\bar{d}(D)$ are independent of light flavor. Now using the experimental values of $D^*_2$, $D_1$, $D^*_{s2}$, $D_{s1}$ and $D^*_{s1}$:

$$m_{D_{s0}} = m_{D^*_{s1}} - \frac{5}{2}(m_{D^*_{s2}} - m_{D_{s1}}) = (2356.2 \pm 1.7)\ \text{MeV} \tag{9.123}$$

$$m_{D_{s1}^*} - m_{D_{s0}} = (103.5 \pm 1.1) \text{ MeV} \tag{9.124}$$

$$m_{D_1^*} - m_{D_0} = (102.0 \pm 1.6) \text{ MeV} \tag{9.125}$$

$$(m_{D_{s1}^*} - m_{D_1^*}) - (m_{D_{s0}} - m_{D_1}) = (1.5 \pm 3.0) \text{ MeV}$$
$$\approx 0 \text{ MeV} \tag{9.126}$$

The mass difference between S-wave $J = 1$ and $J = 0$ mesons:

$$m_{D_s^{*+}} - m_{D_s^+} = (143.8 \pm 0.4) \text{ MeV}$$
$$m_{D^{*0}} - m_{D^0} = (142.12 \pm 0.07) \text{ MeV}$$
$$m_{D^{*+}} - m_{D^+} = (140.65 \pm 0.10) \text{ MeV} \tag{9.127}$$

$$m_{D_s^+} - m_{D^+} = (98.88 \pm 0.301) \text{ MeV}$$
$$m_{D_s^*} - m_{D^*} = (102.05 \pm 0.64) \text{ MeV} \tag{9.128}$$

We conclude from these equations and those proceeding them that mass differences are independent of light flavor and

$$(m_{D_{q1}^*} - m_{D_{q0}}) \approx \frac{1}{\sqrt{2}} \left( m_{D_q^*} - m_{D_q} \right) \tag{9.129}$$

The mass of $D_1^*$ is not experimentally known, since it has not yet been observed. Unlike $D_{s1}^*$ which is a narrow resonance, $D_1^*$ is expected to be broad resonance as its decay to $D^*\pi$ is S-wave decay and it is energetically allowed. However in the potential model its mass can be obtained as follows. From Eqs. (9.107, 9.108):

$$(\bar{\Lambda}_{1s} - \bar{\Lambda}_{1d}) + \frac{1}{2} \left( \bar{\lambda}_{1s} - \bar{\lambda}_{1d} \right) = m_{D_{s2}^*} - m_{D_2^*}$$
$$= m_{D_{s1}} - m_{D_1} \tag{9.130}$$

and from Eqs. (9.109), (9.110)

$$(\bar{\Lambda}_{1s} - \bar{\Lambda}_{1d}) - (\bar{\lambda}_{1s} - \bar{\lambda}_{1d}) = m_{D_{s1}^*} - m_{D_1^*}$$
$$= m_{D_{s0}} - m_{D_0} \tag{9.131}$$

Hence from Eqs. (9.130), (9.131), one gets

$$(m_{D_{s1}} - m_{D_1}) - (m_{D_{s1}^*} - m_{D_1^*}) = \frac{3}{2}(\bar{\lambda}_{1s} - \bar{\lambda}_{1d})$$
$$= (m_{D_{s2}^*} - m_{D_2^*}) - (m_{D_{s1}^*} - m_{D_1^*}) \tag{9.132}$$

Now from the experimental masses of $D_{s2}^*$, $D_2^*$, $D_{s1}$ and $D_1$ :

$$m_{D_{s2}^*} - m_{D_2^*} = (109.8 \pm 1.9) \text{ MeV}$$
$$m_{D_{s1}} - m_{D_1} = (113.0 \pm 0.3) \text{ MeV} \qquad (9.133)$$

Hence from Eqs. (9.126), (9.128) and (9.133), we conclude that in the potential model, the mass difference between the bound states $(c\bar{s})$ and $(c\bar{d})$ is of the order of 100 MeV. This is a general feature of potential model. Thus from Eq. (9.132):

$$\bar{\lambda}_{1s} \approx \bar{\lambda}_{1d}$$

$$(m_{D_{s1}^*} - m_{D_1^*}) \approx (m_{D_{s1}} - m_{D_1}) = (113 \pm 0.36) \text{ MeV} \qquad (9.134)$$

Hence

$$m_{D_1^*} = (m_{D_{s1}^*} - m_{D_{s1}}) - m_{D_1} = (2346.2 \pm 1.4) \text{ MeV} \qquad (9.135)$$

and on using Eq. (9.125):

$$m_{D_0} = m_{D_1^*} - (m_{D_1^*} - m_{D_0}) = (2244 \pm 3) \text{ MeV} \qquad (9.136)$$

However the $J = 0$, P-wave states $0^+ : (D_{s0}^*, D_0^*)$, discovered experimentally have masses $(2317.8 \pm 0.6)$ MeV and $(2318 \pm 29)$ MeV. These states satisfy the following mass relations

$$m_{D_{s0}^*} - m_{D_0^*} \approx 0 \qquad (9.137)$$
$$(m_{D_{s1}^*} - m_{D_{s0}^*}) = (141.7 \pm 1.2) \text{ MeV}$$
$$\approx (m_{D_s^*} - m_{D_s}) \qquad (9.138)$$
$$(m_{D_1^*} - \bar{m}_{D_0^*}) \approx (28 \pm 30) \text{ MeV} \qquad (9.139)$$

The pattern of mass relations for these states given above is hard to understand in the bound state model $(c\bar{q})_{L=1}$. Hence we conclude, the experimentally discovered $0^+$ states $D_0^*$ and $D_{s0}^*$ do not fit as the bound state $(c\bar{q})_{L=1}$. This is because the bound state predicts $0^+$ states $D_{s0}$ and $D_0$ at masses given in Eqs. (9.123) and (9.136), which do not satisfy Eq. (9.137).

We conclude that potential model combined with HQET give masses for S-wave heavy mesons in reasonable agreement with experiment. For P-wave states: $(c\bar{q})_{L=1}$, the potential model gives

$$m_{j=1/2} < m_{j=3/2}$$

$$m_{D_{q1}^*} - m_{D_{q0}} = \frac{5}{2} \left( m_{D_{q2}^*} - m_{D_{q1}} \right)$$

$$m_{D_{q1}^*} - m_{D_{q0}} \approx \frac{1}{\sqrt{2}}(m_{D_q^*} - m_{D_q})$$

Out of the multiplets $\left(m_{D_{q2}^*} - m_{D_{q1}}\right)_{j=3/2}$ and $\left(m_{D_{q1}^*} - m_{D_{q0}}\right)_{j=1/2}$, $D_{s2}^*$, $D_2^*$, $D_{s1}$ and $D_1$ have been experimentally discovered. However experimental discovery of the $D_1^*$, $D_{s0}$, $D_0$ states at the masses predicted by the potential model is still awaiting. The states $D_0^*$ and $D_{s0}^*$ do not follow the general features as bound states of $(c\bar{s})_{L=1}$, $(c\bar{d})_{L=1}$ and $(c\bar{u})_{L=1}$. The states $D_0^*$ and $D_{s0}^*$ are good candidates as tetraquark states (see [8,9]).

## 9.5   Decays of P-wave Mesons

We now discuss the strong decays of P-wave mesons. Parity and angular momentum conservation restricts these decays to the following modes: $D_2^* \to (D\pi)_{l=2}$, $D_2^* \to (D^*\pi)_{l=2}$, $D_1$, $D_1^* \to (D^*\pi)_{l=0,2}$, and $D_0 \to (D\pi)_{l=0}$. Note that $D_1$, $D_1^* \to D\pi$ is forbidden due to parity conservation.

It is convenient to express the decay width in terms of the helicity amplitudes (see Eq. (4.41))

$$\Gamma^J = \frac{|\mathbf{P}_\pi|}{8\pi s} \overline{\sum_\lambda} \left|F_\lambda^J(s)\right|^2 \tag{9.140}$$

In the rest frame of the decaying particle the helicity amplitudes which contribute are $F_0^2$, $F_\pm^2$, $F_0^1$, $F_{\pm1}^1$, $F_0^{\prime1}, F_\pm^{\prime1}$ and $F_0^{\prime0}$. In the heavy quark limit the helicity amplitudes are related as follows:

$j = 3/2$ multiplet:

$$F_0^1 = -2F_\pm^1 = F_0^2 = \frac{2}{\sqrt{3}}F_{\pm1}^2 \tag{9.141}$$

$j = 1/2$ multiplet:

$$\acute{F}_0^1 = -\acute{F}_{\pm1}^1 = \acute{F}_0^0 \tag{9.142}$$

The simplest way to see this is as follows. The emission of pion by $D_J$ would not affect the velocity of heavy quark. Thus it is the operator $\mathbf{S}_q \cdot \mathbf{n}$ which is relevant for these decays. If we select the direction of quantization along z-axis, (i.e. $L_z$ is taken along z-axis) then for the helicity amplitudes, the operator $S_{3q}\sqrt{\frac{4\pi}{3}}\,Y_{10}$ contributes. Then using the wave functions in

Eqs. (9.98) and (9.99), it is easy to derive Eqs. (9.141) and (9.142) by considering the matrix elements of the type

$$F_\lambda^J = f \langle D^* (D), \lambda | S_{3q} Y_{10} | D_J, \lambda \rangle \qquad (9.143)$$

where $f$ is the reduced amplitude. Since hadronic decays of $D_2^*$ are pure D-wave, it follows that $D_1 \to D^* \pi$ is also D-wave. As such for these decays, $F_\lambda^J (s) \sim |\mathbf{p}_\pi|^5$. Similarly since the decay of $D_0$ is pure S-wave, it follows that the decay of $D_1^*$ is also pure S-wave. The above restrictions are consequence of relations (9.141) and (9.142) which hold in the heavy quark spin symmetry limit. Hence for the decays $D_2^* \to D\pi$, $D_2^* \to D^*\pi$, and $D_1 \to D^*\pi$ ,

$$\frac{\Gamma (D_2^* \to D\pi)}{\Gamma (D_2^* \to D^*\pi)} = \frac{2}{3} \frac{|\mathbf{p}_\pi|_{D\pi}^5}{|\mathbf{p}_\pi|_{D^*\pi}^5} \approx 2.46 \qquad (9.144)$$

$$\frac{\Gamma (D_1 \to D^*\pi)}{\Gamma (D_2^* \to D^*\pi)} = \frac{5}{3} \frac{|\mathbf{p}_\pi|_{D_1 D^*\pi}^5}{|\mathbf{p}_\pi|_{D^*\pi}^5} \approx 1.05 \qquad (9.145)$$

where we have used from the experimental data, $|\mathbf{p}_\pi|_{D\pi} = 505$ MeV, $|\mathbf{p}_\pi|_{D^*\pi} = 389$ MeV, $|\mathbf{p}_\pi|_{D_1 D^*\pi} = 355$ MeV. Experimentally

$$\Gamma_{D_2^{0*}} \equiv \Gamma \left( D_2^0 \to D^*\pi^- + D^*\pi^0 + D\pi^- + D^0\pi^0 \right) \qquad (9.146)$$

$$= 43 \pm 4 \text{ MeV}$$

From Eq. (9.145)

$$\frac{\Gamma_{D_1^0}}{\Gamma_{D_2^{0*}}} = \left( \frac{\Gamma \left( D_1^0 \to D^*\pi \right)}{\Gamma \left( D_2^{0*} \to D^*\pi \right)} \right) \frac{1}{1 + \frac{\Gamma(D_2^* \to D\pi)}{\Gamma(D_2^* \to D^*\pi)}} \simeq 0.30 \text{ MeV} \qquad (9.147)$$

which gives

$$\Gamma_{D_1^0} \simeq 13.0 \pm 1.2 \text{ MeV} \quad (20.4 \pm 1.7 \text{MeV}) \qquad (9.148)$$

This is in disagreement with the experimental value given in parentheses. This shows that the decay $D_1 \to D^*\pi$ is not pure D-wave; there may be a component of S-wave. The S-wave widths are usually large, a small component of S-wave may be possible due to symmetry breaking, since heavy quark spin symmetry is not exact. This may be tested for $B_2^*$ and $B_1$ decays where the symmetry breaking effects are expected to be small.

The decays $D_1^* \to D^*\pi$ and $D_0 \to D\pi$ are S-wave decays; thus the decay widths are expected to be large, i.e. in the range of few hundreds of MeV. No experimental data are available even on the masses of $D_1^*$ and $D_0$.

The prediction

$$\frac{\Gamma (D_2^* \to D\pi)}{\Gamma (D_2^* \to D^*\pi)} \approx 2.5$$

and the other predictions on the decay widths of heavy hadrons have to wait for their verification till the experimental data is available.

## 9.6 Problems

(1) Show that $e^{\mp im\psi v \cdot x} h_{\pm v} = e^{-imv \cdot x} h_{\pm v}$.

(2) Using Eq. (9.35), derive Eq. (9.37).

Hint: First show that

$$\left[ h_{+v}(t, \vec{x}), \bar{h}_{+v}(t, \vec{y}) \right] = \frac{1}{v_0} \delta^3(\vec{x} - \vec{y})$$

using the Lagrangian

$$\mathcal{L} = \bar{h}_{+v} iv \cdot D h_{+v}$$

You may use the relation:

$$[AB, C] = A [B, C]_+ - [A, C]_+ B$$

(3) Noting the fact that $J = 1$ mesons $D_1$ and $D_1^*$ both contain the spin singlet component $\chi_-^0$ in their wave functions given in Eqs. (9.98) and (9.99), so that $E_1$ transition of $D_1$ and $D_1^*$ to $D(^1S_0)$ is allowed. Show that

$$A(D_1 \to D^* + \gamma)/A(D_1 \to D\gamma) = \frac{1}{\sqrt{2}}$$

and

$$A(D_1^* \to D^* + \gamma)/A(D_1^* \to D\gamma) = \sqrt{2}$$

## 9.7 References

1. Fayyazuddin and Riazuddin, Phys. Rev. 48, 2224(1993); Modern Phys. Lett. A, 12, 1791(1991).

2. H. Georgi, "Heavy Quark Effective Field Theory" in Proc. Theoretical Advanced Study Institute (1991) Ed. R.K. Ellis, C.T. Hill, and J.D. Lykken (World Scientific Singapore, 1992).

3. Riazuddin and Fayyazuddin, "Heavy Quark Spin Symmetry" in Salamfest, eds. A. Ali, J. Ellis and S. Randjbar-Daemi, World Scientific.

4. Mark B Wise, "Heavy Flavor Theory: overview" in AIP Conference Proceedings 302, Editors P. Drell and D. Rubin (AIP Press, 1993)

5. M Neubert, Phys. Rep. 245, 259 (1994); Int. J. Mod. Phys. A, 4173 (1996).

6. M Neubert, "B decays and the Heavy Quark Expansion" CERN–TH/97-24, hep-ph/9702375, to appear in the second edition of Heavy Flavors, edited by A.J. Buras and M. Linder (World Scientific Singapore)

Heavy Quark Effective Theory

J. D. Jakson, Classical Electrodynamics, Third Edition, Page 190, John Wisley (1998).
8. R.L. Jaffe, Exotica, arXiv:hep-ph/0409065 (2004).
9. F.E. Close, arXiv:hep-ph/0411396 (2004).
10. L. Maiani et al., (arXiv: hep-ph/0412098 (2004); K. Teraski, arXiv:1005.5573 (2010).
11. Particle Data Group, K Nakamura et al., J. of Physics **G 37**, 075021 (2010).

# Chapter 10

# Weak Interaction

## 10.1  $V - A$ Interaction

In analogy with electromagnetic interaction $J_\mu A^\mu$, Fermi proposed for $\beta$-decay the interaction $J^\mu J_\mu$, viz

$$H_{int} = G \left[ \bar{\Psi}_1(x)\gamma_\mu \Psi_2(x) \right] \left[ \bar{\Psi}_3(x)\gamma^\mu \Psi_4(x) \right] + h.c. \qquad (10.1)$$

The above interaction is for the process

$$2 \rightarrow 1 + 3 + \bar{4} \quad (\text{e.g. } n \rightarrow p + e^- + \bar{\nu}_e).$$

The interaction (10.1) can be generalized using five Dirac bilinear covariants. Thus the most general non-derivative four-fermion interaction can be written as

$$H_{int} = \sum_i \left[ \bar{\Psi}_1(x)\Gamma_i \Psi_2(x) \right] \left[ \bar{\Psi}_3(x)\Gamma^i(C_i - C_i'\gamma_5) \Psi_4(x) \right]$$
$$+ h.c. \qquad (10.2)$$

where $\Gamma_i (i = S, V, T, A, P)$ are the five Dirac independent matrices: $1, \gamma_\mu, \sigma_{\mu\nu}, \gamma_\mu\gamma_5, \gamma_5$. In writing Eq. (10.2), we have taken into account the parity violation in $\beta$-decay.

For a massless Dirac particle, if $\Psi$ is a solution of Dirac equation, then $\pm\gamma_5\Psi$ is also its solution. Without loss of generality, we take only negative sign. Suppose particle 4 is massless, then the bilinear

$$\bar{\Psi}_3(x)\Gamma^i \Psi_4(x) \rightarrow -\bar{\Psi}_3(x)\Gamma^i\gamma_5 \Psi_4(x).$$

Hence for this case $C_i = C_i'$. Thus we can write Eq. (10.2) as

$$H_{int} = \sum_i \left[ \bar{\Psi}_1\Gamma_i\Psi_2 \right] \left[ \bar{\Psi}_3 C_i\Gamma^i(1 - \gamma_5)\Psi_4 \right] + h.c. \qquad (10.3)$$

If we identify particle 4 with the neutrino, we have the result that only left handed neutrino takes part in weak processes. This is what is observed experimentally (see below). Thus irrespective of the fact whether neutrino is massless or not, Eq. (10.3) will hold if we take into account the fact that only left handed neutrinos take part in weak processes. Suppose we impose the chiral transformation for the field $\Psi_3$ viz $\Psi_3 \to -\gamma_5 \Psi_3$, then if $H_{int}$ is to be invariant under such a transformation, we have

$$C_S = C_P = C_T = 0.$$

Hence Eq. (10.3) becomes

$$
\begin{aligned}
H_{int} &= \left[ C_V \bar{\Psi}_1 \gamma_\mu \Psi_2 - C_A \bar{\Psi}_1 \gamma_\mu \gamma_5 \Psi_2 \right] \left[ \bar{\Psi}_3 \gamma^\mu (1 - \gamma_5) \Psi_4 \right] \\
&= \frac{G_F}{\sqrt{2}} \left[ \bar{\Psi}_1 \gamma_\mu (1 - \varepsilon \gamma_5) \Psi_2 \right] \left[ \bar{\Psi}_3 \gamma^\mu (1 - \gamma_5) \Psi_4 \right],
\end{aligned}
\tag{10.4}
$$

where we put

$$C_V = \frac{G_F}{\sqrt{2}}, \quad \frac{C_A}{C_V} = \varepsilon. \tag{10.5}$$

Further we note that if we impose the chiral transformation on fields $\Psi_1$ or $\Psi_2$, we have

$$C_V = C_A \tag{10.6}$$

i.e. $\varepsilon = 1$ or V-A theory. We conclude that if one requires invariance of the four- fermion interaction under the chirality transformation of each field separately, we have the V-A theory.

We have written Eq. (10.2) in the order 1 2 3 4. We can go to the order 3 2 1 4 by Fierz reordering theorem:

$$K_i(3214) = \sum_{j=1}^{5} \lambda_{ij} K_j(1234). \tag{10.7}$$

The coefficients $\lambda_{ij}$ are given by the matrix

$$
\lambda_{ij} = -\frac{1}{4}
\begin{pmatrix}
1 & 1 & 1 & 1 & 1 \\
4 & -2 & 0 & 2 & -4 \\
6 & 0 & -2 & 0 & 6 \\
4 & 2 & 0 & -2 & -4 \\
1 & -1 & 1 & -1 & 1
\end{pmatrix},
\tag{10.8}
$$

where

$$K_i(1234) = \left[ \bar{\Psi}_1 \Gamma_i \Psi_2 \right] \left[ \bar{\Psi}_3 \Gamma^i \Psi_4 \right]. \tag{10.9}$$

It is obvious that

$$K_i(3214) = K_i(1432).$$

If we denote by $S$, $V$, $T$, $A$, $P$ the five quadrilinears appearing in the order (1 2 3 4) and $S'$, $V'$, $T'$, $A'$, $P'$ when they appear in the order (3 2 1 4), then from Eqs. (10.7) and (10.8) we get

$$V' - A' = V - A$$
$$S' - T' + P' = S - T + P \qquad (10.10)$$

i.e. these combinations are invariant under Fierz rearrangement.

### 10.1.1 *Helicity of the Neutrino*

To obtain a direct measurement of neutrino helicity, the following reaction was studied

$$^{152}Eu_{(J^P=0^-)} + e^- \rightarrow \, ^{152}Sm^*_{(J^P=1^-)} + \nu_e$$
$$\downarrow$$
$$\rightarrow \left( ^{152}Sm_{(J^P=0^+)} + \gamma \right).$$

The main point of this experiment is that we can select those $\gamma$ rays from the decay of the excited state which go opposite to the $\nu_e$ direction (i.e. in the direction of the recoil nucleus) by having them resonance-scatter from a target of $^{152}Sm$. Balancing the spin along the upward z direction ($\nu_e$ is assumed to be emitted along this direction), one finds that the helicity of the downward $\gamma$-ray will be the same as that for the upward $\nu_e$. By measuring the circular polarization of $\gamma$-ray, the experiment fixed the helicity of the $\gamma$-ray as negative, indicating a left-handed $\nu_e$. Thus it is established that only left-handed neutrinos take part in weak processes.

## 10.2 Classification of Weak Processes

### 10.2.1 *Purely Leptonic Processes*

The well-known example is $\mu$-meson decay

$$\mu^- \rightarrow e^- + \bar{\nu}_e + \nu_\mu.$$

In this process four well-known particles $\mu^-$, $e^-$, $\nu_e$, $\nu_\mu$, called leptons, take part. The decay process is described by V - A interaction [cf. Eqs. (10.4) and (10.6)].

$$-H_{int} = L_W^\ell = -\frac{G_F}{\sqrt{2}} \left\{ \bar{\nu}_\mu \gamma^\mu (1 - \gamma^5) \mu \right\} \left\{ \bar{e} \gamma_\mu (1 - \gamma^5) \nu_e \right\}$$

$$= -\frac{G_F}{\sqrt{2}} L_{(\mu)}^\mu L_{(e)\mu}^\dagger. \tag{10.11a}$$

$L_{(\mu)}^\mu$ and $L_{(e)\mu}$ are lepton currents associated respectively with $\mu$ meson and its associated neutrino $\nu_\mu$ and $e^-$ and $\nu_e$

$$L_{(\mu)}^\mu = \bar{\nu}_\mu \gamma^\mu (1 - \gamma^5) \mu \tag{10.11b}$$

$$L_{(e)\mu} = \bar{\nu}_e \gamma_\mu (1 - \gamma^5) e. \tag{10.11c}$$

The $\gamma^\mu$ and $\gamma^5$ ($\equiv i\gamma^0\gamma^1\gamma^2\gamma^3$) appearing above are the usual Dirac matrices. We write the lepton current as

$$L^\mu = L_{(\mu)}^\mu + L_{(e)}^\mu \tag{10.12}$$

Here $L_\mu^\dagger$ denotes the hermitian conjugate of $L_\mu$. One can also picture the process (1) as being mediated by a vector boson $W_\mu$, the so-called weak vector boson. This is shown in Fig. 10.1 below: Thus all leptonic weak

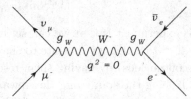

Fig. 10.1   The muon decay mediated by a W-boson.

processes can be described by interaction of the form

$$L_W = -g_W L_\mu W^{-\mu} + h.c. \tag{10.13}$$

where h.c. denotes the hermitian conjugate. Note that Eq. (10.13) is analogous to electromagnetic interaction of say electron which is mediated by photon and is shown in Fig. 10.2.

The interaction responsible for the process shown in Fig. 10.2 is the usual electromagnetic interaction

$$L^{e.m.} = -e j_\mu^{e.m.} a^\mu, \tag{10.14}$$

where $a^\mu$ is the photon field and $j_\mu^{e.m.}$ is the electromagnetic current:

$$j_\mu^{e.m.} = \bar{e} \gamma_\mu e. \tag{10.15}$$

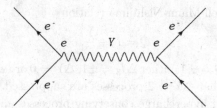

Fig. 10.2   Electromagnetic interaction mediated by a photon.

Note the similarity between Eqs. (10.13), (10.14), (10.11b,c) and (10.15) respectively. Both the electromagnetic and weak currents are vector in character, the appearance of $\gamma_5$ in weak current is due to the fact that parity is not conserved in weak interaction, in fact it is violated maximally. The coupling of electromagnetic current with the photon is characterized by electric charge (related to the fine structure constant $\alpha$ by $\frac{e^2}{4\pi} = \alpha = 1/137$) while that of weak current with the weak vector boson field $W_\mu$ is characterized by $g_W$ (related to the Fermi coupling constant $G_F$ by $\frac{g_W^2}{4\pi} = \frac{G_F}{4\pi\sqrt{2}m_W^2}$).

## 10.2.2   Semileptonic Processes

Some examples of these processes are given below

$$n \to p + e^- + \bar{\nu}_e$$
$$\pi^+ \to e^+ + \nu_e,\ \mu^+ + \nu_\mu$$
$$\pi^- \to e^- + \bar{\nu}_e,\ \mu^- + \bar{\nu}_\mu$$
$$\Sigma^- \to \Lambda^0 + e^- + \bar{\nu}_e$$
$$\Sigma^- \to n + e^- + \bar{\nu}_e$$
$$\Sigma^0 \to p + e^- + \bar{\nu}_e$$
$$K^+ \to \pi^0 + e^+ + \nu_e$$
$$K^- \to \pi^0 + e^- + \bar{\nu}_e. \tag{10.16}$$

From these processes, one notes the following rules:

1. The hadronic charge changes by one unit, i.e. $\Delta Q = \pm 1$
2. In the first three processes, strangeness does not change, in the last five processes it changes by one unit ($\Delta Q/\Delta S = 1$).

For hadrons, Gell-Mann-Nishijima relation

$$Q = I_3 + \frac{Y}{2}$$

implies that for $\Delta Q = \pm 1$, either $\Delta I_3 = \pm 1$, $\Delta Y = 0$ or $\Delta I_3 = \pm 1/2$, $\Delta Y = \pm 1$, if we assume that $\Delta Y = 2$ processes are suppressed. The processes of first kind are called hypercharge conserving processes and those of second kind are called hypercharge changing processes. In all the processes listed above, we see that either $\Delta Y = 0$ or $\Delta Y = \pm 1$; no weak process with $|\Delta Y| > 1$ is seen with the same strength as $|\Delta Y| \geq 1$ transitions. Thus we have the selection rule $\Delta Y = 0, \pm 1$, $\Delta Q = \Delta Y$.

Since there are so many hadrons in nature, therefore to deal with semi-leptonic decays of each of them would be very tedious. Thus we use the simple picture of hadrons made up of quarks. The main thing about the quarks is that they are regarded as truly elementary similar to leptons. Their weak and electromagnetic interactions would then be like those of leptons. Thus in analogy with Eqs. (10.15) and (10.11), their electromagnetic and weak currents are respectively

$$j_\mu^{e.m.} = \frac{2}{3}\bar{u}\gamma_\mu u - \frac{1}{3}\bar{d}\gamma_\mu d - \frac{1}{3}\bar{s}\gamma_\mu s \tag{10.17}$$

while

$$J_\mu^h = \bar{u}\gamma_\mu(1 - \gamma_5)d', \tag{10.18}$$

where

$$d' = \cos\theta_c d + \sin\theta_c s, \, s' = -\sin\theta_c d + \cos\theta_c s \tag{10.19}$$

Since in weak interactions flavor quantum number is not conserved, weak interactions eigenstates $d'$ and $s'$ are not identical with mass eigenstates $d$ and $s$, they are linked as in Eq. (10.19). Here $\theta_c$ is the Cabibbo angle; its value is $\theta_c = 13°$ or $\sin\theta_c = 0.22$. This is introduced since it is seen experimentally that decay rates for $|\Delta Y| = 1$ semi-leptonic decays are suppressed by a factor of about $1/16$ compared to those for $\Delta Y = 0$ processes. We shall deal with $s'$ in Chap. 13. Then in analogy with Eq. (10.11) or (10.13) the interaction responsible for fundamental processes like

$$d \to u + e^- + \bar{\nu}_e$$
$$s \to u + e^- + \bar{\nu}_e \tag{10.20}$$

would be

$$L_W^{h.l.} = -\frac{G_F}{\sqrt{2}}J_\mu^h L_{(e)}^{\mu\dagger}$$

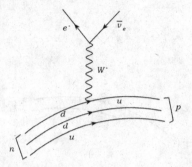

Fig. 10.3  Quark level process for neutron $\beta$-decay.

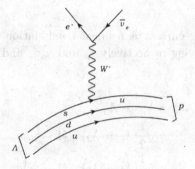

Fig. 10.4  Quark level process for neutron $\Lambda - \beta$-decay.

or

$$L_W = -g_W J_\mu^h W^{-\mu} + h.c. \qquad (10.21)$$

In this picture neutron $\beta$-decay, $\Lambda - \beta$-decay and $\bar{K}^0 \to \pi^+ + e^- + \bar{\nu}_e$, for example, would be pictured as shown in Figs. 10.3, 10.4 and 10.5.

Note the very important fact that both the leptonic and hadronic weak currents in (10.11b, c) and (10.18) are charged, i.e. they carry one unit of charge and the hadronic weak currents (10.18) satisfy the selection rules $|\Delta Y| \le 1$ and $\Delta Q = \Delta Y$. We also note that in terms of flavor SU(3) notation we can write

$$J_\mu^h = \cos\theta_c J_\mu^+ + \sin\theta_c J_\mu^1 \qquad (10.22)$$

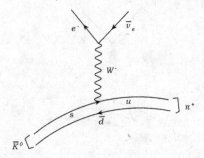

Fig. 10.5   Quark level process for $\bar{K}^0 \to \pi^+ e^- \bar{\nu}_e$.

where weak hadronic current is a linear combination of vector and axial vector currents involving respectively $\gamma_\mu$ and $\gamma_\mu \gamma_5$ and are given by

$$J_\mu^\pm = V_\mu^\pm - A_\mu^\pm$$

$$= \bar{q}\frac{1}{2}(\lambda_1 \pm i\lambda_2)\gamma_\mu(1 - \gamma_5)q \tag{10.23a}$$

$$J_\mu^1 = \bar{q}\frac{1}{2}(\lambda_4 + i\lambda_5)\gamma_\mu(1 - \gamma_5)q. \tag{10.23b}$$

Note also that

$$J_{3\mu} = \bar{q}\frac{1}{2}\lambda_3\gamma_\mu(1 - \gamma_5)q \tag{10.24a}$$

$$J_{8\mu} = \bar{q}\frac{1}{2}\lambda_8\gamma_\mu(1 - \gamma_5)q \tag{10.24b}$$

$$J_\mu^{em} = V_{3\mu} + \frac{1}{\sqrt{3}}V_{8\mu}. \tag{10.24c}$$

Here $q = \begin{pmatrix} u \\ d \\ c \end{pmatrix}$. The heavy quarks and $s'$ will be considered in Chap. 13.

### 10.2.3 Non-Leptonic Processes

Here no leptons are involved. The well-known non-leptonic processes are:

$$\Lambda \to p\pi^- (\Lambda_-^0)$$
$$\Lambda \to p\pi^0 (\Lambda_0^0) \tag{10.25a}$$

$$\Sigma^- \to n\pi^- (\Sigma_-^-)$$
$$\Sigma^+ \to p\pi^0 (\Sigma_0^+) \tag{10.25b}$$
$$\Sigma^+ \to n\pi^+ (\Sigma_+^+)$$

$$\Xi^- \to \Lambda\pi^- (\Xi_-^-)$$
$$\Xi^0 \to \Lambda\pi^0 (\Xi_0^0) \tag{10.25c}$$

or

$$K^0, \bar{K}^0 \to \pi^+\pi^-, \pi^0\pi^0$$
$$K^\pm \to \pi^\pm, \pi^0, \text{ etc.} \tag{10.26}$$

Note that all these decays are strangeness changing ($|\Delta S| = 1$). Let us concentrate on the decays (10.25), the so-called non-leptonic decays of hyperons. If we consider the decaying particle in its rest frame, the conservation of angular momentum $J$ gives

$$J_{in} \equiv \frac{1}{2} = J_{\text{final}} \equiv \ell + s,$$

where $\ell$ is the relative orbital angular momentum of the pion and the baryon in final state. Since spin $s = 1/2$, $\ell$ can be 0 or 1. The pion being pseudoscalar (having odd intrinsic parity), the relative parity of final state with respect to the initial state is

$$P_f = (-1)^0(-1) = -1 \text{ odd for } \ell = 0$$
$$= (-1)^1(-1) = +1 \text{ even for } \ell = 1.$$

The s-wave ($\ell = 0$) decays are parity violating while p-wave ($\ell = 1$) decays are parity conserving. Accordingly decays (10.25) are governed by two amplitudes, parity violating (s-wave) and parity conserving (p-wave). We can write the Lagrangian responsible for non-leptonic decays as

$$L_W^{(h)} = L_W^{h(p.v)} + L_W^{h(p.c)}$$
$$= -\frac{G_F}{\sqrt{2}} J_\mu^h J^{h\mu\dagger} + h.c., \tag{10.27}$$

where $J_\mu^h$ is given in (10.18). The $|\Delta S| = 1$ component of (10.27) behaves as

$$\frac{G_F}{\sqrt{2}} \sin\theta_c \cos\theta_c \{\bar{s}\gamma_\mu(1-\gamma_5)u\} \cdot \{\bar{u}\gamma^\mu(1-\gamma_5)d\}. \tag{10.28}$$

Now $u$ and $d$ belong to isospin doublet $I = 1/2$ while $s$ is isospin singlet $I = 0$. Thus from the combination of angular momentum rules (isospin behaves like angular momentum) first term in curly brackets in Eq. (10.28) has $I = 1/2$ while the second term in curly brackets has $I = 0, 1$. Thus the interaction contains both $\Delta I = 1/2$ and $3/2$ parts. Experimentally $\Delta I = 1/2$ part predominates over $\Delta I = 3/2$ and then decays (10.25a), (10.25b) and (10.25c) respectively get related among themselves. We shall come to these relations later.

### 10.2.4   $\mu$-Decay

Consider the $\mu$-decay

$$\mu \to e^- + \nu_\mu + \bar{\nu}_e$$

From Eq. (10.4), we can write the interaction as

$$H_{int} = \frac{G_F}{\sqrt{2}} \left[ \bar{e}\gamma_\mu(1-\varepsilon\gamma_5)\mu \right] \left[ \bar{\nu}_\mu\gamma^\mu(1-\gamma_5)\nu_e \right]. \tag{10.29}$$

The interaction written in this order is called the charge retention order. It is easier to deal with this order in calculations. Here we have assumed 2-component neutrinos (left-handed $\nu_\mu$ and right-handed $\bar{\nu}_e$) but have allowed $V - \varepsilon A$ interaction, where for $V - A$, $\varepsilon = 1$ and in that case by Fierz rearrangement we get Eq. (10.11a).

From Eq. (10.29), we can write the T-matrix for $\mu^-$ decay:

$$T = \frac{-1}{(2\pi)^6} \sqrt{\frac{m_\mu m_e m_{\nu_\mu} m_{\nu_e}}{p_{10}p_{20}k_{10}k_{20}}} \frac{G_F}{\sqrt{2}}$$
$$\times \left[ \bar{u}(p_2)\gamma_\lambda(1-\varepsilon\gamma_5)u(p_1) \right] \left[ \bar{u}(k_2)\gamma^\lambda(1-\gamma_5)v(k_1) \right] \tag{10.30}$$

where $p_1, p_2, k_1$ and $k_2$ are the four momenta of $\mu^-$, $e^-$, $\nu_\mu$ and $\bar{\nu}_e$ respectively and $u(p_1), u(p_2), u(k_1)$ and $v(k_2)$ are Dirac spinors. From Eq. (10.30), we get

$$d\Gamma = \frac{1}{(2\pi)^5}\delta^4(p_1 - p_2 - k_1 - k_2)\frac{m_\mu m_e m_{\nu_\mu} m_{\nu_e}}{p_{10}p_{20}k_{10}k_{20}}|M|^2 d^3p_2 d^3k_1 d^3k_2 \tag{10.31}$$

where

$$|M|^2 = \overline{\sum_{spin}} |F|^2 = \left(\frac{G_F^2}{2}\right) \overline{\sum_{spin}} |\bar{u}(p_2)\gamma_\lambda(1 - \varepsilon\gamma_5)u(p_1)|^2$$

$$\times |\bar{u}(k_2)\gamma^\lambda(1 - \gamma_5)v(k_1)|^2 \tag{10.32}$$

We can easily calculate $|M|^2$ using the standard trace techniques [see Appendix A]. Neglecting the neutrino masses, we get

$$|M|^2 = \frac{G_F^2}{m_\mu m_e m_{\nu_\mu} m_{\nu_e}} \left[(1 + \varepsilon^2)(p_2 \cdot k_2\, p_1 \cdot k_1 + p_2 \cdot k_1\, p_1 \cdot k_2)\right.$$

$$\left. - (1 - \varepsilon^2) m_\mu m_e k_1 \cdot k_2 + 2\varepsilon(p_1 \cdot k_2\, p_2 \cdot k_1 - p_1 \cdot k_1\, p_2 \cdot k_2)\right] \tag{10.33}$$

Since neutrinos are not observed, we integrate over $d^3k_1 d^3k_2$. Performing these integrations, and writing $d^3p_2 = 4\pi p_e E_e dE_e$, we get

$$d\Gamma = m_\mu \frac{p_e E_e dE_e}{24\pi^3} G_F^2 (1 + |\varepsilon|^2)$$

$$\times \left[3W - 2E_e - \frac{m_e^2}{E_e} + 6\eta \frac{m_e}{E_e}(W - E_e)\right], \tag{10.34}$$

where

$$\eta = \frac{1}{2} \frac{|\varepsilon|^2 - 1}{|\varepsilon|^2 + 1} \tag{10.35}$$

$$W = \frac{m_\mu^2 + m_e^2}{2m_\mu}. \tag{10.36}$$

In evaluating the final result (10.34), we have gone to the rest frame of the muon:

$$E_\mu = m_\mu$$

$$m_\mu = E_e + E_1 + E_2$$

$$O = \mathbf{p}_e + \mathbf{k}_1 + \mathbf{k}_2. \tag{10.37}$$

### 10.2.5 *Remarks*

(1) It is always possible to take $C_V$ as real and take $C_A = \varepsilon C_V$ ($\varepsilon$ complex).

(2) The electron spectrum does not distinguish between $\varepsilon = +1\,(V - A)$ or $\varepsilon = -1\,(V + A)$ interaction.

(3) Any deviation from $\varepsilon = \pm 1$ can be determined by measuring $\eta$ in the electron spectrum. Since $\eta$ is the coefficient of $\left(\frac{m_e(W - E_e)}{E_e}\right)$, it plays a minor role except at low electron energies, where measurements are difficult. The best experimental value of $\eta$ is

$$\eta = -0.007 \pm 0.013 \tag{10.38}$$

which is consistent with zero.

(4) It is instructive to write the electron energy spectrum (10.34) as

$$d\Gamma = \frac{m_\mu p_e E_e dE_e}{12\pi^3} G_F^2 \left(1 + |\varepsilon|^2\right)$$

$$\times \left[3(W - E_e) + 2\rho \left(\frac{4}{3}E_e - W - \frac{1}{3}\frac{m_e^2}{E_e}\right) + 3\eta \frac{m_e}{E_e}(W - E_e)\right],$$

$$(10.39)$$

where $\rho = 3/4$; $\rho$ is called the Michel parameter. In fact the most general interaction without assuming two-component neutrinos gives the electron spectrum of the form within the square brackets. The experimental value of $\rho = 0.7518 \pm 0.0026$ is in excellent agreement with $\rho = 3/4$ as given by $V - \varepsilon A$ theory. We conclude that the two-component neutrino hypothesis is in an excellent agreement with the experimental results. Finally integrating Eq. (10.34), we obtain

$$\Gamma = \tau_\mu^{-1} = G_F^2 P, \tag{10.40a}$$

where

$$P = \left[1 - \frac{8m_e^2}{m_\mu^2}\right] \frac{m_\mu^5}{192\pi^3}. \tag{10.40b}$$

If we include $O(\alpha)$ radiative corrections

$$\tau_\mu^{-1} = G_F^2 P \left[1 + \frac{\alpha}{2\pi}\left(\frac{25}{4} - \pi^2\right) + \left(1 + \frac{2\alpha}{3\pi}\ln\frac{m_\mu}{m_e}\right)\right], \tag{10.40c}$$

where the fine structure constant $\alpha = \frac{1}{137.036}$. The Fermi constant $G_F$ determined from (10.40c), using the experimental value for $\tau_\mu = 2.19703 \times 10^{-6}$ sec, is

$$G_F = 1.16637 \times 10^{-5} \text{ GeV}^{-2}. \tag{10.41}$$

### 10.2.5.1  *Decay of polarized muon*

We have seen that the electron spectrum cannot determine the sign of $\varepsilon$. In order to determine $\varepsilon$, we consider the decay of polarized muon. Let $n_\mu$ be the polarization vector of muon. We note that

$$n^2 = n_\mu n^\mu = -1,$$

$$n \cdot p_1 = 0. \tag{10.42}$$

In the rest frame of the muon $m_\mu n_0 = 0$; thus $n_0 = 0$ and $n \equiv (0, \mathbf{n})$. For this case in taking the trace, we put

$$u(p_1)\bar{u}(p_1) = \left[\left(\frac{\not{p_1} + m_\mu}{2m_\mu}\right)\left(\frac{\gamma_5 \, \gamma \cdot n + 1}{2}\right)\right]. \tag{10.43}$$

Using the standard trace techniques [see Appendix A], and performing the integrations over $d^3k_1 \, d^3k_2$, the differential spectrum in the asymmetry angle for $\mu^-$ decay is

$$d\Gamma = \left(\frac{d\Omega_\gamma}{4\pi}\right) \frac{G_F^2(1 + |\varepsilon|^2)}{12\pi^3} \, m_\mu \, p_e^2 \, dE_e \, \xi \cos\gamma$$

$$\times \left[W - 2E_e + \frac{m_e^2}{m_\mu^2}\right], \tag{10.44}$$

where $\gamma$ is the angle between the electron momentum and the $\mu$–spin direction and

$$\xi = \frac{2 \, R_e \, \varepsilon}{1 + |\varepsilon|^2}. \tag{10.45}$$

It is instructive to write Eq. (10.44) in the form

$$d\Gamma = -\left(\frac{d\Omega_\gamma}{4\pi}\right) \frac{G_F^2}{6\pi^3} \, (1 + |\varepsilon|^2) \, m_\mu p_e^2 dE_e \xi \cos\gamma$$

$$\times \left[(W - E_e) + 2\delta \left(\frac{4}{3}E_e - W - \frac{1}{3}\frac{m_e^2}{m_\mu^2}\right)\right], \tag{10.46}$$

where $\delta = 3/4$ for two-component neutrinos viz for $V - \varepsilon A$ theory. For a general interaction without assuming two-component neutrino, the asymmetry distribution in angle $\gamma$ is of the form given within the square brackets. The experimental value of $\delta$ is $0.749 \pm 0.004$ in excellent agreement with two-component neutrino hypothesis.

The experimental value of $\xi$ is given by

$$\xi P_\mu = 1.003 \pm 0.008. \tag{10.47}$$

## 10.2.6 *Semi-Leptonic Processes*

For a semi-leptonic weak process we can write the interaction Hamiltonian as [cf. Eq. (10.21)].

$$-L_W^{h.l.} = H_{\text{int}} = \frac{G_F}{\sqrt{2}} J_\lambda \left(x\right) \left[\bar{e}(x) \, \gamma^\lambda \left(1 - \gamma_5\right) \nu_e \left(x\right)\right] + h.c. \tag{10.48}$$

To first order in weak interaction, the T-matrix for a semi-leptonic process of the type

$$A \to B + e^- + \bar{\nu}_e$$

is given by

$$T = -\frac{G_F}{\sqrt{2}} \left\langle B \left| J_\lambda \right| A \right\rangle \frac{1}{(2\pi)^3} \sqrt{\frac{m_\nu m_e}{k_0 k_0'}} \left[\bar{u}(k') \, \gamma^\lambda \left(1 - \gamma_5\right) v \left(k\right)\right], \tag{10.49}$$

where $k'$ and $k$ are four momenta of electron and antineutrino. We denote four momenta of $A$ and $B$ by $p$ and $p'$.

## 10.3   Baryon Decays

We consider the case when $A$ and $B$ are spin $1/2$ baryons and $\langle B\,|J_\lambda|\,A\rangle = \langle B\,|V_\lambda - A_\lambda|\,A\rangle$. From Lorentz covariance alone, the most general structure of these matrix elements is given by $[q = p' - p]$

$$\langle B(p')\,|V_\lambda|\,A(p)\rangle = \frac{1}{(2\pi)^3}\sqrt{\frac{m_A m_B}{p_0 p_0'}}\,\bar{u}_B(p')$$
$$\times \left[ g_V\left(q^2\right)\gamma_\lambda + f_V\left(q^2\right)\sigma_{\lambda\nu}\,q_\nu - ih_V\left(q^2\right)q_\lambda\right]u_A(p) \tag{10.50a}$$

$$\langle B(p')\,|A_\lambda|\,A(p)\rangle = \frac{1}{(2\pi)^3}\sqrt{\frac{m_A m_B}{p_0 p_0'}}\,\bar{u}_B(p')$$
$$\times \left[ g_A\left(q^2\right)\gamma_\lambda\gamma_5 + f_A\left(q^2\right)\gamma_5 q_\lambda - \sigma_{\lambda\nu}\,q^\nu\gamma_5 h_A\left(q^2\right)q_\lambda\right]u_A(p). \tag{10.50b}$$

The above equations may be deduced by the following simple argument. If we consider the quantity $\langle B\,|V_\lambda|\,A\rangle\,\sqrt{p_0 p_0'}$, we are dealing with a Lorentz vector [without the square root factor this would not be the case because of the noncovariant orthogonality condition]. Such a Lorentz vector must be expressible as a linear combination of the various vectors that can be constructed from the different tensor quantities associated with the two particles A and B and their Dirac spinors. In actual fact there are five such vectors, namely, $\bar{u}(p')\gamma_\lambda u(p)$, $\bar{u}(p')\sigma_{\lambda\nu}p'^\nu u(p)$, $\bar{u}(p')\sigma_{\lambda\nu}p^\nu u(p)$, $\bar{u}(p')p_\lambda u(p)$, $\bar{u}(p')p'_\lambda u(p)$. Since $u(p)$ and $u(p')$ are free Dirac spinors, satisfying $(\gamma\cdot p - m)u(p) = 0$ and $\bar{u}(p)(\gamma\cdot p' - m) = 0$, only three of these vectors are linearly independent and hence from the five vectors above, it is sufficient to choose three linearly independent ones. We choose $u(p')\gamma_\lambda u(p)$, $u(p')\sigma_{\lambda\nu}q^\nu u(p)$ and $u(p')q_\lambda u(p)$. Then the matrix elements $\langle B\,|V_\lambda|\,A\rangle\,\sqrt{p_0 p_0'}$ must be some linear combination of these three vectors and that is the content of Eq. (10.50a). An entirely parallel arrangement applies in the case of matrix elements $\langle B\,|A_\lambda|\,A\rangle$.

Since the momentum transfer $q = p' - p$ is very small compared to the mass of $A$ or $B$ for the processes we are considering, we can write

$$\langle B(p')\,|J_\lambda|\,A(p)\rangle = \frac{1}{(2\pi)^3}\sqrt{\frac{m_A m_B}{p_0 p_0'}}\,\bar{u}_B(p')$$
$$[g_V\,\gamma_\mu - g_A\,\gamma_\mu\gamma_5]\,u_A(p). \tag{10.51}$$

Now we shall take $A$ and $B$ as members of the spin $1/2$ baryon octet and then

$$J_\lambda = \cos\theta_c\,J_\lambda^+ + \sin\theta_c\,J_\lambda^1, \qquad J_\lambda^\dagger = \cos\theta_c\,J_\lambda^- + \sin\theta_c\,J_\lambda^{1\dagger},$$

$$\tag{10.52}$$

where $J_\lambda^\pm$ and $J_\lambda^1$, $J_\lambda^{1\dagger}$ are $1 \pm i2$ and $4 \pm i5$ components of octet of currents $J_{i\lambda}$ $(i = 1, \cdots, 8)$. As shown in Chap. 5

$$\langle B_k \, |A_{i\lambda}| \, B_j \rangle \equiv \frac{1}{2\pi^3} \sqrt{\frac{m_B m_B}{p_0 p_0'}} \bar{u}(p') \gamma_\lambda \gamma_5 u(p) g_A^{ijk}, \qquad (10.53\text{a})$$

where

$$g_A^{ijk} = i f_{ijk} F + d_{ijk} D. \qquad (10.53\text{b})$$

Since $F_i = -i \int V_{i0}(x, 0) \, d^3x$ is a generator of $SU(3)$, it follows that

$$\langle B_k \, |V_{i\lambda}| \, B_j \rangle \equiv \frac{1}{2\pi^3} \sqrt{\frac{m_B^2}{p_0 p_0'}} \bar{u}(p') \gamma_\lambda u(p) g_V^{ijk}, \qquad (10.54\text{a})$$

where

$$g_V^{ijk} = i f_{ijk}. \qquad (10.54\text{b})$$

Thus if we neglect the momentum transfer $q^2$, $(q^2 \approx 0)$, the matrix elements $\langle B_k \, |J_{i\lambda}| \, B_j \rangle$ are essentially determined in terms of Cabibbo angle $\theta_c$ and the two reduced matrix elements $F$ and $D$. Using Eqs. (10.53a) and (10.54a), the matrix elements of these decays are given below:

| Decay $B \to B' l \nu_e$ | Vector current $g_V$ | Axial vector current $g_A$ | Ratio $g_A/g_V$ | Expt. value of $g_A/g_V$ |
|---|---|---|---|---|
| $n \to p$ | $\cos\theta_c$ | $\cos\theta_c \, (F + D)$ | $F + D$ | $1.2695 \pm 0.0029$ |
| $\Lambda \to p$ | $-\sqrt{\frac{3}{2}}\sin\theta_c$ | $-\sqrt{3/2}\sin\theta_c$ $\times \left(F + \frac{1}{3}D\right)$ | $F + \frac{1}{3}D$ | $0.718 \pm 0.015$ |
| $\Sigma^- \to n$ | $-\sin\theta_c$ | $\sin\theta_c \, (F - D)$ | $F - D$ | $-0.340 \pm 0.017$ |
| $\Xi^- \to \Lambda$ | $\sqrt{\frac{3}{2}}\sin\theta_c$ | $\sqrt{3/2}\sin\theta_c$ $\times \left(F - \frac{1}{3}D\right)$ | $F - \frac{1}{3}D$ | $0.25 \pm 0.05$ |

In order to test the octet hypothesis, we note that if we determine $F$ and $D$ from the first two decays, we find $F - D$ and $F - 1/3D$ for the third and fourth decay in agreement with their experimental values. The parametrization given in Eqs. (10.53a) and (10.54a) is in excellent agreement with experiment. Using the first two entries of the above table, we find $F = 0.444 \pm 0.015$, $D = 0.823 \pm 0.015$.

As an example to show how $g_A/g_V$ is determined, we consider the case of neutron $\beta$-decay $n \to p + e^- + \bar{\nu}_e$ in detail, where from Eqs. (10.51) and (10.52) we have

$$\left\langle p(p') \left| J_\mu^{(+)} \right| n(p) \right\rangle = \frac{1}{2\pi^3} \sqrt{\frac{m_p m_n}{p_0 p_0'}} \bar{u}(p')$$

$$\times [g_V \gamma_\mu - g_A \gamma_\mu \gamma_5] \, u(p) \qquad (10.55)$$

with $g_V = \cos\theta_c$, $g_A = \cos\theta_c (F + D)$. In the rest frame of neutron, we write $k' \equiv (E_e, \mathbf{p}_e)$, $k \equiv (E_\nu, \mathbf{p}_\nu)$, $\mathbf{p} = 0$, $\mathbf{p}' + \mathbf{p}_e + \mathbf{p}_\nu = 0$. Since $q$ is very small as compared with neutron and proton masses, we can treat them non-relativistically. Then

$$\left\langle p(p') \left| J_0^{(+)} \right| n(p) \right\rangle = \frac{1}{(2\pi)^3} \chi_p^+ g_V \chi_n$$

$$\left\langle p(p') \left| J_i^{(+)} \right| n(p) \right\rangle = -\frac{1}{(2\pi)^3} \chi_p^+ g_A \sigma_i \chi_n. \tag{10.56}$$

Let us write the leptonic part as

$$L^\mu = \bar{u}(k') \, \gamma^\mu \, (1 - \gamma_5) \, v(k). \tag{10.57}$$

The amplitude $F$ [cf. Eq. (10.49) and Eq. (2.84)] is given by

$$F = -\frac{G_F}{\sqrt{2}} \, \chi_p^+ \left[ g_V L^0 + g_A \sigma \cdot \mathbf{L} \right] \chi_n. \tag{10.58}$$

We now sum over proton spin and lepton spin and define the neutron spin $\mathbf{S}_n$ as

$$\mathbf{S}_n = \frac{1}{2} \chi_n^+ \sigma \chi_n. \tag{10.59}$$

Using Eq. (2.122), we get for the probability distribution

$$d\Gamma = \frac{G_F^2}{(2\pi)^5} \, p_e^2 (E_{\max} - E_e)^2 A \left[ 1 + \lambda \frac{\mathbf{P}_e \cdot \mathbf{P}_\nu}{E_e E_\nu} \right.$$

$$\left. + \hat{\mathbf{S}}_n \cdot \left( A' \frac{\mathbf{P}_e}{E_e} + B \frac{\mathbf{P}_\nu}{E_\nu} + D \frac{\mathbf{P}_e \times \mathbf{P}_\nu}{E_e E_\nu} \right) \right] dp_e \, d\Omega_e \, d\Omega_\nu \tag{10.60}$$

where $\hat{\mathbf{S}}_n$ is the direction of the neutron spin and

$$A = \left[ |g_V|^2 + 3|g_A|^2 \right]$$

$$\lambda = \frac{1}{A} \left[ |g_V|^2 - |g_A|^2 \right]$$

$$A' = -\frac{2}{A} \left[ |g_A|^2 - \operatorname{Re} g_V \, g_A^* \right]$$

$$B = \frac{2}{A} \left[ |g_A|^2 + \operatorname{Re} g_V \, g_A^* \right]$$

$$D = \frac{2}{A} \operatorname{Im} g_V \, g_A^*. \tag{10.61}$$

The experimental data give the following values of these correlation functions,

$$\lambda = -0.103 \pm 0.004$$

$$A' = -0.1173 \pm 0.0013$$

$$B = 0.9807 \pm 0.0030$$

$$D = (-4 \pm 6) \times 10^{-4} \tag{10.62}$$

If we write $x = |g_A/g_V|$, then the value of $\lambda$ gives

$$x = |g_A/g_V| = 1.261 \pm 0.004. \tag{10.63}$$

The very fact that $B$ is nearly 1 implies the maximum parity violation in $\beta$-decay. The value of $A'$ [assuming $g_V$ and $g_A$ are relatively real, see below] gives $|g_A/g_V| = 1.267 \pm 0.014$, consistent with Eq. (10.63). A non-zero value of $D$ would imply time reversal violation in $\beta$-decay. The experimental value of $D$ is nearly zero and show that time reversal invariance holds. If we write $g_A/g_V = -xe^{i\phi}$, where for $\phi = 0$ or $\pi$, $T$ invariance holds, we obtain

$$\phi = (180.06 \pm 0.07)^\circ. \tag{10.64}$$

Finally, from Eq. (10.60), we obtain for the decay width $\Gamma$:

$$\Gamma = \frac{AG_F^2}{2\pi^3} m_e^5 f(\rho_0), \tag{10.65}$$

where

$$A = |g_V|^2 + 3|g_A|^2$$

and

$$f(\rho_0) = \int_0^{\rho_0} d\rho\, \rho^2 \left( \sqrt{\rho_0^2 + 1} - \sqrt{\rho^2 + 1} \right), \quad \rho_0 = \frac{p_e^{max}}{m_e}. \tag{10.66}$$

Since charged particles are involved, this expression of $f\tau = f\Gamma^{-1}$ is subject to radiative corrections, which are normally incorporated into the factor $f$ along with the first order Coulomb corrections. These corrections change $f$ by about 5%. The average value from direct neutron life-time measurements is

$$\tau = 888.5 \pm 0.8 \text{ sec.} \tag{10.67}$$

Knowing $|g_A/g_V|$, one can determine $G_V = G_F g_V$ from Eq. (10.65). $G_V$ can also be determined from the superallowed $O^+ \to O^+$ pure Fermi decays for which $F\tau$ value is

$$F\tau = \frac{2\pi^3}{m_e^5} \frac{1}{G_V^2 |M_F|^2}, \tag{10.68}$$

where

$$M_F = \left\langle \begin{array}{c} \Psi_f \\ 0^+,\, I = 1 \end{array} \middle| I_\pm \middle| \begin{array}{c} \Psi_i \\ 0^+,\, I = 1 \end{array} \right\rangle. \tag{10.69}$$

$F$ here is different from $f$ for the neutron $\beta$-decay and it must account for the stronger Coulomb effect and for the much more subtle radiative effects associated with the higher electric charge. The quantity $(F\tau)_{AV} =$

3070.6 ± 1.6 sec from the $O^+ \to O^+$decays together with the phase space factor $F$ and the value of $|g_A/g_V|$ given in Eq. (10.63) gives $\tau = 894 \pm 37$ sec, to be compared with the direct neutron life-time measurement given in Eq. (10.67).

Finally the Cabibbo angle

$$\cos\theta_c = \frac{G_V^\beta}{G_V^\mu} \left(1 + \Delta\beta - \Delta\mu\right)^{-1/2} \tag{10.70}$$

where $G_V^\mu = 1.16637\,(13) \times 10^{-5}$ GeV$^{-2}$ while $\Delta\beta$ and $\Delta\mu$ are the "inner" radiative corrections to both nucleon $\beta$-decay and muon decay with $\Delta\beta - \Delta\mu = 0.023\,(2)$. This gives

$$|V_{ud}| = \cos\theta_c = 0.97418 \pm 0.00027. \tag{10.71}$$

$\sin\theta_c$ is determined from $K_{e3}$ decay and its value is $0.2255 \pm 0.0019$.

## 10.4  Pseudoscalar Meson Decays

### 10.4.1  *Pion Decay*

$$\pi^- \to \ell^- + \bar\nu_\ell, \qquad \ell = e, \mu.$$

For this decay, the $T$-matrix is given by

$$T = -\frac{G_F}{\sqrt{2}} \cos\theta_c \left\langle 0 \left| J_\lambda^+ \right| \pi^- \right\rangle$$

$$\times \frac{1}{(2\pi)^3} \sqrt{\frac{m_\ell\, m_\nu}{k_0' k_0}}\, u(k')\, \gamma^\lambda \left(1 - \gamma_5\right) v\,(k). \tag{10.72}$$

Here, we have $p = k' + k$. Now from Lorentz invariance

$$\left\langle 0 \left| J_\lambda^+ \right| \pi^- \right\rangle = -\left\langle 0 \left| A_\lambda^+ \right| \pi^- \right\rangle = -\frac{1}{(2\pi)^{3/2}} \sqrt{\frac{1}{2p_0}}\, i f_\pi\, p_\lambda. \tag{10.73}$$

Using the standard techniques of Chap. 2, the decay rate $\Gamma$ can be easily calculated. We obtain

$$\Gamma\left(\pi^- \to \ell^- + \nu_e\right) = \frac{G_F^2 \cos^2\theta_c}{8\pi}\, f_\pi^2\, m_\ell^2\, m_\pi \left(1 - \frac{m_\ell^2}{m_\pi^2}\right)^2. \tag{10.74}$$

It thus follows that pion decays mainly to muon, its decay to electron is suppressed by a factor $m_e^2/m_\mu^2$ (phase space). In the same way, we can write down the decay rate of $K^- \to \ell^- + \bar\nu_\ell$; it is given by

$$\Gamma\left(K^- \to \ell^- + \bar\nu_\ell\right) = \frac{G_F^2 \sin^2\theta_c}{8\pi}\, f_K^2\, m_\ell^2\, m_K \left(1 - \frac{m_\ell^2}{m_K^2}\right)^2. \tag{10.75}$$

From the experimental values of the decay rates for pion and kaon, we can determine $f_\pi$ and $f_K$. We get $f_\pi \approx 131$ MeV and $f_K/f_\pi \approx 1.22$. From the particle data group: $f_\pi = (130.7 \pm 0.1 \pm 0.36)$ MeV, $f_K = (159.8 \pm 1.4 \pm 0.44)$ MeV.

### 10.4.1.1 Remarks

Suppose pion decay occurs through a vector boson $W$. Then we can write the decay amplitude $F$:

$$F = -g_W \, i \, f_\pi \, p^\mu \, \frac{-g_{\mu\lambda} + \frac{p_\mu p_\lambda}{m_W^2}}{p^2 - m_W^2} \bar{u}(k') \, \gamma^\lambda \, (1 - \gamma_5) \, v\,(k). \qquad (10.76)$$

We write the $W$-propagator in the following form

$$\frac{1}{p^2 - m_W^2} \left[ \left( -g_{\mu\lambda} + \frac{p_\mu p_\lambda}{p^2} \right) + p_\mu p_\lambda \frac{p^2 - m_W^2}{p^2 \, m_W^2} \right]$$

$$= \left( -g_{\mu\lambda} + \frac{p_\mu p_\lambda}{p^2} \right) \frac{1}{p^2 - m_W^2} + \frac{p_\mu p_\lambda}{p^2 m_W^2}. \qquad (10.77)$$

The first part of Eq. (10.77) gives the transverse part of the propagator and second part gives its longitudinal part. If we substitute Eq. (10.77) into Eq. (10.76), we find that the first part of Eq. (10.77) gives zero and the entire contribution comes from the second part. We get

$$F = -\frac{g_W^2}{m_W^2} i \, f_\pi \, p_\lambda \, \bar{u}(k') \, \gamma^\lambda \, (1 - \gamma_5) \, v\,(k)$$

$$= -\frac{g_W^2}{m_W^2} \, i f_\pi \, m_\ell \, \bar{u}(k') \, (1 - \gamma_5) \, v\,(k). \qquad (10.78)$$

Here we have used the Dirac equation $\bar{u}(k') \, (\gamma \cdot k' - m_\ell) = 0$ and $p = k' + k$. Thus we note that the longitudinal part behaves as if the decay has taken place through a scalar particle of zero mass with effective coupling $g_W^2 / m_W^2$. We also note that it gives a contribution proportional to the lepton mass which is reflected in the formula (10.74). This is called helicity suppression.

### 10.4.2 Strangeness Changing Semi-Leptonic Decays

As an example of these decays we consider the decay.

$$K^- \to \pi^0 + \ell^- + \bar{\nu}_\ell, \qquad \ell = e, \mu.$$

We first note the rule: $\Delta Q = \Delta S = 1$. The $T$-matrix is given by

$$T = -\frac{G_F}{\sqrt{2}} \sin \theta_c \, \langle \pi^0 | J_\lambda^1 | K^- \rangle$$

$$\times \frac{1}{(2\pi)^3} \sqrt{\frac{m_\ell \, m_\nu}{k_0' \, k_0}} \, \left[ \bar{u}(k') \, \gamma^\lambda \, (1 - \gamma_5) \, v\,(k) \right]. \qquad (10.79)$$

The Lorentz structure of the hadronic matrix elements is given by

$$\left\langle \pi^0 \left| J_\lambda^1 \right| K^- \right\rangle = \left\langle \pi^0 \left| V_\lambda^1 \right| K^- \right\rangle$$

$$= \frac{1}{(2\pi)^3} \frac{1}{\sqrt{2p_0\, 2p_0'}}$$

$$\times \left[ f_+ \left( q^2 \right) (p + p')_\lambda + f_- \left( q^2 \right) (p - p')_\lambda \right], \quad (10.80)$$

where $p$ and $p'$ are four-momenta of $K^-$ and $\pi^0$, $q = (p' - p)$ and $k'$ and $k$ are four momenta of $\ell^-$ and $\nu_\ell$ respectively. In the rest frame of $K^-$, we have $m_K = \omega + E_\ell + E_\nu$, $\mathbf{p}_\pi + \mathbf{p}_\ell + \mathbf{p}_\nu = 0$. Using the standard techniques of Chap. 2, we get

$$\frac{d\Gamma}{dE_\ell\, d\omega} = \frac{1}{4\pi^3} G_F^2 \sin^2 \theta_c \left| f_+ \left( q^2 \right) \right|^2 \left[ A + B \Re \xi + C \left| \xi \right|^2 \right], \quad (10.81)$$

where

$$A = \left\{ m_K \left[ 2E_\ell\, E_\nu - m_K\, (W - \omega) \right] + \frac{m_\ell^2}{4} (W - \omega) - m_\ell^2\, E_\nu \right\}$$

$$B = \left[ E_\nu - \frac{1}{2} (W - \omega) \right] m_\ell^2$$

$$C = \frac{1}{4} [W - \omega]\, m_\ell^2$$

$$W = \frac{m_K^2 + m_\pi^2 - m_\ell^2}{2\, m_K}$$

$$\xi = f_- \left( q^2 \right) / f_+ \left( q^2 \right) \quad (10.82)$$

For electron, we can neglect its mass, i.e. we put $m_e^2 \approx 0$. Then Eq. (10.81) is much simplified. In this case, we get for the electron spectrum

$$\frac{d\Gamma}{dE_e} = \frac{1}{4\pi^3} G_F^2 \sin^2 \theta_c \left| f_+ \right|^2 m_K\, E_e^2 \frac{(W - E_e)^2}{(m_K - 2E_e)}. \quad (10.83)$$

Here we have put $f_+ \left( q^2 \right) \approx f_+ (0) = f_+$. For this case we obtain

$$\Gamma \left( K_{e3}^\pm \right) = \frac{G_F^2 \sin^2 \theta_c}{768\pi^3} (0.573) \left| f_+ \right|^2 m_K^5. \quad (10.84)$$

In the SU(3) limit $\left\langle \pi^0 \left| V_\lambda^1 \right| K^- \right\rangle \propto i f_{4+i5\,\frac{6-i7}{12}\,3}$ so that $f_+(0) = \frac{1}{\sqrt{2}}$.

Consider the neutral Kaon decays:

$$K^0 \to \pi^- + \ell^+ + \nu_\ell, \qquad \Delta S = \Delta Q$$

$$K^0 \to \pi^+ + \ell^- + \nu_\ell, \qquad \Delta S = -\Delta Q$$

For the first case the hadronic matrix elements are given by

$$\left\langle \pi^- \left| J_\lambda^{1\dagger} \right| K^0 \right\rangle$$

where

$$J_\lambda^1 = J_{4+i5\lambda}, \qquad J_\lambda^{1\dagger} = J_{4-i5\lambda}.$$

$J_\lambda^{1\dagger}$ creates negative charge and $S = -1$. For $\Delta S = -\Delta Q$, no such current can be written down in this conventional theory. For more details for semi-leptonic $K$-decays see Ref. [2].

## 10.5  Hadronic Weak Decays

### 10.5.1  *Non-Leptonic Decays of Hyperons*

Consider the decay

$$B\,(p) \to B'\,(p') + \pi\,(k).$$

The Lorentz structure of the $T$-matrix for this process is given by

$$T = \frac{1}{(2\pi)^{9/2}} \sqrt{\frac{m\,m'}{2k_0\,p_0\,p_0'}}\ \bar{u}(p')\,[\,A - B\gamma_5]\,u\,(p). \qquad (10.85)$$

The amplitudes $A$ and $B$ are functions of scalars: $s = (p'+k)^2$, $t = (p-p')^2$. $A$ is called the parity violating (p.v) [or s-wave] amplitude and $B$ is called the parity conserving (p.c) [or p-wave] amplitude. In the rest frame of baryon $B$

$$\mathbf{p'} = -\mathbf{k}, \quad |\mathbf{p'}| = |\mathbf{k}| = k, \quad \mathbf{p'} = k\mathbf{n},$$

$$m = p_0' + k_0, \quad p_0' = \sqrt{k^2 + m'^2}, \quad k_0 = \sqrt{k^2 + m_\pi^2}$$

$$s = m^2, \quad t = m^2 + m'^2 - 2mp_0'$$

$$k = \frac{1}{2m}\sqrt{\left[m^2 - (m' - m_\pi)^2\right]\left[m^2 - (m' + m_\pi)^2\right]}$$

$$p_0' = \frac{m^2 + m'^2 - m_\pi^2}{2m}. \qquad (10.86)$$

In this frame, the amplitudes $A$ and $B$ are constants. In the rest frame of $B$

$$u\,(p) = \begin{pmatrix} 1 \\ 0 \end{pmatrix} \chi, \quad u\,(p') = \frac{1}{\sqrt{2m'\,(p_0' + m')}} \begin{pmatrix} m' + p_0' \\ \sigma \cdot p' \end{pmatrix} \chi, \qquad (10.87)$$

where $\chi$ is a constant 2-component spinor. Using Eq. (10.87), we may write the $T$-matrix

$$T = \chi^+ M \chi, \tag{10.88a}$$

where

$$M = \frac{1}{(2\pi)^{9/2}} \frac{1}{\sqrt{2\,k_0}} \left[ a_s + a_p\, \sigma \cdot \mathbf{n} \right], \tag{10.88b}$$

$$a_s = \sqrt{\frac{(p_0' + m')}{2\,p_0'}}\, A, \quad a_p = \frac{k}{\sqrt{2\,p_0'\,(p_0' + m')}} B. \tag{10.88c}$$

We note that the p.v. amplitude $A$ is essentially the s-wave amplitude and the p.c. amplitude $B$ accounts for the p-wave amplitude.

The decay width is given by

$$d\Gamma = (2\pi)^7\, \delta^4\, (p - p' - k) \left[ \frac{1}{2} Tr\, \left( MM^\dagger \right) \right] d^3p'\, d^3k. \tag{10.89}$$

Performing the integration, we get the decay width

$$\Gamma = \frac{k\, p_0'}{2\pi m} \left[ |a_s|^2 + |a_p|^2 \right]. \tag{10.90}$$

We now consider the decay of polarized baryon B. Let $\mathbf{S}$ be the polarization (spin) of B. Let $\mathbf{s}$ be the polarization of decayed baryon $B'$. In the rest frame of $B'$, $\mathbf{s}$ gives the spin of $B'$. The decay probability in this case is given by

$$dW = (2\pi)^7\, \delta^4\, (p - p' - k)$$
$$\times \frac{1}{2} \left\{ Tr\, \left[ (1 + \sigma \cdot \mathbf{s})\, M\, (1 + \sigma \cdot \mathbf{S}) \right] M^\dagger \right\} d^3p'd^3k. \tag{10.91}$$

The trace can be easily evaluated and the transition rate is proportional to

$$R = 1 + \alpha \mathbf{S} \cdot \mathbf{n} + \mathbf{s} \cdot \left[ (\alpha + \mathbf{S} \cdot \mathbf{n})\, \mathbf{n} + \beta\, (\mathbf{S} \times \mathbf{n}) + \gamma\, \mathbf{n} \times (\mathbf{S} \times \mathbf{n}) \right], \tag{10.92}$$

where

$$\alpha = \frac{2R_e\, a_s^*\, a_p}{|a_s|^2 + |a_p|^2}, \qquad \beta = \frac{2\Im\, a_s^*\, a_p}{|a_s|^2 + |a_p|^2}$$

$$\gamma = \frac{|a_s|^2 - |a_p|^2}{|a_s|^2 + |a_p|^2}$$

$$\alpha^2 + \beta^2 + \gamma^2 = 1. \tag{10.93}$$

Because of the last constraint, we can write

$$\beta = \left(1 - \alpha^2\right)^{1/2} \sin \phi$$
$$\gamma = \left(1 - \alpha^2\right)^{1/2} \cos \phi$$
$$\phi = \tan^{-1}\left(\beta/\gamma\right). \tag{10.94}$$

One also defines

$$\Delta = -\tan^{-1}\left(\beta/\alpha\right).$$

If we do not observe the polarization of $B'$, we put $\mathbf{s} = 0$ and we get

$$dW/\Gamma = \frac{d\Omega_S}{4\pi}\left[1 + \alpha\, \mathbf{S} \cdot \mathbf{n}\right]. \tag{10.95}$$

Hence we can write the angular distribution

$$I_B\left(\theta\right) = \text{Const}\left[1 + \alpha\, S \cos \theta\right], \tag{10.96}$$

where $\theta$ is the angle between the hyperon spin $S$ and the decayed baryon momentum direction $n$. If $a = 0$, the angular distribution is isotropic. $\alpha = 0$ implies either $a_s = 0$ or $a_p = 0$. For this case parity is conserved. The anisotropy in angular distribution implies nonconservation of parity. From the angular distribution we can determine the product $\alpha S$. Since the polarization $S$ of baryon is not generally known, it is difficult to measure $\alpha$ by this method. Further information about $\alpha$ can be obtained from the polarization of decayed baryon $B'$. From Eq. (10.92), we obtain the polarization of decayed bayron $B'$.

$$\langle \mathbf{s} \rangle = \frac{1}{1 + \alpha\, \mathbf{S} \cdot \mathbf{n}}\left\{(\alpha + \mathbf{S} \cdot \mathbf{n})\,\mathbf{n} + \beta\left(\mathbf{S} \times \mathbf{n}\right) + \gamma\, \mathbf{n} \times \left(\mathbf{S} \times \mathbf{n}\right)\right\}. \tag{10.97}$$

In particular if the original baryon $B$ is unpolarized viz $\mathbf{S} = 0$, we get

$$\langle \mathbf{s} \rangle = \alpha\, \mathbf{n}. \tag{10.98}$$

This equation implies that the baryon $B'$ obtained from the decay of unpolarized baryon $B$ is longitudinally polarized. Thus a measurement of this polarization allowed a direct determination of $\alpha$. The experimental values [8] for $\alpha$, $\beta$ and $\gamma$ are given in Table 10.1.

Now a non-zero value for $\beta$ implies the violation of time reversal invariance in these decays. From Table 10.1, it is clear that $\beta = \left(1 - \alpha^2\right)^{1/2} \sin \phi$ is consistent with zero. Thus the time reversal invariance holds in these decays. $P$ invariance implies either $a_s = 0$ or $a_p = 0$, so that $\alpha = 0$, $\beta = 0$. But Table 10.1 shows that $\alpha$ is non-zero. $C$ invariance implies $\alpha = 0$, $\beta \neq 0$; hence from Table 10.1, it follows that $C$ invariance is also violated. The consequences of $T$ and $C$ invariance quoted above hold if we neglect the final state interactions.

Table 10.1

| Decay | $\alpha$ | $\phi$ | $\gamma$ (derived) | $\Delta$ (derived) |
|---|---|---|---|---|
| $\Lambda^\circ_- : \Lambda$ $\rightarrow p\pi^-$ | $0.642 \pm 0.013$ | $(-6.5 \pm 3.5)^\circ$ | $0.76$ | $(8 \pm 4)^\circ$ |
| $\Lambda^0_0 : \Lambda$ $\rightarrow n\pi^0$ | $0.65 \pm 0.05$ | $-$ | $-$ | $-$ |
| $\Sigma^+_0 : \Sigma^+$ $\rightarrow p\pi^0$ | $-0.980^{+0.017}_{-0.015}$ | $(36 \pm 34)^\circ$ | $0.16$ | $(187 \pm 6)^\circ$ |
| $\Sigma^+_+ : \Sigma^+$ $\rightarrow n\pi^+$ | $0.068 \pm 0.013$ | $(167 \pm 20)^\circ$ | $-0.97$ | $\left(-73^{+133}_{-10}\right)^\circ$ |
| $\Sigma^-_- : \Sigma^-$ $\rightarrow n\pi^-$ | $-0.068 \pm 0.008$ | $(10 \pm 15)^\circ$ | $0.98$ | $\left(249^{+12}_{-120}\right)^\circ$ |
| $\Xi^0_0 : \Xi^0$ $\rightarrow \Lambda + \pi^0$ | $-0.411 \pm 0.022$ | $(21 \pm 12)^\circ$ | $0.85$ | $\left(218^{+12}_{-19}\right)^\circ$ |
| $\Xi^-_- : \Xi^-$ $\rightarrow \Lambda + \pi^-$ | $-0.456 \pm 0.014$ | $(4 \pm 4)^\circ$ | $0.89$ | $(188 \pm 8)^\circ$ |

### 10.5.2 $\Delta I = 1/2$ Rule for Hyperon Decays

The effective weak Hamiltonian responsible for $|\Delta S| = 1$ non-leptonic decays in the conventional theory is given in Eq. (10.28), namely

$$H_W^{\text{eff}} = \frac{G_F}{\sqrt{2}} \sin\theta_c \cos\theta_c \left[ J_\lambda^+ \left( J^{1\lambda} \right)^\dagger + \text{h.c.} \right]$$

$$\equiv \frac{G_F}{\sqrt{2}} \sin\theta_c \cos\theta_c \, H_W, \tag{10.99}$$

where

$$H_W = \left[ J_\lambda^+ \left( J^{1\lambda} \right)^\dagger + \text{h.c.} \right]. \tag{10.100}$$

Now $J_\lambda^+ \sim \bar{u} \, \gamma_\lambda (1 + \gamma_5) \, d$ has $I = 1$, $I_3 = +1$, $J_\lambda^1 \sim \bar{s} \, \gamma_\lambda (1 + \gamma_5) \, u$ has $I = \frac{1}{2}$, $I_3 = +\frac{1}{2}$. Thus in general $H_W$ has a mixture of $\Delta I = 1/2$ and $\Delta I = 3/2$. However, the most striking effect of these decays is the approximate validity of $\Delta I = 1/2$ rule. The decays with $\Delta I = 3/2$ are suppressed. A satisfactory understanding of this rule is still lacking.

We now examine the consequences of $\Delta I = 1/2$ rule in non-leptonic hyperon decays and its approximate experimental validity. Consider first the decays

$$\Lambda^0_- : \Lambda \rightarrow p + \pi^- \qquad \Delta I_3 = 1/2$$
$$\Lambda^0_0 : \Lambda \rightarrow n + \pi^0 \qquad \Delta I_3 = -1/2$$

$$\Delta I = 1/2, \, 3/2, \cdots$$

The simplest possibility is $\Delta I = 1/2$. Assuming this to be the case, the only possible isospinor which one can form is

$$\bar{N} \, \tau \cdot \pi \, \Lambda = \left( \bar{p} \, \pi^0 + \sqrt{2} \, \bar{n} \, \pi^-, \; \sqrt{2} \, \bar{p} \, \pi^+ - \bar{n} \, \pi^0 \right) \Lambda. \tag{10.101}$$

Then for $\Delta Q = 0$, we have

$$\Lambda_-^0 = -\sqrt{2}\Lambda_0^0. \tag{10.102}$$

Hence we get

$$\Gamma\left(\Lambda_-^0\right) = 2\Gamma\left(\Lambda_0^0\right), \tag{10.103a}$$

$$\alpha_{\Lambda_-^0} = \alpha_{\Lambda_0^0}. \tag{10.103b}$$

It is clear from Table 10.1, $\alpha_{\Lambda_-^0} \approx \alpha_{\Lambda_0^0}$ ; experimentally

$$\frac{\Gamma\left(\Lambda_-^0\right)}{\Gamma\left(\Lambda_0^0\right)} \approx \frac{63.9}{35.8} = 1.78. \tag{10.103c}$$

Thus $\Delta I = 1/2$, rule is a good approximation, $\Delta I = 3/2$ amplitude is very much suppressed for $\Lambda$-decays. An exactly similar argument gives

$$\Xi_-^- = -\sqrt{2} \, \Xi_0^0, \tag{10.104a}$$

which implies

$$\Gamma\left(\Xi_-^-\right) / \Gamma\left(\Xi_0^0\right) = 2 \left( Expt : \frac{2.90}{1.639} \approx 1.77 \right) \tag{10.104b}$$

$$\alpha_{\Xi_-^-} / \alpha_{\Xi_0^0} = 1 \left( Expt : \frac{0.456}{0.411} \approx 1.11 \right). \tag{10.104c}$$

For $\Sigma$-decays, assuming $\Delta I = 1/2$, the only isospinors which we can form are

$$a \, \bar{N} \, (\mathbf{\Sigma} \cdot \pi) + i \, b \, \bar{N} \, (\mathbf{\Sigma} \times \pi) \cdot \tau. \tag{10.105a}$$

Writing only the part for which total charge is zero, we have

$$a \, \bar{n} \, \left( \Sigma^- \pi^+ + \Sigma^+ \pi^- + \Sigma^0 \pi^0 \right)$$
$$+ b \, \left( \sqrt{2} \, \bar{p} \, \Sigma^0 \pi^+ - \sqrt{2} \, \bar{p} \, \Sigma^+ \pi^0 - \bar{n} \, \Sigma^+ \pi^- + \bar{n} \, \Sigma^+ \pi^- \right) \tag{10.105b}$$

Thus we get

$$\Sigma_-^- = a + b$$
$$\Sigma_+^+ = a - b$$
$$\Sigma_-^0 = \sqrt{2} \, b \tag{10.106}$$
$$\Sigma_0^+ = -\sqrt{2} \, b.$$

From Eq. (10.106), we get

$$\Sigma_+^+ - \Sigma_-^- = \sqrt{2}\Sigma_0^+. \tag{10.107}$$

The prediction can be tested as follows: In the $(a_s, a_p)$ plane if we regard $\Sigma_+^+, \Sigma_-^-$ and $\sqrt{2}\Sigma_0^+$ as vectors, then they should form a closed triangle.

To sum up, in case the $\Delta I = 1/2$ rule holds, out of 7 decays listed in Eq. (10.25) only four are independent. In the language of flavor SU(3) [cf. Chap. 5], the dominance of $\Delta I = 1/2$ rule is generalized to octet dominance. This can be seen as follows:

   $u$, $d$, $s$, belong to 3 representation of SU(3).

   $\bar{u}$, $\bar{d}$, $\bar{s}$, belonging to $\bar{3}$ representation of SU(3).
Now

$$3 \otimes \bar{3} = 8 \oplus 1.$$

Thus $J_\mu^h$ in Eq. (10.27) belongs to an octet representation of SU(3). Hence $H_{\text{int}}^h$ in Eq. (10.27) or (10.28) contains

$$8 \otimes 8 = 1 \oplus 8 \oplus 8 \oplus 10 \oplus +\overline{10} \oplus 27.$$

It can be seen that only 8 and 27 are relevant for the decays (10.25). Thus $H_{\text{int}}^h$ contains both 8 and 27 where 8 corresponds to $\Delta I = 1/2$ only while 27 contains $\Delta I = 3/2$ as well. Thus in the language of SU(3), generalization of $\Delta I = 1/2$ rule is the octet dominance. The octet dominance for the current-current interaction implies an additional relation (called Lee-Sugawara relation) between s-wave decay amplitudes of (10.25)

$$2A\left(\Xi_-^-\right) + A\left(\Lambda_-^0\right) = +\sqrt{3}\, A\left(\Sigma_0^+\right). \tag{10.108}$$

### 10.5.3  Non-leptonic Hyperon Decays in Non-Relativistic Quark Model

One can recover not only the $\Delta I = 1/2$ rule but also the right order of magnitude of the scale required to reproduce the s- and p-wave fits of non-leptonic hyperon decays. Consider the weak vector boson exchange graph of Fig. 10.6 as the analogue of the gluon exchange quark-quark scattering graph considered in Chap. 7 which quite successfully described the quark spectroscopy.

The matrix elements for the process shown in Fig. 10.6 are of the form

$$M\underset{m_W^2 \gg q^2}{=} \frac{1}{2}\frac{g_W^2}{m_W^2}\sin\theta_c\cos\theta_c\left\{ \bar{u}(p_i')\gamma^\mu\left(1 - \gamma_5\right)\alpha_i^-\, u(p_i)\right.$$

$$\times\ \bar{u}(p_j')\gamma_\mu\left(1 - \gamma_5\right)\beta_i^+ u(p_j) + i \longleftrightarrow j\Big\}, \tag{10.109}$$

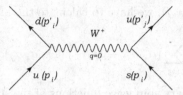

Fig. 10.6   W-boson exchange diagram for $u + s \rightarrow d + u$.

where $q = p_i - p'_i = p'_j - p_j$. $u$'s are Dirac spinors in Dirac space but are column vectors involving $u$, $d$, $s$ quarks in ordinary flavor SU(3) space. $\alpha_i^-$ and $\beta_j^+$ are operators which transform a $u$-like state into a $d$-like state and a $s$-like state into a $u$-like state respectively. We take the leading non-relativistic limit of the above matrix elements. In the leading non-relativistic approximation, only $\gamma_0$ and $\gamma^i \gamma_5$ have non-zero limits. Thus only parity conserving (p.c) part of $M$ survives in the leading non-relativistic approximation and we have in this limit

$$M^{p \cdot c} \sim \frac{1}{\sqrt{2}} G_F \sin \theta_c \cos \theta_c \sum_{i \rangle j} \left( \alpha_i^- \beta_j^+ + \beta_i^+ \alpha_j^- \right) \cdot (1 - \sigma_i \cdot \sigma_j) \quad (10.110)$$

$$M^{p \cdot v} = 0.$$

The latter corresponds to a general result that $\langle B' | (JJ)^{p \cdot v} | B \rangle = 0$ as a consequence of CP and SU(3) invariance. The Fourier transform of Eq. (10.110) gives the effective $H_W$ as

$$H_W^{p \cdot c} = \frac{1}{\sqrt{2}} G_F \sin \theta_c \cos \theta_c \sum_{i > j} \left( \alpha_i^- \beta_j^+ + \beta_i^+ \alpha_j^- \right)$$

$$\times (1 - \sigma_i \cdot \sigma_j) \delta^3 (\mathbf{r}). \quad (10.111)$$

Now it has been shown [see Sec. 11.4.2] that in the current-algebra approach the question of $\Delta I = 1/2$ rule or octet dominance for non-leptonic decays of baryons hinges on the matrix elements

$$\langle B_s | H_W^{p \cdot c} | B_r \rangle \sim a_{rs} \bar{u} u, \quad (10.112)$$

which essentially determine both s- and p-wave amplitudes. Here u is a Dirac spinor for $B_r$ or $B_s$ which denotes a baryon like $\Lambda, \Sigma, \Xi, n$, or $p$. Therefore, we have to take the matrix elements of Eq. (10.111) between the baryon states $B_r$ and $B_s$. We regard the baryon state $B_r$ or $B_s$ as made up of three quarks. We take the spatial wave function for such states to be the same for the octet of baryons $p$, $n$, $\Lambda$, $\Sigma^\pm$, $\Sigma^0$, $\Xi^0$, $\Xi^-$ and denote it by $\Psi_0$. Thus writing

$$d' = \langle \Psi_0 | \delta^3 (\mathbf{r}) | \Psi_0 \rangle = |\Psi_0 (0)|^2, \quad (10.113)$$

where $\mathbf{r} = \mathbf{r}_i - \mathbf{r}_j$ $(i \neq j)$, we have to calculate the matrix elements of the operator

$$\sum_{i)j} \left( \alpha_i^- \beta_j^+ + \beta_i^+ \alpha_j^- \right) (1 - \sigma_i \cdot \sigma_j)$$

between the spin-unitary spin wave functions of the states $p$, $n$, $\Sigma^+$, $\Sigma^0$, $\Lambda$, $\Xi^0$, given in Chap. 6. We obtain

$$a_{\Lambda n} = \frac{G_F}{\sqrt{2}} \sin \theta_c \cos \theta_c d' \left( +\sqrt{6} \right) \qquad (10.114a)$$

$$a_{\Sigma_p^+} = \frac{G_F}{\sqrt{2}} \sin \theta_c \cos \theta_c d' (-6) \qquad (10.114b)$$

$$= -\sqrt{2} a_{\Sigma_n^0} \qquad (10.114c)$$

$$a_{\Xi^0 \Lambda^0} = \frac{G_F}{\sqrt{2}} \sin \theta_c \cos \theta_c d' \left( -2\sqrt{6} \right) \qquad (10.114d)$$

$$a_{\Xi^- \Sigma^-} = 0. \qquad (10.114e)$$

The relation $a_{\Sigma_p^+} = -\sqrt{2} a_{\Sigma_n^0}$ expressed in Eq. (10.114c) ensures the $\Delta I = 1/2$ rule (or octet dominance) and hence $A(\Sigma_+^+) = 0$ (which is good experimentally) in current algebra approach [see Eq. (10.117) below].

Once the octet dominance for $a_{rs}$ is established we can parametrize $a_{rs}$ in the SU(3) limit as

$$a_{rs} = \sqrt{2} \left( 2F' i f_{6rs} + 2D' d_{6rs} \right). \qquad (10.115)$$

Then the relations (10.114) immediately give

$$\frac{D'}{F'} = -1. \qquad (10.116)$$

Now using the current algebra relations [see Sec. 11.4.2] for the s-wave amplitudes one has

$$A\left( \Lambda_-^0 \right) = -\frac{1}{f_\pi} a_{\Lambda n} = -\sqrt{2} \, A \left( \Lambda_0^0 \right)$$

$$A\left( \Xi_-^- \right) = -\frac{1}{f_\pi} \, a_{\Xi^0 \Lambda} = -\sqrt{2} A \left( \Xi_0^0 \right)$$

$$A\left( \Sigma_0^+ \right) = \frac{1}{\sqrt{2} \, f_\pi} \, a_{\Sigma^+ p}$$

$$A\left( \Sigma_+^+ \right) = -\frac{1}{f_\pi} \left( a_{\Sigma^+ p} + \sqrt{2} \, a_{\Sigma^0 n} \right)$$

$$A\left( \Sigma_-^- \right) = \left( \frac{\sqrt{2}}{f_\pi} \right) a_{\Sigma^0 n}. \qquad (10.117)$$

Here $f_\pi$ is the constant which enters in $\pi^- \to \bar{\nu}_\mu + \mu^-$ decay. Then using Eqs. (10.114) and (10.117), we have the relations (10.107) and (10.108). Using the value of $d'$ as determined by the constituent quark spectroscopy [cf. Chap. 7],

$$a_{\Sigma + p} = \frac{-27 \, G_F \, \sin\theta_c \cos\theta_c}{8\sqrt{2}\pi\alpha_s} (m_\Sigma - m_\Lambda) \left(\frac{\hat{m}^2}{1 - \hat{m}/m_s}\right)_{\text{constituent}}$$
$$\approx -105 \text{ eV} \tag{10.118}$$

for the accepted value of $\alpha_s$ $(q^2 \approx 1 \text{ GeV}) \approx 0.5$. This is almost the phenomenological octet dominance scale, which together with $D'/F' \approx -0.86$ [not very far from the prediction (10.116)], are required to fit the s- and p-wave amplitudes of hyperon decays.

## 10.6 Problems

(1) Show that the electron spectrum in the decay of $b$-quark

$$b \to c + e^- + \bar{\nu}_e,$$

using $V - A$ theory is given by (neglecting the electron mass)

$$\frac{d\Gamma}{dy} = \frac{G_F^2}{96\pi^3} m_b^5 \frac{(y_m - y)^2 \, y^2}{(1 - y)^2} \left[(3 - 2y) + \frac{(1 - y_m)(3 - y)}{(1 - y)}\right],$$

where

$$y = \frac{2E_e}{m_b}, \qquad y_m = 1 - \frac{m_c^2}{m_b^2}.$$

Similarly, show that for $c$-quark decay

$$c \to s + e^+ + \nu_e$$

the electron spectrum is given by

$$\frac{d\Gamma}{dy} = \frac{G_F^2}{16\pi^3} m_c^5 \, y^2 \frac{(y_m - y)^2}{(1 - y)}$$

$$y = \frac{2E_e}{m_c}, \qquad y_m = 1 - \frac{m_s^2}{m_c^2}.$$

Hint: For $b \to c + e^- + \bar{\nu}_e$, the matrix elements are

$$T = -\frac{G_F}{\sqrt{2}} \left[\bar{u}(p_2) \, \gamma_\lambda \, (1 - \gamma_5) \, u(p_1)\right] \left[\bar{u}(p_1) \, \gamma^\lambda \, (1 - \gamma_5) \, v(k_2)\right].$$

Use Eqs. (31) and (32) with the replacements $(m_\mu, m_e, m_{\nu m}, m_{\nu e})$ $\to (m_e, m_c, m_e, m_{\nu e})$, $\varepsilon = 1$ so that

$$|M|^2 = \frac{G_F^2}{m_b \, m_c \, m_e \, m_\nu} \, 4 \, p_1 \cdot k_2 \, p_2 \cdot k_1$$

$$= \frac{G_F^2}{m_b \, m_c \, m_e \, m_\nu} \, 2 \, m_b \, E_\nu \left[ m_b^2 - m_c^2 - 2 \, m_b \, E_\nu \right],$$

in the rest frame of $b$. Performing $d^3 p_2$ integration, write $d^3 k_1 d^3 k_2 = k_1^2 \, dk_1 \, dk_2 \, d\Omega$ and use

$$\delta \left( m_b - E_e - E_\nu - \sqrt{k_1^2 + k_2^2 + 2k_1 k_2 \cos\theta + m_c^2} \right)$$

to perform the angular integration to obtain

$$\frac{d\Gamma}{dE_e \, dE_\nu} = \frac{4 G_F^2}{(2\pi)^3} E_\nu \left[ m_b^2 - m_c^2 - 2 \, m_b \, E_\nu \right]$$

where from

$$E_\nu = \frac{m_b^2 - m_c^2 - 2 \, m_b \, E_e}{2 \, (m_b - m_e + E_e \cos\theta)}$$

$$(E_\nu)_{\min} = \frac{m_b^2 - m_c^2 - 2 \, m_b \, E_e}{2 m_b} = \frac{m_b}{2} \, (y_b - y)$$

$$(E_\nu)_{\max} = \frac{(m_b^2 - m_c^2) - 2 \, m_b \, E_e}{2 m_b - 2 E_e} = \frac{m_b \, (y_b - y)}{(1 - y)}.$$

The integration of $E_\nu$ gives the result $\frac{d\Gamma}{dy}$.

For the second problem, the matrix elements are

$$T = -\frac{G_F}{\sqrt{2}} \left[ \bar{u}(p_2) \, \gamma_\lambda \, (1 - \gamma_5) \, u(p_1) \right] \left[ \bar{u}(k_2) \, \gamma^\lambda \, (1 - \gamma_5) \, v(k_1) \right].$$

Results from the first can be obtained by changing $k_2 \longleftrightarrow k_1$, $m_b \to m_c$, $m_c \to m_s$

$$|M|^2 = \frac{G_F^2}{m_c \, m_s \, m_e \, m_\nu} \left[ 4 \, p_1 \cdot k_1 \, p_2 \cdot k_2 \right]$$

$$= \frac{G_F^2}{m_c \, m_s \, m_e \, m_\nu} \, (2 \, m_c \, E_e) \left[ m_c^2 - m_s^2 - 2 \, m_c \, E_e \right]$$

and then follow the same steps as in the first part.

(2) Consider the decay

$$K \to 3\pi.$$

Show that decay rate can be expressed as

$$\Gamma = \frac{1}{2^6 \pi^3 m_K} \frac{Q^2}{6\sqrt{3}} \int dx \, dy \, |A|^2$$

$$0 \le x^2 + y^2 \le 1,$$

where $A$ is the decay amplitude,

$$x = \sqrt{3}\frac{T_2 - T_1}{Q}, \qquad y = \frac{3T_3 - Q}{Q}$$

$T_1, T_2$ and $T_3$ are kinetic energies of pions. Then the energies $\omega_1, \omega_2, \omega_3$ of pions are given by $\omega_i = T_i + m_\pi$ and $Q = T_1 + T_2 + T_3 = \omega_1 + \omega_2 + \omega_3 - 3m_\pi = m_K - 3m_\pi$.

The events in Dalitz plot can be expressed by taking

$$A_j = A_j(0)\left[1 + \frac{2\sigma_j}{3}\frac{m_K}{m_\pi^2}(2\omega_3 - \omega_1 - \omega_2)\right]$$

where $j$ stands for any decay channel of $K$.

(3) Show that if the three pions in the decay of $K \to 3\pi$ are in $I = 1$ states, then

$$\Gamma\left(K_2^0 \to \pi^+\pi^-\pi^0\right) = 2\Gamma\left(K^+ \to \pi^+\pi^0\pi^0\right) \qquad (10.119)$$

$$\Gamma\left(K^+ \to \pi^+\pi^+\pi^-\right) - \Gamma\left(K^+ \to \pi^+\pi^0\pi^0\right)$$
$$= \Gamma\left(K_2^0 \to \pi^0\pi^0\pi^0\right). \qquad (10.120)$$

Equations (10.119) and (10.120) are the necessary conditions for $\Delta I = 1/2$ rule to hold. But they are not sufficient since $I = 1$ state can be reached also by $\Delta I = 3/2$.

Show that for totally symmetric $I = 1$ states

$$\Gamma\left(K^+ \to \pi^+\pi^+\pi^-\right) = 4\Gamma\left(K^+ \to \pi^+\pi^0\pi^0\right),$$
$$\Gamma\left(K_2^0 \to \pi^0\pi^0\pi^0\right) = \frac{3}{2}\Gamma\left(K_2^0 \to \pi^+\pi^-\pi^0\right).$$

(4) Show that if time-reversal invariance holds, the decay amplitudes $A$ and $B$ given in Eq. (10.85) are real, i.e. $\beta = 0$.

## 10.7 References

1. T. D. Lee and C. S. Wu, Weak Interactions. Ann. Rev. Nucl. Sci. 15, 381 (1965).
2. R.E. Marshak, Riazuddin and C. P. Ryan, Theory of Weak Interactions in Particle Physics. Wiley-Interscience, New York (1969).
3. L. B. Okun, Leptons and Quarks, North-Holland Publishing Co., Amsterdam, (1982).
4. E. Commins and P. H. Bucksbaum, Weak Interaction of Leptons and Quarks, Cambridge University Press, Cambridge, England (1983).
5. H. Georgi, Weak Interaction and Modern Particle Theory, Benjamin/Cummings, New York (1984).
6. T. D. Lee, Particle Physics and Introduction to Field Theory, Harwood Academic (revised edition 1988).
7. S. Freedman, Comments on Nuclear and Particle Physics, Part A, Vol. XIX (5), p. 209 (1990).
8. Particle Data Group, K Nakamura et al., Journals of Physics G **37**, 075021 (2010).

# Chapter 11

# Properties of Weak Hadronic Currents and Chiral Symmetry

## 11.1 Introduction

In Chap. 10, we have introduced an octet of vector and axial vector currents

$$V_{i\lambda} = \bar{q}\frac{\lambda_i}{2}\gamma_\lambda q \tag{11.1}$$

$$A_{i\lambda} = \bar{q}\frac{\lambda_i}{2}\gamma_\lambda\gamma_5 q, \tag{11.2}$$

where

$$J_\lambda^\pm = V_{1\pm i2\lambda} - A_{1\pm i2\lambda} \tag{11.3}$$

$$J_\lambda^1, J_\lambda^{1\dagger} = V_{4\pm i5\lambda} - A_{4\pm i5\lambda} \tag{11.4}$$

take part in $|\Delta Y| = 0$ and $|\Delta Y| = 1$ semi-leptonic processes respectively. The electromagnetic current is given by

$$V_\lambda^{em} = V_{3\lambda} + \frac{1}{\sqrt{3}}V_{8\lambda} \tag{11.5}$$

where the first part is the third component of an isovector while the second part is an isoscalar. Now $H_{int}^{em} \sim V_\lambda^{em}a^\lambda$. Since photon field $a^\lambda$ has C-parity $-1$ and the intrinsic parity of the photon is $-1$, we see that CP of $V_\lambda^{em}$ is $+1$. From this we can generalize that CP of vector current $V_\lambda$ is $+1$. The parity of axial-vector current $A_\lambda$ is $+1$ and since the weak Hamiltonian is CP invariant, the C-parity of $A_\lambda$ must be $+1$.

## 11.2 Conserved Vector Current Hypothesis (CVC)

The hypothesis of conserved vector current $(CVC)$ states that $V_\lambda^\pm$ and $V_{3\lambda}(= J_\lambda^{em}, \Delta I = 1)$ are respectively $1 + i2$, $1 - i2$ and $3$ members of an

311

isospin current, which is conserved by strong interaction. The generators of the isospin group $SU_I(2)$ are then given by

$$I_i = \int V_{i0}(\mathbf{x}, t)\, d^3x, \quad i = 1, 2, 3. \tag{11.6}$$

The first consequence of $CVC$ ($\partial^\lambda V_\lambda^\pm = 0$) is that the form factor $h_V(q^2) = 0$ in Eq. (10.50a) where $A$ and $B$ are respectively taken as neutron and proton. [Note: When invariance under $SU(2)$ is assumed, $m_p = m_n = m_N$.]

In order to discuss the other consequences of CVC, we note from Eqs. (11.6) and (10.50a) that

$$\langle p(p') | I_+ | n(p) \rangle$$

$$= \frac{1}{(2\pi)^3} \sqrt{\frac{m_N^2}{p_0 p_0'}} \bar{u}(p') \left[ g_V(q^2)\gamma_0 + i f_V(q^2)\sigma_{0\nu}q^\nu \right] u(p) \int d^3x\, e^{-i\mathbf{q}\cdot\mathbf{x}}. \tag{11.7}$$

Since $I_+$ is conserved in the absence of electromagnetism, $I_+(t)$ is a constant of motion, i.e. $I_+(t) = I_+(0) = I_+$; we can take $t = 0$ and

$$\frac{1}{(2\pi)^3} \int d^3x\, e^{-i\mathbf{q}\cdot\mathbf{x}} = \delta^3(\mathbf{q}). \tag{11.8}$$

Now

$$\bar{u}(p)\gamma_0 u(p) = \frac{p_0}{m_N}$$

$$I_+ | n(p) \rangle = | p(p) \rangle \tag{11.9}$$

and thus

$$\langle p(p') | I_+ | n(p) \rangle = \delta^3(\mathbf{p}' - \mathbf{p}) = \delta^3(\mathbf{q}) \tag{11.10}$$

Hence it follows from Eq. (11.7) that

$$g_V(0) = 1. \tag{11.11}$$

Thus in the absence of electromagnetism, the vector coupling constant in nuclear $\beta$-decay is not renormalized and is equal to its "bare" value. Noting that $[J_\lambda^Y = \frac{1}{\sqrt{3}} V_{8\lambda}$ in SU(3)]

$$[J_\lambda^Y(x), I_+] = 0,$$

$$[V_{3\lambda}, I_+] = V_{1+i2\lambda}(x), \tag{11.12}$$

and

$$I_+ | n \rangle = | p \rangle, \quad \langle p | I_+ = \langle n | \tag{11.13}$$

it follows that

$$\langle p|V_\lambda^+|n\rangle = \langle p|[V_{3\lambda}, I_+]|n\rangle$$
$$= \langle p|[V_\lambda^{em}, I_+]|n\rangle$$
$$= \langle p|V_\lambda^{em}|p\rangle - \langle n|V_\lambda^{em}|n\rangle. \tag{11.14}$$

Now Lorentz invariance gives the electromagnetic form factors of proton and neutron as

$$\langle p(p')|V_\lambda^{em}|p(p)\rangle = \frac{1}{(2\pi)^3}\sqrt{\frac{m_N^2}{p_0 p_0'}}$$
$$\times \bar{u}(p')\left[F_1^p(q^2)\gamma_\lambda + i\frac{F_2^p(q^2)}{2m_N}\sigma_{\lambda\nu}q^\nu\right]u(p) \tag{11.15}$$

$$\langle n(p')|V_\lambda^{em}|n(p)\rangle = \frac{1}{(2\pi)^3}\sqrt{\frac{m_N^2}{p_0 p_0'}}$$
$$\times \bar{u}(p')\left[F_1^n(q^2)\gamma_\lambda + i\frac{F_2^n(q^2)}{2m_N}\sigma_{\lambda\nu}q^\nu\right]u(p) \tag{11.16}$$

where [since $\int d^3x\, V_0^{em}(\mathbf{x}, 0)$ is the electric charge in unit of $e$] it follows, on using Eqs. (11.8)-(11.10) that

$$F_1^p(0) = 1, \; F_1^n(0) = 0. \tag{11.17}$$

Since $\sigma_{\lambda\nu}q^\nu$ gives Pauli type interaction, it also follows that

$$F_2^p(0) = \kappa_p, \; F_2^n(0) = \kappa_n \tag{11.18}$$

where $\kappa_p$ and $\kappa_n$ are the anomalous magnetic moments of proton and neutron respectively. $\kappa_p = 1.792$ and $\kappa_n = -1.913$ in units of nuclear magneton. Hence we get from Eq. (10.50a) and Eqs. (11.12), (11.13) and (11.14) that

$$g_V(q^2) = F_1^p(q^2) - F_1^n(q^2) = F_1^V(q^2)$$
$$f_V(q^2) = \frac{1}{2m_N}\left[F_2^p(q^2) - F_2^n(q^2)\right] = \frac{F_2^V(q^2)}{2m_N}, \tag{11.19}$$

where $F_1^V$ and $F_2^V$ are the isovector electromagnetic nucleon form factors. Their normalization follows from Eqs. (11.17) and (11.18).

$$F_1^V(0) = 1, \; F_2^V(0) = (\kappa_p - \kappa_n). \tag{11.20}$$

Thus in particular

$$g_V(0) = 1,$$
$$f_V(0) = \frac{\kappa_p - \kappa_n}{2m_N} \tag{11.21}$$

Using SU(3), we can write the matrix elements of vector current $V_{i\lambda}$, $i = 1, \cdots, 8$ for an octet of baryons (assuming $q^2 \approx 0$):

$$\langle B_k(p') | V_{i\lambda} | B_j(p) \rangle = \frac{1}{(2\pi)^{3/2}} \sqrt{\frac{m_B^2}{p_0 p_0'}} \, \bar{u}(p')[if_{ijk}\gamma_\lambda]u(p) \tag{11.22}$$

namely the relation (10.54).

## 11.3 Partially Conserved Axial Vector Current Hypothesis (PCAC)

From Eq. (10.73), we have

$$\langle 0 | \partial^\lambda A_\lambda^+(x) | \pi^- \rangle = -ip^\lambda \langle 0 | A_\lambda^+ | \pi^- \rangle e^{-ip \cdot x}$$
$$= \frac{1}{(2\pi)^{3/2}} \frac{1}{\sqrt{2p_0}} f_\pi m_\pi^2 e^{-ip \cdot x}. \tag{11.23}$$

If the axial vector current $A_\lambda^+$ is conserved, then either $f_\pi = 0$ or $m_\pi^2 = 0$. Since for a physical pion $m_\pi^2 \neq 0$, then $f_\pi$ must be zero and pion decay is forbidden. Thus $A_\lambda^+$ is not conserved. Now $\partial^\lambda A_\lambda^+$ has the same quantum numbers as those for a pion. If we now put

$$\partial^\lambda A_\lambda^+ = f_\pi m_\pi^2 \pi^- \tag{11.24}$$

then

$$\langle 0 | \pi^-(x) | \pi^- \rangle = \frac{1}{(2\pi)^{3/2}} \frac{1}{\sqrt{2p_0}} e^{-ip \cdot x} \tag{11.25}$$

Here $\pi^-(x)$ is the pion field operator which creates $\pi^+$ or destroys $\pi^-$. Equation (11.24) is called the PCAC hypothesis. We note from Eq. (11.23), that in the limit $m_\pi^2 \to 0$, the axial vector current is conserved. This implies that strong interactions have an approximate symmetry which is exact in the limit of zero pion mass. Such a symmetry is called chiral symmetry. Chiral symmetry manifests itself in the existence of massless pseudoscalar mesons called Nambu-Goldstone bosons.

We shall come to this point again later. Here we discuss one of the important consequences of PCAC. We apply PCAC to neutron $\beta$-decay. From Eq. (10.50b), we have

$$\langle p(p') | A_\lambda^+ | n(p) \rangle$$
$$= \frac{1}{(2\pi)^3} \sqrt{\frac{m_p m_n}{p_0 p_0'}}$$
$$\times \bar{u}(p') \left[ g_A(q^2)\gamma_\lambda\gamma_5 + f_A(q^2)\gamma_5 q_\lambda - ih_A(q^2)\gamma_5\sigma_{\lambda\nu}q^\nu \right] u(p) \tag{11.26}$$

We note that pion pole contributes to the form factor $f_A(q^2)$ only. It does not contribute to $g_A(q^2)$ nor to $h_A(q^2)$. Separating out the pion pole contribution, we write

$$f_A(q^2) = -\frac{\sqrt{2}g_{\pi NN}f_\pi}{q^2 - m_\pi^2} + \bar{f}_A(q^2) \qquad (11.27)$$

where $\bar{f}_A(q^2)$ is the remaining part of $f_A(q^2)$. From Eqs. (11.26) and (11.27), we get

$$\langle p(p') | \partial^\lambda A_\lambda^+ | n(p) \rangle$$
$$= \frac{1}{(2\pi)^3} \sqrt{\frac{m_p m_n}{p_0 p_0'}}\, \bar{u}(p') i\gamma_5 u(p)$$
$$\times \left[ 2m_N g_A(q^2) - q^2 \frac{\sqrt{2}g_{\pi NN}f_\pi}{q^2 - m_\pi^2} + q^2\, \bar{f}_A(q^2) \right]. \qquad (11.28)$$

Now if we assume that in the limit $m_\pi^2 \to 0$, the axial vector current is conserved, we get,

$$2m_N g_A\left(q^2\right) - \sqrt{2}g_{\pi NN}f_\pi + q^2\, \bar{f}_A(q^2) = 0 \qquad (11.29)$$

At $q^2 = 0$, this gives

$$g_A = \frac{g_{\pi NN}f_\pi}{\sqrt{2}m_N} \qquad (11.30)$$

This is called the Goldberger-Treiman (G-T) relation. Thus G-T relation is exact in the chiral symmetry limit when pion mass is zero and the axial vector current is conserved. This relation can be easily tested as all the quantities in Eq. (11.30) are experimentally known. This relation is valid within 6% agreement with experiment. On the other hand, we note that

$$\langle p(p') | \partial^\lambda A_\lambda^+ | n(p) \rangle = \frac{1}{(2\pi)^3} \sqrt{\frac{m_p m_n}{p_0 p_0'}}\, \bar{u}(p') i\gamma_5 u(p) \left[ 2m_N g_A(q^2) + q^2\, f_A(q^2) \right].$$
$$(11.31)$$

Using PCAC, viz Eq. (11.24),

$$-\frac{\sqrt{2}g_{\pi NN}f_\pi m_\pi^2}{q^2 - m_\pi^2} = 2m_N g_A(q^2) + q^2\, f_A(q^2) \qquad (11.32)$$

Evaluating it at $q^2 = 0$, $m_\pi^2 \neq 0$, we again get the G-T relation. We conclude that the success of the G-T relation implies that deviations from chiral symmetry or equivalently from PCAC are indeed small.

Finally, using $SU(3)$ we can write for $q^2 \approx 0$ for an octet of baryons [cf. Eq. (10.53)].

$$\langle B_k(p') | A_{i\lambda} | B_j(p) \rangle$$
$$= \frac{1}{(2\pi)^{3/2}} \sqrt{\frac{m_B^2}{p_0 p_0'}} \bar{u}(p') \left[ \gamma_\lambda \gamma_5 \left( i f_{ijk} F + d_{ijk} D \right) \right] u(p) \qquad (11.33)$$

In particular for neutron $\beta$-decay,

$$g_A = F + D, \qquad (11.34)$$

where

$$(2\pi)^3 \frac{p_0}{m_N} \langle p | A_{3\lambda} | p \rangle = \frac{1}{2} g_A \bar{u}(p) \gamma_\lambda \gamma_5 u(p) \qquad (11.35)$$

We define a four-vector

$$s^\lambda = \bar{u}(p) \gamma^\lambda \gamma^5 u(p) \qquad (11.36)$$

We note that

$$p \cdot s = 0, \ s^2 = -1. \qquad (11.37)$$

The vector $s^\lambda$ thus gives the spin of the proton. To see it explicitly we go to the rest frame of the proton. In this frame, we get from Eq. (11.37), $s_0 = 0$, $\mathbf{s}^2 = 1$. From Eq. (11.36),

$$\mathbf{s} = \chi^+ \sigma \chi. \qquad (11.38)$$

In quark model, we can write the axial-vector current $A_{i\mu} = \bar{q}\gamma_\mu \gamma_5 \frac{\lambda_i}{2} q$. We define the quantity $\Delta q$ as

$$(2\pi)^3 \frac{p_0}{m} \langle p | \bar{q} \gamma_\lambda \gamma_5 q | p \rangle = \Delta q s_\lambda. \qquad (11.39)$$

In particular for $A_{3\lambda} = \frac{1}{2}(\bar{u}\gamma_\lambda \gamma_5 u - \bar{d}\gamma_\lambda \gamma_5 d)$, we have

$$(2\pi)^3 \frac{p_0}{m} \langle p | A_{3\lambda} | p \rangle = \frac{1}{2}(\Delta u - \Delta d) s_\lambda \qquad (11.40)$$

so that

$$\Delta u - \Delta d = g_A = F + D. \qquad (11.41)$$

## 11.4 Current Algebra and Chiral Symmetry

Isospin conservation implies that strong interactions are invariant under $SU(2)$ group generated by the charges:

$$I_i(t) = \int V_{i0}(\mathbf{x}, t) d^3x, \ i = 1, 2, 3. \tag{11.42}$$

In the same way we can define the axial charges

$$I_i^5(t) = \int A_{i0}(\mathbf{x}, t) d^3x, \ i = 1, 2, 3. \tag{11.43}$$

The generators of the isospin group $SU(2)$ satisfy the commutation relations

$$[I_i(t), I_j(t)] = i\varepsilon_{ijk}I_k(t). \tag{11.44}$$

Since $I_i^5(t)$'s belong to the adjoint representation of $SU(2)$ group, we have

$$[I_i(t), I_j^5(t)] = i\varepsilon_{ijk}I_k^5(t). \tag{11.45}$$

We obtain a closed algebraic system by requiring that

$$[I_i^5(t), I_j^5(t)] = i\varepsilon_{ijk}I_k(t). \tag{11.46}$$

The last relation constitutes a major theoretical assumption. The commutation relations (11.44), (11.45) and (11.46) represent the algebra of the group $SU(2) \times SU(2)$ generated by the vector and axial vector charges. This group is called the chiral $SU(2)$ group.

Let us now write the part of the QCD Lagrangian [cf. Eq. (7.52)] which involves $u$ and $d$ quarks:

$$L_{u,d} = i\bar{q}\gamma^\mu D_\mu q - \frac{m_u + m_d}{2}\bar{q}q - \frac{m_u - m_d}{2}\left(\bar{u}u - \bar{d}d\right), \tag{11.47}$$

where $q = \begin{pmatrix} u \\ d \end{pmatrix}$ is an isodoublet field and we have suppressed color indices. For $m_u = m_d$ this Lagrangian is invariant under the isospin transformation

$$q \to Uq, \tag{11.48}$$

where $U$ is a special unitary matrix, $e^{i\frac{\tau_i}{2}\Lambda_i}$, $\Lambda_i$ being constant. The associated vector current $V_{i\mu} = \bar{q}\frac{\tau_i}{2}\gamma_\mu q$ is conserved. The existence of nearly degenerate isospin multiplets of hadrons shows clearly that $|m_u - m_d|$ is small compared to hadron mass scale ( $\sim$1 GeV). Setting $m_u = m_d = m$, we can write

$$L_{u,d} = i\bar{q}_L\gamma^\mu D_\mu q_L + i\bar{q}_R\gamma^\mu D_\mu q_R - m(\bar{q}_L q_R + \bar{q}_R q_L), \tag{11.49}$$

where we have split $q$ into "left-handed" and "right-handed" components

$$q_{L,R} = \frac{1 \mp \gamma_5}{2} q$$

It is clear that in the limit $m = 0$, the Lagrangian (11.49) would be invariant under independent 'chiral' isospin transformations on $q_L$ and $q_R$:

$$q_L = U_L q_L, \ q_R \to U_R q_R$$

and not only $V_{i\mu}$ but also the axial vector current $\bar{q}\gamma_\mu \gamma_5 \frac{\tau_i}{2} q$ would be conserved. We note that the mass term $m(\bar{q}_L q_R + \bar{q}_R q_L)$ or in general the coupling to scalar and pseudoscalar fields

$$\left[ \phi \bar{q} \begin{pmatrix} 1 \\ \gamma_5 \end{pmatrix} q = \phi (\bar{q}_L q_R \pm \bar{q}_R q_L) \right]$$

would break chiral symmetry. This also demonstrates that the forces between the quarks have to be vector in nature [mediated by spin 1 gluons, cf. the term $\bar{q}\gamma_\mu \lambda \cdot \mathbf{G}^\mu q$ in Eq. (11.47) or Eq. (11.49)]. As we shall see later $m_u \sim 5$ MeV, $m_d \sim 10$ MeV (these are called current quark masses, not to be confused with constituent quark masses of order 300 MeV [cf. Chap. 6]) are small compared to the hadron scale of $O(1 \text{ GeV})$ so that chiral symmetry is nearly exact.

Now if $A_{i\lambda}$ were conserved, the axial charge $I_i^5$ would commute with the Hamiltonian:

$$[I_i^5, H] = 0. \tag{11.50}$$

Hence if we define

$$I_i^5 |X_j\rangle = i\epsilon_{ijk} |Y_k\rangle, \tag{11.51}$$

use of Eq. (11.51) would imply that the states $|Y_k\rangle$ are degenerate in mass with $|X_j\rangle$ even though they have opposite parity. This is because $I_i^5$ has negative parity. This condition can be realized in either of the two ways:

(1) The Wigner-Weyl realization of chiral SU(2) symmetry, in which case $|Y_k\rangle$ would consist of "parity doublets" of $|X_j\rangle$ e.g. if $|X_j\rangle$ were pseudoscalar mesons, $|Y_k\rangle$ would be scalar mesons degenerate in mass with the pseudoscalar mesons. This is not what occurs in nature and therefore chiral symmetry is not realized in nature in this way in contrast to the ordinary isospin symmetry which is realized in this way.

(2) Spontaneously broken symmetry realization of chiral SU(2), in which case $|Y_k\rangle$ would consist of $|X_j\rangle$ plus an odd number of pions with vanishing four-momentum (called soft pions), the pion being a massless "Nambu-Goldstone" boson. In particular

$$I_i^5 |0\rangle = -\frac{i}{2} \frac{f_\pi}{\sqrt{2}} |\pi_i(0)\rangle \neq 0, \tag{11.52}$$

the first part being valid only for single-pion transitions, while

$$I_i |0\rangle = 0. \tag{11.53}$$

As we shall see $m_\pi^2$ would involve $(m_u + m_d)/2$ as a factor and so a measure of explicit chiral symmetry breaking is provided by $m_\pi^2/m_\rho^2 \approx 0.03$, $\rho$ being the non-strange (non Nambu-Goldstone) boson next to pion. The notion of (approximate) spontaneously broken chiral symmetry has been found useful in hadron physics and has given rise to many predictions involving soft pions which are in good agreement with the data [see references]. One such prediction is the Goldberger-Treiman relation (11.30):

$$\left[ \frac{m_A g_A \sqrt{2}}{f_\pi g_{\pi NN}} \right] - 1 = 0 \tag{11.54}$$

to be compared with the experimental value $0.06 \pm 0.01$ of the left-hand side.

The above considerations can be easily generalized to $SU(3)$. Thus the QCD Lagrangian (7.52) shows an approximate global symmetry in the limit $m_q \to 0$, this Lagrangian is invariant under the group $SU(3) \times SU(3)$ generated by the charges associated with the weak currents $J_{i\mu}$. Thus the generators of the group are $(i = 1, \cdots, 8)$.

$$F_i = \int V_{i0}(\mathbf{x}, t) d^3x$$

$$F_i^5 = \int A_{i0}(\mathbf{x}, t) d^3x.$$

They satisfy the commutation relations

$$[F_i, F_j] = i f_{ijk} F_k \tag{11.55}$$

$$\left[ F_i, F_j^5 \right] = i f_{ijk} F_k^5 \tag{11.56}$$

$$\left[ F_i^5, F_j^5 \right] = i f_{ijk} F_k. \tag{11.57}$$

The commutation relations (11.55) and (11.56) follow from flavor $SU(3)$, the commutation relation (11.57) is a new assumption. Equivalently if we define

$$F_i^L = \frac{1}{2} \left( F_i - F_i^5 \right), \quad F_i^R = \frac{1}{2} \left( F_i + F_i^5 \right) \tag{11.58}$$

we get

$$\begin{aligned}
\left[F_i^L, F_j^L\right] &= if_{ijk}F_k^L \\
\left[F_i^R, F_j^R\right] &= if_{ijk}F_k^R \\
\left[F_i^L, F_j^R\right] &= 0.
\end{aligned}$$ (11.59)

Symmetry generated by the above group is called the chiral symmetry. If $(R_1, R_2)$ is a multiplet of group $SU(3) \times SU(3)$, then under parity

$$(R_1, R_2) \rightarrow (R_2, R_1)$$ (11.60)

For example $(8, 1) \rightarrow (1, 8), (3, 3^*) \rightarrow (3^*, 3)$. This means that if this symmetry is realized as a classification symmetry, we must have parity doublets. This is not the case in nature. No parity doublets are found. This implies that the chiral symmetry is realized in the Nambu-Goldstone mode that is to say, there are eight bosons which in the chiral limit have zero mass. As we have already seen, pions are the Nambu-Goldstone bosons which in the chiral $SU(2) \times SU(2)$ limit are massless. The eight pseudoscalar mesons are identified with Nambu-Goldstone bosons of chiral $SU(3) \times SU(3)$ group.

The algebra generated by $F_i$ and $F_i^5$ is called the chiral algebra. This algebra has rather rich physical content because generators of the symmetry group can be identified with observables. The matrix elements can be measured in electroweak interactions. This in fact provides evidence for chiral symmetry [see bibliography].

### 11.4.1 *Explicit Breaking of Chiral Symmetry*

As already seen the chiral symmetry is spontaneously broken [cf. Eq. (11.52)]. Another way of expressing it is that

$$\langle 0 \, |\bar{q}q| \, 0 \rangle \neq 0 \Rightarrow F_i^5 \, |0\rangle \neq 0.$$ (11.61)

To see this, we note that in the quark model, we have the following commutation relations:

$$\begin{aligned}
\left[F_i^5, S_j\right] &= id_{ijk}P_k \qquad i = 0, 1, ..., 8 \\
\left[F_i^5, P_j\right] &= -id_{ijk}S_k \\
\left[F_i, S_j\right] &= if_{ijk}S_k \\
\left[F_i, P_j\right] &= if_{ijk}P_k
\end{aligned}$$ (11.62)

where

$$S_i = \bar{q}\frac{\lambda_i}{2}q, \quad P_i = i\bar{q}\frac{\lambda_i}{2}\gamma_5 q$$ (11.63)

are respectively the scalar and pseudoscalar densities. We note from Eqs. (11.62) that

$$\langle 0 | [P_1 + iP_2, F_{1-i2}^5] | 0 \rangle = i2\sqrt{\frac{2}{3}} \langle 0 | S_0 | 0 \rangle + i\frac{2}{\sqrt{3}} \langle 0 | S_8 | 0 \rangle \qquad (11.64)$$

Now we expect that flavor $SU(3)$ is realized in the usual way and is not spontaneously broken [cf. Eq. (11.53)]. This implies that

$$\langle 0 | S_8 | 0 \rangle = \frac{1}{\sqrt{3}} \langle 0 | [F_{4+i5}, S_{4-i5}] | 0 \rangle = 0 \qquad (11.65a)$$

as

$$F_{4\pm i5} | 0 \rangle = 0. \qquad (11.65b)$$

Thus, if

$$\langle S_0 \rangle_0 = \frac{1}{2}\sqrt{\frac{2}{3}} \langle \bar{u}u + \bar{d}d + \bar{s}s \rangle_0 \neq 0 \qquad (11.66)$$

then we have from Eq. (11.64):

$$F_{1-i2}^5 | 0 \rangle \neq 0. \qquad (11.67)$$

the condition for spontaneously broken symmetry [cf. Eq. (11.52)]. Let us write

$$\langle \bar{u}u \rangle_0 = \langle \bar{d}d \rangle_0 = \langle \bar{s}s \rangle_0 = -v(\text{say}). \qquad (11.68)$$

Hence we have the result that $\langle S_0 \rangle_0 \neq 0$ which implies that chiral symmetry is spontaneously broken and $\langle S_8 \rangle_0 = 0$ implying that flavor $SU(3)$ is not spontaneously broken.

We can write the QCD Hamiltonian density [cf. Eq. (7.52)] as

$$\mathcal{H} = \mathcal{H}_0 + \left( m_u \bar{u}u + m_d \bar{d}d + m_s \bar{s}s \right)$$
$$= \mathcal{H}_0 + \sqrt{\frac{2}{3}} \left( 2\bar{m} + m_s \right) S_0 + \frac{2}{\sqrt{3}} \left( \bar{m} + m_s \right) S_8 + \left( m_u - m_d \right) S_3$$
$$\equiv \mathcal{H}_0 + \mathcal{H}' \qquad (11.69)$$

The Hamiltonian density $\mathcal{H}_0$ is chiral invariant. Here $\bar{m} = (1/2)(m_u + m_d)$. Now

$$\frac{dF_i^5}{dt} = -i \left[ F_i^5, H \right] = -i \left[ F_i^5, H' \right], \qquad (11.70)$$

where

$$H(t) = \int d^3x \mathcal{H}(t, \mathbf{x}).$$

The (charge) continuity equation

$$\frac{dF_i^5}{dt} = \int d^3x \left( \frac{\partial A_{i0}(t, \mathbf{x})}{\partial t} + \nabla \cdot \mathbf{A}_i(t, \mathbf{x}) \right)$$

$$= \int d^3x \partial^\mu A_{i\mu} \tag{11.71}$$

then converts Eq. (11.71) into

$$\partial^\lambda A_{i\lambda} = -i \left[ F_i^5, \mathcal{H}' \right]. \tag{11.72}$$

From Eq. (11.72), we have

$$\langle 0 | \left[ F_j^5, \partial^\lambda A_{i\lambda} \right] | 0 \rangle = -i \langle 0 | \left[ F_j^5, \left[ F_i^5, \mathcal{H}' \right] \right] | 0 \rangle. \tag{11.73}$$

Using Eq. (11.52), namely

$$F_j^5 |0\rangle = -i \frac{f_\pi}{2\sqrt{2}} |\pi_j\rangle$$

$$\langle 0 | F_j^5 = \frac{i}{2\sqrt{2}} \langle \pi_j | f_\pi, \tag{11.74}$$

we obtain

$$\frac{i}{2} \frac{f_\pi}{\sqrt{2}} \left[ \langle \pi_j | \partial^\lambda A_{i\lambda} | 0 \rangle + \langle 0 | \partial^\lambda A_{i\lambda} | \pi_j \rangle \right] = -i \langle 0 | \left[ F_j^5, \left[ F_i^5, \mathcal{H}' \right] \right] | 0 \rangle. \tag{11.75}$$

The use of PCAC relation $\partial^\lambda A_{i\lambda} = (f_\pi/\sqrt{2}) m_i^2 \pi_i$, then gives [$m_{ij}^2$ is symmetric in $i$ and $j$].

$$m_{ij}^2 = -\frac{2}{f_\pi^2} \langle 0 | \left[ F_j^5, \left[ F_i^5, \mathcal{H}' \right] \right] | 0 \rangle \tag{11.76}$$

where $\mathcal{H}'$ is given in Eq. (11.69). Substituting it into Eq. (11.76) and using Eqs. (11.68) and (11.62), one obtains

$$m_{\pi^0}^2 = m_{\pi^+}^2 = \frac{2}{f_\pi^2} (m_u + m_d) v$$

$$m_{K^+}^2 = \frac{2}{f_\pi^2} (m_u + m_s) v, \quad m_{K^0}^2 = \frac{2}{f_\pi^2} (m_d + m_s) v$$

$$m_{\pi^0 \eta}^2 = m_{\eta \pi^0}^2 = \frac{2}{\sqrt{3} f_\pi^2} (m_u - m_d) v$$

$$m_\eta^2 = \frac{2}{3 f_\pi^2} (m_u + m_d + 4 m_s) v \tag{11.77}$$

Let $\Delta$ be the electromagnetic contribution due to photon exchange to $m_{\pi^\pm}^2$. Since $\pi^+$, $K^+$ form a U-spin multiplet the electromagnetic contribution to

$m_{K^\pm}^2$ is also $\Delta$ while it is zero for $m_{\pi^0}^2$, $m_{K^0}^2$, $m_\eta^2$, so that adding $\Delta$ in Eq. (11.77) for $\pi^+$, $K^+$, we get

$$\frac{m_d}{m_u} = \frac{m_{K^0}^2 - m_{K^+}^2 + m_{\pi^+}^2}{2m_{\pi^0}^2 - m_{\pi^+}^2 + m_{K^+}^2 - m_{K^0}^2} \approx 1.8$$

$$\frac{m_s}{m_d} = \frac{m_{K^0}^2 + m_{K^+}^2 - m_{\pi^+}^2}{m_{K^0}^2 - m_{K^+}^2 + m_{\pi^+}^2} \approx 20.1 \tag{11.78}$$

Here we have used the explicit breaking of chiral symmetry in calculating the current quark mass ratios in terms of masses of pseudoscalar mesons. When quark masses go to zero pseudoscalar mesons become zero mass Nambu-Goldstone bosons required by spontaneously broken chiral symmetry.

## 11.4.2 An Application of Chiral Symmetry to Non-Leptonic Decays of Hyperons

Consider the matrix elements [where $B_r$ and $B_s$ are members of the same baryon octet]:

$$\langle B_s(p') | [F_i^5, H_W] | B_r(p) \rangle = \langle B_s(p') | F_i^5 H_W - H_W F_i^5 | B_r(p) \rangle \tag{11.79}$$

where $i = 1, 2, 3$. Using Eq. (11.74) and its hermitian conjugate, we can write it as

$$\langle B_s(p') | [F_i^5, H_W] | B_r(p) \rangle$$
$$= \frac{i}{2} \frac{f_\pi}{\sqrt{2}} [\langle B_s(p') \pi_i(0) | H_W | B_r(p) \rangle + \langle B_s(p') | H_W | \pi_i(0) B_r(p) \rangle]$$
$$= i \frac{f_\pi}{\sqrt{2}} \langle B_s(p') \pi_i(0) | H_W | B_r(p) \rangle. \tag{11.80}$$

In other words in the limit $q_\mu = (p - p')_\mu \to 0$ [called the soft pion limit], if the matrix elements $\langle B_s(p') \pi_i(q) | H_W | B_r(p) \rangle$ are non-singular, then Eq. (11.80) gives

$$\lim_{q \to 0} \langle B_s(p') \pi_i(q) | H_W | B_r(p) \rangle = -i \frac{\sqrt{2}}{f_\pi} \langle B_s(p') | [F_i^5, H_W] | B_r(p) \rangle. \tag{11.81}$$

Now $H_W = H_W^{p.c} + H_W^{p.v}$ [cf. Chap. 10] and it can be shown that for s-waves $[H_W^{p.v}]$, the amplitude on the left-hand side of Eq. (11.81) is non-singular [see below] and we have

$$\lim_{q \to 0} \langle B_s(p') \pi_i(q) | H_W^{p.v} | B_r(p) \rangle = -i \frac{\sqrt{2}}{f_\pi} \langle B_s(p') | [F_i^5, H_W^{p.v}] | B_r(p) \rangle. \tag{11.82}$$

For p-waves $[H_W^{p.c}]$, one can apply the result (11.81) to

$$\lim_{q \to 0} [\langle B_s \, (p') \, \pi_i(q) \, |H_W^{p.c}| \, B_r \, (p) \rangle$$

$$- \langle B_s \, (p') \, \pi_i(q) \, |H_W^{p.c}| \, B_r \, (p) \rangle_{\text{Born}}]$$

$$= -i \frac{\sqrt{2}}{f_\pi} \langle B_s \, (p') \, | \, [F_i^5, H_W^{p.c}] \, | \, B_r \, (p) \rangle \qquad (11.83)$$

where the Born terms are shown in Fig. 11.1.

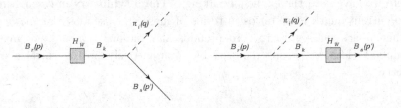

Fig. 11.1   Pole diagram in hyperon decay.

These are singular for $H_W^{p.c}$ in the limit $q_\mu \to 0$ where $m_B = m'_B$ as they behave like $1/|m_B - m'_B|$ but for $H_W^{p.v}$ they behave like $1/|m_B + m'_B|$ and are non-singular. Now as we have seen in Chap. 10 [cf. Eq. (10.28)], the $|\Delta S| = 1$ non-leptonic Hamiltonian is

$$H_W = \frac{G_F}{\sqrt{2}} \sin \theta_c \cos \theta_c \left[ \bar{s} \gamma^\mu (1 - \gamma_5) u \right] \left[ \bar{u} \gamma_\mu (1 - \gamma_5) d \right]. \qquad (11.84)$$

This being the product of two left-handed currents $[F_R = F_i + F_i^5]$ satisfy

$$[F_i^R, H_W] = 0$$

or

$$[F_i^5, H_W] = -[F_i, H_W],$$

i.e.

$$\left[F_i^5, H_W^{p.v,p.c}\right] = -\left[F_i, H_W^{p.c,p.v}\right] \qquad (11.85)$$

Further $F_i$ (being the generator of $SU(3)$ flavor group) acting on $|B_r\rangle$ or $|B_s\rangle$ produces a member of the same octet. To illustrate this point, consider for example, $|B_r\rangle = |\Lambda\rangle$ and $\langle B_s| = \langle p|$, $i = \frac{1+i2}{\sqrt{2}}$. Then

$$F_{1+i2} |\Lambda\rangle = 0 \text{ and } \langle n| = \langle p| F_{1+i2}$$

Thus for s-wave from Eqs. (11.82) and (11.85)

$$\langle p \, (p') \, \pi^- (q) \, |H_w^{p.v}| \, \Lambda^0 \, (p) \rangle = \frac{i}{f_\pi} \langle n \, |H_w^{p.c}| \, \Lambda \rangle \qquad (11.86)$$

Also as shown in Chap. 10, in the exact $SU(3)$ limit $\langle B_s \, |H_W^{p.v}| \, B_r \rangle = 0$. Thus the p-wave non-leptonic decays are given by the Born terms which are also determined by $\langle B_s \, |H_w^{p.c}| \, B_r \rangle$ as far as weak vertices are concerned. These were the results which we employed in Sec. 10.5.3.

## 11.5 Axial Anomaly

As seen in Chap. 7, $\pi^0 \to 2\gamma$ is given by the triangle graph of Fig. 11.2.

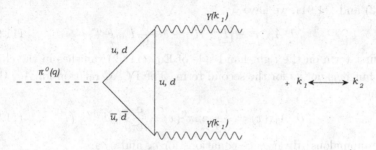

Fig. 11.2  Triangle diagram for $\pi^0 \to 2\gamma$ decay.

In the chiral limit ($m_u = m_d = 0$), this triangle graph gives a finite value for the $\pi^0 \to 2\gamma$ amplitude:

$$M(\pi^0 \to 2\gamma) = \varepsilon^{\mu*}(k_1)\varepsilon^{\nu*}(k_2)\varepsilon_{\mu\nu\alpha\beta}k_1^\alpha k_2^\beta F_{\pi^0\gamma\gamma}(q^2), \qquad (11.87)$$

with

$$F_{\pi^0\gamma\gamma}(0) = N_c[e_u^2 - e_d^2]\left(\frac{-\alpha}{\pi}\right)\frac{g_{\pi qq}}{m_q}, \qquad (11.88)$$

where $N_c$ is the number of colors, e.g. 3, $e_u = 2/3$, $e_d = -1/3$ while the Goldberger-Trieman relation for $\langle q|A_{3\mu}|q\rangle$ with $A_{3\mu} = \frac{1}{2}\left(\bar{u}\gamma_\mu\gamma_5 u - \bar{d}\gamma_\mu d\right)$ gives $(f_\pi/\sqrt{2})g_{\pi qq} = m_q$ so that Eq. (11.87) gives

$$F_{\pi\gamma\gamma}(0) = \frac{-\sqrt{2}\alpha}{\pi f_\pi} \qquad (11.89)$$

It is important to remark that the result (11.89) is unaltered by radiative corrections to the quark triangle and Eq. (11.89) is independent of the masses of fermions in the loop. Equation (11.89) gives

$$\Gamma_{\pi\gamma\gamma} \equiv F_{\pi\gamma\gamma}^2 \frac{m_{\pi^0}^3}{64\pi} = 7.58 \text{ eV} \qquad (11.90)$$

which is remarkably close to experiment with only 2% PCAC correction to the amplitude.

The above result is often stated in terms of contribution to the amplitude due to an axial-vector "anomalous" divergence:

$$\partial^\lambda A_{3\lambda} = \frac{\alpha}{4\pi}F_{\mu\nu}\widetilde{F}^{\mu\nu}, \qquad (11.91)$$

where $F_{\mu\nu} = \partial_\mu a_\nu - \partial_\nu a_\mu$ [$a_\mu$ being electromagnetic potential] and $\widetilde{F}_{\mu\nu} = \frac{1}{2}\varepsilon^{\mu\nu\alpha\beta}F_{\alpha\beta}$. Note that Eq. (11.91) does not arise from equations of motion (11.71). That is why it is called "anomalous" divergence. Combining Eqs. (11.72) and (11.91), we have

$$\partial^\lambda A_{i\lambda} = -\left[F_i^5, \mathcal{H}'\right] + \delta_{i3}\frac{\alpha}{4\pi}F_{\mu\nu}\widetilde{F}^{\mu\nu}. \tag{11.92}$$

The first term on the right-hand side of Eq. (11.92) vanishes in the chiral limit but it is not so for the second term. The PCAC relation for $A_{3\lambda}$ thus becomes

$$\partial^\lambda A_{3\lambda}(x) = \frac{f_\pi}{\sqrt{2}}m_\pi^2\pi^0(x) + \frac{\alpha}{4\pi}F_{\mu\nu}\widetilde{F}^{\mu\nu}. \tag{11.93}$$

The "anomalous" divergence equations for $\eta_8$ and $\eta_0$ are

$$\partial^\lambda A_{k\lambda} = \frac{\alpha}{4\pi}S_P^k F_{\mu\nu}\widetilde{F}^{\mu\nu}, \tag{11.94}$$

where $k = 8$ or $0$ and

$$S_{\eta 8} = N_c\left(\frac{1}{\sqrt{3}}\right)\left[e_u^2 + e_d^2 - 2e_s^2\right] = \frac{1}{\sqrt{3}}$$

$$S_{\eta 0} = N_c\left(\sqrt{\frac{2}{3}}\right)\left[e_u^2 + e_d^2 + e_s^2\right] = 2\frac{2}{\sqrt{3}}. \tag{11.95}$$

Similar considerations show that in $QCD$, the flavor $SU(3)$ singlet current

$$A_{0\mu} = \bar{q}\frac{\lambda_0}{2}\gamma_\mu\gamma_5 q = \frac{1}{2}\sqrt{\frac{2}{3}}\left[\bar{u}\gamma_\mu\gamma_5 u + \bar{d}\gamma_\mu\gamma_5 d + \bar{s}\gamma_\mu\gamma_5 s\right]$$

has "anomalous" divergence

$$\partial^\lambda A_{0\lambda} = \sqrt{\frac{2}{3}}\frac{3\alpha_s}{4\pi}\mathbf{G}_{\mu\nu}\cdot\widetilde{\mathbf{G}}^{\mu\nu} \tag{11.96}$$

where

$$\widetilde{G}^{\mu\nu} = \frac{1}{2}\epsilon^{\mu\nu\alpha\beta}G_{\alpha\beta} \tag{11.97}$$

and $\mathbf{G}_{\mu\nu}$ involving gluon field has been defined in Chap. 7 [cf. Eq. (7.51c)]. Thus

$$\partial^\lambda A_{0\lambda} = \sqrt{\frac{2}{3}}\left[m_u\bar{u}i\gamma_5\bar{u} + m_d\bar{d}i\gamma_5 d + m_s\bar{s}i\gamma_5 s\right]$$

$$+\sqrt{\frac{2}{3}}\frac{3\alpha_s}{4\pi}\mathbf{G}_{\mu\nu}\cdot\widetilde{\mathbf{G}}^{\mu\nu} \tag{11.98}$$

It is clear from Eq. (11.98) that the $SU(3)$ singlet current is not conserved in chiral $SU(3) \times SU(3)$ limit. An application of this will be considered in Chap. 14.

## 11.6 QCD Sum Rules

We have seen in Chap. 7 that the asymptotic freedom property of QCD makes it possible to calculate processes at short distances or for large $q^2$, $q^2$ being the square of the momentum transfer. On the other hand, bound states of quarks and gluons (hadrons or hadron resonances) arise because of large distance confinement effects, i.e. strong coupling effects, which cannot be treated in perturbation theory. The idea of QCD sum rules is to calculate resonance parameters (masses, width) in terms of QCD parameters ($\alpha_s$, quark masses and number of other matrix elements which are introduced to parametrize the non-perturbative effects). We have also seen previously that in the absence of quark masses, the QCD Lagrangian shows a global chiral symmetry, i.e. it is invariant under a global $SU_L(3) \times SU_R(3)$ group. But this chiral symmetry is spontaneously broken, i.e. the ground state is not invariant under this symmetry. This gives rise to [$q = u$, $d$, $s$] [cf. Eq. (11.61)]

$$\langle 0 | \bar{q}q | 0 \rangle \neq 0$$

leading to an octet of zero mass pseudoscalar mesons (so-called Nambu-Goldstone bosons; such bosons acquire masses when QCD Lagrangian is explicitly broken by the quark mass terms). The non-vanishing of the above quark condensate is a non-perturbative effect and gives rise to power corrections to asymptotic freedom effect, which is logarithmic. The essential point of the QCD sum rules, i.e. to relate QCD and non-perturbative parameters of the above type with resonance parameters, is illustrated by the simplest of sum rules, i.e. for a two-point function:

$$\Delta \left( q^2 \right) = \frac{1}{\pi} \int \frac{Im \Delta \left( q^2 \right)}{s - q^2} = \sum_i C_i \left( q^2 \right) \langle 0 | O_i | 0 \rangle \qquad (11.99)$$

The left-hand side is saturated with resonance so that

$$l.h.s. = \sum_i \frac{g_i^2}{m_i^2 - q^2} \qquad (11.100)$$

where ($g_i$, $m_i$) are resonance parameters. The right-hand side is useful only for large $q^2$ in which limit the perturbative QCD allows us to calculate the coefficients $C_i(q^2)$ in the operator product expansion. In practice we want to saturate l.h.s. by a few low lying resonances. Thus we should use some weighting factor to suppress large $s$ contributions on l.h.s. This is done by using Borel transform of the sum rule, which introduces a weighting factor involving a mass parameter $M^2$, which should be sufficiently large to suppress non leading terms on r.h.s. of Eq. (11.100) but not too large in order to suppress contribution from higher hadron states on l.h.s. Thus the problem in practice reduces to finding a region of stability point for $M^2$ so that a small variation in $M^2$ will not affect the physical parameters. In this way from QCD sum rules for two point and three point functions, a large number of constraints on hadron spectrum have been obtained providing not only a consistency check but also a useful phenomenological information on resonance as well as QCD parameters and on $\langle 0|\bar{q}q|0\rangle$. For details see the bibliography.

## 11.7   Problems

(1) Use Dirac Equation, to show that

(a)
$$i\bar{u}(p')\sigma^{\lambda\nu}\,(p'+p)_\nu\,u(p) = \bar{u}(p')\,(p'-p)^\lambda\,u(p)$$

(b)
$$\bar{u}(p')\gamma^\mu u(p) = \bar{u}(p')\left[\frac{(p'+p)^\mu}{2m} + \frac{i}{2m}\sigma^{\mu\nu}\,(p'-p)_\nu\right]u(p).$$

(c) Using Lorentz invariance, show that most of the general matrix elements are of the form:
$$\langle p'\,|\,J_{em}^\lambda\,|\,p\rangle = \frac{1}{(2\pi)^3}\sqrt{\frac{m^2}{p_0 p_0'}}\,\bar{u}(p')\left[F_1\gamma^\lambda + \frac{i}{2m}F_2\sigma^{\lambda\nu}q_\nu + F_3 q^\lambda\right]u(p)$$

(2) Using the conservation of electromagnetic current viz
$$\partial_\lambda J_{em}^\lambda(x) = 0$$
Show that $F_3 = 0$.
Finally $[P = p' + p]$, show that alternatively
$$\langle p'\,|\,J_{em}^\lambda\,|\,p\rangle \sim \bar{u}(\mathbf{p}')\left[\frac{F_1}{2m}P^\lambda + \frac{i}{2m}\,(F_1 + F_2)\,\sigma^{\lambda\nu}q_\nu\right]u(\mathbf{p})$$

$$\sim \bar{u}(\mathbf{p}')\left[(F_1 + F_2)\,\gamma^\lambda - \frac{F_2}{2m}P_\lambda\right]u(\mathbf{p})$$

(3) Determine the commutation relations (11.62).
**Hint:** Use equal-time anticommutation relations of Dirac fields
$$\left[q^i(x), q^{\dagger j}(y)\right]_{x_0=y_0} = \delta^{ij}\delta^3(\mathbf{x}-\mathbf{y})$$
to show that
$$[q^\dagger(x)Oq(x), q^\dagger(y)O'q(y)] = \delta^3(\mathbf{x}-\mathbf{y})q^\dagger(x)\,[O,O']\,q(y)$$
and take operator $O$ and $O'$ approximate to various quantities in (11.62).

## 11.8 References

1. S. L. Adler and R. F. Dashan, Current Algebra and Application to Particle Physics, Benjamin, New York (1968).
2. R. E. Marshak, Riazuddin and C. P. Ryan, Theory of Weak Interaction in Particle Physics, Wiley-Interscience (1969).
3. S. B. Trieman, R. Jackiw and D. J. Gross, Lectures on Current Algebra and its Applications, Princeton University Press, Princeton, New Jersey (1972).
4. V. de Alfaro, S. Fubini, G. Furlan and C. Rossetti, Current in Hadron Physics, North Holland, Amsterdam (1973).
5. M. D. Scadron, Current Algebra, PCAC and the Quark Model, Rep. Prog. Physics, 44, 213 (1981).
6. E. Commins and P. H. Bucksbaum, Weak Interactions of Leptons and Quarks, Cambridge University Press, Cambridge, England (1983).
7. H. Georgi, Weak Interactions and Modern Particle Theory, Benjamin/Cummings, New York (1984).
8. T. D. Lee, Particle Physics and Introduction to Field Theory, Harwood Academic (revised edition 1988).
   For current algebra and chiral symmetry, in addition to the above, see
9. C. H. Llewellyn Smith, Particle Phenomenology: The Standard Model, Proc. of the 1989 Scottish Universities Summer School, Physics of the Early Universe, OUTD-90-160.
10. J. F. Donoghue, Light Quark Masses and Chiral Symmetry, Ann. Rev. Nucl. Part. Sci. 39, 1 (1989).
11. J. F. Donoghue, Chiral Symmetry as an Experimental Science, CERN-TH. 5667/90, Lectures presented at International School of Low-Energy Antiproton, Erice, Jan. 1990.
    **For QCD Sum Rules:**
12. M. A. Shifman, A. I. Vainshtein and V. I. Zakharov, Nucl. Phys. B 147, 385 and 448 (1979).
13. L. J. Reinders, QCD Sum Rules, An Introduction and Some Applications, CERN-TH-3701 (1983): Lectures presented at the 23rd Cracrow School of Theoretical Physics, Zakopane (1983).
14. S. Narison, QCD Spectral Sum Rules, World Scientific Lecture Notes in Physics-Vol. 26, World Scientific, Singapore (1990).
15. S. Narison, SVZ Sum Rules: 30+1 Years later, arXiv:1010.1959 (2010).
16. M. Shifman, arXiv:1101.1122 (2011)

# Chapter 12

# Neutrino

## 12.1 Introduction

Experimental puzzles in the past have led to some important discoveries in Physics. Neutrino, which has spin 1/2, was invented in 1930 by Pauli as the explanation of such a puzzle, namely the conservation of angular momentum and energy in $\beta$-decay

$$n \rightarrow p + e^-,$$

require such a particle, so that

$$n \rightarrow p + e^- + \bar{\nu}_e.$$

Its direct observation was made much later. The electron type antineutrinos are thus produced by the decay of pile neutrons in a fission reactor. These can be captured in hydrogen giving the reaction:

$$\bar{\nu}_e + p \rightarrow e^+ + n,$$

whose cross-section was measured by Reines and Cowan

$$\sigma_{exp} = (11 \pm 2.5) \times 10^{-44} \text{ cm}^2$$

to be compared with the theoretical value

$$\sigma_{th} = (11 \pm 1.6) \times 10^{-44} \text{ cm}^2.$$

Note the extreme smallness of the cross-section. It is a reflection of the fact that neutrino has only weak interaction.

## 12.2   Intrinsic Properties of Neutrinos

(a) Neutrinos are elementary particles with spin $\frac{1}{2}$, electrically neutral and obey Fermi Dirac statistic.

(b) All neutrinos detected are left-handed (see Sec. 10.1.1), i.e. their spin points in the opposite direction from their momenta. All anti-neutrinos are right-handed.

(c) Neutrinos occur in three flavors $\nu_e$, $\nu_\mu$, $\nu_\tau$ associated with the corresponding charged leptons $e^-$, $\mu^-$, $\tau^-$ respectively. Neutrinos "oscillate" from one specie to another with a high probability (see Sec. 12.4). This means that neutrinos produced in a well-defined weak eigenstate $\nu_\alpha$ can be detected in a distinct weak eigenstate $\nu_\beta$ later. This phenomenon of neutrino oscillation is possible if one or more neutrinos have non-vanishing mass.

(d) The discovery of neutrino mass raises the question whether each mass eigenstate $\nu_i$ is identical to its antiparticle $\bar{\nu}_i$ or is distinct from it. If $\bar{\nu}_i = \nu_i$, we call the neutrinos Majorana particles, while if $\bar{\nu}_i \neq \nu_i$ they are called Dirac particles. Usually conserved charges distinguish a particle from its antiparticle. For example electric charge distinguishes electron from positron. But neutrinos carry no electric charge and might not carry any other conserved charge like quantum number. It might be thought that there is a conserved lepton charge (see Sec. 4.1) that distinguish $\nu_e$, $e^-$ from $\bar{\nu}_e$, $e^+$. However conservation of lepton charge is not protected by any gauge symmetry. If such a conserved charge does not exist, then there is nothing to distinguish $\bar{\nu}_i$ and $\nu_i$. Then neutrino may very well be a Majoran particle.

## 12.3   Mass

The question of neutrino mass is one
indexNeutrino Mass of long standing. In the context of the standard model of unified electroweak interactions (Chap. 13), there is no understanding of the origin of masses of elementary fermions. In this category the question of neutrino mass also arises. It has an added importance for the following reasons:

(a) The fact that neutrinos occur asymmetrically in one (left handed) helicity state with lepton number conservation imply that $m_\nu = 0$ (see below). However, there is no local gauge symmetry and no massless gauge boson coupled to the lepton number $L$ to guarantee the masslessness of neutrino and lepton number conservation in contrast to the photon where both the masslessness of photon and charge conservation are consequences of local gauge invariance of Maxwell's equations. One may thus expect a finite mass for neutrino. But the intriguing question is why $m(\nu_e) \ll m(e)$.

(b) Non-vanishing neutrino mass has important implications in Astrophysics. It is a candidate for hot dark matter. It affects history, structure and fate of the universe as we shall seen in Chap. 18.

Experimentally the question of neutrino mass is still open. This is because

(i) $m_\nu$ is small and a smaller quantity is more difficult to measure with high precision than a bigger quantity.

(ii) Neutrino has only weak interaction with matter which implies in practice that no direct measurement of $m_\nu$ is possible.

### 12.3.1  Constraints on Neutrino Mass

#### 12.3.1.1  Direct Limits

We first confine to electron anti-neutrino ($\bar{\nu}_e$). $\bar{\nu}_e$ comes out in $\beta$-decay of Tritium

$$^3H \rightarrow\ ^3He + e^- + \bar{\nu}_e.$$

Electrons from this decay has a very low end-point energy (18.6 keV). As such this process is ideal to look for a possible finite mass of neutrino. If $m_\nu = 0$,

$$\left[\frac{d\Gamma}{p_e^2 dp_e}\right]^{1/2} \propto (E_{max} - E_e).$$

Kurie plot is thus a straight line. If $m_\nu \neq 0$,

$$\left(\frac{d\Gamma}{p_e^2 dp_e}\right)^{1/2} \propto (E_{max} - E_e)^{1/2} \left[(E_{max} - E_e)^2 - m_\nu^2\right]^{1/4}$$

This equation also illustrates why the end point energy range is important for determining $m_\nu = 0$. This gives a distortion at the extreme end of the Kurie plot (see Fig. 2.6). Thus one has to look for such a distortion, but note that the deviation is in fact quite small and the experiment is thus quite difficult. An added complication is the presence of final state ionic

and/or molecular effects that are not well understood. Anyway, the present
limit on the so called effective mass of electron neutrino [for the definition
of $U_{ei}$ see Sec. 12.4] placed

$$m_{\nu_e} = \sqrt{\sum_i |U_{ei}|^2 m_i} < 2 \text{ eV} : m_{\nu_e} \ll m_e$$

The Katrin experiment is expected to improve the limit to $\sim 0.2$ eV. Direct
limits on the other two types of neutrinos are

$$\pi \to \mu\nu_\mu : m_{\nu_\mu} < 190 \text{ keV} : m_{\nu_\mu} \ll m_\mu$$
$$\tau \to 5\pi\nu_\tau : m_{\nu_\tau} < 18.2 \text{ MeV} : m_{\nu_\tau} \ll m_\tau$$

### 12.3.1.2  *Double β-Decay*

The double $\beta$-decay is another way to look for a finite mass of neutrino.
Two kinds of double $\beta$-decay can be considered:

$$(2\nu) \quad (A,\, Z) \to (A,\, Z+2) + 2e^- + 2\bar{\nu}_e \tag{12.1}$$
$$(0\nu) \qquad\qquad \to (A,\, Z+2) + 2e^-.$$

Usually the neutrinos are assumed to be Dirac particles, that is, neutrino
$\nu$ and its anti-neutrino $\bar{\nu}$ are distinct. There is another picture of neutrinos,
called Majorana in which $\nu$ and $\bar{\nu}$ are identical. This implies

$$n \to p + e^- + \bar{\nu}_L \equiv p + e^- + \nu_L$$
$$\nu_L + n \to p + e^-, \tag{12.2}$$

so that $(2n) \to (2p) + 2e^-$ as shown in Fig. 12.1.

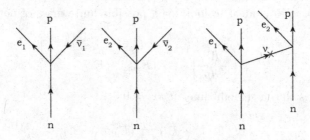

Fig. 12.1   Basic reactions in double $\beta$-decay.

The important physics issues in $(0\nu)$ double $\beta$-decay are:

(i) Lepton number must not be conserved, which is possible if neutrinos
are Majorana particles: $\nu \equiv \bar{\nu}$.

(ii) Helicity of the neutrino cannot be exactly $-1$, this can be satisfied if $m_\nu \neq 0$.

Thus $(0\nu)\beta\beta$ decay is especially interesting in determining $m_\nu$ as half life

$$T_{1/2} \propto Q^{-5} < m_\nu >^{-2}, \qquad (12.3)$$

where $Q$ is the $Q$-value of the reaction involved.

There is now distinct evidence of $(2\nu)\,\beta$-decay:

$$^{82}Se \to\, ^{82}Kr$$
$$T_{1/2} = \left(1.1^{+0.8}_{-0.3}\right) \times 10^{20} \text{ Yrs.} \qquad (12.4)$$

Incidently this is the rarest natural decay process ever observed directly in a laboratory. This would help to provide a standard by which to test the double $\beta$-decay matrix elements of nuclear theory. From the limit on half-life on $(0\nu)\beta\beta$ decay process $^{76}Ge \to\, ^{76}Se + 2e^-$,

$$T_{1/2} > 1.9 \times 10^{25} \text{ Yrs,}$$

the Heidelberg-Moscow experiment gives the best present bound on $m_{\nu_e}$:

$$m_{\nu_e} < (0.35 - 0.5) \text{ eV}$$

Part of the collaboration claims an evidence for the positive signal which would correspond to $m_{\nu_e} \sim 0.4$ eV. Actually, if there is a mixing among neutrinos (see Sec. 12.4 below), then $m_{\nu_e} = \sum_i \lambda_i |U_{ei}|^2 m_{\nu_i}$, where $\lambda_i$ is a possible sign since Majorana neutrinos are $CP$ eigenstates and $U_{ei}$ arises due to two vertices.

### 12.3.1.3  Cosmology

Non-zero neutrino mass alters large scale structure formation within the standard cosmology. The cosmological observations put a bound on the sum of neutrino mass

$$\sum_{i=1}^{3} m_i \leq 0.17 - 1.2 \text{ eV}$$

### 12.3.1.4   *Astrophysical Constraints*

As will be shown in Chap. 18, the mass density of all fairly light ($m_\nu < 1$ MeV) stable neutrinos is

$$\rho_\nu^0 = \sum_i \left( \frac{n_\nu^0}{n_\gamma^0} \right) n_\gamma^0 m_{\nu_i}$$

$$= \frac{3}{11} n_\gamma^0 \sum_i m_{\nu_i}$$

$$= \sum_i 2 m_{\nu_i}(eV) \times 10^{-31}\,\mathrm{gm/cm^3}, \tag{12.5}$$

where $n_\gamma^0 = 400$ cm$^{-3}$ is the present photon number density. Now the average mass density of the universe is

$$\rho_0 = \Omega_0 \rho_{c0},$$

where $\Omega_0 \leq 1$ and $\rho_{c0}$ is the critical density

$$\rho_{c0} = \frac{3 H_0^2}{8 \pi G_N}. \tag{12.6}$$

Here $H_0$ is the Hubble parameter, $H_0 = 3 \times 10^{-18} h_o$ sec$^{-1}$ [100 h km g$^{-1}$ Mpc$^{-1}$], with $h_o = 0.71 \pm 0.01$ and $G_N$ is Newton's gravitational constant. Thus

$$\rho_{c0} = 1.88 \times 10^{-29} h_0^2\,\mathrm{gm/cm^3}$$

$$= 1.05 \times 10^{-5} h_0^2\,\mathrm{GeV/cm^3} \tag{12.7}$$

This together with Eq. (12.5) gives

$$\Omega_\nu h_0^2 = \frac{\rho_\nu^0}{\rho_{c0}} h_0^2 = 1.06 \times 10^{-2} \sum_i m_{\nu_i}\ \mathrm{eV}$$

The sum of the masses of all stable neutrinos is constrained by the WMAP data (see Chap. 18) which gives

$$\Omega_\nu h_0^2 \leq 0.0076$$

Hence

$$\sum_i m_{\nu_i} \leq 0.72\ \mathrm{eV}$$

We may also mention here that the big-bang nucleosynthesis puts constraints on the mass of any metastable (m.s) neutrinos which are

$$m_{\nu_{m.s}}\ (\text{Dirac}) > 32\ \text{MeV or} < 0.95\ \text{MeV}$$

$$m_{\nu_{m.s}}\ (\text{Majorana}) > 25\ \text{MeV or} < 0.37\ \text{MeV}$$

Both the lower limits are in conflict with $m_{\nu_\tau} < 18.2$ MeV, mentioned earlier, implying that $m_{\nu_\tau}$ must actually be below 1 MeV.

### 12.3.2    Dirac and Majorana Masses

It is a general feature of weak interactions that only left handed neutrino $\nu_L$ takes part in it (see Chap. 10). Let us write a Dirac spinor $\psi$ as

$$\psi = \begin{pmatrix} \xi \\ \eta \end{pmatrix}. \tag{12.8}$$

In a representation in which $\gamma_5$ is diagonal,

$$\psi_L = \frac{1-\gamma_5}{2}\psi = \begin{pmatrix} \xi \\ 0 \end{pmatrix}, \; \psi_R = \frac{1+\gamma_5}{2}\psi = \begin{pmatrix} 0 \\ \eta \end{pmatrix}, \tag{12.9}$$

the Dirac equation for the two component spinors $\xi$ and $\eta$ can be written as

$$\left( i\sigma \cdot \nabla - i\frac{\partial}{\partial t} \right)\xi = -m_D\eta$$

$$\left( -i\sigma \cdot \nabla - i\frac{\partial}{\partial t} \right)\eta = -m_D\xi \tag{12.10a}$$

It is the mass which links $\xi$ (or equivalently $\psi_L$) with $\eta$ (or $\psi_R$).

These equations can also be written in the form

$$i\bar{\sigma}^\mu \partial_\mu \xi - m_D\eta = 0$$

$$i\sigma_\mu \partial_\mu \eta - m_D\xi = 0 \tag{12.10b}$$

where

$$\sigma^\mu = (1,\sigma), \; \bar{\sigma}^\mu = (1,-\sigma) \tag{12.10c}$$

Under charge conjugation $C$ (particle $\rightarrow$ antiparticle) $\psi \rightarrow \psi^c = -i\gamma^2\psi^*$ [see Appendix A], so that

$$\xi \rightarrow \xi^c = -i\sigma^2\eta^*$$

$$\eta \rightarrow \eta^c = i\sigma^2\xi^* \tag{12.11}$$

For massless neutrino, $\xi$ and $\eta$ decouple and we have from Eq. (12.10a)

$$\left( i\sigma \cdot \nabla - i\frac{\partial}{\partial t} \right)\xi = 0 \tag{12.12a}$$

$$\left( -i\sigma \cdot \nabla - i\frac{\partial}{\partial t} \right)\eta = 0 \tag{12.12b}$$

These are called the Weyl equations for massless spin $\frac{1}{2}$ particles. It is easy to see that Eqs. (12.12a) and (12.12b) are not disconnected. In fact, using the usual representation of Pauli matrices $\sigma^i$, one verifies that, if $\xi(x)$ is a solution of Eq. (12.12a), $\sigma_2\xi^*(x)$ is a solution of Eq. (12.12b). In order to see physical implications of Weyl Eq. (12.12a), we examine the plane wave solution, given by

$$\xi(x) = w(\mathbf{p})e^{-ip \cdot x} = w(\mathbf{p})e^{i(\mathbf{p} \cdot \mathbf{x} - Et)} \tag{12.13}$$

Then from Eq. (12.12a), we get

$$[\sigma \cdot \mathbf{p} + E]\, w(\mathbf{p}) = 0 \tag{12.14}$$

with $E^2 = \mathbf{p}^2$. Let us denote the positive energy spinor by $u(\mathbf{p})$ and negative energy $(E = -|\mathbf{p}|)$ spinor by $v(\mathbf{p})$. Thus we get

$$\frac{\sigma \cdot \mathbf{P}}{|\mathbf{p}|} u(\mathbf{p}) = -u(\mathbf{p}) \tag{12.15}$$

$$\frac{\sigma \cdot \mathbf{P}}{|\mathbf{p}|} v(\mathbf{p}) = v(\mathbf{p}) \tag{12.16}$$

Hence we get the important result: if neutrino is massless, then 2-component Weyl field $\xi(x)$ satisfying Eq. (12.12a) or equivalently, the chiral projection $\frac{1}{2}[(1 - \gamma_5)\psi]$ of a four component field $\psi(x)$ satisfying massless Dirac equation, give a left-handed (helicity negative) neutrino and a right-handed antineutrino. This is what is realized in nature. If we start with $\eta$-field, then we have opposite case: a right-handed neutrino and left-handed antineutrino. This case is not realized in nature. It is important to remark [see Ref. [2]] that charge conjugation operators given in Eq. (12.11) are not possible in the 2-component case.

If we allow both a finite mass and lepton number non-conservation, then for an electrically neutral lepton, the Lagrangian is

$$\mathcal{L} = \bar{\Psi}\left(i\gamma^{\mu}\partial_{\mu} - m_D\right)\Psi + \frac{m_M}{2}\left(\Psi^T C^{-1}\Psi - \bar{\Psi}C\bar{\Psi}^T\right). \tag{12.17}$$

The second term in Eq. (12.17) is the Majorana mass term and violates lepton number conservation: $\Delta L = 2$. Let us define the new fields $\phi_1$ and $\phi_2$:

$$\phi_1 = \frac{1}{\sqrt{2}}\left(\xi - i\sigma^2\eta^*\right)$$

$$\phi_2 = -\frac{i}{\sqrt{2}}\left(\xi + i\sigma^2\eta^*\right). \tag{12.18}$$

It then follows from Eq. (12.11) that under charge conjugation

$$\phi_{1,2} \overset{C}{\to} \pm\phi_{1,2}, \tag{12.19}$$

i.e. $\phi_{1,2}$ are eigenstates of $C$ with eigenvalues $+1$ and $-1$ respectively. In terms of $\phi_1$ and $\phi_2$, Eq. (12.11) becomes

$$\mathcal{L} = \left[i\phi_1^{\dagger}\bar{\sigma}^{\mu}\partial_{\mu}\phi_1 + \left(\frac{m_D + m_M}{2}\phi_1^T(-i\sigma^2)\phi_1 + h.c.\right)\right.$$
$$\left. + i\phi_2^{\dagger}\bar{\sigma}^{\mu}\partial_{\mu}\phi_2 + \left(\frac{m_D - m_M}{2}\phi_2^T(-i\sigma^2)\phi_2 + h.c.\right)\right]. \tag{12.20}$$

If we start with $\xi$ and $\eta$ or equivalently $\nu_L$ and $\nu_R$, then we can have two Majorana particles of masses $(m_D \pm m_M)/2$. If we start with $\nu_L$ only, $m_D = 0$, we have a Majorana neutrino of mass $m_M$. In this case Eq. (12.20) reduces to

$$\mathcal{L} = i\nu_L^\dagger \bar{\sigma}^\mu \partial_\mu \nu_L + \frac{m_M}{2} \left( \nu_L^T (-i\sigma^2) \nu_L + h.c. \right) \qquad (12.21)$$

We get an important result: a two-component neutrino ($\nu_L$) cannot have a Dirac mass; it can only have Majorana mass, which violates lepton number conservation. Thus one helicity state ($-1$ for neutrino) together with lepton number conservation implies that $m_\nu = 0$. It may be mentioned that if neutrino is massless, there is no distinction between Majorana and Weyl neutrino.

### 12.3.3 *Fermion Masses in the Standard Model (SM) and See-saw Mechanism*

The fermion masses in the standard model are generated through a Yukawa coupling of fermions with a Higgs scalar (see Chap. 13):

$$\mathcal{L} = -g_f \bar{f}_L \phi f_R + h.c. \qquad (12.22a)$$

where $\phi$ develops a vacuum expectation value as shown in Fig. 12.2.

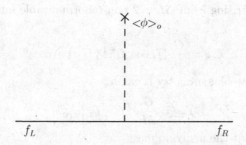

Fig. 12.2   Fermion mass generation.

Here $f_L$ and $\phi$ are doublets while $f_R$ is a singlet under $SU_L(2)$ of the standard model group $SU(2) \otimes U(1)$, e.g.

$$f_L = \begin{pmatrix} \nu_e \\ e^- \end{pmatrix}_L , \quad f_R = e_R. \qquad (12.22b)$$

The SM conserves $L$, nor does it contain any chirally right-handed neutral fields but only the left-handed ones $\nu_L$. If, however, one allows right-handed neutrinos $N_R$ which are $SU(2) \otimes U(1)$ singlets, then one can write the Yukawa interaction

$$\mathcal{L}_{\text{new}} = \mathcal{L}_{\text{SM}} - g_f \bar{f}_L \tilde{\phi} N_R + h.c. \qquad (12.22c)$$

The above mechanism gives $[f_R = e_R$ or $N_R]$

$$\mathcal{L}_{\text{mass}}^{\text{Dirac}} = -g_f \langle \phi \rangle_0 \bar{f}_L f_R + h.c.,$$

leading to the Dirac mass

$$m_f = g_f \langle \phi \rangle_0$$

Thus $m_D(\nu_\ell) = g_\ell \langle \phi \rangle_0$ where $\langle \phi \rangle_o = 175$ GeV and one thus expects $m_D(\nu_\ell) \sim m_D(\ell)$, say within a factor of 10 or so. This does not explain $m(\nu_l) \ll m(l)$. This would require

$$g_\nu \lesssim 10^{-13} - 10^{-12}$$

which looks unnatural if $N_R$ is the same type of field as right-handed components of other fermions. For the neutrino Eq. (12.22a) gives

$$\mathcal{L}_{\text{mass}}^{D} = -m_D \left[ \bar{\nu}_L N_R + \bar{N}_R \nu_L \right] \qquad (12.23)$$

Note that Dirac mass does not mix neutrinos and antineutrinos and as such conserve lepton number. The most economical way to add neutrino mass term to the SM involving only light neutrinos is that the neutrinos have Majorana mass arising from $\Delta L = 2$ non-renormalizable interaction of the form:

$$\mathcal{L}_{\text{eff}} = \frac{G}{M} (f_L \phi)^T C^{-1} (f_L \phi) + h.c. \qquad (12.24a)$$

After the electroweak symmetry breaking

$$\mathcal{L}_{\text{eff}} = \frac{G}{M} \nu_L^T C^{-1} \nu_L \langle \phi \rangle_0^2 \qquad (12.24b)$$

giving nothing but the neutrino mass

$$m_\nu = \frac{G}{M} \langle \phi \rangle_0^2 \qquad (12.24c)$$

and

$$\mathcal{L}_{\text{mass}}^{\text{Majorana}}(\nu) = m_L \nu_L^T C^{-1} \nu_L + h.c. \qquad (12.25)$$

We may remark here that with $G \approx 1$, $\langle \phi \rangle_0 \simeq 175 \,\text{GeV}$, as in the standard model, Eq. (12.24c) gives $m_\nu \approx 10^{-5}$ eV for $M \approx 10^{19}$ GeV (Planck mass

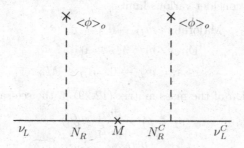

Fig. 12.3   Majorana mass generation.

scale) while for $m_\nu \simeq 0.045$ eV, $M \sim 10^{15}$ GeV not much different from the Grand Unification (GUT) mass scale.

The above mass generation can be pictured as a two-step process shown in Fig. 12.3. This process also gives a Majorana mass to $N_R$

$$\mathcal{L}_{\text{mass}}^{\text{Majorana}}(N) = M_R \left( N_R^T C^{-1} N_R - \overline{N}_R C \overline{N}_R^T \right)$$

$$= M_R \left( N_R^C \right)^T C^{-1} N_R^C + h.c. \tag{12.26}$$

Such a mass term is allowed by the SM, since $N_R$ being the electroweak isospin singlet, all the SM principles including electroweak isospin are preserved. In Eq. (12.26), the second step is obtained by using the charge conjugate of $N_R$

$$\overline{N}_R^C = -N_R^T C^{-1}, \quad N_R^C = C \overline{N}_R^T \tag{12.27}$$

so that $\overline{N}_R C \overline{N}_R^T = -\left( N_R^C \right)^T C^{-1} N_R^C$. Further

$$\overline{N}_R \nu_L = \left( N_R^C \right)^T C^{-1} \nu_L$$

$$= \frac{1}{2} \left[ \left( N_R^C \right)^T C^{-1} \nu_L + \nu_L^T C^{-1} N_R^C \right]$$

Thus we rewrite Eq. (12.24a) in more convenient form

$$\mathcal{L}_{\text{mass}}^{\text{Dirac}}(N) = \frac{1}{2} m_D \left[ \left( \left( N_R^C \right)^T C^{-1} \nu_L + \nu_L^T C^{-1} N_R^C \right) + h.c. \right] \tag{12.28}$$

Referring to Eqs. (12.25), (12.26) and (12.28) the mass matrix in 2-component basis $\nu_L N_R^C$ needs diagonalization. Denoting by prime fields before diagonalization, we have in 2-component basis

$$\mathcal{L}_M = \left( \nu_L'^T \left( N_R'^C \right)^T \right) C^{-1} \begin{pmatrix} m_L & m_{D/2} \\ m_{D/2} & M_R \end{pmatrix} \begin{pmatrix} \nu_L' \\ N_R'^C \end{pmatrix} \tag{12.29}$$

It is useful to consider various limits:

$$\text{Majorana}: m_D \to 0$$
$$\text{Dirac}: m_L, M_R \to 0$$
$$\text{See-saw}: m_L \to 0, \ m_D \ll M_R$$

The diagonalization of the mass matrix (12.29) in the see-saw limit gives

$$\nu_L = \nu'_L - \frac{m_D}{2M_R} N'^C_R$$
$$N^C_R = \frac{m_D}{2M_R} \nu'_L + N'^C_R \tag{12.30}$$

By introducing the charge conjugate left handed component $N_L \equiv (N_R)^C$ we can equivalently write the relation (12.29) in $\nu_L$, $N_L$ basis giving

$$\nu_L = \nu'_L - \frac{m_D}{2M_R} N'_L$$
$$N_L = \frac{m_D}{2M_R} \nu'_L + N'_L$$

Hence we have two Majorana neutrinos $\nu_L$ and $N_R$ with masses

$$m_\nu \simeq m_D^2 / 4M_R \ll m_D$$
$$m_N \simeq M_R \tag{12.31}$$

Depending upon $M_R$, $\nu_L$ could be extremely light and $N_R$ correspondingly heavy. To summarize, in the Dirac case, one must answer the question why

$$(m_{\nu_\ell})_{Dirac} \ll m_\ell \tag{12.32}$$

while in the Majorana case the see-saw mechanism sidesteps this question; here one has

$$(m_{\nu_\ell})_{Majorana} \simeq \frac{m_\ell^2}{4M} \ll m_\ell \tag{12.33}$$

by requiring the existence of a large scale $M$, associated with some new physics. Below we give some typical scales indicative of new physics and the corresponding neutrino masses, which may be relevant for neutrino oscillations (to be discussed below), dark matter and leptogenesis [Chap. 18]:

| $M$ (GeV) | $m_{\nu_e}$ (eV) | $m_{\nu_\mu}$ (eV) | $m_{\nu_\tau}$ (eV) |
|---|---|---|---|
| $M_{Plank}(10^{19})$ | $10^{-14}$ | $4 \times 10^{-10}$ | $10^{-5}$ |
| $M_{GUT}(10^{16})$ | $10^{-11}$ | $4 \times 10^{-7}$ | $10^{-3}$ |
| $M_R(10^{12})$ | $10^{-7}$ | $4 \times 10^{-3}$ | $1$ |
| $10^6$ | $10^{-1}$ | $4 \times 10^3$ | $10^6$ |

## 12.4   Neutrino Oscillations

If neutrinos are massless, then the neutrinos $\nu_e$, $\nu_\mu$, $\nu_\tau$, which enter the weak interaction Lagrangian are also the mass eigenstates. If anyone of them have a mass, then it may be that the mass eigenstates which we denote by $\nu_i (i = 1, 2, 3)$ are different from flavor eigenstates $\nu_w$, $(w = e, \mu, \tau)$. In this case, we can get neutrino oscillations. The phenomenon of neutrino oscillations can provide a mechanism to measure extremely small neutrino masses. We note that two sets of states $|\nu_w\rangle$ and $|\nu_i\rangle$ are connected with each other by a unitary transformation:

$$|\nu_w\rangle = \sum_i U_{wi}|\nu_i\rangle \tag{12.34}$$

$$\sum_w U_{iw} U_{jw}^* = \delta_{ij}. \tag{12.35}$$

Now

$$H(k)|\nu_i\rangle = E_i|\nu_i\rangle \tag{12.36}$$

$$E_i = (k^2 + m_i^2)^{1/2} \approx k + \frac{m_i^2}{2k}, \tag{12.37}$$

since $k \gg m_i$ and we take the extreme relativistic limit. Now at time $t$, $|\nu(t)\rangle$ satisfies the Schrödinger equation:

$$i\frac{d}{dt}|\nu(t)\rangle = H|\nu(t)\rangle.$$

In $\nu_i$ basis, $H$ is a diagonal matrix with eigenvalues $E_1$, $E_2$ and $E_3$. Thus

$$|\nu_i(t)\rangle = e^{-iE_i t}|\nu_i(0)\rangle \equiv e^{-iE_i t}|\nu_i\rangle.$$

Hence from Eq. (12.34), we can write

$$|\nu_w(t)\rangle = \sum_i U_{wi} e^{-iE_i t}|\nu_i\rangle$$

and

$$\langle \nu_{w'}|\nu_w\rangle_t = \sum_i U_{wi} e^{-iE_i t}\langle \nu_{w'}|\nu_i\rangle$$

$$= \sum_i U_{wi} e^{-iE_i t} U_{w'i}^*. \tag{12.38}$$

Thus the probability that at time $t$, the neutrino of type $w$ is converted to the neutrino of type $w'$ is given by

$$P_{w'w} = |\langle \nu_{w'}|\nu_w\rangle_t|^2$$

$$= \sum_i \sum_j (U_{wi} U_{w'i}^*)(U_{wj} U_{w'j}^*)\cos(E_i - E_j)t.$$

$$\tag{12.39a}$$

Neglecting CP-violating phases so that $U$ is real, it is convenient to rewrite it as

$$P_{ww'} = \delta_{w'w} - 4 \sum_{j>i} U_{wi} U_{w'j} U_{wj} U_{w'j} \sin^2(\pi L/\lambda_{ij}) \qquad (12.39b)$$

where $L$ is the distance traveled after which $\nu_w$ is converted to $\nu_{w'}$ and

$$\lambda_{ij} = \frac{4\pi E_\nu}{\Delta_{ij}} = 2.47\, m \left(\frac{E_\nu}{MeV}\right) \frac{eV^2}{\Delta_{ij}} \qquad (12.39c)$$

where we have used the relation $L = ct$,

$$\lambda_{ij} = \frac{2\pi c}{E_i - E_j},$$

$$(E_i - E_j)t = \frac{L(m_i^2 - m_j^2)}{2E_\nu} \qquad (12.39d)$$

$$= \frac{L\Delta_{ij}}{2E_\nu}.$$

The matrix $U$ with matrix elements $U_{wi}$, commonly called PMNS (Pontecorvo-Maki-Nakagawa-Sakata), mixing matrix is given by $U = U_{23}U_{13}U_{12}$

$$U = \begin{pmatrix} 1 & 0 & 0 \\ 0 & c_{23} & s_{23} \\ 0 & -s_{23} & c_{23} \end{pmatrix} \begin{pmatrix} c_{13} & 0 & s_{13}e^{-i\delta} \\ 0 & 1 & 0 \\ -s_{13}e^{i\delta} & 0 & c_{13} \end{pmatrix} \begin{pmatrix} c_{12} & s_{12} & 0 \\ -s_{12} & c_{12} & 0 \\ 0 & 0 & 1 \end{pmatrix} \qquad (12.40)$$

where $\delta$ is known as CP-violating Dirac phase. If neutrinos are Majoran particles, the mixing matrix $U$ is multiplied by $I_\phi$, where $I_\phi = diag\left(1, e^{i\phi_1}, e^{i\phi_2}\right)$ is the diagonal matrix of the Majorana $CP$-violating phases. They can be absorbed with the mass eigenvalues which can then be considered as complex parameter. As a consequence of $CPT$ and $CP$ invariance

$$P_{\nu_{w'}\nu_w} = P_{\bar{\nu}_{w'}\bar{\nu}_w} = P_{\nu_w\nu_{w'}} = P_{\bar{\nu}_w\bar{\nu}_{w'}}$$

The form of transition probability (12.39b) depends on the spectrum of $\Delta m^2$ or $\Delta_{ij}$ chosen and the explicit form of $U$. If $\Delta m^2$ is chosen such that $\lambda \gg L$, then the oscillator term $\sin^2 \frac{\pi L}{\lambda} \to 0$. On the other hand, if $\lambda \ll L$, one has a large number of oscillations and $\sin^2 \frac{\pi L}{\lambda}$ averages out to $\frac{1}{2}$.

For the conversion of $\nu_e$ to $\nu_x$ ($x = \mu$ or $\tau$),

$$U = \begin{pmatrix} \cos\theta & \sin\theta \\ -\sin\theta & \cos\theta \end{pmatrix} \qquad (12.41)$$

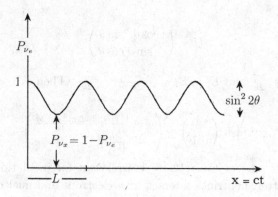

Fig. 12.4   The neutrino oscillations.

and

$$P_{\nu_e \to \nu_x} = \sin^2 2\theta \sin^2 \left[ 1.27 \frac{\Delta m^2}{E_\nu} L \right] \tag{12.42}$$

while the survival probability is $P_{\nu_e \to \nu_e} = 1 - P_{\nu_e \to \nu_x}$. Here $\theta$ is the vacuum mixing angle. $P_{\nu_e \to \nu_e}$ and $P_{\nu_e \to \nu_x}$ oscillate with $L$ as shown in Fig. 12.4.

The amplitude of the oscillations is determined by the mixing angle; the wavelength of the oscillations is $\lambda$. Incidently the above result illustrates the quantum mechanical phenomena of interferometry which provides a sensitive method to probe extremely small effects.

To look for oscillations, one needs factors, which enhance tiny effects: a coherent source (there are many, the sun, cosmic rays, reactors etc.), low energy neutrinos, large base line (size of the sun and that of the earth), large mixing angles and large flux.

## 12.4.1   *Mikheyev-Smirnov-Wolfenstein Effect*

First we write the Hamiltonian in $\nu_e$, $\nu_x$ basis [$x = \mu$ or $\tau$ or $s$ (sterile $\nu$)]:

$$H_\nu(k) = U H U^{-1} \tag{12.43}$$

where $H$ is diagonal in $\nu_1 - \nu_2$ basis:

$$H = \begin{pmatrix} E_1 & 0 \\ 0 & E_2 \end{pmatrix} = k \begin{pmatrix} 1 & 0 \\ 0 & 1 \end{pmatrix} + \frac{1}{4k} \begin{pmatrix} -\Delta m^2 & 0 \\ 0 & \Delta m^2 \end{pmatrix} \tag{12.44}$$

and

$$U = \begin{pmatrix} \cos\theta & \sin\theta \\ -\sin\theta & \cos\theta \end{pmatrix} \tag{12.45}$$

while $E_i \approx k + \frac{m_i^2}{2k}$ and $E_2 - E_1 = \frac{m_2^2 - m_1^2}{2k} = \frac{\Delta m^2}{2k}$. Then

$$H_\nu(k) = \text{const.} \begin{pmatrix} 1 & 0 \\ 0 & 1 \end{pmatrix} + \begin{pmatrix} \frac{-\Delta m^2 \cos 2\theta}{2k} & \frac{\Delta m^2 \sin 2\theta}{4k} \\ \frac{\Delta m^2 \sin 2\theta}{4k} & 0 \end{pmatrix} \tag{12.46}$$

where the first part of Eq. (12.46) is irrelevant for oscillations. Now in traversing matter, neutrinos interact with electrons and nucleons of intervening material and their forward coherent scattering induces an effective potential energy. Such contributions of weak interaction in matter to $H_\nu$ arise due to Feynman diagrams shown in Fig. 12.5.

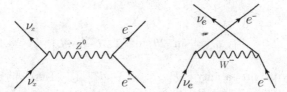

Fig. 12.5   Feynman diagrams for neutral current (n.c.) and charged current (c.c) weak interactions which contribute to $H_\nu$ for oscillations in matter, where $x = e, \mu, \tau$.

The first diagram contributes equally to $\nu_e$, $\nu_\mu$ and $\nu_\tau$ and as such is not relevant $\nu_\mu \leftrightarrow \nu_\mu$ or $\nu_\tau$ oscillations. This gives the effective Hamiltonian [see Chap. 13]:

$$\frac{2G_F}{\sqrt{2}} \left[ \bar{f}_L \gamma^\mu \left( I_{3L} - Q \sin^2 \theta_W \right) f_L \right] [\bar{\nu} \gamma_\mu (1 - \gamma_5)\nu] \tag{12.47}$$

where $f = e^-$, $p$ or $n$ for which respectively $I_{3L} = -\frac{1}{2}, \frac{1}{2}, -\frac{1}{2}$ and $Q = -1, 1, 0$. The second diagram after Fierz rearrangement gives the effective Hamiltonian:

$$\frac{G_F}{\sqrt{2}} \langle \psi_e | e^- \gamma^\mu (1 - \gamma_5)e | \psi_e \rangle \, \bar{\nu}_e \gamma_\mu (1 - \gamma_5)\nu_e \tag{12.48}$$

where $\psi_e$ denotes the state of the medium. These diagrams give the potential energy

$$V_{\nu_e} = \sqrt{2} G_F (n_e^{c.c.} - \frac{1}{2} n_n^{n.c.})$$

$$V_{\nu_{\mu,\tau}} = -\sqrt{2}G_F \frac{1}{2}(n_n^{c.c.})$$

(12.49)

$$V_{\nu_s} = 0$$

where $n_e$ denotes the number of electrons per unit volume and $n_n$ that of neutrons. Then the Hamiltonian in the matter is $[k \simeq E]$

$$H_M(k) = H_\nu(k) + H_W$$
$$= \begin{pmatrix} \frac{-\Delta m^2 \cos 2\theta}{2E} + \sqrt{2}G_F n & \frac{\Delta m^2 \sin 2\theta}{4E} \\ \frac{\Delta m^2 \sin 2\theta}{4E} & 0 \end{pmatrix}$$

(12.50)

where

$$n = n_e \text{ for } \nu_e \leftrightarrow \nu_\mu \text{ or } \nu_\tau$$
$$= n_e - \frac{1}{2}n_n \text{ for } \nu_e \leftrightarrow \nu_s$$

The diagonalization to $\nu_1$, $\nu_2$ basis gives:

$$\begin{pmatrix} \nu_e \\ \nu_x \end{pmatrix} = \begin{pmatrix} \cos\theta_M & \sin\theta_M \\ -\sin\theta_M & \cos\theta_M \end{pmatrix} \begin{pmatrix} \nu_1 \\ \nu_2 \end{pmatrix}$$

(12.51)

with

$$\sin 2\theta_M = \sin 2\theta \frac{l_M}{l_V},$$

$$\cos 2\theta_M = (\cos 2\theta - A)\frac{l_M}{l_V}$$

(12.52)

$$\Delta E = E_2 - E_1 = \frac{1}{2l_M}$$

(12.53)

where

$$A = 2\sqrt{2}G_F n \frac{E}{\Delta m^2}$$

(12.54)

$$l_M = \frac{E}{\Delta m^2}\left[(A - \cos 2\theta)^2 + \sin^2 2\theta\right]^{-1/2},$$

(12.55)

$$l_V = \frac{E}{\Delta m^2}$$

(12.56)

For constant density $n$, the considerations of Sec. 12.3 give the conversion probability

$$P(\nu_e \to \nu_x) = \sin^2 2\theta_M \sin^2\left[1.27\frac{L}{l_M}\right]$$

(12.57)

The following are useful limits:

$$(i) \quad n \to 0, \quad \theta_M = \theta, \quad \Delta E = \frac{\Delta m^2}{2E}$$

$$(ii) \quad n \to \infty, \quad H_M \to \begin{pmatrix} \sqrt{2}G_F n \to 0 \\ \to 0 \quad \quad 0 \end{pmatrix} \tag{12.58}$$

$$\Delta E = 2\sqrt{2}G_F n$$

$$\theta_M = \frac{\pi}{2}, \nu_e = \nu_2, \nu_x = \nu_1$$

$$(iii) \quad n = n_{res} \text{ defined by } \sqrt{2}G_F(n)_{res} = \frac{\Delta m^2 \cos 2\theta}{2E}$$

$$H_M \to \begin{pmatrix} 0 & \frac{\Delta m^2 \sin 2\theta}{4E} \\ \frac{\Delta m^2 \sin 2\theta}{4E} & 0 \end{pmatrix}$$

$$\Delta E_{\text{res}} = \frac{\Delta m^2 \sin 2\theta}{2E} \tag{12.59}$$

$$\theta_M = \frac{\pi}{4}, \nu_2 = \frac{\nu_e + \nu_x}{\sqrt{2}}, \nu_1 = \frac{\nu_e - \nu_x}{\sqrt{2}}$$

Using the above limits, the plot of $E$ versus $n$ is shown in Fig. 12.6.

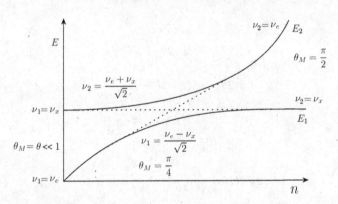

Fig. 12.6   Plot of neutrino energy $E$ versus density $n$ , showing conversion of $\nu_e$, to $\nu_x$ in matter.

Suppose $\nu_e$ is created at $n_0 > n_{res}$ say at the center of the sun, and then it propagates out. If there is no level crossing (shown by dotted lines in Fig. 12.6, then $\nu(n = 0) \simeq \nu_x$ and undetectable. This conversion of $\nu_e$ into $\nu_x$ is the cause of the depletion of observable neutrinos. Now neutrinos of any energy will not go through the resonance. The resonance condition for any given neutrino energy $E$ is:

$$n_{\text{res}}(E) = \cos 2\theta \frac{\Delta m^2}{2\sqrt{2}G_F} \qquad (12.60)$$

We may remark here that for $\nu_e \to \nu_\mu$ or $\nu_\tau$ conversion,

$$n = n_e = \left(\frac{\rho}{m_N}\right) Y$$

where $Y$ denotes the number of electrons per nucleon and is $1/2$ for ordinary matter. Then, the resonance condition (12.59) can be written as

$$\rho_{\text{res}} = \frac{\Delta m^2 \cos 2\theta}{2\sqrt{2}G_F} \frac{m_N}{Y} \frac{1}{E}$$

$$= 1.3 \times 10^7 \, g/cc \frac{1}{2Y} \cos 2\theta \left[\frac{\Delta m^2}{(\text{eV})^2}\right] \left(\frac{\text{MeV}}{E}\right) \qquad (12.61)$$

For $\rho_{res} \geq \rho$ (center of the sun) $= 100 \, g/cc$, we have

$$\left(\frac{E}{\text{MeV}}\right) \leq 1.3 \times 10^5 \frac{\Delta m^2}{(\text{eV})^2} \qquad (12.62)$$

Thus, for example, for $\Delta m^2 \geq 6 \times 10^{-6} \, \text{eV}^2$, we will not have resonance for $E \leq 0.4 \, \text{MeV}$ and the resonance will be at least at $E = 0.8$ MeV. In this case the resonance will not effect $pp$ neutrino for which $E_{max} = 0.44$ MeV but can eliminate $^7B$ neutrinos for which $E_\nu \sim 0.86$ MeV.

### 12.4.2 *Evolution of Flavor Eigenstates in Matter*

The evolution of flavor eigenstates in matter is governed by the equation:

$$i\frac{\partial}{\partial x}\begin{pmatrix} \nu_e(x) \\ \nu_x(x) \end{pmatrix} = H(x)\begin{pmatrix} \nu_e(x) \\ \nu_x(x) \end{pmatrix} \qquad (12.63)$$

where $H(x)$ is given in Eq. (12.50). Note that the $x$ dependence arises due to the $x$ dependence of the density $n$ for varying density case. Using

$$\begin{pmatrix} \nu_e(x) \\ \nu_x(x) \end{pmatrix} = U(x)\begin{pmatrix} \nu_1(x) \\ \nu_2(x) \end{pmatrix} \qquad (12.64)$$

with

$$U(x) = \begin{pmatrix} \cos\theta(x) & \sin\theta(x) \\ -\sin\theta(x) & \cos\theta(x) \end{pmatrix}, \qquad (12.65)$$

we have

$$i\frac{\partial}{\partial x}\begin{pmatrix} \nu_1(x) \\ \nu_2(x) \end{pmatrix} = U^{-1}HU\begin{pmatrix} \nu_1(x) \\ \nu_2(x) \end{pmatrix} - iU^{-1}\frac{\partial U}{\partial x}\begin{pmatrix} \nu_1(x) \\ \nu_2(x) \end{pmatrix}$$

$$(12.66)$$

where

$$U^{-1}HU = \begin{pmatrix} E_1 & 0 \\ 0 & E_2 \end{pmatrix}$$

$$= \frac{E_1 + E_2}{2} \begin{pmatrix} 1 & 0 \\ 0 & 1 \end{pmatrix} + \begin{pmatrix} -\frac{\Delta E}{2} & 0 \\ 0 & \frac{\Delta E}{2} \end{pmatrix} \qquad (12.67)$$

and

$$U^{-1} \frac{\partial U}{\partial x} = \begin{pmatrix} 0 & 1 \\ -1 & 0 \end{pmatrix} \theta'_M(x) \qquad (12.68)$$

where $(')$ means differentiation with respect to $x$. Noting that the first part of Eq. (12.67) is irrelevant for oscillations and using Eq. (12.52) we have

$$i \frac{\partial}{\partial x} \begin{pmatrix} \nu_1(x) \\ \nu_2(x) \end{pmatrix} = \begin{pmatrix} -\frac{1}{4l_M(x)} & i\theta'_M(x) \\ -i\theta'_M(x) & \frac{1}{4l_M(x)} \end{pmatrix} \begin{pmatrix} \nu_1(x) \\ \nu_2(x) \end{pmatrix} \qquad (12.69)$$

For the constant density case, $\theta'_M(x) = 0$ and $l_M$ is independent of $x$, so that Eq. (12.69) has simple solutions

$$\nu_1(x) = \nu_1(0) \exp \left( i \frac{x}{4l_M} \right)$$

$$\nu_2(x) = \nu_2(0) \exp \left( -i \frac{x}{4l_M} \right) \qquad (12.70)$$

where we have taken $x = 0$ as the initial point. Then Eq. (12.64) gives

$$\nu_e(x) = \cos\theta(x)\nu_1(0) \exp \left( i \frac{x}{4l_M} \right) + \sin\theta(x)\nu_2(0) \exp \left( -i \frac{x}{4l_M} \right)$$

$$= \cos\theta(x) \cos\theta^0_M \nu_e(0) \exp \left( i \frac{x}{4l_M} \right)$$

$$- \sin\theta(x) \sin\theta^0_M \nu_e(0) \exp \left( -i \frac{x}{4l_M} \right) \qquad (12.71)$$

where we have used the boundary condition $\nu_x(0) = 0$ [see Eq. (12.64)]. Then the electron neutrino survival probability averaged over the detector position $L$ (from the solar surface) is given by

$$P(\nu_e \longrightarrow \nu_e) = \cos^2\theta_V \cos^2\theta^0_M + \sin^2\theta_V \sin^2\theta^0_M$$

$$= \frac{1}{2} + \frac{1}{2} \cos 2\theta_V \cos 2\theta^0_M \qquad (12.72)$$

where $\theta_V \equiv \theta$ is the vacuum mixing angle. In general when the density $n$ is a function of $x$ one has to solve Eq. (12.69) and as a result $P(\nu_e \to \nu_e)$ is given by the Parke formula:

$$P(\nu_e \to \nu_e) = \frac{1}{2} + \left( \frac{1}{2} - P_j \right) \cos 2\theta \cos 2\theta^0_M \qquad (12.73)$$

where $\theta_M^0$ is the initial mixing angle and $P_j \equiv \exp\left(-\frac{\pi}{2}\gamma\right)$ is the Landau-Zener factor. Here

$$\gamma = \frac{\Delta m^2}{E} \frac{\sin^2 2\theta}{\cos 2\theta} \left(\frac{1}{n}\frac{dn}{dx}\right)^{-1}_{res}$$

and is called the adiabaticity. In the adiabatic limit $\gamma \gg 1$ and $P_j \to 0$ and we recover the relation (12.72). The survival probability $P(\nu_e \to \nu_e)$ as a function of vacuum oscillation length $l_v \propto E/\Delta m$ is displayed for various mixing angles in Fig. 12.7. The transition between the regime of vacuum and matter oscillations is determined by the ratio $l_v/l_M$ (see Eqs. (12.52-12.56). If it is greater than 1 then matter oscillations dominate. If less than $\cos 2\theta$ vacuum oscillations dominate. Generally there is a smooth transition between these two regimes.

## 12.5   Evidence for Neutrino Oscillations

One looks for neutrino oscillations in the following two types of experiments.

### 12.5.1   *Disappearance Experiments*

Reactors are source of $\bar{\nu}_e$ through the $\beta$-decay

$$n \to p + e^- + \bar{\nu}_e$$

and experiment looks for possible decrease in the $\bar{\nu}_e$ flux as a function of distance from the reactor, $\bar{\nu}_e \to X$ [if converted to $\bar{\nu}_\mu$, say, one would see nothing, $\bar{\nu}_\mu$ could have produced $\mu^+$ but does not have sufficient energy to do so].

KamLand experiment confirms that $\bar{\nu}_e$ do indeed disappear when the reactor $\bar{\nu}_e$'s have traveled $\approx 200$km. $\bar{\nu}_e$ flux is only $0.658 \pm 0.044 \pm 0.047$ of what it would be if none of $\bar{\nu}_e$'s were disappearing.

### 12.5.2   *Appearance Experiments*

Here one searches for a new neutrino flavor, absent initially, which can arise from oscillations. Such experiments involve atmospheric neutrino studies, K-2K and MINOS accelerator experiments, which are sensitive to $\Delta m_{23}^2 = m_2^2 - m_3^2$ and $\theta_{23}$ (atmospheric sector), and solar neutrinos. Solar neutrinos and KamLand are sensitive mainly to $\Delta m_{21}^2 = m_2^2 - m_1^2$ and $\theta_{12}$ (solar sector). The $1 - 3$ mixing, if not zero, may give subleading effect. The reactor CHOOZ experiment gives a bound on $\theta_{13}$ as a function of $\Delta m_{31}^2$. The physical effects involved are:

(i) vacuum oscillations for atmospheric neutrinos, K2K, MINOS, CHOOZ and

(ii) MSW effect – see Fig. 12.7 – which illustrates the adiabatical conversion of solar neutrino in the matter of the sun while at low energies solar neutrinos undergo the averaged vacuum oscillations with small matter effects.

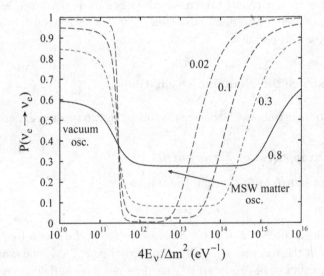

Fig. 12.7   Schematic illustration of the survival probability of $\nu_e$ created at the solar center. The curves are labeled by thesin$^2 2\theta$ values [16].

We now consider some of the experiments and their analysis.

### 12.5.2.1   *Atmospheric neutrino anomaly*

Atmospheric neutrinos are produced in decays of pions (Kaon's) that are produced in the interaction of cosmic rays with the atmosphere:

$$p + A \rightarrow \pi^{\pm} + A'$$
$$\pi^{\pm} \rightarrow \mu^{\pm} \nu_\mu \, (\bar{\nu}_\mu)$$
$$\rightarrow e^{\pm} \nu_e \, (\bar{\nu}_e) \, \bar{\nu}_\mu \, (\nu_\mu)$$

These neutrinos are detected through the reactions $\nu_\mu + n \rightarrow \mu^- + p$, $\bar{\nu}_\mu + p \rightarrow \mu^+ + n$ and $\nu_e + n \rightarrow e^- + p$, $\bar{\nu}_e + p \rightarrow e^+ + n$ and are respectively called $\mu$-like and $e$-like events. One would expect the ratio

$$\frac{N(\nu_\mu)}{N(\nu_e)} \equiv \frac{N(\nu_\mu + \bar{\nu}_\mu)}{N(\nu_e + \bar{\nu}_e)} \simeq 2$$

However this ratio was measured in several detectors and it was found that it is substantially reduced from the value $\sim 2$. Indeed [MC for Monto Carlo]

$$R = \frac{(\mu/e)\, \text{data}}{(\mu/e)\, \text{MC}}$$
$$= 0.658 \pm 0.016 \pm 0.035$$

which should be one in the absence of oscillation.

The results of global analysis gives for the dominant mode of the atmospheric neutrino oscillation ($\nu_\mu \leftrightarrow \nu_\tau$ oscillations)

$$|\Delta m_{32}^2| = (2.39^{+0.11}_{-0.08}) \times 10^{-3} \text{eV}^2$$
$$\sin^2 \theta_{23} = 0.466^{+0.073}_{-0.058} \tag{12.74}$$

### 12.5.2.2 *Solar neutrinos*

Electron type antineutrinos are produced by the decay of pile neutrons in a fission reactor: $n \rightarrow p + e^- + \bar{\nu}_e$, e.g. KamLand. Electron type neutrinos are produced from reactions in the sun called solar neutrinos. The energy of the sun is generated in the reactions of pp and CNO cycles. Energy is generated through nuclear burning involving the transitions of four protons into $^4He$ :

$$4p \rightarrow {}^4He + 2e^+ + 2\nu_e + Q$$

where $Q = 26.7$ MeV is the energy release in the above transition. Thus the generation of the energy of the sun is accompanied by the emission of $\nu_e$'s. The total flux of the neutrinos is connected to the luminosity of the sun $L_O$ by the relation:

$$Q \sum_i \left( l - 2\frac{\bar{E}_i}{Q} \right) \Phi_i = \frac{Lo}{2\pi R^2}$$

where $R$ is the sun-earth distance, $\Phi_i$ is the total flux of neutrinos from the source $i$, and $\bar{E}_i$ is the average energy.

The most important sources of solar neutrinos in the pp cycle, which dominates cooler stars, particularly the sun, are the following reactions:

$$pp \rightarrow {}^2He^+\nu_e : E_\nu < 0.42 \text{ MeV}$$
$$ppe^- \rightarrow {}^2H\nu_e : E_\nu = 1.442 \text{ MeV}$$
$${}^7B_e\, e^- \rightarrow {}^7Li\,\nu_e : E_\nu \sim 0.86 \text{ MeV}$$
$${}^8B \rightarrow {}^8B_e^*e^+\nu_e : E_\nu < 15 \text{ MeV}$$

On the other hand, the CNO cycle dominates hot stars and the following reactions are sources of $\nu_e$'s:

$${}^{13}N \rightarrow {}^{13}Ce^+\nu_e$$
$${}^{15}O \rightarrow {}^{15}Ne^+\nu_e$$

The first reaction in the pp cycle is the main source of solar neutrinos. The third reaction is a source of monochromatic neutrinos. This reaction contributes about 10% to the total flux of solar neutrinos. The fourth reaction contributes only about $10^{-4}$ to the total flux but it is the main source of high energy solar neutrinos (up to 15 MeV).

Due to different detection thresholds, solar neutrinos from different sources can be detected in different reactions. Thus the solar neutrinos with energy $> 0.814$ MeV can be detected in ${}^{37}Cl$ and Super-Kamiokande and those $> 0.233$ MeV in ${}^{71}Ga$. In all experiments the observed event rate is significantly smaller than the rate predicted by the standard solar model: $0.34 \pm 0.04, 0.47 \pm 0.08$ and $0.53 \pm 0.03$ of the expected rate from the standard solar model, respectively for ${}^{37}Cl$, Super Kamiokande and ${}^{71}Ga$.

Particular compelling evidence that the solar neutrinos change flavor has been reported by Sudbury Neutrino Observatory (SNO). SNO measures the high energy part of the solar neutrino flux (${}^8B$ neutrinos). The reactions employed by SNO are

$$\nu d \rightarrow \nu n p$$
$$\rightarrow epp$$
$$\nu e \rightarrow \nu e.$$

SNO measured $\nu_e + \nu_\mu + \nu_\tau$ flux, $\phi_e + \phi_{\mu\tau}$, and the $\nu_e$ flux, $\phi_e$. From the observed rates for the first two reactions, which involve respectively neutral current and charge current, SNO finds that the ratio of two fluxes is

$$\frac{\phi_e}{\phi_e + \phi_{\mu\tau}} = 0.340 \pm 0.023 \tag{12.75}$$

This implies that the flux $\phi_{\mu\tau}$ is not zero. Since all the neutrinos are born in nuclear reaction that produces only electron neutrinos, it is clear that neutrinos change flavor. Corroborating information comes from the detection reaction $\nu e \to \nu e$, studied by both SNO and Super-Kamiokande. Incidently the total neutrino flux $\phi_e + \phi_{\mu\tau}$ measured by SNO is $\left(4.94 \pm 0.21^{+0.38}_{-0.34}\right) \times 10^6$ $\text{cm}^{-2}\text{s}^{-1}$ is in agreement with the Standard Solar Model (SSM) value $\phi_{\text{SSM}} = \left(5.49^{+0.95}_{-0.81}\right) \times 10^6 \text{ cm}^{-2}\text{s}^{-1}$ or $\left(4.34^{+0.71}_{-0.61}\right) \times 10^6 \text{ cm}^{-2}\text{s}^{-1}$ depending on assumption about solar heavy element abundances.

The global analysis of solar neutrino data as well as that of KamLand give best values:

$$\Delta m^2_{12} = \left(7.67^{+0.16}_{-0.19}\right) \times 10^{-5} \text{eV}^2$$
$$\sin^2 \theta_{12} = 0.312^{+0.018}_{-0.049} \tag{12.76}$$

The CHOOZ experiment gives

$$\sin^2 \theta_{13} \leq 0.046 \tag{12.77}$$

One may mention two more experiments, Los Alamos liquid scintillation detector (LSND) and MiniBOONE experiment at Fermilab which search for $\bar{\nu}_\mu \to \bar{\nu}_e$ oscillations, and have found $\bar{\nu}_e$ candidate events. The data is consistent with the $\bar{\nu}_\mu \to \bar{\nu}_e$ oscillations in the 0.1 to $1.0\text{eV}^2$ range for $\Delta m^2$ and small mixing angle. It is not possible to accommodate this mass range in the three known neutrinos picture. If these results are further confirmed, it would require a mechanism to generate a third mass squared difference, involving one or more sterile neutrinos or a new type of flavor transition beyond oscillations.

## 12.6 Neutrino Mass Models and Mixing Matrix and Symmetries

Due to lack of precision in the data, several mass models and mixing patterns are possible which can be limited by some symmetry conditions. From the observed data one sees:

(i) Angle $\theta_{13}$ is suppressed while $\theta_{23}$ and $\theta_{12}$ are large.

(ii) $r = \dfrac{\Delta m^2_{\odot}}{|\Delta m^2_{\text{atm}}|} \equiv \dfrac{\Delta m^2_{21}}{|\Delta m^2_{32}|} = (3.2 \pm 0.2) \times 10^{-2} \ll 1.$

(iii) The sign of mass split $\Delta m^2_{32}$ determines the type of mass hierarchy. It is customary to order the mass eigenstates such that $m^2_1 < m^2_2$, then $\Delta m^2_{32} > 0$ give the normal hierarchy $m^2_3 > m^2_2 > m^2_1$ while $\Delta m^2_{32} < 0$

gives inverted hierarchy $m_2^2 > m_1^2 > m_3^2$. This is depicted in Fig. 12.8.

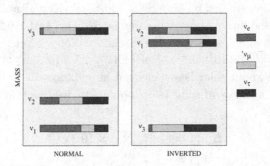

Fig. 12.8   Neutrino mass and flavor spectra for the normal (left) and inverted (right) mass hierarchies. The distribution of flavors (colored parts of boxes) in the mass eigenstates corresponds to the best-fit values of mixing parameters and $\sin^2 \theta_{13} = 0.05$ [21].

The $\left|\Delta m_{32}^2\right|$ given in Eq. (12.74) implies a lower bound on the heaviest of neutrino mass

$$m_h \geq \sqrt{\left|\Delta m_{32}^2\right|} > 0.05 \text{eV} \qquad (12.78)$$

By combining this and all other bounds discussed in Sec. 12.2.1 it is safe to conclude that neutrino weighs less than 1 eV. Neutrino masses may be qualitatively different from charged fermion masses. There is an enormous gap between neutrino masses and the lightest charged fermion $(m_e)$ in contrast to that between $m_e$ and $m_t$ which is populated. Further

$$\frac{(m_\nu)_{\text{mass}}}{m_e} < 2 \times 10^{-6} \qquad (12.79)$$

which need to be understood. This is indication of the new mass scale signifying new physics in nature but it has not yet been pinpointed.

(iv) The leptonic mass Lagrangian in see-saw model can be written as

$$\mathcal{L} = -\bar{L}_i \left(M_l\right)_{ij} e_{Rj} - \bar{L}_i \left(M_D\right)_{ij} N_{Rj} - \frac{1}{2} N_{Ri}^T C^{-1} \left(M_R\right)_{ij} N_{Rj} + h.c. \qquad (12.80)$$

where $i$, $j$ are flavor indices, $L = (e_L, \nu_L)$ are lepton doublets, $e_R$ charged lepton $SU_L(2)$ singlets with non-vanishing hyper charge, $N_R$ are $SU_L(2) \times U(1)$ singlets. It is convenient to have a basis in which

$M_l$ and $M_R$ are simultaneously diagonal

$$M_l \rightarrow \hat{M}_l = U_L^T M_l U_L$$
$$M_R \rightarrow \hat{M}_R = V^T M_R V. \qquad (12.81)$$

Correspondingly $L_i \rightarrow U_L L_i$, $e_{iR} \rightarrow U_R e_{iR}$, $i = e$, $\mu$, $\tau$ is the flavor index. Then the effective Majorana mass matrix for the light neutrino is

$$M_\nu = \hat{M}_D \hat{M}_R^{-1} \hat{M}_D^T \qquad (12.82)$$

where

$$\hat{M}_D = M_D V^*$$

is the Dirac matrix in $\begin{pmatrix} \bar{N}_1 & \bar{N}_2 & \bar{N}_3 \end{pmatrix} \begin{pmatrix} \nu_e \\ \nu_\mu \\ \nu_\tau \end{pmatrix}$ basis.

$M_\nu$ is diagonalized by PMNS mixing matrix given in (12.40)

$$\hat{M}_\nu = U^T M_\nu U \qquad (12.83)$$

As seen from above the data are consistent with having the atmospheric mixing angle $\theta_{23}$ maximal, the reactor angle $\theta_{13}$ zero. This indicates some underlying symmetry. These are the consequences of $\nu_\mu \rightarrow \nu_\tau$ permutation. The general form of such a matrix $U$ is

$$M_\nu = \begin{pmatrix} x & y & y \\ y & z & \omega \\ y & \omega & z \end{pmatrix} \qquad (12.84)$$

which has $\mu - \tau$ (or $2 - 3$ symmetry): $(M_\nu)_{22} = (M_\nu)_{33}$ and $(M_\nu)_{1,2} = (M_\nu)_{1,3}$. Such a matrix is diagonalized by

$$U = \begin{pmatrix} c_{12} & s_{12} & 0 \\ -\frac{s_{13}}{\sqrt{2}} & \frac{c_{12}}{\sqrt{2}} & -\frac{1}{\sqrt{2}} \\ -\frac{s_{12}}{\sqrt{2}} & \frac{c_{12}}{\sqrt{3}} & \frac{1}{\sqrt{2}} \end{pmatrix} I_\phi \qquad (12.85)$$

It must be emphasized that $\mu - \tau$ symmetry cannot be simultaneously imposed for $L_e$ and $L_\nu$. For example in the basis where the charged leptons are diagonal, this would imply $m_\mu = m_\tau$. Deeper origin of $\mu - \tau$ symmetry is not yet known. Ignoring Majorana phases, we have in this case four real parameters, three masses and solar angle.

The diagonalization gives

$$x = c_{12}^2 m_1 + s_{12}^2 m_2$$
$$y = \frac{1}{\sqrt{2}} c_{12} s_{12} (m_2 - m_1)$$
$$z = \frac{1}{2} \left( s_{12}^2 m_1 + c_{12}^2 m_2 + m_3 \right)$$
$$w = \frac{1}{2} \left( s_{12}^2 m_1 + c_{12}^2 m_2 - m_3 \right) \tag{12.86}$$

Furthermore

$$\Delta m_{32}^2 + \Delta m_{13}^2 = \Delta m_{12}^2$$

and

$$r \equiv \frac{\Delta m_{12}^2}{|\Delta m_{32}^2|} = 0.033 \pm 0.004 \tag{12.87}$$

For normal hierarchy ($m_1 \ll m_2 \ll m_3$), referring to Eqs. (12.84), (12.86), $M_\nu$ has the following texture

$$M_\nu = \frac{\sqrt{\Delta m_{31}^2}}{2} \begin{pmatrix} \epsilon & \epsilon & \epsilon \\ \epsilon & 1+\epsilon & -1 \\ \epsilon & -1 & 1+\epsilon \end{pmatrix} \tag{12.88}$$

where $\epsilon \simeq 2\sqrt{r}$. A salient feature of this matrix is the dominant $\mu - \tau$ block.

For inverted mass hierarchy ($m_1 \simeq m_2 \gg m_3$)

$$M_\nu = \sqrt{\Delta m_{13}^2} \begin{pmatrix} 1 & 0 & 0 \\ 0 & 1/2 & 1/2 \\ 0 & 1/2 & 1/2 \end{pmatrix} + \text{ small corrections} \tag{12.89}$$

For degenerate case $m_1 \simeq m_2 \simeq m_3 = m_0$

$$M_\nu = m_0 \begin{pmatrix} 1 & 0 & 0 \\ 0 & 1 & 0 \\ 0 & 0 & 1 \end{pmatrix} + \delta M \tag{12.90}$$

$$\delta M \ll m_0$$

or

$$M_\nu = \begin{pmatrix} 1 & 0 & 0 \\ 0 & 0 & 1 \\ 0 & 1 & 0 \end{pmatrix} + \delta M \tag{12.91}$$

in case of opposite CP-parity of $\nu_2$ and $\nu_3$ ($m_2$ and $m_3$ are of opposite sign).

Some new ingredients are needed to describe correctly the three mixing angles. Two patterns can be considered:

(i) The bimaximal mixing matrix (superscript $m$ for maximal)

$$U_{bm} = U_{23}^m U_{12}^m, \quad U_{13} = 1 \tag{12.92}$$

with maximal $\frac{\pi}{4}$ rotations in $2-3$ and $1-2$ spaces.

$$U_{bm} = \begin{pmatrix} 1 & 0 & 0 \\ 0 & \frac{1}{\sqrt{2}} & \frac{1}{\sqrt{2}} \\ 0 & -\frac{1}{\sqrt{2}} & \frac{1}{\sqrt{2}} \end{pmatrix} \begin{pmatrix} \frac{1}{\sqrt{2}} & \frac{1}{\sqrt{2}} & 0 \\ -\frac{1}{\sqrt{2}} & \frac{1}{\sqrt{2}} & 0 \\ 0 & 0 & 1 \end{pmatrix}$$

$$= \begin{pmatrix} \frac{1}{\sqrt{2}} & \frac{1}{\sqrt{2}} & 0 \\ -\frac{1}{2} & \frac{1}{2} & \frac{1}{\sqrt{2}} \\ \frac{1}{2} & -\frac{1}{2} & \frac{1}{\sqrt{2}} \end{pmatrix} \tag{12.93}$$

Then from Eqs. (12.84), (12.86), $x = \omega + z$, i.e. in addition to $\mu - \tau$ symmetry, there is an additional symmetry $(M_\nu)_{1,1} = (M_\nu)_{2,2} + (M_\nu)_{2,3}$. However identification of $U_{bm}$ with $U_{\text{PMNS}}$ is not possible owing to substantial deviation of $\sin^2 \theta_{12}$ from maximal value $\frac{1}{2}$. Yet $U_{bm}$ can play a dominant role, the correction might originate from charged lepton (mass matrix) so that $U_{PMNS} = U' U_{bm}$, where $U' \simeq U_{12}(\alpha)$ with $\alpha \sim O(\theta_c)$ in analogy to the quark mixing, $\theta_c$ is the Cabbibo angle so that $\theta_{12} + \theta_c = \frac{\pi}{4}$. The $U$ simultaneously generates deviations according to the pattern: $\delta \sin^2 \theta_{12} \simeq \theta_c$ while $\delta \sin^2 \theta_{23} \le \theta_c^2$ and $\delta \sin^2 \theta_{13} \le \theta_c$. It is not trivial to achieve such a pattern.

(ii) The tri-bimaximal [TB] mixing matrix: The data suggest the approximate tri-bimaximal texture of Harrison, Perkin and Scott.

$$U_{bm} = U_{23}^m U_{12}(\theta_{12}),$$

$$= \begin{pmatrix} \frac{\sqrt{2}}{3} & \frac{1}{\sqrt{3}} & 0 \\ -\frac{1}{\sqrt{6}} & \frac{1}{\sqrt{3}} & -\frac{1}{\sqrt{2}} \\ -\frac{1}{\sqrt{2}} & \frac{1}{3} & \frac{1}{\sqrt{2}} \end{pmatrix} I_\phi \tag{12.94}$$

with $\sin^2 \theta_{23} = \frac{1}{2}$. $\sin^2 \theta_{12} = \frac{1}{3}$ and $\sin^2 \theta_{13} = 0$, i.e. mixing parameters are simple numbers $0, \frac{1}{3}, \frac{1}{2}$. This requires $x + y = z + w$ so that

$$M_\nu^{TB} = \begin{pmatrix} x & y & y \\ y & z & x+y-z \\ y & x+y-z & z \end{pmatrix} \tag{12.95}$$

There have been many attempts to understand $\nu_\mu - \nu_\tau$ symmetry and TB. These involve auxiliary symmetries based on $S_3$, $A_4$, $Z_4$, $Z_2$ groups which have been reviewed in recent review articles.

## 12.7   Neutrino Magnetic Moment

With the definition

$$\mu_\nu = \kappa \frac{e\hbar}{2m_e} = \kappa\mu_B, \tag{12.96}$$

where $\mu_B$ is Bohr Magneton, magnetic moment interaction is

$$H_{mag} = \mu_\nu \sigma \cdot \mathbf{B} \tag{12.97}$$

Here $\mathbf{B}$ is the solar magnetic field. The neutrino spin would then precess in the magnetic field, some left handed (LH) neutrinos would become RH and sterile to the detector as shown below in Fig. 12.9.

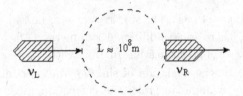

Fig. 12.9   The conversion of $\nu_L$ into $\nu_R$ in the solar magnetic field.

The conversion probability is determined by

$$\kappa\mu_B B\left(\frac{L}{\hbar c}\right). \tag{12.98}$$

Now the solar magnetic field in the convective zone of thickness $L \approx 2 \times 10^8 m$ is $B = (1-5) \times 10^3$ gauss, so that the conversion probability is

$$\kappa(5.79 \times 10^{-9} eV/G)(1-5) \times 10^3 G \frac{2 \times 10^8 m}{[3 \times 10^8 m/s][6.6 \times 10^{-16} eV.s]}$$

$$\approx \kappa(0.6-3)10^{10}. \tag{12.99}$$

This is $O(1)$ if $\kappa = (0.3-1) \times 10^{-10}$ giving $\mu_\nu \approx (0.3-1) \times 10^{-10}\mu_B$.

In the standard model,

$$(\mu_\nu)_{SM} = 3eG_F \frac{m_\nu}{\sqrt{28\pi^2}} \tag{12.100}$$

i.e.

$$(\mu_\nu)_{SM} \sim 3 \times 10^{-19}\mu_B(m_\nu/eV). \tag{12.101}$$

So if $\mu_\nu \approx 10^{-10}\mu_B$, this would definitely indicate physics beyond the standard model. Thus the question of dipole moment of neutrino is very important. What are the other limits on it? The best laboratory limit on $m_\nu$ comes from reactor experiments. In addition to the usual electroweak scattering via $W^\pm$ and $Z^0$ bosons exchange, the process

$$\bar{\nu}_e + e \to \bar{\nu}_e + e$$

could proceed via magnetic scattering which is large in the forward direction and for small $E_\nu$. Consistency with measured cross-section requires

$$\mu_{\nu_e} < 10^{-10} \mu_B. \tag{12.102}$$

More stringent limits have, however, been quoted from astrophysics:

### (a) Nucleosynthesis in the Early Universe

Presence of $\mu_\nu$ mediates $\nu_L e^- \to \nu_R e^-$ scattering. If this occurs frequently in the era before the decoupling of the neutrinos, it doubles the neutrino species and increases the expansion rate of the universe, causing over abundance of helium. To avoid this,

$$\mu_\nu < 8.5 \times 10^{-11} \mu_B. \tag{12.103}$$

### (b) Stellar Cooling

Magnetic scattering of neutrinos produced in thermonuclear reactions may occur, flipping the helicity $[\nu_L \to \nu_R]$ so that the outer regions of the star will no longer be opaque to neutrinos and cooling will proceed much faster. Applied to helium burning star in order that

$$\varepsilon_{\text{exotic}} < \varepsilon_{H_e}$$

where $\varepsilon_{\text{exotic}}$ denotes energy loss due to process of the above types while $\varepsilon_{H_e}$ denotes energy generation rate. This gives

$$\mu < 10^{-11} \mu_B. \tag{12.104}$$

### (c) Limit on $\mu_\nu$ from Supernova 1987A

Neutrinos produced in the initial collapse state have high energies $\sim$ 100 MeV. These high energy neutrinos could escape the following spin-flip magnetic scattering $[\nu_L \to \nu_R]$. Furthermore, a proportion can process back $\nu_R \to \nu_L$ in the galactic magnetic fields and the result on earth could be a signal of high energy ($\sim$ 100 MeV) neutrino interactions in the underground detector with a high rate [note that $\sigma \sim E^2$ in $\bar{\nu}_e + p \to e^+ + n$]. The observance of no signal implies

$$\mu_\nu \leq 10^{-12} \mu_B.$$

In view of the above upper limits on $\mu_\nu$, the neutrino spin precession mechanism does not appear to be a viable solution to the solar neutrino problem.

## 12.8  Problems

(1) Show that

(a)

$$\bar{\nu}_L \gamma^\mu e_L = \frac{1}{2} \bar{\nu} \gamma^\mu (1 - \gamma^5) e$$

(b)

$$\bar{e}_R^c \gamma^\mu \nu_R^c = \frac{1}{2} \bar{e}^c \gamma^\mu \left(1 + \gamma^5\right) \nu^c = -\frac{1}{2} \bar{\nu} \gamma^\mu \left(1 - \gamma^5\right) e = -\bar{\nu}_L \gamma^\mu e_L$$
$$= -\bar{\nu}_L \gamma^\mu e_L$$

(c) The same result follows using,

$$\nu_R^c = C \bar{\nu}_L^T$$
$$e_R^c = C \bar{e}_L^T$$
$$\bar{e}_R^c = -e_L^T C^{-1}$$

$$\nu_R^{cT} C^{-1} N_R = -\bar{\nu}_L N_R.$$

(2) (a) Show that the Dirac mass term

$$m_D \bar{\psi} \psi = \frac{1}{2} m_D \left( \bar{\psi} \psi + \bar{\psi}^c \psi^c \right)$$
$$= \frac{1}{2} m_D \left( \bar{\psi}_L \psi_R + \bar{\psi}_R \psi_L + \bar{\psi}_R^c \psi_L^c + \bar{\psi}_L^c \psi_R^c \right)$$

(b) Show that the Majorana mass term:

$$m_M \left[ \psi^T C^{-1} \psi + h.c \right] = m_M \left[ \psi^T C^{-1} \psi - \bar{\psi} C \bar{\psi}^T \right]$$
$$= -m_M \left[ \bar{\psi}^c \psi + \bar{\psi} \psi^c \right]$$
$$= m_M \left[ \psi_L^T C^{-1} \psi_L + \psi_R^T C^{-1} \psi_R + h.c. \right]$$
$$= -m_M \left[ \bar{\psi}_L^c \psi_R + \bar{\psi}_R^c \psi_L + h.c. \right]$$

For neutrino $\nu$ : $\psi_L = \nu_L$, $\psi_R = 0$. Hence for neutrino $\nu$, the Majorana mass

$$-m_M \left[ \bar{\nu}_R^c \nu_L + h.c. \right] = m_M \left[ \nu_L^T C^{-1} \nu_L + h.c. \right].$$

(3) From equation

$$\frac{\sigma \cdot \mathbf{p}}{|\mathbf{p}|} u(\mathbf{p}) = -u(\mathbf{p})$$

derive

$$\frac{\sigma \cdot \mathbf{p}}{|\mathbf{p}|} v(\mathbf{p}) = v(\mathbf{p}).$$

## 12.9 References

1. T. D. Lee and C. S. Wu, Weak Interactions, Ann. Rev. Nucl. Sci. 15, 381 (1965).
2. R. E. Marshak, Riazuddin and C. P. Ryan, Theory of Weak Interaction in Paticle Physics, Wiley-Interscience, New York (1969).
3. Weak Interaction as Probes of Unification (VPI-1980) AIP Conference Proceedings No. 72 [Editors G. B. Collins, L. N. Chang and J. R. Ficene], AIP, New York (1981), see in particular parts IA and IIA.
4. S.J. Parke, Phys. Rev. Lett. 57, 1275 (1986).
5. F. Boehm and P. Vogel, Physics of Heavy Neutrinos, Cambridge Univ. Press, Cambridge, U.K. (1987).
6. R. Eicher, Nucl. Phys. B (Proc. Supp.) 3, 389 (1988).
7. J. N. Bahcall and R. K. Ulrich, Rev. Mod. Phys. 60, 217 (1988); see also S. Turck-Chiez et al., Astrophys. J. 335, 415 (1988).
8. R. Davis, Jr., A. K. Mann and L. Wolfenstein, Ann. Rev. Nucl. Parti. Sci. 39, 467 (1989).
9. T. K. Huo and J. Pantalone, Rev. Mod. Phys. 61, 937 (1989).
10. J. N. Bahcell, Neutrino Astrophysics, Cambridge University Press, Cambridge, England, (1989).
11. R. Kolb and M. Turner, The Early Universe, Addison and Wesley, California, (1990).
12. S. M. Bilanky, Neutrinos, Past, present, future hep–ph/9710251
13. A Balantekin, exact solutions for matter enhanced neutrino oscillation, hep - ph/9712304
14. H.V. Klapdor-Kleingrothaus *at al.*, Mod. Phys. Lett. 16, 2409 (2001).
15. R.N. Mohapatra *et al.*, hep-ph/0412099
16. R.D. McKeown and p. Vogel, hep-ph/0402025.
17. S. F. King, Rept. prog. Phys. **67**, 107, (2004)
18. Boris Kayser, hep-ph/0504052 and arXiv:1012.4469.
19. A. de Gouvea, hep-ph/0503086 and hep-ph/0401220.
20. Paul Langacker, *Int. J. Mod. Phys* **A (20)**, 5254, (2005).
21. R.N. Mohapatra and A.Y. Smirnov, hep-ph/0603118, Ann. Rev. Nucl. Part. Sci. **59**, 569 (2006) and A.Y. Smirnov, J. Phys.: *Conf. Ser.* **53** 44, (2006).
22. Luca Merlo, arXiv:0909.2760.
23. Carl H. Albright, arXiv:0911.2437.
24. Guido Altarelli, arXiv:1011.5342.
25. M. Zralek, arXiv:1012.2390.

26. Eligio Lisi, Neutrinos: Theory review, ICHEP-Paris (2010).
27. G. Altarelli and F. Feruglio, Rev. Mod. Phys. **82**, 2701, (2010).
28. C.S. Lam *et.al.* Prog. Theor. Phys. Suppl. **183**, 1 (2010)
29. Particle Data Group, K. Nakamura, et.al., Journal of Physics, G **37**, 0750212 (2010).

# Chapter 13

# Electroweak Unification

## 13.1  Introduction

The Fermi theory of $\beta$-decay cannot be the fundamental theory of weak interactions. It leads to many difficulties; it is non-renormalizable theory. In this theory the scattering cross section for the process $\nu_\mu + e^- \rightarrow \nu_e + \mu^-$ is given by Eq. (2.155):

$$\sigma_s = \frac{G_F^2}{\pi} s. \tag{13.1}$$

The above scattering is purely S-wave. Now Eq. (3.177) $[\lambda_1 = \lambda_2 = \pm 1/2]$ gives $\sigma_s = \frac{4\pi}{2} \left| 2F^0 \right|^2$ [the factor 2 in the denominator is average over initial electron spin], where $F^0 = \frac{\eta_0 \, e^{2i\delta_0} - 1}{2ip}$. Now the maximum absorption occurs when $\eta_0 = 0$, so that $\left| F^0 \right|^2 \leq \frac{1}{4p^2} = \frac{1}{s}$. Thus the partial wave unitarity gives

$$\sigma_s = 8\pi \left| F^0 \right|^2 \leq \frac{8\pi}{s} \tag{13.2}$$

so that from Eq. (13.1)

$$\frac{G_F^2}{\pi} s \leq \frac{8\pi}{s}$$

or

$$\frac{G_F \, s}{2\sqrt{2} \, \pi} \leq 1. \tag{13.3}$$

Hence Fermi theory breaks down for $s > (2\sqrt{2}/G_F) = (0.9 \text{ TeV})^2$. Therefore, we need a cut-off $\Lambda_F$ signifying new physics beyond $\Lambda_F$ where from Eq. (13.3)

$$\Lambda_F^{PWU} \leq 0.9 \text{ TeV}. \tag{13.4}$$

Here $PWU$ signifies that this has been obtained from partial wave unitarity. On the other hand, if weak interactions are mediated through vector boson $W$, then instead of Eq. (13.1),

$$\sigma_s = \left(\frac{g_W^2}{4\pi}\right)^2 \frac{32 \pi s}{(s + m_W^2) m_W^2}$$

$$= \frac{G_F^2}{\pi} \frac{s}{\left(1 + \frac{s}{m_W^2}\right)} \tag{13.5}$$

which is finite for all energies, approaching the limiting value

$$\sigma \rightarrow \left(\frac{G_F^2}{\pi}\right) m_W^2.$$

Thus we see from Eq. (13.1) that the W-boson mass $m_W$ provides the cut-off $\Lambda_F$. As we shall see $m_W \approx 80$ GeV, so that $m_W \ll \Lambda_F = 0.9$ TeV.

The charged weak interactions like electromagnetic interaction are vector in character $(V - A)$ and if the mediators of these interactions are vector bosons, then the universality of weak interactions suggests that the underlying theory of these interactions is a gauge theory. Since weak interactions have short range, the vector bosons associated with them must be massive. But the mass term is not gauge invariant. However, if the gauge symmetry is spontaneously broken, then the gauge vector bosons acquire mass. In this way all the desirable features of a gauge theory like universality and renormalizability are preserved.

## 13.2 Spontaneous Symmetry Breaking and Higgs Mechanism

Consider the Lagrangian for the scalar field $\phi(x)$

$$\mathcal{L} = \partial^\mu \bar{\phi} \, \partial_\mu \phi - \mu^2 \, \bar{\phi} \, \phi - \lambda \left(\bar{\phi} \, \phi\right)^2 \tag{13.6}$$

$$V(\phi) = \mu^2 \, \bar{\phi} \, \phi + \lambda \left(\bar{\phi} \, \phi\right)^2$$

The above Lagrangian is invariant under the global gauge transformation

$$\phi(x) = e^{i\Lambda} \phi(x) \tag{13.7}$$

Now

$$\frac{\partial V}{\partial \phi} = \bar{\phi} \left[\mu^2 + 2\lambda \, \bar{\phi} \, \phi\right]$$

It is usual to choose $\lambda > 0$, since for $\lambda < 0$, $V(\phi)$ would have no minimun. For $\mu^2 > 0$, $\frac{\partial V}{\partial \phi} = 0$ at $\phi = 0$. Then we have ordinary field theory of scalar particles of mass $\mu$ and $V(\phi)$ has a local minimum at $\phi = 0$. For $\mu^2 < 0$, $\frac{\partial V}{\partial \phi} = 0$ gives

$$\bar{\phi}\, \phi = |\phi|^2 = -\frac{\mu^2}{2\lambda} = \phi_0^2 \tag{13.8}$$

For this case $V(\phi)$ has a local minimum at $|\phi|^2 = -\frac{\mu^2}{2\lambda}$, $\phi = \pm\sqrt{-\frac{\mu^2}{2\lambda}}$ and the vacuum is degenerate (see Fig. 13.1.). This is a classical approximation to the vacuum expectation value of $\phi$:

$$\langle 0\,|\phi(x)|\,0\rangle = \sqrt{-\frac{\mu^2}{2\lambda}} \equiv \frac{v}{\sqrt{2}}. \tag{13.9}$$

Although the Lagrangian in Eq. (13.6) is invariant under the gauge transformation Eq. (13.7), but the ground state is not. This can be seen as follows.

$$\langle 0\,|\phi(x)|\,0\rangle = \langle 0\,|U^{-1}U\phi(x)U^{-1}U|\,0\rangle$$
$$= \langle 0\,|U^{-1}Ue^{i\Lambda}\phi(x)U^{-1}U|\,0\rangle$$

If $U\,|0\rangle = |0\rangle$, i.e. if the vacuum is invariant under the gauge transformation, then

$$\langle 0\,|\phi(x)|\,0\rangle = e^{i\Lambda}\,\langle 0\,|\phi(x)|\,0\rangle\,,$$

i.e. if $\langle 0\,|\phi(x)|\,0\rangle \neq 0$, $e^{i\Lambda} = 1$, for every $\Lambda$, a contradiction. Hence $U\,|0\rangle \neq |0\rangle$, i.e. the vacuum is not invariant under the gauge transformation. This is the case for spontaneous symmetry breaking. Now $\phi(x)$ can be written in the form

$$\phi(x) = \frac{1}{\sqrt{2}}\,(v + H(x) + i\eta(x)) \tag{13.10}$$

$$\langle 0\,|H + i\eta|\,0\rangle = 0$$

For the case of the spontaneous breaking, $\mathcal{L}$ in terms of the fields $H(x)$ and $\eta(x)$ is given by

$$\mathcal{L} = \frac{1}{2}\partial^\mu H\,\partial_\mu H + \frac{1}{2}\partial^\mu \eta\,\partial_\mu \eta - \frac{1}{2}(2\lambda\, v^2)H^2$$
$$-\frac{\lambda}{4}\left[4vH(H^2 + \eta^2) + (H^2 + \eta^2)^2\right] - \frac{1}{4}\lambda v^4 \tag{13.11}$$

Note that $\frac{1}{4}\lambda v^4$ is the vacuum energy. Thus we have a massive scalar particle $H$ with mass $m_H^2 = 2\lambda\, v^2$ and a massless scalar particle $\eta$, so-called Goldstone boson. Hence the "Spontaneous symmetry breaking" of a global symmetry implies the existence of a massless spin zero particle (Goldstone boson). The symmetry that is spontaneously broken is still a symmetry of the Lagrangian but not of the Hamiltonian.

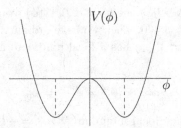

Fig. 13.1   Effective potential $V(\phi)$ for $\mu^2 < 0$, showing local minima.

### 13.2.1   *Higgs Mechanism*

For the local gauge transformation

$$U(x) = e^{i\Lambda(x)}, \phi(x) \rightarrow e^{i\Lambda(x)}\phi(x) \tag{13.12}$$

the situation becomes quite different. Gauge invariance requires a massless vector field $B_\mu$ :

$$B_\mu \rightarrow B_\mu - \frac{1}{g}\partial_\mu \Lambda \tag{13.13}$$

Then we can write the gauge invariant Lagrangian by replacing $\partial_\mu$ by the covariant derivative in Eq. (13.6)

$$\partial_\mu\phi \rightarrow D_\mu\phi = (\partial_\mu - igB_\mu)\phi$$
$$\mathcal{L} = -\frac{1}{4}B^{\mu\nu} B_{\mu\nu} + (\partial_\mu + igB_\mu)\bar{\phi}(\partial_\mu - igB_\mu)\phi - V(\phi) \tag{13.14}$$

where

$$B^{\mu\nu} = \partial^\mu B^\nu - \partial^\nu B^\mu$$

The unwanted zero mass mode due to spontaneous symmetry breaking can be eliminated by means of field dependent gauge transformation:

$$\phi(x) \rightarrow \frac{1}{\sqrt{2}}[v + H(x)]e^{i\eta(x)/v}$$

$$B_\mu(x) \rightarrow B_\mu(x) + \frac{1}{vg}\partial_\mu\eta(x)$$

$$D^\mu\bar{\phi} \equiv (\partial_\mu + igB^\mu)\bar{\phi} \Rightarrow e^{-i\eta/v}\frac{1}{\sqrt{2}}[\partial^\mu H + ig(v + H)B^\mu]$$

$$D_\mu\phi \equiv (\partial_\mu - igB_\mu)\phi \Rightarrow e^{i\eta/v}\frac{1}{\sqrt{2}}[\partial_\mu H - ig(v + H)B_\mu] \tag{13.15}$$

Hence we have

$$\mathcal{L} = -\frac{1}{4}B^{\mu\nu} B_{\mu\nu} + \frac{1}{2}\partial^\mu H \, \partial_\mu \, H + \frac{1}{2}g^2(v^2 + 2vH + H^2)B^\mu B_\mu$$
$$-\frac{1}{2}(\lambda \, v^2)H^2 - \frac{\lambda}{4}H^3(H + 4v) - \frac{1}{4}\lambda v^4 \qquad (13.16)$$

where $m_B^2 = \frac{1}{2}g^2v^2$, $m_H^2 = 2\lambda \, v^2$ and vacuum energy $= \frac{1}{4}\lambda v^4$. The vector boson becomes massive, would be Goldstone field $\eta(x)$ has been transformed away, it has been eaten away by $B_\mu$ to give it a longitudinal component.

## 13.2.2 Gauge Symmetry Breaking for Chiral $U_1 \otimes U_2$ Group

Consider a simple Lagrangian

$$\mathcal{L} = \bar{\Psi} \, i\gamma^\mu \, \partial_\mu \, \Psi + \partial^\mu \, \bar{\phi} \, \partial_\mu \, \phi - h\bar{\Psi}_L \, \Psi_R \, \phi$$
$$-h\bar{\Psi}_R \, \Psi_L \, \bar{\phi} - \mu^2 \, \bar{\phi} \, \phi - \lambda \, (\bar{\phi} \, \phi)^2$$
$$= \bar{\Psi}_L \, i\gamma^\mu \, \partial_\mu \, \Psi_L + \bar{\Psi}_R \, i\gamma^\mu \, \partial_\mu \, \Psi_R + \partial^\mu \, \bar{\phi} \, \partial_\mu \, \phi$$
$$-h\bar{\Psi}_L \, \phi \, \Psi_R - h\bar{\Psi}_R \, \bar{\phi} \, \Psi_L - V(\phi) \qquad (13.17)$$

where

$$\Psi_L = \frac{1}{2}(1 - \gamma_5) \, \Psi \qquad (13.18a)$$

$$\Psi_R = \frac{1}{2}(1 + \gamma_5) \, \Psi \qquad (13.18b)$$

are left-handed and right-handed fermion fields respectively. $\phi$ is a complex scalar field interacting with fermion having a coupling strength $h$. $V(\phi)$ is given by

$$V(\phi) = \mu^2 \, \bar{\phi} \, \phi + \lambda \, (\bar{\phi} \, \phi)^2. \qquad (13.19)$$

Consider the gauge transformations

$$\Psi_L \to e^{i\Lambda_1(x)} \, \Psi_L,$$
$$\Psi_R \to e^{i\Lambda_1(x)} \, \Psi_R, \qquad (13.20a)$$
$$\phi \to \phi$$

$$\Psi_L \to e^{i\Lambda_2(x)} \, \Psi_L,$$
$$\Psi_R \to e^{-i\Lambda_2(x)} \, \Psi_R, \qquad (13.20b)$$
$$\phi \to e^{2i\Lambda_2(x)} \, \phi.$$

Obviously the Lagrangian in Eq. (13.17) is invariant under the gauge transformations (13.20a) if $\Lambda_1$ and $\Lambda_2$ are constants. The gauge group corresponding to gauge transformations (13.20b) is $U_1(1) \otimes U_2(1)$, where we identify $U_1$ with $U_{em}$. If we require the Lagrangian in Eq. (13.17) to be local gauge invariant, then we must introduce two massless gauge fields $A_\mu$ and $B_\mu$, which transform as

$$A_\mu \to A_\mu - \frac{1}{e}\partial_\mu \Lambda_1 \qquad (13.21a)$$

$$B_\mu \to B_\mu + \frac{1}{g}\partial_\mu \Lambda_2. \qquad (13.21b)$$

Then we can write the gauge invariant Lagrangian by replacing $\partial_\mu$ by the covariant derivatives:

$$\partial_\mu \Psi_L \to D_\mu\Psi_L = (\partial_\mu + ieA_\mu - igB_\mu)\,\Psi_L,$$

$$\partial_\mu \Psi_R \to D_\mu\Psi_R = (\partial_\mu + ieA_\mu + igB_\mu)\,\Psi_R,$$

$$\partial_\mu \phi \to D_\mu\phi = (\partial_\mu - 2igB_\mu)\,\phi$$

in Eq. (13.6). Hence the gauge invariant Lagrangian is given by

$$\mathcal{L} = -\frac{1}{4}A^{\mu\nu}\,A_{\mu\nu} - \frac{1}{4}B^{\mu\nu}\,B_{\mu\nu} + \bar{\Psi}_L\,i\gamma^\mu\,(\partial_\mu + ieA_\mu - igB_\mu)\,\Psi_L$$

$$+\bar{\Psi}_R\,i\gamma^\mu\,(\partial_\mu + ieA_\mu + igB_\mu)\,\Psi_R - h\left(\bar{\Psi}_L\phi\Psi_R + \bar{\Psi}_R\bar{\phi}\Psi_L\right)$$

$$+(\partial^\mu + 2igB^\mu)\,\bar{\phi}\,(\partial_\mu - 2igB_\mu)\,\phi - V(\phi). \qquad (13.22)$$

$U(\Lambda_2)\,|0\rangle \neq |0\rangle$ , and the gauge symmetry is spontaneously broken, i.e. $U_1 \times U_2 \to U_1, U_1$ is unbroken.

For spontaneous symmetry breaking we write

$$\phi = \phi' + \frac{v}{\sqrt{2}} = \frac{H + i\eta}{\sqrt{2}} + \frac{v}{\sqrt{2}}$$

$$\langle 0\,|\phi|\,0\rangle = \frac{v}{\sqrt{2}} \qquad (13.23)$$

where $H$ and $\eta$ are hermitian fields with zero expectation values. The Lagrangian (13.17), in terms of the fields $H$ and $\eta$ has the form

$$\mathcal{L} = -\frac{1}{4}A^{\mu\nu}\,A_{\mu\nu} + \bar{\Psi}\left(i\gamma^\mu\partial_\mu - \frac{hv}{\sqrt{2}}\right)\Psi - e\bar{\Psi}\gamma^\mu\Psi A_\mu$$

$$-g\bar{\Psi}\gamma^\mu\gamma_5\Psi B_\mu - \frac{h}{\sqrt{2}}\bar{\Psi}\,(H + i\gamma_5\eta)\,\Psi + \mathcal{L}_B - \tilde{V}(\phi) \qquad (13.24)$$

where

$$\mathcal{L}_B = -\frac{1}{4}B^{\mu\nu}\,B_{\mu\nu} + \frac{1}{2}(\partial_\mu H + 2gB_\mu\eta)^2 + \frac{1}{2}(\partial_\mu\eta - 2gB_\mu(v + H))^2$$

$$(13.25)$$

$$\tilde{V}(\phi) = \frac{1}{2}\left(2\lambda v^2\right)H^2 + \lambda vH(H^2 + \eta^2) + \frac{\lambda}{4}(H^2 + \eta^2)^2 - \frac{\lambda}{4}v^2 \qquad (13.26)$$

If $U_2$ is a global gauge group, then as already seen in previous section we have the Goldstone-Nambu theorem. A spontaneous breakdown of global symmetry leads to a massless scalar particle. But when $U_2$ is a local gauge symmetry, then due to the presence of the term $2gvB^\mu \partial_\mu \eta$ in $\mathcal{L}_B$ a straightforward interpretation of (13.24) is not possible. But we can eliminate this term by a field dependent gauge transformation. Actually what happens is that $\partial_\mu \eta$ combines with $B_\mu$ (which has only transverse components) to form a single massive spin 1 field, $\partial_\mu \eta$ now becomes longitudinal mode of spin 1 field. This can explicitly be seen as follows: Choose the gauge function $\Lambda_2(x)$ to be $\frac{\eta(x)}{2v}$. Then under the gauge transformations

$$\Psi_L = e^{\frac{i\eta(x)}{2v}} \hat{\Psi}_L \quad , \quad \Psi_R = e^{\frac{-i\eta(x)}{2v}} \hat{\Psi}_R$$

$$A_\mu = \hat{A}_\mu \quad , \quad B_\mu = \hat{B}_\mu + \frac{1}{2vg} \partial_\mu \eta(x)$$

$$\phi(x) = \frac{1}{\sqrt{2}} [v + H(x)] e^{\frac{i\eta(x)}{v}} , \tag{13.27}$$

the Lagrangian (13.24) becomes (removing ˆ):

$$\mathcal{L} = -\frac{1}{4} A^{\mu\nu} A_{\mu\nu} - \frac{1}{4} B^{\mu\nu} B_{\mu\nu} + \frac{1}{2} \left(4g^2 v^2\right) B^\mu B_\mu$$

$$+ \bar{\Psi} \left(i\gamma^\mu \partial_\mu - \frac{hv}{\sqrt{2}}\right) \Psi - e\bar{\Psi}\gamma^\mu \Psi A_\mu - g\bar{\Psi}\gamma^\mu \gamma_5 \Psi B_\mu$$

$$- \frac{h}{\sqrt{2}} \bar{\Psi}\Psi H + \frac{1}{2} \left(\partial_\mu H\right)^2 - \frac{1}{2} \left(2v^2\lambda\right) H^2$$

$$+ 2g^2 B^\mu B_\mu \left(H^2 + 2vH\right) - v\lambda H^3 - \frac{1}{4}\lambda H^4. \tag{13.28}$$

It is clear from Eq. (13.28), that the would be Goldstone boson field $\eta(x)$ has been transformed away; it has been eaten away by the field $B_\mu$ to give a longitudinal component. This mechanism is called the Higgs-Kibble mechanism. The massive scalar particle $H$ is called the Higgs particle. To summarize: (i) No massless scalar boson appears. (ii) $A_\mu$ which is associated with unbroken gauge symmetry (electric charge conservation) has zero mass. (iii) The vector boson $B_\mu$ has acquired a mass $m_B = 2gv$. (iv) The fermion field has acquired a mass $m_f = \frac{hv}{\sqrt{2}}$ (v) Both the masses of $B_\mu$ and $\Psi$ arise due to the same symmetry breaking mechanism. (vi) A massive scalar particle with mass $\sqrt{2\lambda v^2}$ appears. This particle is called Higgs particle. Presence of Higgs scalar is an essential feature of spontaneously broken gauge symmetry.

## 13.3   Renormalizability

We give here few remarks about the renormalizability of a gauge theory. Now the fields $A_\mu$ and $B_\mu$ cannot be determined uniquely by field equations. In order to quantize these fields, one has to fix a gauge that is to say break gauge invariance. For the photon field $A_\mu$, a term added to the Lagrangian for this purpose is $-\frac{1}{2}\xi^{-1}\left(\partial^\mu A_\mu\right)^2$. Photon propagator is then given by

$$i\left[-g_{\mu\nu} + (1-\xi)\frac{k_\mu k_\nu}{k^2}\right]\frac{1}{k^2}.$$

For the field $B_\mu$, the gauge fixing term is

$$-\frac{1}{2}\xi^{-1}\left(\partial^\mu B_\mu - \xi m_B \eta_2\right)^2. \tag{13.29}$$

It is so chosen that it cancels awkward looking mixing term $B^\mu \partial_\mu \eta$ in the Lagrangian (13.25). $\xi$ is a parameter which determines the gauge.

The quadratic part of the Lagrangian (13.25) is then given by

$$\mathcal{L}_2 = -\frac{1}{2}B_\mu[-g^{\mu\nu}\partial^2 + \left(1 - \frac{1}{\xi}\right)\partial^\mu\partial^\nu - (2gv)^2 g^{\mu\nu}]B_\nu$$
$$+\frac{1}{2}\left(\partial_\mu H\right)^2 - \frac{1}{2}m_H^2 H^2 + \frac{1}{2}\left(\partial_\mu\eta\right)^2 - \frac{\xi}{2}(2gv)^2\eta^2 \tag{13.30}$$

where $m_B = 2gv, m_\eta^2 = \xi(2gv)^2 = \xi m_B^2$. For this Lagrangian, we have Feynman rules:

(i)   gauge boson propagator

$$i\left[-g_{\mu\nu} + (1-\xi)\frac{k_\mu k_\nu}{k^2 - \xi m_B^2}\right]\frac{1}{k^2 - m_B^2}$$

which is the inverse of

$$-i\left[g^{\mu\nu}(k^2 - m_B^2) + k^\mu k^\nu(1 - \frac{1}{\xi})\right]$$

(ii)  $\eta$ (Goldstone) boson propagator

$$\frac{i}{k^2 - \xi\, m_B^2}$$

(iii) Higgs boson propagator

$$\frac{i}{k^2 - m_H^2}$$

These forms of propagators are expected to give a renormalizable theory for any finite value of $\xi$ since they have a good high $k^2$ behavior, falling like $\frac{1}{k^2}$. This is called R-gauge. The fields $B_\mu$ and $\eta$ separately have no physical significance. In particular the poles at $k^2 = \xi m_B^2$ are unphysical and are canceled out in any S-matrix element, which is also independent of $\xi$. To see this let us consider fermion-fermion scattering through $B - meson$, $\eta - meson$ as represented by the Feynman diagrams, Fig. 13.2: The scattering ($T$) matrix, using the Lagrangian (13.24), for the relevant vertices for the first diagram is given by

$$iT^{(1)} = (ig)^2 \bar{u}(\acute{p}_2)\gamma^\mu\gamma_5 u(p_2)\frac{-i}{k^2 - m_B^2}$$
$$\times \left[g_{\mu\nu} - (1 - \xi)\frac{k_\mu k_\nu}{k^2 - \xi m_B^2}\right]\bar{u}(\acute{p}_1)\gamma^\nu\gamma_5 u(p_1) \qquad (13.31)$$

The expression in square brackets can be written as

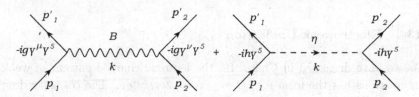

Fig. 13.2   Fermion-fermion scattering through $B$-meson, $\eta$-meson.

$$\left[g_{\mu\nu} - \frac{k_\mu k_\nu}{m_B^2} + k_\mu k_\nu \left(\frac{k^2 - m_B^2}{m_B^2 (k^2 - \xi m_B^2)}\right)\right] \qquad (13.32)$$

Further by using the Dirac Equations

$$(\not{p} - m)\, u(p) = 0, \quad \bar{u}(p)(\not{p} - m) = 0,$$

we can write

$$\bar{u}(\acute{p}_2)\, \not{k}\gamma_5 u(p_2) = 2m_f \bar{u}(p_2')\gamma_5 u(p_2)$$
$$\bar{u}(\acute{p}_1)\, \not{k}\gamma_5 u(p_1) = -2m_f \bar{u}(p_1')\gamma_5 u(p_1),$$

where

$$m_f = \frac{\hbar_f v}{\sqrt{2}}, \ m_B = 2gv$$

and

$$4g^2 m_f^2 = m_B^2 \frac{h_f^2}{2}.$$

Thus the second term in Eq. (13.32) gives

$$-\frac{h_f^2}{2}\bar{u}(\acute{p}_2)\gamma_5 u(p_2)\frac{-i}{k^2 - \xi m_B^2}\bar{u}(\acute{p}_1)\gamma_5 u(p_1) \tag{13.33}$$

which precisely cancels the contribution of the Goldstone boson $\eta$. Thus the fermion-fermion amplitude is independent of $\xi$ and no physical pole at $k^2 = \xi m_B^2$ appears. The result obtained is one which we would have found by neglecting the Goldstone boson and computing the gauge boson exchange by using the uniterilized propagator $\frac{-i}{k^2 - m_B^2}(g_{\mu\nu} - \frac{k_\mu k_\nu}{m_B^2})$. The tensor structure represents a gauge boson polarization sum $\sum \epsilon^\mu(k)\epsilon^{*\nu}(k) = -\left(g_{\mu\nu} - \frac{k_\mu k_\nu}{m_B^2}\right)$ for vector boson on the mass shell having 3 polarization directions. Thus in the cancellation of the $\xi$-dependent part of the gauge propagator, we also find that the Goldstone diagram cancels the contribution of the unphysical time like polarization state of the gauge boson, leaving over the sum required to three physical polarizations.

## 13.4  Electroweak Unification

As we have discussed in Chap. 10, the leptonic charged current of weak interactions has the form $\bar{\nu}_e\gamma_\mu(1 - \gamma_5)e = 2\bar{\nu}_{eL}\gamma^\mu e_L$. The corresponding hadronic charged weak current can be written as $\bar{u}\gamma_\mu(1 - \gamma_5)d' = 2\bar{u}_L\gamma_\mu d'_L$. Here $d'$ means that it is not mass eigenstate. This suggests that we consider

$$\begin{pmatrix} \nu_e \\ e \end{pmatrix}_L, \qquad \begin{pmatrix} u \\ d' \end{pmatrix}_L$$

as left handed doublets in a weak isospin space. The weak currents are then associated with weak isospin raising and lowering operators

$$J_\mu^{(+)} = \bar{\Psi}_L\frac{\tau_+}{2}\gamma_\mu\Psi_L, \qquad J_\mu^{(-)} = \bar{\Psi}_L\frac{\tau_-}{2}\gamma_\mu\Psi_L, \tag{13.34}$$

where $\Psi_L$ is any of the above doublets, $\tau_+ = (\tau_1 + i\tau_2)$ and $\tau_- = (\tau_1 - i\tau_2)$. Let the charges associated with these currents be $Q_+$ and $Q_-$. These charges generate an $SU_L(2)$ algebra

$$[Q_+, Q_-] = 2Q_3. \tag{13.35}$$

The current associated with the charge $Q_3$ is given by

$$J_\mu^3 = \bar{\Psi}_L\frac{1}{2}\tau_3\gamma_\mu\Psi_L. \tag{13.36}$$

The gauge transformation corresponding to the group $SU_L(2)$ is

$$\Psi_L(x) \rightarrow \Psi_L(x) = \exp\left(i\frac{\tau}{2}\cdot\Lambda(x)\right)\Psi_L(x). \tag{13.37}$$

Then the Lagrangian

$$\mathcal{L} = \bar{\Psi}_L i\gamma^\mu D\mu\Psi_L - \frac{1}{4}\mathbf{W}^{\mu\nu} \cdot \mathbf{W}_{\mu\nu}, \tag{13.38}$$

where

$$D_\mu = \partial_\mu + ig\frac{1}{2}\tau \cdot \mathbf{W}_\mu = \partial_\mu + igW_\mu,$$

$$\left(W_\mu = \frac{1}{2}\tau \cdot \mathbf{W}_\mu\right)$$

$$\mathbf{W}_{\mu\nu} = \partial_\mu\mathbf{W}_\nu - \partial_\nu\mathbf{W}_\mu - g\mathbf{W}_\mu \times \mathbf{W}_\nu \tag{13.39a}$$

$$W_{\mu\nu} \equiv \frac{1}{2}\tau \cdot \mathbf{W}_{\mu\nu} = D_\mu W_\nu - D_\nu W_\mu$$

$$= \partial_\mu W_\nu - \partial_\nu W\mu + ig\left[W_\mu, W_\nu\right], \tag{13.39b}$$

is invariant under the gauge transformations [see Chap. 7]:

$$\Psi_L(x) \to U \Psi_L(x)$$

$$W_\mu \to UW_\mu U^\dagger - \frac{i}{g}U\partial_\mu U^\dagger \tag{13.40a}$$

where $U$ is given in Eq. (13.37). For $\Lambda$ infinitesimal, we get

$$\Psi_L(x) \to \left(1 + i\frac{\tau}{2} \cdot \mathbf{\Lambda}(x)\right)\Psi_L(x)$$

$$\mathbf{W}_\mu \to \mathbf{W}_\mu - \mathbf{\Lambda} \times \mathbf{W}_\mu - \frac{1}{g}\partial_\mu\mathbf{\Lambda}. \tag{13.40b}$$

The gauge group $SU_L(2)$ leads to a neutral current $J_\mu^3$ which is neither observed experimentally nor is identical with the electromagnetic current. It is possible to unify weak and electromagnetic forces into a single gauge force, if we extend the gauge group to $SU_L(2) \times U_Y(1)$. For this group we have two gauge couplings $g$ and $g'$ associated with $SU_L(2)$ and $U_Y(1)$ respectively. The weak hypercharge $Y$ is defined by the relation

$$Q = t_3 + \frac{1}{2}Y = \frac{1}{2}\tau_3 + \frac{1}{2}Y.$$

The gauge vector bosons $W^\pm$, $W^0$ belong to the adjoint representation of $SU_L(2)$ and vector boson $B_\mu$ is associated with $U_Y(1)$.

Fermions belong to either fundamental representation [doublet] or trivial representation [singlet]. In view of the structure of charged weak currents given in Eq. (13.34), it is natural to put left-handed fermions into a doublet while the right-handed fermions (except neutrinos, which exist only in left-handed chiral state) are put in the singlet representation of $SU_L(2)$. Thus in the Standard Model, the fermions for each generation belong to the following representations of the group $SU_C(3) \otimes SU_L(2) \otimes U_Y(1)$

$$\Psi_L : \begin{pmatrix} u_i \\ d_i \end{pmatrix}_L : \mathbf{3, 2}, \tfrac{1}{3}$$

$$: \begin{pmatrix} \nu_i \\ e_i \end{pmatrix}_L : \mathbf{1, 2} -1$$

$$\Psi_R : \begin{matrix} u_{iR} & \mathbf{3, 1}, & \tfrac{4}{3} \\ d_{iR} : & \mathbf{3, 1}, & -\tfrac{2}{3} \\ e_{iR} & \mathbf{1, 1}, & -2 \end{matrix} \qquad (13.41)$$

where $i$ is the generation index and hypercharges Y are fixed by $Q = I_3 + \tfrac{1}{2}Y$. The group $SU_C(3)$ is essential as the color plays a crucial role in the cancellation of gauge anomalies (see Sec. 13.12) needed for the renormalizability of the model. Except for this, it does not play any other rule in electroweak interaction and as such we now confine to the electroweak unification group $SU_L(2) \otimes U_Y(1)$. The three generations are

$$Quarks : Leptons$$

$$\begin{pmatrix} u\ c\ t \\ d\ s\ b \end{pmatrix} : \begin{pmatrix} \nu_e\ \nu_\mu\ \nu_\tau \\ e\ \ \mu\ \ \tau \end{pmatrix} \qquad (13.42)$$

The weak eigenstates $d', s', b'$ are not identical with mass eigenstates $d, s, b$ but are related by a unitary matrix $V_d$ and similarly for $u, c, t$.

$$\begin{pmatrix} d' \\ s' \\ b' \end{pmatrix} = V_d \begin{pmatrix} d \\ s \\ c \end{pmatrix}$$

$$\begin{pmatrix} u' \\ c' \\ t' \end{pmatrix} = V_u \begin{pmatrix} u \\ c \\ t \end{pmatrix} \qquad (13.43)$$

The $CKM$ matrix is

$$V_{CKM} = V_u^\dagger V_d = V = \begin{pmatrix} V_{ud}\ V_{us}\ V_{ub} \\ V_{cd}\ V_{cs}\ V_{cb} \\ V_{td}\ V_{ts}\ V_{tb} \end{pmatrix} \qquad (13.44)$$

We can do the same thing for the leptons as was discussed in Chap. 12. However, as neutrino masses are negligible compared to the corresponding charged lepton masses, we would not consider it further in what follows. We can select a basis in which $V_u$ is diagonal. Then there is no need to put primes on $u, c, t$ and $V_{CKM}$ is essentially $V_d$.

In order to break the gauge symmetry spontaneously so that weak vector bosons acquire their masses and fermions also get masses, we need Higgs doublet $\phi$ :

$$\phi = \begin{pmatrix} \phi^+ \\ \phi^0 \end{pmatrix}, \qquad Y = 1. \tag{13.45}$$

The Lagrangian invariant under local gauge transformations

$$\Psi_L \rightarrow \exp\left(\frac{i}{2}\tau \cdot \mathbf{\Lambda} + \frac{i}{2}Y_L\Lambda_0\right)\Psi_L$$

$$\Psi_R \rightarrow \exp\left(\frac{i}{2}Y_R\Lambda_0\right)\Psi_R \tag{13.46}$$

is given by

$$\mathcal{L} = \bar{\Psi}_L i\gamma^\mu \left(\partial_\mu + \frac{i}{2}g\tau \cdot \mathbf{W}_\mu + \frac{i}{2}g'Y_L B_\mu\right)\Psi_L + \bar{\Psi}_{Ra}i\gamma^\mu\left(\partial_\mu + \frac{i}{2}g'Y_{Ra}B_\mu\right)\Psi_{Ra}$$

$$+ \left(\partial^\mu\bar{\phi} - \frac{i}{2}g\bar{\phi}\tau \cdot \mathbf{W}^\mu - \frac{i}{2}g'\bar{\phi}B^\mu\right)\left(\partial_\mu\phi + \frac{i}{2}g\tau \cdot \mathbf{W}_\mu\phi + \frac{i}{2}g'B_\mu\phi\right)$$

$$- h_1\left[\bar{\Psi}_L\phi\Psi_{R1} + \bar{\Psi}_{R1}\bar{\phi}\Psi_L\right] - h_2\left[\bar{\Psi}_L\phi\Psi_{R2} + \bar{\Psi}_{R2}\bar{\phi}\Psi_L\right]$$

$$- \frac{1}{4}\mathbf{W}^{\mu\nu} \cdot \mathbf{W}_{\mu\nu} - \frac{1}{4}B^{\mu\nu} \cdot B_{\mu\nu} - V(\phi) \tag{13.47}$$

where $W_{\mu\nu}$ is given in Eq. (13.39b) and

$$B_{\mu\nu} = \partial_\mu B_\nu - \partial_\nu B_\mu \tag{13.48}$$

$$V(\phi) = \mu^2\bar{\phi}\phi + \lambda\left(\bar{\phi}\phi\right)^2 \tag{13.49}$$

$$\tilde{\phi} = i\tau_2\bar{\phi} = \begin{pmatrix} \bar{\phi}_0^* \\ -\phi^- \end{pmatrix} \tag{13.50}$$

$a = 1,2$ with $\Psi_{R1} = e_R$ or $d_R, \Psi_{R2} = u_R$. Under infinitesimal gauge transformation, vector fields $W_\mu$ transform as given in Eq. (13.40a), but $B_\mu$ transforms as

$$B_\mu \rightarrow B_\mu - \frac{1}{g'}\partial_\mu\Lambda_0. \tag{13.51}$$

In order to break the gauge symmetry spontaneously, assume that

$$\langle\phi\rangle_0 = \begin{pmatrix} 0 \\ \frac{v}{\sqrt{2}} \end{pmatrix}, \tag{13.52}$$

where $v = \sqrt{-\mu^2/\lambda}$, $\langle\phi\rangle_0 = \langle 0|\phi|0\rangle$. In this way, not only $SU_L(2)$ is broken but $U_Y(1)$ is also broken, but it leaves the group $U(1)$ corresponding to electric charge unbroken viz $SU_L(2) \times U_Y(1)$ is broken to $U_Q(1)$. We can now write Eq. (13.45) as

$$\phi = \begin{pmatrix} \phi^+ \\ \frac{(\phi_1 + i\phi_2)}{\sqrt{2}} + \frac{v}{\sqrt{2}} \end{pmatrix}, \tag{13.53}$$

where $\phi^+$ and hermitian fields $\phi_1$ and $\phi_2$ have zero vacuum expectation values. We can select a gauge such that $\phi^+$ and $\phi_2$ disappear from the theory. Instead $\partial_\mu \phi^\pm$ and $\partial_\mu \phi_2$ provide longitudinal components to $W^\pm$ and one of neutral vector bosons respectively. Thus out of the four gauge vector bosons, three become massive and the remaining one remains massless. This massless vector boson is the photon corresponding to unbroken $U_Q(1)$ symmetry. All this amounts to replacing $\phi$ given in Eq. (13.53) by ($\phi_1 = H$)

$$\phi = \begin{pmatrix} 0 \\ \frac{H+v}{\sqrt{2}} \end{pmatrix}. \tag{13.54}$$

With Eq. (13.54), the following term of the Lagrangian (13.47)

$$\left[ \partial^\mu \bar{\phi} - \frac{i}{2} g \bar{\phi} \tau \cdot \mathbf{W}^\mu - \frac{i}{2} g' \bar{\phi} B^\mu \right] \left[ \partial_\mu \phi + \frac{i}{2} g \tau \cdot \mathbf{W}_\mu + \frac{i}{2} g' B_\mu \phi \right]$$

gives

$$\mathcal{L}^{W-H} = \frac{1}{2} \partial^\mu H \partial_\mu H + \frac{g^2}{8} \left( H^2 + 2vH + v^2 \right) \left( 2W_\mu^+ W^{\mu-} + W^{3\mu} W_{3\mu} \right)$$
$$+ \frac{g'^2}{8} \left( H^2 + 2vH + v^2 \right) B^\mu B_\mu - \frac{gg'}{4} \left( H^2 + 2vH + v^2 \right) W^{3\mu} B_\mu, \tag{13.55}$$

where $W_\mu^\pm = (W_{1\mu} \mp iW_{2\mu})/\sqrt{2}$, $\tau \cdot \mathbf{W}_\mu = \begin{pmatrix} W_{3\mu} & \sqrt{2}W_\mu^+ \\ \sqrt{2}W_\mu^- & -W_{3\mu} \end{pmatrix}$. From this equation, it is clear that vector bosons $W_\mu^\pm$ have acquired a mass:

$$m_W^2 = \frac{1}{4} g^2 v^2. \tag{13.56a}$$

For the neutral vector bosons, the mass terms in Eq. (13.55) give the matrix

$$M^2 = \frac{1}{4} \begin{pmatrix} g^2 v^2 & -gg'v^2 \\ -gg'v^2 & g'^2 v^2 \end{pmatrix}. \tag{13.56b}$$

Since $\det (M^2) = 0$, therefore one of the eigenvalues of $M^2$ is zero. The mass matrix (13.56b) can be diagonalize by defining the physical fields $A_\mu$, $Z_\mu$:

$$A_\mu = \cos\theta_W B_\mu + \sin\theta_W W_{3\mu},$$
$$Z_\mu = -\sin\theta_W B_\mu + \cos\theta_W W_{3\mu}. \tag{13.57}$$

Then we get

$$m_A^2 = 0, \qquad A_\mu : \text{photon} \tag{13.58}$$

$$m_Z^2 = \frac{1}{4}\left(g^2 + g'^2\right)v^2$$

$$= \frac{1}{4}g^2v^2\left(\frac{1}{\cos^2\theta_W}\right). \tag{13.59}$$

where

$$\tan\theta_W = \frac{g'}{g} \tag{13.60}$$

and the parameter

$$\rho \equiv \frac{m_W^2}{m_Z^2 \cos^2\theta_W} = 1. \tag{13.61}$$

The fermion masses are given by

$$m_i = h_i \frac{v}{\sqrt{2}} \tag{13.62}$$

Note that each fermion mass has a new coupling, indicating that fermion masses need a more fundamental theory. Further $V(\phi)$ goes into

$$\mu^2(\frac{H+v}{\sqrt{2}})^2 + \lambda(\frac{H+v}{\sqrt{2}})^4$$

giving the Higgs mass

$$m_H^2 = \mu^2 + 3\lambda v^2 = 2\lambda v^2, v = \sqrt{-\mu^2/\lambda} \tag{13.63}$$

From Eq. (13.47), using Eqs. (13.54), (13.57), (13.58) and (13.62), and $Q = T^3 + \frac{1}{2}Y$, the Lagrangian for the fermions can be written as:

$$\mathcal{L}_F = \bar{\Psi}_i\left(i\gamma^\mu\,\partial_\mu - m_i - \frac{gm_i}{2m_W}H\right)\Psi_i$$

$$-\frac{g}{2\sqrt{2}}\bar{\Psi}_i\,\gamma^\mu\left(1-\gamma_5\right)\left(T^+W_\mu^+ + T^-W_\mu^-\right)\Psi_i$$

$$-e\,\bar{\Psi}_i\,\gamma^\mu\,Q_i\,\Psi_i\,A_\mu - \frac{g}{2\cos\theta_W}\bar{\Psi}_i\,\gamma^\mu\left(g_{Vi} - \gamma^5 g_{Ai}\right)\bar{\Psi}_i\,Z_\mu \tag{13.64}$$

where $\Psi_i = \begin{pmatrix} \nu_i \\ l_i \end{pmatrix}$ and $\begin{pmatrix} u_i \\ d_i' \end{pmatrix}$, $e = g'\,\cos\theta_W = g\,\sin\theta_W = \frac{gg'}{\sqrt{g^2+g'^2}}$, $d_i' = V_{ij}\,d_j\,(V : CKM\text{ matrix})$, $m_i$ is the mass of $i$th fermion and $Q_i$ is its charge. $g_{Vi}$ and $g_{Ai}$ are given by

$$g_{Vi} = \left(T_i^3 - 2Q_i\,\sin^2\theta_W\right), g_{Ai} = T_i^3 \tag{13.65a}$$

$$T^\pm = \frac{1}{2}\tau^\pm, \qquad T^3 = \frac{1}{2}\tau^3 \tag{13.65b}$$

We note that the interaction part of the Lagrangian can be written as

$$\mathcal{L}_{int} = -g \sin\theta_W \, J_{em}^{\mu} \, A_\mu - \frac{g}{2\sqrt{2}} \left( J^{+\mu} \, W_\mu^+ + h.\,c. \right) - \frac{g}{\cos\theta_W} J^{Z\mu} \, Z_\mu$$

$$-g \frac{m_i}{2m_w} \bar{\Psi}_i H \Psi_i \tag{13.66a}$$

where

$$J_{em}^\mu = \bar{\Psi}_i \, \gamma^\mu \, Q_i \, \Psi_i$$

$$= \left( -\bar{e} \, \gamma^\mu \, e + \frac{2}{3}\bar{u} \, \gamma^\mu \, u - \frac{1}{3}\bar{d} \, \gamma^\mu \, d \right) + \cdots \tag{13.66b}$$

$$J^{+\mu} = \frac{1}{2} \, \bar{\Psi}_i \, \gamma^\mu \left( 1 - \gamma^5 \right) \tau^+ \Psi_i$$

$$= \left( \bar{\nu}_e \, \gamma^\mu \left( 1 - \gamma^5 \right) e + \bar{u} \, \gamma^\mu \left( 1 - \gamma^5 \right) d' \right) + \cdots \tag{13.66c}$$

$$J^{Z\mu} = \frac{1}{2} \, \bar{\Psi}_i \left( g_{Vi} \, \gamma^\mu - g_A \, \gamma^\mu \, \gamma^5 \right) \Psi_i$$

$$= \frac{1}{2} J^{3\mu} - \sin^2\theta_W \, J_{em}^\mu$$

$$= \frac{1}{2} \left[ \bar{\Psi}_i \left( \gamma^\mu \left( 1 - \gamma^5 \right) \frac{\tau_3}{2} - 2\sin^2\theta_W \, Q_i \, \gamma^\mu \right) \bar{\Psi}_i \right]$$

$$= \frac{1}{4} \left[ \bar{\nu}_e \gamma^\mu \left( 1 - \gamma^5 \right) \nu_e - \bar{e} \, \gamma^\mu \left( 1 - \gamma^5 \right) e \right.$$

$$+ \bar{u} \, \gamma^\mu \left( 1 - \gamma^5 \right) u - \bar{d} \, \gamma^\mu \left( 1 - \gamma^5 \right) d$$

$$\left. - 4\sin^2\theta_W \left( -\bar{e} \, \gamma^\mu \, e + \frac{2}{3}\bar{u} \, \gamma^\mu \, u - \frac{1}{3}\bar{d} \, \gamma^\mu \, d \right) \right] + \cdots \tag{13.66d}$$

where ellipses in Eqs. (13.66) indicate repetition for the second and third generations.

For low momentum transfer phenomena, $q^2 \ll m_W^2$, $m_Z^2$, we can write

$$\frac{g^2}{8m_W^2} = \frac{G_F}{\sqrt{2}} \tag{13.67}$$

$$m_W^2 = \frac{\sqrt{2}e^2}{8G_F \sin^2\theta_W} = \frac{\pi\alpha}{\sqrt{2}G_F \sin^2\theta_W} \tag{13.68}$$

$$m_W = \frac{37.3 \text{ GeV}}{\sin^2\theta_W} \geq 37.3 \text{ GeV} \tag{13.69}$$

$$\rho = 1; \quad m_Z = \frac{m_W}{\cos\theta_W} = \frac{74.6 \text{ GeV}}{\sin^2\theta_W} \geq 74.6 \text{ GeV}. \tag{13.70}$$

Note that $\rho = 1$ is a consequence of the fact that Higgs scalar $\phi$ is an $SU_L(2)$ doublet. The effective neutral current coupling [see Eq. (13.66a)] is

$$\frac{g^2}{m_Z^2 \cos^2\theta_W} = \rho \frac{g^2}{m_W^2} = 8\rho \frac{G_F}{\sqrt{2}}. \tag{13.71}$$

Finally, we note that for the Higgs vacuum expectation value $v$ , using Eqs. (13.56a) and (13.71), we get

$$v^2 = \frac{1}{\sqrt{2}G_F} \approx (246 \text{ GeV})^2 . \tag{13.72}$$

This gives the electroweak unification scale, i.e. the energy scale after which the weak interactions become as strong as electromagnetic interaction.

The fermion masses are given by

$$m_i^2 = h_i^2 \frac{v^2}{2} = h_i^2 \frac{1}{2\sqrt{2}G_F}$$
$$h_i^2 = 2\sqrt{2}\, G_F\, m_i^2, \tag{13.73}$$

i.e. the Yukawa couplings are very weak except for the top quark.

We conclude this section with the following remarks:

 (i) A definite prediction of electroweak unification is the existence of weak neutral current $J_\mu^Z$ with the same effective coupling as charged currents $J_\mu^\pm$ . This current has been found experimentally.
 (ii) The existence of vector bosons $W^\pm$, $Z$, with definite masses given in Eqs. (13.69) and (13.70).
(iii) The theory has one free parameter $\sin^2 \theta_W$.

At low energies $q^2 \ll m_W^2$, one test of the model is to determine $\sin^2 \theta_W$ from different classes of experiments. If $\sin^2 \theta_W$ comes out to be the same in all these experiments, it will support the model. The true test of the model is the existence of vector bosons. This requires much higher energies. We first discuss low energy consequences of the electroweak unification. The vector bosons $W^\pm$ and $Z$ have been found experimentally with masses predicted by the model.

## 13.4.1  Experimental Consequences of the Electroweak Unification

Low energy phenomena $q^2 \ll m_W^2$ : From the Lagrangian (13.66a), for low momentum transfer phenomena $[q^2 \ll m_W^2,\ m_Z^2]$ we can write the effective Lagrangians for charged and neutral currents:

$$\mathcal{L}_{eff}^{CC} = \frac{G_F}{\sqrt{2}} J^{+\mu}\, J_\mu^- \tag{13.74}$$

$$\mathcal{L}_{eff}^{NC} = \rho \frac{G_F}{\sqrt{2}} 8 J^{Z\mu}\, \bar{J}_\mu^Z . \tag{13.75}$$

It is convenient to write $J_\mu^Z$:

$$J_\mu^Z = J_\mu^Z\,(\nu) + J_\mu^Z\,(e) + J_\mu^Z\,(h)\,, \qquad (13.76)$$

where

$$J_\mu^Z\,(\nu) = \frac{1}{4}\,[\bar{\nu}\gamma_\mu\,(1-\gamma_5)\,\nu] \qquad (13.77)$$

$$2J_\mu^Z\,(e) = [\varepsilon_L\,(e)\,\bar{e}\gamma_\mu\,(1-\gamma_5)\,e + \varepsilon_R\,(e)\,\bar{e}\gamma_\mu\,(1+\gamma_5)\,e] \qquad (13.78)$$

$$2J_\mu^Z\,(h) = \sum_{i=u,d,s,\cdots} [\varepsilon_L\,(i)\,\bar{q}_i\gamma_\mu\,(1-\gamma_5)\,q_i + \varepsilon_R\,(i)\,\bar{q}_i\,\gamma_\mu\,(1+\gamma_5)\,q_i] \qquad (13.79)$$

Table 13.1

|            | $e$ | $u$ | $d$ |
|------------|-----|-----|-----|
| $\varepsilon_L$ | $-\frac{1}{2}+\sin^2\theta_W$ | $\frac{1}{2}-\frac{2}{3}\sin^2\theta_W$ | $-\frac{1}{2}+\frac{1}{3}\sin^2\theta_W$ |
| $\varepsilon_R$ | $\sin^2\theta_W$ | $-\frac{2}{3}\sin^2\theta_W$ | $\frac{1}{3}\sin^2\theta_W$ |
| $g_A$ | $-\frac{1}{2}$ | $\frac{1}{2}$ | $-\frac{1}{2}$ |
| $g_V$ | $-\frac{1}{2}+2\sin^2\theta_W$ | $\frac{1}{2}-\frac{4}{3}\sin^2\theta_W$  $C_{1u}=2g_A^e\,g_V^u$  $C_{2u}=2g_V^e\,g_A^u$ | $-\frac{1}{2}+\frac{2}{3}\sin^2\theta_W$  $C_{1d}=2g_A^e\,g_V^d$  $C_{2d}=2g_V^e\,g_A^d$ |

Since the net strangeness of the proton is zero, we will assume that strange quark $s$ and heavy flavor quarks $c$, $b$ etc. make negligible contribution to $J_\mu^Z(h)$ for proton and neutron targets. Then we can write effective Lagrangians for various neutral current processes as follows:

$$\mathcal{L}^{\nu e} = \rho\frac{G_F}{\sqrt{2}}\bar{\nu}\,\gamma^\mu\,(1+\gamma_5)\,\nu\,(2J_\mu^Z\,(e)) \qquad (13.80)$$

$$\mathcal{L}^{\nu h} = \rho\frac{G_F}{\sqrt{2}}\bar{\nu}\,\gamma^\mu\,(1-\gamma_5)\,\nu\,(2J_\mu^Z\,(h)) \qquad (13.81)$$

$$\mathcal{L}^{eh} = -\rho\frac{G_F}{\sqrt{2}}\sum_i - [C_{1i}\bar{e}\gamma^\mu\gamma^5 e\bar{q}_i\gamma_\mu q_i + C_{2i}\bar{e}\gamma^\mu e\bar{q}_i\gamma_\mu\gamma_5 q_i]\,. \qquad (13.82)$$

From Eqs. (13.65a), we can determine the parameters $\varepsilon_L(e)$, $\varepsilon_R(e)$, $\varepsilon_L(i)$, $\varepsilon_R(i)$, $C_{1i}$, and $C_{2i}(e)$, $(i=u,d)$. They are given in Table 13.1.

### 13.4.2   *Need for Radiative Corrections*

Before we discuss the experiments in support of the standard model, let us summarize here the three parameters (not counting Higgs meson mass $m_H$ and the fermion masses) which the minimal model (with $\rho=1$) has: (a) fine structure constant $\alpha = 1/137.035999679(94)$ determined from the

Josephson effect (b) the Fermi coupling constant $G_F = 1.166367(5) \times 10^{-5}$ GeV$^{-2}$ determined from the muon life-time {including lepton mass and $O(\alpha)$ radiative corrections [cf. Eq. (10.40c) ]} and (c) $\sin^2 \theta_W$, determined from neutral current processes or the $W$ and $Z$ masses. Now a best fit to the neutral current neutrino reactions data gives

$$\sin^2 \theta_W = 0.2255 \pm 0.0021 \tag{13.83}$$

This implies that the theory without radiative corrections gives through the relations [cf. Eqs. (13.68) and (13.70)]

$$m_W^2 = \left(\frac{\pi \alpha}{\sqrt{2} G_F}\right) / \sin^2 \theta_W = \frac{A_0^2}{\sin^2 \theta_W}$$

$$m_Z^2 = \frac{m_W^2}{\cos^2 \theta_W}, \tag{13.84a}$$

where

$$A_0 = \left(\frac{\pi \alpha}{\sqrt{2} G_F}\right)^{1/2} = 37.2802 \text{ GeV}, \tag{13.85}$$

$$m_W = 78.42 \text{ GeV}, \qquad m_Z = 89.14 \text{ GeV}. \tag{13.86}$$

These values are to be compared with the experimental ones $m_W = 80.39 \pm 0.06$ GeV and $m_Z = 91.1867 \pm 0.002$ GeV. This shows a need for radiative corrections. First we note that the two coupling constants $g$ and $g'$ which determine the strength of weak interactions are related to $e$ through $e = g\,g'/(g^2+g'^2)$. Since most measurements are made at $Z$ peak, therefore most convenient mass scale for these couplings is at $m_Z$. Thus one should take into consideration the running of $QED$ coupling constant [see Appendix B] which gives the value of $\alpha$ at $m_Z$ :

$$\alpha(m_Z) = \frac{\alpha}{1 - \Delta \alpha} = \frac{\alpha}{1 - (\Pi_{\gamma\gamma}(m_Z) - \Pi_{\gamma\gamma}(0))}$$

$$= \frac{\alpha}{1 - \frac{\alpha}{3\pi} \sum_f Q_f^2 \, N_{cf} \left(-\frac{5}{3} + \ln \frac{m_Z^2}{m_f^2}\right)}, \tag{13.87}$$

where $f = e, \mu, \tau, \ u, d, s, c$ and $b$ and $N_{cf} = 3$ for quarks and 1 for leptons. Equation (13.87) can be directly evaluated for leptons, since their masses are well known. For the light hadronic part, quark masses are not available as reasonable input parameters. The 5-flavor contribution to $\Pi_{\gamma\gamma}$ is extracted from the experimental data on $e^+e^- \rightarrow$ hadrons. The best estimate leads to

$$\alpha(m_Z) = \frac{1}{127.909 \pm 0.019}$$

$$\Delta \alpha = 1 - \frac{\alpha}{\alpha(m_Z)} = 0.0666 \pm 0.0007 \tag{13.88}$$

Thus knowing $G_F$, $\alpha\,(m_Z)$ and $m_Z$, one should be in a position to predict all electroweak observables, including the mixing angle $\sin^2\theta_W = \frac{e^2}{g^2}$. However, the lowest order relation $m_W \,/\, m_Z = \cos\theta_W$ and some other lowest order relations are affected by the fermion loops in gauge bosons propagators. Thus the relation (13.61) is modified to

$$\frac{m_W^2}{m_Z^2\cos^2\theta_W} = \rho = (1+\Delta\rho) \tag{13.89}$$

The leading contribution to $\Delta\rho$ comes from the top quark loop $(m_b \ll m_t)$ to the self energies of $W$ and $Z$ bosons (see Fig. 13.3):

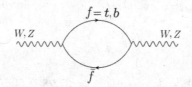

Fig. 13.3    Top and bottom quarks loop contribution to $W$ and $Z$ boson self energies.

$$\Delta\rho = \frac{3G_F}{8\pi^2\sqrt{2}}m_t^2 = 0.0092 \pm 0.0002 \tag{13.90}$$

for $m_t = 172.0 \pm 2.2$ GeV. By contrast, the Higgs boson contribution to $\Delta\rho$ at the one loop level is logarithmic:

$$(\Delta\rho)_{\text{Higgs}} = -\frac{G_F\,m_W^2}{12\pi^2\sqrt{2}}\tan^2\theta_W\left[5\ln\frac{m_H^2}{m_W^2}\right] \tag{13.91}$$

Radiative corrections are scheme dependent, leading to $\sin^2\theta_W$ values which differ by small factors which depend on $m_t$ and $m_H$. A useful scheme [called the on shell scheme] is to take tree level formula $\sin^2\theta_W = 1 - m_W^2 \,/\, m_Z^2$ as the definition of renormalized $\sin^2\theta_W$ to all orders in perturbation theory, i.e. $\sin^2\theta_W \to s_W^2 = 1 - m_W^2 \,/\, m_Z^2$.

Now the tree level expression (13.85) is modified to

$$A_0^2 \to A_0^2\frac{\alpha\,(m_Z)}{\alpha} = A_0^2\frac{1}{1-\Delta\alpha}$$

while using Eq. (13.89),

$$\sin^2\theta_W \to s_W^2 = 1 - \frac{m_W^2}{m_Z^2} = 1 - \frac{m_W^2}{m_Z^2\cos^2\theta_W}\cos^2\theta_W$$

$$= (1-\cos^2\theta_W) + \cos^2\theta_W\left(1 - \frac{m_W^2}{m_Z^2\cos^2\theta_W}\right)$$

$$= \sin^2\theta_W - \Delta\rho\cos^2\theta_W \tag{13.92}$$

Thus the tree level expressions (13.84a) are modified to

$$\left(1 - \frac{m_W^2}{m_Z^2}\right) m_W^2 \equiv m_W^2 \, s_W^2 = m_Z^2 \, s_W^2 \, c_W^2 = \frac{A_0^2}{1 - \Delta r} \tag{13.93}$$

where

$$\Delta r = \Delta\alpha - \Delta\rho \cot^2\theta_W + (\Delta r)_{\text{remainder}}$$

$$= \Delta\alpha - \frac{c_W^2}{s_W^2}\Delta\rho + (\Delta r)_{\text{remainder}} \tag{13.94}$$

where $(\Delta r)_{\text{remainder}}$ contains all possible contributions not included in the fermionic contributions $\Delta\alpha$ and $\Delta\rho$. Likewise, taking into account radiative corrections, the relationship between the $W^{\pm}$-boson mass, $s_W^2$ and $G_F^2$ in the standard model gets modified

$$m_W^2 = \frac{A^2}{s_W^2 \, (1 - \Delta r_W)} \tag{13.95}$$

where $A^2 = A_0^2 \frac{\alpha(Z)}{\alpha} = A_0^2 \frac{1}{1 - \Delta\alpha}$, so that from Eqs. (13.93), (13.94) and (13.95)

$$(1 - \Delta r) = (1 - \Delta\alpha)(1 - \Delta r_W),$$

$$\left(1 - \frac{m_W^2}{m_Z^2}\right) \frac{m_W^2}{m_Z^2} = s_W^2 \, c_W^2 = A_0^2 \frac{\alpha(m_Z)}{\alpha \, m_Z^2} \frac{1}{1 - \Delta r_W}$$

$$= \frac{\pi\alpha(m_Z)}{\sqrt{2}G_F \, m_Z^2 \, (1 - \Delta r_W)}$$

$$= \frac{s_0^2 \, c_0^2}{1 - \Delta r_W} \tag{13.96}$$

where we have used Eq. (13.85) and have defined

$$s_0^2 \, c_0^2 = \frac{\pi\alpha(m_Z)}{\sqrt{2}G_F \, m_Z^2} \tag{13.97}$$

The relation (13.95) defines $\Delta r_W$, which is completely determined by purely weak correction, once $\alpha(m_Z)$ is specified. For $LEP$ physics, $\sin^2\theta_W$ is usually defined from $Z \to \mu^+ \mu^-$ effective vertex. At the tree level we have [cf. Table 13.1]

$$Z \to f \, \bar{f} : \frac{g}{2\cos\theta_W} \bar{f} \, \gamma_\mu \left(g_V^f - g_A^f \, \gamma_5\right) f$$

$$= \left(\sqrt{2}G_F \, m_Z^2\right)^{1/2} \bar{f} \left[\left(I_3^f - 2Q_f \, \sin^2\theta_W\right) \gamma_\mu - I_3^f \, \gamma_\mu \, \gamma_5\right] f \tag{13.98}$$

The weak (non-$QED$) connections to the $Z \to f \bar{f}$ effective vertex can be conveniently written as

$$\left( \sqrt{2} G_F \, m_Z^2 \, \rho_f \right)^{1/2} \bar{f} \left[ \left( I_3^f - 2Q_f \, s_f^2 \right) \gamma_\mu - I_3^f \gamma_\mu \, \gamma_5 \right] f \tag{13.99}$$

where $\rho_f = \frac{1}{1-\Delta\rho}$, $\Delta\rho$ being given in Eq. (13.90) and [cf. Eq. (13.92)]

$$s_W^2 = s_f^2 - c_f^2 \frac{\Delta\rho}{1 - \Delta\rho}$$
$$s_f^2 = s_W^2 + c_W^2 \, \Delta\rho \tag{13.100}$$

Thus $\left[ I_3^\mu = \frac{-1}{2}, \; Q_\mu = -1 \right]$

$$g_V^\mu / g_A^\mu = 1 - 4s_f^2 \tag{13.101}$$

$$g_A^\mu = -\frac{1}{2}\rho_f^{1/2} = -\frac{1}{2}\left( 1 + \frac{\Delta\rho}{2} \right) \tag{13.102}$$

The radiative correction functions $\Delta r_W$, $(\rho_f - 1)$ become the basis on which the corrected standard model and experiments are confronted. These functions involve, among other parameters, the top quark mass $m_t$ whose value due to quadratic dependence of some of these functions on $m_t$ is numerically important and the Higgs boson mass which enters only through logarithm and hence is not effectively bounded.

Now if we use the LEP value $s_f^2 = 0.231119 \pm 0.00014$ for leptons, we can obtain $s_W^2 = 0.22308 \pm 0.00030$ from Eq. (13.100), with $\Delta\rho$ given in Eq. (13.89) and hence $m_W$ through the relation ($m_Z = 91.1876$) $s_W^2 = 1 - m_W^2/m_Z^2$ : $m_W = 80.375 \pm 0.015$ GeV which is consistent with the direct vector boson mass measurements: $m_W = 80.398 \pm 0.025$ GeV and the standard model best fit value: $m_W = 80.376 \pm 0.033$ GeV obtained from Eq. (13.93) (including higher order terms) from $m_Z$, $G_F$, $\alpha$ and $m_t$, $m_H$.

Finally including $m_t$, $m_W$ from the direct experimental measurement, together with $s_W^2$ from neutrino scattering, global fits of the standard model parameters to electroweak precision data give

$$m_t = 173.3 \pm 1.1 \text{ GeV}$$
$$m_H = 87^{+35}_{-26} \text{ GeV}$$
$$\alpha_s(m_Z) = 0.119 \pm 0.003 \text{ GeV} \tag{13.103}$$

The upper limit on $m_H$ at the 95 % CL is $m_H < 185$ GeV after including the information from the 114.4 GeV bound given by LEP, where the theoretical uncertainty is included.

### 13.4.3 Experiments which Determine $\sin^2 \theta_W$

We now discuss three sets of experiments to determine $\sin^2\theta_W$.
1. Consider the process

$$\nu_\mu + e^- \rightarrow \nu_\mu + e^-.$$

This process can occur only through $Z$ exchange (Fig. 13.4). The effective

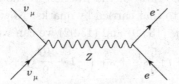

Fig. 13.4 $\nu_\mu$-electron scattering through $Z$-boson exchange.

Lagrangian for this process is given by Eq. (13.80). For this case the laboratory cross-section for $E_\nu \gg m_e$ give

$$\sigma^{\nu_\mu, \bar\nu_\mu} = \frac{G_F^2 \, m_e E_\nu}{2\pi} \left[ (g_V^e \pm g_A^e)^2 + \frac{1}{3} (g_V^e \mp g_A^e)^2 \right], \qquad (13.104)$$

where the upper (lower) sign refers to $\nu_\mu$ ($\bar\nu_\mu$), $E_\nu$ is the incident energy and $G_F^2 \, m_e / 2\pi = 4.31 \times 10^{-42}$ cm$^2$/GeV. The expressions for $g_V^e$ and $g_A^e$ in terms of $\sin^2 \theta_W$ are given in Table 13.1. The most accurate leptonic measurements of $\sin^2\theta_W$ are from the ratio

$$R = \frac{\sigma_{\nu_\mu e}}{\sigma_{\bar\nu_\mu e}} = \frac{3 - 12 \sin^2 \theta_W + 16 \sin^4 \theta_W}{1 - 4 \sin^2 \theta_W + 16 \sin^4 \theta_W}. \qquad (13.105)$$

The most precise experiment (Charm II) determined not only $\sin^2 \theta_W$ but $g_{V,A}^e$ as well. The experimental results are

$$g_V^e = -0.035 \pm 0.017, \ g_A^e = -0.503 \pm 0.017$$
$$\sin^2 \theta_W = 0.2326 \pm 0.0084 \qquad (13.106)$$

2. The deep inelastic neutrino scattering $\nu_\mu + N \rightarrow \nu_\mu + X$ (isoscalar target), gives a precise determination of $\sin^2 \theta_W$ on the mass shell, i.e. $s_W^2$. The relevant Lagrangian is given in Eq. (13.81). The ratio $R_\nu \equiv \sigma_{\nu N}^{NC}/\sigma_{\nu N}^C$ of neutral to charged cross-section has been measured to 1% accuracy. A simple zeroth order approximation gives [see Eqs. (14.79)-(14.80)]

$$R_\nu = g_L^2 + g_R^2 \, r, \qquad R_{\bar\nu} = g_L^2 + g_R^2 \, / \, r, \qquad (13.107a)$$

where

$$g_L^2 = \varepsilon_L \left(u\right)^2 + \varepsilon_L \left(d\right)^2 \approx \frac{1}{2} - \sin^2 \theta_W + \frac{5}{9} \sin^4 \theta_W$$

$$g_R^2 = \varepsilon_R \left(u\right)^2 + \varepsilon_R \left(d\right)^2 \approx \frac{5}{9} \sin^4 \theta_W \tag{13.107b}$$

and $r \equiv \sigma_{\bar{\nu}N}^C / \sigma_{\nu N}^C$ is the ratio of $\bar{\nu}$ and $\nu$ charged current cross-sections, which can be measured directly. In parton model $r \approx \left(\frac{1}{3} + \varepsilon\right) / \left(1 + \frac{1}{3}\varepsilon\right)$, where $\varepsilon \approx 0.125$ is the ratio of the fraction of the nucleon's momentum carried by antiquarks to that carried by quarks. Now from Eqs. (13.107a)-(13.107b) on using Eq. (13.89) and (13.92) we can write

$$R_\nu = \frac{1}{2} - s_W^2 + (1 + r) \frac{5}{9} s_W^4 + s_W^2 \left(-1 + \frac{10}{9}(1 + r)\right) \Delta\rho \tag{13.108a}$$

$$R_{\bar{\nu}} = \frac{1}{2} - s_W^2 + \left(1 + \frac{1}{r}\right) \frac{5}{9} s_W^4 + s_W^2 \left(-1 + \frac{10}{9}\left(1 + \frac{1}{r}\right)\right) \Delta\rho \tag{13.108b}$$

$$R_\nu - r R_{\bar{\nu}} = (1 - r) \left[\frac{1}{2} - s_W^2 (1 + \Delta\rho)\right] \tag{13.109}$$

It is clear from Eq. (13.108a) that dependence of $R_\nu$ on $\Delta\rho$ is weak. Hence this equation is useful to determine $s_W^2$. Using the experimental values $R_\nu = 0.317 \pm 0.003$, $r = 0.440$, and $\Delta\rho = 0.0096$ we obtain from Eq. (13.108a) $s_W^2 = 0.2242 \pm 0.0022$, and (with $m_Z = 91.187$) $m_W = 80.32 \pm 0.11$ GeV. The recent value quoted for $s_W^2$ from $\nu N$ scattering is $0.2255 \pm 0.9021$, which gives $m_W = 80.25 \pm 0.11$ GeV fully consistent with the directly measured value for $m_W = 80.39 \pm 0.06$ GeV. We note from Eq. (13.109), that if we plot $R_\nu$ versus $R_{\bar{\nu}}$ it gives a straight line with a slope determined by $r$. This provides an accurate method to determine $\rho\, s_W^2$ from the experimental data.

   3. Parity violating deep inelastic $eD$ scattering: The relevant Lagrangian for this process through $Z$ exchange, which is parity violating, is given in Eq. (13.82). There is an interference with the parity conserving process through photon exchange. This gives us the parity-violating asymmetry

$$A = \frac{\sigma_R - \sigma_L}{\sigma_R + \sigma_L}, \tag{13.110}$$

where $\sigma_{R,L}$ is the cross-section for the deep inelastic scattering of a right (left)-handed electron $e_{R,L} N \to eX$. In the quark parton model (see Chap. 14 for this model)

$$\frac{A}{q^2} = a_1 + a_2 \frac{1 - (1 - y)^2}{1 + (1 - y)^2}, \tag{13.111a}$$

where $q^2 < 0$ is the momentum transfer and this essentially comes through the photon propagator which appears in the photon exchange process. Here $y$ is the fractional energy transfer from the electron to hadrons. For the deuteron or other isoscalar target neglecting the $s$ quark and antiquarks,

$$a_1 = \frac{3G_F}{5\sqrt{2}\pi\alpha}\left(C_{1u} - \frac{1}{2}C_{1d}\right) \approx \frac{3G_F}{5\sqrt{2}\pi\alpha}\left(-\frac{3}{4} + \frac{5}{3}\sin^2\theta_W\right)$$

$$a_2 = \frac{3G_F}{5\sqrt{2}\pi\alpha}\left(C_{2u} - \frac{1}{2}C_{2d}\right) \approx \frac{9G_F}{5\sqrt{2}\pi\alpha}\left(\sin^2\theta_W - \frac{1}{4}\right) \quad (13.111b)$$

where we have used Table 13.1 in the second step of these formulae. The experimental values for $a_1$ and $a_2$ can be used to determine the mixing angle $\sin^2\theta_W$.

## 13.5   Decay Widths of W and Z Bosons

Consider the decay $W^- \rightarrow e^- + \bar{\nu}_e$:

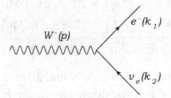

Fig. 13.5   $W$-boson decay.

From Eqs. (13.65b) and (13.66c), the decay amplitude $F$ is given by

$$F = \frac{-g}{2\sqrt{2}}\bar{u}(k_1)\gamma^\lambda\left(1 - \gamma^5\right)v(k_2)\cdot\varepsilon_\lambda, \quad (13.112a)$$

where $\varepsilon_\lambda$ is the polarization of W-boson. From Eq. (13.112a), the decay width can be easily calculated and is given by [ in the limit when we neglect the lepton masses as compared with $m_W$] :

$$\Gamma\left(W^- \rightarrow e^- + \bar{\nu}_e\right) = \left(\frac{g^2}{8}\right)\frac{m_W}{3\left(2\pi\right)^2}\cdot 2\pi$$

$$= \frac{G_F\, m_W^3}{6\sqrt{2}\pi}. \quad (13.112b)$$

We can also calculate the hadronic decays of $W^-$ from the basic processes like $W^- \rightarrow \bar{u}\, d,\ \bar{c}\, s$. Again we get an expression like (13.112b), except that we multiply it by a factor $N_c = 3\left(1 + \frac{\alpha_s(m_W)}{\pi}\right) \approx 3.12$, where the factor 3 is due to color and the factor in the parenthesis is a $QCD$ correction. In this case we also neglect quark masses as compared with $W$-mass. This is a good approximation with the exception of $t$-quark which channel is not open as $m_t = 172.0 \pm 2.2$ GeV. Since in weak interactions a linear combination of mass eigenstates $d$, $s$ and $b$ enters, therefore, we have to multiply the decay rates by square of such factors as $|V_{ud}|^2$, $|V_{cs}|^2$, $|V_{us}|^2$, etc. [see Sec. 13.10]. The link of one generation to succeeding generations is very weak, therefore, we will put $|V_{ud}|^2 \approx \cos^2\theta_c \approx 1$, $|V_{cs}|^2 \approx \cos^2\theta_c \approx 1$, $|V_{us}|^2 \approx \sin^2\theta_c \approx 0$, $|V_{cb}|^2 \approx \sin^4\theta_c \approx 0$. Hence the relative decay widths for three generations are given by:

| $e^-\bar{\nu}_e$ | $\mu^-\bar{\nu}_\mu$ | $\tau^-\bar{\nu}_\tau$ | $\bar{u}\,d$ | $\bar{c}\,s$ |
|---|---|---|---|---|
| 1 | 1 | 1 | $3\left[1+\frac{\alpha_s}{\pi}\right]$ | $3\left[1+\frac{\alpha_s}{\pi}\right]$ |

Thus we get

$$\Gamma\left(W^+ \to l^+\nu_l\right) \approx 227.5 \pm 0.3 \text{ MeV},$$
$$\Gamma\left(W^+ \to u_i d_i\right) \approx (708 \pm 1) \text{ MeV}$$
$$\Gamma_W^{\text{tot}} \approx 2.098 \text{ GeV}, \tag{13.113}$$

to be compared with the experimental value $2.141 \pm 0.041$ GeV.

For the decay $Z \to f + \bar{f}$, the decay amplitude $F$ is given by [from Eqs. (13.65b) and (13.66d)]

$$F = \frac{-g}{\cos\theta_W}\bar{u}(k_1)\left[g_{Vf}\,\gamma^\mu - g_{Af}\,\gamma_\mu\,\gamma^5\right]v(k_2)\,\epsilon_\mu \tag{13.114}$$

and the decay width is given by

$$\Gamma\left(Z \to f\bar{f}\right) = \frac{g^2}{8\cos^2\theta_W m_Z^2}\frac{m_Z^3}{6\pi}\left[g_{Vf}^2 + g_{Af}^2\right]N_c^f$$
$$= \frac{G_F}{\sqrt{2}}\frac{m_Z^2}{6\pi}\left[g_{Vf}^2 + g_{Af}^2\right]N_c^f \tag{13.115}$$

where $N_c^f = 1$ or $3(1+\alpha_s/\pi)$ for $f = $ lepton $(l)$ or quark $q$. First we note that $g_{Af} = -\frac{1}{2}$, $g_{Vf} = \left(-1/2 + |Q_f|2\sin^2\theta_W\right)$. In the presence of radiative corrections, we have [cf. Eqs. (13.99), (13.101) and (13.102)]

$$g_{Af} = -\frac{1}{2}\sqrt{\rho_f} \approx -\frac{1}{2}\left(1 + \frac{1}{2}\Delta\rho\right), \quad x = \frac{g_{Vf}}{g_{Af}} = 1 - 4|Q_f|s_f^2 \tag{13.116}$$

Thus we get

$$\Gamma\left(Z \to l\bar{l}\right) = \frac{G_F}{\sqrt{2}} \frac{m_Z^3}{24\pi} \left[1 + \Delta\rho\right] \left[1 + \left(1 - 4s_f^2\right)^2\right] \tag{13.117a}$$

$$\Gamma\left(Z \to q\bar{q}\right) = \frac{G_F}{\sqrt{2}} \frac{m_Z^3}{24\pi} N_c \left[1 + \Delta\rho\right] \left[1 + \left(1 - 4\left|Q_f\right|s_f^2\right)^2\right] \tag{13.117b}$$

$$\Gamma\left(Z \to \nu\bar{\nu}\right) = \frac{G_F}{\sqrt{2}} \frac{m_Z^3}{24\pi} \left[1 + \rho\right] \times 2 \tag{13.117c}$$

It is convenient to write

$$s_f^2 = (1 + \Delta k) s_o^2 \tag{13.118}$$

where $\Delta k$ signifies non-QED corrections and $s_o^2$ is given in Eq. (13.97) and has the value 0.2311 for $m_Z = 91.187$ GeV. Then

$$\Gamma\left(Z \to f\bar{f}\right) = \frac{G_F}{\sqrt{2}} \frac{m_Z^3}{24\pi} \left[1 + \Delta\rho\right] N_{cf}$$
$$\times \left[1 + \left(1 - 4\left|Q_f\right|\left(1 + \Delta k\right)s_0^2\right)^2\right] \tag{13.119}$$

Let us write [subscript o indicates neglecting $\Delta k$]

$$\Gamma_0\left(Z \to l\bar{l}\right) = \frac{G_F}{\sqrt{2}} \frac{m_Z^3}{24\pi} \left[1 + \left(1 - 4s_0^2\right)^2\right] \tag{13.120a}$$

$$\Gamma_0\left(Z \to q\bar{q}\right) = \frac{G_F}{\sqrt{2}} \frac{m_Z^3}{24\pi} \left[1 + \left(1 - 4\left|Q_q\right|s_0^2\right)^2\right] 3 \left(1 + \frac{\alpha_s}{\pi}\right) \tag{13.120b}$$

From Eqs. (13.120a)-(13.120b), we get on using $s_0^2 = 0.2311$,

$$\Gamma_0\left(Z \to \nu\bar{\nu}\right) = 165.9 \text{ MeV} \tag{13.121a}$$

$$\Gamma_0\left(Z \to e^-e^+\right) = \Gamma_0\left(Z \to \bar{\mu}\mu^+\right) = \Gamma_0\left(Z \to \tau^-\tau^+\right)$$
$$= 83.4 \, (83.984 \pm 0.086) \text{ MeV} \tag{13.121b}$$

$$\Gamma_0\left(Z \to u\bar{u}\right) = \Gamma_0\left(Z \to c\bar{c}\right) = 296.9 \text{ MeV} \tag{13.121c}$$

$$\Gamma_0\left(Z \to d\bar{d}\right) = \Gamma_0\left(Z \to s\bar{s}\right) = \Gamma_0\left(Z \to b\bar{b}\right) = 382.6 \text{ MeV} \tag{13.121d}$$

Thus

$$\Gamma_0\left(Z \to \text{hadrons}\right) 1.742 \, (1.7444 \pm 0.0020) \text{ GeV} \tag{13.122a}$$

$$\Gamma_0\left(Z \to \text{invisible}\right) = \Gamma_Z - \left(\Gamma_{had} + 3\Gamma_{e^+e^-}\right)$$
$$= (2.4952) - (1.742 + 0.250)$$
$$= 502 \, (499.0 \pm 1.5) \text{ MeV} \tag{13.122b}$$

where experimentally $\Gamma_Z = 2.4952 \pm 0.0023$ GeV. The values given in parenthesis are experimental values. Now $3\Gamma_0\left(Z \to \nu\bar{\nu}\right) = 498$ MeV; hence one concludes that $N_\nu = 3$, i.e. there are three generations of neutrinos or three generations of fermions. It may be noted that in calculating the decay widths we have put fermion masses zero. This is a very good approximation; the decay width for $Z \to b\bar{b}$ may need some improvement if $m_b$ is not neglected. Even the theoretical values as given by the Born approximation $\Gamma_0$ are not bad, the small discrepancy can be explained by taking into account the radiative corrections in Eq. (13.121a). We can also use Eq. (13.119) to constrain $\Delta k$ and $\Delta\rho$ by using the exprimental for $\Gamma(Z \to e^+ e^-)$.

Another important observable is forward-backward asymmetry measured at LEP. Consider the process $e^- e^+ \to f\bar{f}$ as depicted in Fig. 13.6.

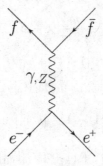

Fig. 13.6   Production of $Z$ in $e^- e^+$ collision and its decay into $f\bar{f}$ pair.

The cross-section for $e^- e^+ \to f\bar{f}$ can be easily calculated by using Eq. (13.114). It is given by (in the limit $s \gg 4m_e^2$, $4m_f^2$, $\beta_e \approx 1$, $\beta_f \approx 1$)

$$
\begin{aligned}
\frac{d\sigma}{d\Omega} = \frac{\alpha^2 N_c^f}{4s} \Bigg\{ & (1 + \cos^2\theta) \left[ Q_f^2 - 2\frac{Q_f\, v_e v_f}{16 \sin^2\theta_W \cos^2\theta_W} \Re\chi(s) \right. \\
& \left. + \frac{(v_e^2 + a_e^2)(v_f^2 + a_f^2)}{(16 \sin^2\theta_W \cos^2\theta_W)^2} |\chi(s)|^2 \right] \\
& + \cos\theta \left[ -\frac{4 Q_f\, a_e a_f}{16 \sin^2\theta_W \cos^2\theta_W} \Re\chi(s) \right. \\
& \left. + \frac{8 v_e v_f a_e a_f}{(16 \sin^2\theta_W \cos^2\theta_W)^2} |\chi(s)|^2 \right] \Bigg\},
\end{aligned} \qquad (13.123a)
$$

where

$$\chi(s) = \frac{s}{s - m_Z^2 + i m_Z \Gamma_Z}, \tag{13.123b}$$

$$v_e = 2g_{Ve} = -1 + 4\sin^2\theta_W, \qquad a_e = 2g_{Ae} = -1$$

$$v_f = 2g_{Vf} = 2T_{3L}^f - 4Q_f \sin^2\theta_W, \qquad a_f = 2g_{Af} = 2T_{3L}^f. \tag{13.123c}$$

$$\sigma = \frac{4\pi\alpha^2}{3s} N_c^f \left[ Q_f^2 - 2Q_f \frac{4g_{Ve}g_{Vf}}{16\sin^2\theta_W \cos^2\theta_W} \frac{s(s - m_Z^2)}{(s - m_Z^2)^2 + m_Z^2\Gamma_Z^2} \right.$$
$$\left. + \frac{16(g_{Ve}^2 + g_{Ae}^2)(g_{Vf}^2 + g_{Af}^2)}{(16\sin^2\theta_W \cos^2\theta_W)^2} \frac{s^2}{(s - m_Z^2)^2 + m_Z^2\Gamma_Z^2} \right] \tag{13.124}$$

Here near and on the peak, integrated cross section is dominated by $Z$-exchange and we get from Eq. (13.115):

$$\sigma_f = \sigma_{\text{peak}}^0 \left| \frac{s\Gamma_Z / m_Z}{s - m_Z^2 + i m_Z \Gamma_Z} \right|^2 (1 + \delta(s)) \tag{13.125a}$$

where

$$\sigma_{\text{peak}}^0 = \frac{12\pi}{m_Z^2} \frac{\Gamma_e \Gamma_f}{\Gamma_Z^2} \tag{13.125b}$$

and the effect of radiative corrections are contained in $\delta(s)$, the large effects due to initial $e^\pm$ bremsstrahlung are represented in $\delta$. The other radiative corrections which lead to improved Born approximation have already been discussed in Secs. 3.2 and 3.4.

The *LEP* data is fitted with an additional modification, i.e. by replacing $s - m_Z^2 + i m_Z \Gamma_Z$ by $s - m_Z^2 + i\frac{s}{m_Z}\Gamma_Z$. Note that the expression for $\Gamma_f$ is given in Eq. (13.115). Thus by measuring $\sigma_{\text{peak}}^0$ for a particular final state e.g. $e^+e^-$ itself, one can directly obtain $\Gamma_{ee}/\Gamma_Z$ or $\frac{\Gamma_{ee} / \Gamma_{f\bar{f}}}{\Gamma_Z}$ and therefore $\Gamma_{f\bar{f}}$. These widths have already been discussed in the beginning of this section.

The forward-backward asymmetry is defined as:

$$A_{FB} = \frac{\int_0^{\pi/2} \left(\frac{d\sigma}{d\Omega}\right) d\Omega - \int_{\pi/2}^{\pi} \left(\frac{d\sigma}{d\Omega}\right) d\Omega}{\int_0^{\pi} \left(\frac{d\sigma}{d\Omega}\right) d\Omega} \tag{13.126}$$

It is clear that this asymmetry is given by $\cos\theta$ term in Eq. (13.123). Near and on the $Z$-peak, we get from Eqs. (13.126) and (13.123c)

$$A_{FB} = 3 \frac{(g_{Ve}g_{Ae})(g_{Vf}g_{Af})}{(g_{Ve}^2 + g_{Ae}^2)(g_{Vf}^2 + g_{Af}^2)}. \tag{13.127}$$

For leptons

$$A^l_{FB} = \frac{3g^2_{Vl} \, / \, g^2_{Al}}{\left(1 + g^2_{Vl} \, / \, g^2_{Al}\right)^2} \tag{13.128}$$

Taking into account the radiative corrections [cf. Eqs. (13.116) and (13.118)], we get from Eq. (13.128)

$$A^l_{FB} = \frac{3\left(1 - 4s^2_f\right)^2}{\left[1 + \left(1 - 4s^2_f\right)\right]^2}$$

$$= \frac{3\left(1 - 4\left(1 + \Delta k\right)s^2_o\right)^2}{\left[1 + 4\left(1 + \Delta k\right)s^2_o\right]^2}, \tag{13.129}$$

where $s^2_o = 0.23116$ and

$$A^l_{FBo} = \frac{3\left(1 - 4s^2_o\right)^2}{\left[1 + \left(1 - 4s^2_o\right)^2\right]^2} = 0.01685 \, (0.01683 \pm 0.00096) \tag{13.130}$$

The value in parentheses is the experimental value. The agreement is quite good.

Finally we discuss the longitudinal polarization of a fermion in the process $e^-e^+ \rightarrow f\bar{f}$. Near and on $Z$-peak, the cross sections for positive and negative helicities are given by

$$\frac{d\sigma^{(+)}}{d\Omega} - \frac{d\sigma^{(-)}}{d\Omega} = \frac{2\alpha^2}{4s} \frac{1}{\left(16\sin^2\theta_W\cos^2\theta_W\right)^2} \left| \frac{s}{s - m^2_Z + i\frac{s}{m_Z}\Gamma_Z} \right|^2$$

$$\times \left[-2\left(v^2_e + a^2_e\right)a_f \, v_f \left(1 + \cos^2\theta\right) - 4\left(v^2_f + a^2_f\right)a_e v_e \cos\theta\right] \tag{13.131a}$$

$$\frac{d\sigma^{(+)}}{d\Omega} + \frac{d\sigma^{(-)}}{d\Omega} = 2\left(\frac{d\sigma}{d\Omega}\right). \tag{13.131b}$$

Hence the polarization at $s = m^2_Z$ is given by

$$-A_f = \frac{\int^\pi_o \left(\frac{d\sigma^{(+)}}{d\Omega} - \frac{d\sigma^{(-)}}{d\Omega}\right) d\Omega}{2\int^\pi_o \left(\frac{d\sigma}{d\Omega}\right) d\Omega}$$

$$= -\frac{2v_f \, a_f}{v^2_f + a^2_f}, = -2\frac{g_{Vf}/g_{Af}}{1 + g^2_{Vf}/g^2_{Af}} \tag{13.132}$$

We can also write Eq. (13.132) in terms of effective mixing angle $s_f$:

$$A_l = 2\frac{1 - 4s^2_f}{1 + \left(1 - 4s^2_f\right)^2} \tag{13.133}$$

Using the value $A_\tau = 0.1431 \pm 0.0046$ we obtain $s^2_f = 0.23201 \pm 0.00057$ to be compared with $s^2_0 = 0.23116 \pm 0.00022$.

## 13.6   Tests of Yang-Mills Character of Gauge Bosons

The vector bosons self-couplings are given by the Lagrangian (13.47)

$$L_W = -\frac{1}{4} \left[ \partial_\mu \mathbf{W}_\nu - \partial_\nu \mathbf{W}_\mu - g \left( \mathbf{W}_\mu \times \mathbf{W}_\nu \right) \right]^2 . \tag{13.134a}$$

This gives the trilinear $W^+ W^- W_3$ coupling as

$$\begin{aligned}
L_W = i\frac{g}{2} \big[ &(\partial_\mu W_{3\nu} - \partial_\nu W_{3\mu}) \left( W^{-\mu} W^{+\nu} - W^{+\mu} W^{-\nu} \right) \\
&+ (\partial_\mu W_\nu^+ - \partial_\nu W_\mu^+) \left( W^{3\mu} W^{-\nu} - W_\nu^3 W^{-\mu} \right) \\
&- (\partial_\mu W_\nu^- - \partial_\nu W_\mu^-) \left( W^{3\mu} W^{+\nu} - W_\nu^3 W^{+\mu} \right) \big] .
\end{aligned} \tag{13.134b}$$

Using $W_\nu^3 = \sin \theta_W A_\nu + \cos \theta_W Z_\nu$, the above equation gives for the $\gamma$, $Z(\lambda, q) \to W^-(k_1, \mu) + W^+(k_2, \mu)$ vertices

$$L = \begin{pmatrix} -g \sin \theta_W \\ -g \cos \theta_W \end{pmatrix} \left[ g^{\mu\nu} (k_1 + k_2)^\lambda + g^{\nu\lambda} (q + k_2)^\mu + g^{\lambda\mu}(-q - k_1) \right] , \tag{13.134c}$$

where $\mu, \nu$ and $\lambda$ are the indices of polarization vectors of $W^-$, $W^+$, $W^3$ respectively. On the other hand, from Eq. (13.55), the Higgs coupling to gauge bosons is given by

$$L_{W-H} = \frac{g^2}{8} \left( H + v \right)^2 \left[ 2 W_\mu^+ W^{-\mu} + \frac{1}{\cos^2 \theta_W} Z_\mu Z^\mu \right] \tag{13.135a}$$

and the Yukawa coupling of Higgs to leptons is given by

$$L_{llH} = \frac{-1}{\sqrt{2}} h_l \bar{l} \, l H, \tag{13.135b}$$

where

$$h_l = \frac{\sqrt{2}}{v} m_l = \left( 2\sqrt{2} G_F \right)^{1/2} m_l. \tag{13.136}$$

One process in which the trilinear couplings can be tested directly is

$$e^+ + e^- \to W^+ + W^-.$$

In the lowest order of $g$, the diagrams shown in Fig. 13.7 contribute to this process.

We are interested in the high energy behavior of the amplitude $M$. The bad behavior comes from the longitudinal polarization of $W$s. The longitudinal polarization vector $\varepsilon_\mu^L$ for a $W$-boson of four-momentum $k_\mu$ is given by

$$\begin{aligned}
\varepsilon_\mu^L &= \frac{1}{m_W}\left(|\mathbf{k}| \;,\; k_o\hat{\mathbf{k}}\right) \\
&= \frac{k_\mu}{m_W} + \frac{m_W}{k_o + |\mathbf{k}|}\left(-1, -\hat{\mathbf{k}}\right) \\
&= \frac{k_\mu}{m_W} + 0\left(\frac{m_W}{E_k}\right)
\end{aligned} \tag{13.137}$$

$$\epsilon^L . k = 0$$

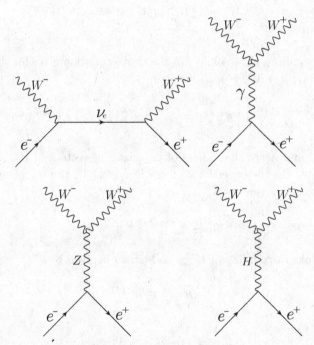

Fig. 13.7  Production of $W^-W^+$ pair in $e^-e^+$ collision through $\nu_e, \gamma$ and $Z$ boson exchange [(a), (b), (c)] and production of $W^-W^+$ through Higgs boson exchange (d).

It is the first term in Eq. (13.137) viz $\frac{k_\mu}{m_W}$ which gives the worst high energy behavior. The amplitude may grow with high energy due to this term, if it is not compensated. The amplitude is given by

$$\begin{aligned}
iT &= iT^{\mu\nu}\epsilon_\mu(k_1)\epsilon_\nu(k_2) \\
&= iT^{\mu\nu}\frac{k_{1\mu}}{m_W}\frac{k_{2\nu}}{m_W} = iT_{LL}
\end{aligned}$$

where $T_{LL}$ for the diagrams of Fig. 13.7 are given by

$$T_{LL}(a) = -\frac{g^2}{4m_W^2}\bar{v}(\acute{p})\left\{(\not{k}_1 - \not{k}_2)\frac{(1-\gamma_5)}{2} + m_e\right\}u(p) \qquad (13.138)$$

$$T_{LL}(b) = \frac{e^2}{2m_W^2}\bar{v}(\acute{p})(\not{k}_1 - \not{k}_2)\,u(p) \qquad (13.139)$$

$$T_{LL}(c) = -\frac{g^2\sin^2\theta_W}{2m_W^2}\bar{v}(\acute{p})(\not{k}_1 - \not{k}_2)$$

$$\times\left[1 - \frac{1}{4\sin^2\theta_W}(1-\gamma_5)\right]\frac{s}{s-m_Z^2}u(p) \qquad (13.140)$$

It is clear from Eqs. (13.138) and (13.139), that there is no possibility of cancellation between $T_{LL}(a)$ and $T_{LL}(b)$ even if $e = g\sin\theta_W$. The third diagram, arises due to trilinear couplings - a feature of gauge theory. Adding all the three contributions with $e = g\sin\theta_W$.

$$T_{LL} = -\frac{e^2}{2m_W^2}\bar{v}(\acute{p})(\not{k}_1 - \not{k}_2)\left[1 - \frac{1}{4\sin^2\theta_W}(1-\gamma_5)\right]u(p)\frac{m_Z^2}{s-m_Z^2}$$

$$-\frac{e^2}{4\sin^2\theta_W}\frac{m_e}{m_W^2}\bar{v}(\acute{p})\,u(p) \qquad (13.141)$$

Thus all the three diagrams together cancel the bad high energy behavior except for the last term in Eq. (13.141), which gives $S$-wave cross-section for $s \gg m_W^2, \sigma_s = \frac{G_F^2}{4\pi}s$. This is in conflict with the unitarity constraint [Eq. (13.2)] $\sigma_s \leq \frac{8\pi}{s}$. This conflict thus starts at

$$s = \frac{4\sqrt{2}}{G_F}\pi = (1.2\text{ TeV})^2. \qquad (13.142)$$

However, even this term is canceled by the diagram (Fig. 13.7d) due to Higgs exchange. This is because this diagram gives

$$T_H = \epsilon_\mu^L(k_1)\epsilon_\nu^L(k_2)T_H^{\mu\nu}$$

with

$$T_H^{\mu\nu} = \frac{2g^2}{g^2v^2}m_W^2\bar{v}(\acute{p})\,u(p)\frac{g^{\mu\nu}}{s-m_Z^2}$$

giving

$$T_H = \frac{e^2}{4\sin^2\theta_W}\frac{m_e}{m_W^2}\bar{v}(\acute{p})\,u(p)\frac{s-2m_W^2}{s-m_H^2} \qquad (13.143)$$

For $s \gg m_H^2$, $T_H = \frac{e^2}{4\sin^2\theta_W} \frac{m_e}{m_W^2} \bar{v}(\not{p})\, u(p)$ which cancels the last term in Eq (13.141). Thus for all $s \to \infty$

$$T_{LL} \to -\frac{e^2 m_Z^2}{2m_W^2} \frac{1}{s} \bar{v}(\not{p})(\not{k}_1 - \not{k}_2)\left[1 - \frac{1}{4\sin^2\theta_W}(1 - \gamma_5)\right] u(p) \quad (13.144)$$

Thus there is no trouble with the high energy behavior in the standard model if $m_H^2 < (1.2\,\text{TeV})^2$. There is similar cancellation for the amplitude $T_{LT}$, which for each individual diagram goes as constant when $s \to \infty$. $\sigma(e^+e^- \to W^-W^+)$ depends crucially on gauge cancellation discussed above. For example, $\sigma(\nu\text{ - exchange})$ for $s \gg m_W^2 \approx \frac{\pi\alpha^2 s}{96\sin^4\theta_W m_W^2}$, this would be the only contribution without $W^-W^+\gamma$ and $W^-W^+Z$ vertices. On the other hand, with the above cancellation

$$\sigma_{s\to\infty}(SM) = \frac{\pi\alpha^2}{2\sin^4\theta_W} \frac{1}{s} \ln\frac{s}{m_W^2}. \quad (13.145)$$

The cross section also contains the threshold factor $\sqrt{1 - \frac{4m_W^2}{s}}$ which tends to 1 as $s \to \infty$. Thus the cross section grows near the threshold and then falls like $\frac{1}{s}$ at large values of $\sqrt{s} \gg m_W$. The cross section is $\sim 10^{-35}\text{cm}^2$ at its maximum which occurs at about 40 GeV above $W^+W^-$ threshold. The situation is shown in Fig. 13.8.

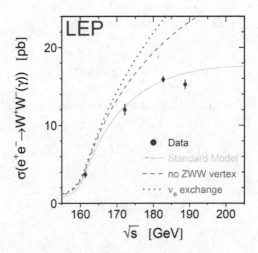

Fig. 13.8  Behavior of $\sigma_{SM}$ with energy $E$ [6,7].

## 13.7   Higgs Boson Mass

The Higgs potential

$$V\left(\phi\right) = \mu^2\phi^2 + \lambda\phi^4, \qquad \phi^2 = \bar{\phi}\phi \tag{13.146}$$

on using Eq. (13.54) goes over to

$$V\left(H\right) = \frac{1}{2}\left(2\lambda v^2\right)H^2 + \lambda v H^4 + \frac{1}{4}\lambda H^4 + \frac{1}{4}\lambda v^4 \tag{13.147a}$$

when the symmetry is spontaneously broken; $\mu^2 = -\lambda v^2\,(\lambda > 0)$. Thus we see that the Higgs boson mass

$$m_H^2 = 2\lambda v^2 \equiv \frac{1}{2}\left|\frac{\partial^2 V}{\partial\phi^2}\right|_{\phi = \frac{v}{\sqrt{2}}} \tag{13.147b}$$

is arbitrary. We now discuss theoretical bounds on the Higgs boson mass:

## 13.8   Upper Bound

### 13.8.1   *Unitarity*

We have seen in Sec. 13.6 that the Higgs boson contribution to the cross section for the process

$$e^- + e^+ \to W_L^+ + W_L^-$$

for $s \gg m_H^2$ is given by

$$\sigma_S = \frac{1}{4\pi}\left(\frac{G_F}{\sqrt{2}}\right)^2 s. \tag{13.148}$$

Then comparing Eq. (13.2), we get

$$\left(\frac{1}{4\pi}\right)\left(\frac{G_F}{\sqrt{2}}\right)^2 s \leq \frac{8\pi}{s} \tag{13.149}$$

This requires a "cut-off" (signaling new physics beyond $\Lambda_{SB}$):

$$\Lambda_{SB}^{PWU} \leq \left(4\sqrt{2}\pi/G_F\right)^{1/2} = (1.2\text{ TeV}). \tag{13.150a}$$

To avoid this conflict, the Higgs mass $m_H$ should be such that

$$m_H < \Lambda_{SB}^{PWU} = (1.2\text{ TeV}). \tag{13.150b}$$

We saw at the beginning of this chapter that $m_W \ll \Lambda_{SB}^{PWU}$. Whether similar thing happens or not for $m_H$ only experiments will tell.

### 13.8.2   *Finiteness of Couplings*

The Higgs-self coupling $\lambda$ is not asymptotically free. In $\lambda$ $\phi^4$ theories, the renormalized group equation gives (see appendix B)

$$\frac{d}{d\ln q^2}\lambda\left(q^2\right) = \frac{3}{4\pi^2}\lambda^2\left(q^2\right) \tag{13.151}$$

This gives

$$\lambda\left(q^2\right) = \frac{\lambda\left(v^2\right)}{1 - \frac{3}{4\pi^2}\lambda\left(v^2\right)\ln\frac{q^2}{v^2}}. \tag{13.152}$$

The minus sign in this equation indicates that the Higgs coupling is not asymptotically free. In fact it implies that regardless of how small $\lambda\left(v^2\right)$ is, $\lambda\left(q^2\right)$ will eventually blow up at some large energy scale $q = \Lambda$. In order to avoid this and to guarantee positivity of $\lambda\left(\Lambda\right) : \lambda\left(\Lambda\right) < \infty$, $\lambda\left(v^2\right) < \frac{4\pi^2}{3}\frac{1}{\ln\frac{\Lambda^2}{v^2}}$, giving $m_H^2 : \left[v^2 = \frac{1}{\sqrt{2}G_F}\right]$

$$m_H^2 \equiv 2\lambda\left(v^2\right)v^2 = \sqrt{2}\frac{\lambda\left(v^2\right)}{G_F} < \frac{4\sqrt{2}\pi^2}{3G_F}\frac{1}{\ln\frac{\Lambda^2}{v^2}} \tag{13.153}$$

The upper bound on $m_H$ is related logarithmically to the scale $\Lambda$ up to which the standard model is assumed to be valid which is very high. For some values of $\Lambda$, the upper bound on $m_H$ is given below

| $\Lambda$ | $m_H$ |
|-----------|-------|
| 1 TeV | 753 GeV |
| $10^{16}$ GeV | 159 GeV |
| $10^{19}$ GeV | 144 GeV |

Thus we see that if we assume the standard model to be valid up to Planck scale, then $m_H \leq 144$ GeV. We also notice that as Higgs mass goes above 800 GeV, $\Lambda$ comes down to Higgs mass. We conclude that as Higgs mass increases, so is the self coupling $\lambda$; the Higgs mass blow up at relatively low energy scale $\Lambda$, heralding the appearance of new non-perturbative physics. We also note that non-perturbative effects as the quartic Higgs coupling becomes large have also been estimated, mostly in the context of the lattice-Higgs model. Again a cut off on the parameter $\lambda\left(v\right)$ provides an upper bound on $m_H : m_H < 700$ GeV.

Finally, we note that the top quark Yukawa coupling $h_t = \sqrt{2}m_t/v$, for $m_t = 172$ GeV, can be of order 1. Top quark coupling modifies the renormalization group equation for the Higgs boson coupling $\lambda$. Top loop corrections reduce $\lambda$ for increasing top-Yukawa coupling. Hence for small

$m_H$ (small $\lambda$), the negative renormalization by $t$-quark coupling derives $\lambda < 0$, leading to instability of electroweak vacuum.

Only a narrow range [11, 12] $130\text{GeV} \leq m_H \leq 180\text{GeV}$ is compatible with the validity of the standard model up to Planck scale.

Finally the precision electroweak data give as previously noted

$$m_H \leq 185\text{GeV} \tag{13.154}$$

The most recent limit, reported at ICHEP 10 is $158 < m_H < 175$ GeV.

## 13.9 Standard Model, Higgs Boson Searches, Production at Decays

The search for Higgs is one of the objectives of new accelerators.

### 13.9.1 *LEP-2*

The dominant production mechanism in LEP2, is $e^- e^+ \to Z \to ZH$. The tree level cross section is given by

$$\sigma \left( e^- e^+ \to ZH \right) = \frac{G_F^2 \, m_Z^4}{96\pi s} \left[ v_e^2 + a_e^2 \right] Q^{1/2} \frac{Q + 12 m_Z^2 \, / \, s}{\left( 1 - m_Z^2 \, / \, s \right)^2}$$

Here the standard model couplings [cf. Eqs.(13.65a) and (13.135a)]

$$g_{Ze^- e^+}^2 = \left( v_e^2 + a_e^2 \right) = \left[ \left( 1 - 4 \sin^2 \theta_W \right)^2 + 1 \right]$$

$$g_{ZZH} = \frac{g^2 v}{2 \cos^2 \theta_W}$$

$$= \frac{g m_Z}{\cos \theta_W} \tag{13.155}$$

have been used. $Q$ is the phase space factor

$$Q \equiv \left[ 1 - \frac{(m_H + m_Z)^2}{s} \right] \left[ 1 - \frac{(m_H - m_Z)^2}{s} \right] \tag{13.156}$$

The LEP has established the lower limit

$$m_H > m_Z + \sqrt{s} \geq 114.4 \text{ GeV} \tag{13.157}$$

for LEP2 $\sqrt{s} = 34$ GeV.

### 13.9.2   *LHC and Tevatron*

The basic processes for the Higgs boson production are [12]:

$$gg \to H$$
$$\to qqV^{(*)}V^{(*)}$$
$$\to qqH$$
$$q\bar{q} \to V^{(*)} \to VH$$
$$gg,\ q\bar{q} \to t\bar{t}H$$

where $V = W$ or $Z$, $V^{(*)}$ is virtual (off-mass shell gauge boson). The corresponding diagrams are shown in Fig. 13.9.

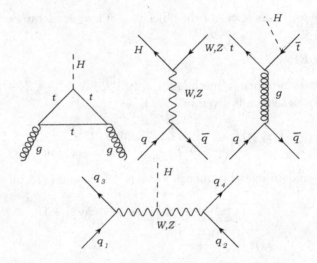

Fig. 13.9   The basic processes for the Higgs boson production.

The SM Higgs production cross-sections at Tevatron for $p\bar{p}$ collision for $\sqrt{s} = 1.96$ TeV are shown in Fig 13.10. The SM Higgs production cross-sections for $\sqrt{s} = 14$ TeV are shown in Fig 13.11. The comparison between Figs. 13.10 and 13.11, shows that SM Higgs production cross-section are significantly larger at LHC than at Tevatron. Thus the chance of SM Higgs discovery at LHC is significant. The SM Higgs being neutral, its detection is only possible through its decay channels.

We now discuss the decays of Higgs, since Higgs searches involve its decay widths.

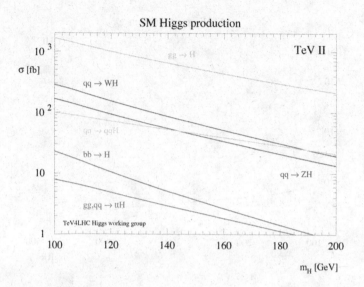

Fig. 13.10   SM Higgs boson production cross sections for $p\bar{p}$ collisions at 1.96 TeV [10].

(i) For $H \to f\bar{f}$, the coupling involved is [cf. (13.73)]

$$h_f^2 = \frac{2m_f^2}{v^2} = 2\sqrt{2}m_f^2 G_F,$$   (13.158)

giving

$$\Gamma\left(H \to f\bar{f}\right) = N_c^f \frac{G_F \, m_f^2}{4\sqrt{2}\pi} m_H \, \beta_f^3$$   (13.159)

where

$$N_c^f = 1 \quad \text{for} \quad f = l^\pm$$
$$= 3 \quad \text{for} \quad f = q$$

$$\beta_f = \left(1 - 4m_f^2 \, / \, m_H^2\right)^{1/2}.$$   (13.160)

It may be noted that there are important QCD corrections for $H \to q\bar{q}$. The bulk of QCD radiative corrections can be mapped into the scale dependence of the quark mass, evaluated at the Higgs mass, i.e. use $m_f$ at $m_H$, i.e. $m_f(m_H)$ in Eq. (13.159) [see Eq. (B. 47)], since $\Gamma(H \to f\bar{f})$ is proportional to $m_f^2$, the decay channels $H = b\bar{b}$ and $H = \tau^+\tau^-$ are more promising for the SM Higgs detected. (ii) For $H \to W^+W^-$ and $Z\bar{Z}$, if $m_H > 2m_W$,

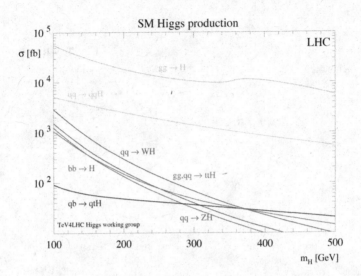

Fig. 13.11   SM Higgs boson production cross sections for $p\bar{p}$ collisions at 14 TeV [10].

$2m_Z$, most promising decay channels are $H = W^+W^-$ and $Z\overline{Z}$ for which the couplings are given by [cf. Eq. (13.135a)]

$$g_{WWH} = g\, m_W, \qquad g_{ZZH} = \frac{g\, m_Z}{\cos\theta_W}, \tag{13.161a}$$

where $\frac{g^2}{8m_W^2} = \frac{G_F}{\sqrt{2}}, g^2 = \frac{e^2}{\sin^2\theta_W}$. These give the widths

$$\Gamma\left(H \to W^+W^-\right) = \frac{G_F}{8\sqrt{2}\pi} m_H^3 \left(1 - x_W\right)^{1/2} r_W \tag{13.161b}$$

$$\Gamma\left(H \to ZZ\right) = \frac{G_F}{16\sqrt{2}\pi} m_H^3 \left(1 - x_Z\right)^{1/2} r_Z, \tag{13.161c}$$

where

$$r_{W,Z} = 1 - x_{W,Z} + \frac{3}{4}x_{W,Z}^2$$

$$x_{W,Z} = \frac{4m_{W,Z}^2}{m_H^2}. \tag{13.162}$$

It is of interest to consider $H \to W_L^+ W_L^-$

$$\Gamma(H \to W_L^+ W_L^-) = \frac{\alpha}{16\sin^2\theta_W} \frac{m_H^3}{m_W^2}(1 - x_W)^{1/2}(1 - \frac{1}{2}x_W)^2 \tag{13.163}$$

Thus we see that for $m_H \gg m_W$ the dominant contribution to the decay comes from the production of the longitudinal polarized W boson. In this

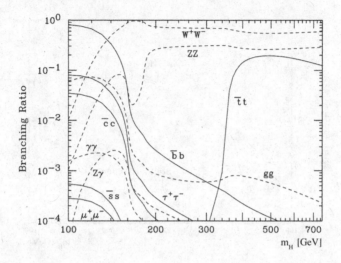

Fig. 13.12  Branching ratios for the main decays of the SM Higgs boson [10].

connection consider the decay of top quark $t \to bW^+$ for which in the unitary gauge

$$\Gamma(t \to bW^+) = \frac{\alpha}{16 \sin^2 \theta_W} \frac{m_t^3}{m_W^2} (1 - \frac{m_W^2}{m_t^2})^2 (1 + \frac{m_W^2}{2m_t^2}),$$

while

$$\Gamma(t \to bW_L^+) = \frac{\alpha}{16 \sin^2 \theta_W} (1 - \frac{m_W^2}{m_t^2})^2 = \Gamma(t \to b + \phi^+),$$

where $\phi^+$ is the Goldstone boson. Thus the top quark decay is expected to be dominated by the emission of longitudinally polarized W-boson and if it turns out to be so, this would be an indirect evidence of Higgs mechanism. One can also show that

$$\Gamma(H \to W_L^+ W_L^-) = \Gamma(H \to \phi^+ \phi^-, W_L^+ \phi^-, \phi^+ W_L^-).$$

It is useful to remember that for $m_H \approx 1.4$ TeV,

$$\Gamma(H \to VV) \approx \frac{1}{2} m_H^3 G_F \sim m_H. \qquad (13.164)$$

(iii) $H \to gg, \gamma\gamma$. The amplitudes for these decays are generated by Feynman diagrams involving triangular quark loops [see Fig. 13.9], while for $\gamma\gamma$ mode the contribution also comes from loop involving W bosons. The decay rates are given by

$$\Gamma(H \to 2g) = \frac{G_F}{4\sqrt{2}\pi} m_H^3 \frac{\alpha_s^2}{9\pi^2} \left| \sum_q I(x) \right|^2 \qquad (13.165)$$

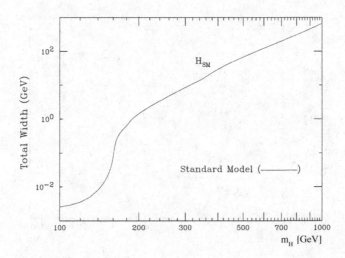

Fig. 13.13   The total decay width of the SM Higgs boson, shown as a function of $m_H$ [10].

where $x = \frac{m_H^2}{m_q^2}$ and $I(x)$ is a form factor that $\to 1$ as $x \to 0$ and $\to 0$ as $x \to \infty$. Thus the dominant contribution comes from very heavy quarks, in particular $t$-quark.

$$\Gamma(H \to 2\gamma) = \frac{G_F}{4\sqrt{2}\pi} m_H^2 \frac{\alpha^2}{18\pi^2} \left| \sum_f Q_f^2 N_C(f) - \frac{1}{4} \right|^2 \tag{13.166}$$

The branching ratios for various decay modes of the SM Higgs are shown in Fig. 13.12 and total decay width in Fig. 13.13.

To sum up the standard model is in very good shape, but Higgs boson $H$ is still a missing link.

## 13.10   Two Higgs Doublet Model (2HDM)

Although the Standard Model (SM) requires a single Higgs for electroweak symmetry breaking, it is interesting to study two Higgs model per se although they may be needed in some extensions of SM. In particular minimal supersymmetric model (MSSM) [see Chap. 17] provides a strong motivation to go beyond one Higgs doublet. In the standard model, Higgs doublet

$$\phi = \begin{pmatrix} \phi^0 \\ \phi^+ \end{pmatrix}, \; Y = 1$$

is coupled to down-quarks and up-quarks (and corresponding leptons)

$$(\bar{\psi}_L \phi d_R + h.c) + (\bar{\psi}_L i\tau_2 \bar{\phi} u_R + h.c)$$

Note that $i\tau_2\bar{\phi}$ has $Y = -1$. However, in MSSM two independent Higgs doublets $\phi_1$ and $\phi_2$ are coupled to down-quarks and up-quarks (and corresponding leptons) respectively;

$$(\bar{\psi}_L \phi_1 d_R + h.c.) + (\bar{\psi}_L i\tau_2 \bar{\phi}_2 u_R + h.c.)$$

Here $\phi_2$ cannot be replaced by $\phi_1$; with this replacement, $L_{\text{Higgs } f\bar{f}}$ will no longer be supersymmetric. A consequence of supersymmetry is that bosonic particles are naturally paired with fermions. Thus corresponding to two different Higgs muliplets $\phi_1$ and $\phi_2$, there are two left-handed fermion multiplets $\tilde{\phi}_1$ and $\tilde{\phi}_2$. The triangular anomaly arising from $\tilde{\phi}_1$ is canceled by $\tilde{\phi}_2$. This cancellation is not possible with one Higgs doublet $\phi_1$ only [4]. In this section, we are interested in the experimental consequences of two Higgs multiplets required by supersymmetry. Two Higgs doublets $(\phi_1, \phi_2)$ potential is given by

$$V(\phi_1, \phi_2) = \mu_1^2 \left(\bar{\phi}_1 \phi_1\right) + \lambda_1 \left(\bar{\phi}_1 \phi_1\right)^2 + \mu_2^2 \left(\bar{\phi}_2 \phi_2\right) + \lambda_2 \left(\bar{\phi}_2 \phi_2\right)^2$$
$$- \mu_3^2 \frac{1}{2} \left(\bar{\phi}_1 \phi_2 + \phi_1 \bar{\phi}_2\right) + \frac{1}{4} \lambda_3 \left(\bar{\phi}_1 \phi_2 + \phi_1 \bar{\phi}_2\right)^2$$
$$- \frac{1}{4} \lambda_4 \left(\bar{\phi}_1 \phi_2 - \phi_1 \bar{\phi}_2\right)^2 + \lambda_5 \left(\bar{\phi}_1 \phi_1 \bar{\phi}_2 \phi_2\right) \qquad (13.167)$$

Minima of $V(\phi_1, \phi_2)$ are given by

$$\frac{1}{\sqrt{2}} v_1 \left[\mu_1^2 + \lambda_1 v_1^2 + \frac{1}{2} (\lambda_3 + \lambda_5) v_2^2\right] - \frac{1}{2\sqrt{2}} v_2 \mu_3^2 = 0 \quad (13.168a)$$

$$\frac{1}{\sqrt{2}} v_2 \left[\mu_2^2 + \lambda_1 v_2^2 + \frac{1}{2} (\lambda_3 + \lambda_5) v_1^2\right] - \frac{1}{2\sqrt{2}} v_1 \mu_3^2 = 0 \quad (13.168b)$$

where $\langle \phi_1 \rangle_0 = v_1, \quad \langle \phi_2 \rangle_0 = v_2$.

After spontaneous symmetry breaking, one can write $\phi_1$ and $\phi_2$ as:

$$\phi_1 = \begin{pmatrix} H_1^+ \\ \frac{1}{\sqrt{2}} (v_1 + h_1 + ia_1) \end{pmatrix} \qquad (13.169a)$$

$$\phi_2 = \begin{pmatrix} H_2^+ \\ \frac{1}{\sqrt{2}} (v_2 + h_2 + ia_2) \end{pmatrix} \qquad (13.169b)$$

It is convenient to introduce an angle $\beta$ such that,

$$\tan \beta = v_2/v_1, \quad v_2 = v \sin \beta, \quad v_1 = v \cos \beta$$
$$v^2 = v_1^2 + v_2^2 \tag{13.170}$$

The physical fields can now be written in the form

$$H^+ = -\sin \beta H_1^+ + \cos \beta H_2^+$$
$$G^+ = \cos \beta H_1^+ + \sin \beta H_2^+ \tag{13.171}$$

The charged Higgs scalars $G^\pm$ (Goldstone-bosons) are absorbed in $W_\mu^\pm$ gauge vector bosons to give them the longitudinal degrees of freedom and to give them masses. For $CP$ odd Higgs scalars $a_1, a_2$, the physical scalars can be written as

$$A^0 = -\sin \beta a_1 + \cos \beta a_2 \tag{13.172a}$$
$$G^0 = \cos \beta a_1 + \sin \beta a_2 \tag{13.172b}$$

where $G^0$ (Goldstone boson) is absorbed in the vector boson $Z_\mu$ to give its mass. The other $CP$-even Higgs scalars $h_1$ and $h_2$ become massive; the physical $CP$-even Higgs scalars are put in the form

$$H^0 = \cos \alpha h_1 + \sin \alpha h_2 \tag{13.173a}$$
$$h^0 = -\sin \alpha h_1 + \cos \alpha h_2 \tag{13.173b}$$

In MSSM: $\phi_1$ is coupled to down quarks and leptons only; $\phi_2$ is coupled to up quarks and leptons only and the Higgs self-couplings $\lambda$s are related to gauge couplings:

$$\lambda_3 = \lambda_4 = -\frac{1}{2}g^2$$
$$\lambda_1 = \lambda_2 = \frac{1}{8}\left(g^2 + g'^2\right)$$
$$\lambda_5 = \frac{1}{4}\left(g^2 - g'^2\right)$$
$$\lambda_3 + \lambda_5 = -\frac{1}{4}\left(g^2 + g'^2\right)$$
$$m_Z^2 = \frac{1}{4}\left(g^2 + g'^2\right)v^2 \tag{13.174}$$
$$m_W^2 = \frac{1}{4}g^2 v^2 \tag{13.175}$$

For MSSM Higgs, we have

$$m_{H^\pm}^2 = \frac{1}{4}gv^2 + m_{A^0}^2 = m_W^2 + m_{A^0}^2 \tag{13.176}$$

$$m_{H^0,h^0}^2 = \frac{1}{2}\left\{(m_Z^2 + m_{A^0}^2) \pm \sqrt{(m_Z^2 + m_{A^0}^2)^2 - 4m_Z^2 m_{A^0}^2 \cos^2 2\beta}\right\} \tag{13.177}$$

$$\tan 2\alpha = \tan 2\beta \frac{m_Z^2 + m_{A^0}^2}{m_{A^0}^2 - m_Z^2} \tag{13.178}$$

These are tree level mass relations subject to radiative corrections. It is clear from the above expressions that two parameters are required to fix all the Higgs masses. A convenient choice is $\tan\beta, m_{A^0}$. Also the following ordering of masses is valid at tree level.

$$m_{h^0} < m_Z, \quad m_{A^0} < m_{H^0}, \quad m_W < m_{H^0}$$

The tree level mass relations are modified by radiative corrections. The largest contribution is a consequence of the incomplete cancellation between $t$ quark and its superpartner (which would be an exact cancellation if supersymmetry were unbroken). After radiative corrections, one gets $m_{h^0} \leq 130$ GeV. This upper bound is reached, when $\tan\beta >> 1$ and $m_{A^0} >> m_Z$ [12]. This is a definite prediction of MSSM which will soon be tested at LHC.

In MSSM, the couplings of Higgs to $W^\pm$ and $Z$ are given by $(V = A, W, Z)$

$$
\begin{aligned}
\mathcal{L}_{HVV} = &\frac{ig}{2\sqrt{2}}\left\{\begin{bmatrix} -\sin(\beta-\alpha)\left(H^0\partial^\mu H^- - H^-\partial^\mu H^0\right) \\ +\cos(\beta-\alpha)\left(h^0\partial^\mu H^- - H^-\partial^\mu h^0\right) \\ +i\left(A^0\partial^\mu H^- - H^-\partial^\mu A^0\right) \end{bmatrix} W_\mu^+ + h.c.\right\} \\
&+ie\left[H^+\partial^\mu H^- - H^-\partial^\mu H^+\right] A_\mu \\
&+\frac{i}{2}\frac{g}{\cos\theta_W}\left\{\begin{bmatrix} (\cos^2\theta_W - \sin^2\theta_W)\left(H^+\partial^\mu H^- - H^-\partial^\mu H^+\right) \\ +i\sin(\beta-\alpha)\left(A^0\partial^\mu H^0 - H^0\partial^\mu A^0\right) \\ -i\cos(\beta-\alpha)\left(A^0\partial^\mu h^0 - h^0\partial^\mu A^0\right) \end{bmatrix} Z_\mu\right\} \\
&+\frac{1}{4}g^2\left\{\left[H^-H^+ + \frac{1}{2}\begin{pmatrix} v^2 + H^{0^2} + h^{0^2} \\ +A^{0^2} + 2v\cos(\beta-\alpha)H^0 \\ +2v\sin(\beta-\alpha)h^0 \end{pmatrix}\right] 2W^{+\mu}W_\mu^-\right\} \\
&+e^2 H^- H^+\left\{\left[A^\mu + \frac{\cos^2\theta_W - \sin^2\theta_W}{2\sin\theta_W\cos\theta_W}Z^\mu\right]^2\right\} \\
&+\frac{1}{8}\frac{g^2}{\cos^2\theta_W}\left\{\left[\begin{matrix} v^2 + H^{0^2} + h^{0^2} \\ +A^{0^2} + 2v\cos(\beta-\alpha)H^0 \\ +2v\sin(\beta-\alpha)h^0 + \cdots \end{matrix}\right] Z^\mu Z_\mu\right\} \tag{13.179}
\end{aligned}
$$

where ellipses denote higher order terms.

From the Lagrangian (13.179), we see that tree level MSSM Higgs $H^0$ and $h^0$ couplings to gauge vector bosons contain the angular factors $\cos(\beta - \alpha)$ and $\sin(\beta - \alpha)$ respectively. The charged Higgs bosons couplings $H^+ H^- \gamma$, $H^+ H^- Z$ and those involving $W^\pm$ and Higgs bosons $H^\mp \left( h^0, H^0, A^0 \right)$ are independent of angular factors.

The MSSM Higgs bosons couplings to fermion pairs are proportional to fermion masses, hence are more relevant for $\tau$, top and bottom quarks. The couplings are given by

$$
L_{Hf\bar{f}} = -\frac{g}{\sqrt{2}m_W} \left\{
\begin{array}{c}
m_\tau \tan\beta \left[ (\bar{\tau}_R \nu_{\tau L}) H^- + h.c. \right] \\
+ \left[ (m_b \tan\beta \bar{b}_R t_L + m_t \cot\beta \bar{b}_L t_R) H^- + h.c \right] \\
\frac{1}{\sqrt{2}} \left[ \begin{array}{c} (m_\tau \bar{\tau}\tau + m_b \bar{b}b) \left( \frac{\sin\alpha}{\cos\beta} h^0 - \frac{\cos\alpha}{\cos\beta} H^0 \right) \\ -m_t \bar{t}t \left( \frac{\cos\alpha}{\sin\beta} h^0 - \frac{\sin\alpha}{\sin\beta} H^0 \right) \end{array} \right] \\
+ \frac{i}{\sqrt{2}} \left[ \left( \tan\beta \left( m_\tau \bar{\tau}\gamma_5\tau + m_b \bar{b}\gamma^5 b \right) + m_t \cot\beta \, \bar{t}\gamma_5 t \right) A^0 \right]
\end{array}
\right\}
$$

$$(13.180)$$

We now discuss the decoupling limit. In the limit $m_{H^0}, m_{H^\pm} \gg m_Z$ from Eqs. (13.176) and (13.177),

$$
m_{H^\pm}^2 = m_{A^0}^2 \left( 1 + m_W^2 / m_{A^0}^2 \right)
$$

$$
m_{H^0}^2 \approx m_{A^0}^2 \left( 1 + \sin^2 2\beta \frac{m_Z^2}{m_{A^0}^2} \right)
$$

$$
m_{h_0}^2 \approx m_Z^2 \cos^2 2\beta \qquad (13.181)
$$

$$
\tan 2\alpha \approx \tan 2\beta
$$

The last equation gives

$$
\beta - \alpha = \pm \frac{n\pi}{2}, \quad n = 0, 1, \cdots
$$

The value $\beta - \alpha = \pi/2$ is interesting, as for this value $\cos(\beta - \alpha) \to 0$, $\sin(\beta - \alpha) \to +1$ and Higgs scalar $h^0$ is decoupled from the heavy Higgs $H^0$, $H^\pm$ and $A^0$ as can be easily seen from Eqs. (13.179) and (13.180). As a consequence $g_{h^0 VV}$ and $g_{h^0 ff}$ reduce to those for the SM Higgs. Hence we conclude that in MSSM, there exists a parameter regime, in which by formally integrating out the heavy scalar state one gets an effective low-energy theory which is precisely that of SM Higgs. In most MSSM models, $1 \lesssim \tan\beta \lesssim m_t/m_b$.

Finally in the decoupling limit, the MSSM Higgs scalar couplings to vector bosons and fermion pairs are given by :

$$g_{H^0WW} = 0, \quad g_{h^0WW} = g_{m_W}$$

$$g_{H^0ZZ} = 0, \quad g_{h^0ZZ} = \frac{g}{\cos\theta_W} m_Z$$

$$g_{H^-\tau^+\nu_\tau} = \frac{g}{\sqrt{2}m_W} m_\tau \tan\beta \,\bar{\tau}\frac{1}{2}\left(1-\gamma_5\right)\nu_\tau$$

$$g_{H^-t\bar{b}} = \frac{g}{\sqrt{2}m_W}\left[\begin{array}{l} m_b \tan\beta \,\bar{b}\frac{1}{2}\left(1-\gamma_5\right)t \\ +m_t \cot\beta \,\bar{b}\frac{1}{2}\left(1+\gamma_5\right)t \end{array}\right]$$

$$g_{h^0\tau\bar{\tau}} = m_\tau\bar{\tau}\tau, \qquad g_{H^0\tau\tau} = 0$$

$$g_{h^0b\bar{b}} = m_b\bar{b}b, \qquad g_{H^0bb} = 0$$

We conclude that the above decay channels are relevant for the detection of MSSM Higgs scalar at LHC.

## 13.11  GIM Mechanism

Since in weak interactions the flavor quantum numbers are not conserved, weak interaction eigenstates of different generations, $d'$, $s'$ and $b'$ are not identical with mass eigenstates $d$, $s$ and $b$. These states are linear combination of $d$, $s$ and $b$. Thus we can write [cf. Eq. (13.44)]

$$d' = V_{ud}\, d + V_{us}\, s + V_{ub}\, b$$
$$s' = V_{cd}\, d + V_{cs}\, s + V_{cb}\, b$$
$$b' = V_{td}\, d + V_{ts}\, s + V_{tb}\, b \tag{13.182}$$

The quarks of one generation are linked to those of the succeeding generations with decreasing strength. Thus for example $V_{ub} \ll V_{us} < V_{ud}$. This is illustrated by the diagram [Fig. 13.14]. If we confine ourselves to ordinary and strange hadrons, then we can safely put $V_{ub} = 0$, but we cannot ignore $V_{cs}$, since charmed quark is linked to strange quark with maximum strength.

As a first approximation, we can ignore the third generation completely and can put $V_{ud} = \cos\theta_c$, $V_{us} = \sin\theta_c$, as given by Cabibbo theory. Thus we can write $d' = d\cos\theta_c + s\sin\theta_c$. In the weak neutral current, we have a term of the form

$$\bar{d}'\,\gamma_\mu\left(1-\gamma_5\right)d'$$
$$= \cos^2\theta_c\,\bar{d}\,\gamma_\mu\left(1-\gamma_5\right)d + \sin^2\theta_c\,\bar{s}\,\gamma_\mu\left(1-\gamma_5\right)s$$
$$+ \sin\theta_c\cos\theta_c\left[\bar{d}\,\gamma_\mu\left(1-\gamma_5\right)s + \bar{s}\,\gamma_\mu\left(1-\gamma_5\right)d\right]$$

$$\tag{13.183}$$

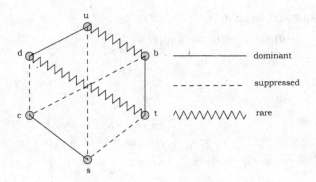

Fig. 13.14   Relative strengths of flavor changing transitions.

which arises from the doublet $\begin{pmatrix} u \\ d' \end{pmatrix}$. The above term can give rise to the following processes (Fig. 13.15).

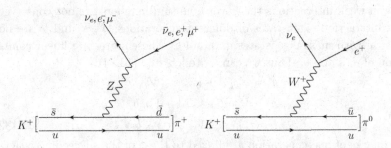

Fig. 13.15   Decay $K^+ \rightarrow \pi^+ + \nu + \bar{\nu}$ through neutral current and $K^+ \rightarrow \pi^0 + e^+ + \nu_e$ through charged current.

It is clear from Fig. 13.15 that both the processes

$$K^+ \rightarrow \pi^+ + \nu + \bar{\nu}$$
$$K^+ \rightarrow \pi^0 + e^+ + \nu_e$$

occur with equal strength. But experimentally

$$\frac{\Gamma\left(K^+ \rightarrow \pi^+ \nu \bar{\nu}\right)}{\Gamma\left(K^+ \rightarrow \pi^0 e^+ \nu_e\right)} < 1.2 \times 10^{-5},$$

i.e. the strangeness changing neutral current is very much suppressed compared with the strangeness changing charged current. Here the charmed quark $c$ comes to the rescue. If we put $V_{cd} = -\sin\theta_c$ and $V_{cs} = \cos\theta_c$, then $s' = -d\sin\theta_c + s\cos\theta_c$ and we get a term

$$\bar{s}' \gamma_\mu (1-\gamma_5) s' = \sin^2\theta_c\, \bar{d}\, \gamma_\mu (1-\gamma_5)\, d + \cos^2\theta_c\, \bar{s}\, \gamma_\mu (1-\gamma_5)\, s$$
$$- \sin\theta_c \cos\theta_c \left[ \bar{d}\, \gamma_\mu (1-\gamma_5)\, s + \bar{s}\, \gamma_\mu (1-\gamma_5)\, d \right] \quad (13.184)$$

from the doublet $\begin{pmatrix} c \\ s' \end{pmatrix}$. From Eqs. (13.183) and (13.184), it is clear that strangeness changing terms are canceled and $J_\mu^Z$ does not contain any strangeness changing term. This mechanism to eliminate the strangeness changing neutral current in tree approximation was suggested by Glashow, Iliapoulas and Maiani ($GIM$) before the experimental discovery of charm.

The $\Delta S = 2$, $K^0 \to \bar{K}^0$ transition shown in Fig. 13.16 is second order in $G_F$. With $GIM$ mechanism, a complete cancellation between $u$ and $c$ couplings occur if $m_c = m_u$. With the known experimental value for this transition, a limit on the mass of $m_c$ can be put and it was predicted that $m_c$ must be less than a few GeV and this is what was found later experimentally.

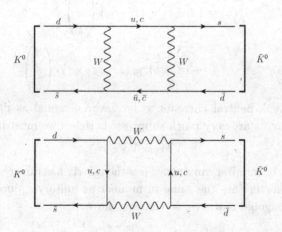

Fig. 13.16   Box diagrams for $\Delta S = 2$, $K^0 - \bar{K}^0$ transitions.

## 13.12   Cabibbo-Kobayashi-Maskawa Matrix

Three generations of fermions are linked with each other by weak interactions. The states $d'$, $s'$, and $b'$ are not mass eigenstates. They are related to mass eigenstates $d$, $s$ and $b$ as follows [cf. Eqs. (13.43) and (13.44)]:

$$\begin{pmatrix} d' \\ s' \\ b' \end{pmatrix} = V \begin{pmatrix} d \\ s \\ b \end{pmatrix} \tag{13.185}$$

where $V$ is a $3 \times 3$ matrix:

$$V = \begin{pmatrix} V_{ud} & V_{us} & V_{ub} \\ V_{cd} & V_{cs} & V_{cb} \\ V_{td} & V_{ts} & V_{tb} \end{pmatrix} \tag{13.186}$$

called 'Cabibbo-Kobayashi-Maskawa' (CMK) matrix.   The hadronic charged weak current can be written as

$$J_\mu^W (h) = (\bar{u}, \bar{c}, \bar{t}) \, \gamma_\mu \, (1 - \gamma_5) \, V \begin{pmatrix} d \\ s \\ b \end{pmatrix} \tag{13.187}$$

and $J_\mu^3 (h)$ which is a part of the neutral current:

$$J_\mu^3(h) = (\bar{u}, \bar{c}, \bar{t}) \, \gamma_\mu \, (1 - \gamma_5) \begin{pmatrix} u \\ c \\ t \end{pmatrix}$$

$$+ (\bar{d}, \bar{s}, \bar{b}) \, \gamma_\mu \, (1 - \gamma_5) \, V^\dagger V \begin{pmatrix} d \\ s \\ b \end{pmatrix}. \tag{13.188}$$

We want weak neutral currents to be flavor diagonal as flavor changing neutral currents are very much suppressed. Hence we must have

$$V^\dagger V = V V^\dagger = 1, \tag{13.189}$$

i.e. $V$ must be a unitary matrix. Thus the matrix has nine real parameters. These parameters are the same in number as unitary group $U_3$. Now $U_3$ has three diagonal matrices, so that we can write

$$V = e^{i\theta \lambda_0} \, e^{i\alpha \lambda_3} \, e^{i\beta \lambda_8} \, C \, e^{i\alpha' \lambda_3} e^{i\beta' \lambda_8}, \tag{13.190}$$

where $C$ is a $3 \times 3$ unitary matrix with 4 real parameters. The five parameters $\theta$, $\alpha$, $\beta$, $\alpha'$ and $\beta'$ can be absorbed into redefinitions of phases of $u$, $c$, $t$ and $d$, $s$, $b$ quarks. Thus we can write

$$V = R_2 R_1 \tilde{C} R_3 \tag{13.191}$$

where $R_1$, $R_2$ and $R_3$ are $3 \times 3$ rotation matrices:

$$R_1 = \begin{pmatrix} c_1 & s_1 & 0 \\ -s_1 & c_1 & 0 \\ 0 & 0 & 1 \end{pmatrix}, \qquad R_2 = \begin{pmatrix} 1 & 0 & 0 \\ 0 & c_2 & s_2 \\ 0 & -s_2 & c_2 \end{pmatrix},$$

$$R_3 = \begin{pmatrix} 1 & 0 & 0 \\ 0 & c_3 & s_3 \\ 0 & -s_3 & c_3 \end{pmatrix}, \tag{13.192}$$

and $\tilde{C}$ is a unitary matrix which can be written as

$$\tilde{C} = \begin{pmatrix} 1 & 0 & 0 \\ 0 & 1 & 0 \\ 0 & 0 & e^{i\delta} \end{pmatrix}. \tag{13.193}$$

Hence we have

$$V = \begin{pmatrix} c_1 & s_1 c_3 & s_1 s_3 \\ -s_1 c_2 & c_1 c_2 c_3 - s_2 s_3 \, e^{i\delta} & c_1 c_2 s_3 + s_2 c_3 \, e^{i\delta} \\ s_1 s_2 & -c_1 s_2 c_3 - c_2 s_3 \, e^{i\delta} & -c_1 s_2 s_3 + c_2 c_3 \, e^{i\delta} \end{pmatrix}, \tag{13.194}$$

where

$$c_i = \cos\theta_i, \qquad s_i = \sin\theta_i \tag{13.195}$$

There is an arbitrary phase $\delta$, which makes the Lagrangian density non-real. Thus the Lagrangian density violates time-reversal invariance. By $CPT$ theorem, it violates $CP$ invariance. Thus there is an attractive possibility of accommodating $CP$ violation in three-generation model; this cannot be done in two-generation model.

If we consider the three generations, the fermion mass matrix for $u$, $c$, $t$ and $d$, $s$, $b$ quarks can be written as

$$\mathcal{L}^q_{\text{mass}} = \frac{v}{\sqrt{2}} \left[ \bar{\Psi}^u_{L_i} \, h_{ij} \, q^u_{R_j} + \bar{\Psi}'^d_{L_i} \, \tilde{h}_{ij} \, q^d_{R_j} \right] + h.c.$$

$$= \left[ (\bar{u}_L, \bar{c}_L, \bar{t}_L) \, M_u \begin{pmatrix} u_R \\ c_R \\ u_R \end{pmatrix} \right]$$

$$+ \left[ (\bar{d}'_L, \bar{s}'_L, \bar{b}'_L) \, \tilde{M} \begin{pmatrix} d'_R \\ s'_R \\ b'_R \end{pmatrix} \right] + h.c. \tag{13.196}$$

Without any loss of generality, we can take $M_u$ to be diagonal matrix viz

$$M_u = \begin{pmatrix} m_u & & \\ & m_c & \\ & & m_t \end{pmatrix}. \tag{13.197}$$

It is clear from Eqs. (13.185) and (13.196) that

$$V^{\dagger} \tilde{M} V = M_d, \qquad (13.198)$$

where $M_d$ is now diagonal matrix.

Below we give the experimental values of CKM matrix elements [10]:

| Matrix element | Experimental value |
|:---:|:---:|
| $|V_{ud}|$ | $0.97425 \pm 0.00022$ |
| $|V_{us}|$ | $0.2252 \pm 0.00109$ |
| $|V_{ub}|$ | $(3.89 \pm 0.44) \times 10^{-3}$ |
| $|V_{cd}|$ | $0.230 \pm 0.011$ |
| $|V_{cs}|$ | $1.023 \pm 0.036$ |
| $|V_{cb}|$ | $(40.6 \pm 1.3) \times 10^{-3}$ |
| $|V_{td}|$ | $(8.4 \pm 0.6) \times 10^{-3}$ |
| $|V_{ts}|$ | $(38.7 \pm 2.1) \times 10^{-3}$ |
| $|V_{tb}|$ | $0.88 \pm 0.07$ |

Note that $|V_{ud}|^2 + |V_{us}|^2 + |V_{ub}|^2$ is consistent with 1 as required by unitarity.

## 13.13    Axial Anomaly

For a theory to be renormalizable, it is essential that vector and axial vector currents are conserved. In electroweak gauge theories, before spontaneous symmetry breaking, fermions are massless and it is, therefore, expected that axial vector current is also conserved. But this is not so, in fact as seen in Chap. 11, axial vector current receives anomalous contribution from the triangle graph: a closed fermion loop with one axial-vector vertex and two vector vertices as shown in Fig. 13.15. This anomalous contribution

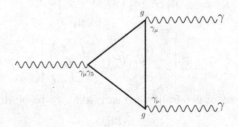

Fig. 13.17   Axial vector current anomaly from the triangle graph.

is equivalent to the statement that in the zero fermion mass limit, the divergence of axial vector current is given by

$$\partial^\mu A_\mu = \frac{g^2}{16\pi^2}\varepsilon^{\mu\nu\alpha\beta}F_{\mu\nu}F_{\alpha\beta} \qquad (13.199)$$

where $F_{\mu\nu}$ is the field tensor of the vector field and $g$ is the coupling constant as shown in Fig. 13.17.

The contribution from $\Delta$ graph arises only if $\Delta$ graph is odd in axial couplings. This contribution is independent of fermion masses and is unaltered by radiative corrections. In $QED$, such graphs do not cause any trouble as photon is not coupled to axial current. Nor does it cause any problem if one or more of the currents is associated with a global symmetry of the theory. In such a case, it can even be useful as for example the case for $\pi^0 \to 2\gamma$, which arises due to the anomaly as discussed in Chap. 11.

In electroweak theory, such graphs are not absent. For example, in the process $e^-e^+ \to \gamma\gamma$ shown in Fig. 13.18, the $\Delta$ graph can cause trouble as it would give bad high energy behavior.

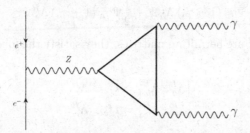

Fig. 13.18   Process $e^-e^+ \to 2\gamma$ through $\Delta$ graph.

To ensure the renormalizability of electroweak theory, it is, therefore, essential to ensure the cancellation of $\Delta$ anomalies. To see this, from Eqs. (13.65a) and (13.65b), we note that current to which $Z$ is coupled is

$$\bar{f}\left[\left(T^3 - 2Q_f \sin^2\theta_W\right)\gamma^\mu + T^3\gamma^\mu\gamma^5\right]f$$

and the anomaly arises from a fermion loop with one axial vector and two vector vertices. Thus for the above diagram the absence of the anomaly requires that

$$\sum_f T_3 Q_f^2, \ \sum_f T_3^3 \propto \sum_f T_3, \ \sum_f T_3^2 Q_f \propto \sum_f Q_f$$

should vanish. However, while $\sum_f T_3$ does vanish but none of the other two do either for leptons or quarks; for the former, $\sum_f T_3 Q_f^2$ is $3\left[-\frac{1}{2}(1)\right] = -3/2$ and for the latter it is $3\cdot3\left[\frac{1}{2}\frac{4}{9} - \frac{1}{2}\frac{1}{9}\right] = 3/2$, where factor 3 is the number of generations and another factor 3 in the quark sum is for color. However, it does vanish when the sum is taken over lepton doublets and quark doublets with a factor of 3 for color. The same is the case for the vanishing of $\sum_q Q_f$. Thus weak interaction gauge theory can be consistently combined with QED if the theory contains equal number of quarks (each having 3 color) and lepton doublets, showing necessity of quark-lepton symmetry. Below we consider cancellation of anomalies in a general way.

Consider a gauge group $G$, where the coupling of the fermions to gauge bosons is given by

$$\mathcal{L}_{\text{int}} = \bar{\Psi}_L i\gamma^\mu \left(\partial_\mu + ig\Lambda_a^L W_{a\mu}\right) \Psi_L + \bar{\Psi}_R i\gamma^\mu \left(\partial_\mu + ig\Lambda_a^R W_{a\mu}\right) \Psi_R. \quad (13.200)$$

The current coupled to gauge bosons is given by

$$J_\mu^a = \frac{1}{2}\bar{\Psi}\gamma_\mu \left(1 - \gamma_5\right) \Lambda_a^L \Psi + \frac{1}{2}\bar{\Psi}\gamma_\mu \left(1 + \gamma_5\right) \Lambda_a^R \Psi. \quad (13.201)$$

Here $\Lambda_a^L$ and $\Lambda_a^R$ are hermitian matrices; they satisfy the following commutation relations

$$\begin{aligned}
\left[\Lambda_a^L, \Lambda_b^L\right] &= if_{abc} \Lambda_c^L \\
\left[\Lambda_a^R, \Lambda_b^R\right] &= if_{abc} \Lambda_c^R.
\end{aligned} \quad (13.202)$$

$\Psi_L$ and $\Psi_R$ need not transform in the same way under $G$ as is the case in electroweak group. $\Lambda_a^L \neq \Lambda_a^R$ in general. The $\Delta$-anomaly (being independent of fermion masses) is proportional to

$$A_{abc}^L - A_{abc}^R$$

where $-$ sign arises since it is odd axial vector vertices which give anomaly and it has to be symmetric in two indices say $a$ and $b$. Thus [ { } denotes anticommutator]

$$A_{abc}^L = Tr\left(\{\Lambda_a^L, \Lambda_b^L\} \Lambda_c^L\right) \quad (13.203a)$$
$$A_{abc}^R = Tr\left(\{\Lambda_a^R, \Lambda_b^R\} \Lambda_c^R\right). \quad (13.203b)$$

Theory is thus anomaly free when

$$Tr\left(\{\Lambda_a^L, \Lambda_b^L\} \Lambda_c^L\right) - Tr\left(\{\Lambda_a^R, \Lambda_b^R\} \Lambda_c^R\right) = 0 \quad (13.204)$$

for all values of $a$, $b$, $c$.

*Examples*

(i) Vector or vector like gauge theory:

$$A^L = A^R \neq 0. \tag{13.205}$$

For such theories either

$$\Lambda_a^L = \Lambda_a^R \tag{13.206}$$

or

$$\Lambda_a^L = U^{-1}\Lambda_a^R U, \tag{13.207}$$

where $U$ is a fixed unitary matrix. The gauge current is given by

$$\begin{aligned}
J_\mu^a &= \bar{\Psi}_L \gamma_\mu \Lambda_a^L \Psi_L + \bar{\Psi}_R \gamma_\mu \Lambda_a^R \Psi_R \\
&= \bar{\Psi}_L \gamma_\mu \Lambda_a^L \Psi_L + \bar{\Psi}_R \gamma_\mu U \Lambda_a^L U^{-1} \Psi_R \\
&= \bar{\Psi} \gamma_\mu \Lambda_a \Psi,
\end{aligned} \tag{13.208}$$

where

$$\Psi = \Psi_L + U^{-1}\Psi_R, \qquad \Lambda_a = \Lambda_a^L, \tag{13.209}$$

is a pure vector. Note that in general the redefinition of $\Psi$ generates $\gamma_5$ terms in the fermion mass matrix. Such a theory is called vector-like.

In $QCD$, the left-handed and right-handed quarks belong to the fundamental representation 3 of $SU_c(3)$. Thus it is a vector theory and is anomaly free.

(ii) $A^L = A^R = 0$.

In this case, fermion representation is such that anomalies cancel separately for left-handed and right-handed fermions. This is the case for example for $SU(2)$. For the fundamental representation 2 of $SU(2)$, $\Lambda_a = \frac{1}{2}\tau_a$ and since $\tau_a \tau_b + \tau_b \tau_a = 2\delta_{ab}$,

$$A_{abc} = Tr\left[\{\tau_a, \tau_b\}\tau_c\right] = 0. \tag{13.210}$$

The representation 2 is a real representation in $SU(2)$. But this is not the case for $SU(n)$, $n > 2$, e.g. representation 3 of $SU(3)$ is not equivalent to $3^*$. Thus for $SU(n)$, $n > 2$ is not safe in general. However, fermions belonging to an octet representation of $SU(3)$ are anomaly free since octet representation is real. This can be seen as follows:

If $\Lambda_a$ form a representation, $-\Lambda_a^*$ also form a representation. The negative sign arises, since matrices $\Lambda_a^*$ satisfy the commutation relation

$$[\Lambda_a^*, \ \Lambda_b^*] = -i \, f_{abc}\Lambda_c^*. \tag{13.211}$$

Hence $-\Lambda_a^*$ form a representation conjugate to $\Lambda_a$. If (as in the case for real representation),

$$\Lambda_a = -U^{-1}\Lambda_a^* U, \tag{13.212}$$

where $U$ is a unitary matrix, then

$$A_{abc} = Tr\left[\{\Lambda_a^*, \ \Lambda_b^*\} \Lambda_c^*\right]$$
$$= -A_{abc}. \tag{13.213}$$

Thus in general real representations are safe. They do not produce axial anomaly. However, a safe representation need not be real.

(iii) The standard model $SU_c(3) \times SU(2) \times U(1)$.

We need to consider $SU(2) \times U(1)$ only as $SU_c(3)$ is anomaly free. The matrices $\Lambda_a^L$ and $\Lambda_a^R$ are given by

$$\Lambda_a^L \ : \ \frac{1}{2}\tau_a^L, \qquad \frac{1}{2}Y_L, \qquad \Lambda_a^R : \frac{1}{2}Y_R$$

$$Q = \frac{1}{2}\tau_3 + \frac{1}{2}Y. \tag{13.214}$$

Now

$$T_R\left(\{\tau_a^L, \ \tau_b^L\} \tau_c^L\right) = 0 \tag{13.215}$$

so that from Eq. (13.204), we have to show that

$$Tr\left(\{\tau_a^L, \tau_b^L\} Y_L\right)$$
$$= 2\delta_{ab} \, Tr \, Y_L$$
$$= 2\delta_{ab} \, Tr\left[2Q - \tau_3\right] = 4\delta_{ab} \, Tr \, Q = 0 \tag{13.216}$$

and

$$Tr\left[Y_L^3\right] - Tr\left[Y_R^3\right] = 0 \tag{13.217}$$

for the cancellation of anomalies. Now

$$Tr\left[Y_R^3\right] = 8Tr\left[Q^3\right]$$
$$Tr\left[Y_L^3\right] = Tr\left[8Q^3 + 6Q \, \tau_3^2 - 6Q^2 \, \tau_3 - \tau_3^3\right]$$
$$= 8TrQ^3 + 6TrQ - 6Tr\left(Q^2 \, \tau_3\right). \tag{13.218}$$

But

$$Tr\left[Q^3\right] \propto TrQ$$
$$Tr\left[Q^2\tau_3\right] \propto Tr\tau_3 = 0 \tag{13.219}$$

Hence for the cancellation of anomaly, we must have

$$Tr \, Q = 0 \tag{13.220}$$

Now

$$Tr \, Q = [0 - 1 + 3(\frac{2}{3} - \frac{1}{3})] = 0 \tag{13.221}$$

Hence in the standard model, lepton anomalies cancel quark anomalies. Note that in the cancellation of anomalies, color plays a crucial role. Left-handed fermions anomalies cancel among themselves and so do the right-handed fermions anomalies.

## 13.14 Problems

(1) In the Standard Model, the symmetry is spontaneously broken by introducing a scalar doublet $\Phi : (I = 1/2, Y = 1)$

$$\Psi_L = \begin{pmatrix} \nu_e \\ e^- \end{pmatrix}_L, e_R.$$

Suppose we introduce a scalar triplet

$$\chi \equiv (\chi_1, \chi_2, \chi_3), \quad I = 1, Y = 2.$$

Write $\chi$ as a $2 \times 2$ matrix

$$\chi = \tau \cdot \chi$$

Write down the Yukawa coupling of $\chi$ for the doublet $\Psi_L$. Show that neutrino acquires a Majorana mass for $\langle \chi^0 \rangle \neq 0$. Write its contribution to $W^\pm$ and $Z$.

(2) Consider high energy neutrino-parton scattering

$$\nu_\mu d \to \mu^- u$$
$$\bar{\nu}_\mu u \to \mu^+ d$$

Show that

$$\frac{d\sigma^{\nu_\mu}}{dQ^2} = \frac{G_F^2}{2\pi}$$
$$\frac{d\sigma^{\bar{\nu}_\mu}}{dQ^2} = \frac{G_F^2}{2\pi} \frac{u^2}{s^2}$$

In the Lab. frame

$$\frac{d\sigma^{\nu_\mu}}{dQ^2} = \frac{G_F^2}{2\pi}, \quad \sigma^{\nu_\mu} = \frac{G_F^2}{\pi} mE$$
$$m_u \approx m_d = m$$
$$\frac{d\sigma^{\bar{\nu}_\mu}}{dQ^2} = \frac{G_F^2}{2\pi} \frac{E'^2}{E^2}, \quad \sigma^{\bar{\nu}_\mu} = \frac{G_F^2}{\pi} m \frac{1}{3} E$$

(3) Show that for high energy neutrino scattering:

$$\nu_\mu(\bar{\nu}_\mu) + e^- \to \nu_\mu(\bar{\nu}_\mu) + e^-$$
$$\frac{d\sigma^{\nu,\bar{\nu}}}{dQ^2} = \frac{G_F^2}{4\pi} \frac{1}{s^2} \left[ (g_{Ve} \pm g_{Ae})^2 s^2 + (g_{V_e} \mp g_{A_e})^2 u^2 \right]$$

In the Lab. frame, show that

$$\sigma^{\nu,\bar{\nu}} = \frac{G_F^2 m_e E_\nu}{2\pi} \left[ (g_{V_e} \pm g_{A_e})^2 \pm \frac{1}{3} (g_{V_e} \mp g_{A_e})^2 \right].$$

(4) For high energy neutrino scattering on parton (quark)

$$\nu\left(\bar{\nu}\right) + q_i \rightarrow \nu\left(\bar{\nu}\right) + q_i.$$

Show that

$$\frac{d\sigma^{\nu,\bar{\nu}}}{dQ^2} = \frac{G_F^2}{4\pi}\frac{1}{s^2}\left[\left(g_{V_i} \pm g_{A_i}\right)^2 s^2 + \left(g_{V_i} \mp g_{A_i}\right)^2 u^2\right]$$

$$= \frac{G_F^2}{4\pi}\left[\left(g_{V_i} \pm g_{A_i}\right)^2 + \left(g_{V_i} \mp g_{A_i}\right)^2\left(1-y\right)^2\right]$$

$$= \frac{G_F^2}{4\pi}\left[\left(g_{V_i}^2 + g_{A_i}^2\right)\left(1 + \left(1-y\right)^2\right) \pm 2g_{V_i}g_{A_i} \pm \left(1 - \left(1-y\right)^2\right)\right]$$

where

$$y = \frac{\nu}{E}, \ \nu = E - E', \ \frac{u}{s} = \left(1-y\right)$$

in the Lab. frame.

(5) From Eq. (13.104), show that

$$\left[\sigma^{\nu_\mu e} + \sigma^{\bar{\nu}_\mu e}\right] \pm 2\left[\sigma^{\nu_\mu e} - \sigma^{\bar{\nu}_\mu e}\right] = \frac{4}{3}\left(g_{V_e} \pm g_{A_e}\right)^2$$

In terms of $\sin^2\theta_W$

$$\left(g_{V_e} + g_{A_e}\right) = \left(-\frac{1}{2} + \sin^2\theta_W\right)$$

$$\left(g_{V_e} - g_{A_e}\right) = \sin^2\theta_W.$$

(6) Show that the decay width for the decay

$$t \rightarrow b + W$$

for the transverse and longitudinal $W$ is given by

$$\Gamma\left(t \rightarrow b + W^\perp\right) = \frac{G_F}{\sqrt{2}}\frac{|\mathbf{k}|^2}{2\pi}m_t\left(\frac{2m_W^2}{m_t^2}\right)$$

$$\Gamma\left(t \rightarrow b + W^\|\right) = \frac{G_F}{\sqrt{2}}\frac{|\mathbf{k}|^2}{2\pi}m_t$$

$$\Gamma\left(t \rightarrow b + W\right) = \frac{G_F}{\sqrt{2}}\frac{|\mathbf{k}|^2}{2\pi}m_t\left(1 + \frac{2m_W^2}{m_t^2}\right)$$

$$= \frac{G_F}{\sqrt{2}}\frac{|\mathbf{k}|^3}{8\pi}\left(1 - \frac{m_W^2}{m_t^2}\right)^2\left(1 + \frac{2m_W^2}{m_t^2}\right).$$

(7) Show that the longitudinal polarization of $f$ in the process $e^- e^+ \to f\bar{f}$ near $Z$-peak is given by

$$A_f = \frac{\int_0^\pi \left( \frac{d\sigma^{(+)}}{d\Omega} - \frac{d\sigma^{(-)}}{d\Omega} \right)}{2 \int_0^\pi \left( \frac{d\sigma}{d\Omega} \right) d\Omega}$$

$$= 2 \frac{g_{V_f}/g_{A_f}}{1 + g_{V_f}^2/g_{A_f}^2}$$

Hint:-

$$\mathcal{L}_e^{\lambda\mu} \propto \left[ \left( g_{V_e}^2 + g_{A_e}^2 \right) \left( k_2^\lambda k_1^\mu + k_2^\mu k_1^\lambda - g^{\mu\lambda} k_1 \cdot k_2 - 2i g_{V_e} g_{A_e} \epsilon^{\lambda\mu\rho\sigma} k_{1\rho} k_{2\sigma} \right) \right]$$

$$\mathcal{L}_{\lambda\mu}^f \propto Tr \left[ \left( \not{p}_1 + m_f \right) \left( \frac{1 + \gamma^5 \not{s}}{2} \right) \gamma_\lambda \left( g_{V_f} - g_{A_f} \gamma_5 \right) \right.$$
$$\left. \times \left( \not{p}_2 - m_f \right) \gamma_\mu \left( g_{V_f} - g_{A_f} \gamma^5 \right) \right]$$

For polarization only terms proportional to $m_f$ contibute. In $L_e^{\lambda\mu} L_{\lambda\mu}^f$, only terms proportional to $\cos\theta$ are relevant and thus only terms containing $\epsilon^{\mu\lambda\rho\sigma} \epsilon_{\mu\lambda\rho'\sigma'}$ will contribute.

## 13.15 References

1. J. C. Taylor, Gauge theories of weak interactions, Cambridge University Press, Cambridge, U. K. (1976).
2. M. A. Beg and A. Sirlin, Gauge theories of weak interactions, Ann. Rev. Nucl. Sci., **24**, 379 (1974); Gauge theories of weak interactions II, Phys. Rep. **88** C, 1 (1982).
3. M. E. Peskin and D. V. Schroeder, An Introduction to Quantum Field Theory, Addison-Wesley (1995).
4. M. E. Peskin, "Beyond standard Model" in proceeding of 1996 European School of High Energy Physics CERN 97-03, Eds: N. Ellis and M. Neubert.
5. G. Altarelli, "The Standard Electroweak Theory and Beyond" CERN-TH / 98-348 hep-ph / 9811456.
6. J. Ellis, "Beyond Standard Model for Hill walkers" CERN-TH / 98-329,hep-ph 9812235.
7. J. L. Rosner, "New developments in precision Electreoweak Physics" Comment Nucl. Phys. 22, **205** (1998).
8. W. Hollik, "Standard Model Theory" CERN-TH / 98-358; KA-TP-18-1998 hep-ph / 9811313, Plenary talk at the XXIX Int. Conf. HEP, Vancouver Canada (1998).

9. M. Spirce and P. M Zerwar, "Electroweak Symmetry Breaking and Higgs Physics" CERL-TH / 97-379, DESY 97-261 hep-ph/9803257.

10. Particle Data Group, N. Nakamura et. al, Journal of Physics G, **37**, 075021 (2010).

11. J. Ellis, arXiv:1004.0648 (2010).

12. H.E. Haber, arXiv:1011.1038 (2010).

13. G. Altarelli, arXiv:1010.5637 (2010).

14. For a review : H.E. Haber, "Present status and future prospects for a Higgs boson discovery at the Tevatron and LHC", arXiv:1011.1038 (2010).

    The references to the original papers can be found in review articles [11, 12].

# Chapter 14

# Deep Inelastic Scattering

## 14.1 Introduction

Lepton-nucleon scattering is an excellent tool to study the structure of nucleon. Electron (muon) scattering clearly shows that nucleon has a structure. Consider for example the scattering

$$e + p \to e' + X.$$

Let $E$ be the energy of the incident electron $e$ and $E'$ be the energy of the scattered electron. Let $q = k - k'$ be the momentum transfer. Then in the lab. frame, the four momenta $P$, $k$ and $k'$ of the target (proton), initial electron and the scattered electron are given by

$$P \equiv (M, \mathbf{0}), \quad k = (E, \mathbf{k})$$
$$k' \equiv (E', \mathbf{k}').$$

Neglecting the mass of the lepton, we have

$$q^2 = (k - k')^2 = -2EE'(1 - \cos\theta) = -4EE' \sin^2 \frac{\theta}{2}$$

$$\mathbf{k} \cdot \mathbf{k}' = EE' \cos\theta \tag{14.1a}$$

We define another invariant $\nu$:

$$M\nu = P \cdot q. \tag{14.1b}$$

In the lab. frame

$$\nu = q_0 = (E - E'). \tag{14.1c}$$

We also define the invariant mass:

$$s = P_X^2 = (q + P)^2 = q^2 + M^2 + 2M\nu. \tag{14.1d}$$

Note that $2M\nu + q^2 \geq 0$; for elastic scattering $2M\nu = -q^2$.

The elastic scattering of electrons on spinless proton can be written in terms of the Mott cross section:

$$\frac{d\sigma}{d\Omega} = \left(\frac{d\sigma}{d\Omega}\right)_M |F(q^2)|^2 \tag{14.2a}$$

where

$$\left(\frac{d\sigma}{d\Omega}\right)_M = \frac{\alpha^2 \cos^2 \frac{\theta}{2}}{4E^2 \sin^4 \frac{\theta}{2}} \tag{14.2b}$$

The structure of the proton manifests itself in terms of the form factor $F(q^2)$. In elastic scattering proton recoils as a whole and the scattering is coherent. The form factor $F(q^2)$ measures the charge distribution of the proton, viz

$$F(q^2) = \int e^{-i\mathbf{q}\cdot\mathbf{r}} \rho(r) d^3 r$$

$$= \int e^{-i\mathbf{q}\cdot\mathbf{r}} \rho(r) r^2 dr \, d\Omega \tag{14.3a}$$

If we expand $F(q^2)$ in powers of $q^2$, we get

$$F(0) = \int \rho(r) d^3 r = 1$$

$$\left.\frac{\partial F(q^2)}{\partial q^2}\right|_{q^2=0} = -2\pi \int r^4 \rho(r) dr \left\langle \cos^2 \theta \right\rangle = -\frac{1}{6} \left\langle r^2 \right\rangle. \tag{14.3b}$$

$\left\langle r^2 \right\rangle$ is called the mean square charge radius.

It is convenient to write the Mott cross section in the form

$$\left(\frac{d\sigma}{dq^2}\right)_M = -\left(\frac{4\pi\alpha^2}{q^4}\right) \frac{E'}{E} \cos^2 \frac{\theta}{2}$$

$$= -\frac{4\pi\alpha^2}{q^4} \left[1 + \frac{q^2}{2mE} + \frac{q^2}{4E^2}\right] \tag{14.4}$$

This is the scattering cross section for the scattering of electrons on spinless (structureless) particles of mass $m$. The scattering cross section for the scattering of electrons on structureless spin $1/2$ particles of mass $m$ and charge $ee_q$ (called parton) can be calculated using the standard trace techniques and is given by $\left[Q^2 = -q^2\right]$ (see Appendix A.7.2)

$$\frac{d\sigma}{dQ^2} = \frac{4\pi\alpha^2 e_q^2}{Q^4} \frac{E'}{E} \cos^2\frac{\theta}{2}\left[1 + \frac{Q^2}{2m^2}\tan^2\frac{\theta}{2}\right]$$

$$= e_q^2\left(\frac{d\sigma}{dQ^2}\right)_{\mathrm{M}}\left[1 + \frac{Q^2}{2m^2}\tan^2\frac{\theta}{2}\right]$$

$$= \frac{4\pi\alpha^2 e_q^2}{Q^4}\left[1 - \frac{Q^2}{2mE} - \frac{Q^2}{4E^2} + \frac{Q^4}{8m^2E^2}\right]$$

$$\left(\frac{d\sigma}{dQ^2}\right)_{\mathrm{M}} = \frac{4\pi\alpha^2}{Q^4}\left[1 - \frac{Q^2}{2mE} - \frac{Q^2}{4E^2}\right] \tag{14.5a}$$

For the longitudinally polarized parton

$$\frac{d\Delta\sigma}{dQ^2} = \frac{4\pi\alpha^2}{Q^4}\frac{Q^2}{2mE^2}e_q^2\left(E + E'\cos\theta\right) \tag{14.5b}$$

## 14.2 Deep-Inelastic Lepton-Nucleon Scattering

We now consider the inelastic scattering of electrons on nucleons (see Fig. 14.1). For this case the matrix elements are

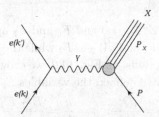

Fig. 14.1   Inelastic charged lepton-proton scattering.

$$T = \frac{e^2}{q^2}\frac{1}{(2\pi^3)}\frac{m_e}{\sqrt{EE'}}\bar{u}_e(k')\gamma_\mu u_e(k)\left\langle X\left|j_{e.m}^\mu\right|P\right\rangle \tag{14.6a}$$

The cross-section is given by [cf. Chap. 2]

$$d\sigma = \frac{1}{v_{in}}\frac{d^3k}{(2\pi)^3}\frac{d^3P_X}{(2\pi)^3}\frac{e^4}{q^4}(2\pi)^4\delta(P_X + k' - k - P)\frac{m_e^2}{EE'}L_{\mu\nu}W^{\mu\nu}, \tag{14.6b}$$

where [see Appendix A.7]

$$v_{in} = \frac{|\mathbf{k}|}{E}$$

$$L_{\mu\nu} = \frac{1}{2m_e^2}\left[k_\mu k_\nu' - g_{\mu\nu}k\cdot k' + k_\nu k_\mu' + im_e\epsilon_{\mu\nu\lambda\rho}q^\lambda n^\rho\right] \tag{14.6c}$$

and $n^\rho$ is the polarization of the electron beam, with $n^\mu n_\mu = -1$, $n^\mu k_\mu = 0$

$$W^{\mu\nu}(q, p, S) = (2\pi)^6 \frac{P_0}{M} \sum_n (2\pi)^4 \delta^4(P+q-P_X) \langle P, S \, |j^\mu_{em}| \, X \rangle \langle X \, |j^\nu_{em}| \, P, S \rangle \, .$$

(14.6d)

Here $S$ denotes the spin of the target and $\sum_n$ denotes the sum over all the quantum numbers of state $X$ and integration over $d^3 P_X$. Then the differential cross-section is given by

$$\frac{d^2\sigma}{d\Omega dE'} = E'(E'^2 - m_e^2)^{1/2} \frac{1}{(2\pi)^3} \frac{e^4}{q^4} \frac{m_e^2}{EE'} L_{\mu\nu} W^{\mu\nu}.$$

(14.6e)

Assuming invariance under C, P and T and conservation of the electromagnetic current $\partial^\mu j^{em}_\mu = 0$, the Lorentz structure of $W^{\mu\nu}$ is

$$\frac{M\dot{W}^{\mu\nu}}{2\pi} = \left(-g^{\mu\nu} + \frac{q^\mu q^\nu}{q^2}\right) F_1(q^2, \nu)$$

$$+ \frac{1}{M\nu} \left(P^\mu - \frac{P \cdot q}{q^2} q^\mu\right) \left(P^\nu - \frac{P \cdot q}{q^2} q^\nu\right) F_2(q^2, \nu)$$

$$+ \frac{i}{\nu} \varepsilon^{\mu\nu\alpha\beta} q_\alpha S_\beta g_1(\nu, q^2) + \frac{i}{\nu} \varepsilon^{\mu\nu\alpha\beta} q_\alpha \left(S_\beta - \frac{q \cdot S}{P \cdot q} P_\beta\right) g_2(\nu, q^2).$$

(14.7)

Here $S^2 = S^\mu S_\mu = -1$, $S^\mu P_\mu = 0$ and $F_1$ and $F_2$ are spin averaged structure functions: $MW_1 \equiv F_1$ and $\nu W_2 \equiv F_2$ while the remaining two are spin dependent structure functions. In Fig. 14.2, we show the plot of $Q^2(= -q^2)$ versus $2M\nu$ where we have defined the variables:

$$x = \frac{Q^2}{2M\nu}, \quad y = \frac{\nu}{E} = \frac{E - E'}{E}$$

$$0 \le y \le 1.$$

(14.8a)

Now

$$(P + q)^2 \ge M^2,$$

so that

$$2P \cdot q - Q^2 \ge 0 \text{ or } 0 \le x \le 1.$$

(14.8b)

If hadron masses are not important, $F$'s dependance on $Q^2$ is unimportant, one might expect that scale invariance holds in the asymptotic (Bjorken) limit $Q^2, \nu \to \infty$ with $x$ fixed. In the "naive" quark model (where the virtual photon interacts with point like constituents), in the

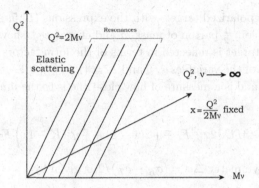

Fig. 14.2   Plot of momentum transfer $Q^2$ versus energy transfer $\nu = E - E'$ in charged lepton-proton scattering, showing various kinematic regions.

limit of quark masses $\to 0$, there are no dimensions and this suggests that in the asymptotic limit the structure functions scale:

$$MW_1(\nu, Q^2) \equiv F_1(\nu, Q^2) \to F_1(x)$$
$$\nu W_2(\nu, Q^2) \equiv F_2(\nu, Q^2) \to F_2(x)$$
$$g_{1,2}(\nu, Q^2) \to g_{1,2}(x). \tag{14.9}$$

In QCD, however, this scaling is broken but only by logarithms of $Q^2/\Lambda_{QCD}^2$.

From Eqs. (14.6) and (14.7), the spin averaged cross-section is given by

$$\frac{d^2\sigma}{d\Omega\, dE'} = \left(\frac{d\sigma}{d\Omega}\right)_M \left[W_2(\nu, Q^2) + 2\tan^2\frac{\theta}{2}W_1(\nu, Q^2)\right]. \tag{14.10a}$$

It is instructive to write this cross-section in the form

$$\frac{d^2\sigma}{dQ^2\, d\nu} = \left(\frac{d\sigma}{dQ^2}\right)_M \left[W_2(\nu, Q^2) + 2\tan^2\frac{\theta}{2}W_1(\nu, Q^2)\right]. \tag{14.10b}$$

We now define right and left polarized cross-sections as

$$\sigma_{R,L} = \sigma \pm \Delta\sigma,$$

where $d^2\sigma/dQ^2\, d\nu$ is given in Eq. (14.10) and

$$\frac{d^2\Delta\sigma}{dQ^2\, d\nu} = \frac{4\pi\alpha^2}{E^2Q^4}\left(\frac{Q^2}{2M\nu}\right)\cos\beta\Bigg\{\left[E + E'\cos\theta\right]g_1(\nu, Q^2)$$
$$- \left[(\frac{E+E'}{\nu})(E - E'\cos\theta) - (E + E'\cos\theta)\right]g_2(\nu, Q^2)\Bigg\} \tag{14.11}$$

where $\beta$ is the angle between $\mathbf{k}$ and the spin quantization direction $\mathbf{S}$. It is instructive to compare Eq. (14.10b) with $\cos\beta = 1$, which corresponds to longitudinally polarized target, with the expressions (14.5a) and (14.5b) for structureless spin $\frac{1}{2}$ parton of mass $m$ and charge $ee_q$, showing that the structure of the target is reflected in terms of the form factors $W_1$, $W_2$, $g_1$, and $g_2$. In terms of the variables $x$, $y$ and $K = (1 - \frac{Q^2}{\nu^2})[= 1 - \frac{2M^2x^2}{Q^2} \to 1$ in the scaling limit and is a measure of how close one is to the limit $Q^2 \to \infty$], we have

$$\frac{d^2\sigma}{dx\,dy} = \frac{4\pi\alpha^2}{Q^4}ME\left[2xy^2F_1 + \left(2(1-y) + \frac{1}{2}y^2(K-1)\right)F_2\right] \quad (14.12)$$

and polarized asymmetry $\Delta\sigma = (\sigma_R - \sigma_L)/2$ is given by

$$\frac{d\Delta\sigma}{dx\,dy} = \frac{4\pi\alpha^2}{Q^4}ME\left[xy\cos\beta\left\{2\left(1 - \frac{y}{2} + \frac{y^2}{4}(K-1)\right)g_1 + y(K-1)g_2\right\}\right].$$
$$(14.13)$$

At high energies $y \to 0$ and $F_2$ and $g_1$ dominate. It may be noted that $g_2$ has never been measured.

The presence of the structure functions in Eq. (14.10) indicates that proton is not a point particle. The structure of the proton can be probed in two ways, one by elastic lepton-nucleon scattering and second by deep inelastic lepton-nucleon scattering. First we discuss the elastic scattering for which $\nu = Q^2/2M$. For this case the structure functions are given by

$$W_2 = \left[F_1^2(Q^2) + \tau F_2^2(Q^2)\right]\delta\left(-\nu + \frac{Q^2}{2M}\right)$$

$$W_1 = \tau\left[F_1^2(Q^2) + F_2^2(Q^2)\right]^2\delta\left(-\nu + \frac{Q^2}{2M}\right) \quad (14.14)$$

where $\tau = Q^2/2M$. Thus from Eq. (14.10), we have

$$\frac{d\sigma}{dQ^2} = \left(\frac{d\sigma}{dQ^2}\right)_M\left\{\left[F_1^2(Q^2) + \tau F_2^2(Q^2)\right] + 2\tau\tan^2\frac{\theta}{2}\left[F_1(Q^2) + F_2(Q^2)\right]^2\right\} \quad (14.15)$$

The form factors for the proton are normalized to $F_1^p(0) = 1$, $F_2^p(0) = \kappa_p$ and for the neutron $F_1^n(0) = 0$, $F_2^n(0) = \kappa_n$ where $\kappa_p = 1.792$ and $\kappa_n = -1.913$ are anomalous magnetic moments of the proton and the neutron respectively. Experimental data is analyzed in terms of Sachs form factors

$$G_E(Q^2) = F_1(Q^2) - \tau F_2(Q^2)$$
$$G_M(Q^2) = F_1(Q^2) + F_2(Q^2) \quad (14.16)$$

These form factors are normalized as follows: $G_E^p(0) = 1$, $G_M^p(0) = \mu_p = 2.792$, $G_E^n(0) = 0$ and $G_M^n(0) = \mu_n$. In terms of $G_E$ and $G_M$, the elastic scattering cross-section is given by

$$\frac{d\sigma}{dQ^2} = \left(\frac{d\sigma}{dQ^2}\right)_{\text{M}} \left\{ \frac{[G_E^2(Q^2) + \tau G_M^2(Q^2)]}{1+\tau} + 2\tau \tan^2 \frac{\theta}{2} G_M^2(Q^2) \right\}.$$
(14.17)

The experimental data is fitted remarkably well by a single form factor

$$G_E^p(Q^2) = \frac{G_M^p(Q^2)}{\mu_p} = \frac{G_M^n(Q^2)}{\mu_n} = \frac{1}{[1 + Q^2/m_V^2]^2}$$

$$G_E^n(q^2) = 0,$$
(14.18)

where $m_V^2 = 0.71 \text{ GeV}^2$. From Eq. (14.15), we get [cf. Eq. (14.3b)]

$$\langle r_E^2 \rangle_p = \frac{12}{m_V^2} = 0.66 \, \text{fm}^2, \quad \langle r_E^2 \rangle_n = 0.$$
(14.19)

Now Eqs. (14.14) and (14.15) clearly show that $\frac{d\sigma}{dQ^2} \to \left(\frac{d\sigma}{dQ^2}\right)_{\text{M}}$ as $Q^2 \to \infty$, i.e. cross section rapidly falls as $Q^2$ become large, clearly showing that the nucleon has a "diffused" structure in the elastic region.

But the behavior of the structure functions $W_2$ and $W_1$ is quite different in the deep inelastic region. The experimental data in this region indicate that the cross section stays large and is of the order of $\left(\frac{d\sigma}{dQ^2}\right)_{\text{M}}$, characteristics of a point particle. This clearly indicates that in this region the scattering is incoherent and is what one would expect if a nucleon consists of non-interacting or weakly interacting point like constituents called partons (quarks). This scattering region thus gives us information about the elementary constituents of nucleon, i.e. about their charges, spin and flavor. Moreover, the structure functions $\nu W_2$ and $M W_1$ show Bjorken scaling, i.e. $\nu W_2$ and $M W_1 \to F_2(x)$ and $F_1(x)$ as $Q^2, \nu \to \infty$ where $x = \frac{2M\nu}{Q^2}$ is fixed. This is clearly indicated in Fig. 14.3 where $F_2(x)$ is plotted against $Q^2$ for various values of $x$. The above characteristics lead to parton model of deep inelastic scattering which we now discuss.

## 14.3   Parton Model

Partons are quarks (spin 1/2), antiquarks (spin 1/2) and gluons (spin 1). Gluons do not contribute here since they carry no electric charge. Thus we shall deal with spin 1/2 partons. If the target is a free quark of flavor $i$, of

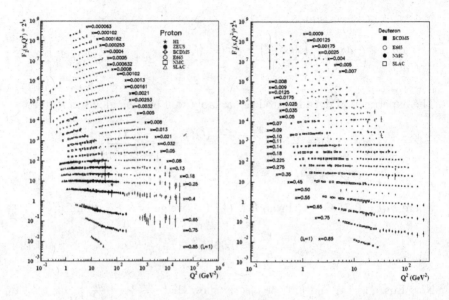

Fig. 14.3   The structure function $F_2$ (a) proton (b) nucleon in deuterium [18].

mass $m$ and charge $e_i$, we have from Eqs. (14.5a) and (14.5b)

$$\frac{d\sigma_i}{dQ^2} = \frac{2\pi\alpha^2 e_i^2}{Q^4}\left[1 + (1-y)^2 + \frac{1}{2}(K-1)y^2\right]$$

$$\frac{d\Delta\sigma_i}{dQ^2} = \frac{4\pi\alpha^2 e_i^2}{Q^4}y\left[1 - \frac{y}{2} + \frac{1}{4}(K-1)y^2\right]  \qquad (14.20)$$

with $y = \nu/E$, $K = (1 - \frac{Q^2}{\nu^2})$

Further from Eq. (14.6d)

$$\frac{mW_i^{\mu\nu}}{2\pi} = \frac{m}{2\pi}(2\pi)^6\frac{p_0}{m}\int(2\pi)^4\frac{d^3p_n}{(2\pi)^3}$$

$$\times \sum \langle p,s\,|j_{em}^\mu|\,p_n\rangle\,\langle p_n\,|j_{em}^\nu|\,p,s\rangle\,\delta^4(p+q-p_n)$$

$$= me_i^2\int m\frac{d^3p_n}{p_{n_0}}\bar{u}(ps)\gamma^\mu\frac{m + \not{p}_n}{2m}\gamma^\nu u(ps)\delta^4(p+q-p_n).$$

$$(14.21a)$$

Now $\frac{d^3p_n}{p_{n_0}} = d^4p_n\delta[p_n^2 - m^2]$ and we obtain from Eq. (14.21a)

$$\frac{m}{2\pi}W_i^{\mu\nu} = me_i^2\bar{u}(ps)\gamma^\mu\left[\not{p} + \not{q} + m\right]\gamma^\nu u(p,s)\delta(2p\cdot q - Q^2).  \qquad (14.21b)$$

To proceed further, we make use of the following identities of Dirac matrices algebra [see Appendix A.2],

$$\gamma^\mu\gamma^\nu = g^{\mu\nu} - i\sigma^{\mu\nu}$$

$$\gamma^\mu\gamma^\rho\gamma^\nu = g^{\mu\rho}\gamma^\nu - g^{\mu\nu}\gamma^\rho + g^{\nu\rho}\gamma^\mu + i\varepsilon^{\mu\rho\nu\sigma}\gamma_5\gamma_\sigma$$

$$\bar{u}(p,s)\gamma_5\gamma^\mu u(p,s) = -s^\mu$$

$$\bar{u}(p,s)\gamma^\mu u(p,s) = \frac{p^\mu}{m}$$

$$\bar{u}(p,s)i\sigma^{\mu\nu}u(p,s) = -\frac{i}{m}\varepsilon^{\mu\nu\alpha\beta}\gamma_5\gamma_\beta p_\alpha \tag{14.22}$$

Then Eq. (14.21b) becomes

$$\frac{m}{2\pi}W_i^{\mu\nu} = e_i^2\delta(2p\cdot q - Q^2)$$
$$\times \left[-g^{\mu\nu}p\cdot q + 2p^\mu p^\nu + q^\mu p^\nu + p^\mu q^\nu + im\varepsilon^{\mu\nu\alpha\beta}q_\alpha s_\beta\right]. \tag{14.23a}$$

Thus the comparison with Eq. (14.7) gives

$$F_{1i} = \frac{1}{2}\delta(x-1)e_i^2, \quad F_{2i} = \delta(x-1)e_i^2$$

$$g_{1i} = \frac{1}{2}\delta(x-1)e_i^2, \quad g_{2i} = 0. \tag{14.23b}$$

Hence from Eqs. (14.12) and (14.13) for a spin $\frac{1}{2}$ parton $i$,

$$\frac{d^2\sigma_i}{dx\,dy} = \frac{4\pi\alpha^2}{Q^4}e_i^2 mE\left[1 + (1-y)^2 + \frac{1}{2}y^2(K-1)\right]\delta(x-1) \tag{14.24a}$$

$$\frac{d^2\Delta\sigma_i}{dx\,dy} = \frac{4\pi\alpha^2}{Q^4}e_i^2 mE\left(\cos\beta\right)y\left[1 - \frac{y}{2} + \frac{y^2}{4}(K-1)\right]\delta(x-1) \tag{14.24b}$$

which are the same as Eq. (14.20) if we integrate over $x$. The comparison of Eq. (14.23) with Eqs. (14.12) and (14.13) clearly shows that if we replace $\delta(1-x)$ in Eq. (14.23) by some distribution functions $F(x)$ and $g(x)$ we get Eqs. (14.12) and (14.13). Hence it follows that in the scaling region, the nucleon is behaving as if it consists of point-like constituents and the structure function $F_{2i}(x)$ or $F_{1i}(x)$ or $g_{1i}(x)$ gives us the $x$-distribution of point-like constituents inside the nucleon. The point-like constituents have been assumed to be free, i.e. interaction between them can be neglected in the scaling region. This is compatible with QCD, as QCD is asymptotically free. More accurately one can write $F_2$ and $F_1$ as $F_2(x, Q^2)$, $F_1(x, Q^2)$; but the dependence on $Q^2$ is very weak (logarithmic). The following physical picture emerges. In the deep inelastic region, the virtual photon interacts in

an incoherent manner and probes roughly the instantaneous construction
of proton. In the center-of-mass frame of electron and proton, we can write
(neglecting lepton mass):

$$k \equiv (P,\, 0,\, 0,\, P),\quad P \equiv \left[ (M^2 + P^2)^{1/2} \simeq P\left(1 + \frac{M^2}{2P}\right),\, 0,\, 0,\, -P \right]$$

$$q_0 = \frac{2M\nu - Q^2}{4P}$$

Let us assume that the target (proton) has point-like constituents called
partons of flavor, $i$. Neglecting any parton momentum transverse to the
target, let us assume that the longitudinal momentum of a parton is given
by $p = xP$. The time of interaction of photon is given by

$$\tau = \frac{1}{q_0} = \frac{4P}{2M\nu - Q^2} = \frac{2P}{M\nu(1 - x)}.$$

The energy of a parton $= \sqrt{p_z^2 + p_\perp^2 + m^2} \approx xP\left(1 + \frac{p_\perp^2 + m^2}{2x^2 P^2}\right)$, so that the
lifetime of virtual parton states is

$$T = \frac{1}{\sum xP\left(1 + \frac{p_\perp^2 + m^2}{2x^2 P^2}\right) - P\left(1 + \frac{M^2}{2P^2}\right)}$$

$$= \frac{2P}{\sum \frac{p_\perp^2 + m^2}{x} - M^2}.$$

For $x$ not going to 0 or 1, $\tau \ll T$ in the deep inelastic region so that one can
consider the partons contained in the proton as free during the interaction.
Hence in the deep inelastic region the photon interacts with the constituents
of proton as depicted in Fig. 14.4.

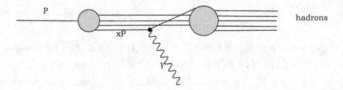

Fig. 14.4   The parton model.

If the target is built from partons of type $i$ and the probability for a
parton $i$ to have momentum fraction $x'$ to $x' + dx'$ is $f_i(x)$, then $p \cdot q =
x'P \cdot q = M\nu x'$,

$$\nu_{\text{parton}} = \frac{p \cdot q}{m} = \frac{M}{m}\nu x',$$

$$\delta(Q^2 - 2p \cdot q) = \delta(Q^2 - 2M\nu x') = \frac{1}{2M\nu}\delta(x - x')$$

and Eq. (14.23a) becomes

$$\frac{m}{2\pi}W_i^{\mu\nu} = e_i^2 \left[-g^{\mu\nu}\frac{x'}{2} + \frac{1}{M\nu}x'^2 P^\mu P^\nu + \cdots + i\varepsilon^{\mu\nu\alpha\beta}q_\alpha s_\beta \frac{x'}{2\nu}\right]\delta(x-x').$$

$$(14.24)$$

Since $\frac{d^2\sigma}{dQ^2\,d\nu} \propto W^{\mu\nu}$, we should write

$$\frac{dW^{\mu\nu}}{d\nu} = \frac{1}{d\nu_{\text{parton}}}\sum_i \int W_i^{\mu\nu}f_i(x')\,dx',$$

or, on using Eq. (14.24),

$$\frac{MW^{\mu\nu}}{2\pi} = M\int\frac{d\nu}{\frac{M}{m}x'\,d\nu}\sum_i\frac{W_i^{\mu\nu}}{2\pi}f_i(x')\,dx'$$

$$= \int\sum_i e_i^2\left[-g^{\mu\nu}\frac{1}{2} + \frac{1}{M\nu}x'P^\mu P^\nu + \cdots\right]\delta(x-x')f_i(x')\,dx'.$$

$$(14.25)$$

Using then the expression (14.7) for $MW^{\mu\nu}/2\pi$, it follows that in the parton model:

$$F_2(x) = \sum_{i,\,\text{spin}} e_i^2 x\, f_i(x)$$

$$= \sum_i e_i^2 x\,(f_{i\uparrow}(x) + f_{i\downarrow}(x))$$

$$= 2xF_1(x), \qquad (14.26)$$

while [cf. Eqs. (14.24) and (14.7)]

$$g_1(x) = \sum_i e_i^2 x\,(f_{i\uparrow}(x) - f_{i\downarrow}(x)).$$

$$g_2(x) = 0, \qquad (14.27)$$

where $\uparrow$ and $\downarrow$ denote respectively parton spin parallel and antiparallel to proton spin $S$.

The relation $F_2(x) = 2xF_1(x)$, which is a consequence of parton having spin $\frac{1}{2}$, is well satisfied experimentally. For the proton target [denoting $f_i(x)$ conveniently by $q(x) + \bar{q}(x)$, $e_i \to e_q$ and with spin sum understood], we have from Eq. (14.20)

$$F_2^{ep} = \sum_{q=u,\,d,\,\cdots} xe_q^2\,[q(x) + \bar{q}(x)]. \qquad (14.28a)$$

In other words,

$$F_2^{ep} = x\left[\frac{4}{9}\,(u(x) + \bar{u}(x)) + \frac{1}{9}\,(d(x) + \bar{d}(x)) + \cdots\right]. \qquad (14.28b)$$

Applying isospin conservation so that the $u(d)$ flavored parton distribution in the proton is the same as $d(u)$ flavored parton distribution in the neutron, whilst the $s$ and $c \cdots$, distributions remain unchanged being isoscalar, the neutron structure function becomes:

$$F_2^{en} = x \left[ \frac{4}{9} \left( d(x) + \bar{d}(x) \right) + \frac{1}{9} \left( u(x) + \bar{u}(x) \right) + \cdots \right]. \quad (14.28c)$$

In the above equations $u(x)$, $d(x)$, $\cdots$, are the probabilities that parton (antiparton) of flavor $u$, $d$, $\cdots$, carries a fraction x of the momentum of the proton or the neutron.

For an isosinglet target $N$, we get

$$F_2^{eN}(x) \equiv \frac{1}{2} \left( F_2^{ep}(x) + F_2^{en}(x) \right)$$

$$= x \left\{ \frac{5}{18} \left[ u(x) + \bar{u}(x) + d(x) + \bar{d}(x) \right] + \frac{1}{9} \left[ s(x) + \bar{s}(x) \right] + \cdots \right\}. \quad (14.28d)$$

Here $\cdots$ means the contributions of other quarks like $c$, $b$, $t$. Note that $\frac{5}{18}$ is just the average squared charge of the $u$, $d$ quarks.

We have thus seen that the parton model leads to the Bjorken scaling of the structure functions in the deep inelastic scattering.

## 14.4   Deep Inelastic Neutrino-Nucleon Scattering

Let us consider the processes

$$\bar{\nu}_\ell + N \to \ell^+ + X$$

$$\nu_\ell + N \to \ell^- + X$$

The matrix elements are given by

$$T_{\bar{\nu}} = -\frac{G_F}{\sqrt{2}} \frac{1}{(2\pi)^3} \sqrt{\frac{m_\ell m_\nu}{EE'}} \bar{v}(k')\gamma_\mu(1 - \gamma_5)v(k') \langle X | J^\mu | N(P) \rangle, (14.29a)$$

$$T_\nu = -\frac{G_F}{\sqrt{2}} \frac{1}{(2\pi)^{3/2}} \sqrt{\frac{m_\ell m_\nu}{EE'}} \bar{u}(k')\gamma_\mu(1 - \gamma_5)u(k') \langle X | J^{\mu\dagger} | N(P) \rangle, \quad (14.29b)$$

where $J^\mu = V^\mu - A^\mu$. Then we have to replace in Eq. (14.6b) $e^4/q^4$ by $G_F^2/2$ and $L_{\mu\nu}$ by

$$L_{\mu\nu}^{\nu,\bar{\nu}} = \frac{2}{m_e m_\nu} \left[ k_\mu k_\nu' - g_{\mu\nu} k \cdot k' k_\nu k_\mu' \pm i\epsilon_{\mu\nu\alpha\beta} k^\alpha k'^\beta \right], \quad (14.29c)$$

while $W_{\mu\nu}$ now contains for the spin averaged case three structure functions $W_1$, $W_2$ and $W_3$. The third function $W_3$ arises due to $V - A$ interference term and appears in Eq. (14.7) as $\frac{1}{2M\nu}\epsilon^{\mu\nu\alpha\beta}P_\alpha q_\beta F_3$, with $\nu W_3 = F_3$. The cross-section is given by

$$\frac{d^2\sigma^{\bar{\nu},\nu}}{dQ^2 d\nu} = \frac{G_F^2}{2}\frac{1}{\pi}\frac{E'}{E}\left[W_2^{\bar{\nu},\nu}(\nu,Q^2)\cos^2\frac{\theta}{2} + 2W_1^{\bar{\nu},\nu}(\nu,Q^2)\sin^2\frac{\theta}{2}\right.$$
$$\left.\mp\frac{E+E'}{M}W_3^{\bar{\nu},\nu}(\nu,Q^2)\sin^2\frac{\theta}{2}\right]. \qquad (14.30)$$

In order to discuss the scaling, we again express the cross-sections in terms of the variables $x$ and $y$. The structure functions show the following scaling behavior in the deep inelastic region:

$$\nu W_2^{\bar{\nu},\nu}(\nu, Q^2) \to F_2^{\bar{\nu},\nu}(x)$$
$$MW_1^{\bar{\nu},\nu}(\nu, Q^2) \to F_1^{\bar{\nu},\nu}(x)$$
$$\nu W_3^{\bar{\nu},\nu}(\nu, Q^2) \to F_3^{\bar{\nu},\nu}(x) \qquad (14.31)$$

The cross section can be written

$$\frac{d^2\sigma^{\bar{\nu},\nu}}{dx\,dy} = \frac{G_F^2 ME}{\pi}\left[\left(1 - y - \frac{M}{2E}xy\right)F_2^{\bar{\nu},\nu}(x)\right.$$
$$\left. + \frac{y^2}{2}2xF_1^{\bar{\nu},\nu}(x) \mp \left(y - \frac{y^2}{2}\right)xF_3^{\bar{\nu},\nu}(x)\right]. \qquad (14.32)$$

For the basic processes

$$\bar{\nu}_\ell + u \to \ell^+ + d, \ \bar{\nu}_\ell + \bar{d} \to \ell^+ + \bar{u}$$
$$\nu_\ell + d \to \ell^- + u, \ \nu_\ell + \bar{u} \to \ell^- + \bar{d}, \qquad (14.33)$$

we have (see problem (13.2)) in the high energy limit, for the first two processes in the first and second line of Eq. (14.33)

$$\frac{d^2\sigma^{\bar{\nu},\nu}}{dQ^2} = \frac{G_F^2}{2\pi}\frac{1}{s^2}\frac{1}{2}\left[(s^2 + u^2) \mp (s^2 - u^2)\right] \qquad (14.34)$$

and similar expression for the other two precesses. They give

$$\sigma^{\bar{\nu}} = \frac{G_F^2}{3\pi}mE, \ \sigma^\nu = \frac{G_F^2}{\pi}mE, \ \frac{\sigma^{\bar{\nu}}}{\sigma^\nu} = \frac{1}{3} \qquad (14.35a)$$

where $m_u \simeq m_d = m$. Now

$$(1 \pm \frac{u^2}{s^2}) = (1 \pm \frac{E'^2}{E^2}) = [1 \pm (1-y)^2] \qquad (14.35b)$$

so that

$$\frac{d\sigma^{\bar{\nu},\nu}}{dQ^2} = \frac{G_F^2}{2\pi}\left[\frac{1+(1-y)^2}{2} \mp \frac{1-(1-y)^2}{2}\right] \qquad (14.35c)$$

Hence we have for large $E$ ($E \gg m$) for the two processes in the first and second line of Eq. (14.33)

$$\frac{d\sigma^{\bar{\nu},\nu}}{dx\,dy} = \frac{G_F^2 mE}{\pi} \left[ \frac{1+(1-y)^2}{2} \mp \frac{1-(1-y)^2}{2} x \right] \delta(1-x) \qquad (14.35d)$$

Thus corresponding to Eq. (14.23b), we have for parton $i$

$$F_{1i} = \frac{1}{2}\delta(x-1) = \frac{1}{2}F_{2i}$$

$$F_{3i} = \delta(x-1) \qquad (14.36)$$

Thus the relation for $F_3$ corresponding to the relation (14.28a) is

$$F_3(x) = 2\sum_q [q(x) - \bar{q}(x)] . \qquad (14.37)$$

In view of Eq. (14.33) [note that the role of $e_q^2$ in Eq. (14.28a) is taken over by the isospin raising and lowering operators, namely $I^{\pm}$], we get

$$F_2(x) = 2xF_1(x)$$

$$F_2^{\bar{\nu}p} = 2x\left[ u(x) + \bar{d}(x) + c(x) + \bar{s}(x) + t(x) + \bar{b}(x) \right]$$

$$F_3^{\bar{\nu}p} = 2\left[ u(x) - \bar{d}(x) + c(x) - \bar{s}(x) + t(x) - \bar{b}(x) \right] , \qquad (14.38a)$$

$$F_2^{\nu p} = 2x\left[ d(x) + \bar{u}(x) + s(x) + \bar{c}(x) + b(x) + \bar{t}(x) \right]$$

$$F_3^{\nu p} = 2\left[ d(x) - \bar{u}(x) + s(x) - \bar{c}(x) + b(x) - \bar{t}(x) \right] \qquad (14.38b)$$

The factor 2 is due to the fact that for weak decays we have both vector and axial vector currents. The corresponding values for neutron are obtained by replacing $u \leftrightarrow d$, $\bar{u} \leftrightarrow \bar{d}$ on the grounds of isospin invariance. Hence for an isosinglet target $N$, we get (suppressing $x$)

$$F_2^{\nu N} = 2x\left[ \frac{1}{2}(u+\bar{u}) + \frac{1}{2}(d+\bar{d}) + s + b + \bar{c} + \bar{t} \right]$$

$$F_3^{\nu N} = 2\left[ \frac{1}{2}(u-\bar{u}) + \frac{1}{2}(d-\bar{d}) + s + b - \bar{c} - \bar{t} \right]$$

$$F_2^{\bar{\nu} N} = 2x\left[ \frac{1}{2}(u+\bar{u}) + \frac{1}{2}(d+\bar{d}) + c + t + \bar{s} + \bar{b} \right]$$

$$F_3^{\bar{\nu} N} = 2\left[ \frac{1}{2}(u-\bar{u}) + \frac{1}{2}(d-\bar{d}) + c + t - \bar{s} - \bar{b} \right] . \qquad (14.39)$$

If we assume that in a nucleon, the probability of having $q$ and $\bar{q}$ ($q = s, c, b, t$) is the same or we neglect $s, \bar{s}, \cdots$, then we can write

$$F_2^{\nu N} = \sum_q x\left[ q + \bar{q} \right] = F_2^{\bar{\nu} N}$$

$$xF_3^{\nu N} = \sum_q x\left[ q - \bar{q} \right] = xF_3^{\bar{\nu} N} . \qquad (14.40)$$

We observe from Eq. (14.28d), neglecting the sea quark contribution of heavy quarks, and Eq. (14.40) that [$\ell = e$ or $\mu$]

$$\frac{F_2^{\ell N}}{\frac{5}{18} F_2^{\nu N}} = \left[1 - \frac{3}{5} \frac{s + \bar{s}}{\sum_q (q + \bar{q})}\right]. \tag{14.41}$$

This ratio has been experimentally tested as the left-hand side is $1.007 \pm 0.063$. This also shows that the strange quark sea contribution is very small. It verifies the charges of $u$ and $d$ valence quarks as their mean square is $\frac{5}{18}$.

## 14.5  Sum Rules

One can write a number of sum rules. First the momentum conservation gives

$$\int_0^1 dx \left[\sum_q x(q + \bar{q})\right] = 1 - \epsilon, \tag{14.42}$$

where $\epsilon$ is the fraction of the momentum carried by the gluon constituents. Hence we get the sum rule

$$\int_0^1 F_2^{\nu N} dx = 1 - \epsilon \tag{14.43}$$

Experimentally, the left-hand side is $0.52 \pm 0.03$ giving the momentum fraction carried by the quarks. Thus the remaining momentum fraction, which is about 50%, is attributed to the gluon constituents.

Since the nucleon has quantum numbers S (strangeness) = 0, C (charm) = 0, B (bottom) = 0 and T (top) = 0, we have

$$0 = \int_0^1 dx \left[q(x) - \bar{q}(x)\right], \tag{14.44}$$

for q = s, c, b and t. On the other hand, the charges of proton and neutron give

$$1 = \int_0^1 dx \left[\frac{2}{3}(u - \bar{u}) - \frac{1}{3}(d - \bar{d})\right],$$

$$0 = \int_0^1 dx \left[\frac{2}{3}(d - \bar{d}) - \frac{1}{3}(u - \bar{u})\right]. \tag{14.45}$$

We can combine them, so that we get

$$1 = \int_0^1 dx \left[(u - \bar{u}) - (d - \bar{d})\right],$$

$$1 = \frac{1}{3}\int_0^1 dx \left[(d - \bar{d}) + (u - \bar{u})\right], \tag{14.46}$$

so that from Eqs. (14.38), (14.44) and (14.46), we have

$$\int_0^1 \left[ F_2^{\bar{\nu}p} - F_2^{\nu p} \right] \frac{dx}{x} = 2, \tag{14.47}$$

and

$$\int_0^1 F_3^{\nu N}(x)\, dx = \int_0^1 dx\, \left[ (u - \bar{u}) + (d - \bar{d}) \right] = 3. \tag{14.48}$$

If we use Eq. (14.38b) and the corresponding equation for the neutron, we get the sum rule (14.44) in the form

$$\int_0^1 \left[ F_2^{\nu n} - F_2^{\nu p} \right] \frac{dx}{x} = 2. \tag{14.49}$$

This is known as the Adler sum rule. It is an exact sum rule obtained from quark structure of electromagnetic and weak hadronic currents and is protected by conservation laws implied by Eqs. (14.44) and (14.45). It is difficult at present to verify it experimentally with good precision as it requires good low $x$ data. On the other hand, the sum rule (14.48), known as the Gross-Llewellyn Smith sum rule, is modified by QCD corrections in the leading order to,

$$\int_0^1 F_3^{\nu N}(x)dx = 3 \left( 1 - \frac{\alpha_s(Q^2)}{\pi} \right). \tag{14.50}$$

The right-hand side of (14.50) for $\alpha_s(Q^2 \approx 3 \text{ GeV}^2) = 0.35 \pm 0.05$ is $2.66 \pm 0.05$ while experimentally the left-hand side is $2.50 \pm 0.018 \pm 0.078$, verifying the sum rule.

Another sum rule which follows from Eqs. (14.28b) and (14.28c) is

$$\int_0^1 \left[ F_2^{ep} - F_2^{en} \right] \frac{dx}{x} = \frac{1}{3} \int_0^1 dx\, \left[ u(x) + \bar{u}(x) - d(x) - \bar{d}(x) \right]$$

$$= \frac{1}{3} \int_0^1 dx\, \left[ u(x) - \bar{u}(x) - d(x) + \bar{d}(x) \right]$$

$$+ \frac{2}{3} \int_0^1 dx\, \left[ \bar{u}(x) - \bar{d}(x) \right]$$

$$= \frac{1}{3} + \frac{2}{3} \int_0^1 dx\, \left[ \bar{u}(x) - \bar{d}(x) \right], \tag{14.51}$$

on using Eq. (14.46). This is known as the Gottfried sum rule. Experimentally the left-hand side is $0.258 \pm 0.017$ implying that the second term on the right-hand side is not zero. Its non-vanishing does not contradict any known principle.

There are two sum rules which involve the spin-dependent structure function $g_1(x)$. We note from Eq. (14.27) that

$$\int_0^1 g_1(x)dx = \frac{1}{2}\sum_q e_q^2 \Delta q, \tag{14.52a}$$

where we have defined (for a nucleon target)

$$\Delta q = \int_0^1 \left\{ [q_\uparrow(x) + \bar{q}_\uparrow(x)] - [q_\downarrow(x) + \bar{q}_\downarrow(x)] \right\} dx. \tag{14.52b}$$

Here $\Delta q$ is the quark contribution to the first moment of the structure function $g_1(x)$. There is also gluon contribution to it, this is due to the short-range interaction of photons with polarized gluons via the quark box diagram, shown in Fig. 14.5. To include this we replace $\Delta q$ by

$$\Delta \tilde{q} = \Delta q - \frac{\alpha_s}{2\pi}\Delta G_q. \tag{14.53}$$

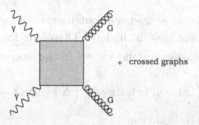

+ crossed graphs

Fig. 14.5   The photon-gluon scattering graph.

This separation is not unambiguous but has been found useful. For the proton target $\Delta \tilde{q}$ has been shown to be related to the matrix elements of the axial vector current $\bar{q}\gamma_\mu\gamma_5 q$

$$\langle p|\bar{q}\gamma_\mu\gamma_5 q| p\rangle = \Delta\tilde{q}(S_\mu), \quad q = u, d, s \tag{14.54}$$

where $S^\mu = \bar{\Psi}\gamma^\mu\gamma_5\Psi$ is the spin of the proton, $\Psi$ being the proton spinor. For the first moment of $g_1^p(x)$, the gluon contribution in relation (14.53) is related to the triangle axial anomaly [cf. Eq. (11.96)] in the divergence of the singlet current $\partial^\mu A_{0\mu}$ :

$$\langle p|\partial^\mu A_{0\mu}|p\rangle = \sqrt{\frac{2}{3}}\frac{3\alpha_s}{4\pi}\left\langle p\left|Tr\left(G^{\mu\nu}\tilde{G}_{\mu\nu}\right)\right|p\right\rangle$$

$$+\sqrt{\frac{2}{3}}\left\langle p\left|\left[m_u\bar{u}i\gamma_5 u + m_d\bar{d}i\gamma_5 d + m_s\bar{s}i\gamma_5 s\right]\right|p\right\rangle$$

$$= \sqrt{\frac{2}{3}}\frac{3\alpha_s}{4\pi}\left(-\Delta G\right)_q 2m_p\bar{\Psi}i\gamma_5\Psi$$

$$+\sqrt{\frac{2}{3}}\left\langle p\left|\left[m_u\bar{u}i\gamma_5 u + m_d\bar{d}i\gamma_5 d + m_s\bar{s}i\gamma_5 s\right]\right|p\right\rangle (14.55)$$

Note that the second term on the right-hand side is not an $SU(3)$ singlet. The first term on the right-hand side also contains a non-singlet part (that is why we have put a subscript q on $\Delta G$ in Eq. (14.52)). For the proton target, Eq. (14.49) gives the sum rule

$$\int_0^1 g_1^p(x)\, dx = \frac{1}{2}\left[\frac{4}{9}\Delta\tilde{u} + \frac{1}{9}\Delta\tilde{d} + \frac{1}{9}\Delta\tilde{s} + \cdots\right]$$

$$= \frac{1}{12}\left\{\left(\Delta\tilde{u} - \Delta\tilde{d}\right) + \frac{1}{3}\left(\Delta\tilde{u} + \Delta\tilde{d} - 2\Delta\tilde{s}\right)\right.$$

$$\left. +\frac{4}{3}\left(\Delta\tilde{u} + \Delta\tilde{d} + \Delta\tilde{s}\right) + \cdots\right\} \qquad (14.56)$$

where $\cdots$ denotes isospin singlet sea contribution of heavy quarks and second and third terms are isospin singlets. Therefore, for the neutron target, only $\left(\Delta\tilde{u} - \Delta\tilde{d}\right)$ changes sign and we get in the isospin conservation limit

$$\int_0^1 [g_1^p(x) - g_1^n(x)]\, dx = \frac{1}{6}\left(\Delta\tilde{u} - \Delta\tilde{d}\right) = \frac{1}{6}g_A \qquad (14.57)$$

since from Eq. (14.54), it is clear that

$$\frac{\Delta\tilde{u} - \Delta\tilde{d}}{2}(S_\mu) = \langle p|A_{3\mu}|p\rangle_{Q^2=0}$$

$$= \frac{1}{2}g_A(S_\mu). \qquad (14.58)$$

Here $g_A$ is the axial vector coupling constant determined from $\beta$-decay of the neutron. The sum rule (14.57) is known as the Bjorken sum rule. If the leading order QCD corrections are included, it then becomes

$$\Gamma_1^p - \Gamma_1^n = \int_0^1 [g_1^p(x) - g_1^n(x)]\, dx$$

$$= \frac{1}{6}g_A\left(1 - \frac{\alpha_s}{\pi}\right). \qquad (14.59)$$

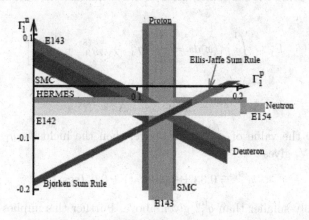

Fig. 14.6    Plot of $\Gamma_1^n$ versus $\Gamma_1^p$. The predictions of the Bjorken and Ellis-Jaffe sum rules are shown on the diagonal band from the lower left to the upper right of the figure. While the data and the Bjorken sum rule overlap within one sigma, the Ellis-Jaffe prediction is roughly two sigma away from the overlap region in the data [16].

This sum rule obtained from quark structure of electromagnetic and weak hadronic currents, is regarded as a fundamental prediction of QCD. For $g_A = 1.270 \pm 0.003$, $\alpha_s = 0.35 \pm 0.05$ one finds for the right hand side of Eq. (14.59), the value $0.187 \pm 0.01$. The experimental situation is best summarized in the $\Gamma_1^n$, $\Gamma_1^p$ plane, Fig. 14.6 which illustrates that Ellis-Jaffe sum rule [see below] is violated by the experimental data whereas Bjorken sum rule is compatible with the data.

One can obtain another sum rule involving only $g_1^p$ if one assumes exact SU(3) flavor symmetry for semi-leptonic decays of baryon octet, so that

$$\left( \Delta\tilde{u} + \Delta\tilde{d} - 2\Delta\tilde{s} \right)(S_\mu)$$
$$= 2\sqrt{3}\, \langle p \,|A_{8\mu}|\, p \rangle_{Q^2=0}$$
$$= g_A^8(S_\mu) = (3F - D)(S_\mu) \tag{14.60}$$

where $g_A$ , $F$ and $D$ have been defined in Chap. 11, namely

$$g_A^{(3)} = \Delta\tilde{u} - \Delta\tilde{d} = g_A = 1.270 \pm 0.003,$$
$$g_A^{(8)} = 0.58 \pm 0.03,$$
$$F = 0.463 \pm 0.023,$$
$$D = 0.803 \pm 0.040. \tag{14.61}$$

Thus neglecting the sea contribution of heavy quarks, one obtains from Eq. (14.56)

$$\Gamma_1^p = \int_0^1 g_1^p(x)dx = \frac{1}{12}\left[g_A^{(3)} + \frac{1}{3}g_A^{(8)} + \frac{4}{3}g_A^{(0)}\right] \tag{14.62}$$

where

$$g_A^{(0)} = \Delta\tilde{u} + \Delta\tilde{d} + \Delta\tilde{s} = \Delta\tilde{\Sigma} \tag{14.63}$$

If we take the value of $g_A^{(8)}$ given above, then the inclusive $g_1$ data with $Q^2 > 1\text{GeV}$, gives

$$g_A^{(0)} = 0.33 \pm 0.03(\text{stat}) \pm 0.05(\text{syst}) \tag{14.64}$$

considerably smaller than $g_A^{(8)}$ given above. Further this implies

$$\Delta\tilde{s} = \frac{1}{3}(g_A^{(0)} - g_A^{(8)}) = -0.08 \pm 0.01(\text{stat}) \pm 0.02(\text{syst}),$$

i.e. strange quark is polarized in the opposite direction to the spin of the particle. If one assumes as is done in naive quark model [OZI rule]

$$\langle p\,|\bar{s}\gamma_\mu\gamma_5 s|\,p\rangle = 0, \tag{14.65}$$

one has

$$g_A^0 \approx g_A^8 = (3F - D) \tag{14.66}$$

which is obviously violated. Nevertheless, if one uses it, then the sum rule (14.62) becomes

$$\Gamma_1^p = \int_0^1 g_1^p(x) = \frac{1}{12}\left[1 + \frac{5}{3}\frac{3F/D - 1}{F/D + 1}\right]. \tag{14.67}$$

This is known as the Ellis-Jaffe sum rule. With $g_A$ and $F/D$ given in Eq. (14.61), the right-hand side of Eq. (14.67) is $0.187 \pm 0.003$ in disagreement with the SMC ($Q^2 = 10\text{ GeV}^2$) data which gives

$$\Gamma_1^p = 0.139 \pm 0.01 \tag{14.68}$$

In view of the above disagreement the assumption (14.65) has been questioned. If one relaxes it, one does not have any prediction. However, one can use the sum rule (14.56), [neglecting $\cdots$] together with (14.68) to determine

$$\frac{1}{2}\left[\frac{4}{9}\Delta\tilde{u} + \frac{1}{9}\Delta\tilde{d} + \frac{1}{9}\Delta\tilde{s}\right] = 0.139 \pm 0.01. \tag{14.69}$$

This together with the values given in Eqs. (14.60) and (14.61) give

$$\Delta\tilde{u} = 0.78 \pm 0.07$$
$$\Delta\tilde{d} = -0.48 \pm 0.08$$
$$\Delta\tilde{s} = -0.14 \pm 0.07. \qquad (14.70)$$

Again $\Delta\tilde{s}$ is not zero as would be the case in naive quark model but is consistent with $\Delta\tilde{s}$ given in Eq. (14.70) so that

$$g_A^0 \equiv \Delta\tilde{\Sigma} = \Delta\Sigma - \frac{3\alpha_s}{2\pi}\Delta\tilde{G} = 0.16 \pm 0.22, \qquad (14.71a)$$

where $\Delta\Sigma = \Delta u + \Delta d + \Delta s$ is the quark contribution to the spin of the proton and $\Delta\tilde{G}$ is the singlet part of $\Delta G$. Again the above value of $g_A^0$ is consistent with the one given in (14.64). Various estimates of $\Delta\Sigma$ indicate that $\Delta\Sigma \approx 0$, which implies that $\frac{\alpha_s}{2\pi}(-\Delta\tilde{G}) = 0.05 \pm 0.07$ which is consistent with present measurements which suggest $(-\frac{3\alpha_s}{2\pi}\Delta\tilde{G}) < 0.06$, this gives $\Delta\tilde{G} < 0.4$ with $\alpha_s \approx 0.3$. Thus one faces with the issue of physical interpretation of $g_A^0$ and possible SU(3) breaking in the estimation of $F$ and $D$ and of $g_A^{(8)}$ from hyperon decays which may change the value of $g_A^0$. Now as is clear from Eq. (14.55) there is the gluonic contribution to $g_A^0$ through the QCD axial anomaly and we have

$$g_A^{(0)} \equiv \left(\sum_q \Delta q - \frac{3\alpha_s}{2\pi}\Delta\tilde{G}\right)_{parton} \qquad (14.71b)$$

where $\left(\Delta\tilde{G}\right)_{parton}$ is the amount of spin carried by polarized gluons in the polarized parton and $(\Delta q)_{partons}$ means the spin carried by the quarks and antiquarks.

To conclude there is a need to understand the underlying dynamics which seems to suppress the $g_A^{(0)}$ relative to the OZI prediction $g_A^{(0)} \simeq g_A^{(8)}$, and the sum rule for the longitudinal spin structure of the proton.

$$\frac{1}{2} = \frac{1}{2}\sum_q \Delta q + \Delta G + L_q + L_G \qquad (14.72)$$

where $L_q$ and $L_G$ denote the orbital angular momentum contributions. It is important to measure both $g_A^0$ and $F_2^0(0)$ [the SU(3) singlet anomalous magnetic moment of the proton] experimentally in order to determine the flavor and spin content of the proton, i.e. to disentangle the different contributions in the above sum rule as well as in (14.71b). There is presently a vigorous experimental program for this [see in particular [17] for the discussion regarding the sum rule for $\Gamma_1^p$ and what follows].

## 14.6   Deep-Inelastic Scattering Involving Neutral Weak Currents

For neutral weak currents mediated by Z-boson (see Chap. 13), the relevant Lagrangian for the processes

$$(\bar{\nu})\nu + N \to (\bar{\nu})\nu + X$$

is given in Eq. (13.107) [see also Table 13.1]. For neutrino-parton scatering

$$(\bar{\nu})\nu + q_i \to (\bar{\nu})\nu + q_i$$

we have [cf. problem (13.4)]

$$\frac{d\sigma_i^{\nu,\bar{\nu}}}{dQ^2} = \frac{G_F^2}{4\pi}\frac{1}{s^2}\left[(g_{Vi} \pm g_{Ai})^2 s^2 + (g_{Vi} \mp g_{Ai})^2 u^2\right]$$

$$= \frac{G_F^2}{4\pi}\left[(g_{Vi} \pm g_{Ai})^2 + (g_{Vi} \mp g_{Ai})^2 (1-y)^2\right] \tag{14.73}$$

$$\sigma_i^{\nu,\bar{\nu}} = \frac{G_F^2}{2\pi}mE\left[(g_{Vi} \pm g_{Ai})^2 + \frac{1}{3}(g_{Vi} \mp g_{Ai})^2\right] \tag{14.74}$$

$$\frac{d^2\sigma_i^{\nu,\bar{\nu}}}{dxdy} = \frac{G_F^2}{2\pi}mEx\left[(g_{Vi} \pm g_{Ai})^2 + (g_{Vi} \mp g_{Ai})^2 (1-y)^2\right]\delta(1-x) \tag{14.75}$$

Hence

$$\frac{d^2\sigma_i^{\nu,\bar{\nu}}}{dxdy}((\nu,\bar{\nu}) + p \to (\nu,\bar{\nu}) + X) = \frac{G_F^2}{4\pi}ME\left(\sum_i x f_i(x)\right)$$

$$\times \left[(g_{Vi} \pm g_{Ai})^2 + (g_{Vi} \mp g_{Ai})^2 (1-y)^2\right] \tag{14.76}$$

Finally, from Eq. (14.76), for the proton target we have

$$\sigma_i^{\nu,\bar{\nu}} = \frac{G_F^2}{2\pi}ME\int_0^1 \sum_i x f_i(x)dx\left[(g_{Vi} \pm g_{Ai})^2 + \frac{1}{3}(g_{Vi} \mp g_{Ai})^2\right]. \tag{14.77}$$

In parton model

$$r = \frac{\sigma_{\nu N}^C}{\sigma_{\bar{\nu}N}^C} = \frac{1}{3} \tag{14.78}$$

$$R_\nu = g_L^2 + \frac{1}{3}g_R^2 \longrightarrow g_L^2 + rg_R^2 \tag{14.79}$$

$$R_{\bar{\nu}} = \frac{\sigma_{\bar{\nu}N}^C}{r\sigma_{\nu N}^C} = \frac{1}{r}\left(g_R^2 + rg_L^2\right) = \frac{g_R^2}{r} + g_L^2 \tag{14.80}$$

which were used in Chap. 13 [cf. Eq. (13.107)]. Finally from Eqs. (14.77), (13.75) and Table 13.1, for the proton target we have

$$F_2^{\nu NC} = 2\rho^2 x \left\{ \left[ (\epsilon_L(u))^2 + (\epsilon_R(u))^2 \right] [u(x) + \bar{u}(x)] \right.$$

$$\left. + \left[ (\epsilon_L(d))^2 + (\epsilon_R(d))^2 \right] [d(x) + \bar{d}(x)] \right\}.$$

$$F_3^{\nu NC} = 2\rho^2 x \left\{ \left[ (\epsilon_L(u))^2 - (\epsilon_R(u))^2 \right] [u(x) - \bar{u}(x)] \right.$$

$$\left. + \left[ (\epsilon_L(d))^2 - (\epsilon_R(d))^2 \right] [d(x) - \bar{d}(x)] \right\}. \tag{14.81}$$

$$F_{2,3}^{\bar{\nu} NC} = F_{2,3}^{\nu NC}.$$

Thus from the experimental data on deep inelastic scattering, we can determine $\epsilon_L(u)$, $\epsilon_L(d)$, $\epsilon_R(u)$ and $\epsilon_R(d)$. This information has been used in Chap. 13. In writing Eqs. (14.81), we have neglected the contribution of strange and heavy quarks. For neutron, we can obtain the structure function by replacing $u(x) \leftrightarrow d(x)$ and $\bar{u}(x) \leftrightarrow \bar{d}(x)$.

We end this chapter by the remarks that the quark-parton model is simple and quite successful. A closer examination of Fig. 14.3 reveals a systematic deviation from exact Bjorken scaling, the structure function increases with increasing $Q^2$ at small $x$ whereas it has opposite behavior for large $x$. The attempts to understand such deviations from the quark-parton model in terms of QCD are beyond the scope of this book.

## 14.7  Problems

(1)

$$\langle p' | J_{em}^\lambda | p \rangle = \sqrt{\frac{m^2}{p_0 p_0'}} \left( \frac{1}{2\pi} \right)^3 \bar{u}(p') \left[ F_1 \gamma^\lambda - \frac{iF_2}{2M} \sigma^{\lambda\nu} q_\nu \right] u(p)$$

$$q = p' - p$$

(a) Express

$$\langle p' | J_{em}^\lambda | p \rangle \sim \bar{u}(p') \left[ (F_1 + F_2) \gamma^\lambda - \frac{F_2}{2M} P^\lambda \right] u(p)$$

(b) Using

$$\bar{u}(p') \gamma^\lambda u(p) = \frac{1}{4M^2 \left( 1 + \frac{Q^2}{4M^2} \right)} \bar{u}(p') \left[ i\epsilon^{\lambda\nu\rho\sigma} P_\rho q_\nu \gamma_\sigma \gamma_5 + 2M P^\lambda \right] u(p)$$

express

$$\langle p' | J_{em}^\lambda | p \rangle$$

$$\sim \frac{1}{1+\tau} \bar{u}(p') \left[ G_E(Q^2) P^\lambda + \frac{i}{4M^2} G_M(Q^2) \epsilon^{\lambda\nu\rho\sigma} P_\rho q_\nu \gamma_\sigma \gamma_5 \right] u(p)$$

where

$$G_E(Q^2) = F_1(Q^2) - \tau F_2(Q^2)$$
$$G_M(Q^2) = F_1(Q^2) + F_2(Q^2)$$

(c) Select the frame

$$\mathbf{p} = -\frac{1}{2}\mathbf{q}, \quad \mathbf{p}' = \frac{1}{2}\mathbf{q}$$
$$p_0 = p'_0 \quad q_0 = 0$$

Show that

$$G_E(0) = 1$$

$$\langle 1/2\mathbf{q} | \mathbf{J}_{em} | -1/2\mathbf{q} \rangle = \frac{1}{1+\tau} \bar{u} \left[ \frac{G_M(q^2)}{2M} i\sigma \times \mathbf{q} \right] u$$

Magnetic Moment :

$$\mu = \int \frac{1}{2} (\mathbf{x} \times \mathbf{J}) \, d^3 x^i$$

Show that

$$\langle 1/2\mathbf{q} | \mu | -1/2\mathbf{q} \rangle = \frac{1}{1+\tau} \delta^3(\mathbf{q}) \frac{G_M}{2M} u^\dagger \sigma u$$

$$\rightarrow \delta^3(q) \frac{G_M(0)}{2M} \chi^\dagger \sigma \chi$$

$$G_M(0) = \mu_p, \quad G_M^n(0) = \mu_n$$

(2) Show that the scattering cross section for elastic scattering of electrons on nucleon is given by (Rosenbluth formula)

$$\frac{d\sigma}{dQ^2} = \left( \frac{d\sigma}{dQ^2} \right)_M \frac{1}{\epsilon(1+\tau)} \left[ \tau G_M^2 + \epsilon G_E^2 \right]$$

$$\left( \frac{d\sigma}{dQ^2} \right)_M = \frac{\pi\alpha^2}{E^2 Q^2} \frac{1}{\tan^2 \theta/2} = \frac{\pi\alpha^2}{4M^2 E^2 \tau} \frac{2\epsilon(1+\tau)}{1-\epsilon}$$

where (for elastic scattering)

$$\epsilon = \frac{1}{1 + 2(1+\nu^2/Q^2)\tan^2\theta/2} = \frac{1}{1 + 2(1+\tau)\tan^2\theta/2}$$

It is clear that $G_M$ can be extracted at $\epsilon = 0$, while $G_E$ is extracted from the $\epsilon$-dependence. $\epsilon$ can be varied at fixed photon energy and momentum by varying electron energy and scattering angle.

(3) The structure functions $W_1$ and $W_2$ are related to the absorptive part for the forward scattering for virtual photon as can be easily seen. Optical theorem gives

$$\left(1 + \frac{\nu^2}{Q^2}\right)\frac{W_2}{W_1} - 1 = \frac{\sigma_L}{\sigma_T} \equiv R$$

where $\sigma_L$ and $\sigma_T$ are the longitudinal and transverse total Compton scattering cross-section respectively

(a) Show that for the elastic scattering

$$\frac{\sigma_L}{\sigma_T} = \frac{G_E^2}{\tau G_M^2}$$

(b) Show that the cross-section $\frac{d^2\sigma}{dQ^2 d\nu}$ can be expressed as

$$\frac{d^2\sigma}{dQ^2 d\nu} = \left(\frac{d\sigma}{dQ^2}\right)_M \frac{2\tan^2\theta/2}{1 - \epsilon} W_1 \left[1 + \epsilon R\right]$$

$$= \left(\frac{d\sigma}{dQ^2}\right)_M \frac{(K-1)F_1}{M(K-2)}\left[1 + \epsilon R\right]$$

$$\frac{d^2\sigma}{dx dy} = \left(\frac{d\sigma}{dQ^2}\right)_M E \frac{2MxF_1(x)}{(2-K)}\left[1 + \epsilon R\right]$$

(4) (a) Express the cross-section $\frac{d^2\Delta^\sigma}{dQ^2 d\nu}$ in the following form

$$\frac{d^2\Delta^\sigma}{dQ^2 d\nu} = \left(\frac{d\sigma}{dQ^2}\right)_M \tan^2\theta/2 \frac{1}{2M\nu}$$

$$\times \left[(g_1 + g_2)(E + E'\cos\theta) - g_2\frac{E + E'}{2}(E - E'\cos\theta)\right]$$

(b) Show that for elastic scattering of longitudinal polarized electrons on polarized protons :

$$\frac{g_1 + g_2}{2M\nu} = \frac{2}{M}G_E G_M \delta\left(\nu - \frac{Q^2}{2M}\right)$$

$$\frac{g_2}{2M\nu} = \frac{2}{M}\left(\frac{\nu}{2M}\right)\frac{1}{1 + \tau}G_M(G_E - G_M)\delta\left(\nu - \frac{Q^2}{2M}\right)$$

$$\frac{d\Delta^\sigma}{dQ^2}\bigg/\frac{d\sigma}{dQ^2} = 2\frac{(1 - \epsilon)G_M}{(1 + \tau)(\tau G_M^2 + \epsilon G_E^2)}\left[\begin{array}{c}\left(E/M - \frac{M\tau}{E}\right)(G_E + \tau G_M)\\ -\tau(2G_E - (1 - \tau)G_M)\end{array}\right]$$

## 14.8   References

1. R. P. Feynman, Photon-Hadron Interactions, Benjamin (1972).
2. P. Roy, Theory of Lepton-Hadron Processes at High Energies, Oxford University Press, Oxford (1975).
3. F. E. Close, An Introduction to Quarks and Partons, Academic Press, New York (1979).
4. G. Altarelli, Partons in Quantum Chromodynamics, Phys. Rep. **81C**, 1 (1982).
5. D. H. Perkins, Introduction to High Energy Physics, Addison-Wesley (Third Edition, 1987).
6. T. D. Lee, Particle Physics and Introduction to Field Theory, Harwood Academic (revised edition 1988).
7. T. Sloan, G. Smadja and R. Voss, The Quark Structure of the Nucleon from the CERN Muon Experiments, Phys. Rep. **162C**, 45 (1988).
8. S. R. Mishra and F. Sciulli, Deep Inelastic Lepton - Nucleon Scattering, Ann. Rev. Nucl. Part. Sci. **39**, 259 (1989).
9. G. Altarelli, Ann. Rev. Nucl. Part. Sci. **39**, 357 (1989).
10. R. Jaffe, Lectures delivered at the "School on High Energy Physics and Cosmology" Quaid-e-Azam University, Islamabad (March 11-25, 1990).
11. R. K. Ellis and W. J. Stirling, QCD and Collider Physics, FERMILAB-Conf.- 90/164-T (1990).
12. Small-x Behavior of Deep Inelastic Structure Function in QCD, Edited by A. Ali and J. Bartels, Nucl. Phys. B **18 C** (1990), Feb. 1991.
13. Riazuddin and Fayyazudin, Flavor and Spin Content of the Proton, M. A. B Beg Memoral Volume (Editors A. Ali and P. Hoodbhoy), World Scientific, Singapore (1991).
14. Proc. of SLAC Summer Institute on Lepton-Hadron Scattering and Topical Conf. Aug. 5-16, (1991).
15. G. Altarelli, R.D. Ball, S.F.Orte and G. Ridolfi, "Theoretical Analysis of Polarized Structure Functions" CERN-TH / 98-61, Talk given by G. Altarelli and G. Riodolfi at Cracow Epiphany Conference On Spin Effects in Particle Physics, Jan 9-11, 1998 Cracow, Poland.
16. M.C. Vetterli, The spin structure of the Nucleon, DESY 98–211, hep-ph/9812420, December 1998.
17. S. D. Bass, Quark spin in the proton, arxiv: 1004.4977 (hep-ph), 28 April 2010.
18. Particle Data Group, N. Nakamura et al., Journal of Physics G, **37**, 075021 (2010).

# Chapter 15

# Weak Decays of Heavy Flavors

In the standard model, three generations of matter replicate themselves with increasing mass scale. We have already discussed the first and second generation leptons $(e, \nu_e)$, $(\mu, \nu_\mu)$. In this chapter, we first discuss the weak decays of $\tau$ lepton (the third generation lepton). Later we study the heavy flavors viz decays of $D$ and $B$ mesons.

The study of heavy flavors provides us an opportunity to discover any deviation from the standard model. However, we will find that the standard model works quite well for heavy flavors. We begin with $\tau$-decays. The mass of $\tau$ lepton is $1776.82 \pm 0.16$ MeV and its mean life is $(290.6 \pm 1.0) \times 10^{-15}$ s. The upper limit on the mass of $\nu_\tau$ is 24 MeV.

## 15.1 Leptonic Decays of $\tau$ Lepton

In the standard model, the third generation leptons $(\tau, \nu_\tau)$ behave exactly in the same manner as $(\mu, \nu_\mu)$. Because $\tau$ lepton mass is 1777 MeV, $\tau$ can decay into light mesons ($\pi$'s and K's). As far as the decay $\tau^- \rightarrow \nu_\tau + e^- + \bar{\nu}_e$ is concerned, in the standard model it should have exactly the same structure as that for the decay $\mu^- \rightarrow \nu_\mu + e^- + \bar{\nu}_e$. Now $e^-$ and $\nu_e$ are common in both these decays. The $e^-$ and $\bar{\nu}_e$ enter in the effective Lagrangian for muon decay in the form $\bar{e}\, \gamma_\mu(1 - \gamma_5)\, \nu_e$, it should occur in this form in the effective Lagrangian for $\tau$-decay. Hence the most general form for the $T$-matrix is given by:

$$T = -\frac{G_F}{\sqrt{2}} \frac{1}{(2\pi)^6} \left( \frac{m_\tau m_e m_{\nu\tau} m_{\nu e}}{p_{10} p_{20} k_{10} k_{20}} \right)^{1/2}$$
$$\times \left[ \bar{u}(\mathbf{p}_2)\gamma^\lambda(1 - \epsilon\gamma^5)u(\mathbf{p}_1) \right] \left[ \bar{u}(\mathbf{k}_1)\gamma_\lambda(1 - \gamma_5)v(\mathbf{k}_2) \right] \quad (15.1)$$

where $p_1$, $p_2$, $k_1$, and $k_2$, are four momenta of $\tau$, $\nu_\tau$, $e$ and $\bar{\nu}_e$ respectively. In the standard model, $\epsilon = +1$. Thus any deviation from the standard model should manifest itself with a value of $\epsilon$ different from 1.

Using the standard techniques, we can easily calculate the electron energy spectrum

$$\frac{d\Gamma}{dx} \approx \frac{G_F^2 m_\tau^5}{384\pi^3} x^2 \left[(1+\epsilon)^2(3-2x) + 6(1-\epsilon)^2(1-x)\right]. \qquad (15.2)$$

In deriving the above expression, we have taken neutrinos to be massless and have put $m_e/m_\tau \approx 0$ and $x = 2E_e/m_\tau$. It is convenient to put Eq. (15.2) in the form

$$\frac{1}{\Gamma}\frac{d\Gamma}{dx} \approx 4x^2 \left[3(1-x) + 2\rho(\frac{4}{3}x - 1)\right], \qquad (15.3)$$

where

$$\Gamma \approx \frac{G_F^2 \, m_\tau^5}{192\pi^3} \frac{(1+\epsilon^2)}{2} \qquad (15.4)$$

and

$$\rho = \frac{3}{8} \frac{(1+\epsilon)^2}{1+\epsilon^2}. \qquad (15.5)$$

Equation (15.4) gives the decay rate for the decay $\tau^- \to e^- + \nu_\tau + \bar{\nu}_e$. Equation (15.5) gives the Michel parameter $\rho$. In the standard model $[V-A$ theory $]$ $\epsilon = +1, \rho = \frac{3}{4}$. The experimental value for $\rho$ is $0.742 \pm 0.027$ in agreement with the theoretical value of $0.75$. This reinforces our assumption that $(\tau, \nu_\tau)$ are sequential leptons.

Using $\epsilon = 1$, we get from Eq. (15.4)

$$\frac{\Gamma(\tau \to \nu_\tau + e + \bar{\nu}_e)}{\Gamma(\mu \to \nu_\mu + e + \bar{\nu}_e)} = \frac{m_\tau^5}{m_\mu^5} \frac{(1 + \delta_{rad}^\tau)}{(1 + \delta_{rad}^\mu)} \qquad (15.6)$$

Since $(1 + \delta_{rad}^\tau)/(1 + \delta_{rad}^\mu) \approx 1$, one can write for the branching ratio

$$BR\,(\tau \to \nu_\tau + e + \bar{\nu}_e) = \left(\frac{\tau_\tau}{\tau_\mu}\right)\left(\frac{m_\tau}{m_\mu}\right)^5 BR\,(\mu \to \nu_\mu + e + \bar{\nu}_e) \qquad (15.7a)$$

Using the experimental values for the masses and decay rates,

$$BR\,(\tau \to \nu_\tau + e + \bar{\nu}_e) = (17.82 \pm 0.09)\% \qquad (15.7b)$$

to be compared with the experimental average $(17.85 \pm 0.05)\%$.

If we neglect $m_\mu/m_\tau$, we get for the decay $\tau \to \nu_\tau + e + \bar{\nu}_\mu$, the same expressions as in Eqs. (15.2)-(15.6). However, taking into account the finite value of $m_\mu/m_\tau$ we get

$$\Gamma\,(\tau \to \nu_\tau + \mu + \bar{\nu}_\mu) = \frac{G_F^2 \, m_\tau^5}{192\pi^3} K\left(\frac{m_\mu^2}{m_\tau^2}\right), \qquad (15.8)$$

where

$$K(y) = 1 - 8y + 8y^3 - y^4 - 12y^2 \ln y. \tag{15.9}$$

Using the experimental values of $m_\mu$ and $m_\tau$, Eqs. (15.8) and (15.4) give

$$BR\left(\tau \to \nu_\tau + \mu + \overline{\nu}_\mu\right) \approx (0.9726 \pm 0.0001) BR(\tau \to \nu_\tau + e + \overline{\nu}_e)$$
$$= (17.33 \pm 0.09)\% \tag{15.10}$$

to be compared with the experimental average $(17.36 \pm 0.5)\%$. Thus we see that $e - \mu - \tau$ universality is satisfied to an excellent degree of accuracy.

## 15.2 Semi-Hadronic Decays of $\tau$ Lepton

We consider a general decay

$$\tau(k) \to X(p_X) + \nu_\tau(k'),$$

where $X$ is any number of hadrons allowed by energy conservation. The $T$-matrix is given by

$$T = -\frac{G'}{\sqrt{2}} \langle 0 | J_\mu^W | X \rangle \ \overline{u}(k') \gamma^\mu (1 - \gamma_5) \ u(k) \frac{1}{(2\pi)^3} \sqrt{\frac{m_\tau m_{\nu_\tau}}{k_0 k_0'}}. \tag{15.11}$$

The decay rate is given by

$$\Gamma = \frac{(2\pi)^3}{(2\pi)^5} \frac{m_{\nu_\tau}}{2} \int d^3 p_X \int \frac{d^3 k'}{k_0' \ p_{X_0}} |F|^2 \ \delta(p_X - k - k') \tag{15.12}$$

where

$$|F|^2 = \frac{G'^2}{\sqrt{2}} (2\pi)^3 (2p_{X_0}) \ \langle 0 | J_\mu^W | X \rangle \ \langle X | J_\lambda^{W\dagger} | 0 \rangle \ L^{\mu\lambda}. \tag{15.13}$$

Note that $G'$ is the effective decay constant, $L_{\mu\lambda}$ is the leptonic part given by

$$L_{\mu\lambda} = \frac{2}{m_\tau m_{\nu_\tau}} \frac{1}{2} \left[ k_\mu' k_\lambda + k_\lambda' k_\mu - g_{\mu\lambda} \ k'.k - i\varepsilon_{\mu\lambda\rho\sigma} \ k^\rho \ k'^\sigma \right]. \tag{15.14}$$

The weak current $J_\mu^W = V_\mu^W - A_\mu^W$. Since the interference term $V^\mu A_\mu$ does not contribute, we can separately consider the vector and axial vector parts. Using the Lorentz invariance and CVC (the spin over final hadrons is summed), we can write quite generally $(q = k - k')$:

$$(2\pi)^3 \int \langle 0 | V_\mu^W | X \rangle \ \langle X | V_\lambda^{W\dagger} | 0 \rangle \ d^3 p_X \delta (p_X - q)$$
$$= \theta (q_0) \left( -q^2 g_{\mu\lambda} + q_\mu q_\lambda \right) \rho_V \left( q^2 \right). \tag{15.15}$$

Similarly we can write

$$(2\pi)^3 \int \langle 0 | A_\mu^W | X \rangle \langle X | A_\lambda^{W\dagger} | 0 \rangle d^3 p_X \delta(p_X - q)$$

$$= \theta(q_0) \left[ (-q^2 g_{\mu\lambda} + q_\mu q_\lambda) \rho_A(q^2) + q_\mu q_\lambda \sigma_A(q^2) \right]. \tag{15.16}$$

In writing Eq. (15.16), we have not used the conservation of axial vector current. The form factor $\sigma_A(q^2)$ arises due to non-conservation of $A_\mu$. From Eqs. (15.12)-(15.16), we get

$$\Gamma_V = \frac{G'^2}{2} \frac{m_\tau^3}{8\pi} \int^{m_\tau^2} \left(1 - \frac{s}{m_\tau^2}\right)^2 \left(1 + \frac{s}{m_\tau^2}\right) \rho_V(s) ds, \tag{15.17}$$

$$\Gamma_A = \frac{G'^2}{2} \frac{m_\tau^3}{8\pi} \int^{m_\tau^2} \left[ \left(1 - \frac{s}{m_\tau^2}\right)^2 \left(1 + \frac{2s}{m_\tau^2}\right) \rho_A(s) \right.$$

$$\left. + \left(1 - \frac{s}{m_\tau^2}\right)^2 \sigma_A(s) \right] ds \tag{15.18}$$

where

$$s = q^2 = (k + k')^2. \tag{15.19}$$

### 15.2.1  *Special Cases*

(1)                                              $\tau^- \to \pi^- + \nu_\tau$

Here

$$\rho_A(s) = \rho_V(s) = 0, \qquad \sigma_A(s) = f_\pi^2 \delta(s - m_\pi^2), \tag{15.20}$$

$G'^2 = G_F^2 \cos^2 \theta_c$. $f_\pi$ is the pion decay constant and is defined by the matrix element

$$\langle 0 | A_\mu^W | \pi \rangle = i f_\pi \frac{1}{(2\pi)^{3/2}} \frac{1}{\sqrt{2q_0}} q_\mu. \tag{15.21}$$

Hence from Eq. (15.18), we get

$$\Gamma_\pi = \frac{G_F^2}{16\pi} \cos^2 \theta_c \, f_\pi^2 \, m_\tau^3 \left(1 - \frac{m_\pi^2}{m_\tau^2}\right)^2. \tag{15.22}$$

(2)                                              $\tau^- \to \rho^- + \nu_\tau$

Here

$$\rho_\nu(s) = f_\rho^2 \delta(s - m_\rho^2), \tag{15.23}$$

where $f_\rho$ is defined by

$$\langle 0 | V_\mu^W | \rho^- \rangle = f_\rho \frac{m_\rho}{(2\pi)^{3/2}} \frac{\varepsilon_\mu}{\sqrt{2q_0}}. \tag{15.24}$$

Hence from Eq. (15.17),

$$\Gamma_\rho = \frac{G_F^2}{16\pi} \cos^2\theta_c \, f_\rho^2 \, m_\tau^3 \left(1 - \frac{m_\rho^2}{m_\tau^2}\right)^2 \left(1 + \frac{2m_\rho^2}{m_\rho^2}\right). \tag{15.25}$$

(3)                              $\tau^- \to a_1^- + \nu_\tau$

Here

$$\rho_A(s) = f_{a_1}^2 \, \delta(s - m_{a_1}^2). \tag{15.26}$$

Hence from Eq. (15.18),

$$\Gamma_{a_1} = \frac{G_F^2 \cos^2\theta_c}{16\pi} f_{a_1}^2 m_\tau^3 \left(1 - \frac{m_{a_1}^2}{m_\tau^2}\right)^2 \left(1 + \frac{2m_{a_1}^2}{m_\tau^2}\right). \tag{15.27}$$

Let us compare these results with their experimental values. From Eqs. (15.22), (15.25) and (15.27), we have

$$\Gamma_\pi/\Gamma_e = (12\pi^2) \left(\frac{f_\pi^2}{m_\tau^2}\right) \cos^2\theta_c \left(1 - \frac{m_\pi^2}{m_\tau^2}\right)^2. \tag{15.28}$$

$$\Gamma_\rho/\Gamma_e = (12\pi^2) \left(\frac{f_\rho^2}{m_\tau^2}\right) \cos^2\theta_c \left(1 - \frac{m_\rho^2}{m_\tau^2}\right)^2 \left(1 + 2\frac{m_\rho^2}{m_\tau^2}\right). \tag{15.29}$$

$$\Gamma_{a_1}/\Gamma_e = (12\pi^2) \left(\frac{f_{a_1}^2}{m_\tau^2}\right) \cos^2\theta_c \left(1 - \frac{m_{a_1}^2}{m_\tau^2}\right)^2 \left(1 + 2\frac{m_{a_1}^2}{m_\tau^2}\right). \tag{15.30}$$

Using $f_\pi = 132$ MeV, $\cos\theta_c = 0.97$ and the experimental values for masses,

$$BR(\tau^- \to \nu_\tau \pi^-) \approx (0.605) \left(\frac{f_\pi}{132}\right)^2 B_e \qquad (10.91 \pm 0.07)\%$$

$$\tag{15.31a}$$

$$BR(\tau^- \to \nu_\tau \rho^-) = (0.557) \left(\frac{f_\rho}{132}\right)^2 B_e \qquad (25.51 \pm 0.09)\%$$

$$\tag{15.31b}$$

$$BR(\tau^- \to \nu_\tau + a_1^-) = (0.329) \left(\frac{f_{a_1}}{132}\right)^2 B_e \qquad (9.32 \pm 0.07)\%$$

$$\tag{15.31c}$$

Using the experimental values given in parenthesis and $B_e = (17.85 \pm 0.05)\%$, we get

$$f_\pi = 132.7 \text{MeV}, \quad f_\rho = 211 \text{MeV} \quad \text{and} \quad f_{a_1} = 166 \text{MeV} \tag{15.31d}$$

Let us compare the above values of the decay constants $f_\pi$ and $f_\rho$ with those determined from light flavor physics. The experimental value of $f_\pi$ is $(130.41 \pm 0.03 \pm 0.20)$MeV and that of $f_\rho$ is $(221 \pm 1)$MeV. The latter value is obtained from the experimental value of decay width $\Gamma_{ee} = (7.04 \pm 0.06)$ keV, using the formula

$$\Gamma\left(\rho \to e^+ e^-\right) = \frac{4\pi\alpha^2}{3} \frac{f_\rho^2}{m_\rho} \left(\frac{1}{2}\right) \tag{15.32}$$

Thus the values of decay constants $f_\pi$ and $f_\rho$ extracted from $\tau$-decays are consistent with those from light flavor physics showing the inner consistency of the standard model. Now the Kawarabayashi-Suzuki-Riazuddin-Fayyazuddin (KSRF) relation and the Weinberg sum rule give respectively

$$f_\rho = \sqrt{2} f_\pi, \, f_{a_1} = \frac{m_\rho}{m_{a_1}} f_\rho = 0.61 f_\rho = 135 \text{ MeV} \tag{15.33}$$

The KSRF value agrees with the experimental value within 16%. The value of $f_{a_1}$, in Eq (15.31d), agrees with $f_{a_1}$ obtained from the $\tau$-decay within 23%.

One can improve the theoretical predictions for $\Gamma(\tau^- \to \rho^- \nu_\tau \to \pi^- \pi^0 \nu_\tau)$ and $\Gamma(\tau^- \to a_1^- \nu_\tau \to \pi^- \rho^0 \nu_\tau)$ by taking into account finite decay widths of $\rho$ and $a_1$ mesons (see problems 1 and 2). However, for hadronic decays $(\tau^- \to f^- + \nu_\tau)$ which proceed through the vector current only, one can use CVC to relate $\Gamma(\tau^- \to f^- + \nu_\tau)$ to the scattering cross section for the process:

$$e^- e^+ \to \gamma \to f^0.$$

The cross section for this process is given by Eq. (A.87)

$$\sigma_f(s) = \frac{16\pi^3 \alpha^2}{s} \rho_\gamma(s). \tag{15.34}$$

Using CVC, we get

$$\rho(s) = 2\rho_\gamma(s) = \frac{2s \, \sigma_f(s)}{16\pi^3 \alpha^2}. \tag{15.35}$$

Hence we have

$$\Gamma(\tau^- \to f^- \nu_\tau) = \frac{G_F^2 \cos^2\theta_c}{128\pi^4 \alpha^2} m_\tau^3$$
$$\times \int^{m_\tau^2} \left(1 - \frac{s}{m_\tau^2}\right)^2 \left(1 + \frac{2s}{m_\tau^2}\right) s \, \sigma_f(s) \, ds. \tag{15.36}$$

Let us apply this to the decay

$$\tau^- \to \pi^- \pi^0 \pi^0 \pi^0 + \nu_\tau$$
$$\to \pi^- \pi^- \pi^+ \pi^0 + \nu_\tau. \tag{15.37}$$

Then we get from Eq. (15.36)

$$\Gamma\left(\tau^- \to \pi^- \pi^0 \pi^0 \pi^0 \nu_\tau\right) + \Gamma\left(\tau^- \to \pi^- \pi^- \pi^+ \pi^0 \nu_\tau\right)$$

$$= \frac{G_F^2 \cos^2 \theta_c}{128 \pi^4 \alpha^2} m_\tau^3 \int^{m_\tau^2} \left(1 - \frac{s}{m_\tau^2}\right)^2 \left(1 + \frac{2s}{m_\tau^2}\right)$$

$$\times s \left[\sigma_{\pi^- \pi^- \pi^+ \pi^+}(s) + \sigma_{\pi^- \pi^+ \pi^0 \pi^0}(s)\right] ds. \tag{15.38}$$

Since the decay proceeds via $I = 1$ weak currents, one also obtains an additional relation

$$\Gamma\left(\tau^- \to \pi^- 3\pi^0 \nu_\tau\right) = \frac{G_F^2 \cos^2 \theta_c}{128 \pi^4 \alpha^2} m_\tau^3 \int^{m_\tau^2} \left(1 - \frac{s}{m_\tau^2}\right)^2 \left(1 + \frac{2s}{m_\tau^2}\right)$$

$$\times s \left[\frac{1}{2} \sigma_{2\pi^- 2\pi^+}(s)\right] ds. \tag{15.39}$$

We have discussed above the dominant decay modes of $\tau^-$ viz $\tau^- \to \nu_\tau e^- \nu_e$, $\tau^- \to \nu_\tau \mu^- \nu_\mu$, $\tau^- \to \nu_\tau (2\pi)^-$, $\tau^- \to \nu_\tau (3\pi)^-$ and $\tau^- \to \nu_\tau (4\pi)^-$. The other small decay modes can also be estimated. All these decay rates occur at the expected rates. The agreement between the theoretical and experimental values is good for each exclusive decay mode.

Finally if we add the decay rates for all exclusive channels for 1 prong events $[\tau^- \to \nu_\tau$ (particle)$^-$ neutrals $(\geq 0)]$, we get the value $(84.72 \pm 0.08)\%$ to be compared with direct inclusive one prong branching ratio $B_1 = (85.36 \pm 0.14)\%$. Thus there is no discrepancy between the two branching ratios. $\tau$-decays are well understood in the standard model.

## 15.3  Weak Decays of Heavy Flavors

In the standard model, the hadronic charged weak current can be written as (see Chap. 13)

$$J_\mu^W = \left(\bar{u}\ \bar{c}\ \bar{t}\right) \gamma_\mu \left(1 - \gamma_5\right) V \begin{pmatrix} d \\ s \\ b \end{pmatrix}, \tag{15.40}$$

where

$$V = \begin{pmatrix} V_{ud} & V_{us} & V_{ub} \\ V_{cd} & V_{cs} & V_{cb} \\ V_{td} & V_{ts} & V_{tb} \end{pmatrix}, \qquad VV^\dagger = 1 \tag{15.41}$$

is the Cabibbo-Kobayashi-Maskawa (CKM) matrix. As we have discussed in Chap. 13, the matrix $V$ has only four real parameters [see Eq. (13.140)].

The weak decays of heavy hadrons are governed by the basic processes (Fig. 15.1). It follows that one has the following selection rules in the

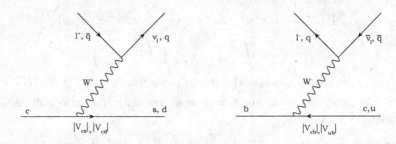

Fig. 15.1   Tree level quark diagram.

standard model for the weak decays of heavy hadrons:

$$\Delta Q = \Delta C = \Delta S = -1, \quad |V_{cs}|, \text{ dominant} \tag{15.42a}$$

$$\Delta Q = \Delta C = -1, \ \Delta S = 0, \ |V_{cd}| \sim \lambda, \text{ suppressed} \tag{15.42b}$$

$$\Delta Q = -\Delta C, \text{ and} \tag{15.42c}$$

$$\Delta Q = \Delta C = -\Delta S, \text{ are strictly forbidden} \tag{15.42d}$$

Thus we expect $D$ mesons ($c\bar{u}$, $c\bar{d}$) to decay predominantly into states with strangeness $S = -1$ ($K^-, \overline{K}^0$+anything) and $D_s$ meson ($c\bar{s}$) to decay predominantly into states with strangeness $S = 0$ ($\phi(\pi's)^+, D_s \rightarrow K^{*+}\overline{K}, K^{*0}\overline{K}^+, (\pi's)^+$).

Similarly

$$\Delta Q = \Delta B = \Delta C = -1, \ |V_{cb}| \sim \lambda^2, \text{ dominant} \tag{15.43a}$$

$$\Delta Q = \Delta B = -1, \ \Delta C = 0, \ |V_{ub}| \sim \lambda^3 \sqrt{\rho^2 + \eta^2}, \text{ suppressed} \tag{15.43b}$$

$$\Delta Q = -\Delta B, \text{ and} \tag{15.43c}$$

$$\Delta Q = \Delta B = -\Delta S, \text{ are strictly forbidden} \tag{15.43d}$$

Thus we expect $B$ mesons ($u\bar{b}$, $d\bar{b}$) to decay predominantly into states with $C = -1$ and $B_s$ meson ($s\bar{b}$) to decay predominantly into states with $S = C = -1$.

### 15.3.1   *Leptonic Decays of D and B Mesons*

The decay constants $f_D$, $f_{D_s}$, $f_B$, $f_{B_s}$ can in principle be determined from the leptonic decays of $D$ and $B$ mesons. Thus for example, using Eq. (15.40), we can write

$$\frac{\Gamma\left(D^+ \to \mu^+ \nu_\mu\right)}{\Gamma\left(\pi^+ \to \mu^+ \nu_\mu\right)} = \frac{f_D^2}{f_\pi^2} \frac{|V_{cd}|^2}{|V_{ud}|^2} \frac{m_\pi}{m_D} \frac{p_D^2}{p_\pi^2}, \tag{15.44a}$$

where

$$p_D = \frac{1}{2} m_D \left(1 - \frac{m_\mu^2}{m_D^2}\right), \qquad p_\pi = \frac{1}{2} m_\pi \left(1 - \frac{m_\mu^2}{m_\pi^2}\right). \tag{15.44b}$$

In order to determine $f_D$, we need $|V_{ud}|^2$ and $|V_{cd}|^2$. Now $|V_{ud}|^2$ can be determined with a great degree of accuracy from nuclear $\beta$-decay. Its value is given by [see Chap. 10].

$$|V_{ud}| = 0.97425(22). \tag{15.45a}$$

Recent analysis of the decays in [13] gives

$$|V_{us}| = 0.2246(12) \tag{15.45b}$$

$$|V_{cd}| = 0.2245(12). \tag{15.45c}$$

$$f_{D^+} = (206.7 \pm 8.5 \pm 2.5)\text{MeV} \tag{15.46}$$

$$|V_{cs}| = 0.97345(22) \tag{15.47}$$

$$f_{D_s^+} = (257.5 \pm 6.1)\text{MeV} \tag{15.48}$$

## 15.3.2 *Semileptonic Decays of D and B Mesons*

The prototype of these decays is $K_{e3}$ decay $\left(K^- \to \pi^0 + e^- + \bar{\nu}_e\right)$. In fact from this decay and hyperon decays, $|V_{us}|$ has been determined:

$$|V_{us}| = 0.2205 \pm 0.0018 \tag{15.49}$$

It is theoretically simplest to begin with semileptonic decays of heavy flavors. We start with the decay

$$D^- \to X^0 + e^- + \bar{\nu}_e : \quad p = p_X + k_1 + k_2$$

where $X^0$ is any number of hadrons consistent with energy conservation and allowed by the selection rules. The T-matrix for this decay is given by

$$T = -\frac{G_F V_{cs}}{\sqrt{2}} \langle X | J_\mu^W | D \rangle \frac{1}{(2\pi)^3} \sqrt{\frac{m_e m_\nu}{k_{10} k_{20}}}$$
$$\times \left[\bar{u}\left(\mathbf{k}_1\right) \gamma^\mu \left(1 - \gamma^5\right) v\left(\mathbf{k}_2\right)\right]. \tag{15.50}$$

Since experimentally, one observes only charged leptons, we sum over hadrons. Thus for the inclusive semileptonic decays, the decay rate is given by:

$$\Gamma = \frac{G_F^2 \; |V_{cs}|^2}{2 \, (2\pi)^6 \; 2 \; p_0} \int \frac{d^3 \, k_1 \, d^3 \, k_2}{k_{10} \; k_{20}} \, (m_e \, m_\nu) \, L^{\mu\lambda} \, A_{\mu\lambda} \qquad (15.51a)$$

where [see Appendix A]

$$L_{\mu\lambda} = \frac{2}{m_e \, m_\nu} \left[ k_{1\mu} \, k_{2\lambda} + k_{1\lambda} \, k_{2\mu} - g_{\mu\lambda} \, k_1 \cdot k_2 - i\varepsilon_{\mu\lambda\rho\sigma} \, k_1^\rho \, k_2^\sigma \right], \quad (15.51b)$$

$$\begin{aligned} A_{\mu\lambda} &= (2\pi)^4 \int d^3 \, p_X \; \delta^4 \, (p - q - \, p_X) \left\langle D \left| J_\lambda^{W\dagger} \right| p_X \right\rangle \left\langle p_X \left| J_\mu^W \right| D \right\rangle \\ &= \int d^4 \, z \; e^{iqz} \left\langle D \left| \left[ J_\lambda^{W\dagger} \, (z) , J_\mu^W \, (0) \right] \right| p \right\rangle \\ &= \frac{1}{(2\pi)^3 \; 2 \; p_0} \left[ \frac{1}{m_D^2} p_\lambda \, p_\mu \; 2\pi \; f_2 \, (\nu, Q^2) - g_{\lambda\mu} \; 2\pi \; f_1 \, (\nu, Q^2) \right. \\ &\quad \left. - \frac{1}{2 \, m_D^2} \; i\varepsilon_{\lambda\mu\alpha\beta} \, p^\alpha \, q^\beta \; 2\pi \; f_3 \, (\nu, Q^2) + \cdots \right]. \end{aligned} \qquad (15.51c)$$

Here

$$q = k_1 + k_2, \qquad q^2 = 2E_e E_\nu \, (1 - \cos\theta)$$

$$\nu = \frac{p \cdot q}{m_D} = E_e + E_\nu. \qquad (15.52)$$

In writing Eq. (15.51c), those form factors which give contribution proportional to lepton mass ($m_e$) are neglected. In Eq. (15.52), we have also put $m_e = m_\nu = 0$, and $k_{10} = E_e$ and $k_{20} = E_\nu$. Thus Eqs. (15.51a) and (15.52) give

$$\begin{aligned} &\frac{d\Gamma}{dE_e \; d\nu \; d\Omega} \\ &= \frac{G_F^2 \; |V_{cs}|^2}{(2\pi)^4} \; \frac{8}{2 \, m_D} E_e^2 E_\nu^2 \\ &\quad \times \left[ \frac{1}{4} \, (1 + \cos\theta) \; f_2 \, (\nu, q^2) + \frac{1}{2} \, (1 - \cos\theta) \; f_1 \, (\nu, q^2) \right. \\ &\qquad \left. + \frac{1}{4 \, m_D} \, (E_e - E_\nu) \frac{1}{2} \, (1 - \cos\theta) \; f_3 \, (\nu, q^2) \right]. \end{aligned} \qquad (15.53)$$

This is a general expression for the semileptonic decay rate of $D$ meson. But this expression is not useful since we do not know the form factors $f_1$, $f_2$ and $f_3$. In order to determine these form factors one has to use some

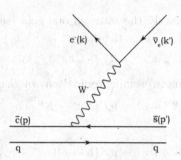

Fig. 15.2   Dominant (spectator) diagram for Cabibbo favored semileptonic decays of D mesons.

models. The simplest model is the spectator quark model. According to this model the decay proceeds as shown in Fig. 15.2. It is assumed in this model that the quarks in the final states fragment into hadrons with unit probability. In this model, one can easily calculate the tensor $A_{\mu\lambda}$ using the usual trace techniques. Noting that $P = xp$, $P' = P - q$, one obtains

$$A_{\mu\lambda} = 2 \frac{(2\pi)^4}{(2\pi)^6} \frac{1}{xp_0} \delta \left(2xp \cdot q - m_c^2 + m_s^2 - q^2\right)$$
$$\times \left[2x^2 \, p_\mu \, p_\lambda + g_{\lambda\mu} \left(-m_c^2 + xp \cdot q\right)\right.$$
$$\left. - x \, i\varepsilon_{\mu\lambda\alpha\beta} \, p^\alpha \, q^\beta\right]. \tag{15.54}$$

Thus comparing it with Eqs. (15.51):

$$f_2\left(\nu, q^2\right) = 8x \, m_D^2 \, \delta \left(2xp \cdot q - m_c^2 + m_s^2 - q^2\right)$$
$$f_1\left(\nu, q^2\right) = -\frac{4}{x} \left(-m_c^2 + xp \cdot q\right) \, \delta \left(2xp \cdot q - m_c^2 + m_s^2 - q^2\right)$$
$$f_3\left(\nu, q^2\right) = -8 \, m_D^2 \, \delta \left(2xp \cdot q - m_c^2 + m_s^2 - q^2\right). \tag{15.55}$$

Hence from Eq. (15.53) and noting here that $x = m_D/m_c \approx 1$,

$$\frac{d\Gamma}{dy} = \frac{G_F^2 \, |V_{cs}|^2}{16\pi^3} m_c^5 \, y^2 \frac{(y_c - y)^2}{(1 - y)}$$
$$\Gamma = \frac{G_F^2 \, m_c^5}{192\pi^3} |V_{cs}|^2 \, F\left(\frac{m_s}{m_c}\right), \tag{15.56}$$

where

$$y = \frac{2 \, E_e}{m_c}, \qquad y_c = 1 - \frac{m_s^2}{m_c^2}$$

$$F\left(\frac{m_s}{m_c}\right) = 1 - 8 \left(\frac{m_s}{m_c}\right)^2 + \left(\frac{m_s}{m_c}\right)^6 - \left(\frac{m_s}{m_c}\right)^8 - 24 \left(\frac{m_s}{m_c}\right)^4 \ln\left(\frac{m_s}{m_c}\right). \tag{15.57}$$

Equation (15.56) is exactly the same as one gets (see problem 11.1 with $G_F^2$ replaced by $G_F^2 \ |V_{cs}|^2$ ) for the decay

$$\bar{c} \to \bar{s} + e^- + \bar{\nu}_e \ (c \to s + e^+ + \nu_e).$$

Similarly for the (inclusive) semileptonic decay of $B$ mesons viz

$$B^+ \to X^0 + e^+ + \nu_e \ (B^- \to X^0 + e^- + \bar{\nu}_e),$$

the basic process is

$$\bar{b} \to \bar{c} + e^+ + \nu_e \ (b \to c + e^- + \bar{\nu}_e),$$

and we get (see problem 11.1 with $G_F^2$ replaced by $G_F^2 \ |V_{cb}|^2$ ).

$$\frac{d\Gamma}{dy} = \frac{G_F^2 \ |V_{cb}|^2}{96\pi^3} m_b^5 \ y^2 \frac{(y_b - y)^2}{(1 - y)^2} \left[ (3 - 2y) + \frac{(1 - y_b)(3 - y)}{(1 - y)} \right]$$

$$\Gamma = \frac{G_F^2 m_b^5 \ |V_{cb}|^2}{192\pi^3} F\left( \frac{m_c}{m_b} \right), \tag{15.58}$$

where

$$y = \frac{2 \ E_e}{m_b}, \qquad y_b = 1 - \frac{m_c^2}{m_b^2}. \tag{15.59}$$

Note that the difference between Eqs. (15.56) and (15.58) which is due to the fact that V - A interference term [the third term in Eq. (15.53)] has opposite sign in the two cases. We conclude this section with the remarks that according to this picture

$$\Gamma\left(D^+ \to X^0 e^+ \nu_e \right) = \Gamma\left(D^0 \to X^- e^+ \nu_e \right)$$
$$\Gamma\left(B^+ \to X^0 e^+ \nu_e \right) = \Gamma\left(B^0 \to X^- e^+ \nu_e \right). \tag{15.60}$$

For the decay

$$D_s^+ \to X^0 + e^+ + \nu_e$$

we get for $\frac{d\Gamma}{dy}$ and $\Gamma$ exactly the same expressions as those given in Eq. (15.56). Similar remarks are applicable to the decay $B_s^0 \to X^- + e^+ + \nu_e$. Hence the quark model predicts

$$\Gamma\left(D^+ \to X^0 e^+ \nu_e \right) = \Gamma\left(D^0 \to X^- e^+ \nu_e \right) = \Gamma\left(D_s^+ \to X^0 e^+ \nu_e\right), \tag{15.61}$$

$$\Gamma\left(B^+ \to X^0 e^+ \nu_e \right) = \Gamma\left(B^0 \to X^- e^+ \nu_e \right) = \Gamma\left(B_s^0 \to X^- e^+ \nu_e\right). \tag{15.62}$$

The experimental branching ratios for these decays are:

$$\left(D^+\right)_{SL} \equiv BR\left(D^+ \to X^0 e^+ \nu_e \right) = (16.07 \pm 0.30) \%$$
$$\left(D^0\right)_{SL} \equiv BR\left(D^0 \to X^- e^+ \nu_e \right) = (6.49 \pm 0.11) \%$$
$$\left(D_s^+\right)_{SL} \equiv BR\left(D_s^+ \to X^0 e^+ \nu_e \right) = (6.5 \pm 0.4)\% \tag{15.63}$$

For $B$ mesons the experimental values are

$$
\begin{aligned}
\left(B^+\right)_{SL} &\equiv BR\left(B^+ \to X^0 e^+ \nu_e\right) = (10.99 \pm 0.28)\,\% \\
\left(B^0\right)_{SL} &\equiv BR\left(B^0 \to X^- e^+ \nu_e\right) = (10.33 \pm 0.28)\,\% \quad (15.64)
\end{aligned}
$$

The experimental values for $\tau_{D^+}$, $\tau_{D^0}$, $\tau_{D_s^+}$, $\tau_{B^+}$ and $\tau_{B^0}$ are respectively $(1040 \pm 7) \times 10^{-15}s$, $(410.1 \pm 1.5) \times 10^{-15}s$, $(500 \pm 7) \times 10^{-15}s$, $(1.638 \pm 0.011) \times 10^{-12}s$, and $(1.525 \pm 0.009) \times 10^{-12}s$. Now

$$
\begin{aligned}
\frac{\Gamma\left(B^0 \to X^- e^+ \nu_e\right)}{\Gamma\left(B^+ \to X^0 e^+ \nu_e\right)} &= (\tau_{B^+}/\tau_{B^0}) \frac{\left(B^0\right)_{SL}}{\left(B^+\right)_{SL}} \\
&\approx 1 ,
\end{aligned}
$$

on using Eq. (15.64) and the experimental value $\tau_{B^+}/\tau_{B^0} = 1$. Comparing it with Eq. (15.60), we conclude that the spectator quark model for inclusive semileptonic decays for $B$ mesons is compatible with the experiment.

However for $D$-mesons $\tau_{D^0} \approx \tau_{D_s^+} \neq \tau_{D^+}$, $\tau_{D^+} \approx 2.5\,\tau_{D^0}$, i.e. isospin is badly broken for $D^+$ and $D^0$.

$$
\begin{aligned}
\frac{\Gamma\left(D^+ \to X^0 e^+ \nu_e\right)}{\Gamma\left(D^0 \to X^- e^+ \nu_e\right)} &= \frac{\left(D^+\right)_{SL}}{\left(D^0\right)_{SL}} \frac{\tau_{D^0}}{\tau_{D^+}} \\
&\approx \frac{2.48 \pm 0.06}{2.5} \\
&\approx 1
\end{aligned}
$$

$$
\begin{aligned}
\frac{\Gamma\left(D_s^+ \to X^0 e^+ \nu_e\right)}{\Gamma\left(D^0 \to X^- e^+ \nu_e\right)} &= \frac{\left(D_s^+\right)_{SL}}{\left(D^0\right)_{SL}} \frac{\tau_{D^0}}{\tau_{D_s^+}} \\
&\approx \frac{1.00 \pm 0.06}{1.22 \pm 0.02} \\
&\approx 0.82 \pm 0.05
\end{aligned}
$$

where the numerical values have been obtained using the experimental values for branching ratios given in Eq. (15.63). Hence except for inclusive semileptonic decays of $D_s^+$ which is about 18% less than the prediction of spectator quark model given in Eq. (15.61), the spectator quark model for inclusive semileptonic decays of $D$-mesons is compatible with experiment.

In the end, it may be noted that one can get an estimate of $|V_{cs}|$, using the $BR\left(D \to X e \nu_e\right)$ from the experiment if we know the quark masses $m_s$ and $m_c$.

### 15.3.3   (Exclusive) Semileptonic Decays of D and B Mesons

The formulation developed in Chap. 9 can be used to calculate the decay rate for semileptonic decays of the type

$$P(p) = M(p') + l + \nu$$

where $P = D$ or $B$. $M$ is a pseudoscalar meson $P'$ or a vector meson $V$. In order to discuss these decays, we first define the form factors

$$\langle P'(p') | \bar{q} \, \gamma_\mu \, Q | P(p) \rangle$$

$$= \frac{1}{(2\pi)^3} \frac{1}{\sqrt{4 \, p_0 \, p_0'}} \left[ f_+(t)(p+p')_\mu + f_-(t)(p-p')_\mu \right] \qquad (15.65a)$$

$$= \frac{1}{(2\pi)^3} \frac{1}{\sqrt{4 \, p_0 \, p_0'}} \left[ (p+p') f_+(t) + \left( \frac{m_P^2 - m_{P'}^2}{t} q_\mu \right) (f_+(t) + f_0(t)) \right]$$

where

$$t = q^2 = (p - p')^2,$$

$$q_\mu = (p - p')_\mu$$

$$f_0(t) = f_+(t) + \frac{t}{m_P^2 - m_{P'}^2} f_-(t) \qquad (15.65b)$$

Here $m_P$ is the mass of $P$ and $m_{P'}$ is that of $P'$. For the vector meson $V$, the form factors are defined:

$$\langle V(p', \epsilon) | \bar{q} \, \gamma_\mu (1 - \gamma_5) \, Q | P(p) \rangle$$

$$= \frac{1}{(2\pi)^3} \frac{-i}{\sqrt{4 \, p_0 \, p_0'}} \left\{ \left[ (m_P + m_V) \epsilon_\mu^* - \frac{(m_P + m_V)}{t} q \cdot \epsilon^* q_\mu \right] A_1(t) \right.$$

$$- \left[ \frac{1}{m_P + m_V} (p + p')_\mu - \frac{(m_P - m_V)}{t} q_\mu \right] q \cdot \epsilon^* A_2(t)$$

$$\left. + \left[ 2 m_V \frac{q \cdot \epsilon^*}{t} q_\mu \right] A_0(t) + \frac{2(-i)}{m_P + m_V} \epsilon_{\mu\nu\lambda\sigma} \, \epsilon^{\nu *} \, p'^\lambda p^\sigma V(t) \right\}$$

$$\hspace{10cm} (15.66)$$

However for the transition $B \to D$ or $B \to D^*$, $q = c$, $Q = B$, it is convenient to write form factors in the following form, which is suitable for taking the HQET limit

$$f_+(t) = \frac{m_B + m_D}{2\sqrt{m_B \, m_D}} h_+(w) \qquad (15.67)$$

$$f_0(t) = \frac{\sqrt{m_B \, m_D}}{(m_B + m_D)} [1 + w] \, h_0(w) \qquad (15.68)$$

$$V(t) = \frac{m_B + m_{D^*}}{2\sqrt{m_B\, m_{D^*}}}\, h_V(w) \tag{15.69}$$

$$A_1(t) = \frac{\sqrt{m_B m_{D^*}}}{m_B + m_{D^*}}\, [1 + w]\, h_{A_1}(w)$$

$$A_2(t) = \frac{m_B + m_{D^*}}{2\sqrt{m_B\, m_{D^*}}}\, h_{A_2}(w)$$

$$A_0(t) = \frac{m_B + m_{D^*}}{2\sqrt{m_B\, m_{D^*}}}\, h_{A_0}(w) \tag{15.70}$$

Note that

$$t = m_B^2 + m_{D^{(*)}}^2 - 2 m_B m_{D^{(*)}} v \cdot v'$$
$$= m_B^2 + m_{D^{(*)}}^2 - 2 m_B m_{D^{(*)}} w \tag{15.71}$$

where

$$w = v \cdot v' \tag{15.72}$$

At $w = 1$, $t = (m_B + m_{D^*})^2 = t_{\max}$, the form factor $\xi(w)$ is normalized as

$$\xi(1) = 1, \qquad \xi(t_{\max}) = 1 \tag{15.73}$$

In HQET limit, the form factors are given by [cf. Eqs. (9.58) and (9.59)]

$$\langle D^0(v') | \bar{c}\, \gamma_\mu b | B^-(v) \rangle = \frac{1}{\sqrt{4\, v_0\, v_0'}} \left[ \xi(w)\, (v + v')_\mu \right] \tag{15.74}$$

$$\langle D^{*0}(v', \epsilon^*) | \bar{c}\, \gamma_\mu (1 - \gamma_5) b | B^-(v) \rangle = -i \frac{1}{\sqrt{4\, v_0\, v_0'}} \xi(w)$$
$$\times \left[ -i\epsilon_{\mu\nu\lambda\sigma}\, \epsilon^{\nu*}\, v'^\lambda v^\sigma + (1 + w)\epsilon_\mu^* - v.\epsilon^* v'_\mu \right] \tag{15.75}$$

Hence in heavy quark limit, we have from Eqs. (15.67)-(15.70) and (15.73)-(15.74)

$$h_+(w) = h_0(w) = h_V(w) = h_{A_1}(w) \tag{15.76}$$
$$= h_{A_2}(w) = h_{A_0}(w) = \xi(w) \tag{15.77}$$

Having defined the matrix elements and form factors, the following table summarize the exclusive semileptonic decays, being considered:

| $Q$ | $q$ | Current | $|V_{qQ}|$ | Decay |
|---|---|---|---|---|
| $c$ | $s$ | $\bar{s}\, \gamma_\mu\, (1 - \gamma_5)\, c$ | $|V_{cs}| = 1.023 \pm 0.036$ | $D \to K l\nu$ <br> $\to K^* l\nu$ |
| $c$ | $d$ | $\bar{d}\, \gamma_\mu\, (1 - \gamma_5)\, c$ | $|V_{cd}| = 0.230 \pm 0.011$ | $D \to \pi l\nu$ <br> $\to \rho l\nu$ |
| $c$ | $s$ | $\bar{s}\, \gamma_\mu\, (1 - \gamma_5)\, c$ | $|V_{cs}| = 1.023 \pm 0.036$ | $D_s \to \eta l\nu$ <br> $\to \phi l\nu$ |
| $b$ | $c$ | $\bar{c}\, \gamma_\mu\, (1 - \gamma_5)\, b$ | $|V_{cb}| = (40.6 \pm 1.3) \times 10^{-3}$ | $B \to D l\nu$ <br> $\to D^* l\nu$ |
| $b$ | $u$ | $\bar{u}\, \gamma_\mu\, (1 - \gamma_5)\, b$ | $|V_{ub}| = (3.89 \pm 0.441) \times 10^{-3}$ | $B \to \pi l\nu$ <br> $\to \rho l\nu$ |

Now for the semileptonic decay $P \to M \, l\nu$, since the meson $M$ is on the mass shell, we can integrate our master equation (15.53) over $\nu$ using the delta function $\delta\left(2m_P\nu - m_P^2 - t + m_M^2\right)$, to obtain

$$\frac{d\,\Gamma}{dt \, dE_\ell} = \frac{G^2}{(2\pi)^3} \frac{1}{8m_P^2} \left\{ \left[ 4\sqrt{K^2 + t} - E_\ell - t \right] f_2\left(t\right) \right.$$
$$\left. + 2t \, f_1\left(t\right) + \frac{t}{m_P} \left[ 2E_\ell - \sqrt{K^2 + t} \right] f_3\left(t\right) \right\}, \quad (15.78)$$

where $K$ is the momentum of meson $M$ in the rest frame of $P$, $t = q^2$ and $G = G_F \left| V_{qQ} \right|$. Integrating Eq. (15.78) over the lepton energy $E_\ell$

$$\frac{d\,\Gamma}{dt} = \frac{G^2}{(2\pi)^3} \frac{1}{8m_P^2} \left[ \frac{2}{3} K^3 \, f_2\left(t\right) + 2tK \, f_1\left(t\right) \right] \quad (15.79)$$

Note that in Eqs. (15.78) and (15.79), those form factors which give contribution proportional to lepton mass $m_l$, have been neglected.

First we consider the case when $M$ is a pseudoscalar meson, i.e. $M = P'$. In this case from Eqs. (15.65a) and (15.51c)

$$f_2\left(t\right) = 4 \, m_p^2 \left| f_+\left(t\right) \right|^2, \qquad f_1\left(t\right) = 0 \quad (15.80)$$

Hence

$$\frac{d\,\Gamma}{dt} = \frac{G^2}{24\pi^3} K^3 \left| f_+\left(t\right) \right|^2 \quad (15.81)$$

For the vector case, i.e. $M = V$, Eqs. (15.66) and (15.51c) after straightforward but some what lengthy calculation give

$$\frac{d\,\Gamma}{dt} = \frac{G^2}{96\pi^3} \, m_V \, \sqrt{w^2 - 1} \, \left| A_1\left(t\right) \right|^2 \, \left(m_P + m_V\right)^2$$
$$\times \left\{ \left(w^2 - 1\right) \left[ 1 + 4 \, m_V \left(m_V - m_P w\right) \frac{r_2\left(t\right)}{\left(m_V + m_P\right)^2} \right. \right.$$
$$\left. + 4m_P^2 \, m_V^2 \, \left(w^2 - 1\right) \frac{r_2^2\left(t\right)}{\left(m_V + m_P\right)^4} \right]$$
$$\left. + 4\frac{t \, m_V^2}{\left(m_V + m_P\right)^4} \left(w + 1\right) \, r_V^2\left(t\right) \left[\left(w + 1\right) + 4w\right] \right\} \quad (15.82)$$

where we have used

$$K_V = m_V \, \sqrt{w^2 - 1}, \qquad E_V = m_V \, w$$
$$t = m_P^2 + m_V^2 - 2m_P m_V \, w \quad (15.83)$$
$$r_2\left(t\right) = \frac{A_2\left(t\right)}{A_1\left(t\right)}, \qquad r_V\left(t\right) = \frac{V\left(t\right)}{A_1\left(t\right)} \quad (15.84)$$

For the transition $B \to D^*$, it is convenient to define

$$R_1(t) = \left[1 - \frac{t}{(m_B + m_{D^*})^2}\right] \frac{V(t)}{A_1(t)}$$

$$R_2(t) = \left[1 - \frac{t}{(m_B + m_{D^*}^2)^2}\right] \frac{A_2(t)}{A_1(t)} \qquad (15.85)$$

Note that in the heavy quark symmetry limit $R_1(t) = R_2(t) = 1$. Using Eqs. (15.84) and (15.85), Eq. (15.82) with $m_P \to m_B$, $m_V \to m_{D^*}$, gives

$$\frac{d\,\Gamma}{d\,w} = \frac{G_F^2 \, |V_{cb}|^2}{48\pi^3} m_{D^*}^3 \sqrt{w^2 - 1} \, (w+1)^2 \left\{ \left[ m_B^2 \left(w^2 - 1\right) \right. \right.$$

$$+ 2m_B \left(m_{D^*} - m_B w\right) (w-1) R_2(t) + m_B^2 (w-1)^2 R_2^2(t) \Big]$$

$$+ \left[ \left(m_B^2 + m_{D^*}^2 - 2m_B m_{D^*}\right) \left(1 + \frac{4w}{w+1}\right)\right] R_1^2(t) \Bigg\} \mathcal{F}^2(w) \quad (15.86)$$

In the symmetry limit, $R_1(t) = R_2(t) = 1$,

$$\frac{d\,\Gamma}{d\,w} = \frac{G_F^2 \, |V_{cb}|^2}{48\pi^3} m_{D^*}^3 \sqrt{w^2 - 1} \, (w+1)^2 \left\{ (m_B - m_{D^*})^2 \right.$$

$$+ \frac{4w \left(m_B^2 + m_{D^*}^2 - 2m_B \, m_{D^*} \, w\right)}{w+1} \Bigg\} \mathcal{F}^2(w) \qquad (15.87)$$

For $B \to D$ transition, Eqs. (15.81) and (15.68) give

$$\frac{d\,\Gamma}{d\,w} = \frac{G_F^2 \, |V_{cb}|^2}{48\pi^3} (m_B + m_D)^2 \, m_D^3 \, \left(w^2 - 1\right)^{3/2} \mathcal{G}^2(w) \qquad (15.88)$$

In the heavy quark symmetry limit

$$\mathcal{G}(w) = \mathcal{F}(w) = \xi(w) \qquad (15.89)$$

For $B \to D$, $D^*$ transitions, the heavy quark spin symmetry gives $R_1(t) = R_2(t) = 1$; but it does not give the form factors $\mathcal{F}(w)$ and $\mathcal{G}(w)$ at any $w$ except at $w = 1$. Lattice calculation including the finite quark mssses corrections gives

$$\mathcal{F}(1) = 0.927 \pm 0.024$$

$$\mathcal{G}(1) = 1.074 \pm 0.018 \pm 0.016 \qquad (15.90)$$

A more refined analysis of symmetry breaking gives

$$R_1 \approx 1 + \frac{4\alpha_s(m_c)}{3\pi} + \frac{\bar{\Lambda}}{2m_c}$$

$$R_2 \approx 1 - \frac{\bar{\Lambda}}{2m_c} \qquad (15.91)$$

Using, $\alpha_s(m_c) \approx 0.34$, $\bar{\Lambda} \approx 0.41$ GeV, [cf. Eq. (9.75b)] $m_c \approx 1.5$ GeV we obtain $R_1 \approx 1.3$, $R_2 \approx 0.9$. The data is fitted using Caprini-Lellauch-Neubert (CLN)[7] parametrization for the form factors $\mathcal{F}(w)$ and $\mathcal{G}(w)$ based on analyticity, unitarity and conformal mapping with variable $z = (\sqrt{w+1} - \sqrt{2})/(\sqrt{w+1} + \sqrt{2})$ :

$$\mathcal{F}(w) = \mathcal{F}(1)\left[1 - 8\rho^2 z + (53\rho^2 - 15)z^2 - (231\rho^2 - 91)z^3\right] \quad (15.92)$$

$$\mathcal{G}(w) = \mathcal{G}(1)\left[1 - 8\rho^2 z + (51\rho^2 - 10)z^2 - (252\rho^2 - 84)z^3\right] \quad (15.93)$$

Note that $\rho^2$ is the slope of the form factor at $w = 1$. For $B \to D^* l\nu$, the fit gives [11]

$$\rho^2 = 1.191 \pm 0.048 \pm 0.028$$

$$R_1(1) = 1.429 \pm 0.061 \pm 0.044$$

$$R_2(1) = 0.827 \pm 0.038 \pm 0.022$$

$$\mathcal{F}(1)V_{cb} = (34.4 \pm 0.3 \pm 1.1) \times 10^{-3} \quad (15.94)$$

The fit also gives [9]

$$h_{A_1}(w^*_{\max}) = 0.52 \pm 0.03 \quad (15.95)$$

The decay $\overline{B}^0 \to D^{*+}\pi^-$, using factorization for the tree graph gives [13]

$$h_{A_0}(w^*_{\max}) = 0.52 \pm 0.03 \quad (15.96)$$

Hence from Eqs. (15.90), (15.104) and (15.105), we conclude that HQET is a good effective theory for heavy to heavy transitions.

For $B \to Dl\nu$, the fit gives [10]

$$\mathcal{G}(1)V_{cb} = (43.4 \pm 1.9 \pm 1.4) \times 10^{-3}$$

$$\rho^2 = 1.20 \pm 0.09 \pm 0.04$$

Let us now consider the semileptonic decays of $D$-mesons. The experimental results on the form factors are as follows [14]:

$$D^+ \to \bar{K}^0 \ell^+ \nu_\ell : \quad f_+(0)|V_{cs}| = 0.707 \pm 0.013 \quad (15.97)$$

$$D^+ \to \bar{K}^{*0} \ell^+ \nu_\ell : \quad V(0) = 1.0 \pm 0.3$$

$$A_1(0) = 0.55 \pm 0.03$$

$$A_2(0) = 0.40 \pm 0.08$$

$$r_2(0) = 0.84 \pm 0.11$$

$$r_V(0) = 1.80 \pm 0.08 \quad (15.98)$$

$$D_s^+ \to \phi \ell^+ \nu_\ell : \quad V(0) = 0.9 \pm 0.3$$

$$A_1(0) = 0.62 \pm 0.06$$
$$A_2(0) = 1.0 \pm 0.3$$
$$r_2(0) = 0.84 \pm 0.11$$
$$r_V(0) = 1.80 \pm 0.08 \tag{15.99}$$

These form factors are of importance for two-body nonleptonic decays of $D$-mesons in the factorization ansatz (see the next section).

There is no model independent way to determine these form factors. First we note that the form factors for various decays are related by the SU(3) as follows

$$-\sqrt{2} f_+ \left(D^+ \to \pi^0\right) = f_+ \left(D^0 \to \pi^-\right)$$
$$= f_+ \left(D^+ \to \bar{K}^0\right)$$
$$= f_+ \left(D^0 \to K^-\right)$$
$$= -\sqrt{3/2}\, f_+ \left(D_s^+ \to \eta_8\right)$$
$$= f_+ \left(D_s^+ \to K^0\right) \tag{15.100}$$

For $D \to V$ transitions, we have similar relations with $\pi \to \rho$ and $K \to K^*$ and since the nonet symmetry holds for vector mesons, we have

$$V\left(D^0 \to \rho^-\right) = V\left(D^+ \to \bar{K}^{*0}\right)$$
$$= V\left(D^0 \to K^{*-}\right)$$
$$= V\left(D_s^+ \to K^{*0}\right)$$
$$= -V\left(D_s^* \to \phi\right)$$

and

$$V\left(D_s^+ \to w\right) = 0 \tag{15.101}$$

Needless to say that similar relations hold for axial vector form factors $A_1$ and $A_2$.

Most of the models agree that the form factors $f_+$ and $V$ are dominated by vector bosons $D^*$ and $D_s^*$. Hence for the Cabbibo favored decays $D^0 \to K^- \ell^+ \nu$, $K^{*-} \ell^+ \nu$ and $D_s^* \to \phi \ell^+ \nu$, the vector mesons dominance shown in Fig. 15.3 gives

$$f_+^{D^0 \to K^-}(t) = g_{D_s^* DK} \frac{f_{D_s^*}\, m_{D_s^*}}{m_{D_s^*}^2 - t} \tag{15.102}$$

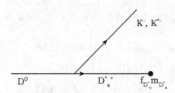

Fig. 15.3   Vector meson dominance for two body decays of $D^0$.

$$V^{D^0 \to K^{*-}}(t) = \left(\frac{m_D + m_{K^*}}{2}\right) 2\frac{g_{D_s^* D^* K}\, f_{D_s^*}\, m_{D_s^*}}{m_{D_s^*}^2 - t} \tag{15.103}$$

$$V^{D_s^* \to \phi}(t) = -(m_D + m_{K^*})\frac{g_{D_s^* D_s \phi} f_{D_s^*} m_{D_s^*}}{m_{D_s^*}^2 - t} \tag{15.104}$$

Now SU(3) gives

$$g_{D_s^* D^0 K^-} = g_{D^{*+} D^0 \pi^-} = g_{D^{*0} D^+ \pi^+} \tag{15.105}$$

The coupling constant $g_{D_s^{*+} D^0 \pi^-}$ can be obtained from the experimental value of the decay width of the $D^{*+} \to D^0\pi^+$ and is given by [cf. Eq. (8.8)]

$$g_{D^{*+} D^0 \pi^+} = 9 \pm 2 \tag{15.106}$$

Heavy quark spin symmetry gives

$$f_{D_s^{*+}} = f_{D_s^+} = (257.5 \pm 6.1)\text{MeV} \tag{15.107}$$

Thus

$$f_+(0) = g_{D_s^{*+} D^0 K^-}\frac{f_{D_s^*}}{m_{D_s^*}} \tag{15.108}$$

$$= 1.10 \pm 0.25 \tag{15.109}$$

(on using SU(3)), to be compared with the experimental value

$$f_+(0)\,|V_{cs}| = 0.707 \pm 0.013 \tag{15.110}$$

From heavy quark spin symmetry and SU(3), one obtains

$$g_{D^* D\rho} = \frac{2}{m_{D^{*+}}} g_{D^* D\pi} \tag{15.111}$$

$$g_{D_s^* D_s \phi} = -\frac{1}{\sqrt{2}} g_{D^* D\rho} = -\frac{\sqrt{2}}{m_{D_s^{*+}}} g_{D^* D\pi} \tag{15.112}$$

$$g_{D_s^* D K^*} = \frac{2}{m_{D_s^*}} g_{D_s^* D K} \tag{15.113}$$

Using these relations, one gets

$$V^{D^0-K^{*-}}(0) = \frac{m_D + m_{K^*}}{2} \frac{2}{m_{D_s^*}^2} g_{D_s^* D K} f_{D_s^*} \tag{15.114}$$

$$\approx 1.43 \pm 0.33 \tag{15.115}$$

$$V^{D_s^0-\phi}(0) = 1.15 \pm 0.26 \tag{15.116}$$

We conclude that single pole approximation for the form factors given in Eqs. (15.102)-(15.104) is only qualitatively valid. However if we take the mass of vector meson $m_{D_s^*}$ slightly larger than its actual value in the pole $(m_{D^{*2}} - t)$, the agreement between the values of the form factors at $t = 0$, with their experimental values will be improved.

### 15.3.4 *Weak Hadronic Decays of B Mesons*

At tree level, the $\Delta B = 1$ weak hadronic decays are described by a single $W$-exchange as shown in the Fig. 15.4, which represents the decay $b \to p + q' + \bar{q}$, $(p = u$ or $c$; $q = u$ or $c$; $q' = d$ or $s)$. The effective Lagrangian

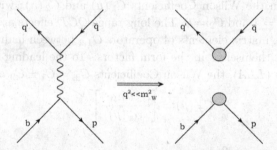

Fig. 15.4   Decay of $b \to p + q' + \bar{q}$ through $W$-exchange.

for this decay is given by

$$\mathcal{L}_{eff} = \frac{G_F}{\sqrt{2}} \left\{ \sum_{q=u,c} V_{pb} V_{qq'}^* \left[ \bar{q}'^\beta \gamma^\mu (1 - \gamma_5) q_\beta \right] \left[ \bar{p}^\alpha \gamma_\mu (1 - \gamma_5) b_\alpha \right] \right\} \tag{15.117}$$

where $\alpha$ and $\beta$ are color indices. Since quarks carry color, the $QCD$ corrections must be taken into account [see Fig. 15.5]. With QCD corrections up to first order

$$\mathcal{L}_{eff} = \frac{G_F}{\sqrt{2}} \sum_{q=u,c} V_{pb} V_{qq'}^* (C_1 O_1^p + C_2 O_2^p) \tag{15.118}$$

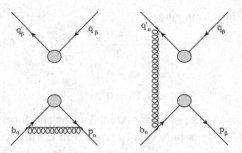

Fig. 15.5   QCD corrections to the Fig. 15.4.

where $C_i$ are Wilson Coefficients evaluated at the renormalization scale $\mu$; the current-current operators $O_{1,2}^p$ are

$$O_1^p = \left(\bar{q}'^\beta q_\beta\right)_{V-A} \left(\bar{p}^\alpha b_\alpha\right)_{V-A} \tag{15.119}$$

$$O_2^p = \left(\bar{q}'^\alpha q_\beta\right)_{V-A} \left(\bar{p}^\beta b_\alpha\right)_{V-A} \tag{15.120}$$

Note that strong interaction due to hard gluon corrections have been taken into account in the Wilson Coefficients $C_1(\mu)$ and $C_2(\mu)$; without these corrections $C_1 = 1$ and $C_2 = 0$. The long range $QCD$ effects are taken into account by the matrix elements of operators $O_{1,2}$ between hadronic states. They manifest themselves in the form factors. To the leading logarithmic approximation ($LLA$), the Wilson Coefficients $C_\pm = C_1 \pm C_2$ are given by

$$C_\pm(\mu) = \left[\frac{\alpha_s(m_W)}{\alpha_s(\mu)}\right]^{\gamma_\pm} \tag{15.121a}$$

where

$$\gamma_+ = \frac{2}{\beta_0} = \frac{2}{11 - \frac{2}{3}n_f}, \quad \gamma_- = -2\,\gamma_+, \quad C_+^2\, C_- = 1 \tag{15.121b}$$

and $n_f$ is the number of active flavors (in the region between $\mu$ and $m_W$). At $\mu = m_b = 4.9$ GeV we get from Eqs. (15.121a), taking $\alpha_s(m_W) \approx 0.12$, $n_f = 5$ and $\alpha_s(m_b) = 0.22$, $\alpha_s(m_c) = 0.34$,

$$C_1(m_b) \approx 1.11, \qquad C_2(m_b) \approx -0.26 \tag{15.122}$$

The operators $O_{1,2}^p$ after Fierz reordering are given by

$$O_1^p = \left(\bar{q}'^\beta q_\beta\right)_{V-A} \left(\bar{p}^\alpha b_\alpha\right)_{V-A}$$

under Fierz reordering $\rightarrow \left(\bar{p}^\alpha q_\beta\right)_{V-A} \left(\bar{q}'^\beta b_\alpha\right)_{V-A} \tag{15.123}$

$$O_2^p = \left(\bar{q}'^\alpha q_\beta\right)_{V-A} \left(\bar{p}^\beta b_\alpha\right)_{V-A}$$

under Fierz reordering $\rightarrow \left(\bar{p}^\beta q_\beta\right)_{V-A} \left(\bar{q}'^\alpha b_\alpha\right)_{V-A} \tag{15.124}$

Now

$$\overline{p}^{\alpha}\, q_{\beta} = \underbrace{(\overline{p}^{\alpha} q_{\beta} - \frac{1}{3}\delta^{\alpha}_{\beta}\overline{p}^{\gamma}q_{\gamma})}_{\text{octet}} + \underbrace{\frac{1}{3}\delta^{\alpha}_{\beta}\overline{p}^{\gamma}q_{\gamma}}_{\text{singlet}} \qquad (15.125)$$

Thus

$$(\overline{p}^{\alpha} q_{\beta})_{V-A}\,(\overline{q}'^{\beta} b_{\alpha})_{V-A} \to \frac{1}{3}\,(\overline{p}^{\gamma} q_{\gamma})_{V-A}\,(\overline{q}'^{\alpha} b_{\alpha})_{V-A}$$

$$= \frac{1}{3} O_2^p \qquad (15.126)$$

Similarly

$$(\overline{q}'^{\alpha} q_{\beta})_{V-A}\,(\overline{p}^{\beta} b_{\alpha})_{V-A} \to \frac{1}{3}\,(\overline{q}'^{\gamma} q_{\gamma})_{V-A}\,(\overline{p}^{\alpha} b_{\alpha})_{V-A}$$

$$= \frac{1}{3} O_1^p \qquad (15.127)$$

Hence with QCD corrections to first order

$$\mathcal{L}_{eff} = \frac{G_F}{\sqrt{2}} \sum_{q=u,c} V_{pb}V_{qq'}^* \left( \begin{array}{l} (C_1 + \frac{1}{3}C_2)\,(\overline{q}'^{\beta} q_{\beta})_{V-A}\,(\overline{p}^{\alpha} b_{\alpha})_{V-A} \\ +(C_2 + \frac{1}{3}C_1)\,(\overline{p}^{\beta} q_{\beta})_{V-A}\,(\overline{q}'^{\alpha} b_{\alpha})_{V-A} \end{array} \right) \qquad (15.128)$$

Usually one writes

$$C_1 + \frac{1}{3}C_2 = a_1: \qquad \text{Tree Diagram } T \qquad (15.129)$$

$$C_2 + \frac{1}{3}C_1 = a_2: \qquad \text{Color suppressed} \qquad (15.130)$$

$$\text{Tree Diagram } C$$

In particular for $p = c$, $q\prime = d$ or $s$

$$\mathcal{L}_{eff} = \frac{G_F}{\sqrt{2}} \sum_{q=u,c} V_{cb}V_{qq'}^* \left( \begin{array}{l} a_1\,(\overline{q}'q)_{V-A}\,(\overline{c}b)_{V-A}\,+ \\ a_2\,(\overline{c}q)_{V-A}\,(\overline{q}'b)_{V-A} \end{array} \right) \qquad (15.131)$$

For $p = u$, $q\prime = d$ or $s$

$$\mathcal{L}_{eff} = \frac{G_F}{\sqrt{2}} \sum_{q=u,c} V_{ub}V_{qq'}^* \left( \begin{array}{l} a_1\,(\overline{q}'q)_{V-A}\,(\overline{u}b)_{V-A}\,+ \\ a_2\,(\overline{u}q)_{V-A}\,(\overline{q}'b)_{V-A} \end{array} \right) \qquad (15.132)$$

For the decays governed by the above Lagrangian, penguin diagrams do not contribute, only tree diagrams are allowed (Fig. 15.6).

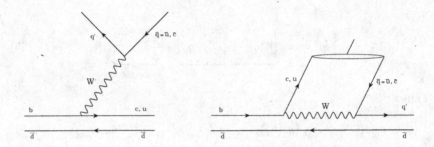

Fig. 15.6   Tree level diagrams (a) color favored (b) color suppressed.

Thus for example

$$\overline{B}^0 \to \pi^- D^+ \quad (K^- D^+) : T$$
$$\overline{B}^0 \to \pi^0 D^0 \quad (\overline{K}^0 D^0) : C$$
$$\overline{B}^0 \to D^- \pi^+ \quad (D^- K^+) : T$$
$$\overline{B}^0 \to \overline{D}^0 \pi^0 \quad (\overline{D}^0 \overline{K}^0) : C$$
$$\overline{B}^0 \to D^+ \overline{D}_s^- \quad : T$$
$$\overline{B}^0 \to J/\psi \overline{K}^0 \quad : C \tag{15.133}$$

We now discuss the case where both tree and penguin diagrams contribute to Figs. 15.7 and 15.8.

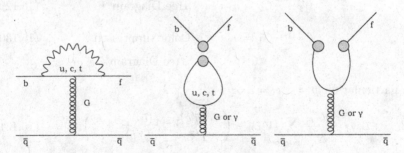

Fig. 15.7   QCD and electromagnetic penguin.

In this case the general effective Lagrangian is given by

$$\mathcal{L}_{eff} = \frac{G_F}{\sqrt{2}} \sum_{p=u,c} \lambda_p^{(f)} \left( C_1 O_1^p + C_2 O_2^p + \sum_{i=3}^{10,7\gamma,8g} C_i O_i \right) \tag{15.134}$$

where

$$\lambda_p^{(f)} = V_{pb} V_{qf}^* , \quad b \to p + f + \bar{q} \tag{15.135}$$

$$O_1^p = (\overline{f}^\beta p_\beta)_{V-A} \, (\overline{p}^\alpha b_\alpha)_{V-A}$$

$$O_2^p = (\overline{f}^\alpha b_\beta)_{V-A} \, (\overline{p}^\beta b_\alpha)_{V-A}$$

$$O_{3,4} = \left\{ (\bar{q}^\beta q_\beta)_{V-A} (\overline{f}^\alpha p_\alpha)_{V-A}, \ (\bar{q}^\alpha q_\beta)_{V-A} (\overline{f}^\beta b_\alpha)_{V-A} \right\}$$

$$O_{5,6} = \left\{ (\bar{q}^\beta q_\beta)_{V+A} (\overline{f}^\alpha p_\alpha)_{V-A}, \ (\bar{q}^\alpha q_\beta)_{V+A} (\overline{f}^\beta b_\alpha)_{V-A} \right\}$$

$$O_{7,8} = \frac{3e_q}{2} \left\{ (\bar{q}^\beta q_\beta)_{V+A} (\overline{f}^\alpha p_\alpha)_{V-A}, \ (\bar{q}^\alpha q_\beta)_{V+A} (\overline{f}^\beta b_\alpha)_{V-A} \right\}$$

$$O_{9,10} = \frac{3e_q}{2} \left\{ (\bar{q}^\beta q_\beta)_{V-A} (\overline{f}^\alpha b_\alpha)_{V-A}, \ (\bar{q}^\alpha q_\beta)_{V-A} (\overline{f}^\beta b_\alpha)_{V-A} \right\}$$

$$O_{7\gamma,8g} = -\frac{m_b}{8\pi^2} \overline{f} \sigma^{\mu\nu} \left[ eF_{\mu\nu}, \ gG_{\mu\nu}^a T_a \right] (1+\gamma_5) b$$

$O_{7\gamma}$ arise from the diagram shown in Fig. 15.8.

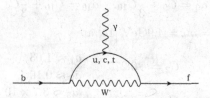

Fig. 15.8 Electromagnetic penguin.

Then following the same arguments discussed above for the tree graphs, the effective Lagrangian (leaving $O_{5,6}$ and $O_{7,8}$; which can be ignored) for $p = u$ is given by

$$\mathcal{L}_{eff}(d) = a_1 (\bar{d}u)_{V-A}((\bar{u}b)_{V-A}) + a_2 (\bar{u}u)_{V-A}((\bar{d}b)_{V-A})$$

$$+ a_3 \left[ (\bar{u}u + \bar{d}d + \bar{s}s)_{V-A} (\bar{d}b)_{V-A} \right]$$

$$+ a_4 \left[ (\bar{d}u)_{V-A}((\bar{u}b)_{V-A}) + (\bar{d}d)_{V-A}((\bar{d}b)_{V-A}) \right.$$

$$\left. + (\bar{d}s)_{V-A}((\bar{s}b)_{V-A}) \right] + a_9 \left[ \left( \frac{2}{3}\bar{u}u - \frac{1}{3}\bar{d}d - \frac{1}{3}\bar{s}s \right)_{V-A} \right] (\bar{d}b)_{V-A}$$

$$+ a_{10} \left[ \frac{2}{3}(\bar{d}u)_{V-A}(\bar{u}b)_{V-A} - \frac{1}{3}(\bar{d}d)_{V-A}(\bar{d}b)_{V-A} \right.$$

$$\left. - \frac{1}{3}(\bar{d}s)_{V-A}(\bar{s}b)_{V-A} \right] \tag{15.136}$$

$$\mathcal{L}_{eff}(s) = a_1(\bar{s}u)_{V-A}((\bar{u}b)_{V-A}) + a_2(\bar{u}u)_{V-A}((\bar{s}b)_{V-A})$$
$$+ a_3\left[(\bar{u}u + \bar{d}d + \bar{s}s)_{V-A}(\bar{s}b)_{V-A}\right]$$
$$+ a_4\left[(\bar{s}u)_{V-A}((\bar{u}b)_{V-A}) + (\bar{s}d)_{V-A}((\bar{d}b)_{V-A}) + (\bar{s}s)_{V-A}((\bar{s}b)_{V-A})\right]$$
$$+ a_9\left[\left(\frac{2}{3}\bar{u}u - \frac{1}{3}\bar{d}d - \frac{1}{3}\bar{s}s\right)_{V-A}\right](\bar{s}b)_{V-A}$$
$$+ a_{10}\left[\frac{2}{3}(\bar{s}u)_{V-A}(\bar{u}b)_{V-A} - \frac{1}{3}(\bar{s}d)_{V-A}(\bar{d}b)_{V-A}\right.$$
$$\left. - \frac{1}{3}(\bar{s}s)_{V-A}(\bar{s}b)_{V-A}\right] \tag{15.137}$$

where $a$'s are given by

$$a_1 = C_1 + \frac{1}{3}C_2, \quad a_2 = C_2 + \frac{1}{3}C_1$$
$$a_3 = C_3 + \frac{1}{3}C_4, \quad a_4 = C_4 + \frac{1}{3}C_3 \tag{15.138}$$
$$a_9 = C_9 + \frac{1}{3}C_{10}, \quad a_{10} = C_{10} + \frac{1}{3}C_9 \tag{15.139}$$

Here $C(\mu)$ with $\mu = m_b = 4.9\text{GeV}$ [14] are

$$
\begin{array}{lll}
C_1 = 1.121 & C_2 = -0.275 & \begin{array}{l} a_1 \approx 1.03 \\ a_2 \approx 0.10 \end{array} \\[2mm]
C_3 = 0.013 & C_4 = -0.028 & \begin{array}{l} a_3 \approx 3.7 \times 10^{-3} \\ a_4 \approx -24 \times 10^{-3} \end{array} \\[2mm]
C_9/\alpha = -1.280 & C_{10}/\alpha = 0.328 & \begin{array}{l} a_9 \approx -8.5 \times 10^{-3} \\ a_{10} \approx -7.1 \times 10^{-4} \end{array}
\end{array}
\tag{15.140}
$$

We denote:

Dominant penguin amplitude: $P$: $a_4$

Color suppressed penguin amplitude: $P^C$: $a_3$

Dominant electroweak penguin amplitude: $P_{EW}$: $a_9$

Non-dominant electroweak penguin amplitude: $P_{EW}^C$: $a_{10}$

### 15.3.5  *Inclusive Hadronic B Decays*

From the Lagrangian (15.128) and Fig. 15.6, it is simple to write the $B \to hadrons$ in the spectator quark model for CKM favored decays

$$\Gamma(\bar{B}_d \to X_{c\bar{u}d}) = \Gamma[\bar{B}_d(b\bar{d}) \to \bar{d}(b \to c + d + \bar{u})] \tag{15.141}$$

$$\Gamma(b \to c + d + \bar{u}) = \frac{G_f^2}{192\pi^2}m_b^5(3C_1^2 + 2C_1C_2 + \frac{1}{3}C_2^2)[|V_{cb}|^2|V_{ud}|^2F(m_c/m_b)]$$

Now using the branching ratio

$$\frac{B(\overline{B}_d \to X_{c\bar{c}s})}{B(\overline{B}_d \to X_{c\bar{u}d})} = 0.12 \tag{15.142}$$

we have

$$\Gamma(\overline{B}_d \to X_{c\bar{c}s}) = \Gamma(b \to c + s + \bar{c}) \tag{15.143}$$

$$= 0.12\ \Gamma(b \to c + d + \bar{u}) \tag{15.144}$$

Thus

$$\Gamma(\overline{B}_d \to X_c) = 1.12\ (b \to c + d + \bar{u}) \tag{15.145}$$

Taking into account

$$\frac{B(b \to X\ \tau\ \bar{\nu}_\tau)}{B(b \to X\ e\ \bar{\nu}_e)} = 0.24 \tag{15.146}$$

we get from Eqs. (15.62) and above equations

$$(\overline{B}^0)_{SL} = (B^-)_{SL} = \frac{1}{2.24 + 1.2(3C_1^2 + 2C_1C_2 + \frac{1}{3}C_2^2)} = \frac{1}{6.06}$$

$$= (B_s)_{SL} \approx 16.5\% \tag{15.147}$$

$$\frac{\Gamma_{\overline{B}^0}}{\Gamma_{B^-}} = 1 \tag{15.148}$$

Experimentally [14],

$$(B^-)_{SL} = (10.99 \pm 0.28)\% \\ (\overline{B}^0)_{SL} = (10.33 \pm 0.29)\% \tag{15.149}$$

Finally

$$\Gamma(\overline{B}^0 \to X_{u\bar{u}d}) = \Gamma(b \to u\ s\ \bar{u}) \tag{15.150}$$

$$= \frac{G_F^2}{192\pi^2} m_b^5 (3C_1^2 + 2C_1C_2 + \frac{1}{3}C_2^2) \times$$

$$|V_{ub}|^2 |V_{ud}|^2 [F(m_d/m_b) \approx 1] \tag{15.151}$$

Thus form Eqs. (15.141), (15.146) and (15.151) one gets

$$\frac{\Gamma(\overline{B}^0 \to X_u)}{\Gamma(\overline{B}^0 \to X_c)} = \frac{|V_{ub}|^2}{|V_{cb}|^2 [F(m_c/m_b) + 0.24]} \tag{15.152}$$

Hence, using the inclusive branching ratios, we can determine both $|V_{cb}|^2$ and $|V_{ub}|^2$.

The spectator quark model predictions $(\overline{B}^0)_{SL} = (B^-)_{SL}$, $\Gamma_{\overline{B}^0}/\Gamma_{B^-} = 1$ are in agreement with the experiments. However, the branching ratios are larger than experimental values by about a factor 1.5. Thus we conclude that the spectator quark model is qualitatively valid although it needs corrections due to the bound state structure.

### 15.3.6 *Radiative Decays of $B_q$ Mesons*

We consider the decays

$$\overline{B}_q \to V + \gamma, \qquad q = \overline{d}, \ \overline{s}, \ \overline{c}$$
$$p = p' + k, \qquad V = K^*, \ D_s^* \tag{15.153}$$

These decays are described by the effective Hamiltonian

$$H_W = \frac{G_F}{\sqrt{2}} V_{cb} \, V_{cs}^* C_{7\gamma} \left[ -\frac{m_b e}{8\pi^2} \overline{s} \sigma^{\mu\nu} (1 + \gamma^5) b F_{\mu\nu} \right] \tag{15.154}$$

The $T$-matrix for these decays is given by

$$T = \frac{-1}{(2\pi)^3} \frac{1}{\sqrt{8k_0 p_0 p_0'}} F = -\frac{G_F}{\sqrt{2}} V_{cb} \, V_{cs}^* C_{7\gamma} \left( -\frac{m_b e}{8\pi^2} \right) \left\langle V \left| \overline{s} \sigma^{\mu\nu} (1 + \gamma^5) b \right| \overline{B}_q \right\rangle$$
$$\times \frac{1}{(2\pi)^{3/2}} \frac{-1}{\sqrt{2k_0}} [k_\mu \epsilon_\nu - k_\nu \epsilon_\mu] \tag{15.155}$$

Now

$$\left\langle V \left| \overline{s} \sigma^{\mu\nu} (1 + \gamma^5) b \right| \overline{B}_q \right\rangle = \frac{1}{(2\pi)^3} \frac{-1}{\sqrt{4p_0 p_0'}} [\eta_\lambda f^{\lambda\mu\nu}] \tag{15.156}$$

where most general $f^{\lambda\mu\nu}$ is given by

$$f^{\lambda\mu\nu} = i\epsilon^{\lambda\mu\nu\rho} k_\rho F'(k^2) + i\epsilon^{\lambda\mu\nu\rho} P_\rho F(k^2)$$
$$+ (g^{\lambda\mu} k^\nu - g^{\lambda\nu} k^\mu) G'(k^2) + (g^{\lambda\mu} P^\nu - g^{\lambda\nu} P^\mu) G(k^2) \tag{15.157}$$

$\epsilon_\mu$ is the polarization vector of $\gamma$ and $\eta_\lambda$ is the polarization vector of $V$, $k = p - p'$ and $P = p + p'$. It is clear that when $f^{\lambda\mu\nu}$ is contracted with $(k_\mu \epsilon_\nu - k_\nu \epsilon_\mu)$, the first and the third terms will give zero. Hence we have

$$\left\langle V \left| \overline{s} \sigma^{\mu\nu} (1 + \gamma^5) b \right| \overline{B}_q \right\rangle = \frac{1}{(2\pi)^3} \frac{-1}{\sqrt{4p_0 p_0'}} \eta_\lambda$$
$$\times [i\epsilon^{\lambda\mu\nu\rho} P_\rho F(k^2) + (g^{\lambda\mu} P^\nu - g^{\lambda\nu} P^\mu) G(k^2)] \tag{15.158}$$

The radiative decay is determined by the two form factors $F(k^2)$ and $G(k^2)$. Now with $\left( g = \frac{G_F}{\sqrt{2}} V_{cb} \, V_{cs}^* C_{7\gamma} \left( \frac{-m_b e}{8\pi^2} \right) \right)$

$$F = g[i\epsilon^{\lambda\mu\nu\rho} \eta_\lambda P_\rho F(k^2) + \eta_\lambda (g^{\lambda\mu} P^\nu - g^{\lambda\nu} P^\mu) G(k^2)](k_\mu \epsilon_\nu - k_\nu \epsilon_\mu)$$
$$= 2g[i\epsilon^{\lambda\mu\nu\rho} \eta_\lambda P_\rho k_\mu \epsilon_\nu F(k^2) + (k.\eta P.\epsilon - \eta.\epsilon P.k) G(k^2)] \tag{15.159}$$

In the rest frame of $B$

$$\mathbf{p} = 0, \quad p = (m_B, 0) \tag{15.160}$$
$$\mathbf{p}' = -\mathbf{k}, \quad m_B = p_0' + k_0, \quad k_0 = |\mathbf{k}| \tag{15.161}$$

$$\epsilon^{\lambda\mu\nu\rho}\eta_\lambda P_\rho k_\mu \epsilon_\nu = 2m_B \eta.(\mathbf{k} \times \epsilon) \tag{15.162}$$

$$k.\eta P.\epsilon - \eta.\epsilon P.k = 2m_B(\eta.\epsilon)\,|\mathbf{k}| \tag{15.163}$$

Thus in the rest frame of $B$

$$F = = 4gm_B\,|\mathbf{k}|\,\left[F(k^2)i\eta.(\mathbf{n} \times \epsilon) + G(k^2)(\eta.\epsilon)\right]$$
$$\tag{15.164}$$

$$|M|^2 = \sum_{pol} |F|^2 = 32g^2 m_B^2\,|\mathbf{k}|^2\,\left[|F(k^2)|^2 + |G(k^2)|^2\right] \tag{15.165}$$

$$\Gamma = \frac{1}{8\pi}\frac{|\mathbf{k}|}{m_B^2}\,|M|^2$$

$$= \frac{g^2}{2\pi}\,|\mathbf{k}|^3\,\left[|F(k^2)|^2 + |G(k^2)|^2\right] \tag{15.166}$$

The decay width is determined by two form factors. For the decay of the type $B \to D_q^*\gamma$, HQET can be used to express these two form factors in terms of only one form factor. The experimental results for the radiative decays $B \to V\gamma$ provide a good framework to test the various models used to calculate these form factors.

## 15.4 Inclusive Hadronic Decays of $D$-Mesons

Corresponding to the Lagrangian (15.128) for the $B$, the Lagrangian for the $D$ mesons is

$$\mathcal{L}_{eff} = \frac{G_F}{\sqrt{2}}V_{cs}\,V_{uq}^*\left(\begin{array}{l}(C_1 + \tfrac{1}{3}C_2)\,(\overline{u}q)_{V-A}\,(\overline{s}c)_{V-A}\\ +(C_2 + \tfrac{1}{3}C_1)\,(\overline{s}q)_{V-A}\,(\overline{u}c)_{V-A}\end{array}\right) \tag{15.167}$$

Thus in the spectator quark model for Cabibbo favored decays $(q = d)$:

$$\Gamma\left[D\,(c\,\overline{q}) \to \overline{q}\,(c \to s + u + \overline{d})\right]$$
$$= \Gamma\,(c \to s + u + \overline{d})$$
$$= \frac{G_F^2\,m_c^5}{192\pi^3}\left[|V_{cs}|^2\,|V_{ud}|^2\,F\left(\frac{m_s}{m_c}\right)\right]\left[3C_1^2 + 2C_1C_2 + \frac{1}{3}C_2^2\right] \tag{15.168}$$

Hence from Eqs. (15.61), (15.168) and (15.62)

$$\left(D^0\right)_{SL} = \left(D^+\right)_{SL} = \frac{1}{2 + \left(3C_1^2 + 2C_1\,C_2 + \frac{1}{3}C_2^2\right)} = \left(D_s^+\right)_{SL}$$

$$\approx \frac{1}{5.5} \approx 18\% \tag{15.169a}$$

$$\frac{\Gamma_{D^0}}{\Gamma_{D^+}} = 1, \text{ where we have used} \tag{15.169b}$$

$$C_1(m_c^2) \approx 1.24, \quad C_2(m_c^2) \approx -0.48 \tag{15.169c}$$

Let us first discuss $(D)_{SL}$, while Eq. (15.169a) is consistent with the experimental value for $(D^+)_{SL}$ but it is about a factor of 2.8 greater than the experimental value for $(D^0)_{SL}$. Also we note that the experimental values, namely

$$\frac{(D^+)_{SL}}{(D^0)_{SL}} \approx 2.48 \pm 0.06 \tag{15.170}$$

$$\frac{\Gamma_{D^0}}{\Gamma_{D^+}} = \frac{\tau_{D^+}}{\tau_{D^0}} \approx 2.54 \tag{15.171}$$

are in disagreement with the predications of spectator quark model given in Eqs. (15.169a).

### 15.4.1  *Scattering and Annihilation Diagrams*

There are two kinds of mechanism for the hadronic decays of heavy mesons which we have not considered. They are depicted for Cabibbo favored decays of $D^0$ and $D_s$ mesons in Figs. 15.9 and 15.10.

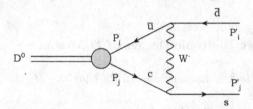

Fig. 15.9   W-exchange diagram for hadronic $D^0$ decay.

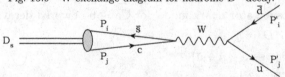

Fig. 15.10   W-annihilation quark level diagram for hadronic $D_s^0$ decay.

The basic processes depicted in these figures are respectively $c+\bar{u} \rightarrow s+\bar{d}$ and $c+\bar{s} \rightarrow u+\bar{d}$ where for the second process, the analogue of Eq. (15.118) gives the effective Lagrangian

$$\mathcal{L}_{eff} = \frac{G_F}{\sqrt{2}}\xi \left[ C_1 \left(\bar{s}^{\alpha} c_{\alpha}\right)_{V-A} \left(\bar{u}^{\beta} d_{\beta}\right)_{V-A} \right.$$
$$\left. + C_2 \left(\bar{s}^{\alpha} c_{\beta}\right)_{V-A} \left(\bar{u}^{\beta} d_{\alpha}\right)_{V-A} \right] \tag{15.172}$$

where $\xi = |V_{cs}| \, |V_{ud}|$ while for the first process the effective Lagrangian is obtained by Fierz rearrangement. Thus the $T$-matrix for the process $c + \bar{u} \to s + \bar{d}$ is given by

$$T_{\text{exch}} \sim \frac{G_F}{\sqrt{2}} \xi \frac{1}{3} \left(C_1 \delta_{\alpha\beta} \, \delta_{\alpha\beta} + C_2 \delta_{\alpha\alpha} \, \delta_{\beta\beta}\right) \left[\bar{u}\left(\mathbf{p}_j'\right) \, \gamma^\mu \left(1 - \gamma^5\right) u\left(\mathbf{p}_j\right)\right]$$

$$\times \left[-\bar{v}\left(\mathbf{p}_i\right) \gamma_\mu \, \left(1 - \gamma_5\right) v\left(\mathbf{p}_i'\right)\right]$$

$$= -\frac{G_F}{\sqrt{2}} \, \xi \, \left(C_1 + 3C_2\right) \left[\bar{u}\left(\mathbf{p}_j'\right) \, \gamma^\mu \left(1 - \gamma^5\right) \, u\left(\mathbf{p}_j\right)\right]$$

$$\times \left[\bar{u}\left(\mathbf{p}_i'\right) \, \gamma_\mu \, \left(1 + \gamma_5\right) \, u\left(\mathbf{p}_i\right)\right] \qquad (15.173a)$$

where we expressed the $v$-spinor in terms of $u$-spinor by the relation $v = C^\dagger \bar{u}^T$. Similarly for the annihilation diagram (Fig. 15.10), Eq. (15.172) gives the $T$-matrix:

$$T_{\text{ann}} \sim \frac{G_F}{\sqrt{2}} \xi \frac{1}{3} \left(C_1 \delta_{\alpha\alpha} \, \delta_{\beta\beta} + C_2 \delta_{\alpha\beta} \, \delta_{\alpha\beta}\right) \left[\bar{v}\left(\mathbf{p}_i\right) \, \gamma^\mu \left(1 - \gamma^5\right) u\left(\mathbf{p}_j\right)\right]$$

$$\times \left[\bar{u}\left(\mathbf{p}_j'\right) \, \gamma_\mu \, \left(1 - \gamma_5\right) v\left(\mathbf{p}_i'\right)\right]$$

$$= -\frac{G_F}{\sqrt{2}} \, \xi \, \left(3C_1 + C_2\right) \left[\bar{u}\left(\mathbf{p}_j'\right) \, \gamma^\mu \left(1 - \gamma^5\right) \, u\left(\mathbf{p}_j\right)\right]$$

$$\times \left[\bar{v}\left(\mathbf{p}_i\right) \, \gamma_\mu \, \left(1 - \gamma_5\right) v\left(\mathbf{p}_i'\right)\right] \qquad (15.173b)$$

where we have used Fierz rearrangement in going from the first line to the second line. Thus we conclude that both diagrams give the same results apart from the color factors. In taking the nonrelativistic limit it is convenient to express $v$-spinor in terms of $u$-spinor and use the relation $\bar{v} \, \Gamma_i \, v = \epsilon_i \, u \, \Gamma_i \, \bar{u}$, with $\epsilon_i = \pm 1$, $\Gamma_i = \gamma_\lambda, \gamma_\lambda \gamma_5$. We note that $m_i'$, $m_j' = m$ (the mass of $u$ and $d$ quark), $m_i = m_c \gg m$, $m_j = m_s$ or $m_d$; $E_i' = E_j' = E$, $m_c + E_j = 2E$, where in the nonrelativistic limit, we have put $E_i = m_c$ and we also put $m_s = m_d$. Using Pauli representation of Dirac matrices, it is a straightforward but long calculation to obtain the cross section $\sigma$ for the scattering or annihilation processes shown in Figs. 15.9 and 15.10. Suppressing the color factor, we get for the singlet and triplet scattering cross sections respectively

$$\sigma_S = \frac{|\xi|^2}{8\pi} G_F^2 \left(8m^2\right) \frac{1}{v}, \qquad (15.174a)$$

$$\sigma_T = \frac{|\xi|^2}{8\pi} G_F^2 \left(\frac{8}{3} m_c^2\right) \frac{1}{v}, \qquad (15.174b)$$

where $v$ is the incoming velocity in the initial state. Now defining the decay width as

$$\Gamma = v \, |\Psi_s(0)|^2 \, \sigma, \qquad (15.175)$$

we get for the triplet state

$$\Gamma\left(^3S_1 \to u\,\bar{d}\right) = \frac{1}{8\pi}\; G_F^2\; \xi^2\; \frac{8}{3}\; m_{D^*}^2\; |\Psi_s(0)|^2, \tag{15.176}$$

where we have put $m_{D^*}^2 = (m_c + m_s)^2 \approx m_c^2$. For the singlet state, we get

$$\Gamma\left(^1S_0 \to u\,\bar{d}\right) = \frac{1}{8\pi}\; G_F^2\; \xi^2\; 8\; m^2\; |\Psi_s(0)|^2. \tag{15.177}$$

Note the important fact that the decay width for the singlet state $(D)$ is proportional to the square of the light quark masses whereas in the spectator quark model it is proportional to $m_c^2$. This is called the helicity suppression.

Inserting back the color factors we have finally

$$\Gamma_{exch}^D = \frac{G_F^2}{8\pi}\; |V_{cs}|^2\; |V_{ud}|^2\; \left(8\;m^2\right) |\Psi_s(0)|^2\; \left(C_1 + 3C_2\right)^2, \tag{15.178}$$

$$\Gamma_{ann}^{D_s} = \frac{G_F^2}{8\pi}\; |V_{cs}|^2\; |V_{ud}|^2\; \left(8\;m^2\right) |\Psi_s(0)|^2\; \left(3C_1 + C_2\right)^2 \tag{15.179}$$

It is clear from Eqs. (15.178) and (15.179), that both the exchange and annihilation diagrams are helicity suppressed, but $\Gamma_{exch}$ is color suppressed as well while $\Gamma_{ann}$ is color enhanced.

It is interesting to see that for the annihilation diagram, one can get the same result just by writing the $T$-matrix for the $D_s \to hadrons$ in the form

$$T = \frac{G_F}{\sqrt{2}}\xi\frac{\delta_{\alpha\alpha}}{\sqrt{3}}\left(C_1 + \frac{1}{3}C_2\right)\langle X\,|J^{W\mu}|\,0\rangle\;\langle 0\,|J_\mu^{W\dagger}|\,D_s\rangle \tag{15.180}$$

where $J_\mu^{W\dagger}$ and $J^{W\mu}$ are color singlet currents with appropriate quantum numbers. Then

$$\Gamma_{ann}^{D_s} = \frac{G_F^2}{2}\; |\xi|^2\; (2\pi)^7 \sum_{spin} \int d^3p_X\; \delta\left(p - p_X\right)$$

$$\times |\langle 0\,|J_\mu^{W\dagger}|\,D_s\rangle|^2\,|\langle X\,|J^{W\mu}|\,0\rangle|^2. \tag{15.181}$$

Now from Lorentz invariance

$$\langle 0\,|J_\mu^{W\dagger}|\,D_s\rangle = i\;\frac{1}{(2\pi)^{3/2}}\frac{1}{\sqrt{2p_0}}\; f_{D_s}\; p_\mu, \tag{15.182}$$

while

$$\sum_{spin} \int d^3p_X\; \delta\left(p_X - p\right)\langle 0\,|J_\mu^{W\dagger}|\,X\rangle\;\langle X\,|J^{W\mu}|\,0\rangle$$

$$= \frac{1}{(2\pi)^3}\theta\left(p_0\right)\left[\left(-p^2\;g_{\mu\lambda} + p_\mu\;p_\lambda\right)\rho\left(p^2\right)\right.$$

$$\left. + p_\mu\;p_\lambda\;\sigma\left(p^2\right)\right].\;. \tag{15.183}$$

Hence

$$\Gamma_{annh}^{D_s} = \frac{G_F^2}{2} |A|^2 \frac{2\pi}{2 m_{D_s}} f_{D_s}^2 m_{D_s}^4 \sigma\left(m_D^2\right). \tag{15.184}$$

Now from dimensional consideration

$$\sigma\left(m_D^2\right) = \frac{2 m^2}{4\pi^2 m_{D_s}^2}. \tag{15.185}$$

Thus

$$\Gamma_{annh}^{D_s} = \frac{G_F^2}{4\pi} |V_{cs}|^2 |V_{ud}|^2 f_{D_s}^2 m_{D_s} \left(m^2\right) \frac{(3C_1 + C_2)^2}{3}. \tag{15.186}$$

One gets exactly the same results if in Eq. (15.180) one replaces $|X\rangle$ by $|u\,\bar{d}\rangle$ (see problem 15.6). Comparing Eq. (15.186) with Eq. (15.179), one obtains

$$f_{D_s}^2 = \frac{12 |\Psi_s(0)|^2}{m_{D_s}}. \tag{15.187}$$

From Eqs. (15.168), (15.177) and (15.178) and (15.186), we get

$$\frac{\Gamma_{annh}^{D_s}}{\Gamma_{sp.}^{D_s}} = 16\pi^2 \frac{m_d^2 f_{D_s}^2 m_{D_s}}{m_c^5 F(m_s/m_c)} \frac{(3C_1 + C_2)^2}{(3C_1^2 + 2C_1 C_2 + \frac{1}{3}C_2^2)}. \tag{15.188}$$

$$\frac{\Gamma_{exch}^{D^0}}{\Gamma_{sp.}^{D^0}} = 8\pi^2 \frac{(m_d^2 + m_s^2) f_D^2 m_D}{m_c^5 F(m_s/m_c)} \frac{(C_1 + 3C_2)^2}{(3C_1^2 + 2C_1 C_2 + \frac{1}{3}C_2^2)}. \tag{15.189}$$

Using $C_1 = 1.24$, $C_2 = -0.48$, $m_c \approx 1.50$ GeV, $F(m_s/m_c) \approx 0.47$, $f_D = 207\,MeV$ and $f_{D_s} = 257$ MeV, we get from Eq. (15.188).

$$\frac{\Gamma_{annh}^{D_s}}{\Gamma_{sp.}^{D_s}} \approx 0.72 \tag{15.190}$$

while $\Gamma_{exch}^{D^0}/\Gamma_{sp.}^{D^0}$ is negligible. The annihilation diagram gives negligible contribution to $D^+$ decays. Taking into account Eq. (15.190)

$$\Gamma(D_s \to hadrons) = (1.7)\,\Gamma_{sp}. \tag{15.191}$$

where $\Gamma_{sp}$ is given in Eq. (15.168) and from Eq. (15.169a)

$$\left(D_s^+\right)_{SL} = \frac{1}{2 + 1.7\left[3C_1^2 + 2C_1 C_2 + \frac{1}{3}C_2^2\right]} \approx 12\% \tag{15.192}$$

$$(6.5 \pm 0.4)\% \quad \text{Experimental}$$

$$\frac{\tau_{D_s^+}}{\tau_{D^0}} \approx \frac{2 + (3C_1^2 + 2C_1 C_2 + \frac{1}{3}C_2^2)}{2 + 1.7\left[3C_1^2 + 2C_1 C_2 + \frac{1}{3}C_2^2\right]} \approx 0.69 \tag{15.193}$$

$$(1.22 \pm 0.02) \quad \text{Experimental} \quad .$$

Both these predictions are not in agreement with their experimental values especially the first one which is by a factor of 2 larger than the experimental value.

We conclude that the contribution of the annihilation diagram is although helicity suppressed, but enhancement by a factor of $192\pi^2$ due to phase space and that due to color factor more than compensate the helicity suppression. However, there are still problems between the predictions of spectator quark model and the experimental values. We conclude that the spectator quark model is not adequate for the inclusive $D$ decays.

Finally the annihilation diagram for $B$ decays are Cabibbo suppressed and they may be neglected.

## 15.5   Problems

(1) Taking into account finite width for $\rho$ meson and using Eq. (15.17), show that

$$
\Gamma\left(\tau^- \to \rho^- \nu_\tau \to \pi^- \pi^0 \nu_\tau\right) = \frac{G_F^2 \cos^2 \theta_c}{384\pi^3} \, m_\tau^3 \int_{2m_\pi^2}^{m_\tau^2} |F_\pi(s)|^2
$$

$$
\times \left(1 - \frac{s}{m_\tau^2}\right)^2 \left(1 + \frac{2s}{m_\tau^2}\right) \left(1 - \frac{4\,m_\pi^2}{s}\right)^{3/2} ds, \qquad (15.194)
$$

where

$$
F_\pi(s) = \frac{f_{\rho\pi\pi}\, f_\rho}{\left[(s - m_\rho^2) + i\, m_\rho\, \Gamma\right]}
$$

Hint:

$$
\left[2\pi\delta\left(s - m_\rho^2\right) \to \frac{2\,m_\rho\, \Gamma}{(s - m_\rho^2) + m_\rho^2\, \Gamma^2}\right].
$$

Considering the process

$$
e^- e^+ \to \gamma \to e^+ e^-,
$$

show that the cross-section is given by ($s \gg 4m_e^2$)

$$
\sigma_{\pi^+ \pi^-}(s) = \frac{\pi}{3\,s}\, \alpha^2\, |F_\pi(s)|^2 \left(1 - \frac{4\,m_\pi^2}{s}\right)^{3/2}
$$

where $F_\pi(s)$ is the electromagnetic form factor of pion:

$$
\langle \pi^+ \pi^- | J_\lambda^{em} | 0\rangle \propto F(q^2)(p_1 - p_2)_\lambda
$$

$$
s = q^2 = (p_1 + p_2)^2.
$$

Using Eq. (15.36), show that we get back Eq. (15.194). From Eq. (15.194), find the decay rate for $\tau^- \to \pi^- \pi^0 \, \nu_\tau$ through $\rho$-resonance.

(2) Taking into account finite width of $a_1$ meson, and Eq. (15.18), show that (taking $m_\pi = 0$)

$$\Gamma\left(\tau^- \to a_1^- \nu_\tau \to \pi^- \rho^0 \nu_\tau\right)$$
$$= \Gamma\left(\tau^- \to \pi^0 \rho^- \nu_\tau\right)$$
$$= \frac{G_F^2 \cos^2\theta_c}{96\pi^3} m_\tau^3 \int_{m_\rho^2}^{m_\tau^2} ds \, |F_{\rho\pi}(s)|^2 \left(1 - \frac{s}{m_\tau^2}\right)^2 \left(1 + \frac{2s}{m_\tau^2}\right) \left\{1 + \frac{s}{8\,m_\rho^2}\right.$$
$$\times \left. \left[\left(1 + \frac{m_\rho^2}{s}\right)^2 + 2r\left(1 - \frac{m_\rho^2}{s}\right)^2 + r^2 \frac{\left(1 - \frac{m_\rho^2}{s}\right)^4}{\left(1 + \frac{m_\rho^2}{s}\right)^2}\right]\right\} \tag{15.195}$$

where

$$F_{\rho\pi}(s) = \frac{f_{a_1} F_{a_1\rho\pi}}{\left[(s - m_{a_1}^2) + i\, m_{a_1} \Gamma\right]}.$$

The $a_1\rho\pi$ couplings are defined by the decay amplitude $T$:

$$T \propto 2\, m_{a_1} F_{a_1\rho\pi} \left[\eta \cdot \varepsilon + \frac{r}{k \cdot q}(\eta \cdot k)(\varepsilon \cdot q)\right]$$

where $\eta_\mu$, $\varepsilon_\mu$ are polarization vectors of $a_1$ and $\rho$, $q$ and $k$ are their four momenta and $r$ is the ratio of $D$ to $S$ waves couplings. In order to derive (15.195), first show that

$$\Gamma(a_1 \to \rho\,\pi)$$
$$= \frac{(F_{a_1\rho\pi})^2}{6\pi} m_{a_1} \left(1 - \frac{m_\rho^2}{m_{a_1}^2}\right) \left\{1 + \frac{m_{a_1}^2}{8\,m_\rho^2}\left[\left(1 + \frac{m_\rho^2}{m_{a_1}^2}\right)^2\right.\right.$$
$$\left.\left. +2r\left(1 - \frac{m_\rho^2}{m_{a_1}^2}\right)^2 + r^2 \frac{\left(1 - \frac{m_\rho^2}{m_{a_1}^2}\right)^4}{\left(1 + \frac{m_\rho^2}{m_{a_1}^2}\right)^2}\right]\right\}.$$

Using the experimental numbers for $\Gamma\left(\tau^- \to \pi^- \rho^0 \nu_\tau\right)$ and $\Gamma(a_1 \to \rho\,\pi)$, determine $F_{a_1\rho\pi}$ and $r$, using $f_{a_1}^2 = f_\rho^2 = 2f_\pi^2$.

(3) Show that for the decays

$$\overline{B}^0 \to \pi^0\pi^0 : \overline{A}_{00} = \frac{1}{\sqrt{2}}(P - C)$$
$$\overline{B}^0 \to \pi^+\pi^- : \overline{A}_{+-} = (T + P)$$
$$B^- \to \pi^-\pi^0 : \overline{A}_{-0} = \frac{1}{\sqrt{2}}(T + C)$$

Hence show that

$$\frac{1}{\sqrt{2}}\overline{A}_{+-} - \overline{A}_{00} = \overline{A}_{-0}$$

Derive the above relation also from the isospin analysis, noting that Bose statistics exclude $I = 1$; as the two pions in the final state are in s-state($l = 0$): $I = 0$, $I = 2$.

(4) Neglecting $P^C$ and $P^C_{EW}$, show that for

$$\overline{B}^0 \rightarrow \overline{K}^0\pi^0 : \overline{A}_{00} = \frac{1}{\sqrt{2}}(C - P + P_{EW})$$

$$\overline{B}^0 \rightarrow K^-\pi^+ : \overline{A}_{-+} = (T + P)$$

$$B^- \rightarrow K^-\pi^0 : \overline{A}_{-0} = \frac{1}{\sqrt{2}}(T + C + P + P_{EW})$$

$$B^- \rightarrow \overline{K}^0\pi^- : \overline{A}_{0-} = P \qquad\qquad (15.196)$$

(5) For decays

$$\overline{B}^0 \rightarrow \overline{K}^0\rho^0$$
$$\overline{B}^0 \rightarrow \overline{K}^0 w$$
$$\overline{B}^0 \rightarrow \overline{K}^0 \phi, q$$

show that

$$\left\langle \overline{K}^0\rho^0 \left| H_w(s) \right| B^0 \right\rangle = \frac{1}{2}(C - P + P_{EW})$$

$$\left\langle \overline{K}^0 w \left| H_w(s) \right| \overline{B}^0 \right\rangle = \frac{1}{2}(C + P + \frac{1}{3}P_{EW})$$

$$\left\langle \overline{K}^0\phi \left| H_w(s) \right| \overline{B}^0 \right\rangle = (-P + \frac{1}{3}P_{EW})$$

Assuming factorization for the electroweak penguin, show that

$$f_\rho\, F_1^{B-K}(m_\rho^2) - \frac{1}{3}f_\varpi\, F_1^{B-K}(m_\rho^2) - \frac{2}{3}f_\phi\, F_1^{B-K}(m_\phi^2) = 0$$

$$\Rightarrow f_\rho - \frac{1}{3}f_\varpi - \frac{2}{3}f_\phi = 0$$

giving a sum rule, with excellent agreement with experimental values for (cf. Problem 8.6)

$$f_\rho = 221,\ \text{MeV} \quad f_w = 194,\ \text{MeV} \quad f_\phi = 228\ \text{MeV}$$

(6) Using Eqs. (15.180) and (15.181) and writing

$$\langle X \,|\, J_\mu^W \,|\, 0\rangle = \langle \mu\bar{d} \,|\, J_\mu^W \,|\, 0\rangle$$

$$= \frac{1}{(2\,\pi)^3}\sqrt{\frac{m^2}{p_{10}\,p_{20}}}\ [\bar{u}\,(p_1)\ \gamma_\mu\ (1-\gamma_5)\ v\,(p_2)]$$

$$\int d^3\,p_X\ \delta\,(p-p_X) \to \int d^3\,p_1\ d^3\,p_2\ \delta\,(p-p_1-p_2)\,,$$

show that

$$\Gamma\,(D_s \to u\bar{d}) = \frac{(3C_1+C_2)^2}{3}\,\frac{G_F^2}{4\,\pi}\ |V_{cs}|^2\ |V_{ud}|^2\,f_{D_s}^2\ m_{D_s}\ (m^2)\,,$$

(7) Writing

$$\langle 0 \,|\, J_\mu^{W\dagger} \,|\, D_s^*\rangle \sim f_{D_s^*}\,\varepsilon_\mu$$

where $\varepsilon_\mu$ is the polarization of $D_s^*$, show that

$$\Gamma\,(D_s^* \to u\bar{d}) = \frac{(3C_1+C_2)^2}{3}\,\frac{G_F^2}{4\,\pi}\ |V_{cs}|^2\ |V_{ud}|^2\,\frac{1}{3}f_{D_s^*}^2\ m_{D_s^*}.$$

Comparing it with Eq. (15.187) when multiplied by the color factor $(3C_1+C_2)^2$, show that

$$f_{D_s^*}^2 = 12\ |\Psi_s\,(0)|^2\,m_{D^*}.$$

Hence show that

$$f_{D_s^*} = (m_{D_s}\,m_{D_s^*})^{1/2}\,f_{D_s}.$$

## 15.6   References

1. M. L. Perl, Rep. Prog. Phys. **55**, 653 (1992).

2. A. S. Schwarz, $\tau$ physics in "Lepton and Photon Interaction" XVI Int. symposium, Ithaca NY (1993) (eds P. Drell and D. Rubin) AIP, page 671; M. S. Witherell, Charm decay physics, ibid Page 198; M. B. Wise, Heavy flavor theory; ibid Page 253.

3. G. Buchalla and A. J. Buras, and M. E. Lautenbacher, Rev. Mod. Phys. **68**, 1125 (1996).

4. M. Neubrat, "B Decays and the Heavy-Quark Expansion" CERN-TH/97–24 hep-ph/ 9702375 [ To appear in the second edition of "Heavy Flavors" edited by A. J. Buras and M.Lindner; World Scientific] "Heavy-Quark Effective Theory and Weak Matrix Elements" CERN-TH/98–2 hep-ph/ 980 1269 [Invited talk presented at Int. Europhysics Conf. on High Energy Physics Jerusalem], Israel 19-26 Aug. 1997.

5. M. Neubrat, and B. Stech, "Non-Leptonic Weak Decays of B Mesons" CERN-TH/97–99 hep-ph/ 9705292 [ To appear in the second edition of "Heavy Flavors" edited by A. J. Buras and M.Lindner; World Scientific]

6. J. L. Rosner "B Physics - A Theoretical Overview" Nuclear Instruments and Methods in Physics Research A **408**, 308 (1998).

7. I. Caprini, L Lellauch and M. Neubert, "Dispersive Bounds on the shape of $\overline{B} \rightarrow D^{(*)} l\overline{\nu}$ Form Factors", CERN-TH/97-91, arxiv: 9712417v1, Dec 1997, Nuclear Physics B530, 153 (1998).

8. Thomas Mannel, Effective Field Theories in Flavour Physics, Springer. (2004).

9. A. Aubert et al., [Babar Collaboration] Phys. Rev. D **77**, 032002 (2008).

10. A. Aubert et al., [Babar Collaboration] arxiv: 0807.4978v1 [hep-ex] July (2008).

11. S. Faller, et al., "$B \rightarrow D^{(*)}$ Form factors from QCD Light-Cone Sum Rules", arxiv: 08090.0222v2 [hep-ph] Feb (2009).

12. Fayyazuddin, Phys. Rev. D**80**, 094015 (2009)

13. J. L. Rosner and S. Stone, "Leptonic decays of Charged Pseudoscalar Mesons" arxiv: 1002.1635v2 [hep-ex]

14. Particle Data Group, N. Nakamura et. al, Journal of Physics G, **37**, 075021 (2010).

# Chapter 16

# Particle Mixing and $CP$-Violation

## 16.1 Introduction

Symmetries have played an important role in particle physics. In quantum mechanics a symmetry is associated with a group of transformations under which a Lagrangian remains invariant. Symmetries limit the possible terms in a Lagrangian and are associated with conservation laws. Here we will be concerned with the role of discrete symmetries: Space Reflection (Parity) $P$: $\mathbf{x} \to -\mathbf{x}$, Time Reversal $T$: $t \to -t$ and Charge Conjugation $C$: *particle* $\to$ *antiparticle*.

Quantum Electrodynamics (QED) and Quantum Chromodynamics (QCD) respect all these symmetries. Also, all Lorentz invariant local quantum field theories are $CPT$ invariant. However, in weak interactions $C$ and $P$ are maximally violated separately but what about $CP$? This is the main topic of this chapter.

First indication of parity violation was revealed in the decay of a particle with spin parity $J^P = 0^-$, called $K$-meson into two modes $K^0 \to \pi^+\pi^-$ (parity violating), and $K^0 \to \pi^+\pi^-$ $\pi^0$(parity conserving).

Lee and Yang in 1956, suggested that there is no experimental evidence for parity conservation in weak interaction. They suggested a number of experiments to test the validity of space reflection invariance in weak decays. One way to test this is to measure the helicity of outgoing muon in the decay:

$$\pi^+ \to \mu^+ + \nu_\mu$$

The helicity of muon comes out to be negative, showing that parity conservation does not hold in this decay. In the rest frame of the pion, since $\mu^+$ comes out with negative helicity, the neutrino must also come out with negative helicity because of the spin conservation, confirming the fact that neutrino is left-handed. Thus

$$\pi^+ \to \mu^+(-) + \nu_\mu$$

Under charge conjugation,

$$\pi^+ \xrightarrow{C} \pi^- \qquad \mu^+ \xrightarrow{C} \mu^- \qquad \nu_\mu \xrightarrow{C} \bar\nu_\mu$$

Helicity $\mathcal{H} = \frac{\sigma \cdot \mathbf{p}}{|\vec{p}|}$ under $C$ and $P$ transforms as,

$$\mathcal{H} \xrightarrow{C} \mathcal{H}, \qquad \mathcal{H} \xrightarrow{P} -\mathcal{H}$$

Invariance under $C$ gives,

$$\Gamma_{\pi^+ \to \mu^+(-)\nu_\mu} = \Gamma_{\pi^- \to \mu^-(-)\bar\nu_\mu}$$

Experimentally,

$$\Gamma_{\pi^+ \to \mu^+(-)\nu_\mu} >> \Gamma_{\pi^- \to \mu^-(-)\bar\nu_\mu}$$

showing that $C$ is also violated in weak interactions. However, under $CP$,

$$\Gamma_{\pi^+ \to \mu^+(-)\nu_\mu} \xrightarrow{CP} \Gamma_{\pi^- \to \mu^-(+)\bar\nu_\mu}$$

which is seen experimentally. Thus, $CP$ conservation holds for this decay.

Let us now consider $K^0 - \bar{K}^0$ complex, which as mentioned above gave the first hint of parity violation and again it is here that $CP$ violation was first discovered experimentally. In hadronic and electromagnetic interactions, the hypercharge $Y$ is conserved so that $K^0(Y = 1) \longleftrightarrow \bar{K}^0(Y = -1)$ transitions are not possible. In a production process involving hadronic (or electromagnetic) interaction, $K^0$ and $\bar{K}^0$ appear as two distinctly different particles. In the presence of weak interaction, $Y$ is no longer conserved and transitions between $K^0$ and $\bar{K}^0$ can occur, for example.

$$K^0 \underset{\text{weak}}{\to} \pi^+\pi^- \underset{\text{weak}}{\to} \bar{K}^0, |\Delta Y| = 2$$

Thus if we write $H = H_0 + H_W$, where $H_0 = H_{had} + H_{e.m}$, $K^0$ and $\bar{K}^0$, which are eigenstates of $H_0$, are no longer eigenstates of $H$. A linear combination of $K^0$ and $\bar{K}^0$ will be eigenstates of $H$. Such states cannot be eigenstates of $C$ or $P$ since neither is conserved in weak interaction; $CP$ is a better choice. Choosing the $CP$ phase

$$CP\,|K^0\rangle = -\,|\bar{K}^0\rangle$$

so that

$$CP\,|\bar{K}^0\rangle = -\,|K^0\rangle, \tag{16.1}$$

it is easy to see that

$$|K_{1,2}^0\rangle = \frac{1}{\sqrt{2}}\left[|K^0\rangle \mp |\bar{K}^0\rangle\right] \tag{16.2}$$

are eigenstates of $CP$ with eigenvalues $\pm 1$. Further if $CP$ is conserved so that $[H, CP] = 0$, then

$$\begin{aligned}\langle K_2^0 |H| K_1^0\rangle &= \langle K_2^0 |(CP)^{-1}H\,CP| K_1^0\rangle \\ &= -\langle K_2^0 |H| K_1^0\rangle\end{aligned} \tag{16.3}$$

so that $\langle K_2^0 |H| K_1^0\rangle = 0 = \langle K_1^0 |H| K_2^0\rangle$, showing that $H$ is diagonal in the basis provided by $|K_1^0\rangle$ and $|K_2^0\rangle$. Thus eigenstates of $H$ can be chosen to be eigenstates of $CP$.

Now $\bar{K}^0$ is the antiparticle of $K^0$; they should have the same mass. But $K_1^0$ is not the antiparticle of $K_2^0$ and so they can have different properties. In fact due to weak interaction, $K_1^0$ and $K_2^0$ should have slightly different rest energies; experimentally $(m_{K_2} - m_{K_1})/m_K \sim 10^{-14}$ and it is remarkable that such a small quantity is measured. What about their life times? Energetically kaons can decay into two or three pions. Consider $2\pi$ final state. As seen in Sec. 4.4, $C$ parity of $2\pi$ state is $(-1)^\ell$ where $\ell$ is the relative orbital angular of $2\pi$ system. Thus

$$\begin{aligned}CP|\pi^+\pi^-\rangle &= (-1)^\ell(-1)^2(-1)^\ell|\pi^+\pi^-\rangle \\ &= (-1)^{2\ell}|\pi^+\pi^-\rangle = |\pi^+\pi^-\rangle\end{aligned}$$

Similarly

$$CP|\pi^0\pi^0\rangle = |\pi^0\pi^0\rangle \tag{16.4}$$

Thus only $K_1^0$ can decay into $2\pi$ if $CP$ is conserved in weak interaction and $K_2 \to 2\pi$ is forbidden. $K_2^0$ will have other modes, e.g. three pionic which can have $CP = -1$. Now decay energy available for $2\pi$ mode is about 220 MeV and for 3 pionic model it is about 90 MeV. Thus the phase space available for decay into three pions is considerably smaller than that for two pions, implying

$$\tau_1 \equiv \tau(K_1^0) \ll \tau(K_2^0) \equiv \tau_2$$

Experimentally $\tau(K_1^0) = 0.893 \times 10^{-10}$ sec. and $\tau(K_2^0) = 0.517 \times 10^{-7}$ sec. so that $\tau_1/\tau_2 = 1/580$.

As seen above, if $CP$ is conserved, $K_2^0 \to \pi^+\pi^-$ is forbidden. But $K_2 \to \pi^+\pi^-$ occurs, showing that $CP$ is not conserved. Numerically it is not a big effect.

$$\frac{A(K_2^0 \to \pi^+\pi^-)}{A(K_1^0 \to \pi^+\pi^-)} = 2.269 \times 10^{-3} \tag{16.5}$$

As such it is harder to understand in contrast to separate $P$ and $C$ violation in weak interaction, which is maximal. This is built into V-A interaction since as is well known, vector and axial vector currents transform oppositely under $P$ and $C$ separately. This is not so for $CP$ and in fact vector and axial vector currents transform in the same way under $CP$. What then is the cause of $CP$ violation? This question is answered in the next section.

## 16.2   *CPT* and *CP* Invariance

It is instructive to discuss the restrictions imposed by $CPT$ invariance. $CPT$ invariance implies,

$$
\begin{aligned}
{}_{\text{out}}\langle f\,|\mathcal{L}|\,X\rangle &= {}_{\text{out}}\left\langle f\,\Big|(CPT)^{-1}\,\mathcal{L}\,CPT\,\Big|\,X\right\rangle \\
&= \eta_T^{X*}\eta_T^{f}\,{}_{\text{in}}\left\langle \tilde{f}\,\Big|(CP)^{\dagger}\,\mathcal{L}^{\dagger}\,(CP)^{-1\dagger}\Big|\,\tilde{X}\right\rangle^{*} \\
&= \eta_T^{X*}\eta_T^{f}\left\langle \tilde{X}\,\Big|(CP)^{-1}\,\mathcal{L}\,(CP)\Big|\,\tilde{f}\right\rangle_{\text{in}}
\end{aligned}
\qquad (16.6)
$$

where

$$
T\,|X^0\rangle = -\eta_T^{X}\,\Big|\tilde{X}^0\Big\rangle
$$

$$
\langle f_{out}|\,T^{-1} = \left\langle \tilde{f}_{in}\,\Big|\,\eta_T^{f*}\right.
\qquad (16.7)
$$

and $\tilde{\ }$ means momentum and spin of the states are reversed. Since we are in the rest frame of $X$, $T$ will reverse only the magnetic quantum number and so we can drop $\tilde{\ }$. Further, we may choose the $CP$ phase such that

$$
CP\,|X\rangle = -\,|\bar{X}\rangle
\qquad (16.8)
$$

$$
CP\,|f\rangle = \eta_{CP}^{f}\,|\bar{f}\rangle
\qquad (16.9)
$$

Thus we have

$$
{}_{\text{out}}\langle f\,|\mathcal{L}|\,X\rangle = \eta_f\left\langle \overline{X}\,|\mathcal{L}|\,\bar{f}\right\rangle_{\text{in}}
\qquad (16.10)
$$

where

$$
\eta_f = \eta_{CP}^{f}\eta_T^{*}\eta_T^{X*}
\qquad (16.11)
$$

Hence on using

$$
|f\rangle_{\text{in}} = S_f\,|f\rangle_{\text{out}} = \exp(2i\delta_f)\,|f\rangle_{\text{in}}
\qquad (16.12)
$$

we get from Eq. (16.6)

$$
{}_{\text{out}}\langle f\,|\mathcal{L}|\,X\rangle = \eta_f\,e^{2i\delta_f}\left\langle \overline{X}\,|\mathcal{L}|\,\bar{f}\right\rangle_{\text{out}}
\qquad (16.13)
$$

$$
= \eta_f\,e^{2i\delta_f}\,{}_{\text{out}}\langle f\,|\mathcal{L}|\,X\rangle^{*}
\qquad (16.14)
$$

where $\delta_f$ is the strong interaction phase for the state $|f\rangle$. Hence finally we have

$$\bar{A}_{\bar{f}} = \eta_f e^{2i\delta_f} A_f^* \qquad (16.15)$$

If $CP$-invariance holds, then

$$_{out}\langle f |\mathcal{L}| X\rangle = \eta_{CP}^{f*} {}_{out}\langle \bar{f} |\mathcal{L}| \bar{X}\rangle \qquad (16.16)$$

implying

$$\bar{A}_{\bar{f}} = -\eta_{CP}^f A_f. \qquad (16.17)$$

We now discuss the implication of $CPT$ constraint with respect to $CP$ violation of weak decays. The weak amplitude is in general complex; it contains the final state strong phase $\delta_f$ and in addition it may also contain a weak phase $\phi$. Taking out both these phases,

$$A_f = e^{i\phi} e^{i\delta_f} |A_f| \qquad (16.18)$$

$CPT$ [Eq. (16.15)] gives,

$$\bar{A}_{\bar{f}} = \eta_f e^{2i\delta_f} A_f^* = \eta_f e^{-i\phi} e^{i\delta_f} |A_f| \qquad (16.19)$$

while CP invariance [Eq. (16.17)] gives

$$\bar{A}_{\bar{f}} = -\eta_{CP}^f e^{i\phi} e^{i\delta_f} |A_f| \qquad (16.20)$$

Thus for $CP$ violation, at least two amplitudes with different weak phases are required

$$A_f = A_{1f} + A_{2f} \qquad (16.21)$$

$CPT$ gives

$$\bar{A}_{\bar{f}} = e^{2i\delta_{1f}} A_{1f}^* + e^{2i\delta_{1f}} A_{2f}^*$$
$$A_{if} = e^{i\phi_i} e^{i\delta_{if}} |A_{if}|$$

where $(\delta_{1f}, \delta_{2f})$, $(\phi_1, \phi_2)$ are strong final state phases and the weak phases respectively. Thus the $CP$ violation asymmetry is given by

$$A_{CP} = \frac{\overline{\Gamma}(\overline{X} \to \overline{f}) - \Gamma(X \to f)}{\overline{\Gamma}(\overline{X} \to \overline{f}) + \Gamma(X \to f)}$$
$$= \frac{2|A_{1f}||A_{2f}| \sin\phi \sin\delta_f}{|A_{1f}|^2 + |A_{2f}|^2 + 2|A_{1f}||A_{2f}| \cos\phi \cos\delta_f} \qquad (16.22)$$

where $\delta_f = \delta_{2f} - \delta_{1f}$, $\phi = \phi_1 - \phi_2$. Hence the necessary condition for non-zero $CP$ violation is $\delta_f \neq 0$ and $\phi \neq 0$. As we shall see, the weak phase is a consequence of phases in CKM matrix. Four parameters $|A_{1f}|$, $|A_{2f}|$, $\phi$ and $\delta_f$ have been introduced here. Can these parameters be determined from the decay rates? The answer is in general positive as we see in the subsequent sections.

## 16.3   CP-Violation in the Standard Model

The current in the standard model is of the form, $\bar{\Psi}_i \gamma^\mu (1 - \gamma^5) \Psi_j$. Under $CP$ and time reversal it transforms as [see Appendix A]

$$2\bar{\Psi}_{iL}\gamma^\mu \Psi_{jL} \equiv \bar{\Psi}_i \gamma^\mu (1 - \gamma^5)\Psi_j$$
$$\xrightarrow{CP} -\eta(\mu)\,\bar{\Psi}_j \gamma^\mu (1 - \gamma^5)\Psi_i$$
$$\xrightarrow{T} \eta(\mu)\,\bar{\Psi}_i \gamma^\mu (1 - \gamma^5)\Psi_j, \qquad (16.23)$$
$$W_\mu \xrightarrow{CP} \eta_W W_\mu^\dagger$$

where

$$\eta(\mu) = \eta_i \eta_j^* = \begin{cases} +1, & \text{if } \mu = 0 \\ -1, & \text{if } \mu = 1,2,3 \end{cases} \qquad (16.24)$$

Note that under $CP$, $\bar{\Psi}_i \gamma^\mu (1 - \gamma^5)\Psi_j$ goes over to its hermitian conjugate. Thus if a Lagrangian $\mathcal{L}$ has a structure

$$\mathcal{L} = aO + a^* O^\dagger$$

where $O$ is an operator of type $\bar{\Psi}_j \gamma^\mu (1 - \gamma^5)\Psi_i$, then using Eq. (16.24)

$$\mathcal{L} \xrightarrow{CP} aO^\dagger + a^* O$$

Hence CP violation requires $a^* \neq a$.

The charged current interaction Lagrangian which is flavor changing given in Eq. (13.64) is

$$\mathcal{L}_{CC} = -\frac{g}{\sqrt{2}} \bar{\Psi}_{iL} \gamma^\mu (T^+ W_\mu^+ + T^- W_\mu^-)\Psi_{jL} \qquad (16.25)$$

where $\Psi_{iL} = \begin{pmatrix} u_i \\ d_i' \end{pmatrix}_L$ is doublet under weak isospin subgroup. It can be rewritten in more convenient form

$$\mathcal{L}_{CC} = -\frac{g}{2\sqrt{2}} \sum_{\substack{i=u,c,t \\ q=d,s,b}} \left( \bar{i}_L \gamma^\mu q' W_\mu + h.c. \right) \qquad (16.26)$$

Before symmetry breaking all masses are zero and $\mathcal{L}_{CC}$ given in Eq. (16.26) seems to be CP invariant in view of Eq. (16.23). However due to spontaneous symmetry breaking of the weak isospin, there arises a miss alignment between the weak isospin subgroup and quark mass matrix, expressed by

$$\bar{i}_L \gamma_\mu q'_L = V_{iq} \bar{i}_L \gamma_\mu q_L \qquad (16.27)$$

Thus the charged current interaction Lagrangian (16.26) becomes

$$\mathcal{L}_{cc} = -\frac{g}{\sqrt{2}} \{ W^\mu \sum_{iq} V_{iq} \bar{i}_L \gamma_\mu q_L + W^{\mu\dagger} \sum_{iq} V_{iq}^* \bar{q}_L \gamma_\mu i_L \} \qquad (16.28)$$

Then using Eqs. (16.23) and (16.26), we obtain

$$CPL_{cc}(CP)^{-1} = -\frac{g}{\sqrt{2}} \left\{ W^{\mu\dagger} \sum_{iq} V_{iq} \bar{q}_L \gamma_\mu i_L + W^\mu \sum_{\substack{i \\ q}} V_{iq}^* \bar{i}_L \gamma_\mu q_L \right\}$$

(16.29)

where phase $\eta(W)$ $\eta(q)$ $\eta^*(i)$ can be chosen to be $+1$. This is identical with (16.28) except that

$$V_{iq} \to V_{iq^*}$$  (16.30)

On the other hand

$$(CP)\mathcal{L}_{NC}(CP)^{-1} = \mathcal{L}_{NC}$$  (16.31)

where $\mathcal{L}_{NC}$ is the neutral current interaction Lagrangian, involving only diagonal couplings,

$$\mathcal{L}_{NC} = -\frac{g}{\cos\theta_W} Z^\mu \sum_{\substack{q=u,c,t \\ d,s,b}} \left\{ \left[ I_3(q) - Q(q)\sin^2\theta_W \right] \bar{q}_L \gamma_\mu q_L \right.$$

$$\left. -Q(q)\sin^2\theta_W\ \bar{q}_R \gamma_\mu q_R \right\}$$  (16.32)

Thus the neutral current in the interaction Lagrangian is necessarily $CP$ invariant. On the other hand from Eqs. (16.28) and (16.29), it is clear that

$$(CP)\mathcal{L}_{cc}(CP)^{-1} = \mathcal{L}_{cc}$$  (16.33)

if and only if $V$ is real $[V_{iq} = V_{iq}^*]$ or can be made real. Thus the standard model of electroweak interaction is capable of $CP$-violation.

Suppose we have $N$ generations so that $V$ is an $N \times N$ matrix and as such has $N^2$ complex elements or $2N^2$ real parameters. But since $V$ has to be unitary, it has $N^2$ real parameters. Then there is freedom to define any quark field by a phase, for example, $d \to e^{i\theta}d$, $W^\mu V_{id}\bar{i}_L \gamma_\mu d_L \to W^\mu V_{id}\bar{i}_L \gamma_\mu (e^{i\theta}d_L) = W^\mu (V_{id}e^{i\theta})(\bar{i}_L \gamma_\mu d_L)$ and $e^{i\theta}$ can be absorbed in the redefinition of $V_{id}$ without changing physics. Thus phase of any individual CKM matrix has no physical meaning [what counts is the relative phase]. Hence the number of phases which have no physical meaning [remember there are $2N$ fields] are $(2N - 1)$. Therefore, number of independent parameters in $V_{N \times N}$ are

$$N^2 - (2N - 1) = (N - 1)^2$$

$$= \begin{cases} 0 & N = 1 \\ 1 & N = 2 \\ 4 & N = 3 \end{cases}$$  (16.34)

One way of choosing the parameters is mixing angles and complex phases. Now if $V_{N \times N}$ were orthogonal matrix, then the number of independent parameters would be $N^2 - N - \frac{N(N-1)}{2} = \frac{N(N-1)}{2}$, which give the number of mixing angles. Then the number of phases are

$$(N-1)^2 - \frac{N(N-1)}{2} = \frac{(N-1)(N-2)}{2}$$

$$= \begin{cases} 0 & N=1 \\ 0 & N=2 \\ 1 & N=3 \end{cases} \qquad (16.35)$$

Thus the standard model of electroweak interaction is capable of $CP$ violation provided that $V$ is at least $3 \times 3$, i.e. three mixing angles and one phase. In other words $CP$ violation can be accommodated if the number - of generations is at least three. This observation was made before the third generation was discovered.

It is convenient to express the CKM matrix, as parameterized in the the Maiani-Wolfenstien way

$$V = \begin{pmatrix} V_{ud} & V_{us} & V_{ub} \\ V_{cd} & V_{cs} & V_{cb} \\ V_{td} & V_{ts} & V_{tb} \end{pmatrix}$$

$$\simeq \begin{pmatrix} 1 - \frac{1}{2}\lambda^2 & \lambda & A\lambda^3(\rho - i\eta) \\ -\lambda & 1 - \frac{1}{2}\lambda^2 & A\lambda^2 \\ A\lambda^3(1 - \rho - i\eta) & -A\lambda^2 & 1 \end{pmatrix} + O\left(\lambda^4\right) \quad (16.36)$$

where $\lambda \simeq \sin\theta_c \simeq 0.2253 \pm 0.0007$, is the Cabibbo angle and $|A| = 0.808 \pm^{0.022}_{0.015}$ is determined from semileptonic B-decays. $\eta \neq 0$ if $CP$ is not conserved. The unitarity of $V$, $VV^\dagger = 1$, gives

$$V_{ud}^* V_{ub} + V_{cb}^* V_{cd} + V_{td}^* V_{tb} = 0 \qquad (16.37a)$$

To leading order in $\lambda$, this relation can be written using Eq. (16.36) as

$$V_{ub}^* + V_{td} - \lambda V_{cb} = 0 \qquad (16.37b)$$

The relations (16.37a) and (16.37b) can be represented by a triangle in the complex plane (Fig. 16.1)

$$V_{ub} = |V_{ub}| e^{-i\gamma}$$
$$V_{cb} = A\lambda^2 \qquad (16.38)$$
$$V_{td} = |V_{td}| e^{-i\beta}$$

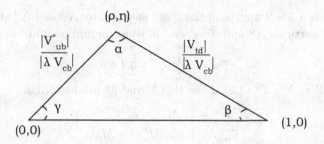

Fig. 16.1 The CKM-unitarity triangle in the Wolfenstein parametrization.

where,

$$\tan\gamma = \frac{\eta}{\rho} = \frac{\bar{\eta}}{\bar{\rho}}, \quad \tan\beta = \frac{\bar{\eta}}{1-\bar{\rho}}, \tag{16.39a}$$

$$\bar{\rho} = \rho(1 - \frac{\lambda^2}{2}), \quad \bar{\eta} = \eta(1 - \frac{\lambda^2}{2}), \tag{16.39b}$$

$$\text{with } \bar{\rho} = 0.132 \pm^{0.022}_{0.014}, \quad \bar{\eta} = 0.341 \pm 0.013. \tag{16.39c}$$

The weak phases $\gamma$ and $\beta$ play a leading role in $CP$ violation. However, these weak phases are in $V_{ub}$ and $V_{td}$, which connect the first generation with the third generation. Hence the role of $\beta$ and $\gamma$ in $K$ and $D$ decays is peripheral as both $K$ and $D$ are bound states of the first and second generation quarks.

## 16.4 Particle Mixing

As seen in Sec 16.1, $K^0$ and $\bar{K}^0$ can mix. We now develop a general formalism for particle mixing. Let $X^0$ and $\bar{X}^0$ be two pseudoscalar particles ($X = K, B$ or $D$; $\bar{X}$ being the antiparticle of $X$). Let $|\Psi(t)\rangle$ be a state at time $t$. It is a coherent mixture of $|X^0\rangle$ and $|\bar{X}^0\rangle$

$$|\Psi(t)\rangle = a(t)|X^0\rangle + \bar{a}(t)|\bar{X}^0\rangle \tag{16.40}$$

where $t$ is measured in the rest system of the particle $X^0$. Then the time evolution of the state

$$\Psi(t) = \begin{pmatrix} a(t) \\ \bar{a}(t) \end{pmatrix}$$

is given by

$$i\frac{d\Psi}{dt} = \mathfrak{m}\Psi \tag{16.41}$$

where $m$ is a $2 \times 2$ matrix in the space spanned by $X^0$ and $\bar{X}^0$ states and since the particles $X^0$ and $\bar{X}^0$ decay, $\mathfrak{m}$ is not hermitian and has the form

$$\mathfrak{m}_{\alpha'\alpha} = M_{\alpha'\alpha} - \frac{i}{2} \Gamma_{\alpha'\alpha} \tag{16.42}$$

with $\alpha, \alpha' = X^0, \bar{X}^0$ (1,2). Note that $\Gamma$ and $M$ are hermitian

$$\Gamma^\dagger = \Gamma, M^\dagger = M$$
$$\Gamma^*_{\alpha\alpha'} = \Gamma_{\alpha'\alpha}, M^*_{\alpha\alpha'} = M_{\alpha'\alpha} \tag{16.43}$$

If one now assumes $CPT$ invariance, then Eq. (16.12) gives

$$\left\langle X^0 \left| \mathfrak{m} \right| X^0 \right\rangle = \left\langle \bar{X}^0 \left| \mathfrak{m} \right| \bar{X}^0 \right\rangle \tag{16.44}$$

where we have used the fact that for a single particle state $|f\rangle = \left| X^0 \right\rangle_{out}$, $\delta_f = 0$, $\eta^X_{CP} = -1$ according to the choice of phase (16.7). Thus Eq. (16.44) gives $m_{11} = m_{22}$ or

$$M_{11} = M_{22}, \ \Gamma_{11} = \Gamma_{22} \tag{16.45}$$

that is particle-antiparticle have identical mass and the same total width. Note that if we take $f = \bar{X}^0$ in Eq. (16.10), we get an identity so that with $CPT$ invariance alone $m_{12}$ and $m_{21}$ are not related. However, if we assume $CP$ invariance, then Eq. (16.15) for $f = \bar{X}^0 [\eta^X_{CP} = -1]$ gives

$$\left\langle \bar{X}^0 |\mathfrak{m}| X^0 \right\rangle = \left\langle X^0 |\mathfrak{m}| \bar{X}^0 \right\rangle,$$

and thus $CP$ invariance implies

$$\mathfrak{m}_{21} = \mathfrak{m}_{12} \tag{16.46}$$

We have the result that in the $X^0 - \bar{X}^0$ space $\mathfrak{m}$ is a $2 \times 2$ matrix of the form

$$\mathfrak{m} = \begin{pmatrix} A & B \\ C & A' \end{pmatrix} = \begin{pmatrix} M_{11} - i\frac{\Gamma_{11}}{2} & M_{12} - i\frac{\Gamma_{12}}{2} \\ M_{21} - i\frac{\Gamma_{21}}{2} & M_{22} - i\frac{\Gamma_{22}}{2} \end{pmatrix} \tag{16.47}$$

where $CPT$ invariance alone (which we now assume) requires

$$A = A' \tag{16.48}$$

or

$$M_{11} = M_{22} \qquad\qquad \Gamma_{11} = \Gamma_{22}$$

But hermiticity of the matrices $M$ and $\Gamma$ [see Eqs. (16.43)] gives

$$M_{12} = M^*_{21}, \Gamma_{12} = \Gamma^*_{21}$$
$$M_{11} = M^*_{11}, \Gamma_{11} = \Gamma^*_{11} \tag{16.49}$$
$$M_{22} = M^*_{22}, \Gamma_{22} = \Gamma^*_{22}$$

Then the diagonalization of the matrix (16.47) gives the eigenvalues

$$\gamma_{1,2} = A \mp \sqrt{BC} = A \mp pq \qquad (16.50)$$

where

$$p^2 = B = M_{12} - \frac{i}{2}\Gamma_{12}$$

$$q^2 = C = M_{21} - \frac{i}{2}\Gamma_{21} = M_{12}^* - \frac{i}{2}\Gamma_{12}^* \qquad (16.51)$$

and using Eq. (16.48)

$$A = M_{11} - \frac{i}{2}\Gamma_{11} = M_{22} - \frac{i}{2}\Gamma_{22}$$

Then the corresponding eigenstates are

$$|X_\mp\rangle = \frac{1}{\sqrt{|p|^2 + |q|^2}} \left[ p\,|X^0\rangle \mp q\,|\bar{X}^0\rangle \right] \qquad (16.52)$$

Hence we have the result [cf. Eq. (16.50)]

$$M_{11} - \frac{i}{2}\Gamma_{11} - pq = \gamma_1 = m_1 - \frac{i}{2}\Gamma_1$$

$$M_{11} - \frac{i}{2}\Gamma_{11} + pq = \gamma_2 = m_2 - \frac{i}{2}\Gamma_2 \qquad (16.53)$$

so that taking real and imaginary parts

$$m_1 = M_{11} - \Re pq$$

$$m_2 = M_{11} + \Re pq$$

$$\Gamma_1 = \Gamma_{11} + 2\Im pq$$

$$\Gamma_2 = \Gamma_{11} - 2\Im pq \qquad (16.54)$$

Thus finally we have

$$\Delta m = m_2 - m_1 = 2\Re pq$$

$$m = \frac{m_1 + m_2}{2} = M_{11}$$

$$\Delta\Gamma = \Gamma_2 - \Gamma_1 = -4\Im pq$$

$$\Gamma = \frac{1}{2}(\Gamma_1 + \Gamma_2) = \Gamma_{11} \qquad (16.55)$$

Let us define

$$\frac{1 - \epsilon}{1 + \epsilon} = \frac{q}{p} = \sqrt{\frac{C}{B}} = \sqrt{\frac{M_{12}^* - \frac{i}{2}\Gamma_{12}^*}{M_{12} - \frac{i}{2}\Gamma_{12}}} \qquad (16.56)$$

If $CP$ is conserved, then $B = C$, $q = p(\epsilon = 0)$ so that the mass eigenstates given in Eq. (16.52) become

$$|X_{1,2}\rangle = \frac{1}{\sqrt{2}} \left[ |X^0\rangle \mp |\bar{X}^0\rangle \right] \qquad (16.57)$$

which are now also the eigenstates of $CP$:

$$CP |X_1\rangle = |X_1\rangle$$
$$CP |X_2\rangle = - |X_2\rangle \qquad (16.58)$$

It follows that $CP$-violation is determined by the parameter

$$\epsilon = \frac{p - q}{p + q} \qquad (16.59)$$

Since the particles $X^0$ and $\bar{X}^0$ are unstable, it is the particles $X_1 \equiv X_-$ and $X_2 \equiv X_+$ defined in Eq. (16.52) which have definite masses $m_1$ and $m_2$ and decay widths $\Gamma_1$ and $\Gamma_2$ respectively. Let $|\Psi(t)\rangle$ be a state at time $t$. In the $X_1$ and $X_2$ basis, we can write

$$|\Psi(t)\rangle = a(t) |X_1\rangle + b(t) |X_2\rangle \qquad (16.60)$$

$$i\frac{d}{dt} |\Psi(t)\rangle = \begin{pmatrix} m_1 - \frac{i}{2}\Gamma_1 & 0 \\ 0 & m_2 - \frac{i}{2}\Gamma_2 \end{pmatrix} |\Psi(t)\rangle. \qquad (16.61)$$

The solution is

$$a(t) = a(0) \exp\left[ -i\left( (m_1 - \frac{i}{2}\Gamma_1)t \right) \right]$$

$$b(t) = b(0) \exp\left[ -i\left( m_2 - \frac{i}{2}\Gamma_2 \right)t \right] \qquad (16.62)$$

Suppose we start with $X^0$, viz $|\Psi(0)\rangle = |X^0\rangle$, then from Eq. (16.52), we get

$$a(0) = b(0) = \frac{\sqrt{|p|^2 + |q|^2}}{2p}. \qquad (16.63)$$

Hence from Eqs. (16.60), (16.62) and (16.63)

$$\begin{aligned}
|\psi(t)\rangle &= \frac{\sqrt{|p|^2 + |q|^2}}{2p} \left[ \exp\left( -im_1 t - \frac{1}{2}\Gamma_1 t \right) |X_1\rangle \right. \\
&\quad \left. + \exp\left( -im_2 t - \frac{1}{2}\Gamma_2 t \right) |X_2\rangle \right] \\
&= \frac{1}{2} \left\{ \left[ \exp\left( -im_1 t - \frac{1}{2}\Gamma_1 t \right) + \exp\left( -im_2 t - \frac{1}{2}\Gamma_2 t \right) \right] |X^0\rangle \right. \\
&\quad \left. - \frac{q}{p} \left[ \exp\left( -im_1 t - \frac{1}{2}\Gamma_1 t \right) - \exp\left( -im_2 t - \frac{1}{2}\Gamma_2 t \right) \right] |\bar{X}^0\rangle \right\}
\end{aligned}$$

$$(16.64)$$

Equation (16.64) clearly shows the particle mixing. Similarly if we start with $|\bar{X}^0\rangle$ we get after time $t$

$$|\psi(t)\rangle = \frac{1}{2}\left\{\frac{p}{q}\left[\exp\left(-im_1t - \frac{1}{2}\Gamma_1t\right) - \exp\left(-im_2t - \frac{1}{2}\Gamma_2t\right)\right]|X^0\rangle\right.$$
$$\left. - \left[\exp\left(-im_1t - \frac{1}{2}\Gamma_1t\right) + \exp\left(-im_2t - \frac{1}{2}\Gamma_2t\right)\right]|\bar{X}^0\rangle\right\} \quad (16.65)$$

From Eqs. (16.64) and (16.65), we can determine $X^0$ and $\bar{X}^0$ mixing. It is clear that if we start with $X^0$, then at time $t$, the probability of finding the particles $X^0$ or $\bar{X}^0$ is given by [using Eq. (16.64)]

$$\left|\langle X^0|\psi(t)\rangle\right|^2 = \frac{1}{4}\left[e^{-\Gamma_1t} + e^{-\Gamma_2t} + 2e^{-\Gamma t}\cos\Delta mt\right] \quad (16.66)$$

$$\left|\langle \bar{X}^0|\psi(t)\rangle\right|^2 = \frac{1}{4}\left|\frac{1-\epsilon}{1+\epsilon}\right|^2\left[e^{-\Gamma_1t} + e^{-\Gamma_2t} - 2e^{-\Gamma t}\cos\Delta mt\right] \quad (16.67)$$

We define the mixing parameter $r$ as

$$r = \frac{\int\limits_0^T\left|\langle\bar{X}^0|\psi(t)\rangle\right|^2 dt}{\int\limits_0^T\left|\langle X^0|\psi(t)\rangle\right|^2 dt} \quad (16.68)$$

where $T$ is a sufficiently long time. In the limit $T \to \infty$, using Eqs. (16.66) and (16.67), we get

$$r = \left|\frac{1-\epsilon}{1+\epsilon}\right|^2\frac{x^2+y^2}{2+x^2-y^2} \quad (16.69)$$

where $x = \frac{\Delta m}{\Gamma}$ and $y = \frac{\Delta\Gamma}{2\Gamma}$. If we start with $\bar{X}^0$, we can use Eq. (16.65) then

$$\bar{r} = \frac{\int\limits_0^T\left|\langle X^0|\psi(t)\rangle\right|^2 dt}{\int\limits_0^T\left|\langle\bar{X}^0|\psi(t)\rangle\right|^2 dt} \xrightarrow{T\to\infty} \left|\frac{1+\epsilon}{1-\epsilon}\right|^2\frac{x^2+y^2}{2+x^2-y^2} \quad (16.70)$$

When $CP$-violation effects are neglected, then

$$r = \bar{r} = \frac{x^2+y^2}{2+x^2-y^2} \quad (16.71)$$

The asymmetry parameter $a$

$$a = \frac{\bar{r}-r}{\bar{r}+r} = \frac{4\Re\epsilon}{1+|\epsilon|^2} \quad (16.72)$$

is a measure of $CP$-violation. We define another parameter $\chi$ which is also a measure of particle mixing. Let $\chi$ be the probability for $X^0 \to \bar{X}^0$, then

$$\chi = \int_0^T \left| \langle \bar{X}^0 | \psi(t) \rangle \right|^2 dt$$

$$1 - \chi = \int_0^T \left| \langle X^0 | \psi(t) \rangle \right|^2 dt$$

Thus

$$r = \frac{\chi}{1-\chi}, \chi = \frac{r}{1+r} \tag{16.73}$$

Similarly, we get

$$\bar{r} = \frac{\bar{\chi}}{1-\bar{\chi}}, \bar{\chi} = \frac{\bar{r}}{1+\bar{r}} \tag{16.74}$$

We note from definitions, $x = \Delta m / \Gamma$, $y = \Delta \Gamma / 2\Gamma$

$$0 \le x^2 \le \infty$$
$$0 \le y^2 \le 1$$

Obviously

$$0 \le r \le 1$$

We now discuss how the mixing parameter $r$ can be measured experimentally. Suppose that $X^0$ and $\bar{X}^0$ are produced in the reaction

$$e^- e^+ \to X^0 \, \bar{X}^0.$$

Taking into account the particle mixing, we have four possible final states $X^0 \, \bar{X}^0$, $\bar{X}^0 X^0$, $X^0 X^0$, $\bar{X}^0 \bar{X}^0$. Experimentally $X^0 \bar{X}^0$ and $\bar{X}^0 X^0$ are indistinguishable. We can define a parameter

$$R = \frac{N(X^0 X^0) + N(\bar{X}^0 \bar{X}^0)}{N(X^0 \bar{X}^0) + N(\bar{X}^0 X^0)} \tag{16.75}$$

which can be measured experimentally. $N(X^0 X^0)$ can be identified by some convenient final states (e.g. two charged leptons $l^- l^-$). If $X \bar{X}^0$ pair is produced incoherently (for example not through a resonance of definite spin and parity and $C$-parity), then

$$R = \frac{\bar{\chi}(1-\chi) + \chi(1-\bar{\chi})}{(1-\bar{\chi})(1-\chi) + \chi\bar{\chi}}. \tag{16.76}$$

Neglecting $CP$-violation effects, i.e. using $\overline{\chi} = \chi$, we get

$$R = \frac{2\chi(1 - \chi)}{(1 - \chi)^2 + \chi^2}$$
$$= \frac{2r}{1 + r^2} \tag{16.77}$$

Now suppose that $X^0\bar{X}^0$ are produced through a resonance with $J^{PC} = 1^{--}$, for example

$$e^-e^+ \rightarrow \Upsilon \rightarrow B^0\bar{B}^0.$$

For this case we have to consider a state with $C = -1$ viz

$$[|\Psi(t)\rangle \,|\bar{\Psi}(t)\rangle - |\bar{\Psi}(t)\rangle \,|\Psi(t)\rangle].$$

If the two decays take place at $t_1$ and $t_2$, then neglecting $CP$-violation, we have from Eqs. (16.64) and (16.65):

$$|\Psi(t_1)\rangle \,|\bar{\Psi}(t_2)\rangle - |\bar{\Psi}(t_1)\rangle \,|\Psi(t_2)\rangle$$
$$= (g_+(t_1)g_-(t_2) - g_-(t_1)g_+(t_2)) \,|X^0 X^0\rangle$$
$$+ (g_-(t_1)g_-(t_2) - g_+(t_1)g_+(t_2)) \,|X^0 \bar{X}^0\rangle$$
$$+ (g_+(t_1)g_+(t_2) - g_-(t_1)g_-(t_2)) \,|\bar{X}^0 X^0\rangle$$
$$+ (g_-(t_1)g_+(t_2) - g_+(t_1)g_-(t_2)) \,|\bar{X}^0 \bar{X}^0\rangle \tag{16.78}$$

where

$$g_\pm(t) = \left[e^{-im_1 t} \exp\left(-\frac{1}{2}\Gamma_1 t\right) \pm e^{-im_2 t} \exp\left(-\frac{1}{2}\Gamma_2 t\right)\right]$$
$$= e^{-im_1 t} \exp\left(\frac{-1}{2}\Gamma t\right) \left[\exp(\frac{i}{2}\Delta m t) \exp(\frac{1}{4}\Delta\Gamma t)\right.$$
$$\left. \pm \exp(\frac{-i}{2}\Delta m t) \exp\left(-\frac{1}{4}\Delta\Gamma t\right)\right] \tag{16.79}$$

Hence we have

$$\left.\begin{array}{l} N\left(X^0 X^0\right) \\ N\left(X^0 \bar{X}^0\right) \end{array}\right\} = \int_0^\infty dt_1 \int_0^\infty dt_2 \,|g_\pm(t_1)g_\mp(t_2) - g_\mp(t_1)g_\pm(t_2)|^2$$
$$= 4 \int_0^\infty dt_1 \int_0^\infty dt_2 \, e^{-\Gamma(t_1+t_2)} \left[\exp\left(\frac{1}{2}\Delta\Gamma(t_2 - t_1)\right)\right.$$
$$\left. + \exp\left(-\frac{1}{2}\Delta\Gamma(t_2 - t_1)\right) \pm 2\Re \exp\left(-i\Delta m(t_2 - t_1)\right)\right]$$
$$= 4 \left[\frac{2}{\Gamma^2 - \frac{1}{4}(\Delta\Gamma)^2} \pm \frac{2}{\Gamma^2 + (\Delta m)^2}\right] \tag{16.80}$$

Noting from Eq. (16.78) that $N(X^0 X^0) = N(\bar{X}^0 \bar{X}^0)$ and $N(X^0 \bar{X}^0) = N(\bar{X}^0 X^0)$, we get [cf. Eq. (16.69) with $\epsilon = 0$]

$$R = \frac{N(\bar{X}^0 X^0)}{N(X^0 X^0)} = \frac{(\Delta m)^2 + \frac{1}{4}(\Delta \Gamma)^2}{2\Gamma^2 + (\Delta m)^2 - \frac{1}{4}(\Delta \Gamma)^2} = r \qquad (16.81)$$

## 16.5   $K^0 - \bar{K}^0$ Complex and $CP$-Violation in $K$-Decay

We now apply the general formalism developed in Sec. 16.4 to the $K^0 \bar{K}^0$ system. Here we denote $K_1$ and $K_2$ as $K_S$ and $K_L$. First we discuss hypercharge oscillations. Suppose that at $t = 0, K^0$ $(Y = 1)$ is produced by the reaction $\pi^- p \rightarrow K^0 \Lambda^0$. The initial state is then pure $Y = 1$. It is clear from Eq. (16.64) [with $X = K$] that a kaon beam which has been produced in a pure $Y = 1$ state has changed into one containing both the parts with $Y = 1$ and $Y = -1$. Experimentally $\bar{K}^0$ can be verified through the observation of hadronic signature such as $\bar{K}^0 p \rightarrow \pi^+ \Lambda^0$ since $\pi^+ \Lambda^0$ can only be produced by $\bar{K}^0$ and not by $K^0$. The probability of finding $Y = -1$ component at time $t$ in the kaon produced at $t = 0$ in a pure $Y = 1$ state is given by Eq. (16.67) $[|\varepsilon| << 1]$.

$$P(K^0 \rightarrow \bar{K}^0, t) \equiv |\langle \bar{K}^0 | \psi(t) \rangle|^2 \simeq \frac{1}{4} \left\{ \exp\left(-\frac{t}{\tau_S}\right) + \exp\left(-\frac{t}{\tau_L}\right) \right.$$
$$\left. -2 \exp\left[-\frac{1}{2}\left(\frac{t}{\tau_S} + \frac{t}{\tau_L}\right)\right] \cos \Delta mt \right\} \qquad (16.82)$$

where $\Delta m = m_L - m_S$ since $\tau_L = (5.17 \pm 0.14)10^{-8}s$ is much larger than $\tau_S = (0.8935 \pm 0.0008)10^{-10}s$,

$$P\left(K^0 \rightarrow \bar{K}^0, t\right) \simeq \frac{1}{4}\left(1 + e^{-t/\tau_S} - 2e^{-\frac{1}{2}t/\tau_S}\cos\left(\Delta m\right)t\right) \qquad (16.83)$$

If kaons were stable $(\tau_L, \tau_S \rightarrow \infty)$, then,

$$P\left(K^0 \rightarrow \bar{K}^0, t\right) = \frac{1}{2}\left[1 - \cos\left(\Delta m\right)t\right] \qquad (16.84)$$

which showed that a state produced as pure $Y = 1$ state at $t = 0$ continuously oscillates between $Y = 1$ and $Y = -1$ state with frequency $\omega = \frac{\Delta m}{\hbar}$ and period of oscillation,

$$\tau = \frac{2\pi}{(\Delta m/\hbar)}. \qquad (16.85)$$

Kaons, however, decay and their oscillations are damped.

By measuring the period of oscillation, $\Delta m$ can be determined:

$$\Delta m = m_L - m_S = (3.489 \pm 0.008) \times 10^{-12} \text{ MeV.} \qquad (16.86a)$$

Such a small number is measured as a consequence of quantum mechanical phenomena of interferometry. On the other hand

$$\Delta \Gamma = \Gamma_L - \Gamma_S \simeq -\Gamma_S \simeq -2\Delta m \qquad (16.86b)$$

We now discuss $CP$ violation in $K^0 - \bar{K}^0$ mixing. As seen in Sec. 16.1, experimentally it was found that long lived $K_2^0$ does decay to $\pi^+ \pi^-$ showing that $CP$ is not conserved; but the probability is quite small [cf. Eq. (16.5)]. Small $CP$ non-conservation can be taken into account by defining,

$$|K_S\rangle = |K_1^0\rangle + \epsilon |K_2^0\rangle$$
$$|K_L\rangle = |K_2^0\rangle + \epsilon |K_1^0\rangle \qquad (16.87)$$

where $\epsilon$ is a small number. Thus $CP$ non-conservation manifests itself by the ratio:

$$\eta_{+-} = \frac{A(K_L \to \pi^+ \pi^-)}{A(K_S \to \pi^+ \pi^-)} \qquad (16.88)$$

Now $CP$ non-conservation implies,

$$M_{12} \neq M_{12}^*, \qquad \Gamma_{12} \neq \Gamma_{12}^*. \qquad (16.89)$$

Since $CP$ violation is a small effect, therefore,

$$\Im M_{12} \ll \Re M_{12} \qquad \Im \Gamma_{12} \ll \Re \Gamma_{12}. \qquad (16.90)$$

Now from Eq. (16.51)

$$p^2, q^2 \approx \left( \Re M_{12} - \frac{i}{2} \Re \Gamma_{12} \right) \left[ 1 \pm \frac{i \Im M_{12} + \Im \Gamma_{12}}{\Re M_{12} - \frac{i}{2} \Re \Gamma_{12}} \right] \qquad (16.91)$$

Hence from Eq. (16.59)

$$\frac{2\epsilon}{1+\epsilon} = 1 - \frac{q}{p} = \frac{i \Im M_{12} + \frac{1}{2} \Im \Gamma_{12}}{\Re M_{12} - \frac{i}{2} \Re \Gamma_{12}} \qquad (16.92)$$

Now we get from Eqs. (16.55) and (16.91), on using the approximation Eq. (16.90)

$$\Delta m = m_L - m_S = 2\Re pq = 2\Re M_{12}$$
$$\Delta \Gamma = \Gamma_L - \Gamma_S = -4\Im pq = 2\Re \Gamma_{12} \qquad (16.93)$$

Hence we get

$$\epsilon = \frac{i \Im M_{12} + \frac{1}{2} \Im \Gamma_{12}}{\Delta M - i \Delta \Gamma / 2} \qquad (16.94)$$

The parameter $\epsilon$ determines $CP$-violation due to $K^0 - \overline{K}^0$ mixing. It is important to detect the $CP$-violation, if any, in the decay amplitudes

$$A\left(K^0 \to \pi\pi(I)\right) = A_I e^{i\delta_I} \qquad (16.95)$$

and

$$A\left(\overline{K}^0 \to \pi\pi(I)\right) = A_I^* e^{i\delta_I} \qquad (16.96)$$

where we have used the relation (16.19) obtained from $CPT$ invariance. Due to Bose statistics $I = 0$ or $2$. $CP$ violation is detected by looking for a difference between $CP$-violation for the final $\pi^0\pi^0$ state and that for $\pi^+\pi^-$. Using Clebsch-Gordon (CG) coefficients,

$$A\left(K^0 \to \pi^+\pi^-\right) = \frac{1}{\sqrt{3}}\left[\sqrt{2}A_0 e^{i\delta_0} + A_2 e^{i\delta_2}\right]$$

$$A\left(K^0 \to \pi^0\pi^0\right) = \frac{1}{\sqrt{3}}\left[A_0 e^{i\delta_0} - \sqrt{2}A_2 e^{i\delta_2}\right] \qquad (16.97)$$

and the corresponding ones for $\bar{K}^0$ obtained from (16.96). The dominant decay amplitude is $A_0$ due to $\Delta I = 1/2$ rule, $|A_2/A_0| \simeq 1/22$. Neglecting terms of order $\epsilon\Re\frac{A_2}{A_0}$, $\epsilon\frac{\Im A_0}{\Re A_0}$ and $\epsilon\frac{\Im A_2}{\Re A_0}$, and ignoring the over all phase factor $e^{i\delta_0}$ we get,

$$\eta_{+-} \equiv |\eta_{+-}| e^{i\phi_{+-}} \simeq \widetilde{\epsilon} + \epsilon'$$

$$\eta_{00} \equiv |\eta_{00}| e^{i\phi_{00}} \simeq \widetilde{\epsilon} - 2\epsilon' \qquad (16.98)$$

where,

$$\widetilde{\epsilon} = \epsilon + i\frac{\Im A_0}{\Re A_0} \qquad (16.99)$$

$$\epsilon' = \frac{i}{\sqrt{2}} e^{i(\delta_2 - \delta_0)}\left(\frac{\Re A_2}{\Re A_0}\right)\left(\frac{\Im A_2}{\Re A_2} - \frac{\Im A_0}{\Re A_2}\right) \qquad (16.100)$$

The quantities $\Im A_0$, $\Im A_2$ and $\Im\widetilde{\epsilon}$ depend on the choice of phase convention. The choice $\Im A_0 = 0$ called the Wu-Yang phase convention, gives $\widetilde{\epsilon} = \epsilon$. The value of $\epsilon'$ is independent of phase convention. It follows from Eq. (16.94) and (16.86b)

$$\tan\phi_\epsilon = -2\Delta m/\Delta\Gamma \simeq \frac{\pi}{4} \qquad (16.101)$$

where we have used that $\Im(\Gamma_{12}/\Delta\Gamma)$ is negligible. Experimentally $\phi_\epsilon$ is determined to be

$$\phi_\epsilon = (43.5 \pm 0.7)^\circ \qquad (16.102)$$

Clearly $\epsilon'$ measures the $CP$-violation in the decay amplitude, since $CP$-invariance implies $A_2$ to be real. This is referred to as direct $CP$ violation. The measured quantity $\left|\frac{\eta_{00}}{\eta_{+-}}\right|^2$ is very close to unity so that one can write

$$R = \left|\frac{\eta_{00}}{\eta_{+-}}\right|^2 = \left|\frac{\epsilon - 2\epsilon'}{\epsilon + \epsilon'}\right|^2, \qquad \epsilon' \ll \epsilon$$

$$\simeq \left|1 - \frac{3\epsilon'}{\epsilon}\right|^2 \simeq 1 - 6\mathrm{Re}\,(\epsilon'/\epsilon)$$

After 35 years of experiments at Fermilab and CERN, results have converged on a definitive non-zero result for $\epsilon'$,

$$\Re\,(\epsilon'/\epsilon) = \frac{1 - R}{6} \tag{16.103}$$

$$= (1.65 \pm 0.26) \times 10^{-3}. \tag{16.104}$$

$$\sin(\phi_{00} - \phi_{+-}) = 3\Re(\epsilon'/\epsilon)\tan(\phi_\epsilon - \phi_{\epsilon'}) \tag{16.105}$$

The measurement of $R$ provides a test for $CP$ violation since in case $\Re(\epsilon'/\epsilon)$ is not zero, $CP$ violation must occur in a decay amplitude and its present value is an evidence for it. Further we note from Eq. (16.100),

$$\phi_{\epsilon'} = \delta_2 - \delta_0 + \frac{\pi}{2} \approx 42.3 \pm 1.5° \tag{16.106}$$

where numerical value is based on an analysis of $\pi\pi$ scattering. The experimental values of $CP$-violation parameters are as follows

$$|\epsilon| = (2.228 \pm 0.011) \times 10^{-3}$$

$$|\eta_{+-}| = (2.233 \pm 0.010) \times 10^{-3}$$

$$|\eta_{00}| = (2.222 \pm 0.010) \times 10^{-3} \tag{16.107}$$

$$\phi_{+-} = (43.4 \pm 0.7)°$$

$$\phi_{00} - \phi_{+-} = (0.2 \pm 0.4)° \tag{16.108}$$

Finally $[\phi_{00} - \phi_{+-}]$ can be used to test $CPT$ symmetry, which gives the relation (16.105). Using Eqs. (16.102), (16.104) and (16.106), one can limit the right-hand side of Eq. (16.108) to be under 0.02° showing the experimental value for the phase $\Delta\phi$ in Eq. (16.108) to be consistent with $CPT$, although further accuracy will be desired.

Finally what are theoretical expectations for the ratio $\epsilon'/\epsilon$. First we note that in the standard model, the tree level diagrams shown in Fig. 16.3 involve CKM elements $\lambda_u = V_{ud}^* V_{us}$ and $\lambda_u^*$ so that $A(K_L \to \pi^+\pi^-)$ involves $(\lambda_u - \lambda_u^*) = \Im \lambda_u = 0$ [cf. Eq. (16.36)]. Thus $\frac{\epsilon'}{\epsilon}$ arises from the

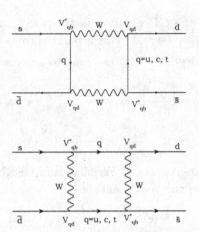

Fig. 16.2   Box diagrams for $|\Delta S| = 2$ transition.

ratio of so called "penguin" diagram shown in Fig. 16.4 to the box diagram shown in Fig. 16.2.

The $CP$-violation is determined by $\sum_i \Im \lambda_i = \Im \lambda_t$, [cf. Eq. (16.36)] where $\lambda_i = V_{id} V_{is}^*$, $i = u, c, t$. This involves t quark which belongs to the third generation and which has very small mixing with the first and the second generation. This also explains why $CP-$violation is so small in kaon decays. The theoretical prediction for $\Re (\epsilon'/\epsilon)$ is not precise [for $m_t > m_W$] but most theoretical calculations give

$$\Re\left(\frac{\epsilon'}{\epsilon}\right) < 3 \times 10^{-3} \qquad (16.109)$$

since this ratio depends on various parameters which are not as yet well determined. This is consistent with its experimental value given in Eq. (16.104).

We now discuss the $CP$-asymmetry in leptonic decays of kaon. Let us define the decay amplitudes ($l = e, \mu$)

$$K^0 \to \pi^- + l^+ + \nu_l : f \qquad \frac{\Delta S}{\Delta Q} = 1$$

The $CPT$ invariance gives

$$\overline{K}^0 \to \pi^+ + l^- + \overline{\nu}_l : f^*$$

Similarly for

$$K^0 \to \pi^+ + l^- + \overline{\nu}_l : g^*$$

$$\overline{K}^0 \to \pi^- + l^+ + \nu_l : g \qquad \frac{\Delta S}{\Delta Q} = -1 \qquad (16.110)$$

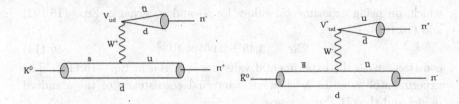

Fig. 16.3    Tree level diagrams for $K \to 2\pi$ decays.

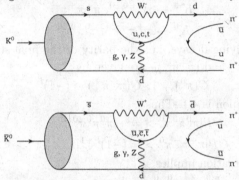

Fig. 16.4    Penguin diagrams for $K \to 2\pi$ decays.

Hence using Eqs. (16.52) and (16.59) $[X_+ = K_L, X_- = K_S]$

$$A(K_L^0 \to \pi^- + l^+ + \nu_l) = \frac{1}{\sqrt{2}}[(1+\epsilon)f + (1-\epsilon)g] \qquad (16.111)$$

$$A(K_L^0 \to \pi^+ + l^- + \bar{\nu}_l) = \frac{1}{\sqrt{2}}[(1+\epsilon)g^* + (1-\epsilon)f*]$$

The $CP$-asymmetry parameter $\delta_l$ can be written as

$$\delta_l = \frac{\Gamma(K_L^0 \to \pi^- l^+ \nu_l) - \Gamma(K_L^0 \to \pi^+ l^- \bar{\nu}_l)}{\Gamma(K_L^0 \to \pi^- l^+ \nu_l) + \Gamma(K_L^0 \to \pi^+ l^- \bar{\nu}_l)}$$

$$= \frac{2\mathrm{Re}\epsilon[|f|^2 - |g|^2]}{|f|^2 + |g|^2 + (fg^* + f^*g) + O(\epsilon^2)} \qquad (16.112)$$

In the standard model $\frac{\Delta S}{\Delta Q} = -1$ transitions are not allowed, thus $g = 0$. Hence

$$\delta_l \approx 2\mathrm{Re}\epsilon = (3.32 \pm 0.06)10^{-3}[\text{Expt. value}] \qquad (16.113)$$

Now

$$2\mathrm{Re}\epsilon = 2\,|\epsilon|\cos\phi_\epsilon$$

which, on using experimental values for $|\epsilon|$ and $\phi_\epsilon$ given in Eqs. (16.102) and (16.107)

$$2\Re\epsilon = (3.45 \pm 0.05) \times 10^{-3} \qquad (16.114)$$

consistent with the experimental value for $\delta_l$ given in Eq. (16.113). The experimental value of $\delta_l$ shows the internal consistency of the standard model and the $CPT$ invariance.

Finally we discuss $CP$-asymmetries for $K \to 3\pi$ decays. The decays

$$K^+ \to \pi^+\pi^0\pi^0,\ \pi^+\pi^+\pi^-$$

$$K^0 \to \pi^+\pi^-\pi^0,\ \pi^0\pi^0\pi^0$$

are parity conserving decays, i.e. the parity of the final state is $-1$. Now the $C$-parity of $\pi^0$ and $(\pi^+\pi^-)_{l'}$ are given by

$$C(\pi^0) = 1,\ C(\pi^+\pi^-) = (-1)^{l'}$$

and G-parity of pion is $-1$. Thus

$$CP|\pi^0\pi^0\pi^0> = -|\pi^0\pi^0\pi^0>$$

$$CP|\pi^+\pi^-\pi^0> = (-1)^{l'+1}|\pi^+\pi^-\pi^0>$$

Hence $CP$-conservation implies

$$K_2^0 \to \pi^0\pi^0\pi^0 \text{ allowed.}$$

$$K_1^0 \to \pi^0\pi^0\pi^0 \text{ is forbidden.}$$

$$K_1^0 \to \pi^+\pi^-\pi^0 \text{ allowed if } l' \text{ is odd.}$$

$$K_2^0 \to \pi^+\pi^-\pi^0 \text{ allowed if } l' \text{ is even.} \qquad (16.115)$$

Now G-parity of three pions $\pi^+\pi^-\pi^0$ :

$$G = C(-1)^I = (-1)^{l'+I} = -1$$

Hence

$$l' = \text{even},\ I(\text{odd});\ I = 1,3$$

$$l' = \text{odd},\ I(\text{even});\ I = 0,2 \qquad (16.116)$$

Only $l' = 0$ decays are favored as the decays for $l' > 0$ are highly suppressed due to centrifugal barrier. Hence $K_1^0 \to \pi^+\pi^-\pi^0$ is highly suppressed. Thus we have to take into account $I = 1,3$ amplitudes viz $a_1$ and $a_3$. $I = 3$ contribution is expected to be suppressed as it requires $\Delta I = \frac{5}{2}$ transition.

Hence $CP$-asymmetries of $K^0 \to 3\pi$ decays are given by

$$\eta_{000} = \frac{A(K_S \to \pi^0\pi^0\pi^0)}{A(K_L \to \pi^0\pi^0\pi^0)} = \frac{(a_1 - a_1^*) + \epsilon(a_1 + a_1^*)}{(a_1 + a_1^*) + \epsilon(a_1 - a_1^*)} = \frac{[i\Im a_1 + \epsilon\Re a_1]}{\Re a_1 + i\epsilon\Im a_1}$$

$$\approx \epsilon + i\frac{\Im a_1}{\Re a_1}$$

$$\eta_{+-0} = \frac{A(K_S \to \pi^+\pi^-\pi^0)}{A(K_L \to \pi^+\pi^-\pi^0)} \approx \epsilon + i\frac{\Im a_1}{\Re a_1} = \eta_{000} \qquad (16.117)$$

## 16.6  $B^0 - \bar{B}^0$ Complex

Here the general formalism developed in Sec. 16.5 is applied to $B^0 - \bar{B}^0$
complex. For $B_q^0$ ($q = d$ or $s$) we show below that both $m_{12}$ and $\Gamma_{12}$ have
the same phase. This follows from the following consideration. We can
write

$$M_{12} - i\Gamma_{12} = \langle \bar{B}_q^0 | H_{eff}^{\Delta B=2} | B_q^0 \rangle \qquad (16.118)$$

$H_{eff}^{\Delta B=2}$ induces particle-antiparticle transition. For $\Delta m_{12}$, $H_{eff}^{\Delta B=2}$ arises
from the box diagram as shown in Fig. 16.2 with $s$ replaced by $b$, where
the dominant contribution comes out from the $t$-quark. Thus,

$$M_{12} \propto (V_{tb})^2 (V_{tq}^*)^2 m_t^2 \qquad (16.119)$$

Now

$$\Gamma_{12} \propto \sum_f \langle \bar{B}^0 | H_W | f \rangle \langle f | H_W | B^0 \rangle \qquad (16.120)$$

where the sum is over all the final states which contribute to both $B^0$ and
$\bar{B}^0$ decays. The common final states for the $B_d^0$ and $\bar{B}_d^0$ decays are shown
in Fig. 16.5 while for $B_s^0$ and $\bar{B}_s^0$ can be obtained by changing $d$ to $s$.

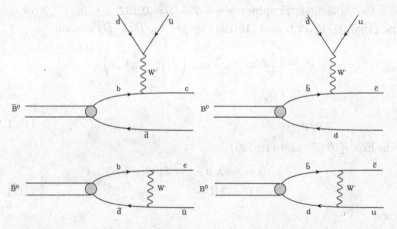

Fig. 16.5   Diagrams showing the common final states for the $B^0$ and $\bar{B}^0$ decays.

Thus,

$$\Gamma_{12} \propto \left( V_{cb} V_{cq}^* + V_{ub} V_{uq}^* \right)^2 m_b^2 \qquad (16.121)$$

Hence we have the result that,

$$\frac{|\Gamma_{12}|}{|M_{12}|} \sim \frac{m_b^2}{m_t^2} \qquad (16.122)$$

Now for $B_d^0 \to \bar{B}_d^0$ transition, using Eq. (16.36)

$$(V_{tb})^2 (V_{td}^*)^2 = A^2 \lambda^6 \left[(1-\rho)^2 + \eta^2\right] e^{2i\beta}$$

$$= \left[A\lambda^3 (1 - \rho + i\eta)\right]^2 = \left(V_{cb} V_{cq}^* + V_{ub} V_{uq}^*\right)^2 \quad (16.123)$$

Hence $M_{12} >> \Gamma_{12}$ and both $M_{12}$ and $\Gamma_{12}$ have the same phase

$$M_{12} = |M_{12}| \, e^{2i\beta}, \qquad \Gamma_{12} = |\Gamma_{12}| \, e^{2i\beta}, \qquad \phi_M = -\beta \quad (16.124)$$

On the other hand, for $B_s^0 \to \bar{B}_s^0$ transition:

$$(V_{tb})^2 (V_{ts}^*)^2 = |V_{ts}|^2 \approx A^2 \lambda^4 \quad (16.125)$$

$$M_{12} = |M_{12}|, \qquad \Gamma_{12} = |\Gamma_{12}| \quad (16.126)$$

$$\phi_M = 0 \quad (16.127)$$

Also we have,

$$\frac{\Delta m_{B_s}}{\Delta m_{B_d}} = \frac{|M_{12}|_s}{|M_{12}|_d} = \frac{1}{\lambda^2 \left[(1-\bar{\rho})^2 + \bar{\eta}^2\right]} \xi \approx 34\xi \quad (16.128)$$

where $\xi$ is $SU(3)$ breaking parameter. The numerical value is obtained using the experimental values $\lambda = 0.225$, $\bar{\rho} = 0.132$, $\bar{\eta} = 0.341$. Now using Eqs. (16.51), (16.122) and (16.124) we get for $B_d^0 - \bar{B}_d^0$ system

$$pq = \left[\left(M_{12} - \frac{i}{2}\Gamma_{12}\right) \left(M_{12}^* - \frac{i}{2}\Gamma_{12}^*\right)\right]^{1/2}$$

$$= |M_{12}| - \frac{i}{2} |\Gamma_{12}|$$

$$q/p = e^{-2i\beta} \quad (16.129)$$

From Eqs. (16.55) and (16.129)

$$\Delta m = \Delta m_{B_D} = 2 |M_{12}|$$

$$\Delta \Gamma = 2 |\Gamma_{12}| \quad (16.130)$$

Hence

$$\Delta\Gamma / \Delta m_{B_D} << 1$$

Now from Eqs. (16.51), (16.52), (16.124) and that $|\Gamma_{12}| << |M_{12}|$ the mass eigenstates $B_L^0$ and $B_H^0$ can be written as:

$$|B_L^0\rangle = \frac{1}{\sqrt{2}} \left[|B^0\rangle - e^{-2i\beta} \, |\bar{B}^0\rangle\right] \quad (16.131)$$

$$|B_H^0\rangle = \frac{1}{\sqrt{2}} \left[|\bar{B}^0\rangle + e^{-2i\beta} \, |B^0\rangle\right] \quad (16.132)$$

In this case, $CP$ violation occurs due to phase factor $e^{-2i\beta}$ in the mass matrix.

One gets (from Eq. (16.64)), using Eqs. (16.130), (16.131) and (16.132),

$$|B^0(t)\rangle = e^{-imt}e^{-\frac{1}{2}\Gamma t}\left\{\cos\left(\frac{\Delta m}{2}t\right)|B^0\rangle - ie^{-2i\beta}\sin\left(\frac{\Delta m}{2}t\right)|\bar{B}^0\rangle\right\}$$
(16.133)

Similarly we get,

$$|\bar{B}^0(t)\rangle = -e^{-imt}e^{-\frac{1}{2}\Gamma t}\left\{\cos\left(\frac{\Delta m}{2}t\right)|\bar{B}^0\rangle - ie^{2i\beta}\sin\left(\frac{\Delta m}{2}t\right)|B^0\rangle\right\}$$
(16.134)

Suppose we start with $B^0$ viz $|B^0(0)\rangle = |B^0\rangle$, the probabilities of finding $\bar{B}^0$ and $B^0$ at time $t$ are given by,

$$P(B^0 \to \bar{B}^0, t) = \left|\langle\bar{B}^0|B^0(t)\rangle\right|^2$$
$$= \frac{1}{2}e^{-\Gamma t}(1 - \cos(\Delta m)t)$$
$$P(B^0 \to B^0, t) = \left|\langle B^0|B^0(t)\rangle\right|^2$$
$$= \frac{1}{2}e^{-\Gamma t}(1 + \cos(\Delta m)t)$$
(16.135)

These are equations of a damped harmonic oscillator, the angular frequency of which is,

$$\omega = \frac{\Delta m}{\hbar}$$

From this by measuring the frequency of the oscillation one can determine $\Delta m$. Now the mixing parameter,

$$r = \frac{\chi}{1-\chi} = \frac{\int_0^T \left|\langle\bar{B}^0|B^0(t)\rangle\right|^2 dt}{\int_0^T \left|\langle B^0|B^0(t)\rangle\right|^2 dt}$$
$$\xrightarrow{T\to\infty} \frac{x^2}{2+x^2} = \frac{(\Delta m/\Gamma)^2}{2+(\Delta m/\Gamma)^2}$$
(16.136)

where we have used Eq. (16.69) and have neglected $(\Delta\Gamma/\Delta m_B)^2$, in view of Eq. (16.130).

Experimentally, for $B_d^0$ and $B_s^0$,

$$\Delta m_{B_d^0} = (0.507 \pm 0.005) \times 10^{-12} s^{-1} = (3.337 \pm 0.033) \times 10^{-10} \text{MeV}$$

$$\tau_{B_d^0} = (1.525 \pm 0.009) \times 10^{-12} s \qquad (16.137a)$$

$$\Delta m_{B_s^0} = (117 \pm 0.8) \times 10^{-12} s^{-1} = (1.17 \pm 0.01) \times 10^{-10} \text{MeV}$$

$$\tau_{B_s^0} = (1.472 \pm^{0.0024}_{0.0026}) \times 10^{-12} s \qquad (16.137b)$$

$$x_d = \left( \frac{\Delta m_{B_d^0}}{\Gamma_{B_d^0}} \right) = 0.77 \pm 0.008 \qquad (16.137c)$$

$$x_s = \left( \frac{\Delta m_{B_s^0}}{\Gamma_{B_s^0}} \right) = 26.02 \pm 0.5 \qquad (16.137d)$$

We also note from Eq. (16.119) that in the framework of standard model

$$\frac{\Delta m_{B_s^0}}{\Delta m_{B_d^0}} = \frac{m_{B_s}}{m_{B_d}} \xi^2 \left| \frac{V_{ts}}{V_{td}} \right|^2 \qquad (16.138)$$

where $\xi$ is an $SU(3)$ flavor symmetry breaking factor, extracted from Lattice QCD to be $\xi = 1.20 \pm^{0.047}_{0.035}$. From the above relation $\left| \frac{V_{ts}}{V_{td}} \right|$ can be determined. Non zero values of $x_d$ and $x_s$ clearly show mixing between $B_q$, $B_{\bar{q}}(q = s, d)$. The large value of the $x_s$ compared to $x_d$ is in conformity with Eq. (16.128).

From Eqs. (16.133) and (16.134), the decay amplitudes for $B^0(t) \to f$ and $\bar{B}^0(t) \to \bar{f}$

$$A_f(t) = \langle f | H_w | B^0(t) \rangle$$
$$\bar{A}_{\bar{f}}(t) = \langle \bar{f} | H_w | \bar{B}^0(t) \rangle \qquad (16.139)$$

are given by,

$$A_f(t) = e^{-imt} e^{-\frac{1}{2}\Gamma t} \left\{ \cos\left( \frac{\Delta m}{2} t \right) A_f - i e^{-2i\beta} \sin\left( \frac{\Delta m}{2} t \right) \bar{A}_f \right\} \quad (16.140)$$

$$\bar{A}_{\bar{f}}(t) = -e^{-imt} e^{-\frac{1}{2}\Gamma t} \left\{ \cos\left( \frac{\Delta m}{2} t \right) \bar{A}_{\bar{f}} - i e^{+2i\beta} \sin\left( \frac{\Delta m}{2} t \right) A_{\bar{f}} \right\}$$

$$\qquad (16.141)$$

These equations give the decay rates,

$$\Gamma_f(t) = e^{-\Gamma t} \left\{ \frac{1}{2} \left( |A_f|^2 + |\bar{A}_f|^2 \right) + \frac{1}{2} \left( |A_f|^2 - |\bar{A}_f|^2 \right) \cos \Delta m t \right.$$

$$\left. - \frac{i}{2} \left( 2i\Im e^{-2i\beta} A_f^* \bar{A}_f \right) \sin \Delta m t \right\} \qquad (16.142)$$

$$\bar{\Gamma}_{\bar{f}}(t) = e^{-\Gamma t} \left\{ \frac{1}{2} \left( |A_{\bar{f}}|^2 + |\bar{A}_{\bar{f}}|^2 \right) - \frac{1}{2} \left( |A_{\bar{f}}|^2 - |\bar{A}_{\bar{f}}|^2 \right) \cos \Delta m t \right.$$

$$\left. + \frac{i}{2} \left( 2i\Im e^{-2i\beta} A_{\bar{f}}^* \bar{A}_{\bar{f}} \right) \sin \Delta m t \right\} \qquad (16.143)$$

For $\Gamma_{\bar{f}}$ and $\bar{\Gamma}_f$ change $f \to \bar{f}$ and $\bar{f} \to f$ in $\Gamma_f$ and $\bar{\Gamma}_{\bar{f}}$ respectively.

As a simple application of the above equations, consider the semi-leptonic decays of $B^0$,

$$B^0 \to l^+ \nu X^- : f \quad \text{for example } X^- = D^-$$
$$\bar{B}^0 \to l^- \bar{\nu} X^+ : \bar{f} \quad \text{for example } X^+ = D^+$$

In the standard model, $\bar{B}^0$ decay into $l^+\nu X^-$ and $B^0$ decay into $l^-\bar{\nu}X^+$ are forbidden. Thus, $\bar{A}_f = 0$, $A_{\bar{f}} = 0$ and we have

$$\Gamma_f(t) = e^{-\Gamma t} \frac{1}{2} |A_f|^2 \left(1 + \cos \Delta mt\right)$$

$$\Gamma_{\bar{f}}(t) = e^{-\Gamma t} \frac{1}{2} |\bar{A}_{\bar{f}}|^2 \left(1 - \cos \Delta mt\right) \tag{16.144}$$

And since $|\bar{A}_{\bar{f}}| = |A_f|$ because of $CPT$ invariance [cf. Eq. (16.19)] we have

$$\delta = \frac{\int_0^\infty \Gamma_{\bar{f}}(t)dt}{\int_0^\infty \Gamma_f(t)dt} = \frac{x_d^2}{2 + x_d^2} = r_d \tag{16.145}$$

Non-zero value of $\delta$ would indicate mixing. If, however, $\bar{A}_f \neq 0$ and $A_{\bar{f}} \neq 0$ due to some exotic mechanism, then $\delta \neq 0$ even without mixing. Now

$$\frac{\Gamma\left(\mu^- X^+\right)}{\Gamma\left(\mu^+ X^-\right) + \Gamma\left(\mu^- X^+\right)} = \frac{r_d}{1 + r_d} = \chi_d \tag{16.146}$$

$$= 0.172 \pm 0.010 \ (Expt \ value)$$

which gives,

$$x_d = 0.723 \pm 0.032 \tag{16.147}$$

in agreement with $x_d$ given in Eq. (16.136c).

## 16.7 *CP*-Violation in *B*-Decays

In this section, we discuss the $CP$-violation for $B \to f, \bar{f}$ when $f$ and $\bar{f}$ are eigenstates of $CP$

$$|\bar{f}\rangle = CP|f\rangle = \eta_{CP}^f |f\rangle$$

$$\eta_{CP}^f = \pm 1 \tag{16.148}$$

For this case we get, from Eqs. (16.142) and (16.143),

$$\mathcal{A}_f(t) = \frac{\Gamma_f(t) - \bar{\Gamma}_f(t)}{\Gamma_f(t) + \bar{\Gamma}_f(t)} = \left\{ \cos\left(\Delta mt\right) \left(|A_f|^2 - |\bar{A}_f|^2\right) \right.$$

$$\left. -i\sin\left(\Delta mt\right) \left(i2\eta_{CP}^f \Im e^{-2i\beta} A_f^* \bar{A}_f\right) \right\} / \left(|A_f|^2 + |\bar{A}_f|^2\right)$$

$$= \cos\left(\Delta mt\right) C_f + \eta_{CP}^f \sin\left(\Delta mt\right) S_f \tag{16.149}$$

where,

$$C_f = \frac{1 - \left|\bar{A}_f\right|^2 / \left|A_f\right|^2}{1 + \left|\bar{A}_f\right|^2 / \left|A_f\right|^2} = \frac{1 - |\lambda_f|^2}{1 + |\lambda_f|^2} \qquad \lambda_f = \frac{\bar{A}_f}{A_f} \qquad (16.150)$$

This is the direct $CP$ violation and,

$$S_f = \frac{2\Im \left(e^{-2i\beta} \lambda_f\right)}{1 + |\lambda_f|^2} \qquad (16.151)$$

is the mixing induced $CP$-violation.

If the amplitude with one CKM phase dominates, then Eq. (16.18) and $CPT$ relation (16.19) $(f = \bar{f})$ give,

$$\lambda_f = \frac{\bar{A}_f}{A_f} = \frac{e^{i(\phi + \delta_f)}}{e^{i(-\phi + \delta_f)}} = e^{2i\phi} \qquad (16.152)$$

where $2\phi$ is the weak phase difference between the decays $B^0 \to f$ and $B^0 \to \bar{B}^0 \to f$. Hence from Eqs. (16.149) and (16.152), we obtain, $C_f = 0$ and

$$\mathcal{A}_f(t) = \eta_{CP}^f \sin(\Delta m t) \sin(-2\beta + 2\phi) \qquad (16.153)$$

In particular for the decay, $B^0 \to J/\psi K_s$, $\eta_{CP}^f = 1$ and in the standard model from the transition $b \to c\bar{c}s$, it is easy to see that $\phi = 0$ since the CKM elements involved are $V_{cb}^* V_{cs}$ which do not involve any CKM phases [cf. Eq. (16.36)]. Thus $S_f = -\sin 2\beta$ and from Eq. (16.149)

$$\mathcal{A}_{\psi K_s} = \frac{\int_0^\infty \left[\Gamma_f(t) - \bar{\Gamma}_f(t)\right] dt}{\int_0^\infty \left[\Gamma_f(t) + \bar{\Gamma}_f(t)\right] dt}$$

$$\mathcal{A}_{\psi K_s} = -\sin(2\beta) \frac{(\Delta m/\Gamma)}{1 + (\Delta m/\Gamma)^2} \qquad (16.154)$$

$\mathcal{A}_{\psi K_s}$ has been experimentally measured. It gives,

$$\sin 2\beta = 0.673 \pm 0.023 \qquad (16.155)$$

Theoretically this is the cleanest way of obtaining $\sin 2\beta$. One may remark that final states in decays $B^0 \to \phi K_s$ and $B^0 \to \eta' K_s$ are also eigenstates of the CP and as such they also provide $\sin 2\beta$ measurements. This is because although these decays are penguin dominated, they have the same phase (namely $\phi = 0$ in the standard model) as that for the $b \to c\bar{c}s$ at the tree level decays, since $V_{tb}^* V_{ts} = -V_{cb}^* V_{cs}(1 + O(\lambda^2))$ [cf. Eq. (16.36)]. Thus such modes may provide a way to look for new physics which contribute to the amplitude a different weak phase which makes $S_f \neq -\sin 2\beta$ and $C_f \neq 0$. Experimentally

$$S_{\eta' K_s} = 0.6 \pm 0.07 \qquad (16.156)$$

We now discuss the direct $CP$-violation in $B$ decays. As discussed in Sec. 16.2, this would require two contributions with different CKM phases. Consider for example $B^0 \to \pi\pi$ decay where final states are $CP$ eigenstates with $\eta_{CP} = 1$ and involve $b \to u\bar{u}d$ decay [see Fig. 16.6]. At the tree level CKM elements involved are $V_{ub}^* V_{us}$ which contain phase $\gamma$ [cf. Eq. (16.36)] so that from, Eqs. (16.151) and (16.152) are

$$S_{\pi^+\pi^-} = -\sin(-2\beta - 2\gamma) = \sin 2\alpha$$

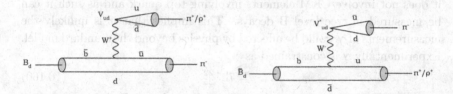

Fig. 16.6   Quark level diagrams for $B \to \pi^+(\rho^+)\pi^-$ decays.

However, there is a sizable contribution of $b$-$d$ penguin amplitudes, which have different CKM phase than $b \to u\bar{u}d$ tree amplitudes in $B^0 \to \pi\pi$ decays. Thus $S_{\pi^+\pi^-}$ does not measure $\sin 2\alpha$, but

$$S_{\pi^+\pi^-} = \sqrt{1 - C_{\pi^+\pi^-}^2}\,\sin(2\alpha - \delta_{+-})$$

where $\delta_{+-}$ is the phase difference between $e^{2i\gamma} A_{\pi^+\pi^-}$ and $A_{\pi^+\pi^-}$.

An isospin analysis gives a sum rule [see problem 15.3]

$$\frac{1}{\sqrt{2}} A_{\pi^+\pi^-} = A_{\pi^0\pi^0} + A_{\pi^+\pi^0} \qquad (16.157)$$

similar to the one for $K \to 2\pi$ decays. The sum rule (16.157) follows from the fact that $B^0 \to \pi\pi$ decay are s-wave (orbital $A.M = 0$) and Bose statistics require $I = 0, 2$ and $A_{\pi^+\pi^0}$ gets contribution from $I = 2$ only. A global isospin analysis, using the branching fraction of all these modes give

$$C_{\pi^+\pi^-} = 0.38 \pm 0.06, \quad S_{\pi^+\pi^-} = -0.65 \pm 0.07$$
$$C_{\pi^0\pi^0} = 0.43^{+0.25}_{-0.24} \qquad (16.158)$$

The analysis however puts on a loose constraint on $\alpha$.

Finally we discuss $B \to \rho\pi$, which is an example where final state is, e.g. $B^0 \to \rho^+\pi^-$ is not a $CP$ eigenstate, but this decay proceeds via the same quark level diagram [see Fig. 16.6] as $B^0 \to \pi^+\pi$ and likewise $B^0$ and $\overline{B}^0$ decay to $\rho^+\pi^-$ and thus involve the CKM phase $\alpha$. These are for amplitudes $B^0 \to \rho^\pm\pi^\mp$ and $\overline{B}^0 \to \rho^\mp\pi^\pm$ with decay rates given in Eqs. (16.142) and (16.143). The analysis of these decays allows the extraction of $\alpha$ with a single discrete ambiguity $\alpha \to \alpha + \pi$. $\alpha$ is constrained to

$$\alpha = 89.0 \pm^{4.4}_{4.2} \tag{16.159}$$

Finally, since [see unitarity triangle, Fig. 16.1]

$$\gamma = -\arg\left(-\frac{V_{ud}V_{ub}^*}{V_{cd}V_{cb}^*}\right)$$

it does not involve CKM elements involving top quark and as such it can be measured in tree level B decays. This implies that it is unlikely the measurements of $\gamma$ would be affected by physics beyond the standard model. Experimentally $\gamma$ is constrained as

$$\gamma = 73^{+22}_{-25} \tag{16.160}$$

To conclude we have discussed only typical modes of $B$ to illustrate how various $CP$ violating parameters can be extracted. There are many other modes, for which the reader is referred to the literature, some of which are listed at the end of this chapter. [See in particular Secs. 11 and 12 of Ref. [27] on which this section is mainly based.]

## 16.8   *CP*-Violation in Hadronic Weak Decays of Baryons

So far we have discussed the $CP$ violation in $K^0 - \overline{K}^0, B_q^0 - \overline{B}_q^0$ systems. There is a need to study $CP$ violation outside these systems. The hadronic weak decays of baryons and antibaryons provide another framework to study $CP$ violation.

The hadronic weak decay

$$N(p) \to N(p') + \pi(q)$$

is described by the amplitude $[\mathbf{n} = \mathbf{p}'/|\mathbf{p}'|]$

$$M_S = \overline{u}(\mathbf{p}')[A - \gamma_5 B]u(\mathbf{p}) \sim \chi^\dagger[a_s + a_p\sigma.\mathbf{n}]\chi \tag{16.161}$$

(Note here we have designated a baryon by N, not to confuse with a B-meson and $\pi$ is any pseudoscalar meson.) The relationships between $A(B)$ and $a_s(a_p)$ are given in Sec. 11.2.4.

Under charge conjugation $(C)$:

$$u(\mathbf{p}) \to C\overline{v}^T(\mathbf{p}), \quad C = i\gamma^0\gamma^2$$

Under space reflection $(P)$:

$$u^{(r)}(\mathbf{p}) \to u^{(r)}(-\mathbf{p}) = \gamma^0 u(\mathbf{p})$$

Under time reversal $(T)$:

$$u^{(r)}(\mathbf{p}) \to u^{*(-r)}(-\mathbf{p}) = B u^{(r)}(\mathbf{p}), \quad B = \gamma^1\gamma^3$$

Thus, under these transformations

$$M_f \overset{CP}{\to} -\overline{v}(\mathbf{p})[A + \gamma_5 B]v(\mathbf{p}') = \overline{M_{\overline{f}}} \sim \chi^\dagger\,(-a_s + a_p\sigma.\mathbf{n}) \qquad (16.162a)$$

$$M_f \overset{T}{\to} \overline{u}(\mathbf{p}')[A^* - \gamma_5 B^*]u(\mathbf{p}) \qquad (16.162b)$$

$$M_f \overset{CPT}{\to} -\overline{v}(\mathbf{p})[A^* + \gamma_5 B^*]v(\mathbf{p}') = \overline{M_{\overline{f}}} \qquad (16.162c)$$

When final state interactions are taken into account, the partial wave amplitudes $a_s$ and $a_p$ acquire strong final state phases $e^{i\delta_f^s}$ and $e^{i\delta_f^p}$ respectively. Thus with final state interactions

$$_{out}\langle f\,|H|\,B\rangle \overset{CPT}{\to} _{in}\langle\overline{f}\,|H|\,\overline{B}\rangle^* = e^{2i\delta_f}\,_{out}\langle\overline{f}\,|H|\,\overline{B}\rangle^* \qquad (16.163)$$

Hence under $CPT$

$$M_f \overset{CPT}{\to} -\overline{v}(\mathbf{p}')[e^{2i\delta_f^s}A^* - \gamma_5 e^{2i\delta_f^p}B^*]v(\mathbf{p})$$
$$= \overline{M_{\overline{f}}} \sim \chi^\dagger[-e^{2i\delta_f^s}a_s^* + e^{2i\delta_f^p}a_p^*\sigma.\mathbf{n}]\chi \qquad (16.164)$$

Hence from Eq. (16.162a) and the definition of decay rate $\Gamma$, asymmetry parameter $\alpha$, the transverse polarization of final baryon $\beta$ and the longitudinal polarization of the final baryon $\gamma$ given in Sec. 10.2.4, we conclude that $CP$ symmetry gives

$$\overline{\Gamma} = \Gamma, \qquad \overline{\alpha} = -\alpha, \qquad \overline{\beta} = -\beta \qquad (16.165)$$

Again from the definition of these parameters and Eq. (16.164), we note that both $CP$ and $CPT$ invariance give the same result given in Eq. (16.165), unless the S-wave amplitude A and P-wave amplitude B have different weak phases. Hence to leading order, $CP$-odd observables are

$$\delta\Gamma = \frac{\Gamma - \overline{\Gamma}}{\Gamma + \overline{\Gamma}}$$

$$\delta\alpha = \frac{\alpha + \overline{\alpha}}{\alpha - \overline{\alpha}}$$

$$\delta\beta = \frac{\beta + \overline{\beta}}{\beta - \overline{\beta}} \qquad (16.166)$$

Such asymmetries can be measured in the proposed Super-Lear accelerator in

$$p\bar{p} \;\to\; \Lambda\overline{\Lambda} \;\to\; p_f\pi^- \; \bar{p}_f\pi^+ \tag{16.167}$$

where one studies the asymmetry

$$\bar{A} = \frac{N_p^+ - N_p^- + N_{\bar{p}}^+ - N_{\bar{p}}^-}{N_{\text{total}}} \;=\; \mathcal{P}_\Lambda \alpha_\Lambda \delta\alpha_\Lambda \tag{16.168}$$

Here $N_p^\pm$ is the number of protons with $(\mathbf{p}_i \times \mathbf{p}_\Lambda) \cdot \mathbf{p}_f$ greater than or less than zero. $\mathcal{P}_\Lambda$ denotes the polarization of $\Lambda$. Similarly in the reaction

$$p\bar{p} \to \; \Xi\bar{\Xi}$$
$$\to \Lambda\pi^- \; \bar{\Lambda}\pi^+$$
$$\to p_f \; \pi^-\pi^- \; \bar{p}_f \; \pi^+\pi^+ \tag{16.169}$$

the relevant asymmetry is

$$\bar{B} = \frac{\bar{N}_p^+ - \bar{N}_p^- + \bar{N}_{\bar{p}}^+ - \bar{N}_{\bar{p}}^-}{N_{\text{total}}}$$
$$= \frac{\pi}{8}\mathcal{P}_\Xi \alpha_\Lambda \beta_\Xi (\delta\alpha_\Lambda + \delta\beta_\Xi) \tag{16.170}$$

where $\bar{N}_p^\pm$ denotes number of events with $\mathcal{P}_\Xi \cdot (\mathbf{p}_f \times \mathbf{p}_\Lambda)$ greater than or less than zero.

We now discuss the isospin analysis. First we note that assuming $CPT$ only, Eq. (16.163) gives

$$a_\ell(I) \equiv \langle f_{\ell I}^{out}|H_W|B\rangle = \eta_f \, e^{2i\delta_\ell(I)} \langle \bar{f}_{\ell I}^{out} \,|H_W|\bar{B}\rangle^*$$
$$= \eta_f \, e^{2i\delta_\ell(I)}\bar{a}_\ell^* \,(I) \tag{16.171}$$

where $\delta_\ell \,(I)$ are strong phases. Selecting the phase $\eta_f$ as $(-1)^{\ell+1}$, Eq. (16.163) gives

$$a_\ell(I) = (-1)^{\ell+1}e^{2i\delta_\ell(I)} \, \bar{a}_\ell^* \,(I) \tag{16.172}$$

Denoting by $\phi_\ell(I) \; CP$ odd phases, we define

$$a_s = \sum_I e^{i(\delta_s^I + \phi_s^I)}S_I$$
$$a_p = \sum_I e^{i(\delta_p^I + \phi_p^I)}P_I \tag{16.173}$$

Then from Eq. (16.164)

$$\bar{a}_s = -\sum_I e^{i(\delta_s^I - \phi_s^I)}S_I$$
$$\bar{a}_p = \sum_I e^{i(\delta_p^I - \phi_p^I)}P_I \tag{16.174}$$

To leading order, we obtain

$$\delta\Gamma = \sqrt{2}\,\frac{S_{33}}{S_{11}}\,\sin\,(\delta_3^s - \delta_1^s)\,\sin\,(\phi_3^s - \phi_1^s)$$
$$= 0, \text{ for } \Xi, \tag{16.175}$$

since there is only one isospin final state in $\Xi$ decay and neglecting $\frac{S_{33}}{S_{11}}$ and $\frac{P_{33}}{P_{11}}$, which are very small $(\sim \frac{1}{25})$ if $\Delta I = \frac{1}{2}$ rule dominates,

$$\delta\alpha = -\tan\,(\delta_1^p - \delta_1^s)\tan\,(\phi_1^p - \phi_1^s)$$
$$\delta\beta = \cot\,(\delta_1^p - \delta_1^s)\tan\,(\phi_1^p - \phi_1^s) \tag{16.176}$$

Thus in order to get non-vanishing $\delta\Gamma$, $\delta\alpha$ and $\delta\beta$, the following conditions must be satisfied: (i) the amplitudes must have $CP$ violating phases, (ii) there must be final state phases, (iii) there must be two or more decay channels, (iv) the $CP$ phases and final state phases must be different in different channels. The expectations in the standard model are

|  | $\delta\Gamma$ | $\delta\alpha$ | $\delta\beta$ |
|---|---|---|---|
| $\Lambda \to p\pi^-$ | $10^{-6}$ | $10^{-4}$ | $3 \times 10^{-3}$ |
| $\Xi \to p\pi^-$ | $0$ | $10^{-4}$ | $10^{-3}$ |

These estimates have considerable uncertainty and are model dependent. The above observables can also be used to study the effects of extensions of the standard model.

The decays of $B(\bar{B})$ mesons to baryon-antibaryon pair $N_1 \bar{N}_2$ $(\bar{N}_1 N_2)$ and subsequent decays of $N_2, \bar{N}_2$ or $(N_1, \bar{N}_1)$ to a lighter hyperon (antihyperon) plus a meson provide a means to study $CP$-odd observables as for example in the process,

$$e^-e^+ \to B, \bar{B} \to N_1\bar{N}_2 \to N_1\bar{N}_2'\bar{\pi}, \qquad \bar{N}_1 N_2 \to \bar{N}_1 N_2'\pi$$

The decay $B \to N_1\bar{N}_2(f)$ is described by the matrix element,

$$M_f = F_q e^{i\phi}\,[\bar{u}(\mathbf{p}_1)(A_f + \gamma_5 B_f)v(\mathbf{p}_2)] \tag{16.177}$$

whereas $B \to \overline{N}_1 N_2(\bar{f})$ is described by the matrix elements

$$M'_f = F'_q e^{i\phi'}\,\left[\bar{u}(\mathbf{p}_2)(A'_{\overline{f}} + \gamma_5 B'_{\overline{f}})v(\mathbf{p}_1)\right] \tag{16.178}$$

where $F_q$ is a constant containing CKM factor, $\phi$ is the weak phase. The amplitude $A_f$ and $B_f$ are in general complex in the sense that they incorporate the final state phases $\delta_p^f$ and $\delta_s^f$ and they may also contain weak phases $\phi_s$ and $\phi_p$. Note that $A_f$ is the parity violating amplitude ($p$-wave) whereas $B_f$ is parity conserving amplitude (s-wave). The $CPT$ invariance gives the matrix elements for the decay $\bar{B} \to \bar{N}_1 N_2(\bar{f})$ :

$$\bar{M}_{\bar{f}} = F_q e^{-i\phi} \left[ \bar{u}(\mathbf{p}_2)(-A_f^* e^{2i\delta_p^f} + \gamma_5 B_f^* e^{2i\delta_s^f})v(\mathbf{p}_1) \right] \qquad (16.179)$$

if the decays are described by a single matrix element $M_f$. If $\phi_s = 0 = \phi_p$ then $CPT$ and $CP$ invariance give the same predictions viz

$$\bar{\Gamma}_{\bar{f}} = \Gamma_f, \qquad \bar{\alpha}_{\bar{f}} = -\alpha_f, \qquad \bar{\beta}_{\bar{f}} = -\beta_f, \qquad \bar{\gamma}_{\bar{f}} = \gamma_f \qquad (16.180)$$

In order to test these predictions, consider for example the decay

$$B_d^0 \to p\bar{\Lambda}_c^- \to p\bar{p}K^0 \qquad (16.181)$$

$$\bar{B}_d^0 \to \bar{p}\Lambda_c^+ \to \bar{p}p\overline{K}^0 \qquad (16.182)$$

By analyzing the final states $p\bar{p}K^0$, $\bar{p}p\overline{K}^0$ one may test $\bar{\alpha}_{\bar{f}} = -\alpha_f$ for the chamed hyperon (antihyperon) decays.

## 16.9   Problems

(1) For the decay
$$\bar{B}^0 \to K^- \pi^+ \quad : \quad \bar{A}_{-+} = (T + P) \text{ c.f. problem 15.4}$$
where
$$T = |V_{ub}||V_{cs}|e^{-i\gamma}e^{i\delta_T}|T|$$
$$P = |V_{cb}||V_{cs}|e^{i\delta_P}|P|$$
$$\bar{A}_{-+} = |V_{cb}||V_{cs}||P|e^{i\delta_P} \left[ 1 + re^{-i\gamma}e^{-i\delta_{+-}} \right]$$
$$r = \frac{|V_{ub}||V_{cs}|}{|V_{cb}||V_{cs}|} \frac{|T|}{|P|}, \qquad \delta_{+-} = \delta_P - \delta_T,$$
show that
$$A_{CP}\left(B^0 \to K^+\pi^-\right) = \frac{-2r \sin\gamma \sin\delta_{+-}}{1 + 2r\cos\gamma\cos\delta_{+-} + r^2}$$

(2) Using the result of problem 15.5, show that
$$\mathcal{C}(\phi K_s) = 0, \qquad \mathcal{S} = -\sin 2\beta$$
Neglecting the electroweak contribution, show that
$$\mathcal{C}(\rho^0 K_s) = -\mathcal{C}(\omega^0 K_s).$$

(3) The effective weak Lagrangian for the decay $\bar{B}_0 \to \pi^+\pi^-$ is
$$\mathcal{L}_W = (V_{ub}V_{ud}^*) \left[ \bar{d}\gamma^\mu \left(1 - \gamma^5\right) u \right] \left[ \bar{u}\gamma_\mu \left(1 - \gamma_5\right) b \right]$$
$$\equiv (V_{ub}V_{ud}^*) O,$$
Show that
$$\mathcal{L}_W \xrightarrow{CP} (V_{ub}V_{ud}^*) O^\dagger$$
$$\mathcal{L}_W \xrightarrow{CPT} (V_{ub}^* V_{ud}) O^\dagger.$$

## 16.10    References

1. J. F. Donoghue, X. G. He, S. Pakvsa, Phys. Rev. D 34 833 (1986); J. F. Donoghue, B. R. Holstein, and G. Valencia, Phys. Rev. B 178 319 (1986).

2. For a review, see for example $CP$ violation edited by C. Jarlskog, World Scientific (1989).

3. J. D. Bjorken, Topics in B-physics, Nucl. Phys. 11 (proc.suppl.) 325 (1989);

4. N. Isgur and M. B. Wise Phys. Lett B 232, 113 (1989) Phys. Lett. B 237, 527 (1990).

5. S. Balk, J. G. Korner, G. Thompson, F. Hussain J. Phys. $C$ 59, 283-293 (1993).

6. J. F. Donohue et.al. Phys. Rev. Lett. 77, 2187 (1996).

7. Y. Grossman and H. R. Quinn. Phys. Rev. D 58 017504 (1998);

8. M. Suzuki and L. Wofenstein, Phys. Rev. D 60, 074019 (1999).

9. M. Beneke, G. Buchalla, M. Neubart and C. T. Sachrajda, Phys. Rev. Lett, 83, 1914 (1999)

10. J.Charles. Phys Rev. D 59 054007 (1999);

11. M. Beneke, G. Buchalla, M. Neubart and C. T. Sachrajda, Nucl. Phys. B591 313 (2000);C. W. Bauer, D. Pirjol and I. W. Stewart, Phys. Rev. Lett. 87, 201806 (2001) hep-ph/0107002.

12. H. Quinn. B Physics and $CP$ violation. hep-ph/0111177 v1.

13. M. Gronau et. al. Phys. Lett B 514 315 (2001).

14. M. Beneke and M. Neuebert, Nucl. Phys. B675, 338 (2003).

15. R. D. Peccei. Thoughts on $CP$ violation. hep-ph/0209245.

16. L. Wolfenstein. $CP$ violation: The past as prologue. hep-ph/0210025.

17. Fayyazuddin, JHEP 09, 055 (2002).

18. Fayyazuddin. Phys. Rev. D 70 114018 (2004).

19. V. Page and D. London, Phys. Rev. D 70, 017501 (2004).

20. M. Gronau and J. Zupan: hep-ph/0407002, 2004 Refernces to earlier literature can be found in this reference.

21. P. Ball, R. Zweicky and W. I. Fine. hep-ph/0412079 v1.

22. M.Gronau and J.L. Rosner, hep-ph/0807.3080 v3

23. Fayyazuddin, arXiv: hep-ph/0709.3364, Phys. Rev. D 77 014007 (2008).

24. S. Faller et al., hep-ph/0809.0222 v1.

25. G. Duplancic et al., hep-ph/0801.1796 v2.

26. Fayyazuddin, arXiv: hep-ph/0909.2085, Phys. Rev. D 80,094015

(2009).

27. K. Nakamura et al., Particle Data Group, Journal of Physics, G **37**, 0750212 (2010).

# Chapter 17

# Grand Unification, Supersymmetry and Strings

## 17.1 Grand Unification

As we have seen all fundamental forces are of gauge nature. Thus they may be deduced from some generalized gauge principle. Ingredients of gauge models are

(i) Choice of gauge group
(ii) Choice of fundamental representations
(iii) If gauge symmetry is spontaneously broken, choice of Higgs sector generates mass parameters.

Gauge principle restricts the form of interaction. Also gauge model may be renormalizable if its fermion content is such that the model is anomaly free. At low energies we have a spontaneously broken $SU(2) \times U(1)$ gauge group for electroweak forces and an exact $SU_c(3)$ gauge group for the strong quark–gluon forces. Thus the standard model involves

$$G_1 \equiv \underset{g_2}{SU(2)} \times \underset{g'}{U(1)} \times \underset{g_s > g_2 > g'}{S_cU(3)}$$

The fermion content of $G_1$ for the first generation is

$$\begin{pmatrix} u \\ d \end{pmatrix}_L^a \qquad u_R^a \qquad d_R^a \qquad a = r, y, b$$
$$\;\;(2,3) \qquad (1,3) \qquad (1,3)$$
$$\begin{pmatrix} \nu_e \\ e^- \end{pmatrix}_L$$
$$\;\;(2,1) \qquad e_R$$
$$\qquad\qquad (1,1)$$

Thus we have 15 two-component fermion states per generation. The electroweak part of $G_1$ is spontaneously broken

$$G_1 \equiv SU(2)_L \times U(1) \times SU_c(3) \rightarrow G_2 \equiv U_{em}(1) \times SU_c(3).$$

Also the experimental data show that

$$\rho \equiv \left( \frac{m_W}{m_z \cos \theta_W} \right)^2 \approx 1$$

which implies that $SU_L(2) \times U(1)$ breaking predominantly occurs only through a $SU_L(2)$ Higgs doublet or doublets. Despite the fact that the above picture is capable of providing a current phenomenological description of all the observed "low energy physics", many questions given below remain:

(i) 3 independent coupling constants
(ii) no charge quatization because of $U(1)$ factor
(iii) no relation between lepton and quark masses
(iv) why are 3 generations identical in representation content but vastly different in mass ?
(v) why is the intergeneration mixing small ?
(vi) no principle limiting the number of $SU(2)$ generations $- e, \mu, \tau, \cdots$

Could the situation be improved ? Grand unification of electroweak and strong quark–gluon forces answers some of these questions but say nothing about the generation problem. The basic hypothesis is that there exists a simple group $G$

$$G \supset G_1 \equiv SU_L(2) \times U(1) \times SU_c(3)$$

which is characterized by a single coupling constant and that all interactions are generated by $G$. Quarks and leptons are in general members of the same multiplets of the group $G$. Then at some energy scale, $G$ suffers a breakdown to $G_1$:

$$G \rightarrow G_1 \rightarrow G_2 \equiv U_{em}(1) \times SU_c(3)$$
$$M_X \gg m_W \qquad \approx 100 \text{ GeV}$$

The simple groups relevant for grand unification fall into three categories (i) The unitary groups $SU_{l+1} = SU(n)$ ($l$ is the rank of the group) with $l(l+2)$ parameters or generators. (ii) The orthogonal groups $SO_{2l}$ with $l(2l-1)$ generators. (iii) The exceptional groups, in particular $E_6$ with 78 parameters.

The rank $l$ specifies the number of generators (set of Hermitian matrices) which commute among themselves and can be simultaneously diagonalized. The vector bosons associated with a gauge group belong to the adjoint representation of the group with dimensionality equal to number of parameters.

The unitary groups $SU(4)$ and $SU(5)$ of rank 3 and 4 with 15 and 24 generators have been constructed for grand unification. The orthogonal groups $SO(4)$ and $SO(6)$ with 6 and 15 generators are also of interest. In fact $SO(4)$ and $SO(6)$ are isomorphic to $SU(2) \times SU(2)$ and $SU(4)$ respectively.

For grand unification, the orthogonal group $SO_{2l} \equiv SO(4n+2)$, for $n = 2$ viz $SO(10)$ is of special interest.

In the choice of a gauge group, the fermion representation of gauge group must be anomaly free. The representation of orthogonal groups (except $SO(6)$ which is isomorphic to $SU(4)$) are all anomaly free. Only the groups $SU(n \geq 3)$ and $SO(4n+2 \geq 10)$ have complex spinor representations. The orthogonal group $SO(4n+2 \geq 10)$ have two complex spinor representations of dimensions $2^{2n}$. Thus for $SO(10)$, the spinor representations are 16 and $\overline{16}$. Hence all the two-component 15 fermion states, per generation for the standard model plus one right-handed Majorana neutrino are accommodated in the spinor representation 16 of $SO(10)$. The following chain of subgroups of $SO(10)$ are of interest.

(i)

$$SO(10) \supset SU(5) \times U(1)$$
$$SU(5) \underset{M_X}{\to} SU(3) \times SU(2) \times U(1)$$
$$16 \to \overline{5} + 10 + 1$$
$$45 \to 24 + 1 + 10 + \overline{10}$$

Hence the representation $\overline{5} + 10$ of $SU(5)$ is anomaly free

$$\overline{5} \to (\overline{3}, 1) + (1, 2)$$
$$10 \to (3, 2) + (\overline{3}, 1) + (1, 1)$$

$$24 \to \quad (3, 2) \quad + \quad (\overline{3}, 2) \quad + \quad (8, 2) \quad + \underbrace{(1, 3) + (1, 1)}_{W^{\pm},\ Z^0,\ \gamma}$$

$$\underbrace{\begin{matrix} 6 \text{ X's} & \quad 6 \text{ Y's} & \quad 8 \text{ Gluons} \\ Q = \pm 4/3 & \quad Q = \pm 1/3 & \quad Q = 0 \end{matrix}}_{\text{lepto-quarks}}$$

Baryon number and lepton number violating processes through exchange of lepto-quarks are generated. However in $SU(5)$; $\Delta(B-L) = 0$ and as such for proton decay, the allowed and forbidden channels are:

$$p \to e^+ \pi^0, p \to \nu_e^c \pi^+$$
$$p \nrightarrow e^- + (anything)^{++}, p \nrightarrow \nu_e + \pi^+$$

Charge operator

$$Q = \frac{2}{\sqrt{6}}\lambda_{15} = \frac{4}{\sqrt{6}}F_{15}$$

(ii)

$$SO(10) \supset SO(6) \times SO(4)$$
$$\supset SU_c(4) \times SU_L(2) \times SU_R(2) \quad \text{Pati-Salam group}$$

$$SU_c(4) \times SU_L(2) \times SU_R(2) \xrightarrow[M_X]{} SU_c(3) \times U(1) \times SU_L(2) \times SU_R(2)$$

The exceptional groups $E_4$ and $E_6$ are isomorphic to $SU(5)$ and $SO(10)$ respectively. The exceptional groups $E_7$ and $E_8$ have only real spinor representations. The only relevant group is $E_6$ which has **78** parameters. The adjoint representation of $E_6$ has dimensionality **78**. The spinor representation has dimensionality **27**. The following chains are of interest.

$$E_6 \supset SO(10) \times U(1)$$
$$27 \to 16 + 10 + 1$$
$$78 \to 45 + 16 + \overline{16} + 1$$

$$E_6 \supset SU(5)$$
$$27 \to (\overline{5}, 10) + (\overline{5}, 5) + (1, 1)$$

$$E_6 \supset SU_c(3) \times SU(3) \times SU(3)$$
$$27 \to (1^c, 3, \overline{3}) + (3^c, \overline{3}, 1) + \quad (\overline{3}^c, 1, 3)$$
$$\text{Leptons} \quad\quad \text{quarks} \quad\quad \text{antiquarks}$$
$$78 \to (8^c, 1, 1) + (1^c, 8, 1) + (1^c, 1, 8) + (3^c, 3, 3) + (\overline{3}^c, \overline{3}, \overline{3})$$

### 17.1.1 $q^2$ Evolution of Gauge Coupling Constants and the Grand Unification Mass Scale

At presently available energies $g_s$, $g_2$ and $g'$ are very different. How then can we have $G$ with a single coupling constant ? This is possible since due to quantum radiative corrections $g$'s are $q^2$ dependent. Thus if we have a grand unification theory (GUT), there must be a point $q^2$ where $g_s$, $g_2$ and $g'$ coincide. To see how this comes about, let us consider the $q^2$ evolution equation for the effective coupling constant in a general gauge theory [see Appendix B for more details].

$$\frac{d\alpha_g^{-1}}{d \ln q^2} = b + \ldots, \tag{17.1}$$

for $q^2 >$ masses of fermions and gauge bosons but $q^2 < M_X^2$ and

$$b = \frac{1}{4}\left(\frac{11}{3}C_2(G) - \frac{4}{3}T_f\right). \tag{17.2}$$

For $SU_c(3), C_2(G) = 3$,

$$T_f \delta_{AB} = Tr\left(\frac{\lambda^A}{2}\frac{\lambda^B}{2}\right) \text{[number of } SU_c(3) \text{ triplets]}$$

$$= \frac{1}{2}\delta_{AB}n_f, \tag{17.3}$$

where $n_f$ is the number of quark flavors, known to be six. For $SU_L(2)$ in the electroweak group,

$$C_2(G) = 2, \ T_f\delta_{rs} = Tr\left(\tau_r/2 \ \tau_s/2\right)$$

$$\times \left[\frac{1}{2} \text{ number of left-handed doublets}\right], \tag{17.4a}$$

where $\frac{1}{2}$ comes from the fact that we have only left-handed couplings. Thus

$$T_f\delta_{rs} = \frac{1}{2}\delta_{rs}\frac{1}{2}\left(2n_f\right). \tag{17.4b}$$

Here $2n_f = 12$ appears since each generation has one lepton doublet and 3 quark doublets (one for each color). For $U(1)$ group of electroweak

$$C_2 = 0, \ T_f = \frac{1}{2}\sum_f\left(\frac{1}{2}Y\right)^2, \tag{17.5a}$$

because each fermion has either left-handed or right-handed coupling. Thus

$$T_f = \frac{1}{2}\left\{\frac{n_f}{2}\frac{1}{4}\left[3\cdot\frac{1}{9} + 3\cdot\frac{1}{9} + 1 + 1 + 3\cdot\frac{16}{9} + 3\cdot\frac{4}{9} + 4\right]\right\}$$

$$= \frac{5}{6}n_f. \tag{17.5b}$$

It is convenient to introduce $g_1 = \sqrt{5/3}\,g'$. Thus for $q^2 \gg m_W^2, m_Z^2, m_t^2$

$$\frac{d\alpha_s^{-1}}{d \ln q^2} = b_s, \ b_s = \frac{1}{4\pi}\left(\frac{33}{3} - \frac{2}{3}n_f\right) > 0 \qquad (17.6)$$

$$\frac{d\alpha_2^{-1}}{d \ln q^2} = b_2, \ b_2 = \frac{1}{4\pi}\left(\frac{22}{3} - \frac{2}{3}n_f\right) > 0 \qquad (17.7)$$

$$\frac{d\alpha_1^{-1}}{d \ln q^2} = b_1, \ b_1 = \frac{1}{4\pi}\left(-\frac{2}{3}n_f\right) < 0 \qquad (17.8)$$

These renormalization group (RG) equations have the solution

$$\alpha_i^{-1}\left(q^2\right) = \alpha_i^{-1}\left(m_Z^2\right) + b_i \ln \frac{q^2}{m_Z^2} \qquad (17.9)$$

Hence as $q^2$ increases

(1) $\alpha_s\left(q^2\right)$ decreases
(2) $\alpha_2\left(q^2\right)$ also decreases but less rapidly than $\alpha_s\left(q^2\right)$
(3) $\alpha_1\left(q^2\right)$ increases

Thus, since $\alpha_1 < \alpha_2 < \alpha_s$ at available energies, at some $q^2 = m_X^2, \alpha_s, \alpha_2$ and $\alpha_1$ should coincide [see Fig. 17.1]

$$C_3^2 \alpha_s\left(M_X^2\right) = C_2^2 \alpha_2\left(M_X^2\right) = C_1^2 \alpha_1\left(M_X^2\right) = \alpha_G, \qquad (17.10)$$

where $C_3, C_2$ and $C_1$ are group theory numbers (so that the generators of the group are properly normalized) and are of order 1. For example for SU(5), $C_3^2 = C_2^2 = C_1^2 = 1$. $M_X$ is called the grand unification mass scale at which one has only one free coupling constant $\alpha_G$. Since the gauge coupling constants are supposed to merge into one in GUT, the value of $\sin^2 \theta_W$, which measures the relative strengths of $\alpha_1$ and $\alpha_2$ at $q^2 = m_Z^2$, namely

$$\frac{\alpha_1\left(m_Z^2\right)}{\alpha_2\left(m_Z^2\right)} = \frac{5}{3}\tan^2\theta_W, \ \alpha_2\left(m_Z^2\right) = \frac{\alpha\left(m_Z^2\right)}{\sin^2\theta_W}, \qquad (17.11)$$

enters into the determination of $\alpha_G$ and $M_X$. Whether the three coupling constants meet at a single point $q^2 = M_X^2$ depends on the gauge group $G$. It may be noted that to include the contribution of Higgs doublet one adds $-\frac{1}{6}n_H$ in the expression given in Eqs. (17.7) and (17.8), where $n_H$ is the number of Higgs doublets.

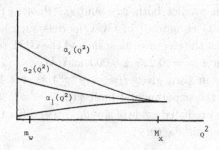

Fig. 17.1   Behavior of $\alpha_s\left(q^2\right)$, $\alpha_2\left(q^2\right)$ and $\alpha_1\left(q^2\right)$ versus $q^2$.

## 17.1.2   General Consequences of GUTS

The general consequences which one would expect from GUTS are

(1) $G$ being simple, the charge operator will be a generator of the group and traceless. So if it acts on any representation of $G$ containing quarks and leptons, it would give some relation between quark and lepton charges (sum of charges in each multiplet $= 0$), i.e. we would have charge quantization.

(2) The fact that quarks and leptons share the same representation(s) of $G$, there would be relationship between quark and lepton masses.

(3) Since quarks and leptons share the same representation(s) of $G$ and since gauge theories contain vector bosons linking all particles in a multiplet, there would in general be some interaction changing quarks into leptons, thereby violating baryon charge $(B)$ and lepton charge $(L)$ conservation. At present energy scale $E \ll E_{GUT} \approx M_X$, we have effective $B$ and $L$ conservation but this conservation cannot be exact.

In general $B$ violating forces will make proton unstable and so one has to watch that protons do not decay too quickly, the present experimental limit on proton decay is

$$\tau_p \geq 2.1 \times 10^{29} \text{ years (independent of modes)}$$
$$> 10^{31} \text{ to } 5 \times 10^{33} \text{ years (mode dependent)} \qquad (17.12)$$

Using $\alpha_3\left(m_Z\right)$ and $\alpha\left(m_Z\right)$ in $\tilde{M}_S$ renormalization scheme adopted for the definition of the coupling constants:

$$\alpha_3\left(m_Z\right) = \alpha_s\left(m_Z\right) = 0.1214 \pm 0.0031$$
$$\alpha^{-1}\left(m_Z\right) = 127.88 \pm 0.09 \qquad (17.13)$$

as inputs, one can predict both $M_X$ and $\sin^2\theta_W (m_Z)$ [cf. Eqs.(17.10) and (17.11) and RG equations] in GUT models such as $SU(5)$ with no extra scales between the electroweak scale and the GUT scale $M_X$. Typical predictions are $\sin^2\theta_W = 0.215 \pm 0.003$ and $M_X \approx \left(2^{+2}_{-1}\right) \times 10^{14}$ GeV. This value of $M_X$ in turn gives $\tau(p \rightarrow e^+\pi^0) \approx 4 \times 10^{29\pm0.7\pm1.2}$ years which contradicts the experimental limit $\tau(p \rightarrow e^+\pi^0) \geq 5 \times 10^{32}$ years. Likewise the above predicted value of $\sin^2\theta_W (m_Z)$ differ from the presently determined value of $\sin^2\theta_W (m_Z)$.

$$\sin^2\theta_W (m_Z) = 0.23124 \pm 0.00017 \qquad (17.14)$$

by six standard derivations. The same mismatch between theory (single – breaking GUT models) and low energy measurements given in Eqs. (17.13) and (17.14) is observed if one uses the three effective coupling constants from their measured values to the GUT scale and above. This is shown in Fig. 17.2, which shows that the three couplings evolved to the GUT scale do not meet at a point. This observation and the others discussed in Sec. 1.6 perhaps point to the presence of new physics between the electroweak scale and the GUT scale.

A related problem, which also requires new physics, is the so-called "hierarchy problem". The point is that the Higgs mass is subject to large quantum radiative loop corrections:

$$\delta m_H^2 = O\left(\alpha/\pi\right)\Lambda^2,$$

which is unnatural for $m_H^2 \ll \delta m_H^2$ if the cut-off $\Lambda$ used to regularize the divergent loop integrals is of the order of Planck scale ($10^{19}$ GeV), the only natural scale available in nature, unless there is some new physics at a scale $\ll$ Planck scale. Then one can interpret $\Lambda$ as a cut-off representing the energy scale at which new physics beyond the SM appears. In fact loop corrections give

$$\delta m_H^2 = \frac{G_F}{\sqrt{2}}\frac{\Lambda^2}{4\pi^2}\left[6m_W^2 + 3m_Z^2 - m_H^2 - 12m_t^2\right],$$

where

$$\frac{G_F}{\sqrt{2}} = \frac{\pi\alpha}{2m_W^2\sin^2\theta_W}.$$

Putting the masses of $W$, $Z$ bosons and $t$ quark,

$$\delta m_H^2 = -\left[\left(\frac{\Lambda}{0.7}\text{ TeV}\right)200\text{ GeV}\right]^2$$

which implies $\Lambda < 1$ TeV, if $m_H \leq 200$ GeV. Thus it is unnatural to have the masses of $W$ and $Z$ bosons at 80 and 90 GeV, respectively, and the Higgs boson mass below 200 GeV, unless the standard model is somehow "cut-off" and embedded in a richer structure that tames the ultraviolet divergence in the Higgs mass at an energy no higher than about 1 TeV. One such an alternative is provided by suppersymmetry, where there exists a symmetry between fermions and bosons; for any fermion there exists a corresponding boson of the same mass and vice versa. As a result there is a cancellation of the divergences in the radiative loop corrections since fermions and bosons contribute with opposite sign. As a result there is a scale associated with it.

Thus as seen above one consequence of suppersymmetry (see next section) is that bosonic particles are naturally paired with fermionic ones. Each minimal pairing is called a supermultiplet. For example: a left-handed fermion, its right-handed antiparticle, a complex boson and its conjugate form a chiral supermultiplet. On the other hand, a massless vector field and a left-handed fermion form a vector super-multiplet – two transversely polarized vector boson states, plus the left-handed fermion and its antiparticle. Thus for $\mathcal{N} = 1$ supersymmetry one has the following helicity states

$$\text{chiral: } (1/2, 0), \text{ gauge: } (1, 1/2), \text{ graviton: } (2, 3/2)$$

Thus in the minimal supersymmetric extension of the standard model, we have the following particles

| Particle | Spin | Spartner | Spin |
|---|---|---|---|
| quark: $q$ | 1/2 | squark: $\tilde{q}$ | 0 |
| lepton: $l$ | 1/2 | slepton: $\tilde{l}$ | 0 |
| photon: $\gamma$ | 1 | photino: $\tilde{\gamma}$ | 1/2 |
| weak vector boson: $W$ | 1 | wino: $\tilde{W}$ | 1/2 |
| weak vector boson: $Z$ | 1 | zino: $\tilde{Z}$ | 1/2 |
| Higgs: $H$ | 0 | higgsino: $\tilde{H}$ | 1/2 |
| gluon: $G$ | 1 | gluino: $\tilde{G}$ | 1/2 |

Due to the presence of supersymmetric particles the RG coefficients $b$'s given in Eqs. (17.6), (17.7) and (17.8) are modified. This modification leads to a solution such that the couplings do meet at a point [see Fig. 17.3]. The unification scale in such extensions is higher than the value of $M_X$ discussed above in the context of $SU(5)$ model. This would imply a longer life time for the proton, evading the present experimental bound.

Supersymmetry is needed from another point of view, which is discussed in the next section.

Before we end this section, we may mention that another popular GUT model, $SO(10)$ [rotation group in ten dimensions in internal space with spinor representations], when broken in a single decent to $SU_L(2) \times U(1) \times SU_C(3)$ is also in conflict with the limit (17.12). However, in contrast to $SU(5)$, $SO(10)$ admits various symmetry breaking patterns, some containing new intermediate mass scales. One such chain of symmetry breaking is

$$SO(10) \underset{M_X}{\rightarrow} SU_L(2) \times SU_R(2) \times SU_C(4)$$

$$\underset{m_R}{\rightarrow} SU_L(2) \times U(1) \times SU_C(3)$$

$$\underset{m_L}{\rightarrow} U_{em}(1) \times SU_C(3)$$

where $SU_C(4)$ is Pati-Salam group. Here it is possible to avoid the conflict with the limit (17.12). However, there is no prediction for $\sin^2 \theta_W$; in fact its value is used to fix the intermediate mass scale $m_R$ which is of the order of $10^{13}$ GeV and being so large has no observable consequences.

To conclude GUTS have several attractive features mentioned above, but their predictive power is limited. However, the idea that quarks and leptons can be treated on an equal footing, and that both lepton and baryon number violations are possible in such unified theories, is now an integral part of GUT models and their extension.

## 17.2    Poincaré Group and Supersymmetry

### 17.2.1    *Introduction*

Even though there is no conclusive evidence at present time that supersymmetry is a symmetry of the world, supersymmetry is a favored way of reducing some problems in phenomenology beyond the standard model such as unification of gauge coupling constants of standard model at GUT scale (see last section). Issues such as the fine-tuning problem due to a fundamental Higgs are naturally avoided in the supersymmetric theories since the normally large radiative corrections are softened due to the presence of the fermionic partner Higgsino. Similarly the hierarchy problem can also be resolved in this framework. Supersymmetry is seen as a positive feature of string theory since the theory requires it and one does not have to introduce it by hand.

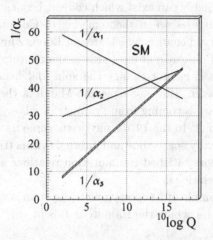

Fig. 17.2  Running of the three gauge couplings in minimal SU(5) GUT showing disagreement with a single unification point [5].

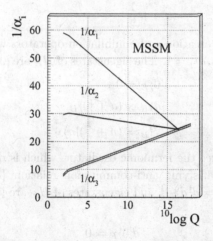

Fig. 17.3  Running of the three gauge couplings in minimal supersymmetric extension of the standard model [5].

Supersymmetry (SUSY) pair the bosons with fermions. But bosons and fermions behave very differently. Consider rotating a spin-$j$ particle by an angle $\theta$ around an axis, say $z$-axis. While bosons come back to themselves after a rotation by $2\pi$ but fermions do not! Pauli's exclusion principle hold for fermions but not for bosons. Because they behave so differently it seems unlikely that a symmetry can exist which convert bosons into fermions and vice versa. Also there is no-go theorem due to Coleman and Mandula, which forbids conserved charges which are not Lorentz invariant other than those belonging to Poincaré's group (such as momentum). But if a charge is Lorentz invariant it cannot change the spin and hence cannot change a fermion into a boson. However Coleman-Mandula theorem only talked about conserved charges coming from symmetries whose generators satisfy commutation relations. In the 1970's physicists come up with the idea of a supersymmetry to unify space-time and internal symmetries, a symmetry in which some generators satisfied commutation relations and some satisfied anti-commutation relations.

The simplest example of a supersymmetric system is provided by simple harmonic oscillator for which the Hamiltonian is $[\hbar = 1, \ m = 1, \ \omega = 1]$

$$H_B = \frac{1}{2}(p^2 + q^2)$$
$$= \frac{1}{\sqrt{2}}(q + ip)(q - ip) - \frac{1}{2}[q, p]$$
$$= a^\dagger a + \frac{1}{2}$$

where $a^\dagger$ and $a$ are creation and annihilation operators which satisfy commutation relation $[a, a^\dagger] = 1$. The eigenstates of $H_B$ are $|0\rangle_B, |1\rangle_B, |2\rangle_B, \cdots$.

$$a|0\rangle_B = 0$$
$$|n\rangle_B \propto (a^\dagger)^n |0\rangle_B$$
$$H_B = (n + \frac{1}{2})|n\rangle_B$$

We now introduce the fermionic oscillator, which is defined by the operators $d$ and $d^\dagger$ satisfying anti-commutaion relation $\{d, d^\dagger\} = 1$. The Hamiltonian is $H_F = d^\dagger d - \frac{1}{2}$ and eigenstates of $H_F$ are $|0\rangle_F$ and $d^\dagger|0\rangle_F$ so that

$$d|0\rangle_F = 0$$

For the combined system, the total Hamiltonian is

$$H = H_B + H_F$$
$$= a^\dagger a + d^\dagger d$$

and eigenstates are

$$|0\rangle, \quad |1\rangle = a^\dagger|0\rangle \quad |2\rangle = (a^\dagger)^2|0\rangle \quad \cdots$$
$$|1\rangle = d^\dagger|0\rangle \quad |2\rangle = d^\dagger|1\rangle \quad \cdots$$

Note the degeneracy of states in supersymmetric harmonic oscillator. The degeneracy in energy indicates that there is a symmetry.

Define the operator $Q = a^\dagger d$, $Q^\dagger = a d^\dagger$. $Q$ is no longer a pure bosonic or fermionic object, and $Q$ and $Q^\dagger$ provide the simplest SUSY algebra

$$\{Q, Q\} = \{Q^\dagger, Q^\dagger\} = 0$$
$$\{Q, Q^\dagger\} = H$$
$$[Q, H] = [Q^\dagger, H] = 0 \tag{17.15}$$

These relations, which can be easily derived, explain degeneracy mentioned above. The above considerations show that any enlargement of the space-time group require generators which anti-commute. A brief review of the Poincaré group, whose generators satisfy the commutation relations will be of help in introducing the spinor generators, which anti-commute.

### 17.2.2  Poincaré Group

Poincaré transformation is

$$x'^\mu = a^\mu + \Lambda^\mu_\nu x^\nu$$
$$a^\mu \ : \ \text{translation}$$
$$\Lambda^\mu_\nu \ : \ \text{Lorentz transformation} \tag{17.16}$$

The corresponding infinitesimal transformation is

$$x'^\mu = x^\mu + \delta x^\mu = x^\mu + a^\mu + \epsilon^\mu_\nu x^\nu \tag{17.17}$$

Associated with Poincaré transformation is a unitary operator:

$$U = e^{-ia^\mu P_\mu + \frac{i}{2} \epsilon^{\mu\nu} M_{\mu\nu}} \tag{17.18}$$

As discussed in Chap. 3, the ten generators of the Poincaré group, are identified as:

$P_\mu = -i\partial_\mu$ generators of translation, energy-momentum

$M_{\mu\nu} = L_{\mu\nu} + S_{\mu\nu}$ generators of the Lorentz group

where

$L_{\mu\nu} = i(x_\mu \partial_\nu - x_\nu \partial_\mu)$,

$M^{ij} = -\epsilon^{ijk} J_k = \epsilon^{ijk} J^k = M_{ij} = \epsilon_{ijk} J_k$ : generators of the rotation group

$M^{0i} = K^i = -M^{i0}$ : generators of the Lorentz boost

$$e^{\frac{i}{2} \epsilon^{\mu\nu} M_{\mu\nu}} = e^{-i\omega \cdot \mathbf{J} - i\zeta \cdot \mathbf{K}} \tag{17.19}$$

These generators should satisfy the same commutation relations which the corresponding differential generators $P_\mu$ and $L_{\mu\nu}$ satisfy. They are

$$[P_\mu, P_\nu] = 0$$
$$[M_{\mu\nu}, P_\lambda] = i(g_{\nu\lambda}P_\mu - g_{\mu\lambda}P_\nu)$$
$$[M_{\mu\nu}, M_{\lambda\rho}] = i(g_{\nu\lambda}M_{\mu\rho} - g_{\mu\lambda}M_{\nu\rho} + g_{\mu\rho}M_{\nu\lambda} - g_{\nu\rho}M_{\mu\lambda}) \quad (17.20)$$
$$M_{\mu\nu} = -M_{\nu\mu}$$

For a Dirac spinor field $\psi(x)$:

$$[M_{\mu\nu}, \psi(x)] = -(L_{\mu\nu} + \Sigma_{\mu\nu})\psi(x) \quad (17.21)$$

where

$$\Sigma_{\mu\nu} = \frac{i}{4}(\gamma_\mu\gamma_\nu - \gamma_\nu\gamma_\mu)$$
$$= \frac{i}{4}\begin{pmatrix} \sigma_\mu\bar{\sigma}_\nu - \sigma_\nu\bar{\sigma}_\mu & 0 \\ 0 & \bar{\sigma}_\mu\sigma_\nu - \bar{\sigma}_\nu\sigma_\mu \end{pmatrix}$$
$$\equiv \begin{pmatrix} \sigma_{\mu\nu} & 0 \\ 0 & \bar{\sigma}_{\mu\nu} \end{pmatrix} \quad (17.22)$$

where $\sigma_\mu = (\sigma_0, -\sigma^i)$, $\bar{\sigma}_\mu = (\sigma_0, \sigma^i)$ and $\sigma^i = -\sigma_i$.

We close the section on Poincaré group by introducing the Pauli-Lubanski operator

$$W_\lambda = -\frac{1}{2}\varepsilon_{\lambda\alpha\beta\sigma}M^{\alpha\beta}P^\sigma \quad (17.23)$$

It is a vector and plays the role of covariant spin operator. Under the Poincaré transformation $W_\lambda$ transforms as

$$[P_\mu, W_\lambda] = 0 \quad (17.24)$$
$$[M_{\mu\nu}, W_\lambda] = i(g_{\nu\lambda}W_\mu - g_{\mu\lambda}W_\nu) \quad (17.25)$$

This follows from the fact that $W_\lambda$ is a vector under the Lorentz group. It is easy to show that

$$[W_\mu, W_\nu] = i\varepsilon_{\mu\nu\lambda\sigma}W^\lambda P^\sigma \quad (17.26)$$

The Lie algebra of the Poincaré group has two invariants

$$P^2 = P^\mu P_\mu \quad (17.27)$$
$$W^2 = W^\mu W_\mu \quad (17.28)$$
$$P^\mu W_\mu = 0 \quad (17.29)$$

which gives

$$\frac{\mathbf{P}\cdot\mathbf{W}}{P_0} = W_0 = \mathbf{J}\cdot\mathbf{P} \quad (17.30)$$

For massless particle

$$W^\mu = \lambda P^\mu \qquad (17.31)$$

$$\lambda = \frac{\mathbf{J} \cdot \mathbf{P}}{|\mathbf{P}|} \qquad (17.32)$$

Thus for a massless particle, $\lambda$ is the helicity with eigenvalues $\pm s$, where for a spinor $s = \frac{1}{2}$ and for a vector particle $s = 1$.

### 17.2.3  Two-Component Weyl Spinors

The Weyl spinors $\xi$ and $\eta$ or equivalently the Chiral projections of 4-component Dirac Spinor $\psi$,

$$\psi_L = \frac{1 - \gamma_5}{2} \psi \qquad (17.33)$$

$$\psi_R = \frac{1 + \gamma_5}{2} \psi, \qquad (17.34)$$

have already been introduced in Chap. 12 and Appendix A, where it is shown that 4-component Dirac spinor can be written as

$$\psi = \begin{pmatrix} \xi \\ i\sigma^2 \chi^* \end{pmatrix} \qquad (17.35)$$

where $\chi = \xi^C = -i\sigma^2 \eta^*$ and $\eta = i\sigma^2 \xi^{C*}$.

On the other hand for the Majorana spinor, which is self conjugate

$$\psi_M = \psi_M^C = C\bar{\psi}_M \qquad (17.36)$$

$$\psi_{M\alpha} = C_{\alpha\beta} \bar{\psi}_M^\beta \qquad (17.37)$$

Here charge conjugation matrix $C$ is

$$C = \begin{pmatrix} \epsilon & 0 \\ 0 & -\epsilon \end{pmatrix} \qquad (17.38)$$

$$\epsilon = i\sigma^2$$

The fundamental spinor representation of the Lorentz group are

$$\xi = \begin{pmatrix} \xi^1 \\ \xi^2 \end{pmatrix}, \qquad \xi^* = \begin{pmatrix} \xi^{\dot{1}} \\ \xi^{\dot{2}} \end{pmatrix} \qquad (17.39)$$

designated as

$$D(\tfrac{1}{2}, 0), \; D(0, \tfrac{1}{2})$$

so that.

$$\left(\frac{1}{2}, 0\right) \otimes \left(0, \frac{1}{2}\right) = \left(\frac{1}{2}, \frac{1}{2}\right)$$

gives Lorentz vector.

It is convenient to introduce the following lowering and raising spinor metric

$$\epsilon^{\alpha\beta} = i\sigma^2 = \begin{pmatrix} 0 & 1 \\ -1 & 0 \end{pmatrix} = \epsilon^{\dot{\alpha}\dot{\beta}}$$

$$\epsilon_{\alpha\beta} = -i\sigma^2 = \begin{pmatrix} 0 & -1 \\ 1 & 0 \end{pmatrix} = \epsilon_{\dot{\alpha}\dot{\beta}} \tag{17.40}$$

$$\xi_\alpha = \epsilon_{\alpha\beta}\xi^\beta, \qquad \xi_{\dot{\alpha}} = \epsilon_{\dot{\alpha}\dot{\beta}}\xi^{\dot{\beta}} \tag{17.41}$$

Define

$$\overline{\xi} = i\sigma^2\xi^*, \qquad \overline{\chi} = i\sigma^2\chi^* \tag{17.42}$$

Hence the Dirac 4-spinor can be expressed in terms of the undotted and dotted spinors as

$$\psi = \begin{pmatrix} \xi^\alpha \\ \overline{\chi}_{\dot{\alpha}} \end{pmatrix} \tag{17.43}$$

Finally from Eqs. (17.21), (17.22) and (17.43)

$$[M_{\mu\nu}, \xi^\alpha] = -i(\sigma_{\mu\nu})^\alpha_\beta \xi^\beta$$

$$[M_{\mu\nu}, \overline{\chi}_{\dot{\alpha}}] = -i(\overline{\sigma}_{\mu\nu})^{\dot{\beta}}_{\dot{\alpha}} \overline{\chi}_{\dot{\beta}} \tag{17.44}$$

where

$$(\sigma_{\mu\nu})^\alpha_\beta = \frac{1}{4}\left[(\sigma_\mu)^{\alpha\dot{\alpha}}(\overline{\sigma}_\nu)_{\dot{\alpha}\beta} - \mu \leftrightarrow \nu\right]$$

$$(\overline{\sigma}_{\mu\nu})^{\dot{\beta}}_{\dot{\alpha}} = \frac{1}{4}\left[(\overline{\sigma}_\mu)_{\alpha\dot{\alpha}}(\sigma_\nu)^{\alpha\dot{\beta}} - \mu \leftrightarrow \nu\right] \tag{17.45}$$

### 17.2.4 *Spinor Algebra, Supersymmetry*

Introduce two Weyl spinor generators

$$Q^\alpha, \quad \overline{Q}_{\dot{\alpha}} \equiv i\sigma^2 Q^{*\alpha} \tag{17.46}$$

It follows from Eq. (17.44):

$$[M_{\mu\nu}, Q^\alpha] = -i(\sigma_{\mu\nu})^\alpha_\beta Q^\beta$$

$$[M_{\mu\nu}, \overline{Q}_{\dot{\alpha}}] = -i(\overline{\sigma}_{\mu\nu})^{\dot{\beta}}_{\dot{\alpha}} \overline{Q}_{\dot{\beta}} \tag{17.47}$$

$$[P_\mu, Q^\alpha] = 0 = [P_\mu, \overline{Q}_{\dot{\alpha}}] \tag{17.48}$$

This is not obvious (see problem 17.2).

Now $Q^\alpha$ and $\overline{Q}_{\dot\alpha}$ respectively transform as $(1/2, 0)$ and $(0, 1/2)$ under the Lorentz group. Thus $\{Q^\alpha, \overline{Q}^{\dot\beta}\}$ transforms as $(1/2, 1/2)$ under the Lorentz group and $P^\mu$ is the only vector generator of the Poincaré group. Thus to close the algebra we require $\{Q^\alpha, \overline{Q}^{\dot\beta}\}$ and $\{Q^\alpha, Q_\beta\}$ both of which are to be bosons and linear in $P_\mu$ and $M_{\mu\nu}$:

$$\{Q^\alpha, \overline{Q}^{\dot\beta}\} = t(\sigma^\mu)^{\alpha\dot\beta} P_\mu \tag{17.49}$$

$$\{Q^\alpha, Q_\beta\} = s(\sigma^{\mu\nu})^\alpha_\beta M_{\mu\nu} \tag{17.50}$$

Thus, on using Eq. (17.20)

$$[P_\lambda, \{Q^\alpha, Q_\beta\}] = s(\sigma^{\mu\nu})^\alpha_\beta [P_\lambda, M_{\mu\nu}]$$
$$= s(\sigma^{\mu\nu})^\alpha_\beta (-i)(g_{\nu\lambda} P_\mu - g_{\mu\lambda} P_\nu)$$

$$[P_\lambda, \{Q^\alpha, \overline{Q}^{\dot\beta}\}] = t(\sigma^\mu)^{\alpha\dot\beta} [P_\lambda, P_\mu] = 0$$

Since $Q^\alpha$, $Q_\beta$, $\overline{Q}^{\dot\beta}$ all commute with $P_\lambda$, the left-hand side is zero. Thus we have $s = 0$, but the second equation is identically satisfied and as such does not fix $t$. Hence we have

$$\{Q^\alpha, Q_\beta\} = 0$$
$$\{Q_\alpha, Q_\beta\} = 0 = \{Q_{\dot\alpha}, Q_{\dot\beta}\} \tag{17.51}$$

$$\{Q^\alpha, \overline{Q}^{\dot\beta}\} = 2(\sigma^\mu)^{\alpha\dot\beta} P_\mu$$
$$= 2(\overline{\sigma}^\mu)^{\dot\beta\alpha} P_\mu \tag{17.52}$$

where we have taken $t = 2$, as the normalization condition. Equations (17.47), (17.48), (17.51) and (17.52) give the supersymmetry algebra.

Finally for the 4-component Majorana spinor $Q_M = Q$ from Eqs. (17.21) and (17.22),

$$[P_\mu, Q_\alpha] = 0 \tag{17.53}$$
$$[M_{\mu\nu}, Q_\alpha] = -(\Sigma_{\mu\nu})_{\alpha\beta} Q^\beta \tag{17.54}$$
$$\{Q_\alpha, \overline{Q}_\beta\} = (\gamma^\mu)_{\alpha\beta} P_\mu \tag{17.55}$$

or equivalently

$$\{Q_\alpha, Q_\beta\} = -(\gamma^\mu C)_{\alpha\beta} P_\mu$$

since [cf. Eq. (17.37)]

$$Q_\alpha = C_{\alpha\beta} \overline{Q}^\beta.$$

### 17.2.5  *Supersymmetric Multiplets*

First we note that by taking the trace of Eq. (17.52), where only $\sigma^0$ has non-zero trace equal to 2.

$$\left( Q^1 \bar{Q}^{\dot{1}} + Q^2 \bar{Q}^{\dot{2}} + \bar{Q}^{\dot{1}} Q^1 + \bar{Q}^{\dot{2}} Q^2 \right) = 4P_0 \qquad (17.56)$$

that is

$$4 \left\langle \Psi \left| P^0 \right| \Psi \right\rangle = \sum_{\alpha=1}^{2} \left\langle \Psi \left| Q^\alpha (Q^\alpha)^* + (Q^\alpha)^* Q^\alpha \right| \Psi \right\rangle \geq 0 \qquad (17.57)$$

implying that the spectrum of $H = P_0$ is semi-positive definite. In particular for vacuum state $|0\rangle$, $E_{vac} = 0$ implies

$$\langle 0 | P_0 | 0 \rangle = 0 \Leftrightarrow Q^\alpha |0\rangle = 0 \qquad (17.58)$$

Thus the vanishing of vacuum energy is a necessary and sufficient condition for the existence of a unique vacuum.

Now the Poincaré group has two invariants

$$P^2 = P_\mu P^\mu, \ W^2 = W_\mu W^\mu = -m^2 \mathbf{J}^2 \qquad (17.59)$$

and $\mathbf{J}^2$ has eigenvalues $j(j+1)$. Now while

$$[P^2, Q^\alpha] = 0 = \left[ P^2, \bar{Q}^{\dot{\alpha}} \right] \qquad (17.60)$$

but

$$[W^2, Q^\alpha] \neq 0 \qquad (17.61)$$

Thus the massive irreducible representations of the SUSY algebra will certainly contain different spins: $j = 1/2$ super charge $Q^\alpha$ acting on a state of spin $j$ results in a state of spin $j \pm 1/2$, thereby mixing fermions and bosons.

Since $Q^\alpha$ changes fermion number by one unit, we may write

$$(-1)^{N_F} Q^\alpha = -Q^\alpha (-1)^{N_F}$$

where $N_F$ is the fermion number operator.

Now consider a finite dimensional representation $R$ of the super algebra. Then since

$$Tr \left[ (-1)^{N_F} \left\{ Q^\alpha, \bar{Q}^{\dot{\beta}} \right\} \right] \qquad (17.62)$$

$$= Tr \left[ -Q^\alpha (-1)^{N_F} \bar{Q}^{\dot{\beta}} + Q^\alpha (-1)^{N_F} \bar{Q}^{\dot{\beta}} \right] = 0, \qquad (17.63)$$

using Eq. (17.49)

$$2\,(\sigma^{\mu})^{\alpha\dot{\beta}}\,tr\left[(-1)^{N_F}\,P_{\mu}\right] = 0 \tag{17.64}$$

For a fixed non-zero $P_{\mu}$, $tr\left[(-1)^{N_F}\right] \equiv \sum_m \left\langle m\left|(-1)^{N_F}\right|m\right\rangle = 0$. Since $(-1)^{N_F}$ has value $+1$ on a bosonic state and $-1$ on a fermionic state, this means that

$$n_B\,(R) - n_F\,(R) = 0 \tag{17.65}$$

where $n_B$ is the number of bosons in the representation $R$ and $n_F$ is the number of fermions in the representation $R$.

We now consider the representations of the SUSY algebra that can be realized by one particle states. We start with the massless case, since in most of the phenomenologically interesting scenarios the non-zero masses of the particles that we observe are generated by SUSY breaking effects. As already seen in Sec. 17.2.2 for massless particle

$$\lambda = \frac{\mathbf{J}\cdot\mathbf{P}}{P_0} = \frac{\mathbf{J}\cdot\mathbf{P}}{|\mathbf{P}|} \tag{17.66}$$

is the helicity and $W^0 = \mathbf{J}\cdot\mathbf{P}$. Now consider a massless state $|p,\lambda\rangle$ with momentum $p$. Then

$$P^{\mu}|p,\lambda\rangle = p^{\mu}|p,\lambda\rangle \tag{17.67}$$

$$P_0|p,\lambda\rangle = E|p,\lambda\rangle \tag{17.68}$$

$$\frac{\mathbf{J}\cdot\mathbf{P}}{P_0}P_0|p,\lambda\rangle = E\frac{\mathbf{J}\cdot\mathbf{P}}{\mathbf{P}}|p,\lambda\rangle = E\lambda|p,\lambda\rangle \tag{17.69}$$

$$\tag{17.70}$$

i.e.

$$W^0|p,\lambda\rangle = E\lambda|p,\lambda\rangle \tag{17.71}$$

Further we note that

$$[W^0,Q^{\alpha}] = -\frac{1}{2}(\sigma\cdot\mathbf{P})^{\alpha}_{\beta}P^{\beta} \tag{17.72}$$

$$[W^0,\overline{Q}_{\dot{\alpha}}] = -\frac{1}{2}(\sigma\cdot\mathbf{P})^{\dot{\beta}}_{\dot{\alpha}}P_{\dot{\beta}} \tag{17.73}$$

From Eqs. (17.74) and (17.75):

$$W^0Q^{\alpha}\,|p,\lambda\rangle = \lambda E Q^{\alpha}\,|p,\lambda\rangle - \frac{1}{2}(\sigma.\mathbf{P})^{\alpha}_{\beta}Q^{\beta}\,|p,\lambda\rangle \tag{17.74}$$

Selecting

$$Q^1\,|p,\lambda\rangle = 0 = Q^2\,|p,\lambda\rangle$$

it is easy to derive the following result for massless state

$$Q^2 |p, \lambda\rangle = \text{constant } |p, \lambda - 1/2\rangle$$
$$= \sqrt{4E} |p, \lambda - 1/2\rangle$$
$$Q^2 |p, \lambda - 1/2\rangle = \sqrt{4E} |p, \lambda\rangle \qquad (17.75)$$

Hence there are just two states with helicity $\lambda$ and $\lambda - 1/2$. The most common of them, which we encounter are

$$\lambda = 1/2, 1, 2$$

$\lambda = 1/2$ : Chiral supermultiplet: a Weyl spinor with helicity $1/2$, and a scalar; plus $CTP$ conjugate Weyl fermion of helicity $-1/2$ and another scalar.

$\lambda = 1$ : Vector supermultiplet: a massless vector particle, a fermion of helicity $1/2$ and then $CTP$ conjugate viz $\lambda = -1$ and $\lambda = -1/2$.

$\lambda = 2$ : graviton supermuliplet $(2, 3/2)$ and $CPT$ conjugate $\lambda = -2$, $\lambda = -3/2$.

To conclude: bosonic particles are naturally paired with fermionic ones. Each minimal pairing is called supermultiplet. As we have seen above: a left-handed fermion ($\lambda = 1/2$), its right-handed antiparticle ($\lambda = -1/2$), a complex boson ($\lambda = 0$) and its conjugate form a supermultiplet. A massless vector field ($\lambda = 1$) and a left-handed fermion form a vector multiplet, two transversely polarized vector boson states, plus left-handed and right-handed fermion and its antiparticle. Thus for $\mathcal{N} = 1$ supersymmetry one has the helicity states, given in Sec. 17.1.2.

## 17.3  Supersymmetry and Strings

### 17.3.1  *Introduction*

One of the main puzzles in quantum theory is how to reconcile General Relativity with quantum mechanics. The usual method of taking the classical Lagrangian and quantizing it fails because of insurmountable difficulties in making sense of the renormalization program, which has been so successful in other quantum field theories.

In most situations the domains in which quantum field theories are interesting and the domains in which General Relativity is relevant have no overlap. General Relativity is used when dealing with massive bodies of interest at large distance scales in astrophysics and cosmology and quantum mechanics is used at short distance scales. However, there are situations where both theories become relevant. For instance, close to a black hole quantum effects become relevant as evidenced by Hawking radiation. When one begins to probe distances of the order of the Planck scale one expects that quantum gravitational effects will become important.

The impasse in the field theoretic approach to gravity can be circumvented by using string theory. String theory is a novel program which replaces the plethora of particles that exist by a single string! In this approach the vibrational modes of the string correspond to different particles. Whereas in field theory it seems virtually impossible to include dynamical gravity, in string theory quite the opposite situation prevails: one cannot have string theory without gravity! This is because in the spectrum of string theory there is always a massless spin 2 field, which is naturally identified as the graviton.

Another feature of string theory is that it requires supersymmetry. Even though there is no conclusive evidence at the present time that supersymmetry is a symmetry of the world, supersymmetry is a favored way of resolving some problems in phenomenology beyond the Standard Model as discussed in Sec. 17.2.1. Issues such as the fine-tuning problem due to a fundamental Higgs are naturally avoided in supersymmetric theories since the normally large radiative corrections due to a fundamental scalar Higgs are suppressed due to the presence of its fermionic partner, the Higgsino. Similarly the hierarchy problem can also be resolved in this framework. Supersymmetry is thus seen by many as a positive feature of string theory since the theory requires it and one does not have to introduce it by hand.

## 17.3.2 *Supersymmetry*

Space-time supersymmetry is a symmetry which generalizes ordinary Poincaré symmetry by augmenting the usual generators with fermionic generators. They satisfy certain commutation relations with the bosonic generators and anti-commutation relations with the remaining fermionic ones:

$$[Q_{\alpha i}, P_\mu] = 0,$$
$$[Q_{\alpha i}, M_{\mu\nu}] = \frac{1}{2}(\Sigma_{\mu\nu})^\alpha_\beta Q_{\beta i}, \qquad (17.76)$$
$$\{Q_{\alpha i}, Q_{\beta j}\} = -\delta_{ij}(\gamma_\mu C)_{\alpha\beta} P_\mu + C_{\alpha\beta} Z_{ij} + (\gamma_5 C)_{\alpha\beta} Z'_{ij}.$$

$P_\mu$ are generators of translations and $M_{\mu\nu}$ are Lorentz generators. Together they generate the Poincaré group. The fermionic generators $Q$ are in the

Majorana representation and $C$ is the charge conjugation matrix so that:
$$Q_{\alpha i} = C_{\alpha\beta}\overline{Q}_i^{\beta}. \qquad (17.77)$$
The index $i$ runs over the number of supersymmetries $i = 1, ..., \mathcal{N}$. In the simplest case $\mathcal{N} = 1$, the other cases are known as extended supersymmetries. The $Z$ and $Z'$ are so-called central charges, they are anti-symmetric in the indices $i, j$ and commute with everything. They only exist when one has extended supersymmetry.

One of the consequences of supersymmetry is that bosonic particles are naturally paired with fermionic ones so that the number of on-shell degrees of freedom of fermions and bosons are the same. Each minimal pairing consistent with a certain amount of supersymmetry is called a "multiplet". For instance, in four dimensions the smallest amount of supersymmetry has four real fermionic generators and is referred to as $\mathcal{N} = 1$ supersymmetry. In this case one can have an $\mathcal{N} = 1$ "vector multiplet" which consists of a spin 1 gauge boson along with its supersymmetric partner, a Majorana fermion. The fermions and bosons both have two on-shell degrees of freedom. In addition to the vector multiplet one can have a "chiral multiplet" consisting of a complex scalar and its partner, a Weyl fermion. Again the degrees of freedom are the same, i.e. two. One can have upto 16 real supersymmetries (usually referred to as $\mathcal{N} = 4$ supersymmetry) without introducing anything above spin 1 in four dimensions. Beyond that one has to include higher spin degrees of freedom. Another useful limit to remember is that if one restricts the highest spin of the fields to 2, corresponding to the graviton, the maximum amount of supersymmetry is generated by 32 real fermionic generators (often referred to as $\mathcal{N} = 8$ supergravity). The highest space-time dimension in which a supersymmetric theory can be written down with fields with highest spin equal to 2, is 11 dimensions. This is why eleven dimensional supergravity plays a distinguished role in supersymmetric physics.

When supersymmetry is an exact symmetry, the bosonic and fermionic partners in a multiplet have the same mass. Clearly, this is not seen in nature. For instance, there is no experimentally observed scalar with the same mass as the electron which would qualify as the electron's supersymmetric partner. Phenomenological models then have to break supersymmetry. The mechanism of supersymmetry breaking is not well understood, however, once one assumes that superymmetry is broken at some high energy scale, one can incorporate in low energy models the breaking by simply introducing terms which break it. The number of such terms can be restricted to soft-breaking terms which are relevant in the infrared. These terms push the masses of the (as yet) unobserved supersymmetric partners of the known fields up, to account for their unobserved status while carefully avoiding contradictions with well measured data.

Supersymmetry is a vast area of research which deserves and has received book-length accounts[1]. In the next subsection we will content ourselves with a simple example to illustrate the ideas touched on in our exposition.

### 17.3.2.1 Supersymmetric Yang-Mills: An Example

To illustrate the basic ideas of supersymmetry we analyze a toy model: $\mathcal{N} = 1$ supersymmetric Yang-Mills theory. As mentioned earlier, in a minimally supersymmetric model containing a vector field we need to introduce fermions with as many on-shell degrees of freedom as the vector field. A vector meson in $d$ dimensions has $d-2$ physical degrees of freedom, whereas a fermion field with $n$ components has $n/2$ on-shell degrees of freedom. In four dimensions we need to find a fermion field with 2 on-shell degrees of freedom to match the vector field's physical polarizations. Both Weyl and Majorana fermion have 2 real on-shell degrees of freedom. Consider the following Lagrangian:

$$\mathcal{L} = -\frac{1}{4}F_{\mu\nu}^a F^{a\mu\nu} + \frac{i}{2}\overline{\psi}^a \gamma^\mu (D_\mu \psi)^a, \tag{17.78}$$

where a sum over repeated indices is implied. $a$ is a group theory index and runs over the generators of the gauge group since all fields transform in the adjoint representation of the gauge group:

$$F_{\mu\nu}^a = \partial_\mu A_\nu^a - \partial_\nu A_\mu^a + g f^{abc} A_\mu^b A_\nu^c$$
$$(D_\mu \psi)^a = \partial_\mu + g f^{abc} A_\mu^b \psi^c. \tag{17.79}$$

The fermionic field $\psi$ is taken to be a Majorana field:

$$\psi_\alpha^a = C_{\alpha\beta}\overline{\psi}^{\beta a}. \tag{17.80}$$

This Lagrangian is invariant under the Poincaré group and local gauge transformation, in addition it enjoys a fermionic symmetry:

$$\delta A_\mu^a = \frac{i}{2}\overline{\epsilon}\gamma_\mu \psi^a$$
$$\delta \psi^a = -\frac{1}{4}F_{\mu\nu}^a \gamma^{\mu\nu}\epsilon. \tag{17.81}$$

---

[1] See for instance, J. Wess and J. Bagger, "Supersymmetry and Supergravity" Princeton University Press (1992).

$\epsilon$ is an "infinitesimal" spinor which anti-commutes with fermionic fields and commutes with bosonic fields. And $\gamma_{\mu\nu} = i\sigma_{\mu\nu}$ as defined in Appendix A. This fermionic symmetry combined with the Poincaré symmetry is known as $\mathcal{N} = 1$ supersymmetry.

We can derive equal-time (anti-)commutation relations for the fields $\psi$ and $A_\mu$. There is a subtlety which needs to be mentioned here. Since the field $A_0$ does not have a conjugate momentum one cannot quantize it in the usual way, more sophisticated methods are called for. In the following we pick the gauge $A_0 = 0$ and agree to impose the equation of motion of the $A_0$ field (Gauss' law) by hand on all physical states. In this gauge we can write down the following equal-time commutation relations:

$$\{\psi_\alpha^a(x), \psi^{*\beta b}(y)\} = \delta^{(3)}(x - y)\delta_\alpha^\beta \delta^{ab}$$
$$\left[F_{0i}^a(x), A_j^b(y)\right] = i\delta_{ij}\delta^{ab}\delta^{(3)}(x - y). \tag{17.82}$$

Using these commutation relations and using the Majorana condition, we can write down the generators of supersymmetry in terms of the fields:

$$Q_\alpha = -\frac{1}{4} \int d^3x F_{\mu\nu}^a (\gamma^{\mu\nu}\gamma^0)_\alpha^\beta \psi^{\beta a}. \tag{17.83}$$

One can easily verify that these generators generate the above supersymmetry transformations in the gauge $A_0 = 0$:

$$[\bar\epsilon Q, \psi] = \bar\epsilon\{Q, \psi\} = -\frac{1}{4}F_{\mu\nu}^a \gamma^{\mu\nu}\epsilon$$
$$[\bar\epsilon Q, A_i] = \bar\epsilon[Q, A_i] = \frac{i}{2}\bar\epsilon\gamma_i\psi^a \tag{17.84}$$

$\mathcal{N} = 1$ Super Yang-Mills (SYM) has some properties in common with ordinary QCD. For instance, the one-loop beta function of this theory is given by:

$$\Lambda\frac{dg}{d\Lambda} = -\frac{3}{16\pi^2}(f^{acd}f^{acd})g^3. \tag{17.85}$$

The beta function is negative implying that the theory is asymptotically free just like ordinary QCD. Also like QCD, it is believed that SYM is confining and develops a mass gap. In addition, SYM has instantons which contribute to correlation functions.

## 17.4   String Theory and Duality

There are five known string theories, which are called the Type I, Type IIA, Type IIB, Heterotic $SO(32)$, and Heterotic $E_8 \times E_8$ string theories.

They are at first sight very different. For instance, the Type I and the two Heterotic theories have half the supersymmetries of the Type II theories. Similarly, the Type I and Heterotic theories have non-abelian gauge groups while the others don't. One key feature that they do have in common is that they are all formulated in 10 dimensions.

In 1995, the ground breaking work of Hull, Townsend and Witten unified these theories. They argued that, while naively the theories had distinct properties, in many cases they were non-perturbatively the same. Many of these properties can be understood by thinking of these theories as limits of a single theory: "M-theory".

The key concept unifying the string theories is called "duality". The basic idea is simple. Consider a physical system which has two distinct descriptions A and B, say. A is then said to be dual to B, and vice versa. If the two descriptions are different, as they must for duality to be non-trivial, there must be mechanisms by which their apparent disparity can be overcome. Also, their region of validity must be such that one doesn't find any obvious contradiction. There are many different dualities. We list a few to illustrate the concept.

Strong-weak coupling duality. This is a very powerful type of duality which relates a theory A, say, at strong coupling to another theory B at weak coupling. An example of this duality is provided by the Type I and $SO(32)$ Heterotic theories in 10 dimensions. Their couplings are inversely related. Thus when one of them is strongly coupled, the other is weakly coupled. Another example is that of the Type IIB theory which is self-dual under strong-weak duality. This means that the weakly coupled theory is the same as the strongly coupled theory with some fields interchanged.

T-duality. In its most general form T-duality relates string theories on different manifolds to each other. An example is of Type IIA on $R^9 \times S^1$ (where $S^1$ is a circle) which is dual to Type IIB on $R^9 \times S^1$. The radii of the two circles are related by $R_A = \alpha'/R_B$ ($\alpha'$ is the string tension which is the same as the 10 dimensional Planck length squared). Here we find that two distinct string theories on different manifolds (different because of their radii) are dual. Similarly, we have that Heterotic string theory on $R^6 \times T^4$ ($T^4$ is the four dimensional torus) is dual to Type IIA on $R^6 \times K3$ ($K3$ is a Ricci flat manifold of complex dimension 2).

### 17.4.1  *M-theory*

Perhaps the most amazing dualities involve M-theory. Very little is known about M-theory and yet it is a powerful tool in string theory. The defining feature of M-theory is that at low energies it is accurately described by 11 dimensional supergravity. One duality states that M-theory on a circle of radius $R$ is the same as type IIA string theory in 10 dimensions with coupling constant $g_s = (R/l_p)^{3/2}$ (where $l_p$ is the 11 dimensional Planck length). A surprising consequence of this identification is that strongly coupled type IIA string theory develops a new dimension (since in that limit $R$ becomes large)! Another, similar, duality states that M-theory on a line segment is equivalent to $E_8 \times E_8$ Heterotic string theory.

One of the appeals of duality is that it allows one to formulate the notion of non-perturbative string theory by changing the description. A key method used in establishing duality is to work with the various supergravities which capture the low-energy dynamics of string theories. The field content of supergravity consists of the massless modes of the string theory in question. For instance, the Type IIA supergravity describes the low-energy dynamics of Type IIA string theory. It has a number of massless fields of which the bosonic fields are as follows:

$$\begin{aligned}
&\phi && \text{scalar dilaton} \\
&g_{\mu\nu} && \text{graviton} \\
&B_{\mu\nu} && \text{anti-symmetric 2-tensor} && (17.86) \\
&A_\mu && \text{abelian gauge field} \\
&A_{\mu\nu\rho} && \text{anti-symmetric 3-tensor} && (17.87)
\end{aligned}$$

The anti-symmetric fields all couple to extended objects known as p-branes. Just as a gauge field couples to a point particle, an antisymmetric (p+1)-tensor couples to a p-brane. An important example is $B_{\mu\nu}$ which couples to the Type IIA fundamental string.

We can compare the above field content to that of 11 dimensional supergravity. The massless bosonic content of 11 dimensional supergravity is:

$$\begin{aligned}
&G_{\mu\nu} && \text{graviton} \\
&C_{\mu\nu\rho} && \text{anti-symmetric 3-tensor} && (17.88)
\end{aligned}$$

At first sight it seems to bare little resemblance to the type IIA field content. Recall, however, that M-theory on $R^9 \times S^1$ is supposed to be equivalent to Type IIA string theory. When we compactify on $S^1$ and take the radius to be small we can ignore the dependence of the fields on the compact coordinate, as is usual when one performs dimensional reduction. From the ten dimensional point of view we can make the following identifications:

$$e^{4\phi/3} = G_{11,11}$$
$$A_\mu = G_{11,\mu}$$
$$g_{\mu\nu} = G_{\mu\nu} \tag{17.89}$$
$$B_{\mu\nu} = C_{\mu\nu,11}$$
$$A_{\mu\nu\rho} = C_{\mu\nu\rho} \tag{17.90}$$

Thus we see that all the fields are accounted for. The dilaton serves as a coupling constant in type IIA supergravity. The usual string-frame dilaton is related. We see immediately that when the dilaton is large the radius of the circle becomes large and type IIA supergravity becomes a poor approximation for 11 dimensional supergravity. We understand this to mean that Type IIA is a perturbative theory which is non-perturbatively equivalent to M-theory on $S^1$.

The spectrum of p-branes is different in the two theories, but they too are related as above. We illustrate this identification with a few examples. Type IIA string theory has 0-branes which couple to the gauge field $A_\mu$, in M-theory they correspond to momentum modes along $S^1$. Since momentum is quantized in the $S^1$ direction in integer units of $2\pi/R$, where $R$ is the radius of the compact direction, the number of units is naturally identified with the number of 0-branes. A striking difference is that M-theory contains no strings. It does, however, have a 2-brane (membrane) which when wrapped on the $S^1$ appears as a string in 10 dimensions as long as one is justified in ignoring scales smaller than the radius of the compact direction.

All string dualities have to satisfy consistency checks of the above kind. Fortunately there are many tests one can perform. Here the importance of a distinguished set of states known as BPS states are particularly useful. BPS states preserve some fraction of the total space-time supersymmetry, by virtue of which they are the lowest mass states in their class and are guaranteed to be stable. Many of their properties can be established exactly, even when the theory is strongly coupled.

## 17.5   Some Important Results

Many new insights have been gained using duality. Although these areas do not directly touch on finding phenomenologically viable models some do demonstrate the ability to study phenomena which generically exist in realistic models. We briefly discuss some of these below.

In the last few years, using duality, considerable progress has been made in our understanding of gauge theories, particularly supersymmetric gauge theories. Significant results include the demonstration of confinement and chiral symmetry breaking in four dimensional gauge theories.

String theories have been used to study black holes. One of the most exciting new results concerns the problem of black hole entropy. The Beckenstein-Hawking entropy is a thermodynamic quantity which satisfies a generalized version of the second law of thermodynamics. It has recently been given a statistical mechanical basis by relating it to microscopic states of a black hole.

Recently, progress has been made in finding a connection between gravity and field theory. One manifestation of this has been a proposal that a quantum mechanics model known as Matrix theory captures the dynamics of M-theory. Many checks have been performed to test the ability of Matrix theory to reproduce supergravity calculations with success. Another approach known as the Maldacena conjecture has led to a radically new connection between conformal field theories and supergravity in AdS backgrounds.

## 17.6   Conclusions

We have given just a flavor of the vast and rapidly growing area of supersymmetry and string theory dualities. The interested reader should consult review articles and books for a thorough introduction to the subject. A good place to start is the recent book by Polchinski (J. Polchinski, "String Theory" Vols. 1 and 2, Cambridge University Press (1998)).

## 17.7   Problems

(1) Show that

(a)
$$\sigma^2 \sigma^\mu \sigma^2 = (\bar{\sigma}^\mu)^T$$

(b)
$$(\sigma^\mu)_{\alpha\dot\beta} = (\bar{\sigma}^\mu)_{\dot\beta\alpha}$$

(c)
$$\bar{\psi}_D \psi_D = \chi_\alpha \xi^\alpha + \bar{\xi}^{\dot\alpha} \bar{\chi}_{\dot\alpha}$$

(d)
$$\bar{\psi}_M \gamma^\mu \psi_M = 0$$

(e)
$$\bar{\psi}_D \gamma^\mu \psi_D = \bar{\xi}^{\dot\alpha} (\bar{\sigma}^\mu)_{\dot\alpha\beta} \xi^\beta + \chi_\alpha (\sigma^\mu)^{\alpha\dot\beta} \bar{\chi}_{\dot\beta}$$
$$= -\xi_\alpha (\sigma^\mu)^{\alpha\dot\beta} \bar{\xi}_{\dot\beta} + \chi_\alpha (\sigma^\mu)^{\alpha\dot\beta} \bar{\chi}_{\dot\beta}$$

(2) Show that
$$[P^\mu, Q^\alpha] = 0$$

**Hint:** $P^\mu$ is 4-vector, only other 4-vector available is $\sigma^\mu$. It follows that
$$[P^\mu, Q^\alpha] = c(\sigma^\mu)^{\alpha\dot\beta} \bar{Q}_{\dot\beta}$$

where $c$ is a complex number. Take the complex conjugate of the above equation, $(P^{\mu*} = -P^\mu)$:
$$[P^\mu, \bar{Q}^{\dot\alpha}] = -c^* (\bar{\sigma}^\mu)^{\dot\alpha\beta} Q_\beta$$
$$\left[P^\mu, \bar{Q}_{\dot\beta}\right] = -c^* (\bar{\sigma}^\mu)_{\dot\beta\alpha} Q^\alpha$$

Then use the Jacobi identity
$$[P^\mu, [P^\nu, Q^\alpha]] + [P^\nu, [Q^\alpha, P^\mu]] + [Q^\alpha, [P^\mu, P^\nu]] = 0$$

(3) For a chiral supermultiplet, the Lagrangian is given by
$$\mathcal{L} = i\xi^\dagger \bar{\sigma}^\mu \xi + \frac{1}{2} \partial^\mu \phi^* \partial_\mu \phi$$

Show that the above Lagrangian is invariant under the infinitesimal supersymmetric transformations
$$\delta_\epsilon \phi = \sqrt{2} \epsilon^T C \xi \tag{17.91a}$$
$$\delta_\epsilon \xi = \sqrt{2} i \sigma^\mu \partial_\mu \phi C \epsilon^* \tag{17.91b}$$

where $\epsilon$ is a parameter which transforms as a left handed chiral spinor and $C$ is charge conjugation matrix.

(4) $Q_\alpha$ and $Q_{\dot\alpha}$ are generators of SUSY transformations. For chiral super-multiplets $\xi$ and $\phi$, the SUSY transformations is given by Eqs. (17.91a and 17.91b)

(a) Show that

$$[\delta_{\epsilon_1}, \delta_{\epsilon_2}]\,\phi = 2i\left[\epsilon_1^\alpha(\sigma^\mu)_{\alpha\dot\beta}\bar\epsilon_2^{\dot\beta} - \epsilon_2^\alpha(\sigma^\mu)_{\alpha\dot\beta}\bar\epsilon_1^{\dot\beta}\right]\partial_\mu\phi \qquad (17.92)$$

(b) The transformation in terms of supersymmetric generators $Q$ and $\bar Q$ are:

$$\delta_\epsilon = \epsilon\cdot Q + \bar\epsilon\cdot\bar Q = -\epsilon^\alpha Q_\alpha + \bar\epsilon_{\dot\beta}\bar Q^{\dot\beta}$$

Show that

$$[\delta_{\epsilon_1}, \delta_{\epsilon_2}] = \epsilon_1^\alpha\{Q_\alpha, \bar Q_{\dot\beta}\}\bar\epsilon_2^{\dot\beta} - \epsilon_2^\alpha\{Q_\alpha, \bar Q_{\dot\beta}\}\bar\epsilon_1^{\dot\beta}$$

Hence show that $[\delta_{\epsilon_1}, \delta_{\epsilon_2}]\,\phi$ satisfy Eq. (17.92), which shows that the transformations defined in Eqs. (17.91a) and (17.91b) are SUSY transformations.

## 17.8   References

1. M. K. Gaillard and L. Miani "New quarks and leptons"Quarks and leptons cargees 1979, p. 443 (Ed. M. Levy et.al.) Plenum press, New York.

2. P. Langacker, "Grand Unified Theories and Proton Decay"Phys. Rep. 72C, 185 (1981).

3. A. Zee, The unity of forces in the universe, Vol. 1 World Scientific (1982).

4. J. Wess and J. Bagger, "Supersymmetry and supergravity" Princeton University Press (1992).

5. R.E. Marshak, Conceptual Foundations of Modern Particle Physics, World Scientific (1993).

6. D. Bailin, A. Love, Supersymmetric Gauge Field Theory and String Theory, Taylor & Francis; 1st edition (1994)

7. M.E. Peskin, "Beyond standard model" in proceeding of 1996 European School of High Energy Physics CERN 97-03, Eds. N. Ellis and M. Neubert.

8. S.P. Martin, A supersymmetry primer, Perspective in supersymmetry Ed. G.L. Kane, World Scientific hep-ph/9709356.

9. J. Ellis, "Beyond Standard Model for Hillwalker" CERN-TH/98-329, hep-ph 9812235.

10. J. Polchinski, "String Theory" Vols. 1 and 2, Cambridge university press (1998).
11. R. Slansky: Appendix A, "Instant review of group theory", p. 338.
12. Adel Bilal, Introduction to Supersymmetry, arXiv: hep-th/0101055 (2001).
13. B. Zwiebach, "A first course in String Theory", Cambridge University Press (2004)
14. Particle Data Group, K. Nakamura et al., Journal of Physics, G **37**, 0750212 (2010).

# Chapter 18

# Cosmology and Astroparticle Physics

## 18.1 Cosmological Principle and Expansion of the Universe

On a sufficiently large scale, universe is homogeneous and isotropic. This is called the cosmological principle. In such a universe, a coordinate system in which matter is at rest at any moment is called a co-moving coordinate system. An observer in this coordinate system is called a co-moving observer. Any co-moving observer will see around himself a uniform and isotropic universe. Cosmological principle implies the existence of a universal cosmic time, since all observers see the same sequence of events[1] with which to synchronize their clocks. In particular they all start their clocks with big bang.

A homogeneous and isotropic universe is described by the Friedmann-Robertson-Walker (FRW) metric which describes the geometry of the universe

$$ds^2 = c^2 dt^2 - R^2(t) \left[ \frac{dr^2}{1 - kr^2} + r^2 \left( d\theta^2 + \sin^2\theta \ d\phi^2 \right) \right]. \quad (18.1)$$

$t, r, \theta, \phi$ are co-moving 4-coordinates and $R(t)$ describes the expansion of the universe and is appropriately known as the scale factor. Notice that in a homogeneous and isotropic world it is necessarily a function only of time.

$k$ is related to the 3-space curvature or geometry at a particular time. With suitable rescaling of $R(t)$, $k$ can be made to have values +1, 0, or −1 corresponding to the positively curved, flat or negatively curved (spatial sections of the) universe respectively[2]. One can write Eq. (18.1) as

$$ds^2 = c^2 dt^2 - R^2(t) d\sigma^2$$

---

[1]For example, the evolution of matter density.
[2]Also known as closed, flat or open universe in literature.

where $d\sigma^2$ is the line element of a Riemannian 3-space of constant curvature, independent of time. By change of variables

$$\frac{dr^2}{1 - kr^2} = d\chi^2$$

so that

$$\chi = \int \frac{dr}{(1 - kr)^{\frac{1}{2}}}$$

$$= \begin{cases} \sin^{-1} r & \text{for } k = +1 \\ r & \text{for } k = 0 \\ \sinh^{-1} r & \text{for } k = -1 \end{cases}$$

we can write the line element $d\sigma^2$ as

$$d\sigma^2 = d\chi^2 + f_k^2(\chi) \left( d\theta^2 + \sin^2 \theta d\phi^2 \right)$$

where

$$f_k(\chi) = \begin{cases} \sin \chi & \text{for } k = +1 \\ \chi & \text{for } k = 0 \\ \sinh \chi & \text{for } k = -1 \end{cases}$$

FRW metric can thus be written in a more convenient form

$$ds^2 = c^2 dt^2 - R^2(t) \left[ d\chi^2 + f_k^2(\chi) \left( d\theta^2 + \sin^2 \theta d\phi^2 \right) \right] \tag{18.2}$$

Now $r$ (or equivalently $\chi$) being a co-moving coordinate is a label for a particular galaxy and does not change as the universe expands. Furthermore for an isotropic universe $\theta$ and $\phi$ also remain fixed for a particular galaxy. This means that in a given direction [3] the physical distance between two points (or galaxies), say, at $r = 0$ and $r = r'$ is given by

$$\ell = R(t) \ r', \tag{18.3}$$

and the velocity of expansion is given by

$$v = \frac{dl}{dt} = \dot{R}(t) \ r'$$

$$= \frac{\dot{R}(t)}{R(t)} R(t) \ r' = H\ell, \tag{18.4}$$

where

$$H = \frac{\dot{R}(t)}{R(t)} \tag{18.5}$$

---

[3] That is, at the same values of $\theta$ and $\phi$.

is called the Hubble parameter and determines the expansion rate of universe. Let us denote by $t_o$ the present time and by $t_e$ the time at which light was emitted from a distant galaxy. Correspondingly we denote the detected wavelength by $\lambda_o$ and emitted (laboratory) wavelength by $\lambda_e$ of some electromagnetic spectral line. We define the redshift

$$z \equiv \frac{\Delta \lambda}{\lambda_e} = \frac{\lambda_0 - \lambda_e}{\lambda_e}$$

$$1 + z = \frac{\lambda_0}{\lambda_e} = \frac{R(t_0)}{R(t_e)}. \qquad (18.6a)$$

Then

$$\frac{\dot{R}}{R_0} = \frac{H(z)}{1+z} \qquad (18.6b)$$

Equation (18.6) shows that the redshift directly indicates the relative linear size of the universe when the photon was emitted. This redshift is experimentally observed and it clearly shows that the universe is expanding.

The highest redshift so far discovered $z = 6.96$ so that the Lyman-alpha line appears in the red part of the spectrum around 7200 Å. This implies that $\frac{R(t_0)}{R(t_e)} = (1 + z) = 7.96$. In the matter dominated universe $R \sim t^{2/3}$ (see below). This gives [with $t_0 = 1.5 \times 10^{10}$ yrs, the present age of the universe] $t_e \simeq \frac{1}{14} \left( 1.4 \times 10^{10} \text{ yrs} \right) \simeq 10^9$ yrs. The existence of these high $z$-objects implies that by the time the universe was about $10^9$ yrs old, some galaxies (or at least their inner region) had already been formed.

For small time intervals since emission compared to $H_0^{-1}$, Eq. (18.6a-18.6b) takes the form

$$z \simeq \Delta t \frac{\dot{R}_0}{R_0} \simeq \frac{l}{c} \frac{\dot{R}_0}{R_0} = \frac{l}{c} H_0, \qquad (18.6c)$$

where $l$ is the distance to the source.

## 18.2  The Standard Model of Cosmology

The framework for the Standard Model of Cosmology is provided by Einstein's theory of gravity

$$R_{\mu\nu} - \frac{1}{2} g_{\mu\nu} R = \frac{8\pi G_N}{c^4} T_{\mu\nu} + g_{\mu\nu} \Lambda \qquad (18.7)$$

where $R_{\mu\nu}$ is the curvature tensor, $R = g^{\mu\nu} R_{\mu\nu}$, $T_{\mu\nu}$ is the energy-momentum tensor, $\Lambda$ is the cosmological constant and $G_N$ of course is the Newton's gravitational constant. The above equation relates how matter (the R.H.S.) influence the geometry of space-time (the L.H.S.) and the expansion of the Universe and vice versa. It is assumed that the matter in the universe behaves like an ideal fluid which has

$$T^{\mu\nu} = \left(\rho c^2 + p\right) U^\mu U^\nu - g^{\mu\nu} p, \tag{18.8}$$

where $p$ is the isotropic pressure, $\rho$ is the energy density and $U^\mu = (1, 0, 0, 0)$ is the velocity vector for the isotropic fluid in a co-moving frame. Then

$$T_{00} = \rho c^2$$
$$T_{ij} = -p g_{ij} \tag{18.9}$$

Using the FRW metric, one can calculate components of $R_{\mu\nu}$ [only the calculation of 00 and 11 component is sufficient][4] and then a substitution in Eq. (18.7) gives

$$\frac{\ddot{R}}{R} = \frac{1}{3} \Lambda c^2 - \frac{4\pi G_N}{3c^4} \left(\rho c^2 + 3p\right), \tag{18.10}$$

and

$$\frac{\ddot{R}}{R} + 2\frac{\dot{R}^2 + kc^2}{R^2} = \Lambda c^2 + \frac{4\pi G_N}{c^4} \left(\rho c^2 - p\right), \tag{18.11}$$

where we have used Eq. (18.8). We can use Eq. (18.10) to get rid of $\ddot{R}$ in Eq. (18.11)

$$H^2 = \left(\frac{\dot{R}}{R}\right)^2 = \frac{8\pi G_N}{3} \rho - \frac{kc^2}{R^2} + \frac{1}{3} \Lambda c^2. \tag{18.12}$$

Equations (18.10) and (18.12) are known as Friedmann-Lemaitre equation. Further the energy-momentum conservation gives

$$\dot{\rho} = -3H \left(\rho c^2 + p\right) \tag{18.13}$$

This equation can also be derived from the first law of thermodynamics, which assuming adiabatic expansion gives

$$d\left(\rho R^3 c^2\right) + p d\left(R^3\right) = 0. \tag{18.14}$$

From Eq. (18.12), one can obtain the expression for the critical density $\rho_c$ [total density required for the flat, $k = 0$, universe for a given expansion rate] with $\Lambda = 0$,

$$\rho_c = \frac{3}{8\pi G_N} \left(\frac{\dot{R}(t)}{R(t)}\right)^2 = \frac{3H^2(t)}{8\pi G_N} \tag{18.15}$$

---

[4]Of course, the inverse metric is also required.

The cosmological density parameter is then defined by

$$\Omega = \frac{\rho}{\rho_c} \qquad (18.16)$$

In addition we need the equation of state. We take this to be that of the ideal gas

$$p = nk_BT \qquad (18.17)$$

where $n$ is the particle density and $k_B$ is the Boltzmann constant ($k_B = 0.86 \times 10^{-10}$ MeV/K) (where K: Kelvin). If we take $k_B = 1$, then the temperature is measured in MeV. In particular 0.86 MeV$= 10^{10}$K. We note that for a non-relativistic (NR) gas, $k_BT \ll mc^2$, so that

$$p \ll mnc^2, \qquad (18.18)$$

i.e.

$$p \ll \rho c^2,$$

where

$$\rho = mn.$$

Then we say that the universe is matter dominated. For extreme relativistic gas (ER)

$$p = \frac{1}{3}\rho c^2$$

$$\rho c^2 = 3nk_BT \qquad (18.19)$$

and we say that Universe is radiation dominated. Present Universe is matter dominated, i.e., $p \approx 0$. Thus from Eq. (18.14), we have

$$d\left(\rho R^3 c^2\right) = 0$$

or

$$\rho R^3 = \text{constant} = \frac{3}{4\pi}M \,, \qquad (18.20)$$

where $M$ is a constant of the dimension of mass[5]. At this stage it is convenient to introduce the conformal time $\eta$ : $dt = Rd\eta$,

$$\dot{R} = \frac{1}{R}\frac{dR}{d\eta}. \qquad (18.21)$$

Using this relation with Eq. (18.12) we have then [$\Lambda = 0$, $c = 1$]

$$\frac{1}{R^2}\left(\frac{dR}{d\eta}\right)^2 = \frac{8\pi G}{3}\,\rho\,R^2 - k. \qquad (18.22)$$

---

[5]In other words the mass of a comoving region remains a constant.

Integrating Eq. (18.22) with the help of Eq. (18.20), we get for $k = 1$ (positively curved Universe)

$$R = MG\,(1 - \cos\eta)$$
$$t = MG\,(\eta - \sin\eta),\tag{18.23}$$

and for $k = -1$ (negatively curved Universe),

$$R = MG\,(\cosh\eta - 1)$$
$$t = MG\,(\sinh\eta - \eta).\tag{18.24}$$

For $k = 0$ (flat Universe)

$$R = (2MG_N)^{1/2}\,\eta,\qquad\qquad t = \frac{1}{2}\,(2MG_N)^{1/2}\,\eta.\tag{18.25}$$

All the three cases are shown in Fig. 18.1.

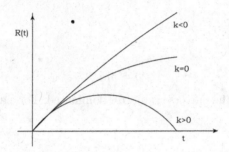

Fig. 18.1  Plot of scale factor $R(t)$ versus time t for closed $(k > 0)$, open $(k < 0)$ and flat $(k = 0)$ Universe.

For $k = 1$, the Universe will recollapse in a finite time, whereas for $k = 0, -1$ the Universe will expand indefinitely. These simple conclusions can be altered when $\Lambda \neq 0$ [see Fig. 18.2]. Thus in the presence of $\Lambda$ there is no link between geometry (curvature) and the fate of the Universe. A so-called closed Universe $(k = 1)$ can expand forever.

## 18.3   Cosmological Parameters and the Standard Model Solutions

The essential feature of the standard model of cosmology, that the Universe "started" in a hot and dense phase and has been undergoing expansion since then, is based on two directly observed facts:

(a): The distant cosmological objects were found to be moving away from the observers with velocities proportional to their distances: $v = Hl$.

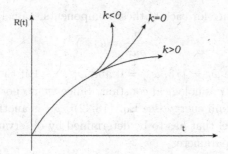

Fig. 18.2   Plot of scale factor $R(t)$ in the presence of $\Lambda$.

(b): The universe is filled with a gas of photons with temperature $T_0 \simeq$ 2.7K called the Cosmic Microwave Background (CMB) radiation. This is supposed to be relic of the early Universe. The observed isotropy of the CMB radiation $\left(\Delta T/T \sim 10^{-5}\right)$ provides the strongest direct support of the cosmological principle.

Another indirect evidence that validates this picture is the successful prediction of Helium and other light element abundances created in the early universe. These calculations are done within the framework of the big bang model.

Some of the important parameters within this framework are as follows.
(i) Hubble parameter

$$H(t) = \frac{\dot{R}(t)}{R(t)} = (1+z)\frac{\dot{R}(t)}{R_0} \tag{18.26}$$

This measures the expansion rate at a given time in the history of the Universe. Its present value is $H(t_0) = H_0 = 100h$ kms$^{-1}$ Mpc$^{-1}$, $h \approx 0.71 \pm 0.01$. $H_0^{-1} = 9.78\ h^{-1}$ Gyr$= 2998h^{-1}$ Mpc.

(ii) Density parameters
The energy density $\rho$ can have several sub-components $\rho_i$ and we can define a density parameter for each of these.

$$\Omega_i(t) = \frac{\rho_i}{\rho_c(t)}; \qquad \rho_c(t) = \frac{3H^2(t)}{8\pi G} \tag{18.27}$$

$\rho_{co} = 1.88 \times 10^{-29}h^2$ gcm$^{-3} = 1.05 \times 10^{-5}h^2$ GeVcm$^{-3}$. Although dominant in the early universe $\Omega_r$ and $\Omega_\nu$ are completely negligible compared to $\Omega_m$ and $\Omega_V$ today[6]. Also in order to make use of Eq. (18.12), we need

---

[6]From now on we will use subscript "$r$" for radiation, "$m$" for matter, "$\nu$" for neutrinos and "$V$" for the "vacuum energy". So for example $\Omega_r$ means $\Omega_{radiation}$ and so on.

the equation of state for each of these components, namely,

$$\frac{p_i}{\rho_i c^2} = \omega_i, \qquad (18.28)$$

where for example $\omega_r = 1/3$, $\omega_m = 0$ and $\omega_V = -1$ if the vacuum energy happens to be the cosmological constant. Since we do not know about the nature of the vacuum energy [see Eq. (18.32)], $\omega_V$ is another parameter in the standard model that has to be determined by observations.

(iii) Deceleration parameter

$$q = -\frac{R(t)\ddot{R}(t)}{\dot{R}^2(t)} = -\frac{\ddot{R}(t)}{R(t)}\frac{1}{H^2(t)} \qquad (18.29)$$

We can use Eq. (18.13) with the equation of state (18.28) to give

$$\dot{\rho}_i = -3\frac{\dot{R}}{R}\rho_i c^2(1 + \omega_i)$$

which has the solution:

$$\frac{\rho_i}{\rho_{i0}} = (\frac{R}{R_0})^{-3(1+\omega_i)} = (1+z)^{3(1+\omega_i)} \qquad (18.30)$$

If the first term on the right of Eq. (18.12) dominates, then its integration on using Eq. (18.30), gives

$$R \sim t^{\frac{2}{3(1+\omega_i)}}, \ \omega_i \neq -1 \qquad (18.31)$$

Instead of using the cosmological constant $\Lambda$, we can equivalently use the vacuum energy (which is more general)

$$\Lambda c^2 \rightarrow 8\pi G\rho_V \qquad (18.32)$$

Then the Friedmann equation (18.12) gives the sum rule

$$\frac{H^2(z)}{H_0^2} = \sum_i \Omega_i(1+z)^{3(1+\omega_i)} - \frac{k}{H_0^2 R_0^2}(1+z)^2 \qquad (18.33)$$

where of course $\Omega_i = \frac{\rho_{i0}}{\rho_{c0}}$ [cf. Eq. (18.16)]. For a universe with radiation, matter and vacuum energy, Eq. (18.33) takes the form

$$\frac{kc^2}{H_0^2 R_0^2} = (\Omega_m + \Omega_r + \Omega_V - 1) \qquad (18.34)$$

where the subscript "0" as usual indicates the present day values. Especially for a flat ($k = 0$) universe it means that $\Omega_m + \Omega_r + \Omega_V = 1$ or $\Omega_m + \Omega_V \simeq 1$ as $\Omega_r$ is negligible.

Using the Friedmann equation (18.10) the deceleration parameter $q = -\frac{\ddot{R}}{RH^2}$, can be written as

$$q(z) \equiv -\frac{\ddot{R}(t)}{R(t)}\frac{1}{H_0^2}\frac{H_0^2}{H^2(t)}$$

$$= (\frac{H_0}{H(z)})^2[\frac{1}{2}\sum_i \Omega_i(1 + 3\omega_i)(1 + z)^{3(1+\omega_i)}]$$

$$= (\frac{H_0}{H(z)})^2[\frac{1}{2}\Omega_m(1 + z)^3 - \Omega_V] \qquad (18.35)$$

The last step follows since $\Omega_r$ is negligible and $\omega_V = -1$. Further

$$q_0 = \frac{1}{2}\sum_i (1 + 3\omega_i)\Omega_i$$

$$= \frac{1}{2}\Omega_m + \Omega_r + \frac{1 + 3\omega_V}{2}\Omega_V \qquad (18.36)$$

We can conclude that

(i) For $\omega_V < -\frac{1}{3}$, the vacuum energy may lead to accelerating expansion. Current data support it and gives $\Omega_V = 0.72 \pm 0.05$ and $\Omega_m = 0.28 \pm 0.05$ if $k = 0$ [see Sec. 18.4]. We see that $\Lambda$ acts as "repulsive" (anti) gravity tending to speed up the expansion. It should be noted that it is the negative pressure of $\Lambda$ rather than its energy density that does that.

(ii) The presence of vacuum energy implies that there is no link between geometry (curvature) and the fate of the Universe: (a) A closed Universe ($k > 0$ which implies $\Omega_V > (1 - \Omega_m)$) can expand for ever if $\Lambda$ is positive (b) An open Universe ($k < 0$ which implies $\Omega_V \leq (1 - \Omega_m)$ ) must recollapse if $\Lambda$ is negative.

(iii) From the relation (18.6),

$$\frac{dR}{R_0} = \frac{H(z)}{1 + z}dt, \ \frac{dR}{R_0} = -\frac{1}{(1 + z)^2}dz$$

so that

$$dt = -\frac{dz}{(1 + z)H(z)} \qquad (18.37)$$

For $t = t_0$, $R = R_0$, $z = 0$ and $t \to 0$, $R \to 0$, $z \to \infty$. Thus the age of the Universe may be written as

$$t_0 = \int_0^\infty \frac{dz}{(1 + z)H(z)} \qquad (18.38)$$

On using Eqs. (18.33) and (18.34), with $\omega_V = -1$ and neglecting $\Omega_r$,

$$H_0 t_0 = \int_0^\infty \frac{dz}{(1+z)\left[(1+z)^2\left(\Omega_m z + 1\right) - z\left(2+z\right)\Omega_v\right]^{1/2}} \quad (18.39)$$

For the special case $\Omega_m + \Omega_V = 1$, $\Omega_m < 1$, by a change of variables $X = 1 + \frac{1-\Omega_V}{\Omega_V}\left(1+z\right)^3$, the above integral simplifies to

$$H_0 t_0 \simeq \frac{1}{\frac{2}{3}\frac{1}{\sqrt{\Omega_V}}\displaystyle\int_{1/\sqrt{\Omega_V}}^\infty \frac{dX}{X^2-1}}$$

giving

$$H_0 t_0 = \frac{1}{3}\frac{1}{\sqrt{\Omega_V}}\ln\frac{1+\sqrt{\Omega_V}}{1-\sqrt{\Omega_V}} \quad (18.40)$$

For $\Omega_V \approx 0.72$, $H_0 t_0 = 0.98$ gives $t_0 = 0.98 H_0^{-1} = 9.6 h^{-1}\text{Gyr} = 13.5\text{Gyr}$, the age of the Universe.

Finally from Eqs. (18.30) and (18.31) we have, for

(i) matter dominated Universe, $\omega = 0$, $\rho \sim R^{-3}$, $R \sim t^{\frac{2}{3}}$

(ii) radiation dominated Universe, $\omega = \frac{1}{3}$, $\rho \sim R^{-4}$, $R \sim t^{\frac{1}{3}}$

(iii) curvature dominated Universe, $H^2 = \frac{\dot{R}^2}{R^2} = \frac{|k|c^2}{R^2}$, $R \sim t$

(iv) vacuum dominated Universe, $\frac{\dot{R}}{R} = \sqrt{\frac{1}{3}\Lambda c^2}$, i.e. $R \propto e^{\sqrt{\frac{1}{3}\Lambda c^2}\,t} = e^{Ht}$ with $H = \frac{4\pi}{3}G\rho_V$. This gives what is now known as the inflationary phase of the Universe.

Thus we see that the curvature dominates at rather late time if a Cosmological term does not dominate sooner. We note that $\frac{\Omega_V}{\Omega_m} = \frac{\rho_V}{\rho_m} \propto R^3$. Thus at early times we have that the vacuum energy much more suppressed compared to that of matter or radiation while it dominates at late times. Hence we would expect a transition from deceleration to acceleration at some $z$. As will be discussed in Sec. 18.4. there is evidence for this in the present data. This seems to occur at $z = 0.5$, i.e. about 5 billion years ago.

## 18.4 Accelerating Universe and Dark Energy

The discovery that the universe is accelerating its expansion was made in 1998 by two independent groups.

### 18.4.1 Evidence from Supernovae

One of the key evidences for accelerating universe comes from supernovae of type Ia which are generally thought to be thermonuclear explosions of white dwarfs and serve as effective standard candles[7] ; thus one can deduce their distances from their apparent brightness. One defines the luminosity distance as follows

$$d_L \equiv \sqrt{\frac{L_s}{4\pi F}},$$

where $L_s$ is the intrinsic luminosity of the source and $F$ is the flux. It can be shown that the observed luminosity of the source, $L_o$ is related to $L_s$ by $L_0(\text{redshift}) = \frac{1}{(1+z)^2} L_s$, so that

$$d_L = (1+z)\sqrt{\frac{L_o}{4\pi F}}. \tag{18.41}$$

Now

$$d_L = (1+z)\, r_e R_0$$

where $r_e$ is the coordinate distance

$$r_e = \int_{t_e}^{t} \frac{dt}{R(t)} = \frac{1}{R_0} \int_{t_e}^{t} (1+z)\, dt = \frac{1}{R_0} \int_0^z \frac{dz}{H(z)}.$$

Here we have put $c = 1$ and used the relation $\frac{\dot{R}}{R_0} = \frac{H(z)}{1+z}$. Thus

$$d_L = (1+z) \int_0^z \frac{dz}{H(z)}. \tag{18.42}$$

This gives

$$H(z) = \frac{1}{\frac{d}{dz}\left(\frac{d_L(z)}{1+z}\right)} \tag{18.43}$$

The quantity in the denominator can be measured. $H(z)$ on the other hand knows the history of the Universe and determines the expansion of the Universe. It depends on $\Omega_m$ and $\Omega_V$ and thus on equation of state. Supernova had actually been detected to be systematically 25% fainter than expected i.e. more distant than one would expect for a cosmic whose expansion has recently been showing down, or even coasting. The situation is summarized in Fig. 18.3.

They thus provide the first direct evidence of the transition from the earlier decelerating phase to the present accelerating expansion. Note, in particular the transition from decelerating to accelerating at $z = 0.5$.

---

[7]Standard candles are astronomical objects that have standard or known intrinsic luminosity that can be inferred from a certain characteristic possessed by the objects in that class. For example, SN Ia have a characteristic graph between the observed light intensity and time (called the "light curve").

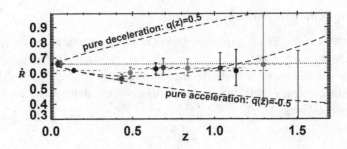

Fig. 18.3   Cosmic expansion rate $\dot{R}$ is plotted against the redshift $z$. Positive and negative slope indicate decelerating and accelerating expansions respectively [16].

## 18.4.2   Evidence from CMB Data

We can see the universe 380,000 years after Big Bang by studying the microwave background radiation, which is a direct relic of the universe when it becomes transparent to electromagnetic radiation. Fluctuations in the CMB radiation[8] have been detected with angular resolutions from 7 degree to few arc minutes in the sky. As the matter is tightly coupled with radiation early on, these fluctuations indicate the first clumping of matter particles into cosmic structures. Overdense regions have a higher temperature and thus photon pressure tends to oppose this clumping. The net result is gravity driven acoustic-like oscillations. These oscillations left their signature in the anisotropy of the CMB.

After seven years of running, Wilkinson Microwave Anisotropy Probe (WMAP) provides a detailed picture of the temperature fluctuations which one can analyze by a Gaussian model which fit the WMAP data. These fluctuations are usually expressed by spherical harmonic expansion

$$\frac{\Delta T}{T} \equiv \frac{T(\theta, \phi) - T}{T} = \sum_{l,m} a_{lm} Y_{lm}(\theta, \phi) \tag{18.44}$$

Most of the cosmological information is contained in the two-point temperature correlation function equivalently expressed by the angular power spectrum

$$l(l+1)C_l = \sum_m \frac{|a_{lm}|^2}{4\pi} \tag{18.45}$$

which is plotted against multipole moment $l$ [see Fig. 18.4]. $C_l$ depends on all the cosmological parameters.

---

[8]Typically at the level of a few parts in $10^5$.

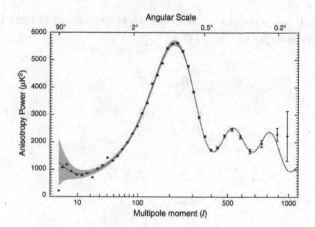

Fig. 18.4   Angular power spectrum of CMB temperature fluctuations [21,25].

The position of first peak tells us that the biggest CMB hot spots make an angle of $1° \sim 180°/l$, implying $l \sim 200$ on the sky if CMB photons are not distorted by the large scale cosmic curvature between us and the horizon of the last scattering surface. This in turns implies that geometry of the universe is flat: $k = 0$ [for $k < 0$, angle would be smaller implying $l > 200$, and for $k > 0$, angle would be larger implying $l < 200$]. Since the non-baryonic matter is impervious to radiation pressure, the sound amplitudes depend sensitively on the baryonic density. This has the important consequence that the height of the second peak relative to the first peak is a sensitive measure of baryon density, giving

$$\Omega_b = 0.04 \tag{18.46}$$

in agreement with that inferred from Big Bang nuclear synthesis of primordial $H$ and $H_e$ [see Sec. 18.8].

Figure 18.5 summarizes the three independent data sets, namely supernovae, CMB and the structure formation. Comparing the results with the estimates about $\Omega_b$ mentioned in (18.46), there is evidence for the existence of Dark Matter (DM). DM by definition is non-luminous and its presence can only be inferred from the gravitational effects on the visible matter. Its composition is unknown, although there are many candidates for it. Historically the earliest results came from the rotational speeds of galaxies and orbital velocities of galactic clusters. If all matter were luminous, the rotational speed of the galactic disc, $V(r) \sim \sqrt{\frac{GM(r)}{r}}$, $M(r)$ being the mass inside the orbit, $r$ is the distance from the orbit. But the observations shown that $V(r) \sim$ constant. This implies the existence of dark halo with mass density $\rho \propto \frac{1}{r^2}$, i.e. $M(r) \propto r$.

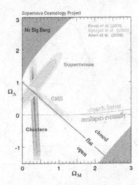

Fig. 18.5  Evidence for dark energy. Shown are a combination of observations of the cosmic microwave background (CMB), supernovae (SNe) and baryon acoustic oscillations (BAO) [17,25].

Other observations consistent with the existence of DM are:

(i) gravitational lensing of background orbits by galaxy clusters
(ii) the temperature distribution of hot gas in galaxies and cluster of galaxies.

The data shown in Fig. 18.5 gives us the following values for the basic set of cosmological parameters.

$$\Omega = 1.006 \pm 0.006, \text{ consistent with spatially flat universe}$$
$$h = 0.71 \pm 0.01$$
$$\Omega_m h^2 = 0.133 \pm 0.006 \tag{18.47}$$
$$\Omega_b h^2 = 0.0227 \pm 0.006$$
$$\Omega_\nu h^2 < 0.0076$$

Further analysis of structure formation in the universe indicates that most of the DM should be "cold" i.e. should have been non-relativistic at the onset of galaxy formation. This is consistent with the upper bound on the contributions of light neutrinos[9] given above.

From Eq. (18.32), with $\Omega_r$ negligible and $\omega_V = -1$, we see that the present data implies the deceleration parameter

$$q_0 = \frac{1}{2}\Omega_m - \Omega_\Lambda < 0,$$

---

[9]Which are not "cold" at that epoch.

i.e. the universe is accelerating in its expansion. Further, as has already been mentioned, there is evidence in the present data for a transition from deceleration to acceleration at some $z$.

There have been a number of attempts to explain this phenomenon. Each approach is plagued by its own problems like fine tuning, instabilities, etc. There are four broad class of models that have been put forward to explain this issue.

(1) Cosmological Constant: The energy density of the vacuum.
(2) Quintessence: Scalar Fields of unknown origin with negative pressure.
(3) Modified Gravity: General Relativity has to be replaced by another theory at the cosmological scales.
(4) Backreaction: General Relativity is correct but we do not know how to apply it in cosmology.

We briefly discuss them one by one.

The standard model of cosmology with cosmological constant $\Lambda$, also known as $\Lambda CDM$ is the simplest model which is compatible with all the data till now and there is no other model which gives better statistical fit. But present theory, namely the Standard Model of particle interactions, cannot explain it since

$$\rho_\Lambda|_{\text{obs}} = \frac{\Lambda}{8\pi G} \sim \left(10^{-3}\text{eV}\right)^4$$

while

$$\rho_\Lambda|_{\text{theory}} \sim M_{\text{fundamental}}^4 \geq M_{\text{susy}}^4 \sim (1\text{TeV})^4 >> \rho_\Lambda|_{\text{obs}}.$$

To summarize, the cosmological constant $\Lambda$ explains the late time acceleration of the Universe. But we do not know what really is CDM and we cannot give a theoretical justification to the Cosmological Constant.

### 18.4.3  *Quintessence*

Just as we can think of cosmological constant as an effective stress energy term, we can also construct other stress energy terms. This can be most easily done by the use of scalar fields. Since energy density is always positive, this means that we need to have a sufficiently negative pressure. Such a system can be constructed out of scalar fields. Consider a real scalar field $\phi$, with potential $V(\phi)$, with Lagrangian density

$$\mathcal{L} = \frac{1}{2}g^{\mu\nu}\partial_\mu\phi\partial_\nu\phi - V(\phi) \tag{18.48}$$

The stress energy tensor for this can be calculated as

$$T_{\mu\nu} = \partial_\mu\phi\partial_\nu\phi - g_{\mu\nu}\left[\frac{1}{2}\partial^\lambda\phi\partial_\lambda\phi + V(\phi)\right] \qquad (18.49)$$

If the scalar field is homogeneously distributed in space, then the space derivatives of it are zero and thus we get the energy density and pressure $[T^{ij} = p\delta^{ij}]$:

$$\rho = T^{00} = \frac{1}{2}\dot{\phi}^2 + V(\phi) \qquad (18.50)$$

$$p = \frac{1}{2}\dot{\phi}^2 - V(\phi)$$

and from this we can calculate the equation of state as

$$w_\phi = \frac{\dot{\phi}^2 - 2V(\phi)}{\dot{\phi}^2 + 2V(\phi)} \qquad (18.51)$$

Using the conservation of energy condition $\nabla_\mu T^{\mu\nu} = 0$, we get the equation of motion for the scalar field

$$\ddot{\phi} + 3H\dot{\phi} + \frac{dV}{d\phi} = 0 \qquad (18.52)$$

From Eq. (18.51), we see that for constant fields, we get $w = -1$ just as for the cosmological constant. From these equations, we see that with a suitable choice of potential, we can explain the accelerated expansion of the universe.

There are, however, some problems with scalar fields. For example, consider the potential $V(\phi) = \frac{1}{2}m_\phi^2\phi^2$. Now assuming that $<\phi> \sim M_P$ and noting that the observed value for energy density is about $(10^{-12}\text{GeV})^4$, we can estimate $m_\phi$ as

$$\rho \sim V(\phi) \sim m_\phi^2\phi^2 \Rightarrow m_\phi \sim 10^{-33}\text{eV} \qquad (18.53)$$

In the standard model of particle physics, the masses are of the order of MeVs or GeVs. So this extremely small value constitutes a **fine tuning** problem for the scalar fields, i.e. how do we construct a model from known particle physics that leads to such a small mass?

On top of that we have a new cosmological constant problem to solve. Since the accelerated expansion is now explained completely by scalar fields, it is hard to understand why $\Lambda = 0$. So we see that while quintessence solves some problems, it creates new ones. Before ending this section, it should be emphasized that there have been a large number of proposals on scalar field models as the source of cosmic acceleration. To date we do not have any scalar field model that evades the fine tuning problem.

### 18.4.4 *Modified Gravity*

The Einstein-Hilbert action

$$S = \frac{1}{2K} \int d^4x \sqrt{-g} R + S_M \qquad (18.54)$$

leads to the Einstein field equations. It has been shown that the simplest classes of Modified Gravity Theory is the $f(R)$ gravity, i.e. replace $R$ by $f(R)$, the equation of motion is then

$$R_{\mu\nu} \frac{df}{dR} - \frac{1}{2} g_{\mu\nu} f(R) + g_{\mu\nu} \Box \frac{df}{dR} - \nabla_\mu \nabla_\nu \frac{df}{dR} = K T_{\mu\nu} \qquad (18.55)$$

It is important to note here that for general relativity, $f(R) = R$ and Eq. (18.55) reduces to Einstein's field equation:

$$R_{\mu\nu} - \frac{1}{2} g_{\mu\nu} R = K T_{\mu\nu}$$

We can write Eq. (18.55) in a more familiar form:

$$G_{\mu\nu} \equiv R_{\mu\nu} - \frac{1}{2} g_{\mu\nu} R$$

$$= \frac{K}{f'(R)} + g_{\mu\nu} \frac{[f(R) - R f'(R)]}{2 f'(R)} + \frac{[\nabla_\mu \nabla_\nu f'(R) - g_{\mu\nu} \Box f'(R)]}{f'(R)} \qquad (18.56)$$

$$= \frac{K}{f'(R)} \left( T_{\mu\nu} + T_{\mu\nu}^{(eff)} \right) \qquad (18.57)$$

where we wrote $df/dR = f'(R)$. It is interesting to note that Eq. (18.55) is fourth order in the metric, while general relativity is only second order because the covariant derivative terms vanish in this case. Taking trace of this equation gives

$$f'(R) R - 2 f(R) + 3 \Box f'(R) = K T. \qquad (18.58)$$

This clearly shows that this equation relates $R$ with $T$ differentially, not algebraically as in the case of general relativity, where $R = -KT$. This already indicates that the field equations of $f(R)$ theories will admit a larger variety of solutions than Einstein's theory. One very important result is that $T = 0$ no longer implies $R = 0$. There are constraints on $f(R)$ Models: General relativity has been tested extensively at the solar system level and at earth bound laboratories. There are also some astrophysical tests that support general relativity. So it is evident that any modified gravity theory has to pass these tests and possibly reduce to general relativity at the solar system level. Thus one cannot be sure that modified gravity is a viable alternative to general relativity.

## 18.5 Hot Big Bang: Thermal History of the Universe

### 18.5.1 *Thermal Equilibrium*

Consider an arbitrary volume $V$ in thermal equilibrium with a heat bath at temperature $T$. The particle density $n_i$ ($i$, particle index) at temperature $T$ is given by

$$n_i = \frac{N_i}{V} = \frac{g_i}{2\pi^2} \left( \frac{k_B T}{\hbar c} \right)^3 \int_0^\infty \left[ \exp\left( \frac{E}{k_B T} \right) \pm 1 \right]^{-1} z^2 dz. \quad (18.59)$$

The energy density is given by

$$\rho_i \, c^2 = \frac{g_i}{2\pi^2} \left( \frac{k_B T}{\hbar c} \right)^3 (k_B \, T) \int_0^\infty \left[ \exp\left( \frac{E}{k_B T} \right) \pm 1 \right]^{-1} \left( \frac{E}{k_B T} \right) z^2 dz, \quad (18.60)$$

where

$$z = \frac{qc}{k_B T}, \qquad E = \left[ (q_i c)^2 + \left( m_i c^2 \right)^2 \right]^{1/2} \quad (18.61)$$

and $g_i$ are the number of spin states, $q_i$ is the momentum of the particle and $m_i$ is its mass. The + sign is for the fermions (F) and − sign is for the bosons (B). In particular for $i$ = photon, $m = 0$, $g = 2$. In writing Eqs. (18.60) and (18.61), we have put the chemical potential $\mu_i = 0$ for photons. Since particles and antiparticles are in equilibrium with photons $\mu_i = -\mu_{\bar{\imath}}$, if there is no asymmetry between the number of particles and antiparticles, $\mu_i = \mu_{\bar{\imath}} = 0$. Even if that is not true, the difference between the number of particles and antiparticles is small compared with the number of photons,

$$\left| \frac{\mu_i}{k_B T} \right| = \left| \frac{\mu_{\bar{\imath}}}{k_B T} \right| \ll 1 \quad (18.62)$$

and the chemical potential can be neglected. For the photon gas, we get from Eqs. (18.60) and (18.61)

$$n_\gamma = 2 \frac{\zeta(3)}{\pi^2} \left( \frac{k_B T}{\hbar c} \right)^3 = 2 \frac{1.2}{\pi^2} \left( \frac{1}{\hbar c} \right)^3 (k_B T)^3 \quad (18.63)$$

$$\rho_\gamma \, c^2 = 6 \frac{\zeta(4)}{\pi^2} \left( \frac{1}{\hbar c} \right)^3 (k_B T)^4$$

$$= \frac{\pi^2}{15} \left( \frac{1}{\hbar c} \right)^3 (k_B T)^4 \approx 2.7 \, n_\gamma \, (k_B T). \quad (18.64)$$

In Eqs. (18.63) and (18.64) $\zeta(r)$, $r = 3, 4$ is the Riemann zeta function. For a gas of extreme relativistic particles (ER), $k_B T \gg m_i c^2$, $qc \gg m_i c^2$, we thus get

$$n_B = \left(\frac{g_B}{2}\right) n_\gamma, \qquad \rho_B = \left(\frac{g_B}{2}\right) \rho_\gamma \tag{18.65}$$

$$n_F = \frac{3}{4}\left(\frac{g_F}{2}\right) n_\gamma, \qquad \rho_F = \frac{7}{8}\left(\frac{g_F}{2}\right) \rho_\gamma. \tag{18.66}$$

The entropy $S$ for the photon gas is given by

$$S = \frac{R^3}{T}\frac{4}{3}\,\rho_\gamma(T). \tag{18.67}$$

For any relativistic gas

$$S = \frac{R^3}{T}\frac{4}{3}\,\rho(T). \tag{18.68}$$

Thus for a gas consisting of extreme relativistic particles (bosons and fermions): ($\hbar = c = 1$)

$$n(T) = \frac{1}{2}\,g'(T)\,n_\gamma(T)$$

$$= \frac{1.2}{\pi^2}\,g'(T)\,(k_B T)^3 \tag{18.69}$$

$$\rho(T) = \frac{1}{2}\,g_*(T)\,\rho_\gamma(T)$$

$$= \frac{\pi^2}{30}\,g_*(T)\,(k_B T)^4 \tag{18.70}$$

$$S = \frac{R^3}{T}\frac{2}{3}\,g_*(T)\,\rho_\gamma(T), \tag{18.71}$$

where

$$g'(T) = \sum_B g_B + \frac{3}{4}\sum_F g_F \tag{18.72}$$

$$g_*(T) = \sum_B g_B + \frac{7}{8}\sum_F g_F \tag{18.73}$$

are called the "effective" degrees of freedom. We note that entropy per unit volume is given by

$$\frac{s}{k_B} \equiv \frac{1}{k_B}\frac{S}{R^3} = \frac{2\pi^2}{45}\,g_*(T)\,(k_B T)^3. \tag{18.74}$$

For non-relativistic gas $k_B T \ll m_i c^2$, we use the Boltzmann distribution

$$n_i = \frac{g_i}{2\pi^2}\left(\frac{k_B T}{\hbar c}\right)^3 \int_0^\infty \exp\left(-\frac{E}{k_B T}\right) z^2\,dz \tag{18.75}$$

$$E \approx m_i c^2 \left[ 1 + \frac{1}{2} \frac{q^2 c^2}{(m_i c^2)^2} \right]. \tag{18.76}$$

From Eq. (18.75), we get

$$n_i = \left[ \frac{g_i}{(2\pi)^{3/2}} \right] \left( \frac{k_B T}{\hbar c} \right)^3 \left[ \left( \frac{m_i c^2}{k_B T} \right)^{3/2} e^{-m_i c^2 / k_B T} \right] \tag{18.77}$$

$$\rho_i = n_i \, m_i. \tag{18.78}$$

### 18.5.2  *The Radiation Era*

For extreme relativistic gas, $\rho = \frac{1}{3} \rho c^2$, we get from Eq. (18.14)

$$R^3 \frac{d\rho}{dR} + 4\rho R^2 = 0. \tag{18.79}$$

Thus, we have

$$\rho = A^2 R^{-4} \qquad \text{(A : constant)}. \tag{18.80}$$

Hence

$$\frac{\rho_{N \cdot R}}{\rho_{E \cdot R}} \propto R \to 0 \qquad \text{as} \quad R \to 0.$$

Therefore, we have the important result. Early universe is dominated by extreme relativistic particles, i.e. the universe is radiation dominated in early stages. Since $\rho \to \frac{1}{R^4}$ for the early universe, we can neglect the second and third terms on the left-hand side of Eq. (18.12) as compared with the first term. Thus we get

$$H = \frac{\dot{R}}{R} = \sqrt{\frac{8\pi}{3}} \, (G_N \, \rho)^{1/2}. \tag{18.81}$$

Now using Eq. (18.70),

$$H = \sqrt{\frac{4\pi^3}{45}} \, [g_* \, (T)]^{1/2} \frac{(k_B T)^2}{M_P} \approx 0.21 \, g_*^{1/2} \left( \frac{k_B T}{\text{MeV}} \right)^2 \text{sec}^{-1}. \tag{18.82}$$

where $M_P$ is the Planck mass $= 1/\sqrt{G_N}$. Further

$$R\dot{R} = A \sqrt{\frac{8\pi G_N}{3}}. \tag{18.83}$$

Hence

$$\begin{aligned} t &= \sqrt{\frac{3}{32\pi G_N}} \frac{R^2}{A} = \sqrt{\frac{3}{32\pi}} \frac{1}{\sqrt{G_N \, \rho}} \\ &= \sqrt{\frac{45}{16\pi^3}} \, g_*^{-1/2} \frac{\hbar M_P}{(k_B T)^2} \\ &= 2.42 \, g_*^{-1/2} \left( \frac{\text{MeV}}{k_B T} \right)^2 \text{sec}. \end{aligned} \tag{18.84}$$

Thus

$$Ht \approx 0.5. \tag{18.85}$$

We consider two examples:

(i) For $g_* = g_\gamma = 2$,

$$t \approx 1.7 \left(\frac{MeV}{k_B T}\right)^2 \text{ sec.} \tag{18.86}$$

Thus for $k_B T = m_e c^2 \approx 0.51$ MeV, $t = 6.5$ sec and $H \approx 0.08$ sec$^{-1}$.

(ii) For $m_\mu > k_B T > m_e$,

$$g_* = g_\gamma + \frac{7}{8}(g_e + 3g_\nu)$$

$$= 2 + \frac{7}{8}(4+6) = \frac{43}{4} \tag{18.87}$$

and for $m_\pi > k_B T > m_\mu$,

$$g_* = 2 + \frac{7}{8}\left(g_e + g_\mu + 3g_\nu\right)$$

$$= \frac{57}{4}. \tag{18.88}$$

Here we have taken the number of neutrinos $N_\nu = 3$. Now we get from Eqs. (18.81) and (18.83) at $k_B T = 1$ MeV

$$H \approx 0.67 \left(\frac{k_B T}{MeV}\right)^2 \text{ sec}^{-1} \approx 0.67 \text{ sec}^{-1} \tag{18.89}$$

and

$$t \approx 0.74 \left(\frac{MeV}{k_B T}\right)^2 \text{ sec} \approx 0.74 \text{ sec} \tag{18.90}$$

Now Eq. (18.90) gives the time evolution of the universe in radiation era. From Eq. (18.81), we have the important result that $H \propto \sqrt{\rho}$, i.e. the higher the energy density in the early universe, the faster will be the expansion rate.

As we have seen, the radiation density falls off as $R^{-4}$ and the energy density in non-relativistic matter falls of as $R^{-3}$. The universe eventually becomes matter dominated. At $t = t_{eq}$, matter density becomes equal to radiation density, i.e.

$$\rho_m(t_{eq}) = \rho_r(t_{eq}), \tag{18.91}$$

where on using Eq. (18.20)

$$\rho_m \left(t_{eq}\right) = \Omega_{mo} \, \rho_{co} \left(\frac{R_o}{R_{eq}}\right)^3 \tag{18.92}$$

and from Eqs. (18.64), (18.66) and (18.80) we get

$$\rho_r \left(t_{eq}\right) = \frac{\pi^2}{30} \left[g_\gamma \, (k_B T_{\gamma o})^4 + \frac{7}{8} 3 \, g_\nu \, (k_B T_{\nu o})^4\right]$$

$$\times \left(\frac{R_o}{R_{eq}}\right)^4$$

$$= \frac{\pi^2}{30} \left[2 + \frac{21}{4} \left(\frac{4}{11}\right)^{4/3}\right] (k_B T_o)^4 \left(\frac{R_o}{R_{eq}}\right)^4 \tag{18.93}$$

where we have used $\left(\frac{T_{\nu_o}}{T_{\gamma_o}}\right)^3 = \frac{4}{11}$, [see Eq. (18.135)] and $T_{\gamma_o} = T_o$. Hence from Eqs. (18.92) and (18.93), we obtain for $T_o = 2.725 K$,

$$1 + z_{eq} = \frac{R_o}{R_{eq}} = \Omega_m \, \rho_c \, \left(2.27 \times 10^6 \text{ MeV}^{-1} \text{cm}^3\right)$$

$$= 2.4 \times 10^4 \, \Omega_m \, h^2,$$

i.e. with

$$\Omega_m \, h^2 = 0.113$$

$$z_{eq} \simeq 3200 \tag{18.94}$$

From Eqs. (18.93) and (18.94), we get

$$k_B T_{eq} = (k_B T_0) \frac{R_0}{R_{eq}} = (k_B T_0) \left(1 + z_{eq}\right)$$

$$= 5.6 \times 10^{-6} \, \Omega_m \, h^2 \text{ MeV}$$

$$\simeq 0.75 \text{ eV} \tag{18.95}$$

and from Eqs. (18.84) and (18.95), we obtain

$$t_{eq} \approx 3.0 \times 10^{10} \, \left(\Omega_m \, h^2\right)^{-2} \text{ sec} \simeq 1.7 \times 10^{12} \text{ sec} \tag{18.96}$$

In the dense early universe the radiation would have been held in thermal equilibrium with matter and would have scattered repeatedly off free electrons. But when the expansion had cooled the matter below 3000 K ($k_B T \simeq 0.26$ eV), so that from Eq. (18.84) with $g_* = 2 + \frac{7}{8}(6) = \frac{29}{4}$, $t \simeq 1.3 \times 10^{13}$ sec $\simeq 4 \times 10^5$ yrs, the primordial plasma would have recombined with atoms, the universe thereafter becoming transparent to light.

Table 18.1  Cosmic History (some critical phases)

| Era | Age (in seconds) | Temperature K | Remarks |
|---|---|---|---|
| Big Bang | 0 | | Vacuum to matter transition |
| End of Inflation | $10^{-35}$ | $10^{27}$ | |
| Electro-weak | $10^{-10}$ | $10^{18}$ | $W^{\pm}, Z^0$: Electroweak transition |
| Quark | $10^{-5}$ | $3 \times 10^{12}$ | Hadronization |
| Lepton | $8 \times 10^{-5}$ | $1.2 \times 10^{12}$ | $\mu^{\pm}$ annihilation |
| | $8 \times 10^{-3}$ | $1.2 \times 10^{11}$ | $\nu_{\mu}$'s decouple |
| | 0.7 | $10^{10}$ | $\nu_e$'s decouple |
| | 6 | $6 \times 10^9$ | $e^+ e^-$ annihilation |
| Particle | 100 | $[1.3 - 0.8] \times 10^9$ | Nucleosynthesis |
| Photon | $2 \times 10^{12}$ | $2 \times 10^{14}$ | Radiation era ends, CMB decouples |
| | $10^{13}$ | $3 \times 10^3$ | from plasma |
| Atoms | $10^{14}$ | $10^3$ | Plasma to atom; |
| Now | $4 \times 10^{17}$ | 2.75 | Present |

The experimentally detected microwave photons are therefore direct messengers from an era when the universe had an age of about $4 \times 10^5$ yrs. But photons are still around and they fill the universe and have nowhere else to go. The thermal radiation last scattered at this epoch is now detected as the cosmic background radiation. This epic, which corresponds $1 + z_{dec} \simeq 1100$, defines a "surface" known as "surface of last scattering".

We summarize the cosmic history of the universe (see Table 18.1): In particular

$$\rho_{\text{radiation}} = \rho_{\text{matter}}$$
$$z_{eq} = 3230; \ t_u = 56000 \text{ yrs}$$

CMB decouples from Plasma

$$z_{\text{dec}} = 1089, \ t_u = 380,000 \text{ yrs}$$
$$T_{CMB} = 2970 \text{ K}$$

First stars are thought to form at

$$z_r = 20, \ t_u = 2 \times 10^8 \text{yrs}$$

Now

$$z = 0, \ t_u = 13.7 \times 10^9 \text{ yrs}$$
$$T_{CMB} = 2.725 \text{ K}$$

We end this section by writing some useful numbers. From Eqs. (18.63) and (18.64), using the present temperature $T_0 = 2.735\text{K}$, we get

$$n_{\gamma_0} \approx \frac{2.4}{\pi^2} \left( \frac{k_B T_0}{\hbar c} \right)^3 \approx 415 \text{ cm}^{-3} \qquad (18.97)$$

$$\rho_{\gamma_0} \approx 2.6 \times 10^{-10} \text{ GeV cm}^{-3}. \qquad (18.98)$$

Thus $n_\gamma$ at temperature $T$ is given by

$$n_\gamma \approx 415 \left( \frac{T}{2.725} \right)^3 \text{ cm}^{-3}. \qquad (18.99)$$

In addition we write down the following estimates. There are about $10^{57}$ nucleons in a typical star. There are about $10^{11}$ galaxies in the universe, each galaxy has about $10^{11}$ stars. Thus there are about $10^{79}$ baryons in the universe. This is to be compared with $10^{89}$ photons within the part of the universe we can observe; this number is obtained by thermodynamical arguments. Thus number of baryons/number of photons $\approx 10^{-10}$. The present size of the observable universe is $10^{28}$ cm. Further the baryon number density $n_b$ is given by $n_b \sim \left( \frac{3}{4\pi} \frac{10^{79}}{(10^{28})^3} \right) \sim 10^{-6} \text{ cm}^{-3}$.

Another quantity of interest is baryon number density. First we note from Eq. (18.20) that it scales as $R^{-3}$. It is convenient to define the baryon density in terms of parameter $\eta$ viz

$$\eta \equiv \frac{n_B}{n_\gamma} \qquad (18.100)$$

where $n_B = n_b - n_{\bar{b}}$. The baryon number density, is given by

$$n_B = \frac{\rho_B}{m_B} = \frac{\rho_B}{\rho_c} \frac{\rho_c}{m_B}$$

$$= \frac{\Omega_B}{m_B} \rho_c \qquad (18.101)$$

where $\rho_B$ is the baryon energy density and $\Omega_b \equiv \frac{\rho_b}{\rho_c}$. Now using [cf. Eq. (18.27)]

$$\rho_c \approx 1.05 \times 10^{-25} h^2 \text{ GeV cm}^{-3} \qquad (18.102)$$

and taking $m_B = 1$ GeV, we obtain

$$n_B \approx \Omega_b \left( 1.05 \times 10^{-5} h^2 \right) \text{ cm}^{-3} \qquad (18.103)$$

$$\eta = \frac{n_B}{n_\gamma} = 2.65 \times 10^{-8} \Omega_b h^2$$

$$= (6.06 \pm 0.16) \times 10^{-10} \qquad (18.104)$$

if we use the value of $\Omega_b h^2$ given in Eq. (18.47) and

$$\rho_B = \eta \, n_{\gamma_0} (1 \text{ GeV}) = \eta \, (412) (1 \text{ GeV}) \text{ cm}^{-3}$$

$$= 7.0 \times 10^{-22} \eta \text{ gm cm}^{-3} \qquad (18.105)$$

## 18.6  Freeze Out

At high temperatures ($k_B T \gg m$), thermodynamic equilibrium is maintained through the processes of decays, inverse decays and scattering. As the universe cools and expands, the reaction rates will fail to keep up with the expansion rate and there will come a time when equilibrium will no longer be maintained. At various stages then, depending on masses and interaction strengths, different particles will decouple with a "freeze out" surviving abundance. We now determine conditions under which the statistical equilibrium is established.

From dimensional analysis, the reaction rate for a typical process can be written as follows. For the decay of an X-particle, the decay rate is given by

$$\Gamma_X \sim g_d\, \alpha_X\, m_X\, \frac{m_X}{\left[(k_B T)^2 + m_X^2\right]^{1/2}}, \tag{18.106}$$

where $m_X$ is the mass of the X-particle, $\alpha_X = \frac{f_X^2}{4\pi}$ is the measure of coupling strength of X-particle to the decay products, and $g_d$ are number of spin states for the decay channels. Note that

$$\Gamma_X \quad \approx \quad \begin{array}{ll} g_d\, \alpha_X\, m_X & k_B T \ll m_X \\ g_d\, \alpha_X\, \frac{m_X^2}{k_B T} & k_B T \gg m_X. \end{array} \tag{18.107}$$

The reaction rate for the scattering processes is given by

$$\Gamma = \langle n\sigma\, v \rangle \tag{18.108}$$

where $v$ is the velocity and $n$ is the number of target particles per unit volume. For a weak scattering process

$$\langle \sigma\, v \rangle = g_W^4\, \frac{(k_B T)^2}{\left[(k_B T)^2 + m_W^2\right]^2}. \tag{18.109}$$

Since $n \sim (k_B T)^3$, we can write the reaction rate for a weak process

$$\Gamma \sim g_W^4\, \frac{(k_B T)^5}{\left[(k_B T)^2 + m_W^2\right]^2}. \tag{18.110}$$

For $k_B T \ll m_W$, we get

$$\Gamma \sim \frac{g_W^4}{m_W^4}\, (k_B T)^5 \approx G_F^2\, (k_B T)^5. \tag{18.111}$$

The condition for thermal equilibrium is

$$\Gamma \geq H, \tag{18.112}$$

i.e. the reaction rate $\Gamma$ must be greater than the expansion rate to maintain the thermodynamic equilibrium.

We now consider a specific example. At about a temperature of 10 MeV, the universe is made up of neutrons, protons, $\nu$'s, $\bar{\nu}$'s, $e^{\pm}$ and $\gamma$'s in thermodynamic equilibrium. At about a few MeV, the neutrinos decouple. To see this consider the processes

$$\nu_e + e^- \leftrightarrow \nu_e + e^-, \qquad \nu\bar{\nu} \leftrightarrow e^+ e^-.$$

For these processes $\langle \sigma v \rangle = \frac{G_F^2}{\pi} s$ and $\frac{2}{3\pi} G_F^2 s$ respectively, where $s$ is the centre of mass energy. Thus the reaction rate

$$\Gamma \approx \frac{2}{3\pi} G_F^2 \ (k_B T)^5. \tag{18.113}$$

Now using Eq. (18.89), we find from Eq. (18.113) $[G_F = 1.166 \times 10^{-5}$ GeV$^{-2}]$

$$0.7 \left( \frac{k_B T}{\text{MeV}} \right)^2 \text{sec}^{-1} = \frac{2}{3\pi} G_F^2 \ (1 \ \text{GeV})^4 \left( \frac{k_B T}{\text{MeV}} \right)^5 \frac{1}{\hbar} \text{sec}^{-1}$$

$$= 0.04 \left( \frac{k_B T}{\text{MeV}} \right)^5 \text{sec}^{-1}. \tag{18.114}$$

Thus the decoupling temperature for neutrinos is given by

$$k_B T_D = 2.6 \text{ MeV}. \tag{18.115}$$

Hence for $k_B T < 2.6$ MeV, neutrinos are decoupled. The neutrinos are extreme relativistic particles. For (ER) particles

$$N_{\text{Eq}}^{\text{ER}} = n_{\text{Eq}}^{\text{ER}} \ V \propto T^3 \ R^3. \tag{18.116}$$

Now the entropy for ER gas is given by [cf. Eq. (18.67)]

$$S = \frac{R^3}{T} \frac{4}{3} \ \rho(T). \tag{18.117}$$

But $\rho(T) \propto R^{-4}$, therefore, $S \sim (1/RT)$. Thus for the entropy to remain constant $T \propto R^{-1}$. Hence from Eq. (18.116), we have the important result. In equilibrium ER particles are conserved. This can also be seen as follows:

As we have discussed in the beginning of this section, at high temperatures all interacting species $i$, $j$, $l$, $m$ are in thermodynamic equilibrium through the reactions of the type

$$i \ j \leftrightarrow l \ m.$$

As the reaction rate $\Gamma < H$ (the expansion rate), the species involved decouple and their abundance is frozen out. Consider an arbitrary volume $V$ and let $N_i$ be the number of particles of type $i$ in this volume. Thus for the reaction $i\, j \leftrightarrow l\, m$, we have

$$\frac{dN_i}{dt} = \Gamma_{\text{prod}}\, N_l - \Gamma_{\text{ann}}\, N_i \qquad (18.118)$$

where

$$\Gamma_{\text{prod}} = \langle v\, \sigma_{lm \to ij} \rangle\, n_m \qquad (18.119)$$
$$\Gamma_{\text{ann}} = \langle v\, \sigma_{ij \to lm} \rangle\, n_j. \qquad (18.120)$$

Thus

$$\frac{dN_i}{dt} = \langle v\, \sigma_{lm \to ij} \rangle\, n_m\, N_l - \langle v\, \sigma_{ij \to lm} \rangle\, n_j\, N_i. \qquad (18.121)$$

Now $N_i = n_i V$, therefore

$$\frac{dN_i}{dt} = V\, \frac{dn_i}{dt} + n_i\, \frac{dV}{dt} \propto \left[ R^3 \frac{dn_i}{dt} + 3\, n_i\, R^2 \frac{dR}{dt} \right]. \qquad (18.122)$$

Hence we have from Eq. (18.121)

$$\frac{dn_i}{dt} = -3\, n_i\, \frac{\dot{R}}{R} + \langle v\, \sigma_{lm \to ij} \rangle\, n_m\, n_l - \langle v\, \sigma_{ij \to lm} \rangle\, n_j\, n_i. \qquad (18.123)$$

The principle of detailed balance gives

$$\frac{\langle v\, \sigma_{lm \to ij} \rangle}{\langle v\, \sigma_{ij \to lm} \rangle} = \frac{n_l^{Eq}\, n_m^{Eq}}{n_i^{Eq}\, n_j^{Eq}}. \qquad (18.124)$$

Thus from Eqs. (18.121) and (18.123),

$$\frac{dN_i^{Eq}}{dt} = \frac{V\, \langle v\, \sigma_{ij \to lm} \rangle}{n_i^{Eq}\, n_j^{Eq}} \left[ \left( n_l^{Eq}\, n_m^{Eq} \right)^2 - \left( n_i^{Eq}\, n_j^{Eq} \right)^2 \right] \qquad (18.125)$$

and

$$\frac{d\, n_i^{Eq}}{dt} = -3\, n_i^{Eq}\, \frac{\dot{R}}{R} + \frac{\langle v\, \sigma_{ij \to lm} \rangle\, Eq}{n_i^{Eq}\, n_j^{Eq}} \left[ \left( n_l^{Eq}\, n_m^{Eq} \right)^2 - \left( n_i^{Eq}\, n_j^{Eq} \right)^2 \right]. \qquad (18.126)$$

Note here that $n_i^{Eq}$ etc. are given by Eq. (18.59). Now

$$\frac{d\, N_i^{Eq}}{dt} = 0 \qquad (18.127)$$

implies $\left[ \left( n_l^{Eq}\, n_m^{Eq} \right)^2 = \left( n_i^{Eq}\, n_j^{Eq} \right)^2 \right]$, which means that we must have

$$\frac{d\, n_i^{Eq}}{dt} = -3\, n_i^{Eq}\, \frac{\dot{R}}{R}. \qquad (18.128)$$

From this equation, we get

$$n_i^{Eq} \propto \frac{1}{R^3}, \qquad (18.129)$$

and from Eq. (18.125), we get

$$(n_i \, n_j)_{\text{Eq}} = (n_l \, n_m)_{\text{Eq}}$$

i.e. in equilibrium extreme relativistic particles are conserved. The condition (18.129) is always satisfied for the extreme relativistic particle.

Weakly interacting particles may decouple when they are ER, massless particles are always ER. For massive particles whose interactions are sufficiently strong to be capable of maintaining equilibrium when $k_B \, T < m$ : [cf. Eq. (18.77)].

$$N_{\text{Eq}}^{\text{NR}} = n_{\text{Eq}}^{\text{NR}} \, V \propto (m \, k_B \, T)^{3/2} \exp\left(-\frac{m}{k_B \, T}\right) \frac{1}{T^3}. \qquad (18.130)$$

## 18.7   Limit on Neutrino Mass

We now use the result that neutrinos decouple at a temperature of a few MeV (i.e. they go out of equilibrium before $e^- e^+$ annihilation heated up the photon background radiation). Thus $T_{\nu 0}$ will be less than $T_{\gamma 0}$. Using Eqs. (18.69) and (18.72), we get

$$\frac{n_{\nu 0}}{n_{\gamma 0}} = \frac{3}{4}\left(\frac{T_{\nu 0}}{T_{\gamma 0}}\right)^3. \qquad (18.131)$$

Now using Eq. (18.68) or (18.117), the entropy before $e^- e^+$ annihilation is given by

$$S = \frac{4}{3}\left(\rho_{e^-} + \rho_{e^+} + \rho_\gamma\right)\frac{R^3}{T_{\text{before}}}, \qquad (18.132)$$

and the entropy after $e^- e^+$ annihilation is given by

$$S = \frac{4}{3}\rho_\gamma \frac{R^3}{T_{\text{after}}}. \qquad (18.133)$$

Thus we have from Eqs. (18.132), (18.133), (18.64) and (18.66)

$$\left(\frac{7}{8} \times 2 + \frac{7}{8} \times 2 + 2\right) T_{\text{before}}^3 = 2 \, T_{\text{after}}^3. \qquad (18.134)$$

Noting that $T_{\text{before}} = T_{\nu 0}$ and $T_{\text{after}} = T_{\gamma 0}$ , we get

$$\left(\frac{T_{\nu 0}}{T_{\gamma 0}}\right)^3 = \frac{4}{11}. \qquad (18.135)$$

Hence we have from Eq. (18.131)

$$\frac{n_{\nu0}}{n_{\gamma0}} = \frac{3}{11} \qquad (18.136)$$

and [cf. Eq. (18.97)]

$$n_{\nu0} = \frac{3}{11}(412) \text{ cm}^{-3}. \qquad (18.137)$$

The present neutrinos density in GeV cm$^{-3}$ can be written as

$$\rho_{\nu0} \approx n_{\nu0}\left(\sum_i m_{\nu i} \text{ in eV}\right) \times 10^{-9} \text{ GeV cm}^{-3}. \qquad (18.138)$$

Using Eq. (18.137), we get

$$\rho_{\nu0} \approx (112)\left(\sum_i m_{\nu i} \text{ eV}\right) \times 10^{-9} \text{ GeV cm}^{-3} \qquad (18.139)$$

On using Eq. (18.27)

$$\Omega_\nu h^2 = \frac{\sum_i m_{\nu i} \text{ eV}}{94} \qquad (18.140)$$

WMAP [c.f. Eq. (18.47)] gives

$$\Omega_\nu h^2 < 0.0076$$

Thus

$$\sum_i m_{\nu i} < 0.71 \text{ eV} \qquad (18.141)$$

## 18.8 Primordial Nucleosynthesis

At temperatures $\geq 1$ MeV, the weak reactions such as

$$\bar{\nu}_e + p \leftrightarrow e^+ + n$$
$$e + p \leftrightarrow \nu_e + n \qquad (18.142)$$

are still fast compared with the expansion rate of the Universe to maintain thermodynamic equilibrium between $p$ and $n$. The abundance ratio at equilibrium is given by

$$\frac{n}{p} \sim e^{-\Delta m/(k_B T)}, \qquad k_B T > k_B T_D \sim 1 \text{ MeV}. \qquad (18.143)$$

Using $\Delta m = (m_n - m_p) = 1.3$ MeV and $k_B T = k_B T_D = 1$ MeV, we find $n/p = 0.27$. The decoupling temperature $T_D$ is estimated as follows. The reaction rate for reactions given in Eq. (18.142) is given by

$$\Gamma = \frac{7\pi}{30} G_F^2 \left(1 + 3g_A^2\right) (k_B T)^5$$

$$\approx 4.22 \, G_F^2 \, (k_B T)^5 = 0.8 \left(\frac{k_B T}{\text{MeV}}\right)^5 \sec^{-1}. \tag{18.144}$$

The decoupling temperature is given by $\Gamma = H$ viz [cf. Eq. (18.89), where we have taken $N_\nu = 3$]

$$0.8 \left(\frac{k_B T}{\text{MeV}}\right)^5 = 0.7 \left(\frac{k_B T}{\text{MeV}}\right)^2. \tag{18.145}$$

Thus

$$k_B T = k_B T_D \simeq 1 \text{ MeV}. \tag{18.146}$$

As the temperature cools and past the decoupling temperature $k_B T_D \approx 1$ MeV, it is no longer possible to maintain the thermal equilibrium. The ratio $n/p$ thereafter is frozen out and is approximately constant (it decreases slowly due to weak decay of neutron). The freeze out $n/p$ ratio is given by

$$n/p \approx e^{-Q/k_D T_D \text{ MeV}} \approx 0.16, \tag{18.147}$$

where we have used the $Q$-value (the energy released when various particles are emitted in $\beta$ decay) $Q = (m_n - m_p) + m_e = 1.8$ MeV. For $T > T_S$, the deuteron formed is knocked out by photo dissociation

$$\gamma + D \rightarrow p + n,$$

since the binding energy $\Delta B$ for the deuteron is only 2.2 MeV. The formation of deuteron actually starts after $T_S \approx 0.1$ MeV; $T_S$ is called nucleosynthesis temperature. The estimate that $T_S \approx 0.1$ MeV can be obtained as follows:

$$\frac{n_\gamma^{\text{diss}}}{n_B} \sim \frac{1}{\eta} e^{-\Delta B/k_B T} \leq 1. \tag{18.148}$$

Thus

$$-\frac{\Delta B}{T_S} \approx \ln \eta. \tag{18.149}$$

Using $\Delta B \approx 2.2$ MeV , and $\eta \approx 6 \times 10^{-10}$, we find $T_S \approx 0.1$ MeV.

For $T > T_S$, photodissociation is so rapid that deuteron abundance is negligibly small and this provides a bottleneck to further nucleosynthesis.

The deuteron "bottleneck" thus delays nucleosynthesis till $T \leq 0.1$ MeV. But once the bottleneck is passed, nucleosynthesis proceeds rapidly and essentially all neutrons are incorporated into $^4He$ :

$$n + p \to D + \gamma$$
$$D + D \to \; ^3H + p, \qquad ^3He + n$$
$$^3H + D \to \; ^4He + n$$
$$^3H + \; ^4He \to \; ^7Li$$

It is clear from the above reactions that $^4He$ abundance is given by

$$Y = \frac{2\,(n \;/\; p)}{1 + n \;/\; p} = \frac{0.32}{1.16} = 0.27. \tag{18.150}$$

The ratio $Y$ changes from $T_D$ to $T_S$ due to the neutron decay $n \to p + e^- + \bar{\nu}_e$. During this time $n/p$ changes from 0.16 to 0.14. Thus at $T = T_S$,

$$Y = \frac{0.28}{1.14} = 0.25. \tag{18.151}$$

We conclude that $n/p$ ratio or $Y$ depends on three parameters:

(i) decoupling temperature $T_D$, which in turn depends on the number of light particles, e.g. number of neutrino flavors $N_\nu$.

(ii) neutron decay in between $T_D$ and $T_S$, i.e. on the decay rate of neutron or neutron half-life $\tau_{1/2}$.

(iii) $\eta = n_B/n_\gamma$.

In fact $Y$ is arguably the most sensitive function of $\Gamma/H$. Now $T_D$ depends on the expansion rate $H$; the expansion rate depends upon the effective degrees of freedom $g_*$, the higher the $g_*$, the faster the expansion rate. This implies higher $T_D$ and hence higher $n/p$ freeze out abundance. Thus the higher the $g_*$, the higher will be $Y$. But $g_* = 2 + \frac{7}{8}(4 + 2N_\nu)$, where $N_\nu$ are the number of neutrino species. For $N_\nu = 3$, $g_* = \frac{43}{4}$ and we obtained $T_D \approx 1$ MeV and $Y \approx 0.25$. The observed primordial abundance of $^4He$ gives $Y = 0.234 \pm 0.002\,(\pm0.005)$. The half-life for neutron decay, the parameter needed in the above analysis, is $\tau_{1/2} = 885.7 \pm 0.8$ s.

Taking $N\nu = 3$ as given by LEP data [cf. Sec. 18.13]: $N\nu = 2.999 \pm 0.016$, one can use the observed primordial abundances of D, $^4$He. It turns out that small amount of deuterium D which remains unburned in the nucleosyntheis reaction is very sensitive to $\eta$. Now D/H ratio in primary samples of the universe has been measured

$$D/H = (2.84 \pm 0.26) \times 10^{-5} \tag{18.152}$$

pinning

$$\eta = (5.80 \pm 0.27) \times 10^{-10}$$

consistent with the value given in Eq. (18.104), obtained from WMAP data, although the physics (Nuclear Physics) involved here is different from the one involved in WMAP (Gravitational). Table 18.1 summarizes the various phase transitions the Universe is thought to have undergone.

## 18.9   Inflation

There are several problems in the standard model of Cosmology. We now discuss two of these problems and their proposed solutions in the inflationary scenario.

### 18.9.1   *Horizon Problem*

In terms of the conformal time $\eta$

$$cdt = R(t)\, d\eta, \tag{18.153}$$

the FRW metric factorizes

$$ds^2 = R(\eta)^2 \left\{ d\eta^2 - d\chi^2 - f_k^2 \left( d\theta^2 + \sin^2\theta d\phi^2 \right) \right\} \tag{18.154}$$

It can be seen that the radial (i.e. $d\theta = 0 = d\phi$) propagation of light ($ds^2 = 0$) is particularly simple in these coordinates.

$$d\eta = d\chi \tag{18.155}$$

[for the flat universe $\chi = r$]. Thus in a given time, a photon can go as far as

$$\Delta\chi = \Delta\eta = \int_{t_1}^{t_2} \frac{cdt}{R(t)}, \tag{18.156}$$

i.e. just the interval of the conformal time. Using Eqs. (18.37) and (18.5), we can rewrite the integral as

$$\Delta\chi = -\frac{c}{R_0} \int \frac{dz}{H(z)} \tag{18.157}$$

where from Eq. (18.33) [$k = 0$]

$$\frac{H(z)}{H_0} = \left[ \sum_i \Omega_i (1+z)^{3(1+\omega_i)} \right]^{1/2} \tag{18.158}$$

Thus we see that the integral converges if $\omega > -\frac{1}{3}$ for the dominant component in the above sum as $z \to \infty$ (corresponding to $k = 0$, $R = 0$). Thus light signals can only propagate a finite distance between the Big bang and the present; there is then said to be particle horizon and space-time points outside the particle horizon are causally disconnected as shown in the Fig. 18.6.

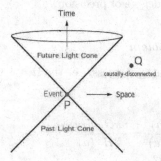

Fig. 18.6    Light cones and causality. The interior of the light cone, consisting of all null and time-like geodesics, defined the region of space-time causally related to that event. Causally disconnected regions of space-time are outside the light cone.

Thus in the FRW cosmology where radiation dominates at early times there exists such a horizon. At later times, horizon is largely determined by matter dominated phase, for which ($\omega = 0$),

$$D_H = R_o \chi$$
$$= \frac{1}{H_0 \sqrt{\Omega}} \int_z^\infty (1 + z)^{-3/2}$$
$$= \frac{1}{H_0 \sqrt{\Omega}} \frac{2}{(1 + z)^{1/2}} \tag{18.159}$$

where in the units with $c = 1$,

$$H_0^{-1} \simeq 3000 h^{-1} \; Mpc. \tag{18.160}$$

Thus the horizon at the time of formation of CMB (last scattering surface ($z \simeq 1100$)) was of size

$$D_H \simeq \frac{18}{\sqrt{\Omega h^2}} \; \text{Mpc} \simeq 120 \; \text{Mpc} \tag{18.161}$$

for

$$\Omega h^2 = \Omega_b h^2 \simeq 0.023. \tag{18.162}$$

This subtends an angel of about $1°$. The question arises "Why then the large number of causally disconnected regions we see on the microwave sky are all at the same temperature". This is called the Horizon problem. Thus it is difficult to explain the high degree of isotropy in the present universe. The uniformity of the temperature of the background microwave radiation $\left(\frac{\Delta T}{T} \leq 10^{-5}\right)$ provides a strong evidence that the universe is isotropic and homogeneous to a high degree of precision.

## 18.9.2   *Flatness Problem*

A related problem is that of flatness, i.e. the $\Omega = 1$ Universe is unstable[10]. Now $\left[\text{with } a\left(t\right) = \frac{R(t)}{R_0}\right]$

$$\Omega\left(a\right) \equiv \frac{\rho\left(a\right)}{\rho_c\left(a\right)} = \frac{8\pi G_N \rho\left(a\right)}{3H^2\left(a\right)}$$

$$= \frac{H_0^2}{H^2\left(a\right)} \frac{\rho\left(a\right)}{\rho_c}$$

$$= \frac{\left\{\Omega_V + \Omega_m a^{-3} + \Omega_r a^{-4}\right\}}{\left\{\left(\Omega_V + \Omega_m a^{-3} + \Omega_r a^{-4}\right) - \left(\Omega_0 - 1\right) a^{-2}\right\}} \qquad (18.163)$$

where we have used the scaling of various components of density with respect to $R$ given in Eq. (18.31), Friedmann equation (18.12) and the equation (18.34)

$$\frac{k}{H_0^2 R_0^2} = \Omega_V + \Omega_m + \Omega_r - 1$$

$$= \Omega - 1 \qquad (18.164)$$

Thus since the early part of the Universe is radiation dominated

$$\Omega\left(a_{in}\right) \simeq 1 + \frac{\Omega - 1}{\Omega_r} a_{in}^2 \qquad (18.165)$$

Thus as $a_{in} \to 0$, this requires that $\Omega\left(t\right)$ to be unity to arbitrary precision as the initial time tend to zero; a universe of non-zero curvature today requires very finely tuned initial conditions which looks unnatural. The natural solution is either $\rho = \rho_c$, $\Omega = 1$ (for a reason to be discovered) for the whole history of the universe or some non standard mechanism intervened to drive $\rho_0 \to \rho_{c0}$, $\Omega_0 \to 1$.

---

[10]In the sense that if one is slightly away from this special value $\Omega - 1$ grows with time.

### 18.9.3    *Realization of Inflation*

A possible solution to both the problems lies in that if dominant energy component in the early universe did not satisfy $\omega > -\frac{1}{3}$, which is indeed the case if vacuum energy $\Omega_V$ dominates in the beginning. Then the conformal time $\to -\infty$ [cf. integral in Eq. (18.156)], i.e. the Big Bang singularity is pushed to negative time.' This implies that there was much more conformal time between the singularity and decoupling than we thought, allowing the microphysical processes to equalize the temperature over the last scattering surface as shown in Fig. 18.7.

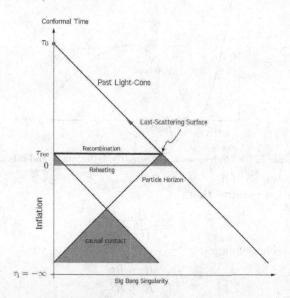

Fig. 18.7    Conformal diagram of inflationary cosmology. Inflation extends conformal time to negative values! The end of inflation creates an *apparent* Big Bang at $\eta \equiv \tau = 0$. There is, however, no singularity at $\tau = 0$ and the light cones intersect at an earlier time [21].

The basic idea of this scenario is that there was an epoch when the vacuum energy density dominated the energy density of the Universe. Then we write

$$\rho = \rho_V + \rho_r = \rho_V + \frac{\pi^2}{30}\ g_*(T)\ (k_B\ T)^4 . \qquad (18.166)$$

The radiation era density $\rho_r \sim \frac{1}{R^4}$ , but $\rho_V$ is constant independent of $R$. Suppose $\rho_V \gg \rho_r$ in the early universe. Thus from Friedmann equation

(18.12), we have

$$\frac{\dot{R}}{R} = H = \sqrt{\frac{8\pi\, G_N\, \rho_V}{3}} = \text{const.} \qquad (18.167)$$

Thus $\rho_V$ acts like an effective cosmological constant. We get [see Fig. 18.8]

$$R(t) = e^{Ht}, \qquad t \le 10^{-35}\text{s} \qquad (18.168)$$

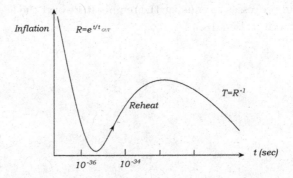

Fig. 18.8   Scale factor $R(t)$ verses t, showing the inflationary and reheating phase.

The inflation requires that

$$\epsilon = -\frac{\dot{H}}{H^2} < 1 \qquad (18.169)$$

This can be seen as follows:

$$\dot{H} = \frac{\ddot{R}}{R} - \frac{\dot{R}^2}{R^2} \qquad (18.170)$$

Using the Friedmann Eqs. (18.12) and (18.10), it is easy to see that

$$\epsilon = \frac{3}{2}\left(1 + \frac{p}{\rho}\right) \qquad (18.171)$$

The condition for inflation is

$$\frac{p}{\rho} = \omega < -\frac{1}{3} \qquad (18.172)$$

which means $\epsilon < 1$. It is very likely that the Universe has undergone one or more phase transitions caused by a symmetry breaking involving a scalar field, the inflaton, $\phi$. Such a scalar field as already seen satisfy the equation of motion given in Eq. (18.52)

$$\ddot{\phi} + 3H\dot{\phi} = -\frac{dV}{d\phi} = -V' \qquad (18.173)$$

Further using $p$ and $\rho$ as given in Eq. (18.50), we have from the Friedmann equation (18.12)

$$H^2 = \frac{8\pi G_N}{3}\left[\frac{1}{2}\dot{\phi}^2 + V(\phi)\right]$$

$$p + \rho = \dot{\phi}^2 \tag{18.174}$$

$$\epsilon = \frac{3}{2}\frac{\dot{\phi}^2}{V + \frac{1}{2}\dot{\phi}^2} = 4\pi G_N\frac{\dot{\phi}^2}{H^2} \tag{18.175}$$

We see that $\epsilon \ll 1$, implies $\frac{1}{2}\dot{\phi}^2 \ll V$, which on using first of Eqs. (18.174) and (18.173) leads to the condition given in Eq. (18.169).

## 18.9.4  *Slow-roll Inflation*

As seen above the inflation requires $\epsilon \ll 1$, which implies

$$\frac{1}{2}\dot{\phi}^2 \ll V \tag{18.176}$$

We now discuss how this condition is implemented in slow-roll approximation. For slow-roll approximation

$$\left|\ddot{\phi}\right| \ll \left[3H\dot{\phi}\right] \tag{18.177}$$

so that from the equation of motion (18.173)

$$3H\dot{\phi} = -V' \tag{18.178}$$

Thus

$$\frac{1}{2}\dot{\phi}^2 = \frac{1}{2}\left(\frac{V'}{3H}\right)^2$$

$$H^2 = \frac{8\pi G_N}{3}V \tag{18.179}$$

Consistency then requires $[G_N^{1/2} = \frac{1}{M_{Pl}}]$

$$\frac{1}{2}\dot{\phi}^2 = \frac{1}{2}\left(\frac{V'}{3H}\right)^2$$

$$= \frac{M_{Pl}^2}{48\pi}\left(\frac{V'}{V}\right)^2 V \ll V \tag{18.180}$$

i.e.

$$\left(\frac{V'}{V}\right)^2 \ll \frac{48\pi}{M_{Pl}^2} \tag{18.181}$$

or

$$\epsilon_v \equiv \frac{M_{Pl}^2}{16\pi} \left(\frac{V'}{V}\right)^2 \ll 1 \qquad (18.182)$$

By differentiating (18.180) and using Eqs. (18.178) and (18.182), it is easy to see that

$$\delta \equiv \frac{\ddot{\phi}}{3H\dot{\phi}}$$

$$= \left[-\frac{M_{Pl}^2}{8\pi}\frac{V''}{V} + \epsilon_v\right]$$

$$= \epsilon_v - \eta_v \qquad (18.183)$$

where

$$\eta_v = \frac{M_{Pl}^2}{8\pi}\frac{V''}{V} \qquad (18.184)$$

The slow-roll condition (18.177), namely $|\delta| \ll 1$ then gives

$$|\eta_v| \ll 1 \qquad (18.185)$$

Before we proceed further with the discussion of the above equation, it is instructive to make the following remarks:

In addition to scalar (density perturbations) contribution to CMB anisotropies there is a tensor contribution arising presumably from primordial gravitational waves. In this respect a useful parameter is the ratio $r$ between the power spectrum of tensors to scalars. Its importance lies in the fact that it is the only non-generic prediction of inflation. In fact $r \simeq 16\epsilon_v$, where $\epsilon_v$ is defined in Eq. (18.182). WMAP has placed bound $r < 0.24$ at 95% confidence level, whereas PLANCK Satellite would be able to go up to $r \sim 0.03$.

Coming back to Eq. (18.182), define $dN = Hdt$, which measures the number of e-folds N of inflationary expression $R(t) \sim e^{\int Hdt} = e^N$. Thus

$$N = \int Hdt = \int \frac{H}{\dot{\phi}}d\phi$$

$$= \int \frac{3H^2}{3H\dot{\phi}}d\phi$$

$$= \int 8\pi G_N \frac{V}{V'}d\phi$$

$$= 2\sqrt{\pi}G_N^{1/2} \int \frac{d\phi}{\sqrt{\epsilon_v}}$$

$$= \sqrt{4\pi} \int \frac{d\phi}{\sqrt{\epsilon_v}M_{Pl}} \qquad (18.186)$$

where we have used Eq. (18.182). Now from Eqs. (18.184) and (18.185), it is clear that $\sqrt{\epsilon_v}$ is a slowly varying function so that

$$\Delta N \sim \frac{\sqrt{4\pi}}{\sqrt{\epsilon_v}} \frac{\Delta\phi}{M_{\text{Pl}}} \tag{18.187}$$

Now we want inflation to last for a sufficiently long time ($N \sim 60$ e-fold)[11] giving

$$\frac{\Delta\phi}{M_{\text{Pl}}} \sim 60\sqrt{\epsilon_v} \sim \mathcal{O}(1) \left(\frac{r}{0.01}\right)^{1/2}. \tag{18.188}$$

This is worrisome since the range of scalar field is almost Planck scale if one has to use effective field theory. In other words, higher dimensional Planck suppressed operators give corrections

$$\Delta V = \frac{O_6}{M_{\text{Pl}}^2} \sim V_0 \frac{\Delta\phi^2}{M_{\text{Pl}}^2} \tag{18.189}$$

of order 1. Therefore, the effective field description might not be useful if primordial gravitational waves are detected by PLANCK. This problem can be avoided if underlying ultraviolet theory has a symmetry which forbids the appearance of these higher order terms. PLANCK Satellite data expected in 2012 will be of great interest in order to constrain different models of inflation.

## 18.10 Baryogenesis

Barayogenesis[12] is a term used to understand the origin of matter-antimatter asymmetry; there is only matter in the universe, but no antimatter. Quantitatively it is expressed as

$$\Delta B = \frac{n_B - n_{\bar{B}}}{s} \sim O\left(10^{-10}\right) \tag{18.190}$$

where $s$ is the entropy and $n_B$ and $n_{\bar{B}}$ are number of baryons and anti-baryons in the Universe. A measure of smallness of the baryon density is provided by the ratio of baryons to photons in the Cosmic Microwave Background Radiation (CMBR):

$$\eta_B = \frac{n_B}{n_\gamma}$$

---

[11] This much expansion under inflation is required to at least explain the homogeneity of the presently observed Universe.

[12] See [14,19].

$\left(n_\gamma = 410.50 \text{ cm}^{-3}\right)$ which does not change as universe expands. As seen in Secs. 18.5 and 18.8 two independent determinations of this $\eta_B$, one from WMAP and the other from primordial nuclear synthesis, give $\eta_B = 6 \times 10^{-10}$.

Further there is good evidence that there are no large regions of anti-matter at any but cosmic distance scales. For example anti-proton $\bar{p}$ to proton $p$ ratio in cosmic rays is

$$\frac{\bar{p}}{p} \sim 10^{-4}$$

In general $p$ and $\bar{p}$ may annihilate if they are brought together, e.g.

$$p + \bar{p} \to \gamma + \gamma$$

How does it affect the nuclear density $n_B$ or $n_{\bar{B}}$? One can show that if we start with $n_B = n_{\bar{B}}$, then for $T \leq 1$ GeV the freeze out temperature comes out to be $T = T^* = 20$ MeV, at which

$$\eta = \frac{n_B}{n_\gamma} = \frac{n_{\bar{B}}}{n_\gamma} \simeq \left(\frac{m_N}{K_B T}\right)^{3/2} e^{-\frac{m_N}{K_B T}} = 2 \times 10^{-18} \tag{18.191}$$

This contradicts $\eta = 6 \times 10^{-10}$, which then reflects some primordial asymmetry in the Universe

$$n_B \to n_B - n_{\bar{B}}$$

$$\eta = \frac{n_B - n_{\bar{B}}}{n_\gamma} = 6 \times 10^{-10}$$

Understanding the nature of $\eta$, namely that of Baryogenesis is the second big problem of Cosmology, the first one being Dark matter and Dark energy.

In the early universe

$$n_B - n_{\bar{B}} = \frac{1}{3}\left(n_q - n_{\bar{q}}\right)$$
$$s \approx n_q + n_{\bar{q}}$$
$$\Delta B \sim \frac{n_q - n_{\bar{q}}}{n_q + n_{\bar{q}}} \approx 10^{-10} \tag{18.192}$$

Thus there is one extra quark per $10^{10}$ quark-antiquark pairs. Where did it come from?

### 18.10.1 Sakharov's Conditions

Sakharov was the first to suggest that $\eta$ might be understood in terms of microphysical laws rather than some sort of boundary condition provided that the following three conditions are satisfied.

(1) Baryon number violation must occur in the fundamental laws. Firstly if at very early universe, $\Delta B = 0$ interactions were in equilibrium, then one can say that the universe "started" with net zero baryon number. As such if the universe is to end up with a non-zero asymmetry, $\Delta B \neq 0$ interactions are obviously necessary.

(2) CP symmetry must be violated; otherwise every reaction which produces a particle will be necessarily accompanied by a reaction which produces its antiparticle at exactly the same rate, so

$$n_B \underset{CP}{\rightarrow} n_{\bar{B}}$$

and even if $B$ is violated, we can never establish baryon-antibaryon asymmetry.

(3) Departure from thermal equilibrium, if through its early history, the universe was in thermal equilibrium, then even $B$ and CP violating interactions could not produce a net asymmetry. This is a direct consequence of $\theta \equiv CPT$ invariance. The density matrix at time $t$ is: [$\beta = \frac{1}{kT}$, $H$ is Hamiltonian]

$$\rho(t) = e^{-\beta(t)H(t)}$$

Equilibrium average of baryon number operator $\hat{B}$ is

$$\langle B \rangle_T = Tr\left(e^{-\beta H}\hat{B}\right)$$
$$= Tr\left(\theta^{-1}\theta e^{-\beta H}\hat{B}\right)$$
$$= Tr\left(\theta e^{-\beta H}\hat{B}\theta^{-1}\right)$$
$$= Tr\left(\theta e^{-\beta H}\theta^{-1}\theta\hat{B}\theta^{-1}\right)$$
$$= Tr\left(e^{-\beta H}\left(-\hat{B}\right)\right)$$
$$= -\langle B \rangle_T$$

where we have used the fact that $H$ commutes with $\theta$. Thus $\langle B \rangle_T$ vanishes. In other words if there is thermal equilibrium, all asymmetries vanished even if all quantum numbers are not conserved. Thus all the three conditions must be satisfied simultaneously.

## 18.10.2   *Various Scenarios for Baryogenesis*

### 18.10.2.1   *Baryogenesis at GUT (Grand Unification theories) Level*

In a typical GUT, quarks and leptons are assigned in one representations and as such there are gauge bosons, called leptoquarks $X$ and $\bar{X}$ giving rise to $B$ violating processes

$$X \to ql \qquad X \to \bar{Q}\bar{q}$$

and similarly for $\bar{X}$. Thus we see that the decays of $X$ bosons violate baryon number; they also violate CP, meeting the first two conditions. The third condition is supplied by the expansion of the universe. It is usually assumed that at temperature well above the GUT scale, the universe was in thermal equilibrium. As the temperature drops below the mass of $X$-boson, the reactions which produce the X-bosons are not sufficiently rapid to keep up with the expansion rate to maintain equilibrium. The condition for departure from equilibrium gives

$$KT_D \geq m_X \qquad \left(\approx 10^{15} - 10^{16} \text{ GeV}\right)$$

However, a reheating temperature after inflation greater than $10^9$ GeV leads to cosmological difficulties. Thus it is very unlikely that GUT Baryogenesis is the origin of observed baryon asymmetry.

### 18.10.2.2   *Electroweak Baryogenesis*

In the SM, both baryon number, B and lepton number, L, symmetries hold at classical level. However, because of chiral nature of electroweak theory, at quantum level $(J_B^\mu)$ and $(J_L^\mu)$ are not conserved, so-called electroweak anomaly.

$$\partial_\mu J_B^\mu = \left(\frac{1}{3}\right)_{B_q} (3 \text{ colors}) \, N_g \left(\frac{\alpha^2}{\pi} W_a^{\mu\nu} \tilde{W}_{a\mu\nu}\right) \qquad (18.193)$$

$$= \partial_\mu J_L^\mu$$

where, $N_g$ denotes number of generations, $W_a^{\mu\nu}$ are $SU(2)$ field strengths, $\frac{\alpha_2}{\pi}$ coupling corresponding to $SU_L(2)$ and $\tilde{W}_a^{\mu\nu} = \frac{1}{2}\varepsilon^{\mu\nu\alpha\beta}W_{a\alpha\beta}$. Thus

$$\partial_\mu \left(J_B^\mu - J_L^\mu\right) = \partial_\mu J_{B-L}^\mu = 0$$

$$\partial_\mu \left(J_B^\mu + J_L^\mu\right) \neq 0$$

implying

$$\Delta B = B(+\infty) - B(-\infty)$$
$$= \int_{-\infty}^{+\infty} dt \partial_0 \int d^3x J_B^0(t, \mathbf{x})$$
$$= \int_{-\infty}^{+\infty} dt \int d^3x \partial_\mu J_B^\mu(t, \mathbf{x}) \qquad (18.194)$$

Similarly for $\Delta L$. Note that $\Delta(B - L) = 0$, the electroweak anomaly preserves $B - L$. But

$$\Delta(B + L) \neq 0$$
$$= 2N_g \nu$$

where

$$\nu = \frac{\alpha^2}{8\pi} \int d^4x W_a^{\mu\nu} \tilde{W}_{a\mu\nu} \qquad (18.195)$$

Physically one can understand how baryon number is not conserved without explicit $B$-violating terms in the Lagrangian. Let us recall Dirac equation for massless fermions

$$i\frac{\partial}{\partial t}\psi = i\gamma^0 \gamma (\partial - ig\mathbf{A}) \psi = H_{Dirac}(t) \psi \qquad (18.196)$$

and Dirac concept of vacuum, an infinite sea of electrons occupying negative energy as shown in Fig. 18.9.

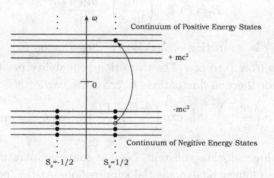

Fig. 18.9  Dirac picture of vacuum.

Non-Abelian theories (QCD or $SU_L(2)$) have a non-trivial vacuum structure with an infinite number of ground states (topological charges). Now we have a background field, energy levels change, which produces a fermion number. This is pictorially depicted in Fig. 18.10. While in QCD, baryon number is conserved, not so for $SU_L(2)$. Now for $N_g = 3$, the distance between two ground states is

$$\Delta B = \Delta L = 3$$

and at zero temperature, the transition (through tunneling) probability is

$$\sigma(B + L \neq 0) \sim e^{-\frac{4\pi}{\alpha_2}} \sim e^{-165}$$

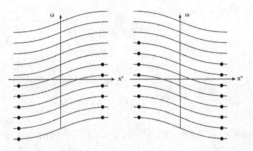

Fig. 18.10   (a) and (b) for QCD case while for electroweak case we have only (a) i.e. only left handed fermions [19].

The situation is different at high temperatures when large thermal fluctuations called "sphalerons" are possible. The sphaleron energy is

$$E_{sp}(T) \equiv c\left(\frac{m_H}{m_W}\right)\frac{\pi m_W(T)}{\alpha_2} \qquad (18.197)$$

where $c\left(\frac{m_H}{m_W}\right)$ is a function of $\lambda$, $\lambda \sim \frac{m_H^2}{m_W^2}$ [for $T = 0$, $E_{sp} \approx 7 - 14$ TeV as $\lambda$ increases from 0 to $\infty$]. The transition probability per unit time per unit volume for thermal fluctuations to cross the barrier is

$$P_{sp}(T) = \mu\left(\frac{m_W}{\alpha_2 T}\right)^3 m_W^4 \exp\left(\frac{-E_{sp}(T)}{T}\right), \quad \text{for } T < T_{EW} \text{ (barrier)}$$

$$(18.198)$$

where $\mu$ is a dimensionless constant. Now at some temperature $m_W(T) \rightarrow 0$, there will no longer be exponential suppression. At this point the computation of the transition rate is a difficult problem - there is no small parameter but noting that the only important scale in the symmetric phase is $\alpha_2 T$, scaling arguments suggest the form

$$P_{sp}(T) = \kappa(\alpha_2 T)^4, \quad \text{for } T > T_{EW} \qquad (18.199)$$

where numerical estimates suggest $\kappa \sim 0.1 - 1$.

Now $B$ and $L$ violating processes are in thermal equilibrium for [$H$ is the Hubble parameter which determines the expansion rate]

$$\Gamma_{sp} \equiv \frac{P_{sp}(T)}{T^3} > H \approx 1.66 \frac{g^{*\frac{1}{2}} T^2}{M_{\text{Pl}}} \tag{18.200}$$

This gives [$\alpha_2 \approx 0.029$]

$$T < \left(\kappa g^{*\frac{1}{2}}\right) \alpha_2^4 M_{\text{Pl}} \approx 10^{12} \text{ GeV} \tag{18.201}$$

The $B$ and $L$ violating processes are in thermal equilibrium for temperature in the range

$$T_{EW} \approx 100 \text{ GeV} < T < T_{sp} \approx 10^{12} \text{ GeV} \tag{18.202}$$

This implies that even if there were both $B$ and $L$, but no net $B - L$ established above $T_{\max} \sim \alpha_2^4 M_{\text{Pl}} = 10^{12}$ GeV, they will get washed out down to $T_{EW}$.

Now B-violation is rapid as compared to cosmological expansion at $\langle\phi\rangle_T < T$, where $\langle\phi\rangle_T$ is Higgs expectation value at temperature $T$. But the Universe expands slowly. Expansion time

$$H^{-1} = \frac{M_{\text{Pl}}}{T_{EW}^2} \sim 10^{14} \text{ GeV}^{-1} \sim 10^{-10} \text{ sec.} \tag{18.203}$$

This is too large to have deviations from thermal equilibrium. One way out is if the EW phase transition is of first order since then the co-existence of broken $\langle\phi\rangle \neq 0$ and unbroken phase $\langle\phi\rangle = 0$ at the phase transition at $T_c \sim T_{EW}$, [see Fig. 18.11], where there is condensation of the Higgs field, is a departure from equilibrium. After the phase transition baryon asymmetry should not be washed out, $\langle\phi\rangle_T > T$ after the phase transition. However, one cannot get the first order transition unless ($m_H < 73$ GeV), excluded by the LEP limit ($m_H > 114$ GeV). Another problem is the size of CP violation in the SM: $\lambda = 0.2$

$$\eta \sim \alpha_2^4 \varepsilon_{CP} = 10^{-16} \lambda^6 \sin\delta = 6 \times 10^{-12}\delta$$
$$\sim 10^{-18} \tag{18.204}$$

This requires SM extensions so as to have new bosons which strongly couple with Higgs plus a new source of CP-violation.

### 18.10.3 Leptogenesis

The second possibility is the Leptogenesis[13] where one tries to generate $L \neq 0$ but not $B$ from neutrino physics giving a net $B - L$. This is an

---

[13]See [11,13,14].

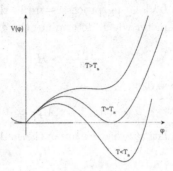

Fig. 18.11    Electroweak phase transition.

attractive possibility due to the fact that, as seen in Chap. 12 neutrino physics has entered a flourishing era.

As was discussed in Chap. 12, light neutrino masses arise in the see-saw mechanism, which involves 3 heavy right-handed Majorana neutrinos $N_i$, whose mass matrix breaks the lepton number (cf. Eq. (12.80)) needed for Leptogenesis. One starts from a thermal distribution of heavy Majorana neutrinos which have CP violating decay modes into standard leptons (CP violation can enter through phases in the Yukawa couplings and the mass matrices of N's). The asymmetry is then generated by the out of equilibrium CP violating decays in the early Universe of $N_i \to l_i H$ versus $N_i \to \bar{l}_i H$ at the temperature $T \sim M \equiv M_i \ll M_j$.

The crucial ingredients in leptogenesis scenario is CP asymmetry generated through the interference between tree level and one-loop Majorana neutrino decay diagrams. In the simplest extension of SM, these are shown in Fig. 18.12.

Then the CP asymmetry is caused by interference between the above diagrams and in a basis where M (mass matrix for heavy right-handed neutrinos) is diagonal and real,

$$\varepsilon_i = \frac{\Gamma\left(N_i \to l_i H\right) - \Gamma\left(N_i \to \bar{l}_i H\right)}{\Gamma\left(N_i \to l_i H\right) + \Gamma\left(N_i \to \bar{l}_i H\right)}$$

$$= \frac{1}{8\pi} \sum_{j \neq i} I_{kl}^{ij} \left[ f\left(\frac{|M_j|^2}{|M_i|^2}\right)\right] \tag{18.205}$$

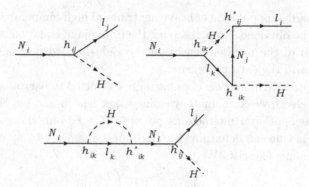

Fig. 18.12 Feynman diagrams of Majorana neutrino decay for tree level and one loop level.

where

$$f(x) = \sqrt{x} \left[ \frac{2-x}{1-x} - (1+x) \ln \left( \frac{1+x}{x} \right) \right]$$

$$\rightarrow \frac{1}{\sqrt{x}} \left( -\frac{3}{2} \right) \quad \text{as} \quad \frac{1}{x} \rightarrow 0 \ (x \rightarrow \infty) \quad (18.206)$$

$$f(x) \approx -\frac{1}{x} \quad \text{for x} \ll 1$$

$$I_{kl}^{ij} = \frac{1}{|h_{ij}|^2} \Im \left[ h_{ik} h_{il} h_{jk}^* h_{jl}^* \right] \quad (18.207)$$

Using

$$(M_D)_{ij} = h_{ij} v$$

we can write

$$I_{kl}^{ij} = \frac{1}{v^2 (m_D m_D^*)_{ii}} \left[ \Im \left( (m_D m_D^*)_{ij} \right)^2 \right] \quad (18.208)$$

Once the right-handed neutrinos are produced in early universe and as the universe cools through temperatures of the order of the masses of $N$'s, their long life time would allow them to decay out of equilibrium, thereby generating an asymmetry between leptons and anti-leptons.

The lepton asymmetry $Y_L$ is related to CP asymmetry through the relation

$$Y_L = \frac{n_L - n_{\bar{L}}}{s} = \kappa \frac{n}{s} \epsilon_i \quad (18.209)$$

where $\kappa \leq 1$ is efficient factor which accounts for the wash out processes [inverse decay and lepton number violating scattering; such processes can

create a thermal population of heavy neutrinos of high temperature $T > M$]
and it can be obtained through solving the Boltzmann equations. $\frac{n}{s} \sim 10^{-3}$
is the ratio of the number density of the right-handed neutrinos $(N_i)$ in
equilibrium to the entropy density.

The lepton asymmetry is then partially converted to baryon asymmetry
thanks to electroweak anomaly, yielding a net lepton and baryon number
(recall that sphaleron interactions preserve $B - L$, but violate $B$ and $L$
separately). One can determine the resulting asymmetry by an elementary
thermodynamic exercise. We have

$$n_i - n_{\bar{i}} = \frac{2}{\pi^2} g' T^3 \left( \frac{2\mu_i}{T} \right) \tag{18.210}$$

where $\mu_i$ are chemical potential. This implies

$$n_B = B \left( \frac{4}{\pi^2} g' T^2 \right) \tag{18.211}$$

$$n_L = L \left( \frac{4}{\pi^2} g' T^2 \right) \tag{18.212}$$

where $B$ and $L$ are baryon and lepton asymmetries respectively.

Note that in SM

$$q_{L_i} = \begin{pmatrix} u_{L_i} \\ d_{L_i} \end{pmatrix} \qquad B = \frac{1}{3}, \ L = 0$$

$$u_{R_i}, \ d_{R_i}$$

$$l_{L_i} = \begin{pmatrix} \nu_{L_i} \\ e_{L_i} \end{pmatrix} \qquad B = 0, \ L = 1$$

$$e_{R_i}$$

Thus in the above equation

$$B = 3 \times \frac{1}{3} \sum_i (2\mu_{q_i} + 2\mu_{u_i} + 2\mu_{d_i}) \tag{18.213}$$

$$L = \sum_i (2\mu_{l_i} + 2\mu_{e_i}) \tag{18.214}$$

In high temperature plasma, quarks, leptons and Higgs interact via Yukawa
and gauge couplings and in addition, via the non-perturbative sphaleron
processes.

In thermal equilibrium all these processes yield constraints between various chemical potentials. The effective interaction

$$O_{B+L} = \prod_i (q_{L_i} q_{L_i} q_{L_i} l_{L_i}) \tag{18.215}$$

yields

$$\sum_i (3\mu_{q_i} + \mu_{l_i}) = 0 \qquad (18.216)$$

Another constraint is provided by vanishing of total hypercharge of plasma

$$\sum_i \left( \mu_{q_i} + 2\mu_{u_i} + \mu_{d_i} - \mu_{l_i} - \mu_{e_i} + \frac{2}{N}\mu_\phi \right) = 0 \qquad (18.217)$$

Further invariance of Yukawa couplings $\bar{q}_{L_i}\phi d_{R_i}$ gives

$$\mu_{q_i} - \mu_\phi - \mu_{d_i} = 0$$
$$\mu_{q_i} - \mu_\phi - \mu_{u_i} = 0$$
$$\mu_{l_i} - \mu_\phi - \mu_{e_i} = 0$$

When all Yukawa interactions are in equilibrium, these interaction establish equilibrium in different generations

$$\mu_{l_i} = \mu_l, \qquad \mu_{q_i} = \mu_q \quad \text{etc.}$$

Combining these relations, one can solve for the chemical potentials in terms of the lepton and baryon chemical potentials and finally in terms of the initial $B - L$ .

$$B = a(B - L)$$
$$B(1 - a) = -aL \qquad (18.218)$$

where

$$a = \frac{8N + 4}{22N + 13} = \frac{28}{79} \approx \frac{1}{3}$$

$N$ being the number of generations. Thus finally we obtain

$$Y_B \left( \equiv \frac{n_B - n_{\bar{B}}}{s} \right) = aY_{B-L}$$

$$= \frac{a}{a - 1} Y_L$$

$$= \frac{a}{a - 1} \kappa \frac{n}{s} \varepsilon_i$$

The observed $Y_B$ is

$$Y_B = \eta \left( \frac{n_\gamma}{s} \right)$$

$$\simeq \frac{1}{7}\eta$$

$$= \frac{1}{7} \left( 6 \times 10^{-10} \right)$$

can be obtained if

$$\varepsilon_i \geq 10^{-6}$$

There exist many models of leptogenesis where $\varepsilon$ is usually expressed in terms of $CP$-violating Majorana phases and experimentally known mass parameters and $\Delta m^2$ for light neutrino matrix, giving $\varepsilon_i \geq 10^{-6}$.

## 18.11  Problems

(1) Calculate the Ricci tensor $R_{\mu\nu}$ and Ricci scalar $R$ by using the FRW metric (18.1), where

$$R_{\mu\nu} = \partial_\rho \Gamma^\rho_{\mu\nu} - \partial_\nu \Gamma^\rho_{\mu\rho} + \Gamma^\rho_{\sigma\rho}\Gamma^\sigma_{\mu\nu} - \Gamma^\rho_{\sigma\mu}\Gamma^\sigma_{\nu\rho}$$

with

$$\Gamma^\rho_{\mu\nu} = \frac{1}{2}g^{\rho\sigma}\left[\partial_\mu g_{\nu\sigma} + \partial_\nu g_{\mu\sigma} - \partial_\sigma g_{\mu\nu}\right].$$

**Hint:** First calculate $R_{00}$ and $R_{11}$.

(2) Verify that the 00-component of Einstein equation (18.7) for $\Lambda = 0$ gives the Friedmann equation

$$H^2 \equiv \left(\frac{\dot{R}}{R}\right)^2 = \frac{8\pi G_N}{3}\rho - \frac{kc^2}{R^2}.$$

(3) Show that the trace of Einstein equation (18.7) for $\Lambda = 0$ gives the Friedmann 2nd equation (acceleration equation)

$$\frac{\ddot{R}}{R} = -\frac{4\pi G_N}{3c^4}\left(\rho c^2 + 3p\right).$$

(4) Show that the two Friedmann equations implies the continuity equation

$$\dot{\rho} = -3H\left(\rho c^2 + p\right)$$

(5) Show that conservation of energy condition $\nabla_\mu T^{\mu\nu} = 0$ leads to continuity equation

$$\dot{\rho} = -3H\left(\rho c^2 + p\right)$$

where $T^{\mu\nu}$ is defined in Eq. (18.8).

(6) For a scalar field $\phi$, with potential $V(\phi)$ and Lagrangian density

$$\mathcal{L} = \frac{1}{2}g^{\mu\nu}\partial_\mu\phi\partial_\nu\phi - V(\phi)$$

Calculate the stress energy tensor $T_{\mu\nu}$ (18.49).

(7) Using the conservation of energy condition $\nabla_\mu T^{\mu\nu} = 0$, verify that the equation of motion for the scalar field is

$$\ddot{\phi} + 3H\dot{\phi} + \frac{dV}{d\phi} = 0$$

(8) Find the slow roll parameters $\epsilon_v(\phi)$ and $\eta_v(\phi)$ of inflation for $V(\phi) = \frac{1}{2}m^2\phi^2$ and $V(\phi) = \lambda\phi^4$.

(9) By using the Eq. (18.186), calculate the number of $e$-folds for $V(\phi) = \frac{1}{2}m^2\phi^2$ and $V(\phi) = \lambda\phi^4$.

(10) By using the relation (18.188), calculate the value of parameter $r$ for the scalar potentials $\frac{1}{2}m^2\phi^2$ and $\lambda\phi^4$.

## 18.12  References

1. P. J. E. Peebles, Physical Cosmology, Princeton University Press, Princeton, N. J. (1971).
2. D. W. Sciama, Modern Cosmology, Cambridge University Press, Cambridge (1972).
3. S. Weinberg, Gravitation and Cosmology, Wiley NY, (1972).
4. D. Denegri, The Number of Neutrino Species CERN-EP/89-72. Rev. Mod. Phys.
5. G. Steigman, Ann. Rev. Nucl. Part. Sci. 29, 313 (1979).
6. F. Wilczek, Erice Lecture on Cosmology, Proc. 1981 Int. Sch. of Subnucl. Physics, "Ettore Majorana".
7. A. H. Guth, Phys. Rev. D **23**, 347 (1981).
8. A. Zee, Unity of Forces in the Universe Vol. II (World Scientific, Singapore 1982). A collection of original papers relevant to this chapter can be found in this book.
9. L. D. Landau and E. M. Lifshitz, The Classical Theory of Fields (4th Edition) and Statistical Mechanics (3rd Edition), Part I, Pergamon Press 1985.
10. M. S. Turner, Cosmology and Particle Physics, Lectures at the NATO Advanced Study Inst. Edited by T. Ferbal, Plenum Press (1985).
11. M. Fukugita and T. Yanagida, Phys. Lett. B **174**, 45 (1986).
12. A. D. Linde, Particle Physics and Cosmology, in Proc. XXIV Int. Conf. on High Energy Physics (Editors R. Kotthaus and J. H. Khn), Springer-Verlag, Heidelberg, Germany (1989).
13. W. Buchmuller, hep-ph/0101102; hep-ph/0107153; W. Buchmuller, P. Di Bari and M. Plumacher, Phys. Lett. B **547**, 128 (2002); Annals.

Phys. **315**, 305 (2005).

14. Riazuddin, arXiv, hep-ph/0302020 (2003) and references within it.
15. A. Liddle, An Introduction to Modern Cosmology, 2nd Edition, Wiley (2003).
16. A. G. Riess et al., astro-ph/0611572, Astrophys. J. **659**, 98 (2007).
17. N.E. Mavromatos, arXiv:0708.0134 (2007).
18. S. Weinberg, Cosmology, Oxford University Press, (2008).
19. V. Rubakov, ICTP Lectures on Astro-Particle Phsics, AS-ICTP Summer School on Cosmology, Trieste, Italy (2008).
20. D. Wands, ICTP Lectures on Inftlation and Cosmological Perturbations, AS-ICTP Summer School on Cosmology, Trieste, Italy (2008).
21. D. Baumann, TASI Lectures on Inflation, arXiv:0907.5424 (2009).
22. D. Baumann, ICTP Lectures on Inftlation, AS-ICTP Summer School on Cosmology, Trieste, Italy (2010).
23. A.J. Tolley, ICTP Lectures on Dark Energy and Modified Gravity, AS-ICTP Summer School on Cosmology, Trieste, Italy (2010).
24. G. Steigman, arXiv:1008.4765 (2010).
25. K. Nakamura et al. Particle Data Group, Journal of Physics, G **37**, 0750212 (2010) [see in perticular secs. 19 and 21].
26. K. A. Olive, arXiv:1001.5014 (2010).

# Appendix A

# Quantum Field Theory

## A.1  Spin 0 Field

Spin zero particle of mass $m$ is described by a field $\phi(x)$ which in the absence of interactions, satisfies the Klein-Gordon equation.

$$\left(\Box^2 + m^2\right)\phi(x) = 0. \tag{A.1}$$

In quantum mechanics, $\phi(x)$ is regarded as a c-number. In quantum field theory, $\phi(x)$ is a field operator which can create and annihilate the field quantum.

The Fourier decomposition of $\phi(x)$ is

$$\phi(x) = \frac{1}{(2\pi)^{3/2}} \int \frac{d^3k}{\sqrt{2k_0}} \left[a(k)e^{-ik.x} + b^\dagger(k)e^{ik.x}\right] \tag{A.2a}$$

$$\phi^\dagger(x) = \frac{1}{(2\pi)^{3/2}} \int \frac{d^3k}{\sqrt{2k_0}} \left[a^\dagger(k)e^{ik.x} + b(k)e^{-ik.x}\right] \tag{A.2b}$$

where $\phi^\dagger(x)$ is hermitian conjugate of $\phi(x)$ and $k \cdot x = k_0 x_0 - \mathbf{k}.\mathbf{x}, k_0 = \sqrt{\mathbf{k}^2 + m^2} > 0$. In Eq. (A.2), $a(k)$ and $b(k)$ are interpreted as follows:

$a^\dagger(k):$    creation operator for the particle (spin 0 and mass $m$)

$a(k):$    annihilation operator for the particle (spin 0 and mass $m$)

$b^\dagger(k):$    creation operator for the antiparticle (spin 0 and mass $m$)

$b(k):$    annihilation operator for the antiparticle (spin 0 and mass $m$).

$a(k)$ and $b(k)$ satisfy the following commutation relations

$$[a(k), a^\dagger(k')] = \delta^3(\mathbf{k} - \mathbf{k}') \tag{A.3a}$$

$$[b(k), b^\dagger(k')] = \delta^3(\mathbf{k} - \mathbf{k}') \tag{A.3b}$$

$$[a(k), b(k')] = [a(k), b^\dagger(k')] = 0. \tag{A.3c}$$

If $|0\rangle$ denotes the vacuum state then one particle state of 4-momentum $k$ is given by

$$|k\rangle = a^\dagger(k)|0\rangle. \tag{A.4}$$

States of two or more particles are constructed by operating on the vacuum with appropriate two or more operators. Note that

$$a^\dagger(x) = \frac{-i}{(2\pi)^{3/2}} \int \frac{d^3x}{\sqrt{2k_0}} e^{-ik.x} \frac{\overleftrightarrow{\partial}}{\partial x^0} \phi^\dagger(x) \tag{A.5a}$$

$$b^\dagger(x) = \frac{-i}{(2\pi)^{3/2}} \int \frac{d^3x}{\sqrt{2k_0}} e^{ik.x} \frac{\overleftrightarrow{\partial}}{\partial x^0} \phi(x) \tag{A.5b}$$

where

$$\frac{\overleftrightarrow{\partial}}{\partial x^0} = \frac{\overrightarrow{\partial}}{\partial x^0} - \frac{\overleftarrow{\partial}}{\partial x^0}$$

Define

$$N_+(k) = a^\dagger(k)a(k)$$
$$N_-(k) = b^\dagger(k)b(k). \tag{A.6}$$

It follows from the commutation relations (A.3) that $N_+(k)$ and $N_-(k)$ have the eigenvalues $0, 1, 2, \cdots$ and are known as number operators for the particles and antiparticles. Then

$$n = \sum_k N_+(k) = \text{Total number of particles} \tag{A.7a}$$

$$\bar{n} = \sum_k N_-(k) = \text{Total number of antiparticles.} \tag{A.7b}$$

It may also be noted that for free fields

$$[\phi(x), \phi^\dagger(x')] = i\Delta(x - x'), \tag{A.8a}$$

$$\Delta(x) = -\frac{i}{(2\pi)^3} \int d^4k \varepsilon(k_0) e^{-ik.x} \delta(k^2 - m^2) \tag{A.8b}$$

$$\varepsilon(k_0) = \begin{cases} +1 & k_0 > 0 \\ -1 & k_0 < 0 \end{cases}. \tag{A.8c}$$

We note that

$$\Delta(x - x') = 0 \qquad \text{for } (x - x')^2 < 0 \qquad \text{(A.8d)}$$

viz the space-like distances. Then from Eq. (A.8), it follows that the commutator is zero for space-like separation. This is the statement of the micro causality. Also

$$\left| \frac{\partial \Delta(x - y)}{\partial x_0} \right|_{x_0 = y_0} = -\delta^3(\mathbf{x} - \mathbf{y}) \qquad \text{(A.9a)}$$

and from Eq. (A.8d), we get

$$\Delta(0, \mathbf{x} - \mathbf{y}) = 0. \qquad \text{(A.9b)}$$

If we assume that $\phi(x)$ is not a free field but an interacting (Heisenberg) field, then the creation operator defined in (A.5) is not time-independent. We define, however, the operators $a^\dagger_{\substack{in \\ out}}(k)$ by the relation

$$a^\dagger_{\substack{in \\ out}}(k) = \lim_{\substack{t \to -\infty \\ +\infty}} \frac{-i}{(2\pi)^{3/2}\sqrt{2k_0}} \int e^{-ik.x} \overleftrightarrow{\partial} \phi^\dagger(x) d^3x \qquad \text{(A.10a)}$$

Under time reversal these operators transform as

$$a^\dagger_{\substack{in \\ out}}(k_0, \mathbf{k}) \xrightarrow{T} a^\dagger_{\substack{out \\ in}}(k_0, -\mathbf{k}) \qquad \text{(A.10b)}$$

$a^\dagger_{in}(k)$ and $a^\dagger_{out}(k)$, acting on the vacuum, create incoming and outgoing states respectively and so from (A.10b), incoming (outgoing) states are transformed into outgoing (incoming) states by the time reversal transformation.

## A.2 Spin 1/2 Particle

Spin 1/2 particle of mass $m$ is described by a field $\Psi(x)$, which in the absence of interactions, satisfies the Dirac equation $\left[\partial_\mu = \frac{\partial}{\partial x^\mu} = \left(\frac{\partial}{\partial t}, \nabla\right)\right]$

$$(i\gamma^\mu \partial_\mu - m) \Psi(x) = 0. \qquad \text{(A.11a)}$$

The adjoint of $\Psi(x)$, $\bar{\Psi}(x) = \Psi^\dagger(x) \gamma^0$ satisfies the equation

$$\bar{\Psi}(x) \left(-i\gamma^\mu \overleftarrow{\partial}_\mu - m\right) = 0. \qquad \text{(A.11b)}$$

$\gamma^\mu$ are Dirac matrices. We choose $\gamma^\mu$ :

$$\gamma^{0\dagger} = \gamma^0, \gamma^{i\dagger} = -\gamma^i \qquad i = 1, 2, 3, \qquad \text{(A.12)}$$

$$\gamma^{\mu\dagger} = \gamma^0\gamma^\mu\gamma^0$$

$\gamma^\mu$'s satisfy the anticommutation relation

$$[\gamma^\mu, \gamma^\nu]_+ \equiv \gamma^\mu\gamma^\nu + \gamma^\nu\gamma^\mu = 2g^{\mu\nu}. \tag{A.13}$$

There are 16 independent Dirac matrices:

| Matrices | Components |
|---|---|
| $1$ | $1$ |
| $\gamma^\mu$ | $4$ |
| $\sigma^{\mu\nu} = \frac{i}{2}\left(\gamma^\mu\gamma^\nu - \gamma^\nu\gamma^\mu\right)$ | $6$ |
| $\gamma^5 = i\gamma^0\gamma^1\gamma^2\gamma^3$ | $1$ |
| $i\,\gamma^\mu\gamma^5$ | $4$ |

$$\gamma^5 = i\gamma^0\gamma^1\gamma^2\gamma^3 = \gamma_5 \tag{A.14a}$$

$$\gamma^5 = -\frac{i}{4!}\epsilon_{\alpha\beta\rho\lambda}\gamma^\alpha\gamma^\beta\gamma^\rho\gamma^\lambda \tag{A.14b}$$

$$\epsilon_{\alpha\beta\rho\lambda} = -\epsilon^{\alpha\beta\rho\lambda} \tag{A.14c}$$

$$\epsilon_{0123} = -1 = -\epsilon^{0123} \tag{A.14d}$$

$$\epsilon_{ijk} = -\epsilon^{ijk} \tag{A.14e}$$

$\gamma^5$ is also hermitian

$$\gamma^{5\dagger} = \gamma^5 \tag{A.15}$$

$\gamma^5$ anticommutes with $\gamma^\mu$ viz

$$\gamma^5\gamma^\mu = -\gamma^\mu\gamma^5 \tag{A.16}$$

Two representations of $\gamma$-matrices are useful.

### A.2.1   *Pauli Representation of $\gamma$ Matrices*

$$\gamma^i = \begin{pmatrix} 0 & \sigma^i \\ -\sigma^i & 0 \end{pmatrix}, \quad \gamma_i = -\gamma^i \tag{A.17a}$$

$$\gamma^0 = \begin{pmatrix} 1 & 0 \\ 0 & -1 \end{pmatrix} = \gamma_0 \tag{A.17b}$$

$$\gamma = \begin{pmatrix} 0 & \sigma \\ -\sigma & 0 \end{pmatrix}. \quad \gamma^5 = \gamma_5 = \begin{pmatrix} 0 & 1 \\ 1 & 0 \end{pmatrix} \tag{A.17c}$$

## A.2.2 Weyl Representation of $\gamma$ Matrices

$$\gamma^0 = \begin{pmatrix} 0 & 1 \\ 1 & 0 \end{pmatrix} = \gamma_0; \qquad \gamma^i = \begin{pmatrix} 0 & \sigma^i \\ -\sigma^i & 0 \end{pmatrix} \tag{A.18a}$$

$$\gamma^5 = \begin{pmatrix} -1 & 0 \\ 0 & 1 \end{pmatrix} = \gamma_5 \tag{A.18b}$$

Define

$$\sigma^\mu = (1, \sigma^i) = (1, \sigma); \qquad \sigma_\mu = (1, -\sigma)$$
$$\overline{\sigma}^\mu = (1, -\sigma^i) = (1, -\sigma); \qquad \overline{\sigma}_\mu = (1, \sigma)$$

$$\gamma^\mu = \begin{pmatrix} 0 & \sigma_\mu \\ \overline{\sigma}_\mu & 0 \end{pmatrix} \tag{A.18c}$$

The Fourier decomposition of $\Psi(x)$ is

$$\Psi(x) = \frac{1}{(2\pi)^{3/2}} \int d^3 p \sqrt{\frac{m}{p_0}}$$

$$\times \sum_{r=1}^{2} \left[ a_r(p) \, u_r(p) \, e^{-ip.x} + b_r^\dagger(p) \, v_r(p) \, e^{ip.x} \right] \tag{A.19a}$$

$$\overline{\Psi}(x) = \frac{1}{(2\pi)^{3/2}} \int d^3 p \sqrt{\frac{m}{p_0}}$$

$$\times \sum_{r=1}^{2} \left[ a_r^\dagger(p) \, \overline{u}_r(p) \, e^{ip.x} + b_r(p) \, \overline{v}_r(p) \, e^{-ip.x} \right] \tag{A.19b}$$

where

$$\overline{u} = u^\dagger \gamma^0, \qquad \overline{v} = v^\dagger \gamma^0 \tag{A.20}$$

and $u$ and $v$ satisfy the equations

$$(\gamma.p - m) \, u_r(p) = 0 \tag{A.21a}$$

$$(\gamma.p + m) \, v_r(p) = 0 \tag{A.21b}$$

$$\overline{u}_r(p) \, (\gamma.p - m) = 0 \tag{A.21c}$$

$$\overline{v}_r(p) \, (\gamma.p + m) = 0 \tag{A.21d}$$

$a(p)$ and $b(p)$ are interpreted as follows:

$a_r^\dagger(p)$ :    creation operator of the particle with momentum $\mathbf{p}$ and spin component $r$

$a_r(p)$ :    annihilation operator of the particle with momentum $\mathbf{p}$ and spin component $r$

$b_r^\dagger(p)$ :    creation operator of the antiparticle with momentum $\mathbf{p}$ and spin component $r$

$b_r(p)$ :    annihilation operator of the particle with momentum $\mathbf{p}$ and spin component $r$.

The operators $a$ and $b$ satisfy the anticommutation relation

$$\left[a_r\left(p\right), a_{r'}^\dagger\left(p'\right)\right]_+ = \delta_{rr'}\delta^3\left(\mathbf{p} - \mathbf{p}'\right) \qquad (A.22a)$$

$$\left[b_r\left(p\right), b_{r'}^\dagger\left(p\right)\right]_+ = \delta_{rr'}\delta^3\left(\mathbf{p} - \mathbf{p}'\right) \qquad (A.22b)$$

and all other anticommutation relations give zero. Define number operators:

$$N_r^{(+)}\left(p\right) = a_r^\dagger\left(p\right) a_r\left(p\right) \qquad (A.23a)$$

$$N_r^{(-)}\left(p\right) = b_r^\dagger\left(p\right) b_r\left(p\right). \qquad (A.23b)$$

Then from the anticommutation relations (A.22), we have

$$\left[N_r^{(\pm)}\left(p\right)\right]^2 = N_r^{(\pm)}\left(p\right). \qquad (A.24)$$

Thus, we see that $N_r^{(\pm)}\left(p\right)$ have eigenvalue 0 or 1. This means that each state is either empty or has a single particle of definite spin and momentum. Thus the anticommutation relations lead to description of a system of particles which obey the Pauli exclusion principle or in other words obey the Fermi-Dirac statistics.

Note that

$$a^\dagger(x) = \frac{1}{(2\pi)^{3/2}} \sqrt{\frac{m}{p_0}} \int d^3x \overline{\psi}(x)\gamma^0 u_r(p)e^{-ip.x} \qquad (A.25)$$

with similarly expressions for other creation and annhilation operators.

The spinors $u$ and $v$ satisfy the following orthogonality relations:

$$\overline{u}_r\left(p\right) u_{r'}\left(p\right) = \delta_{rr'} = -\overline{v}_r\left(p\right) v_{r'}\left(p\right) \qquad (A.26a)$$

$$u_r^\dagger\left(p\right) u_{r'}\left(p\right) = \frac{E_p}{m}\delta_{rr'} = v_r^\dagger\left(p\right) v_{r'}\left(p\right) \qquad (A.26b)$$

$$\overline{v}_r\left(p\right) u_{r'}\left(p\right) = \overline{u}_r\left(p\right) v_{r'}\left(p\right) = 0. \qquad (A.26c)$$

They also satisfy the completeness relations

$$\sum_{r=1}^{2}\left[u_\alpha^r\left(p\right)\overline{u}_\beta^r\left(p\right) - v_\alpha^r\left(p\right)\overline{v}_\beta^r\left(p\right)\right] = \delta_{\alpha\beta}, \qquad (A.27)$$

where $\alpha$ and $\beta$ are spinor indices; $\alpha$, $\beta = 1, 2, 3, 4$.

$$\sum_{r=1}^{2}u_\alpha^r\left(p\right)\overline{u}_\beta^r\left(p\right) = \left(\frac{\not{p}+m}{2m}\right)_{\alpha\beta} \equiv \left(\Lambda_+\left(p\right)\right)_{\alpha\beta} \qquad (A.28a)$$

$$-\sum_{r=1}^{2}v_\alpha^r\left(p\right)\overline{v}_\beta^r\left(p\right) = \left(\frac{-\not{p}+m}{2m}\right)_{\alpha\beta} \equiv \left(\Lambda_-\left(p\right)\right)_{\alpha\beta}. \qquad (A.28b)$$

$\Lambda_+ (p)$ and $\Lambda_- (p)$ are the projection operators for particles and antiparticles respectively. One also writes $\gamma^\mu \, p_\mu = \gamma \cdot p = \slashed{p}$.

Using the Pauli representation of $\gamma$-matrices, we can write

$$u_r (p) = R w^{(r)} \tag{A.29a}$$

where

$$R = \frac{1}{\sqrt{2m \, (p_0 + m)}} \begin{pmatrix} (p_0 + m) \ I \\ \sigma \cdot \mathbf{p} \end{pmatrix} \tag{A.29b}$$

$$p_0 \equiv E_p = \sqrt{p^2 + m^2} \tag{A.29c}$$

$$w^{(1)} = \begin{pmatrix} 1 \\ 0 \end{pmatrix} \qquad \text{and} \qquad w^{(2)} = \begin{pmatrix} 0 \\ 1 \end{pmatrix}. \tag{A.29d}$$

$\bar{u}_r (p)$ is given by

$$\bar{u}_r (p) = w^{(r)\dagger} R^\dagger \gamma^0 = w^{(r)\dagger} \overline{R}, \tag{A.30a}$$

where

$$\overline{R} = \frac{1}{\sqrt{2m \, (p_0 + m)}} \left( (p_0 + m) \, I, -\sigma \cdot \mathbf{p} \right). \tag{A.30b}$$

$v_r (p)$ is given by

$$v_r (p) = -i\gamma^2 u_r^* (p). \tag{A.31}$$

Finally, we note that for free fields $[\slashed{\partial}_x = \gamma^\mu \partial_\mu]$

$$\begin{aligned}
\left[ \Psi_\alpha (x), \, \overline{\Psi}_\beta (x') \right]_+ &= i \, (i \, \slashed{\partial}_x + m)_{\alpha\beta} \, \Delta (x - x') \\
&= -i S_{\alpha\beta} (x - x'),
\end{aligned} \tag{A.32a}$$

where

$$S (x - x') = (-i \, \slashed{\partial}_x - m) \, \Delta (x - x').$$

$$\begin{aligned}
-i S (x - x')|_{x_0 = x_0'} &= -i \left( -i\gamma^0 \frac{\partial}{\partial x^0} - \gamma \cdot \nabla - m \right) \Delta (x - x')|_{x_0 \to x_0'} \\
&= \gamma^0 \delta^3 (\mathbf{x} - \mathbf{x}').
\end{aligned} \tag{A.32b}$$

$$\left[ \Psi (x), \, \overline{\Psi} (x') \right]_{+_{x_0 = x_0'}} = \gamma^0 \delta^3 (\mathbf{x} - \mathbf{x}') \tag{A.32c}$$

## A.3   Trace of $\gamma$ Matrices

We note that $\gamma$-matrices are traceless

$$Tr\left(\gamma^{\mu}\right) = 0, \quad \mu = 0, 1, 2, 3$$
$$Tr\left(\gamma^{5}\right) = 0. \tag{A.33}$$

Now

$$Tr\left(\gamma^{\mu}\gamma^{\nu}\right) = Tr\left(\gamma^{\nu}\gamma^{\mu}\right). \tag{A.34}$$

Therefore, from Eq. (A.13), we have

$$Tr\left(\gamma^{\mu}\gamma^{\nu}\right) = g^{\mu\nu}Tr\left(\widehat{1}\right) = 4g^{\mu\nu} \tag{A.35}$$

and

$$Tr\left(\not{k}\,\not{p}\right) = 4p \cdot k \tag{A.36}$$

where

$$\not{k} = \gamma^{\mu}k_{\mu} \tag{A.37}$$

Further, we have

$$\gamma^{\mu}\gamma^{\nu}\gamma^{\rho}\gamma^{\sigma} + \gamma^{\sigma}\gamma^{\mu}\gamma^{\nu}\gamma^{\rho} = 2g^{\rho\sigma}\gamma^{\mu}\gamma^{\nu} - 2g^{\nu\sigma}\gamma^{\mu}\gamma^{\rho} + 2g^{\mu\sigma}\gamma^{\nu}\gamma^{\rho} \tag{A.38}$$

Therefore,

$$Tr\left(\gamma^{\mu}\gamma^{\nu}\gamma^{\rho}\gamma^{\sigma}\right) = 4\left[g^{\sigma\rho}g^{\mu\nu} - g^{\nu\sigma}g^{\mu\rho} + g^{\mu\sigma}g^{\nu\rho}\right] \tag{A.39}$$

We now show that trace of odd number of $\gamma$-matrices is zero

$$\begin{aligned} Tr\left(\gamma^{\mu}\gamma^{\nu}...\gamma^{\lambda}\right) &= Tr\left(\left(\gamma^{5}\right)^{2}\gamma^{\mu}\gamma^{\nu}...\gamma^{\lambda}\right) \\ &= (-1)^{n} Tr\left(\gamma^{5}\gamma^{\mu}\gamma^{\nu}...\gamma^{\lambda}\gamma^{5}\right) \\ &= (-1)^{n} Tr\left(\gamma^{5}\gamma^{5}\gamma^{\mu}\gamma^{\nu}...\gamma^{\lambda}\right) \text{ (cyclic property)} \\ &= (-1)^{n} Tr\left(\gamma^{\mu}\gamma^{\nu}...\gamma^{\lambda}\right) \end{aligned} \tag{A.40}$$

Hence for odd $n$

$$Tr\left(\gamma^{\mu}\gamma^{\nu}...\gamma^{\lambda}\right) = 0. \tag{A.41}$$

Now

$$\gamma^{\mu}\gamma^{\nu}\gamma^{\rho} = i\varepsilon^{\mu\nu\rho\lambda}\gamma_{5}\gamma_{\lambda} + g^{\nu\rho}\gamma^{\mu} - g^{\mu\rho}\gamma^{\nu} + g^{\mu\nu}\gamma^{\rho}. \tag{A.42}$$

Noting that we can write

$$\gamma^{5} = \frac{-i}{4!}\varepsilon_{\alpha\beta\sigma\rho}\gamma^{\alpha}\gamma^{\beta}\gamma^{\nu}\gamma^{\rho}, \tag{A.43}$$

we have

$$Tr\left(\gamma^5\gamma^\mu\right) = 0 \tag{A.44}$$

Then from Eq. (A.42)

$$Tr\left(\gamma^5\gamma^\mu\gamma^\nu\right) = 0 \tag{A.45}$$

$$Tr\left(\gamma^5\gamma^\mu\gamma^\nu\gamma^\rho\right) = 0 \tag{A.46}$$

and

$$Tr\left(\gamma^5\gamma^\mu\gamma^\nu\gamma^\rho\gamma^\sigma\right) = 4i\varepsilon^{\mu\nu\rho\sigma} \tag{A.47}$$

with the definition $\varepsilon_{0123} = -1$ and $\varepsilon_{0ijk} = \varepsilon_{ijk}$ while $\varepsilon^{0123} = 1$. In calculations, we usually come across the matrix elements of the form

$$\mathcal{L}_{\mu\nu} = \sum_{\text{spin}} \left[\overline{u}\left(k_2\right)\gamma_\mu\left(1 + a\gamma_5\right)u\left(k_1\right)\right]\left[\overline{u}\left(k_1\right)\gamma_\nu\left(1 + a\gamma_5\right)u\left(k_1\right)\right]^* \tag{A.48}$$

$$= \sum_{\text{spin}} \left[\overline{u}\left(k_2\right)\gamma_\mu\left(1 + a\gamma_5\right)u\left(k_1\right)\right]\left[u^\dagger\left(k_1\right)\left(1 + a\gamma_5\right)\gamma_\nu^\dagger\gamma^0 u\left(k_2\right)\right]. \tag{A.49}$$

Now $[\gamma^0\gamma_\nu^\dagger\gamma^0 = \gamma_\nu]$

$$\gamma^0\left(1 + a\gamma_5\right)\gamma_\nu^\dagger\gamma^0 = \left(1 - a\gamma_5\right)\gamma_\nu$$

$$= \gamma_\nu\left(1 + a\gamma_5\right). \tag{A.50}$$

Therefore,

$$\mathcal{L}_{\mu\nu} = \sum_{\text{spin}} \overline{u}\left(k_2\right)\gamma_\mu\left(1 + a\gamma_5\right)u\left(k_1\right)\overline{u}\left(k_1\right)\gamma_\nu\left(1 + a\gamma_5\right)u\left(k_2\right)$$

$$= \sum_{\text{spin}} \overline{u}_\alpha\left(k_2\right)\left[\gamma_\mu\left(1 + a\gamma_5\right)\right]_{\alpha\alpha'}u_{\alpha'}\left(k_1\right)\overline{u}_{\beta'}\left(k_1\right)\left[\gamma_\nu\left(1 + a\gamma_5\right)\right]_{\beta'\delta}u_\delta\left(k_2\right)$$

$$= \sum_{\text{spin}} \left(\frac{\not{k}_2 + m_2}{2m_2}\right)_{\delta\alpha}\left[\gamma_\mu\left(1 + a\gamma_5\right)\right]_{\alpha\alpha'}\left(\frac{\not{k}_1 + m_1}{2m_1}\right)_{\alpha'\beta'}\left[\gamma_\nu\left(1 + a\gamma_5\right)\right]_{\beta'\delta}$$

$$= \frac{1}{4m_1m_2}Tr\left[\left(\not{k}_2 + m_2\right)\gamma_\mu\left(1 + a\gamma_5\right)\left(\not{k}_1 + m_1\right)\gamma_\nu\left(1 + a\gamma_5\right)\right]. \tag{A.51}$$

Here

$$\not{k}_2 = \gamma\cdot k_2, \qquad \not{k}_1 = \gamma\cdot k_1. \tag{A.52}$$

Using the formulae for the traces of $\gamma$-matrices given previously, we get

$$\mathcal{L}_{\mu\nu} = \frac{4}{4m_1m_2}\left\{\left(1 + a^2\right)\left(k_{2\mu}k_{1\nu} + k_{2\nu}k_{1\mu} - k_1\cdot k_2 g_{\mu\nu}\right)\right.$$

$$\left. + m_1m_2\left(1 - a^2\right)g_{\mu\nu} + 2ia\varepsilon_{\mu\nu\rho\sigma}k_1^\rho k_2^\sigma\right\}. \tag{A.53}$$

Similarly for

$$\mathcal{L}_{\mu\nu} = \sum_{\text{spin}} \left[\bar{v}\left(k_2\right)\gamma_\mu\left(1 + a\gamma_5\right)u\left(k_1\right)\right]\left[\bar{v}\left(k_2\right)\gamma_\nu\left(1 + a\gamma_5\right)u\left(k_1\right)\right]^*, \quad (A.54)$$

we get

$$\mathcal{L}_{\mu\nu} = \frac{4}{4m_1 m_2}\left\{\left(1 + a^2\right)\left(k_{2\mu}k_{1\nu} + k_{2\nu}k_{1\mu} - k_1 \cdot k_2 g_{\mu\nu}\right)\right.$$
$$\left. - m_1 m_2\left(1 - a^2\right)g_{\mu\nu} + 2ia\varepsilon_{\mu\nu\rho\sigma}k_1^\rho k_2^\sigma\right\}. \quad (A.55)$$

For

$$\mathcal{L}_{\mu\nu} = \sum \left[\bar{u}\left(k_2\right)\gamma_\mu\left(1 + a\gamma_5\right)v\left(k_1\right)\right]\left[\bar{u}\left(k_2\right)\gamma_\nu\left(1 + a\gamma_5\right)v\left(k_1\right)\right]^*, \quad (A.56)$$

we get the same value as given in Eq. (A.55).

## A.4   Spin 1 Field

Electromagnetic field (photon) with mass $m = 0$.

In the absence of interactions, the electromagnetic field $A_\mu\left(x\right)$ satisfies the field equation

$$\Box^2 A_\mu\left(x\right) = 0. \quad (A.57)$$

There is an additional condition

$$\partial^\mu A_\mu\left(x\right) = 0. \quad (A.58)$$

The Fourier decomposition of $A_\mu\left(x\right)$:

$$A_\mu\left(x\right) = \frac{1}{\left(2\pi\right)^{3/2}}\int \frac{d^3 k}{\sqrt{2k_0}}\sum_\lambda \varepsilon_\mu^\lambda\left(k\right)\left[a_\lambda\left(k\right)e^{-ik.x} + a_\lambda^\dagger\left(k\right)e^{ik.x}\right], \quad (A.59)$$

where $\varepsilon_\mu^\lambda\left(k\right)$, are four vectors called polarization vectors. $a_\lambda\left(k\right)$ and $a_\lambda^\dagger\left(k\right)$ are interpreted respectively as the annihilation and creation operator of the photon with momentum $k$ and polarization $\varepsilon_\mu^\lambda\left(k\right)$. They satisfy the following commutation relations

$$\left[a_\lambda\left(k\right), a_{\lambda'}^\dagger\left(k'\right)\right] = \delta_{\lambda\lambda'}\delta^3\left(\mathbf{k} - \mathbf{k}'\right), \quad (A.60a)$$

$$\left[a_\lambda\left(k\right), a_{\lambda'}\left(k'\right)\right] = \left[a_\lambda^\dagger\left(k\right), a_{\lambda'}^\dagger\left(k'\right)\right] = 0. \quad (A.60b)$$

The poarization vector $\varepsilon_\mu^\lambda\left(k\right)$ satisfies the following relations

$$\varepsilon^\lambda\left(k\right) \cdot \varepsilon^{\lambda'}\left(k\right) = \delta_{\lambda\lambda'} \quad (A.61)$$

$$\sum_{\lambda=0} k \cdot \varepsilon^\lambda = 0. \quad (A.62)$$

For transverse photon polarization, the four-vector $\varepsilon_\mu^\lambda(k)$ can be chosen as

$$\varepsilon_\mu^\lambda(k) \equiv \left(0,\ \varepsilon^\lambda(k)\right), \tag{A.63}$$

so that we have

$$k \cdot \varepsilon^\lambda = \mathbf{k} \cdot \varepsilon^\lambda = 0 \tag{A.64}$$

and

$$\sum_{\lambda=1}^{2} \varepsilon_\mu^\lambda(k)\ \varepsilon_\nu^\lambda(k) = -g_{\mu\nu} + \frac{k_\mu k_\nu}{k^2 - (k \cdot \eta)^2} + \frac{k^2 \eta_\mu \eta_\nu}{k^2 - (k \cdot \eta)^2}$$
$$- \frac{(k \cdot \eta)\, k_\mu \eta_\nu + k_\nu \eta_\mu}{k^2 - (k \cdot \eta)^2} \tag{A.65}$$

where $\eta = (1, 0, 0, 0)$.

## A.5    Massive Spin 1 Particle

A spin 1 particle of mass $m$ is described by a vector field $\phi_\mu(x)$, which in the absence of interactions satisfies the equation

$$\left(\Box^2 + m^2\right) \phi_\mu(x) = 0 \tag{A.66a}$$

with the subsidiary condition

$$\partial^\mu \phi_\mu(x) = 0. \tag{A.66b}$$

The Fourier decomposition of $\phi_\mu(x)$ is given by

$$\phi_\mu(x) = \frac{1}{(2\pi)^{3/2}} \int \frac{d^3k}{\sqrt{2k_0}} \sum_{\lambda=1}^{3} \varepsilon_\mu^\lambda(k) \left[a_\lambda(k)\, e^{-ik.x} + b_\lambda^\dagger(k)\, e^{ik.x}\right] \tag{A.67a}$$

$$\phi_\mu^\dagger(x) = \frac{1}{(2\pi)^{3/2}} \int \frac{d^3k}{\sqrt{2k_0}} \sum_{\lambda=1}^{3} \varepsilon_\mu^{\lambda*}(k) \left[a_\lambda^\dagger(k)\, e^{ik.x} + b_\lambda(k)\, e^{-ik.x}\right] \tag{A.67b}$$

$a_\lambda(k)$ and $b_\lambda(k)$ satisfy the following commutation relations:

$$\left[a_\lambda(k), a_{\lambda'}^\dagger(k')\right] = \delta_{\lambda\lambda'}\delta^3(\mathbf{k} - \mathbf{k}'), \tag{A.68a}$$

$$\left[b_\lambda(k), b_{\lambda'}^\dagger(k')\right] = \delta_{\lambda\lambda'}\delta^3(\mathbf{k} - \mathbf{k}'). \tag{A.68b}$$

$a_\lambda^\dagger(k)$ $(a_\lambda(k))$ are creation (annihilation) operators for the particle with polarization $\lambda$ and momentum $k$. $b_\lambda^\dagger(k)$ $(b_\lambda(k))$ are creation (annihilation) operators for the antiparticle with polarization $\lambda$ and momentum $k$.

The polarization vector $\varepsilon_\mu^\lambda$ satisfies the relation

$$\varepsilon^\lambda \cdot \varepsilon^{\lambda'} = \delta_{\lambda\lambda'}, \qquad k \cdot \varepsilon^\lambda = 0 \tag{A.69a}$$

$$\sum_{\lambda=1}^{3} \varepsilon_\mu^\lambda \varepsilon_\nu^\lambda = -g_{\mu\nu} + \frac{k_\mu k_\nu}{m^2}. \tag{A.69b}$$

## A.6 Feynman Rules for S-Matrix in Momentum Space

| | | |
|---|---|---|
| For each internal photon line: | | $\frac{-i}{(2\pi)^4} \frac{g_{\mu\nu}}{k^2+i\varepsilon}$ |
| For each internal fermion line: | | $\frac{i}{(2\pi)^4} \frac{\not{p}+m}{p^2-m^2+i\varepsilon}$ |
| For each internal pion line: | | $\frac{i}{(2\pi)^4} \frac{1}{k^2-m_\pi^2+i\varepsilon}$ |
| For each external fermion line entering the graph, depending upon whether the line is in the initial or final state | | $\frac{1}{(2\pi)^{3/2}} \sqrt{\frac{m}{p_0}} u_r(p)$ or $\frac{1}{(2\pi)^{3/2}} \sqrt{\frac{m}{p_0}} v_r(p)$ |
| For each external fermion line leaving the graph, depending upon whether the line is in the final or initial state | | $\frac{1}{(2\pi)^{3/2}} \sqrt{\frac{m}{p_0}} \bar{u}_r(p)$ or $\frac{1}{(2\pi)^{3/2}} \sqrt{\frac{m}{p_0}} \bar{v}_r(p)$ |
| For each external photon line: | | $\frac{1}{(2\pi)^{3/2}} \frac{1}{\sqrt{2k_0}} \varepsilon_\mu^\lambda(k)$ |
| For each external spin 0 meson line: | | $\frac{1}{(2\pi)^{3/2}} \frac{1}{\sqrt{2k_0}}$ |
| For photon-fermion vertex: | $-ie\gamma^\mu$ | $H_I = e\bar{\Psi}\gamma^\mu\Psi A_\mu$ |
| For pion-fermion vertex: | $g\gamma^5$ | $H_I = ig\bar{\Psi}\gamma^5\Psi\phi$ |
| For photon-meson vertex: | $-ie(q+q')_\mu$ | |
| For a massive vector boson of mass $m_W$ (in unitary gauge) | | $\frac{i}{(2\pi)^4} \frac{-g_{\mu\nu}+\frac{k_\mu k_\nu}{m_W^2}}{k^2-m_W^2}$ |

A factor $(2\pi)^4 \delta^4 (p - p' \pm k)$ at each vertex. A factor $(-1)$ for each closed fermion loop.

Further one has $\frac{\int d^4 l}{(2\pi)^4}$ for each loop integral where the four momentum $l$ is not fixed by energy-momentum conservation. Multiply by $\delta_p = 1\ (-1)$ and $-1\ (1)$ respectively for the direct and exchange term of fermion (antifermion)-fermion (antifermion) scattering.

Feynman rules for a hermitian self-interacting spin 0 boson with the Lagrangian

$$\mathcal{L} = \frac{1}{2} \left[ (\partial_\mu \phi)^2 - \mu^2 \phi^2 \right] - \frac{\lambda}{4!} \phi^4$$

are as follows

| | |
|---|---|
| For each external line | $\dfrac{1}{(2\pi)^{3/2}} \dfrac{1}{\sqrt{2k_0}}$ |
| For each internal line | $\dfrac{i}{(2\pi)^4} \dfrac{1}{\sqrt{2k_0}} \dfrac{1}{k^2 - \mu^2 + i\varepsilon}$ |
| For vertex | $-i\lambda$ |
| A factor $(2\pi)^4 \delta^4 (k_1 + k_2 - k_3 - k_4)$ | |

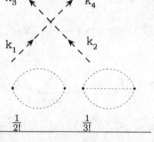

| at each vertex | |
|---|---|
| For each loop integral $\frac{\int d^4 l}{(2\pi)^4}$ | |
| statistical factors | $\dfrac{1}{2!} \qquad \dfrac{1}{3!}$ |

## A.7 Application of Feynman Rules

As a simple application of Feynman rules, we consider the process

$$e^- (p_1) + e^+ (p_2) = \mu^- (p_1') + \mu^+ (p_2')$$

$$S = \int d^4 k\, \frac{m_\mu}{\sqrt{p_{10}'\, p_{20}'}}\, \frac{1}{(2\pi)^3}\, \overline{u}\,(p_1')\,(-ie)\,\gamma^\lambda v\,(p_2')$$

$$\times \frac{-i}{(2\pi)^4}\, \frac{g_{\lambda\nu}}{k^2}\, \overline{v}\,(p_2)\,(-ie\gamma^\nu)\,u\,(p_1)\, \frac{1}{(2\pi)^3}\, \frac{m_e}{\sqrt{p_{10} p_{20}}}$$

$$\times (2\pi)^4 \delta^4 (p_1 + p_2 - k)\,(2\pi)^4 \delta^4 (k - p_1' - p_2'). \tag{A.70}$$

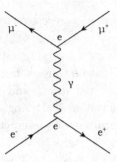

Fig. A.1   One photon exchange Feynman diagram for the process $e^- + e^+ \to \mu^- + \mu^+$.

Therefore, using the relation $S = 1 + i\,(2\pi)^4\,\delta^4\,(P_i - P_f)\,T$ :

$$T = \frac{1}{(2\pi)^6}\frac{m_\mu\,m_e g_{\lambda\nu}}{\sqrt{p_{10}\,p_{20}\,p'_{10}\,p'_{20}}}\frac{e^2}{k^2}\left[\bar{u}\,(p'_1)\,\gamma^\lambda v\,(p'_2)\right]\left[\bar{v}\,(p_2)\,\gamma^\nu u\,(p_1)\right], \quad (A.71)$$

Defining the amplitude:

$$F = \frac{e^2}{k^2}\left[\bar{u}\,(p'_1)\,\gamma^\lambda v\,(p'_2)\right]\left[\bar{v}\,(p_2)\,\gamma_\lambda u\,(p_1)\right]. \qquad (A.72)$$

$$|F|^2 = \frac{e^4}{k^4}\overline{\sum_{\text{spin}}\sum_{\text{spin}}}\left|\bar{u}\,(p'_1)\,\gamma^\lambda v\,(p'_2)\right|^2\left|\bar{v}\,(p_2)\,\gamma_\lambda u\,(p_1)\right|^2. \qquad (A.73)$$

Using Eqs. (A.55) and (A.56) $(a = 0)$,

$$|F|^2 = \frac{e^4}{k^4}\frac{1}{4}\frac{1}{m_e^2}\frac{1}{m_\mu^2}\left[\begin{array}{c} p'_1\cdot p_2\,p'_2\cdot p_1 + p'_2\cdot p_2\,p'_1\cdot p_1 + m_e^2\,p'_1\cdot p'_2 \\ + m_\mu^2\,p_2\cdot p_1 + 2m_e^2 m_\mu^2 \end{array}\right]. \quad (A.74)$$

$$s = (p_1 + p_2)^2 = (p'_1 + p'_2)^2$$
$$t = (p_1 - p'_1)^2 = (p'_2 - p_2)^2$$
$$u = (p_1 - p'_2)^2 = (p'_1 - p_2)^2$$
$$s + t + u = 2(m_\mu^2 + m_e^2)$$

From Eq. (A.74),

$$|F|^2 = \frac{1}{4}\frac{e^4}{s^2}\frac{2}{m_\mu^2 m_e^2}\left[\frac{(t - (m_\mu^2 + m_e^2))^2}{4} + \frac{(u - (m_\mu^2 + m_e^2))^2}{4} + \frac{1}{2}s(m_\mu^2 + m_e^2)\right]$$

$$(A.75)$$

Now in the c.m. frame

$$\mathbf{p}_1 = -\mathbf{p}_2 = \mathbf{p}, \quad \mathbf{p}_1' = \mathbf{p}_2' = \mathbf{p}'$$

$$E = E'$$

$$s = 4E^2 = 4E'^2 = E_{cm}^2$$

$$t = -\frac{s}{2}\left[1 - \frac{2(m_\mu^2 + m_e^2)}{s} - \beta_e\beta_\mu\cos\theta\right]$$

$$u = -\frac{s}{2}\left[1 - \frac{2(m_\mu^2 + m_e^2)}{s} + \beta_e\beta_\mu\cos\theta\right]$$

$$\beta_e = \frac{2\,|\mathbf{p}|}{\sqrt{s}}, \qquad \beta_\mu = \frac{2\,|\mathbf{p}'|}{\sqrt{s}}$$

Hence in the c.m. frame [cf. Eq. (2.100)]

$$\frac{d\sigma}{d\Omega} = \frac{\alpha^2}{s}\frac{\beta_\mu}{\beta_e}\frac{1}{2}\left[\frac{(t-(m_\mu^2+m_e^2))^2}{s^2} + \frac{(u-(m_\mu^2+m_e^2))^2}{s^2} + \frac{2(m_\mu^2+m_e^2)}{s}\right]$$

$$= \frac{\alpha^2}{4s}\frac{\beta_\mu}{\beta_e}\left[1 + (2 - \beta_e^2 - \beta_\mu^2) + \beta_e^2\beta_\mu^2\cos^2\theta\right] \tag{A.76}$$

$$\sigma = \frac{4\pi\alpha^2}{3}\frac{\beta_\mu}{\beta_e}\left[1 + \frac{2-\beta_e^2-\beta_\mu^2}{2} + \frac{(1-\beta_e^2)(1-\beta_\mu^2)}{4}\right] \tag{A.77}$$

In the extreme relativistic limit $\beta_e \to 1$, $\beta_\mu \to 1$

$$\frac{d\sigma}{d\Omega} = \frac{\alpha^2}{4s}(1 + \cos^2\theta)$$

$$\sigma(e^-e^+ \to \mu^-\mu^+) = \frac{4\pi\alpha^2}{3}\frac{1}{s} \tag{A.78}$$

Similarly

$$\sigma(e^-e^+ \to q\bar{q}) = \frac{4\pi\alpha^2}{3s}3e_q^2$$

where $ee_q$ is the charge of the quark $q$ and factor 3 is due to that each quark carries 3 colors. If the quark fragment into hadrons with 100% probability, then

$$R = \frac{\sigma(e^-e^+ \to hadrons)}{\sigma(e^-e^+ \to \mu^-\mu^+)} = 3\sum_q e_q^2, \quad s \gg m_e^2, m_\mu^2, m_q^2 \tag{A.79}$$

Finally ($Q^2 = -t$)

$$\frac{d\sigma}{dQ^2} = \frac{2\pi\alpha^2}{s^2}\frac{1}{\beta_e^2}\left[\frac{(t-(m_\mu^2+m_e^2))^2}{s^2} + \frac{(u-(m_\mu^2+m_e^2))^2}{s^2} + \frac{2(m_\mu^2+m_e^2)}{s}\right]$$

$$\tag{A.80}$$

In the high energy limit

$$\frac{d\sigma}{dQ^2} = \frac{2\pi\alpha^2}{s^2}\left[\frac{t^2+u^2}{s^2}\right] \tag{A.81}$$

The above equation is Lorentz invariant and can be evaluated in any frame.

### A.7.1   $e^+e^- \rightarrow$ *Hadrons*

In this section we evaluate the total cross-section for the process $e^+e^- \rightarrow$ *hadrons* in the lowest order in electromagnetism, i.e. through one photon exchange gives [see Fig. A.2]

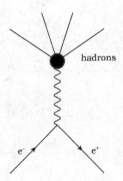

Fig. A.2   One photon exchange Feynman diagram for the process $e^+e^- \rightarrow$ hadrons.

$$T = \frac{1}{(2\pi)^3}\frac{m_e}{\sqrt{E_1 E_2}}\frac{(-i)g_{\lambda\mu}}{q^2 - i\epsilon} < n|J_{em}^\lambda|0 > (-e^2) \times \bar{v}(k_2)\gamma^\mu u(k_1)$$

$$\sigma_{\text{hadron}}^{e^+e^-} = \frac{16\pi^2\alpha^2}{v_{in}}(2\pi)^4\frac{(2\pi)^6}{(2\pi)^6}\int\frac{d^3p_n}{(2\pi)^3}\delta(p_n - q)\frac{m_e^2}{E_1 E_2}|F|^2 \tag{A.82}$$

Now

$$|F|^2 = \sum_{\substack{\text{lepton spin}\\\text{spin}}}\sum\left|< n|J_{em}^\lambda|0 >\right|^2 (2\pi)^3\left|\bar{v}(k_2)\gamma_\lambda u(k_1)\right|^2\frac{m_e^2}{E_1 E_2}$$

$$= J^{\mu\lambda}\mathcal{L}_{\lambda\mu} \tag{A.83}$$

where

$$\mathcal{L}_{\lambda\mu} = \frac{1}{m_e^2}\left(k_{2\lambda}k_{1\mu} + k_{2\mu}k_{1\lambda} - (k_1 \cdot k_2)g_{\lambda\mu}\right)$$

$$J^{\mu\lambda} = (2\pi)^3\sum_{\text{spin}}< 0|J_{em}^\mu|n >< n|J_{em}^\lambda|0 > \tag{A.84}$$

Introduce

$$A^{\mu\lambda} = \int \frac{d^3p_n}{(2\pi)^3} \delta(p_n - q) J^{\mu\lambda}$$

$$= \int \frac{d^3p_n}{(2\pi)^3} \sum_{\text{spin}} \langle 0|J_{em}^\mu|n\rangle \langle n|J_{em}^\lambda|0\rangle \delta(p_n - q) \qquad (A.85)$$

$$= \sum_n <0|J_{em}^\mu|n><n|J_{em}^\lambda|0> \delta(p_n - q)$$

$$= \frac{1}{(2\pi)^3} \Theta(q_0)(-q^2 g^{\mu\lambda} + q^\mu q^\lambda)\rho(q^2) \qquad (A.86)$$

Here $\Theta(q_0)$ appears since $q_0$ is positive, as $q^2$ is time-like. Using $v_{in} = \left|\frac{\mathbf{k_1}}{E_1} - \frac{\mathbf{k_2}}{E_2}\right| = |\mathbf{k}| \frac{2E}{E^2} \approx 2$ and contracting $A^{\mu\lambda}$ with $\mathcal{L}_{\lambda\mu}$, we obtain

$$\sigma_{\text{hadron}}^{e^+e^-} = \frac{16\pi^3\alpha^2}{s}\rho(s) \qquad (A.87)$$

where $\rho(s)$ is the structure function which represents the blob. Hence

$$\sigma_{\text{hadron}}^{e^+e^-} = \frac{4\pi\alpha^2}{3s}\sigma_\gamma(s) \qquad (A.88)$$

where

$$\rho(s) = \frac{1}{12\pi^2}\sigma_\gamma(s). \qquad (A.89)$$

## A.7.2   *Electron Scattering and Structureless Spin 1/2 Target*

First we consider electron-muon scattering, $e\mu \to e\mu$ through one photon exchange (see Fig. A.3).

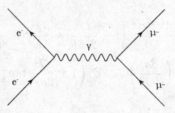

Fig. A.3   One photon exchange Feynman diagram for the process $e^-\mu^- \to e^-\mu^-$.

We can obtain $|F|^2$ from Eq. (A.75) by interchanging $s \leftrightarrow t$ (known as crossing symmetry). Hence we have

$$|F|^2 = \frac{1}{4}\frac{e^2}{t^2}\frac{2}{m_\mu^2 m_e^2}\left[\frac{(s - m_\mu^2)^2}{4} + \frac{(u - m_\mu^2)^2}{4} + \frac{1}{2}tm_\mu^2\right] \qquad (A.90)$$

Thus for high energy electron scattering on any spin $1/2$ target of mass $m$ (for example parton of charge $e_q$), we have

$$|F|^2 = \frac{1}{4} \frac{e^4 e_q^2}{t^2} \frac{2}{m^2 m_e^2} \left[ \frac{(s-m^2)^2}{4} + \frac{(u-m^2)^2}{4} + \frac{1}{2} t m^2 \right] \tag{A.91}$$

In the lab frame

$$k = (E_L, -\mathbf{k}_L), \qquad k' = (E_L', \mathbf{k}_L')$$
$$p = (m, 0), \qquad p' = (p_0', \mathbf{p}_L').$$

From now on we will drop the subscript "$L$"

$$s = (k+p)^2 = m^2 + 2mE \tag{A.92}$$

$$t = (k - k')^2 = (p' - p)^2 = q^2 = -Q^2$$
$$\simeq -2EE'(1 - \cos\theta) = 2m^2 - 2mp_0' = -2m(E - E')$$
$$u = (p - k')^2 = m^2 - 2mE' = (k - p')^2 \tag{A.93}$$

Hence in the lab frame

$$\frac{d\sigma}{dQ^2} = \frac{\pi}{2} \frac{e_{q^2} \alpha^2}{m^2 E^2 Q^4} \left[ (s-m^2)^2 + (u-m^2)^2 + 2m^2 t \right] \tag{A.94}$$

$$\frac{d\sigma}{d\Omega} = \frac{e_q^2 \alpha^2}{2m^2} \frac{1}{Q^4} \frac{E'}{E} \left[ (s-m^2)^2 + (u-m^2)^2 + 2m^2 t \right] \tag{A.95}$$

Now, Eqs. (A.92)-(A.93) gives

$$\left[ (s-m^2)^2 + (u-m^2)^2 + 2m^2 t \right] = 4m^2 \left[ E^2 + E'^2 - m(E - E') \right]$$
$$= 8m^2 E^2 \left[ 1 - \frac{Q^2}{2mE} - \frac{Q^2}{4E^2} + \frac{Q^4}{8m^2 E^2} \right]$$

Thus

$$\frac{d\sigma}{dQ^2} = \frac{4\pi e_q^2 \alpha^2}{Q^4} \left[ 1 - \frac{Q^2}{2mE} - \frac{Q^2}{4E^2} + \frac{Q^4}{8m^2 E^2} \right]$$
$$= \left( \frac{d\sigma}{dQ^2} \right)_{Mott} e_q^2 \left( 1 + \frac{Q^2}{2m^2} \tan^2 \frac{\theta}{2} \right) \tag{A.96}$$

where

$$\left( \frac{d\sigma}{dQ^2} \right)_{Mott} = \frac{4\pi \alpha^2}{Q^2} \frac{E'^2}{E^2} \cos^2 \frac{\theta}{2}$$
$$= \frac{4\pi \alpha^2}{Q^4} \left[ 1 - \frac{Q^2}{2mE} - \frac{Q^2}{4E^2} \right] \tag{A.97}$$

and Eq. (A.93) has been used in going from one form to the other. Now $\left( \frac{d\sigma}{dQ^2} \right)_{Mott}$ gives the elastic scattering of electrons on spinless, structureless proton, where as $\frac{d\sigma}{dQ^2}$ in Eq. (A.96) or Eq. (A.97) ($e_q^2 = 1$), gives the scattering of electrons on structureless spin 1/2 proton. The second term in Eq. (A.96) is due to spin. For particles with structure like proton; the above cross-section is modified by introducing form factors. Thus for spin 1/2 particle we will have two form factors $F_1(Q^2)$ and $F_2(Q^2)$ multiplying the first and second terms.

Finally for high energy electron scattering neglecting the mass term, in Eq. (A.94)

$$\frac{d\sigma}{dQ^2} = 2\pi e_q^2 \alpha^2 \frac{\alpha^2}{s^2}\left(\frac{s^2 + u^2}{t^2}\right) \tag{A.98}$$

### A.7.2.1 *Electron Scattering on a Polarized Parton*

For this purpose, polarized electron is required. Introduce a four-vector $n^\mu$ :

$$n^\mu n_\mu = -1; \qquad k \cdot n = 0$$

where $n^\mu$ is the polarization vector of the electron. For polarized parton, introduce the four-vector $s^\mu$ :

$$s^\mu s_\mu = -1, \qquad p \cdot s = 0$$

For polarized electron:

$$\mathcal{L}^e_{\lambda\mu} = \sum_{\text{spin}} \left| \bar{u}(k')\gamma_\lambda \frac{1 + \gamma_5\gamma.n}{2} u(k) \right|^2$$

$$= \frac{1}{2m_e^2} \left[ k'_\lambda k_\mu + k'_\mu k_\lambda - g_{\mu\lambda} k.k' - im_e \epsilon_{\lambda\mu\nu\rho} q^\nu n^\rho \right]$$

For polarized parton:

$$\mathcal{L}^{\lambda\mu}_P = \sum_{\text{spin}} \frac{1}{2m^2} \left[ p^\lambda p'^\mu + p^\mu p'^\lambda - g^{\mu\lambda} p.p' + m^2 g^{\mu\lambda} + im\epsilon^{\mu\lambda\nu'\rho'} q_{\nu'} s_{\rho'} \right]$$

$$|F|^2 = \frac{e^4}{t^2} e_q^2 \mathcal{L}^{\lambda\mu}_P \mathcal{L}^l_{\mu\lambda}$$

Hence for polarized scattering, the relevant $|F|^2$ is

$$|F|^2_{pol} = \frac{e_q^2 (4\pi)^2 \alpha^2}{t^2} \frac{1}{4m^2 m_e^2} (2mm_e) \left[ \delta^{\nu'}_\nu \delta^{\rho'}_\rho - \delta^{\nu'}_\rho \delta^{\rho'}_\nu \right] q^\nu n^\rho q_{\nu'} s_{\rho'}$$

$$= \frac{e_q^2 (4\pi)^2 \alpha^2}{t^2} \frac{1}{2mm_e} \left[ q^2 n.s - q.nq.s \right]$$

For longitudinal polarization of electron

$$k = (E, 0, 0, E), \qquad \mathbf{k}' = E' \left[ \cos \phi \sin \theta, \sin \phi \sin \theta, \cos \theta \right]$$

$$n = \lambda_e \left[ \frac{E}{m_e}, 0, 0, \frac{E}{m_e} \right], \qquad \lambda_e = \pm \begin{matrix} \text{(right handed)} \\ \text{(left handed)} \end{matrix}$$

$$q^2 = t = (k - k')^2 = -2EE'(1 - \cos \theta)$$

$$q_z = k_z - k_z' = E - E' \cos \theta$$

$$p = (m, 0)$$

$$p.s = 0 \Rightarrow s^0 = 0, \qquad s = (0, \mathbf{s}).$$

We consider the longitudinally polarized parton:

$$\mathbf{s} = (0, 0, 1)$$

Thus

$$\left[ q^2 n.s - q.nq.s \right] = -\frac{\lambda_e}{m_e} t \left[ E + E' \cos \theta \right]$$

$$= \frac{\lambda_e t}{4mm_e} \left[ (s - u) - t \frac{m}{E} \right]$$

$$|F|^2 = \frac{e_q^2 (4\pi)^2 \alpha^2}{t^2} \frac{1}{2mm_e} \left[ q^2 n.s - q.nq.s \right]$$

$$= \frac{e_q^2 (4\pi)^2 \alpha^2}{t^2} \frac{t}{2mm_e} E \left[ \frac{-\lambda_e}{m} (E + E' \cos \theta) \right]$$

$$\frac{d\sigma}{dQ^2} = \frac{4\pi e_q^2 \alpha^2}{Q^4} \frac{1}{2mE^2} Q^2 (E + E' \cos \theta)$$

$$\frac{d\sigma}{dQ^2} = \frac{8\pi e_q^2 \alpha^2}{Q^4} \frac{Q^2}{4mE^2} E \left( 2 - \frac{\nu}{E} - \frac{Q^2}{2E^2} \right) \qquad \text{(A.99)}$$

## A.8   Discrete Symmetries

### A.8.1   *Charge Conjugation*

Dirac equation in the presence of electromagnetic field is given by

$$\left[ i\gamma^\mu \left( \partial_\mu + ieA_\mu \right) - m \right] \Psi(x) = 0 \qquad \text{(A.100)}$$

For the adjoint field $\overline{\Psi}$, Eq. (A.11b) can be written:

$$\left[ -i \left( \gamma^\mu \right)^T \left( \partial_\mu - ieA_\mu \right) - m \right] \overline{\Psi}^T(x) = 0 \qquad \text{(A.101)}$$

Under the charge conjugation

$$\Psi(x) \rightarrow \Psi^c(x) = U_c \Psi(x) u_c^{-1} \tag{A.102}$$

$$A_\mu(x) \rightarrow A_\mu^c(x) = U_c A_\mu(x) u_c^{-1}. \tag{A.103}$$

If the Dirac equation is invariant under charge conjugation then:

$$\left[ i\gamma^\mu \left( \partial_\mu + ieA_\mu^c \right) - m \right] \Psi^c(x) = 0. \tag{A.104}$$

Now we can write Eq. (A.101) as

$$C \left[ -i \left( \gamma^\mu \right)^T \left( \partial_\mu - ieA_\mu \right) - m \right] C^{-1} C \overline{\Psi}^T(x) = 0, \tag{A.105}$$

where $C$ is a unitary matrix, called the charge conjugation matrix. Equation (A.105) is identical to Eq. (A.104), provided that

$$\gamma^\mu = -C \left( \gamma^\mu \right)^T C^{-1} \tag{A.106}$$

$$A_\mu^c = -A_\mu \tag{A.107}$$

$$\Psi^c(x) = C \overline{\Psi}^T(x) \tag{A.108}$$

Also one can write

$$\Psi^c(x) = C \overline{\Psi}^T(x) = -\gamma^0 C \Psi^* \tag{A.109}$$

$$\overline{\Psi}^c(x) = -\Psi^T(x) C^{-1} \tag{A.110}$$

Both in Pauli and Weyl representations for $\gamma$-matrices

$$\left( \gamma^\mu \right)^T = \begin{cases} \gamma^\mu & \text{for} \quad \mu = 0, 2 \\ -\gamma^\mu & \text{for} \quad \mu = 1, 3 \end{cases} \tag{A.111}$$

Therefore we have from Eq. (A.106):

$$C = -i\gamma^2 \gamma^0$$

$$C^\dagger = -C = C^T$$

$$C^2 = -1 \tag{A.112}$$

$$\Psi^c(x) = -i\gamma^2 \Psi^* \tag{A.113}$$

Under charge conjugation

$$u \rightarrow v = C\overline{u}^T \tag{A.114}$$

$$\overline{u} \rightarrow \overline{v} = -u^T C^{-1} \tag{A.115}$$

We define the left-handed and the right-handed fields

$$\psi_L = \frac{(1 - \gamma_5)}{2} \psi; \qquad \overline{\psi}_L = \overline{\psi} \frac{(1 + \gamma_5)}{2}$$

$$\psi_R = \frac{(1 + \gamma_5)}{2} \psi; \qquad \overline{\psi}_R = \overline{\psi} \frac{(1 - \gamma_5)}{2}$$

$$\psi_L^c = \frac{(1-\gamma_5)}{2}\psi^c = C\left(\overline{\psi}\frac{(1-\gamma_5)}{2}\right)^T = C\overline{\psi}_R^T$$

$$\overline{\psi}_L^c = -\psi_R^T C^{-1}$$

Similarly

$$\psi_R^c = C\overline{\psi}_L^T, \qquad \overline{\psi}_R^c = -\psi_L^T C^{-1}$$

For a spin 1/2 particle described by a spinor $\psi$, we have $\psi_L$ : left-handed particle; $\psi_L^c$ : left-handed antiparticle, $\psi_R$ : right-handed particle, $\psi_R^c$ : right-handed antiparticle. However, the neutrino which takes part in weak interaction exists only in one chirality state viz left-handed. Thus for a neutrino we have left-handed neutrino $\nu_L$ and right-handed antineutrino $\nu_R^c$.

In the Weyl representation, one can write

$$\Psi = \left(\frac{1-\gamma^5}{2}\right)\Psi + \left(\frac{1+\gamma^5}{2}\right)\Psi$$

$$= \begin{pmatrix} \Psi_L \\ 0 \end{pmatrix} + \begin{pmatrix} 0 \\ \Psi_R \end{pmatrix}$$

$$= \begin{pmatrix} \xi \\ 0 \end{pmatrix} + \begin{pmatrix} 0 \\ \eta \end{pmatrix} = \begin{pmatrix} \xi \\ \eta \end{pmatrix} \tag{A.116}$$

where

$$\Psi_L = \left(\frac{1-\gamma^5}{2}\right)\Psi \equiv \xi, \quad \Psi_R = \left(\frac{1+\gamma^5}{2}\right)\Psi \equiv \eta \tag{A.117}$$

are two component left-handed and right-handed spinors. Hence from Eq. (A.110), we get

$$\xi^c = -i\sigma^2\eta^*$$

$$\eta^c = i\sigma^2\xi^* \tag{A.118}$$

Sometimes it is convenient to write a right-handed field in terms of a left-handed antiparticle field (cf. Eq. (A.118)):

$$\eta = i\sigma^2\xi^{c*} \tag{A.119}$$

so that Eq. (A.116) becomes

$$\Psi = \begin{pmatrix} \xi \\ i\sigma^2\xi^{c*} \end{pmatrix} \tag{A.120}$$

The Majorana spinor $\Psi_M$ is defined as

$$\Psi_M^C = \Psi_M = C\overline{\Psi}_M^T$$

$$\Psi_{M\alpha} = C_{\alpha\beta}\overline{\Psi}_M^\beta$$

Now the Dirac Lagrangian

$$\mathcal{L} = \bar{\Psi}\left(i\gamma^\mu\partial_\mu - m_D\right)\Psi + \frac{m_M}{2}\left(\Psi^T C^{-1}\Psi - \bar{\Psi}C\bar{\Psi}^T\right) \tag{A.121}$$

(where the second term in Eq. (A.121) is the Majorana mass term and violates lepton number conservation) can be written in terms of two component chiral fields using Eqs. (A.118), (A.112) and (A.120):

$$\mathcal{L} = i\left[\xi^\dagger\bar{\sigma}^\mu\partial_\mu\xi + \xi^{c\dagger}\bar{\sigma}^\mu\partial_\mu\xi^c\right] - m_D\left[\xi^{*T}i\sigma^2\xi^{c*} + \xi^{c^T}\left(-i\sigma^2\right)\xi\right]$$
$$+\frac{m_M}{2}\left[\xi^T i\sigma^2\xi - \xi^{c*^T}i\sigma^2\xi^{c*} + \xi^{c^T}i\sigma^2\xi^c - \xi^{*^T}i\sigma^2\xi^*\right] \tag{A.122}$$

where we have used

$$\sigma^2\,\sigma^\mu\,\sigma^2 = (\bar{\sigma}^\mu)^T$$

$$\left(\sigma^2\xi^{c*}\right)^\dagger\sigma^\mu\partial_\mu\left(\sigma^2\xi^{c*}\right) = \xi^{c^T}(\bar{\sigma}^\mu)^T\partial_\mu\xi^{c*}$$

$$= -\partial_\mu\xi^{c\dagger}\bar{\sigma}^\mu\xi^c \quad \text{(fermion fields anticommute)}$$

$$= \xi^{c\dagger}\bar{\sigma}^\mu\partial_\mu\xi^c \quad \text{(partial integration)}$$

Equation (A.122) can be put in the compact form

$$\mathcal{L} = i\xi_i^\dagger\bar{\sigma}^\mu\partial_\mu\xi_i - \frac{1}{2}\left(m_{ij}\xi_i^T\left(-i\sigma^2\right)\xi_j + h.c.\right) \tag{A.123}$$

where $i$, $j = 1$, 2 and $m_{ij}$ is the symmetric mass matrix

$$\begin{pmatrix} m_M & m_D \\ m_D & m_M \end{pmatrix}$$

and $\xi_1 = \xi = \Psi_L$ and $\xi_2 = \xi^c = -i\sigma^2\eta^* = -i\sigma^2\Psi_R^*$.

## A.8.2 Space Reflection

$$\mathbf{x} \rightarrow -\mathbf{x}, \qquad t \rightarrow t$$

$$x^\mu \rightarrow x'^\mu = \Lambda^\mu{}_\nu x^\nu$$

where

$$\Lambda^\mu{}_\nu = \begin{pmatrix} 1 & 0 & 0 & 0 \\ 0 & -1 & 0 & 0 \\ 0 & 0 & -1 & 0 \\ 0 & 0 & 0 & -1 \end{pmatrix}$$

$$x^\mu \rightarrow x'^\mu \equiv (x^0, -\mathbf{x})$$

Dirac Equation:

$$\left[ i\gamma^\mu \frac{\partial}{\partial x^\mu} - m \right] \psi(x^0, \mathbf{x}) = 0 \qquad (A.124)$$

in the reflected coordinate system becomes

$$\left[ i\gamma^0 \frac{\partial}{\partial x^0} - i\gamma^i \frac{\partial}{\partial x^i} - m \right] \psi'(x^0, -\mathbf{x}) = 0$$

$$\gamma^0 \left[ i\gamma^0 \frac{\partial}{\partial x^0} - i\gamma^i \frac{\partial}{\partial x^i} - m \right] \gamma^0 \gamma^0 \psi'(x') = 0$$

Now using $\gamma^0 \gamma^i \gamma^0 = -\gamma^i$, we have

$$\left[ i\gamma^\mu \frac{\partial}{\partial x^\mu} - m \right] \gamma^0 \psi'(x') = 0$$

It reduces to Eq. (A.124), if

$$\gamma^0 \psi'(x') = \psi(x)$$
$$\psi'(x') = \gamma^0 \psi(x)$$

Changing $x' \to x$, we have

$$\psi'(x) = \hat{P} \psi(x) \hat{P}^{-1} = \eta_P \gamma^0 \psi(x')$$

Finally under space reflection

$$\psi'(x') = \eta_P \gamma^0 \psi(x)$$
$$\bar{\psi}'(x') = \eta_P^* \bar{\psi}(x) \gamma^0$$

Under space reflection

$$u^{(r)}(\mathbf{p}) \to u^{(r)}(-\mathbf{p}) = \gamma^0 u^{(r)}(\mathbf{p})$$
$$v^{(r)}(\mathbf{p}) \to v^{(r)}(-\mathbf{p}) = -\gamma^0 v^{(r)}(\mathbf{p}) \text{ (cf. Problem A.4)}$$

### A.8.3  *Time Reversal*

$$t \to -t, \mathbf{x} \to \mathbf{x} \qquad x^\mu \to x'^\mu = (-x^0, \mathbf{x})$$

The Dirac equation

$$[i\gamma^\mu \partial_\mu - m] \psi(x^0, \mathbf{x}) = 0 \qquad (A.125)$$

under time reversal $t \to -t$,

$$\left[ -i\gamma^{0*}(-\partial_0) - i\gamma^{i*}\partial_i - m \right] \psi'(-x^0, \mathbf{x}) = 0$$

Multiply the above equation on the left by a matrix $B^{-1}$ :

$$B^{-1} \left[ -i\gamma^{0*}(-\partial_0) - i\gamma^{i*}\partial_i - m \right] B B^{-1} \psi'(-x^0, \mathbf{x}) = 0 \qquad (A.126)$$

For the Dirac equation to be invariant under time reversal, Eq. (A.126), should reduce to Eq. (A.125). This requires:

$$B^{-1}\psi'(-x^0, \mathbf{x}) = B^{-1}\psi'(x') = \eta_T \psi(x)$$
$$\psi'(x') = \eta_T B\psi(x)$$

and (note that $\gamma^0$ is hermitian while $\gamma^i$ is antihermitian)

$$B^{-1}\gamma^{0*}B = \gamma^0, \; B^{-1}\left(\gamma^0\right)^T B = \gamma^0$$
$$B^{-1}\gamma^{i*}B = -\gamma^i, \; B^{-1}\left(\gamma^i\right)^T B = \gamma^i \qquad \text{(A.127)}$$

or

$$B^{-1}\left(\gamma^\mu\right)^T B = \gamma^\mu$$

$$B^T B = 1, \; B^T = -B$$

Thus

$$\psi'(x) = \Pi\psi(x)\Pi^{-1} = \eta_T B\psi(x')$$
$$\text{or} \quad \psi'(x') = \eta_T B\psi(x)$$
$$\overline{\psi}'(x') = \eta_T^* \overline{\psi}(x)B^\dagger$$

From Eq. (A.127), it is clear we can select

$$B = \gamma^1\gamma^3$$
$$B^\dagger = \gamma^3\gamma^1 \qquad \gamma^0 B^\dagger \gamma^0 = B^\dagger$$
$$BB^\dagger = 1; \qquad B^\dagger = B^{-1}$$

$B$ is a unitary matrix. Under time reversal

$$u^{(r)}(\mathbf{p}) \to u^{*(-r)}(-\mathbf{p}) = Bu^{(r)}(\mathbf{p})$$
$$v^{(r)}(\mathbf{p}) \to v^{*(-r)}(-\mathbf{p}) = Bv^{(r)}(\mathbf{p}) \qquad \text{(cf. problem A.4)}$$

## A.9 Problems

(1) Under Lorentz transformation, the Dirac spinor $\psi(x)$ transforms as

$$\psi'(x') = S\psi(x)$$

where for infinitesimal transformation

$$S = 1 - \frac{i}{4}\epsilon_{\mu\nu}\sigma^{\mu\nu}$$

Show that

$$S^{-1}\gamma^5 S = \gamma^5$$
$$S^{-1}\gamma^\mu S = \gamma^\mu + \epsilon^\mu_\nu \gamma^\nu$$

Using the above result, show that Dirac bilinears, under proper Lorentz transformation, transform:

$$S = \overline{\Psi}\Psi, \ P = \overline{\Psi}\gamma^5\Psi \qquad \text{as a scalar}$$
$$V^\mu = \overline{\Psi}\gamma^\mu\Psi, \ A^\mu = \overline{\Psi}\gamma^\mu\gamma^5\Psi \qquad \text{as vector}$$
$$T^{\mu\nu} = \overline{\Psi}\sigma^{\mu\nu}\Psi \qquad \text{as tensor}$$

Show that under space reflection $S$ is a scalar, $P$ is a pseudoscalar (i.e. changes sign), $V^\mu$ is a vector (i.e. its space components change sign) while $A^\mu$ is an axial vector (i.e. its space components do not change sign).

(2) Show that

(a)

$$\gamma_5\gamma_\lambda = \frac{i}{3!}\epsilon_{\lambda\mu\nu\rho}\gamma^\mu\gamma^\nu\gamma^\rho$$

(b) Using this result and

$$\epsilon^{\mu\nu\rho\lambda}\epsilon_{\mu\nu'\rho'\lambda'}$$
$$= -\left[\delta^\nu_{\nu'}\delta^\rho_{\rho'}\delta^\lambda_{\lambda'} + \delta^\nu_{\rho'}\delta^\rho_{\lambda'}\delta^\lambda_{\nu'} + \delta^\nu_{\lambda'}\delta^\rho_{\nu'}\delta^\lambda_{\rho'} - \delta^\nu_{\nu'}\delta^\rho_{\lambda'}\delta^\lambda_{\rho'} - \delta^\nu_{\rho'}\delta^\rho_{\nu'}\delta^\lambda_{\lambda''} - \delta^\nu_{\lambda'}\delta^\rho_{\rho'}\delta^\lambda_{\nu'}\right]$$

show that

$$\gamma^\mu\gamma^\nu\gamma^\lambda = g^{\mu\nu}\gamma^\lambda - g^{\mu\lambda}\gamma^\nu + g^{\nu\lambda}\gamma^\mu + i\epsilon^{\mu\nu\lambda\rho}\gamma^5\gamma_\rho$$

(c) Show that using the above result

$$[\gamma^\lambda, \ \sigma^{\mu\nu}] = 2i\left(g^{\mu\lambda}\gamma^\nu - g^{\nu\lambda}\gamma^\mu\right)$$
$$\{\gamma^\lambda, \ \sigma^{\mu\nu}\} = -2\epsilon^{\mu\nu\lambda\rho}\gamma^5\gamma_\rho$$

(3) Show under space reflection $(P)$, time reversal $(T)$ and charge conjugation $(C)$

$$\overline{\Psi}_i\gamma^\mu\Psi_j \text{ and } \overline{\Psi}_i\gamma^\mu\gamma^5\Psi_j$$

transform as

$$P: \begin{cases} \overline{\Psi}_i\gamma^\mu\Psi_j \to (-1)^\mu\overline{\Psi}_i\gamma^\mu\Psi_j \\ \overline{\Psi}_i\gamma^\mu\gamma^5\Psi_j \to (-1)(-1)^\mu\overline{\Psi}_i\gamma^\mu\gamma^5\Psi_j \end{cases}$$

$$T: \begin{cases} \overline{\Psi}_i\gamma^\mu\Psi_j \to (-1)^\mu\overline{\Psi}_i\gamma^\mu\Psi_j \\ \overline{\Psi}_i\gamma^\mu\gamma^5\Psi_j \to (-1)^\mu\overline{\Psi}_i\gamma^\mu\gamma^5\Psi_j \end{cases}$$

$$C: \begin{cases} \overline{\Psi}_i\gamma^\mu\Psi_j \to -1\left(\overline{\Psi}_i\gamma^\mu\Psi_j\right)^T = -\overline{\Psi}_j\gamma^\mu\Psi_{ij} \\ \overline{\Psi}_i\gamma^\mu\gamma^5\Psi_j \to \left(\overline{\Psi}_i\gamma^\mu\gamma^5\Psi_j\right)^T = \overline{\Psi}_j\gamma^\mu\gamma^5\Psi_i \end{cases}$$

$$\text{where } (-1)^\mu = \begin{cases} +1 & \mu = 0 \\ -1 & \mu = i \end{cases}$$

(4) Show that

$$u(-\mathbf{p}) = \gamma^0 u(\mathbf{p})$$
$$v(-\mathbf{p}) = -\gamma^0 v(-\mathbf{p})$$
$$u^{(-r)*}(-\mathbf{p}) = B u^{(r)}(\mathbf{p})$$

One may use

$$u(\mathbf{p}) = \sqrt{\frac{E+m}{2m}} \begin{pmatrix} \chi^{(r)} \\ \frac{\sigma \cdot \mathbf{p}}{E+m}\chi^{(r)} \end{pmatrix}$$

(5) For a particle of mass $m$ and momentum $p^\mu$, introduce a four-vector $s^\mu$

$$s^\mu s_\mu = -1, \; p \cdot s = 0$$

$s^\mu$ is called the polarization vector. Introduce a (spin) projection operator

$$\frac{1 + \gamma_5 \gamma \cdot s}{2}$$

Show that

$$\left(\frac{1 + \gamma_5 \gamma \cdot s}{2}\right)^2 = \frac{1}{2}(1 + \gamma_5 \gamma \cdot s)$$
$$[1 + \gamma_5 \gamma \cdot s, \; \gamma \cdot p] = 0$$

Show that in the rest frame of particle

$$\frac{1 + \gamma_5 \gamma \cdot s}{2} u^{(r)}(\mathbf{p}) = \frac{1}{2}(1 + \sigma \cdot \mathbf{s})\chi^{(r)}$$

To take into account polarization of spin $\frac{1}{2}$ particle

$$u(\mathbf{p}) \to \frac{1 + \gamma_5 \gamma \cdot s}{2} u(\mathbf{p})$$
$$v(\mathbf{p}) \to \frac{1 + \gamma_5 \gamma \cdot s}{2} v(\mathbf{p})$$

Hence show that

$$u(\mathbf{p})\bar{u}(\mathbf{p}) \to \frac{\not{p} + m}{2m} \frac{1 + \gamma_5 \not{s}}{2}$$
$$v(\mathbf{p})\bar{v}(\mathbf{p}) \to \frac{\not{p} - m}{2m} \frac{1 + \gamma_5 \not{s}}{2}$$

(6) (a) Show that for the operators $A$, $B$, $C$ and $D$

$$[AB, CD] = \{A, \; C\} DB - AC \{B, \; D\} + A \{C, \; B\} D - C \{A, \; D\} B$$

(b) For the spinor field operator $\Psi(x)$, we have

$$\{\Psi_\alpha(x),\ \overline{\Psi}_\beta(x)\} = (i\gamma^\mu\partial_\mu + m)_{\alpha\beta}\, i\Delta(x-y)$$
$$\equiv iS_{\alpha\beta}(x-y)$$

Using (a), show that

$$[\overline{\Psi}_\alpha(x)\Psi_\beta(x),\ \overline{\Psi}_{\beta'}(y)\Psi_{\alpha'}(y)] = \overline{\Psi}_\alpha(x)(i)S_{\beta\beta'}(x-y)\Psi_\alpha(y)$$
$$+\overline{\Psi}_{\beta'}(y)(-i)S_{\alpha\alpha'}(y-x)\Psi_{\beta'}$$

(c) The matrices $\Gamma_a, \Gamma_b$ satisfy anticommutation relation

$$\{\Gamma_a,\ \Gamma_b\} = 2\delta_{ab}(a,\ b = 1,\ 2,\ 3)$$

Using (a), show that the matrices

$$R_{ab} = -\frac{i}{4}(\Gamma_a\Gamma_b - \Gamma_b\Gamma_a)$$

satisfy the commutation relation

$$[R_{ab},\ R_{cd}] = i\,[\delta_{ac}R_{db} - \delta_{bd}R_{ac} + \delta_{cb}R_{ad} - \delta_{ad}R_{cb}],$$

i.e. the algebra of group O(4).

(7) Show that bilinear $\psi^T C^{-1}\psi$ is also a Lorentz scalar. Further show that

$$\psi^T C^{-1}\psi = -\overline{\psi}^c\psi$$

(8) A positive energy Dirac spinor is given in (Pauli rep. of $\gamma$-matrices):

$$u^{(r)}(p) = \begin{pmatrix} \chi^{(r)} \\ \frac{\sigma\cdot\mathbf{P}}{E+m}\chi^{(r)} \end{pmatrix}$$

As $m \to 0$, write

$$\omega^{(r)}(p) = \frac{1-\gamma^5}{2}u^{(r)}(p)$$

show that

$$\mathcal{H}\omega^{(r)}(p) = -\omega^{(r)}(p)$$

where $\mathcal{H}$ is the helicity operator $\frac{\sigma\cdot\mathbf{P}}{|p|}$, i.e. $\omega^{(r)}(p)$ is an eigenstate of $\mathcal{H}$ with eigenvalue $-1$. Interpret your result.

(9) For spin 1/2 baryon, $J_\lambda = V_\lambda - A_\lambda$, the form factors are given by

$$\langle B(p')|J_\lambda|A(p)\rangle = \frac{1}{(2\pi)^3}\sqrt{\frac{m_A m_B}{p_0 p_0'}}\overline{u}(p')$$
$$\times \begin{bmatrix} (g_V\gamma_\lambda - g_A\gamma_\lambda\gamma_5) + if_V(q^2)\sigma_{\lambda\nu}q^\nu \\ +ih_A(q^2)\sigma_{\lambda\nu}q^\nu\gamma_5 + h_V(q^2)q_\lambda - f_A(q^2)q_\lambda\gamma_5 \end{bmatrix} u(p)$$

(a) Show that if time reversal invariance holds, then all form factors are real.

(b) In quark model

$$J_\lambda = \overline{\Psi}_i \gamma_\lambda (1 - \gamma_5) \Psi_j$$

Using CP invariance,

$$\langle B \,|J_\lambda|\, A \rangle = \langle B \,\big|\overline{\Psi}_i \gamma_\lambda (1 - \gamma_5) \Psi_j\big|\, A \rangle$$
$$\overset{CP}{\rightarrow} = -(-1)^\lambda \langle \bar{B} \,\big|\overline{\Psi}_j \gamma_\lambda (1 - \gamma_5) \Psi_i\big|\, \bar{A} \rangle$$

Show that the form factors for $A \rightarrow B$ and $\bar{A} \rightarrow \bar{B}$ are related as follows

$$\bar{g}_V = g_V, \bar{f}_V = f_V, \ \bar{h}_V = -h_V$$
$$\bar{g}_A = g_A, \ \bar{h}_A = -h_A, \ \bar{f}_A = f_A.$$

# Appendix B

# Renormalization Group and Running Coupling Constant

## B.1 Feynman Rules for Quantum Chromodynamics

For canonical covariant quantization, the QCD Lagrangian given in Eq. (7.32) is written as [repeated indices imply summation]

$$\mathcal{L} = -\frac{1}{4}G_A^{\mu\nu}g_{A\mu\nu} + \bar{q}^a i\gamma^\mu \left(\partial_\mu - ig_s T_A\right)_a^b q_b - \frac{1}{2\xi}\left(\partial^\mu g_{A\mu}\right)^2 + \text{Ghosts} \quad (B.1)$$

where

$$G_A^{\mu\nu} = \partial^\mu g_A^\nu - \partial^\nu g_A^\mu + g_s \, f_{ABC} g_B^\mu G_C^\nu$$

$$[T_A, T_B] = i \, f_{ABC} \, T_C, \quad T_A = \frac{1}{2}\lambda_A$$

$$Tr \, (T_A, T_B) = \frac{1}{2}\delta_{AB}, \text{ for the fundamental representation.}$$

$$\sum f_{ACD} \, f_{BCD} = C_2\left(G\right)\delta_{AB}$$

$$= N\delta_{AB}, \quad \text{for } SU(N) \text{ gauge group} \quad (B.2)$$

$$(T_A)_c^a \, (T_A)_b^c = C_F \delta_b^a = \frac{N^2-1}{2N}\delta_b^a, \quad \text{for } SU(N).$$

In the Lagrangian (B.1), $-\frac{1}{2\xi}\left(\partial^\mu g_{A\mu}\right)^2$ is the gauge fixing term, $\xi$ being the fixing parameter. The supplementary non-physical fields, called ghosts are needed for covariant quantization in order to cancel the probabilities of observing scalar (or time-like) and longitudinal gluons.

Quantizing in a renormalizable gauge leads to the following Feynman rules:

$A,\mu \quad k \quad B,\nu$    $\delta_{AB}\frac{-i}{k^2}\left[g_{\mu\nu} - \frac{(1-\xi)k_\mu k_\nu}{k^2}\right]$, gluon propagator

$A \quad k \quad B$    $-i\delta_{AB}/k^2$, ghost propagator

$a \quad k \quad b$    $i\delta_b^a \frac{1}{\not{k}-m}$, quark propagator

$$-g_s f_{ABC} \left[ \begin{array}{c} (p-q)_\nu \, g_{\lambda\mu} + (q-r)_\lambda \, g_{\mu\nu} \\ + (r-p)_\mu \, g_{\nu\lambda} \end{array} \right]$$

$$-i g_s^2 \, f_{ABE} \, f_{CDE} \, (g_{\lambda\nu} g_{\mu\sigma} - g_{\lambda\sigma} g_{\mu\nu})$$
$$-i g_s^2 \, f_{ACE} \, f_{BDE} \, (g_{\lambda\mu} g_{\nu\sigma} - g_{\lambda\sigma} g_{\mu\nu})$$
$$-i g_s^2 \, f_{ADE} \, f_{CBE} \, (g_{\lambda\nu} g_{\mu\sigma} - g_{\lambda\mu} g_{\sigma\nu})$$

$$g_s f_{ABC} \, p_\mu$$

$$i g_s \gamma_\mu \, (T_A)_{ab}$$

The other factors are

(i)     $\int \frac{d^4 l}{(2\pi)^4}$ for each loop integral

(ii)    $(-1)$ · for closed fermion (ghost) loop

(iii)   Statistical factors like

$\frac{1}{2!}$     $\frac{1}{3!}$     etc.

## B.2   Renormalization Group, Coupling Constant and Asymptotic Freedom

We now show that the self-coupling of gluons envisaged in the first term of the Lagrangian (B.1) has the consequences that QCD has a remarkable property of being asymptotically free, i.e. the quark - quark force becomes weak at large momentum transfer or short distances, such as probed in deep-inelastic collision [cf. Chap. 14]. In other words, the coupling constant $\alpha_s$ depends on the momentum transfer in such a way that $\alpha_s \left(Q^2\right) \to 0$ as $Q^2 \to \infty$.

Consider the radiative corrections to quark-quark-gluon ($qqG$) vertex, where at one-loop level these corrections are shown in Fig. B.1, $[Q^2 = -q^2]$.

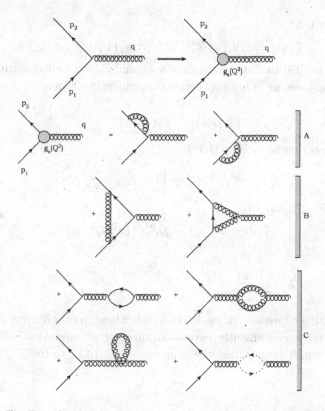

Fig. B.1   One-loop corrections to quark-quark-gluon $qqG$ vertex.

The one-loop corrections to $qqG$ vertex shown above are infinite. One must define a high $l^2$- cut off ($l$ bing loop momentum) $\lambda^2$, so that the loop integrals converge. We have then

$$\Gamma_{A\mu} = -iT_A\gamma_\mu\Gamma_s\left(Q^2,\lambda,g_s\right) \qquad \text{(B.3a)}$$

where

$$\Gamma_s\left(Q^2,\lambda,g_s\right) = g_s\left[1 + g_s^2\left(a_0 + b_0\ln\frac{\lambda^2}{Q^2}\right)\cdots\right], \qquad \lambda^2 \gg Q^2 \quad \text{(B.3b)}$$

where $\cdots$ denotes the corrections from higher order loops. Note here that the cut-off dependent logarithmic contributions from diagrams A [involving quark self-energy diagram] and the first of diagrams B [involving the quark gluon vertex function] cancel due to gauge invariance as is also the case in quantum electrodynamics. Since the theory is renormalizable, we must be able to write it as

$$\Gamma_s\left(Q^2/\lambda^2, g_s\right) = Z_s^{1/2}\left(\lambda^2/\mu^2, g_s\right)\Gamma_s^R\left(Q^2/\mu^2, g_s\right) \qquad (B.4)$$

where $\mu$ is called the renormalization scale and $Z_s$ is a multiplicative renormalization constant. One may define the renormalization scale through the relation

$$\Gamma_s^R\left(Q^2/\mu^2, g_s\right)\big|_{Q^2=\mu^2} = g_s. \qquad (B.5)$$

Then neglecting $a_0$ in Eq. (B.3b)

$$Z_s^{1/2}\left(\lambda^2/\mu^2, g_s\right) = \left[1 + g_s^2 b_0 \ln\frac{\lambda^2}{\mu^2}\cdots\right] \qquad (B.6a)$$

and [cf. Eq. (B.3b)]

$$\Gamma_s\left(Q^2/\lambda^2, g_s\right) = g_s Z_s^{1/2}\left(\lambda^2/Q^2, g_s\right). \qquad (B.6b)$$

Thus

$$g_s\left(Q^2\right) \equiv \Gamma_s^R\left(Q^2/\mu^2, g_s\right)$$
$$= g_s Z_s^{-1/2}\left(\lambda^2/\mu^2, g_s\right) Z_s^{1/2}\left(\lambda^2/Q^2, g_s\right). \qquad (B.7)$$

This relation expresses the basic renormalization group property.

It is more conveniently expressed through an equivalent differential equation which follows from the $\mu$-independence of $\Gamma_s$ so that

$$\frac{d\Gamma_s}{d\mu} = 0, \qquad (B.8a)$$

or, using Eq. (B.4),

$$\frac{1}{\Gamma_s^R\left(\mu\right)}\frac{d\Gamma_s^R\left(\mu\right)}{d\mu} = -\frac{1}{Z_s^{1/2}}\frac{dZ_s^{1/2}}{d\mu} \qquad (B.8b)$$

This can be rewritten as $[\Gamma_s^R\left(\mu\right) = g_s\left(\mu\right)]$

$$\frac{dg_s\left(\mu\right)}{d\ln\mu} = -g_s\frac{1}{Z_s^{1/2}}\frac{dZ_s^{1/2}}{d\ln\mu} = \beta\left(g_s\right) \qquad (B.9a)$$

so that

$$\beta\left(g_s\right) = -g_s\frac{1}{Z_s^{1/2}}\frac{dZ_s^{1/2}}{d\ln\mu} \qquad (B.9b)$$

where $Z_s^{1/2}$ is given in Eq. (B.6a). Equations (B.9) are known as the renormalization group equations for the effective coupling constant $g_s\left(\mu\right)$. Writing

$$Z_s^{1/2} = 1 + \sum_{k=1}^{\infty}\left(\ln\frac{\lambda^2}{\mu^2}\right)^k Z_{s,k}^{1/2}\left(g_s\right), \qquad (B.10a)$$

Eq. (B.9b) gives

$$\beta\left(g_s\right) = -2g_s^3 \frac{dZ_{s,1}^{1/2}}{dg_s^2}$$

$$= -2g_s^3 \left[b_0 + b_1 g_s^2 + \cdots\right], \tag{B.10b}$$

where we have used Eq. (B.6a).

To integrate Eq. (B.9a), it is convenient to write it, on using Eq. (B.10b), as [putting $\mu^2 = Q^2$]

$$\frac{d\left(1/g_s^2\right)}{d\ln Q^2} = 2\left[b_0 + b_1 g_s^2 + \cdots\right]$$

or

$$d\ln Q^2 = d\left(1/g_s^2\right) \frac{1}{2b_0} \left[1 + \frac{b_1}{b_0}\left(1/g_s^2\right)^{-1} + \cdots\right]^{-1}$$

$$= d\left(1/g_s^2\right) \frac{1}{2b_0} \left[1 - \frac{b_1}{b_0}\left(1/g_s^2\right)^{-1} + \cdots\right]. \tag{B.11}$$

If we keep only the lowest order term, we have

$$\frac{d\left(\alpha_s^{-1}\right)}{d\ln Q^2} = b \tag{B.12a}$$

where $\alpha_s = \frac{g_s^2}{4\pi}, b = 8\pi b_0$. Integration of Eq. (B.12a) gives

$$\alpha_s^{-1}\left(Q^2\right) = \alpha_s^{-1}\left(\mu^2\right) + b\ln\frac{Q^2}{\mu^2} \tag{B.12b}$$

Note that what renormalization group does is to relate the coupling constant at two different scales. We may also write Eq. (B.12b) as

$$\alpha_s^{-1}\left(Q^2\right) = b\ln\frac{Q^2}{\Lambda_{QCD}^2} \tag{B.12c}$$

where $\left[\alpha_s^{-1}\left(\mu^2\right) \equiv \alpha_\mu^{-1}\right]$

$$\frac{1}{b}\alpha_\mu^{-1} - \ln\mu^2 = -\ln\Lambda_{QCD}^2$$

or

$$\frac{\Lambda_{QCD}^2}{\mu^2} = \exp\left(-\frac{1}{b\alpha_\mu}\right). \tag{B.12d}$$

$\Lambda_{QCD}$ is one parameter which determines the size of $\alpha_s\left(Q^2\right)$. It must be determined from experiment. Thus finally we have from Eq. (B.12)

$$\alpha_s\left(Q^2\right) = \frac{1}{\alpha_\mu^{-1} + b\ln\frac{Q^2}{\mu^2}} + O\left(\alpha_s^2\left(Q^2\right)\right)$$

$$= \frac{1}{b\ln\frac{Q^2}{\Lambda_{QCD}^2}} + O\left(\alpha_s^2\left(Q^2\right)\right). \tag{B.13}$$

| $Z_3$ | $1 + \frac{g_s^2}{32\pi^2} \left[ \left( \frac{13}{3} - \xi \right) C_2 - \frac{4}{3} n_f \right] \ln \frac{\lambda^2}{\mu^2}$ |
|---|---|
| $Z_1$ | $1 + \frac{g_s^2}{32\pi^2} \left[ \left( \frac{17}{6} - \frac{3}{2}\xi \right) C_2 - \frac{4}{3} n_f \right] \ln \frac{\lambda^2}{\mu^2}$ |
| $Z_{3F}$ | $1 - \frac{g_s^2}{16\pi^2} \xi \, C_F \, \ln \frac{\lambda^2}{\mu^2}$ |
| $Z_{1F}$ | $1 - \frac{g_s^2}{16\pi^2} \xi \, C_F \, \ln \frac{\lambda^2}{\mu^2} - \frac{g_s^2}{32\pi^2} \left( \frac{3}{2} + \frac{1}{2}\xi \right) C_2 \, \ln \frac{\lambda^2}{\mu^2}.$ |

Note that we have been able to sum the leading logs here [compare (B.13) with $\alpha_\mu \left( 1 - \alpha_\mu b \, \ln \frac{Q^2}{\mu^2} + \cdots \right)$ in the ordinary perturbation theory]. Thus Eq. (B.13) goes beyond the ordinary perturbation theory. The perturbation now is with respect to $\alpha_s \left( Q^2 \right)$.

We now determine $b$. For this purpose, we need $Z_s$ [cf. Eqs. (B.8) and (B.9)]. But we note from Fig. B.1 that $Z_s$ is given by

$$Z_s^{1/2} = Z_{3F} \, Z_{1F}^{-1} \, Z_3^{1/2} \qquad (B.14)$$

where the renormalization constants $Z_{3F}$, $Z_{1F}$ and $Z_3$ arise respectively from diagrams A [self-energy part of the fermions (quarks) propagator], B [vertex part for the fermion] and C [the vacuum polarization or the self-energy part of the gluon propagator]. The values of these constants are summarized in Table B.1, which also includes $Z_1$ which corresponds to the triple gluon vertex [i.e. the first of diagrams (B) with the quark lines replaced by the gluon lines while the second is replaced by ⟍⟋ ]. In this table $C_2$ and $C_F$ are defined in Eq. (B.1) and $n_f$ denote the number of fermion flavors.

From Eq. (B.14) and Table B.1, we have to order $g_s^2$

$$Z_s^{1/2} = 1 + \frac{g_s^2}{32\pi^2} \left( \frac{11}{3} C_2 - \frac{2}{3} n_f \right) \ln \frac{\lambda^2}{\mu^2}. \qquad (B.15)$$

Thus from Eq. (B.10b) [note that the gauge fixing parameter $\xi$ is canceled out]

$$\beta \left( g_s \right) = -2g_s^3 \left[ b_0 + O \left( g_s^2 \right) \right], \qquad (B.16a)$$

so that

$$b_0 = \frac{1}{32\pi^2} \left( \frac{11}{3} C_2 - \frac{2}{3} n_f \right) \qquad (B.16b)$$

$$b = 8\pi b_0 = \frac{1}{4\pi} \left( \frac{11}{3} C_2 - \frac{2}{3} n_f \right). \qquad (B.16c)$$

Hence in summary, we have from Eq. (B.13)

$$\alpha_s\left(Q^2\right) = \frac{1}{\alpha_\mu^{-1} + \frac{1}{4\pi}\left(\frac{11}{3}C_2 - \frac{2}{3}n_f\right)\ln\frac{Q^2}{\mu^2}} + O\left(\alpha_s^2\left(Q^2\right)\right)$$

(B.17a)

$$= \frac{1}{\left(\frac{11}{3}C_2 - \frac{2}{3}n_f\right)\ln\frac{Q^2}{\Lambda_{QCD}^2}} + O\left(\alpha_s^2\left(Q^2\right)\right).$$

(B.17b)

It is a very useful equation and the single parameter $\Lambda_{QCD}$ becomes the QCD scale which effectively defines the energy scale at which the running coupling constant attains its maximum. $\Lambda_{QCD}$ can be determined from experiments and turns out to be [see Chap. 7]

$$\Lambda_{QCD} = 140 \pm 60 \text{ MeV.} \tag{B.18}$$

Note that for $SU_c(3)$ $[C_2 = 3]$, $\left(11 - \frac{2}{3}n_f\right)$ is positive for $\frac{2}{3}n_f < 11$ (which is certainly true for known six quark flavors $n_f = 6$) and then $\alpha_s\left(Q^2\right)$ decreases as $Q^2$ increases. This is made possible because of coefficient 11 which comes from the self coupling of gluons, non-Abelian nature of QCD. The logarithmic deviation from asymptotic freedom is a characteristic of QCD and the tests of the theory have to be sought to detect logarithmic scaling violations.

## B.3  Running Coupling Constant in Quantum Electrodynamics (QED)

For QED, only fermion loops (i.e. the first of diagrams C in Fig. B.1 with gluon replaced by photon and $g_s^2$ by $e^2$) contribute to electric charge renormalization so that in Table B.1 only $Z_3$ without $C_2$ is relevant. Note, however, that the contributing charged fermions are $e, u, d, \mu, c, s, \tau, b$ and $t$ so that $e^2\left(n_f/2\right)$ in the expression for $Z_3$ is replaced by

$$e^2\sum_f Q_f^2 = e^2\frac{n_f}{2}\left[1 + 3\left(\frac{4}{9} + \frac{1}{9}\right)\right]$$

where $\left(n_f/2\right)$ are the number of generations [3 in our case] and the factor 3 outside the parenthesis is due to the color. Thus

$$Z_{em}^{1/2} = 1 + \frac{e^2}{32\pi^2}\frac{1}{2}\left[-\frac{4}{3}\left(8\,\frac{n_f}{3}\right)\right]\ln\frac{\lambda^2}{\mu^2} \tag{B.19a}$$

giving

$$\beta_{em} = -\frac{2e^3}{32\pi^2}\left(-\frac{16n_f}{9}\right) \qquad (B.19b)$$

The equation analogues to (B.12) is then

$$\frac{d\alpha_e^{-1}(Q^2)}{dQ^2} = b_{em} = \frac{1}{4\pi}\left(-\frac{16n_f}{9}\right) \qquad (B.20a)$$

giving

$$\alpha_e(Q^2) = \frac{1}{\alpha_e^{-1}(Q^2) - \frac{4n_f}{9\pi}\ln\frac{Q^2}{\mu^2}} \qquad (B.20b)$$

and increases with $Q^2$ in contrast to $\alpha_s(Q^2)$ which decreases with $Q^2$.

Let us apply Eq. (B.20b) for $\mu^2 = m_e^2$, where $\alpha_e(m_e^2)$ is determined from Thomson scattering, for example, $\left[\alpha \equiv \alpha_e(m_e^2) = \frac{1}{137}\right]$. No matter how small $\alpha$ one has, one can always increase $Q^2$ to a point where $\alpha_e(Q^2)$ which was given in Eq. (B.20b) becomes infinite [Landau ghost]. This, however, occurs [for six flavors] at

$$Q^2 = m_e^2 \exp\left(\frac{3\pi}{8}\alpha^{-1}\right) \approx 10^{63}\ \text{GeV}^2 \qquad (B.21)$$

which is even larger than $M_P^2 \approx 10^{38}\ \text{GeV}^2$ by several orders of magnitude.

Finally, we wish to remark that the formula (B.20b) holds for $m_e^2 \leq Q^2 < m_W^2$. For $Q^2 \geq m_W^2$, we have to consider the contribution of charged $W^\pm$ bosons to $\beta_{em}$. In this case

$$b_{em} = \frac{1}{4\pi}\left(-\frac{16}{9}n_f + \frac{22}{3}\right) \qquad (B.22)$$

and $\alpha_e(Q^2)$ still increases with $Q^2$ for $n_f = 6$ (or $> 6$).

## B.4    Running Coupling Constant for SU(2) Gauge Group

For SU(2) group $C_2 = 2$ and therefore from Eq. (B.16c)

$$b_{SU(2)} = \frac{1}{4\pi}\left(\frac{22}{3} - \frac{2}{3}n_f\right) \qquad (B.23)$$

and correspondingly Eq. (B.17a) becomes

$$\alpha_2(Q^2) = \frac{1}{\alpha_\mu^{-1} + \frac{1}{4\pi}\left(\frac{22}{3} - \frac{2}{3}n_f\right)\ln\frac{Q^2}{\mu^2}} \qquad (B.24)$$

where $\alpha_2 = \frac{g_2^2}{4\pi}$, $g_2$ being the coupling constant associated with $qqW^\pm$ vertex, $W^\pm$, $W_3$ being the gauge bosons associated with SU(2) gauge group. Note that for six quark flavors $(n_f = 6)$, $\left(\frac{22}{3} > 4\right)$ and $\alpha_2(Q^2)$ is falling with $Q^2$, although at a rate less than $\alpha_s(Q^2)$ for the SU$_C$(3) group.

## B.5   Renormalization Group and High $Q^2$ Behavior of Green's Function

Consider now in general a renormalized Green's function (propagator or vertex function or a related quantity) in QCD denoted by

$$\Gamma_R\left(p_i, \alpha_s, \mu, \xi\right) = Z^{-1}\left(\lambda^2/\mu^2, \alpha_s, \xi\right) \Gamma\left(p_i, \alpha_{s0}, \xi_0\right), \tag{B.25}$$

where $Z$ is a multiplicative renormalization factor and $\Gamma$ on the right-hand side knows nothing about $\mu$ so that $\frac{d\Gamma}{d\mu} = 0$. This implies that $\Gamma_R$ satisfies the renormalization group equation

$$\frac{1}{\Gamma_R}\frac{d\Gamma_R}{d\mu} = -\frac{1}{Z}\frac{dZ}{d\mu}$$

$$\left[\frac{\partial}{\partial\mu} + \frac{d\alpha_s}{d\mu}\frac{\partial}{\partial\alpha_s} + \frac{d\xi}{d\mu}\frac{\partial}{\partial\xi}\right]\Gamma_R = \left[-\frac{1}{Z}\frac{dZ}{d\mu}\right]\Gamma_R$$

or

$$\left[\mu\frac{\partial}{\partial\mu} + 2\overline{\beta}\left(\alpha_s\right)\frac{\partial}{\partial\alpha_s} + \delta\left(\alpha_s,\xi\right)\frac{\partial}{\partial\xi} - 2\gamma\left(\alpha_s\right)\right]\Gamma_R\left(p_i,\alpha_s,\mu,\xi\right) = 0 \tag{B.26a}$$

where [cf. Eqs. (B.9) and (B.10)]

$$\delta\left(\alpha_s,\xi\right) = \mu\frac{d\xi}{d\mu}$$

$$\gamma = -\frac{1}{2Z}\frac{dZ}{d\ln\mu} = -g_s^2\frac{dZ}{dg_s^2}$$

$$\overline{\beta}\left(\alpha_s\right) = \frac{1}{2}\mu\frac{d\alpha_s}{d\mu} = \frac{1}{4\pi}g_s\,\beta\left(g_s\right). \tag{B.26b}$$

To simplify matters, let us work in the Landau gauge $\xi = 0$, then

$$\left[\mu\frac{\partial}{\partial\mu} + 2\overline{\beta}\left(\alpha_s\right)\frac{\partial}{\partial\alpha_s} - 2\gamma\left(\alpha_s\right)\right]\Gamma_R\left(p_i,\alpha_s,\mu\right) = 0. \tag{B.27}$$

The above equation also determines the high $Q^2$ behavior of $\Gamma_R$. To see this, we first note that there is another constraint on $\Gamma$ which comes from dimensional analysis. Assume we scale all momenta in $\Gamma_R\left(p\right)$, $p_i \to \lambda\, p_i$

$$\Gamma_R\left(\lambda p_i, \alpha_s, \mu\right) = \mu^D\, F\left(\lambda^2\frac{p_i \cdot p_j}{\mu^2}, \alpha_s\right), \tag{B.28}$$

$D$ is the dimension of the Green's function (e.g. for inverse gluon propagator $\Gamma \sim p^2$ and we have $D = 2$). $F$ is a dimensionless function of dimensionless variables. From Euler's theorem for homogeneous function

$$\left[\lambda\frac{\partial}{\partial\lambda} + \mu\frac{\partial}{\partial\mu} - D\right]\Gamma_R\left(\lambda\, p_i,\alpha_s,\mu\right) = 0. \tag{B.29}$$

Put $t = \ln \lambda$ and combine the naive scaling equation (B.29) with the renormalization group equation (B.27) [which gives the dynamical constraint] to eliminate $\mu \frac{\partial}{\partial \mu}$ and obtain

$$\left[ -\frac{\partial}{\partial t} + 2\bar{\beta}\left(\alpha_s\right) \frac{\partial}{\partial \alpha_s} + D - 2\gamma\left(\alpha_s\right) \right] \Gamma_R \left(\lambda p_i, \alpha_s, \mu\right) = 0. \qquad (B.30)$$

Its general solution can be obtained by the method of characteristics. First one solves [cf. Eq. (B.26b) with $t = \ln \mu$]

$$\frac{d\,\bar{\alpha}_s\,(t, \alpha_s)}{dt} = 2\bar{\beta}\left(\bar{\alpha}_s\,(t)\right) \qquad (B.31)$$

with the condition $\bar{\alpha}_s\,(0, \alpha_s) = \alpha_s$. The general solution of Eq. (B.30) can then be expressed in terms of that of the above differential equation. In this way one obtains

$$\Gamma_R\left(\lambda p_i, \alpha_s, \mu\right) = \lambda^D \Gamma_R\left(p_i, \bar{\alpha}_s\,(t), \mu\right) \exp\left[-2 \int_0^t dt' \gamma\left(\bar{\alpha}_s\,(t')\right)\right]. \qquad (B.32)$$

What we learn from this general solution is that the behavior of Green's functions when all momenta are scaled up is governed by $\bar{\alpha}_s\,(t)$. Now as already seen in Sec. B.2

$$\bar{\beta}\left[\left(\bar{\alpha}_s\,(t)\right)^2\right] = -\left(\bar{\alpha}_s\,(t)\right)^2 b + \cdots \qquad (B.33a)$$

and similarly we can expand

$$\gamma\left(\bar{\alpha}_s\,(t)\right) = \gamma_0 \bar{\alpha}_s\,(t) + \cdots \qquad (B.33b)$$
$$\Gamma_R\left(p_i, \bar{\alpha}_s\,(t), \mu\right) = \Gamma_{R_0}\left(p_i, \mu\right) + \Gamma_{R_1}\left(p_i, \mu\right) \bar{\alpha}_s\,(t) + \cdots . \qquad (B.33c)$$

Thus to solve Eq. (B.31) in the lowest order, we make use of Eq. (B.33a) and rewrite it as

$$dt = -\frac{d\bar{\alpha}_s}{2b\bar{\alpha}_s^2 \left[1 + \cdots\right]}$$

giving

$$\bar{\alpha}_s\,(t) = \frac{1}{\alpha_s^{-1} + 2bt} + 0\left(\bar{\alpha}_s^2\right). \qquad (B.34)$$

Remember that $t \sim \ln \lambda$ and this is the same functional dependence for $\bar{\alpha}_s$ as before for $\alpha_s\left(Q^2\right)$ in Sec. B.2. Thus noting that for large $t$, $\bar{\alpha}_s\,(t) \sim \frac{1}{2bt}$, the use of Eqs. (B.33a) and (B.34) in the first order enable us to write Eq. (B.32) for large $t$ or $\lambda$ as

$$\Gamma_R(\lambda\, p_i, \alpha_s, \mu)$$

$$\sim \lambda^D \Gamma_{R0}(p_i, \mu) \exp\left(-2\int_0^t \frac{\gamma_0}{2bt'}dt'\right)$$

$$= \lambda^D \Gamma_{R0}(p_i, \mu)\, t^{-\gamma_0/b}$$

$$= \lambda^D \Gamma_{R0}(p_i, \mu)\, (\ln \lambda)^{-\gamma_0/b} \tag{B.35}$$

where $\gamma$ or $\gamma_0$ is called the anomalous dimension of $\Gamma_R$, which can be determined from Eq. (B.26b). If it were zero, we would have obtained canonical scaling behavior $\lambda^D$ as in the traditional parton model [cf. Chap. 14]. Noting that $\bar{\alpha}_s(t) \sim \frac{1}{2bt}$ $[t \sim \ln \lambda]$, we can say from Eq. (B.35) that the large $Q^2$ behavior of $\Gamma_R(Q^2)$ is

$$\frac{\Gamma_R(Q^2)}{\Gamma_R(Q^2)} = \left(\frac{Q^2}{Q_0}\right)^D \left[\frac{\bar{\alpha}_s(Q^2)}{\bar{\alpha}_s(Q_0^2)}\right]^{\gamma_0/b} \tag{B.36}$$

where the second factor can be written as

$$\left[\frac{\bar{\alpha}_s(Q^2)}{\bar{\alpha}_s(Q^2)}\right]^{\gamma_0/b} = \left\{\bar{\alpha}_s(Q^2)\left[\alpha_\mu^{-1} + b\ln\frac{Q_0^2}{\mu^2}\right]\right\}^{\gamma_0/b}$$

$$= \left\{\bar{\alpha}_s(Q^2)\left[\alpha_\mu^{-1} + b\ln\frac{Q^2}{\mu^2} - b\ln\frac{Q^2}{Q_0^2}\right]\right\}^{\gamma_0/b}$$

$$= \left\{1 - b\,\bar{\alpha}_s(Q^2)\ln\frac{Q^2}{Q_0^2}\right\}^{\gamma_0/b}, \tag{B.37}$$

with $b = \frac{1}{4\pi}\left[\frac{11}{3}C_2 - \frac{2}{3}n_f\right]$ [cf. Eq. (B.16c)]. Thus it is clear that the renormalization group equation has enabled us to sum up terms of the form $\left[\alpha_s(Q^2)\ln Q^2\right]^N$ whereas in ordinary perturbation theory we would have to deal with a power series in $\alpha_s(Q^2)\ln Q^2$.

Analogous logarithmic violation of the scaling will hold in the deep inelastic structure functions and similar physical quantities.

Let us now consider some simple applications:

## B.5.1 *Gluon Propagator*

From Table B.1,

$$Z_3 = 1 + \frac{2\alpha_s}{8\pi}\left[\left(\frac{13}{3} - \xi\right)C_2 - \frac{4}{3}n_f\right]\ln\frac{\lambda}{\mu} + O\left(\alpha_s^2\right) \tag{B.38}$$

$$\gamma_v = \frac{\alpha_s}{8\pi} \left[ \left( \frac{13}{3} - \xi \right) C_2 - \frac{4}{3} n_f \right] + O\left(\alpha_s^2\right)$$

$$\gamma_{v0} = \frac{1}{8\pi} \left[ \left( \frac{13}{3} - \xi \right) C_2 - \frac{4}{3} n_f \right]. \tag{B.39}$$

$$D_{AB\mu\nu} = -i\delta_{AB} \left[ \left( g_{\mu\nu} - \frac{k_\mu k_\nu}{k^2} \right) + \xi \frac{k_\mu k_\nu}{k^2} \right] \frac{1}{k^2} d\left(-k^2\right) \tag{B.40a}$$

where

$$d\left(-k^2\right) \sim \left[\alpha_s\left(-k^2\right)\right]^{\gamma_{v0}/b}. \tag{B.40b}$$

### B.5.2    *Fermion Propagator*

$$S_F^{-1}(p) = \left( \not{p} - m_\mu - \sum(p) \right), \tag{B.41a}$$

where

$$\sum(p) = m_\mu \sum\nolimits_1 \left(p^2\right) + \left( \not{p} - m_\mu \right) \sum\nolimits_2 \left(p^2\right). \tag{B.41b}$$

Let us define an effective or running mass through the following equations

$$S_F(p) = s_c\left(p^2\right) \frac{\not{p} + m\left(p^2\right)}{p^2 - m^2\left(p^2\right)} \tag{B.40c}$$

$$s_c\left(p^2\right) = \left( 1 + \sum\nolimits_2 \left(p^2\right) \right)^{-1} \tag{B.40d}$$

$$m\left(p^2\right) = m_\mu \left[ 1 + \frac{\sum_1\left(p^2\right)}{1 + \sum_2\left(p^2\right)} \right]. \tag{B.40e}$$

The fermion self-energy diagram is given in Fig. B.2.

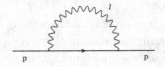

Fig. B.2   Fermion self-energy at one-loop level.

This determines $\sum(p)$ at one-loop level. The renormalization mass $m_R$ is defined by

$$m_R = Z_m\, m_B \tag{B.42}$$

where $Z_m$ is the multiplicative mass renormalization constant and is given by

$$Z_m = 1 + \sum_1$$
$$= 1 + 2\frac{3\alpha_s}{4\pi}C_F \ln\frac{\lambda}{\mu} \tag{B.43}$$

while from Table B.1:

$$Z_{3F} = \left(1 - \sum_2\right) - 1$$
$$= 1 - 2\frac{\alpha_s}{4\pi}(C_F \, \xi)\ln\frac{\lambda}{\mu}. \tag{B.44}$$

Thus to the leading order

$$\gamma_{m0} = \frac{3\,C_F}{4\pi} \tag{B.45}$$

$$\gamma_{F0} = \frac{C_F}{4\pi}(-\xi), \tag{B.46}$$

where for $SU_c(3)$ $C_F = \frac{4}{3}$ [cf. Eq. (B.2)]. Hence

$$s_c\left(p^2\right) \sim \left[\alpha_s\left(-p^2\right)\right]^{\gamma_{F0}/b} \tag{B.47a}$$

while

$$S_F\left(p^2\right) \sim \frac{1}{\not{p}}\left[\alpha_s\left(-p^2\right)\right]^{\gamma_{F0}/b}. \tag{B.47b}$$

For large $p^2$ we note from Eq. (B.36) that

$$\frac{m\left(p^2\right)}{m\left(p_0^2\right)} = \left[\frac{\alpha_s\left(-p^2\right)}{\alpha_s\left(-p_0^2\right)}\right]^{\gamma_{m0}/b}$$
$$= \left[\frac{\alpha_s\left(-p^2\right)}{\alpha_s\left(-p_0^2\right)}\right]^{3C_F/\left(\frac{11}{3}C_2 - \frac{2}{3}n_f\right)}$$
$$= \left[\frac{\alpha_s\left(-p^2\right)}{\alpha_s\left(-p_0^2\right)}\right]^{4/\left(11 - \frac{2}{3}n_f\right)}. \tag{B.48}$$

## B.6    References for Appendices

1. J. J. Sakurai, Advanced Quantum Mechanics, Addison-Wesley, Reading, Massachusetts (1967).
2. J. D. Bjorken and S. D. Drell, Relativistic Quantum Fields, McGraw-Hill, New York (1964).
3. C. Itzykson and J. B. Zuber, Quantum Field Theory, McGraw-Hill, New York (1980).
4. M.E. Peskin and D.V. Schroeder, An Introduction to Quantum Field Theory, Addison-Wesley, Reading, Massachusetts (1995).

# Index

Accelerating Universe, 566
  Dark Energy, 566
  Evidence from CMB Data, 568
  Evidence from Supernovae, 567
Aharanov and Bohm Experiment, 186
Astroparticle Physics, 557
Asymptotic freedom, 640
  property of QCD, 640
Axial Anomaly, 325, 416

Baryogenesis, 595
  at GUT Level, 598
  Electroweak Baryogenesis, 598
  Sakharov's Conditions, 597
Baryon, 9, 128
  decays, 292
  heavy, 232
Baryon States, 130
Beta ($\beta$) decay, 44, 46
Big Bang Theory, 568
Bottom Quark, 228
Bottomonium, 233

C-parity, 107, 234
Cabibbo angle, 284, 296
Charge Conjugation, 107, 628
  hadronic interactions, 109
  invariance, 109, 110
  matrix, 629
  parity, 108
Charges, 91, 94
  strong, 10

weak color, 11
Charm, 221, 224
  Discovery, 221
  isospin, 223
  SU(3) classification, 223
Charmonium, 233
Chiral Symmetry, 311, 317
  eplicit breaking, 320
  non-leptonic decays of hyperons,
    323
CKM Matrix, 414, 458
Color, 9, 10, 16–18, 181
  confining potential, 17
  confinement, 10
  electric potential, 17
Conservation
  baryon charge, 94
  electric charge, 91
  Energy and momentum, 26
  hypercharge, 96
  lepton charge, 94
  muon number, 95
  parity, 68
  probability, 34
  strangeness, 96
Conserved Vector Current, 311
  CVC, 311
Cornell potential, 208
Cosmic Microwave Background, 568
Cosmology, 557
  Accelerating Universe, 566
  Baryogenesis, 595

Cosmological Principle, 557
Hot Big Bang, 574
Inflation, 588
Leptogenesis, 601
Limit on Neutrino Mass, 584
Modified Gravity, 573
Standard Model, 559
Standard Model Solutions, 562
CP-violation, 497, 503, 508
in $B^0 - \bar{B}^0$ system, 511
in $K^0 \bar{K}^0$ system, 504
Cross-section, 55
Current Algebra, 317

Dalitz plot, 43
Dark Energy, 566
Accelerating Universe, 566
Decay of polarized Muon, 290
Decay widths of $W$ and $Z$ bosons, 389
Deep Inelastic Scattering, 425
involving neutral weak currents,
446
Lepton-Nucleon Scattering, 427
Neutrino-Nucleon Scattering, 436
Detailed balance principle, 76
diquark-diantiquark, 246
Dirac Equation, 611
Dirac gamma matrices, 611, 612
Dirac Lagrangian, 631
Duality, 549
and String Theory, 548

Effective Lagrangian of HQET, 255
Electromagnetic Interaction, 50, 100
Electroweak Baryogenesis, 598
Electroweak Unification, 365, 374
Experimental Consequences, 381
Equation of continuity, 185
Euler's theorem, 647
Expansion of the Universe, 557

Fermi plot, 45
Fermi-Dirac statistics, 614
Fermion, 10
propagator, 650
self energy diagram, 650

Fermion flavor, 644
Feynman Rules, 620
application to $e^+ e^- \rightarrow \mu^+ \mu^-$, 621
for QCD, 639
Fifth Quark Flavor, 228
Flatness Problem, 590
Freeze Out, 581
Friedmann-Robertson-Walker metric,
557

G-Parity, 112
G-T relation, 315
Gamma $(\gamma)$ matrices trace, 616
Gauge principle, 184, 188
Gauge transformation, 93
$U_Q(1)$, 92
Gell-Mann-Nishijima relation, 98,
124, 224, 284
Gell-Mann-Okubo mass formulae, 150
Ghosts, 639
GIM Mechanism, 411
Globally conserved quantum
numbers, 91, 94, 95
Gluon, 16, 197
exchange potential, 16, 17
longitudnal, 639
Goldberger-Treiman (G-T) relation,
315
Grand Unification, 525
mass scale, 529
Green's function
in QCD, 647
GUTS, 531
General Consequences, 531

Hadron Spectroscopy, 202
Hadronic cross-section, 55
Hadronic Decay Width, 238
Hadrons, 206
Heavy Baryons, 232, 267
Heavy Flavors, 221, 457
Weak decays, 451
Heavy Mesons, 224
Heavy Quark, 259, 265
spin symmetry, 259
Helicity, 281

Higgs Boson Mass, 399
  Upper bound, 399
Horizon Problem, 588
Hot Big Bang, 574
  Freeze Out, 581
  Radiation Era, 576
  Thermal Equilibrium, 574
HQET, 255
Hypercharge, 96
Hyperons
  chiral symmetry, 323
  non-leptonic decays, 299

Inflation, 588
  Flatness Problem, 590
  Horizon Problem, 588
  Realization, 591
  Slow-roll Inflation, 593
Interaction picture, 31
Internal Symmetries, 91, 97
  Charge Conjugation, 107
  G-Parity, 112
Invariance Principle, 65
Invariant BBP Couplings SU(3), 145
Irreducible Representation of SU(3), 134
Isospin, 97, 100, 101

Kurie plot, 45

Leptogenesis, 601
Leptonic decays
  of $\tau$ lepton, 451
  of D and B mesons, 458

M-theory, 549
Magnetic moment of Baryons, 164
Mass spectrum, 209, 213
Mass Splitting in Flavor SU(3), 148
Meson, 126, 275
Mikheyev-Smirnov-Wolfenstein
  Effect, 345
Modified Gravity, 573
Muon $(\mu)$ Decay, 288

Nambu-Goldstone bosons, 327

Neutrino, 331
  Helicity, 281
  Magnetic Moment, 360
  Mass, 332
  Mass limit, 584
  Oscillations, 343
Neutrino Mass, 584
  Astrophysical Constraints, 336
  Constraints, 333
  Dirac and Majorana, 337
Neutrino Oscillations, 343
  Evidence, 351
  Resonant Matter, 345
Non-Relativistic Quark Model
  non-leptonic Hyperon Decays, 304
Nucleon Magneton, 168

Octet, 311
One Gluon Exchange potential, 202
Optical theorem, 78
OZI rule, 153

Parity, 66, 68
Partial wave unitarity of two-particle, 79
Partially Conserved Axial Vector
  Current, 314
  PCAC, 314
Parton Model, 431
Phase space, 36, 41, 101
  two-body Lorentz invariant, 79
Photon, 1, 5, 11, 17
  exchange, 17
Pion, 70, 76
  exchange, 55
  intrinsic parity, 70
Positronium, 110, 111
Primordial Nucleosynthesis, 585
Pseudoscalar Meson Decays, 296

Quantum Chromodynamics, 17, 194, 197, 199
  asymptotic freedom, 18, 640
  Feynman rules, 639
  Lagrangian, 639
  long range potential, 16

scale factor, 645
Quantum Electrodynamics, 17, 645
  running coupling constant, 645
Quantum Field Theory, 609
  application of Feynman rules, 621
  charge conjugation, 628
  Feynman rules, 620
  massive spin 1, 619
  spin 0, 609
  spin 1, 618
  spin 1/2, 611
Quark, 8, 9
  confinement, 16, 18
  flavor, 10
Quarkonium, 233
  Leptonic Decay Width, 237
  mass spectrum, 245
  non-relativistic Treatment, 240
Quintessence, 571

Radiation Era, 576
Radiative Corrections, 382
  quark triangle, 325
Radiative Decays, 176, 177
  vector mesons, 170
Redshift, 559
Relativistic quantum mechanics, 188
Renormalizability, 372
Renormalization factor
  multiplicative, 647
Renormalization group, 639
  asymptotic freedom of QCD, 640
  effective coupling constant, 640
  high $Q^2$ behaviour of Green's
        function, 647
  property, 642
  renormalization constant, 642
  renormalization scale, 642
  running coupling constant, 639
  running coupling constant for
        SU(2) gauge group, 646
  running coupling constant in QCD,
        639
  running coupling constant in QED,
        645
  running mass, 650

Resonance $\Delta$ (delta), 103
Resonance $\rho$ (rho), 102
Resonance production, 101
Running coupling constant
  for SU(2) gauge group, 646
  in QCD, 639
  in QED, 645

S-Matrix, 34, 71, 77
Sakharov's Conditions, 597
Nucleon-Nucleon scattering process,
        54
Scattering process, 26–28, 39, 41
  Neutrino-electron, 53
See-saw Mechanism, 339
Selection rules, 91
Semi detailed balance principle, 76
Semileptonic decays, 459
  Exclusive, 464
  of D and B mesons, 459
Sixth Quark Flavor, 228
Slow-roll Inflation, 593
S-Matrix, 32
Solar Neutrinos, 353
Space-Time Symmetries, 66, 68
Spectrum, 209
Spin of $\Delta$, 103
Spin of pion, 76
Spin projection operators, 615
Spin-Spin Interaction, 209
Standard Model of Cosmology, 559
Strangeness, 96
String Theory, 525, 548
  and duality, 548
  Heterotic, 549
  Supersymmetry, 544
SU(3), 119
  sextet representation, 233
SU(6), 159
SU(6) and Quark Model, 159
SU(6) wave function for Mesons, 160
SU(N), 141
Sum Rules, 327, 439
  QCD, 327
Supergravity, 550
Supersymmetry, 525, 545

and Strings, 544
Symmetries, 65, 91
   time reversal, 73

T-duality, 549
T-Matrix, 77, 296
Tau ($\tau$) Lepton, 451
   Semi-Hadronic decays, 453
Tetraquark, 246
   diquark-diantiquark, 246
Time reversal, 73
Top Quark, 228
Transition rate, 31

U-Spin, 132
Unitarity constraints, 77, 80, 84
Unitary Group, 119

V-A Interactions, 279
Virial Theorem, 242
VPP Couplings SU(3), 146

Weak Decays
   of Heavy Flavors, 451, 457
Weak Hadronic Currents
   Properties, 311
Weak Interaction, 52, 101, 279
Weak Processes, 281
   non-leptonic, 287
   purely leptonic, 281
   semileptonic, 283, 291

Yang-Mills fields, 197
Young's Tableaux, 135
Young's tableaux for SU(N), 141